Commercial Satellite Launch Vehicle Attitude Control Systems Design and Analysis

(H-infinity, Loop Shaping, and Coprime Factorization Approach)

Chong H. Kim, Ph.D.
The Boeing Company (Retiree)

CHK

ISBN : 978-0-6151-6511-0
Library of Congress Control Number: 2007908290
Printed in the United States of America

To my mother, Chung Yul Kahn,
my wife, Chung Kang,
and children, Jennifer and David

My life has been so much blessed and enriched by these remarkable individuals.

PREFACE

This book is written for aerospace engineers in industry who have completed their undergraduate work and are interested in the design and analysis of rocket attitude control systems. It attempts to include every step in the equation mathematical derivation for easy understanding.

Designing a Guidance, Navigation, and Control (GN&C) system for a satellite launch vehicle is no simple matter: It would take many years of multi-disciplinary efforts if a successful design is to be achieved. The attitude control system design, in particular, requires a complex synergy of physics, chemistry, engineering, and computing technology.

The primary contribution of this book is an introduction of a new technical approach to the design of robust-performance attitude control systems for launch vehicles (LV). Designs of current LV attitude control systems are based on the classical Single Input Single Output (SISO) control theory, which lacks robustness because the SISO system math model cannot represent the real system as closely as the Multi-Input Multi-Output (MIMO) system math model can. Thus, it is unable to deal effectively with disturbances such as wind because it cannot compute appropriate feedback signals that reflect the wind effect. However, adopting MIMO control theory, we can deal with more than one input, e.g., attitude, AOA (Angle of Attack), and wind at the same time. And it becomes possible to compute appropriate feedback signals that reflect the attitude, wind, and AOA behaviors. It will be shown that the theory used here truly offers a technique that enables us to design control systems that are reasonably insensitive to math modeling errors and can withstand disturbances such as wind effectively. That is, the control system thus designed can contain strong wind and also reduce the AOA to some extent. However, it may cause excessive LV drift from the nominal trajectory. The guidance system must incorporate proper trajectory design in order to compensate for the excessive drift. It is to be noted that the new approach offers a tremendous advantage in the design process in that it does not need state estimators, such as Kalman filter. Extensive simulation results, which demonstrate the effectiveness of this approach, are presented in this book. The theory has been used in many other applications, but not in launch vehicle attitude control systems design as comprehensively and practically as described in this book.

Chapter 1 explains the basic rocket theory that provides some insight as to how the rocket works. Chapter 2 introduces an aspect of H-infinity theory starting with the well-known Riccati equation, which is used to solve optimization problem in H-infinity. In this chapter, the core of the H-infinity pertinent to optimal system design is briefly introduced. Recent development of the H-infinity design technique has opened up a new avenue in which this powerful method has been applied to various control system designs. H-infinity is basically a method or a technique by which optimization-based control system design can be accomplished. The major advantages of H-infinity are that it guarantees robustness and stability, and it can be applied to multi-input multi-output control system design. It was named for an eminent mathematician, G. H. Hardy, who investigated the behavior of all stable transfer functions. The name H-infinity actually refers to a mathematical framework (ref. 37).

In Chapter 3, an LV state space system equation is derived.

In Chapter 4, based on the math model derived in Chapter 3, an attitude control system is designed using H-infinity. The design effectiveness is verified by simulation. It should be noted that this new control system may require a new set of control surfaces such as fins, fin actuators, on-and-off reaction jets, and/or Vernier engines to implement feedback loop. The math modeling error sensitivity analysis is also presented in this chapter. In practice, it is difficult to determine system parameter values exactly. It is shown here that the control system designed by this method demonstrates reasonable insensitivity to parameter estimation errors. Finally, the software implementation issue is discussed. The dynamic feedback block, designed by use of H-infinity, usually is a higher order system that requires a great deal of computation time, which is not recommended for a flight control system application. It is desirable to have a lower order feedback block, and in this chapter, an order reduction technique is demonstrated.

In Chapter 5, the influence of wind on the attitude control system is analyzed. Since the wind affects the rocket's behavior significantly when the rocket is flying through the atmosphere, it is necessary to analyze and predict the wind effect before rocket launch in order to ensure attitude stability and optimal performance during the first stage of the flight. The rocket should be able to pass through the atmosphere without structural or thermal damage. Attitude control system performance through the atmosphere is one of the serious concerns in the current flight system. The reason is that the current launch vehicle control systems were designed using classical control theories such as Bode, Nyquist, and Root Locus, it is not possible to deal with attitude input command and the wind together at the same time. More specifically, classical control theory can be used only for a system with a single input. Therefore, since the LV attitude control system designed in Chapter 4 has three inputs and two outputs, the MIMO control theory is used for our design. It is verified that a feedback signal that reflects the wind behavior can be computed in H-infinity MIMO optimization process such that the control system can manage the wind disturbance dynamically regardless how strong the wind is. Single input single output classical control theory does not allow us to compute the feedback signal that reflects the wind dynamics. In addition, even AOA can be controlled. It is shown in the later part of this Chapter that the AOA can be reduced to some extent by applying an appropriately designed AOA input command. Once a stable control system is designed for multiple inputs, e.g., attitude, AOA command, and wind, the control system will maintain its stability effectively irrespective of the wind strength. It is shown that the H-infinity MIMO theory used here truly offers a technique that enables us to design a control system that can withstand strong wind disturbances as well as some extent of the math modeling errors.

Those who are acquainted with the current rocket configuration are familiar with the SISO (single-input single-output) control system. But when it comes to MIMO (multi-input multi-output) control (i.e., inputs: attitude command, angle of attack, and wind, and outputs: attitude and attitude rate), it could cause a confusion. In our current system, we can have only one input, that is, the attitude command. Now the question is how we can implement the angle of attack input and wind input. In order to apply these unconventional inputs, we need to reconfigure the current rocket structure by adding 1) special control surface or on-and-off reaction jets that controls the angle of attack and 2) a combination of wind sensor and motor that response to the wind feedback (read Chapter 5 for detail)..

The way that this book is written is probably different from other technical books. Prior to the computer era, book publication was not as simple as it is today; therefore, books were published without detailed derivation of equations. In this book, every step of the math derivation is presented to make it easier for readers to follow.

I wish to express my thanks to those professors who guided me through the years in my undergraduate and graduate studies in the United States, including Professors Henry M. Black and George M. Junkhan at Iowa State University; Professor Rufus Oldenburger at Purdue University; Professor Kumpati S. Narendra at Yale University; and Professor David P. Lindorff at the University of Connecticut. Dr. Lindorff inspired and encouraged me greatly as I was working on my doctorate at the University of Connecticut.

Without a doubt, the Almighty God has been so gracious to me throughout my life and kept me in His plan, even in the plan of salvation.

Chong H. Kim, Ph.D.
Aug. 25, 2007
Fountain Valley, CA 92708

Contents

Notation

[t = to, t=tf]∫ F(t) dt = integration of F(t) from t = to to t = tf.

( )d = d( )/dt

( )dd = d[d( )/dt]/dt
= d^2( )/(dt)^2

( )^2 = square of ( ).

sqrt = square root

wrt = with respect to

( )* = transpose of ( )

Inv ( ) = inverse of ( )

Symbols

(See Appendix I)

Chapter 1 Basic Rocket Theory

Rocket theory is well known and has been presented in numerous publications in various forms. In this chapter, important terms and equations are explained and derived in such a manner that those who are new to this field of study can understand the theoretical concepts of rocket theory with minimal effort (ref. 32).

1.1 Rocket Propulsion

Rocket propulsion can be described mathematically as a function of the following five elements:

1) propellant mass flow rate (mdot = dm/dt)
2) propellant exhaust velocity (Ve)
3) rocket nozzle exit area (Ae)
4) gas pressure at the rocket exit area (Pe)
5) ambient pressure (Pa)

The propulsion thrust equation is:

$$F = \text{mdot}\ Ve + Ae(Pe - Pa) \qquad (1.1\text{-}1)$$

Equation (1.1-1) can be rewritten as,

$$\begin{aligned} F &= \text{mdot}\ [\ Ve + (Ae/\text{mdot})\ (Pe - Pa)] \\ &= \text{mdot}\ C \end{aligned}$$

where

$$C = [\ Ve + (Ae/\text{mdot})\ (Pe - Pa)] \qquad (1.1\text{-}2)$$

The "C" in Eqn. (1.1-2) is the Effective Exhaust Velocity and Pa is close to zero in space.

There are four more basic terms related to the rocket propulsion.

1) <u>Specific Impulse (Isp):</u>

$$\begin{aligned} Isp &= F/(g\ \text{mdot}) \qquad (1.1\text{-}3) \\ &= K\ \text{sqrt}(\ Tc/M)\ \ \text{for chemical rockets} \end{aligned}$$

where

g = gravity force
K = proportionality constant
Tc = chamber temperature
M = molecular weight of combustion products

Specific Impulse is a measure of energy content embedded in propellant, representing propellant efficiency. Liquid propellants have higher Isp than solid propellants. For example, the Isp of Liquid Fluorine Hydrogen is 300 - 385 sec, while Potassium Perchlorite Thikol has an Isp of 170 - 210 sec. Notice that the Isp unit is time. Isp basically represents the length of time during which 1 lb of the fuel produces 1 lb of force consistently from the beginning of the fuel burning to the end. An interpretation of " Isp = 210 sec." therefore would mean that it takes 210 seconds to burn 1.00 lb (weight) of fuel while the thrust force (F) remains consistently 1.00 lb from the beginning to the end. This interpretation comes from the unit analysis as shown below:

Isp = 210.0 lbf / [1 lbf(fuel weight) /sec]
= 210.0 sec. 1 lbf / 1 lb (fuel weight)
= 210.0 sec.

Derivation of the Isp equation is shown in Section. 1.2.1. From Eqn. (1.1-2), we see that high efficiency can be achieved by maximizing the term, "(Pe – Pa)," which implies maximization of Pe.

2) Nozzle Area Expansion Ratio (ε)

Nozzle Area Expansion Ratio (ε) plays an important role in efficiency maximization. It is defined as,

$$\varepsilon = Ae/At \qquad (1.1\text{-}4)$$

where

Ae = nozzle exit area
At = nozzle throat area

3)Characteristic Velocity (C*)

Characteristic velocity is a measure of the energy available from the combustion process and is defined as,

$$C^* = Pc\ At/mdot \qquad (1.1\text{-}5)$$

where

Pc = chamber pressure.

Example:

Monopropellant Hydrogen	= 1,333 m/sec
Bipropellants	= 1,640 m/sec
Cryogenic LO2/LH2	= 2,360 m/sec

Most launch vehicles nowadays use cryogenic LO2/LH2.

4) Thrust Coefficient (Ct)

Thrust Coefficient is a measure of efficiency of converting the energy to exhaust velocity and characterizes the nozzle performance.

$$Ct = F/(Pc\ At) \qquad (1.1\text{-}6)$$

Typical values of Ct are 1.6 (ε=30:1) and 1.86 (ε=200:1). From C* and Ct, Isp can be computed.

$$Isp = Ct\ \ C^* / g \qquad (1.1\text{-}7)$$

1.2 Rocket Equation Derivation

The rocket equation that provides a means of computing ΔV (Vf – Vo) is derived. The ΔV represents the primary measure of performance capability of a propulsion system. The rocket equation is presented below and the derivation follows.

$$\begin{aligned} \Delta V &= Vf - Vo \\ &= v \ln(mo/mf) - g(tf - to) \qquad (1.2\text{-}1) \end{aligned}$$

where

Vo = initial velocity
Vf = final velocity
v = velocity difference between rocket and a particle in engine thrust

(note that $v>0$)
mo = initial mass
mf = final mass
g = gravity
tf = final time
to = initial time

Derivation

Linear momentum is defined as the product of mass and velocity.

$$P = mV \qquad (1.2\text{-}2)$$

where

P = linear momentum
m = mass
V = velocity

Force is defined as the time derivative of linear momentum,

$$F = dP/dt$$
$$= (dm/dt)V + m(dV/dt) \qquad (1.2\text{-}3)$$

When there is no engine thrust, there is no rocket acceleration, and its velocity remains constant assuming that there are no other disturbances and neglecting the gravity. Then the linear momentum of rocket is,

$$Po = moVo \qquad (1.2\text{-}4)$$

After the engine is activated for a short period of time, the momentum of the rocket and the engine thrust is,

$$Pf = (mo - \Delta m)(Vo + \Delta V) + u\,\Delta m \qquad (1.2\text{-}5)$$

where

$(mo - \Delta m)$ = reduction of mass where Δm is fuel consumption
$(Vo + \Delta V)$ = rocket velocity gain
Δm = mass leaving from the rocket ($\Delta m > 0$)
u = relative velocity (see Fig. 1.2-1, $u>0$)

Here the term "(mo – Δm)" is the rocket mass after fuel of mass, Δm, is consumed, and (Vo + ΔV) is the rocket velocity after the use of the fuel. The term u Δm accounts for the momentum loss from the rocket as the thrust expels the particles at the velocity of "u," which is the difference between the rocket velocity and the particle's velocity at the nozzle exit. From the conservation of linear momentum (of the whole system),

$$Po = Pf$$

And from Eqn. (1.2-4) and Eqn. (1.2-5),

$$\begin{aligned} moVo &= (mo - \Delta m)(Vo + \Delta V) + u\,\Delta m \\ &= moVo - \Delta m Vo + mo\,\Delta V - \Delta m\,\Delta V + u\,\Delta m \end{aligned}$$

Canceling the term, moVo, and ignoring the higher order term, Δm ΔV ,

$$mo\,\Delta V - \Delta m Vo = -u\,\Delta m \qquad (1.2\text{-}6)$$

Now dividing both sides of the Eqn. (1.2-6) by Δt and taking the time limit,

$$mo\,V' - m'Vo = -u\,m' \qquad (1.2\text{-}7)$$

where

( )' stands for d( )/dt.

Notice that "mo" can be written as "m," and "Vo," as "V" <u>in the limit</u>. Eqn. (1.2-7) is valid only in vacuum and gravity-free environment. Including the gravity and ignoring the aerodynamic force,

$$mV' - m'V = -u\,m' - mg \qquad (1.2\text{-}8)$$

In Eqn. (1.2-8), (- u m') is the thrust. Rearranging the equation above,

$$\begin{aligned} mV' + mg &= (-u + V)\,m' \\ m(dV/dt) + mg &= (-u + V)(dm/dt) \\ m\,dV &= (-u + V)\,dm - mg\,dt \end{aligned}$$

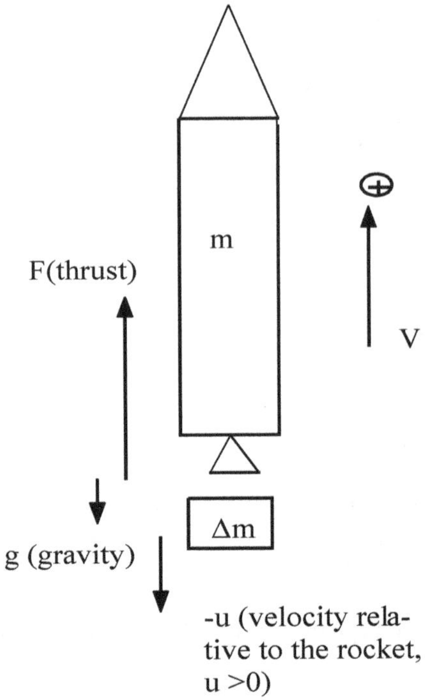

Fig. 1.2-1 Relative Velocity of Δm

Let $v = -(-u + V)$, then

$$m\, dV = -v\, dm - mg\, dt \quad (1.2\text{-}9)$$

In propulsion, $u > V$, $(-u + V) < 0$, and thus $v > 0$.

Dividing both sides of Eqn. (1.2-9) by "m,"

$$dV = -v\,(dm/m) - g\, dt$$

Integrating both sides of the equation above, Eqn. (1.2-1) is obtained. Rewriting the equation below,

$$(Vf - Vo) = v \ln(mo/mf) - g\,(tf - to)$$

where

Vf = final rocket velocity
Vo = initial rocket velocity
mo = initial rocket mass
mf = final rocket mass
tf = time when the rocket mass is equal to mf
to = initial time
v = relative velocity ($v > 0$ for propulsion)

1.2.1 Derivation of Isp (Specific Impulse)

The unit used for Specific Impulse is second. It is basically burning duration of a pound of fuel while exerting 1 pound of thrust force. It ranges from 170 sec. to 400 sec. The longer the Isp is, the better the quality of the fuel is. The Isp is defined as an integration of the thrust, F, from 1 lbm to 0 lbm. The Isp, which is time, must be positive.

Dividing Eqn. (1.2-9) by "dt,"

$$m\,(dV/dt) + mg = -v\,(dm/dt) \qquad (1.2.1\text{-}1)$$

The first term represents the product of rocket mass and its acceleration (the rocket force), and the second term is the weight of the rocket. The sum of these two terms is equal to the thrust of the rocket, -v (dm/dt); here, $v>0$ and $(dm/dt) <0$. So the rocket thrust, F, is,

$$F\ (\text{rocket thrust}) = -v\,(dm/dt)$$

And

Isp = [mi=1lbm, mf=0 lbm] ∫ F dt (see below for the definition of the symbol of [to=0, tf=t] ∫ F dt)
= [mi=1lbm, mf=0 lbm] ∫ -v(dm/dt) dt
= - [mi=1lbm, mf=0 lbm]∫ v(dm)
= - [mi=1/g, mf=0 lbm] ∫ v(dm) (see below for derivation of mi=1/g)

---

1) [ti =A, t2=B] ∫ f(t) dt means integration of f(t) from A to B

2) F = (m) (a) lbf = m slugs * a ft/sec^2

Now we can write,
1.0 lbf = 1 slugs * 1 ft/sec^2
= (1/32.174)* 32.174 slugs * 1 ft/sec^2
= (1/32.174) slugs * 32.174 ft/sec^2 (1.2.1-1)

In other expression,

1.0 lbf = 1.0 lbm * 32.174 ft/sec^2 (1.2.1-2)

From Eqn. (1.2.1-1) and Eqn. (1.2.1-2),

1.0 lbm = (1/g) slugs

Isp = - v [mi=1/g, mf=0] (m), v: constant
= - v (0.0 - 1/g)
= v/g (1.2.1-3)

The equation shows that the longer the Isp is, the larger the relative velocity will be, which means that fuel of longer Isp achieves better performance. For example, if Isp = 200 sec., it means that 1 lbm fuel produces 1.0 lbf thrust for 200 seconds, which implies that if we burn the same fuel of 1.0 lbm for 1 second, then we can generate 200 lbf thrust for 1.0 second.

1.2.2 Specific Impulse Parametric Study

In order to express the final time in terms of the specific impulse, the rocket thrust term is integrated from "to" to "tf."

[t =to, t =tf] ∫ F dt = -v [t =to, t =tf] ∫ (dm/dt)dt
F [t =to, t =tf] ∫ dt = -v [m1=mo, m2=mf] ∫ (dm)
F (tf - to) = -v (mf – mo)

Assuming to=0,

tf = -(v/F) (mf – mo)
= -(vmo/F)[(mf/mo) – 1]
= - [(g)(v/g)mo/F] [(mf/mo)-1]

Using Eqn. (1.2.1-1),

$$tf = -[(Isp)gmo/F] [(mf/mo)-1]$$

The term, "gmo/F," is the ratio of the rocket weight and the rocket thrust at the beginning of a flight. Defining

$$\Re = F/gmo,$$

"tf" can be written as,

$$tf = (Isp)(1/\Re)[1-(mf/mo)] \qquad (1.2.2\text{-}2)$$

Substituting Eqn. (1.2.2-2) into Eqn. (1.2-1).

$$\begin{aligned} Vf - Vo &= v \ln(mo/mf) - g(Isp)(1/\Re)[1-(mf/mo)] \text{ , to=0} \\ &= g(v/g) \ln(mo/mf) - g(Isp)(1/\Re)[1-(mf/mo)] \\ &= g \, Isp \ln(mo/mf) - g \, Isp \, (1/\Re)[1-(mf/mo)] \\ &= g \, Isp\{ \ln(mo/mf) - (1/\Re)[1-(mf/mo)]\} \end{aligned} \qquad (1.2.2\text{-}3)$$

Eqn. (1.2.2-3) relates several interesting rocket parameters: Isp, $\Re$, mf/mo, and Vf. The usefulness of this equation will be demonstrated below.

The initial velocity of a rocket at a launch pad at the equator can be computed as shown below.

Mean equatorial radius	= 2.092567257e+07 ft
Earth rate of rotation	= 0.0000729 rad/sec
Initial velocity	= 1,525.5 ft/sec

Throughout the case study in the following, Vo (initial velocity) is assumed to be 1,525.5 ft/sec.

1) $\Re$ =1.0 (thrust to weight ratio at t=0)

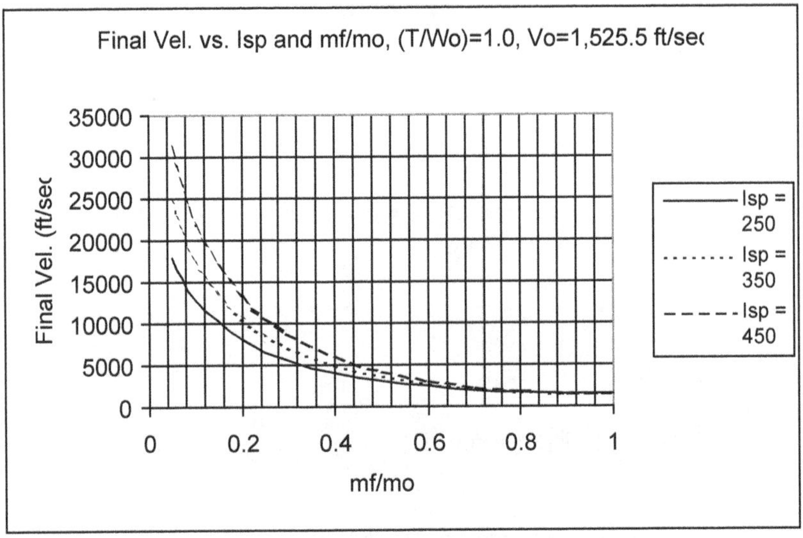

Fig, 1.2.2-1 Rocket Final Velocity as a Function of Isp and Mass Ratio, $\Re$ =1.0

This is the case in which the initial thrust to rocket weight ratio is equal to 1.0. In this case, the rocket does not move from the launch pad at t=0.0+, but does move in inertial space. However, as the fuel is being consumed, the ratio of mf/mo decreases and gradually the rocket takes off and the velocity increases. The final velocities as the mass ratio(mf/mo) reduces to 0.1 (almost the end of the flight) from 1.0 (at launch pad) is tabulated for three ISPs when $\Re$ =1.0.

Isp (s)	mf/mo	Vf (ft/sec)
250	0.1	12,807.19
350	0.1	17,319.87
450	0.1	21,832.55

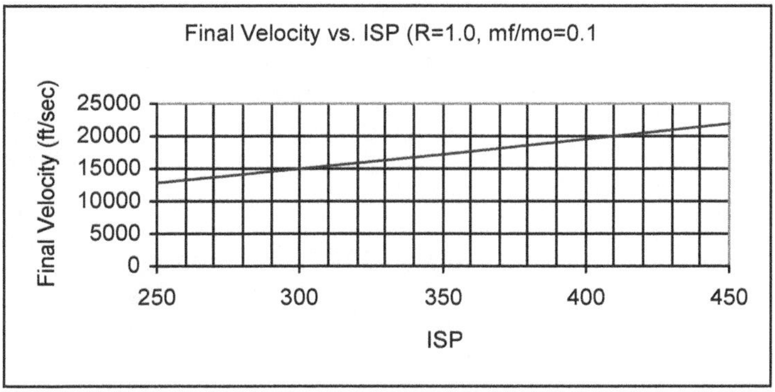

Fig. 1.2.2-2 Final Velocity vs. ISP (mf/mo=1)

2) $\Re$ =1.2 (thrust to weight ratio at t=0)

The initial ratio of the thrust and the rocket weight of the heavy vehicles currently being flown is about 1.2. For a heavy vehicle, it would be difficult for the ratio of mf/mo to reach less than 0.2 at the end of the flight. For Isp = 450, the final velocity that can be obtained when mf/mo=0.2 at the end of the flight is 15,175.22 ft/sec.

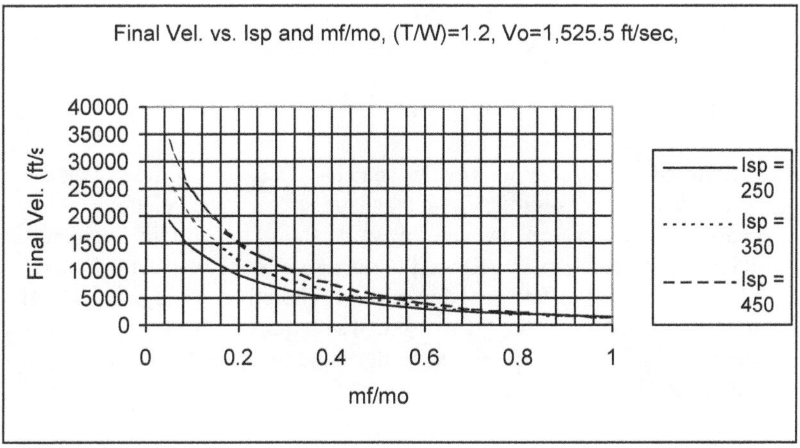

Fig, 1.2.2-3 Rocket Final Velocity as a Function of Isp and Mass Ratio, $\Re$ =1.2

The velocity when the mass ratio(mf/mo) decreases to 0.2 (almost at the end of the flight) from 1.0 (prior to launch) is tabulated for three ISPs when $\Re$ =1.2.

Isp (sec)	mf/mo	Vf (ft/sec)
250	0.2	9,108.68
350	0.2	12,141.95
450	0.2	15,175.22

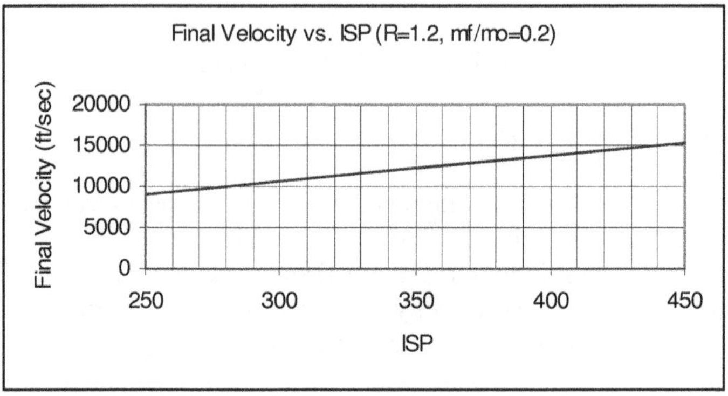

Fig. 1.2.2-4 Final Velocity vs. ISP (R=1.2, mf/mo=0.2)

3) $\Re$ =2.0 (thrust to weight ratio at t=0)

This high ratio is possible for smaller vehicles. Fig. 1.2.2-5 shows that the rocket velocity increases quickly, and as Isp increases, the velocity increases more quickly. However, this study shows that the final velocity could not reach the velocity required to put a satellite into an earth-bound orbit even when the mf/mo gets down to 0.2. Therefore, a two-stage rocket is needed to place a satellite on an orbit. The required velocity is approximately 25,000 ft/sec. for an earth-bound orbit and approximately 35,000 ft/sec. for a lunar mission.

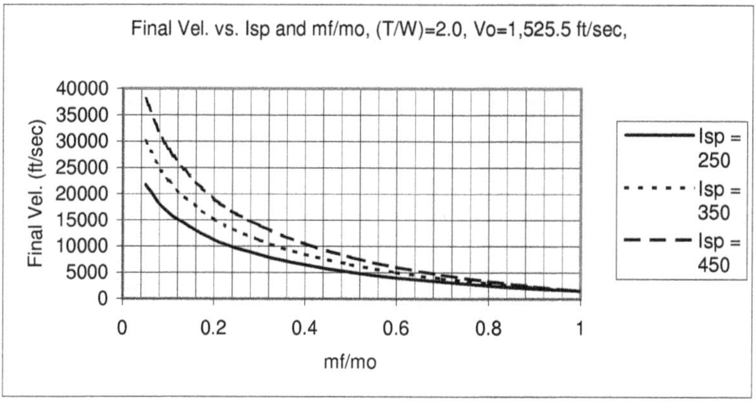

Fig, 1.2.2-5 Rocket Final Velocity as a Function of Isp and Mass Ratio, $\Re$ =2.0

The final velocities as the mass ratio (mf/mo) decreases to 0.2 (almost the end of the flight) from 1.0 (before launch) are tabulated for three ISPs when $\Re$ =2.0.

Isp (s)	mf/mo	Vf (ft/sec)
250	0.2	11,253.61
350	0.2	15,144.86
450	0.2	19,036.10

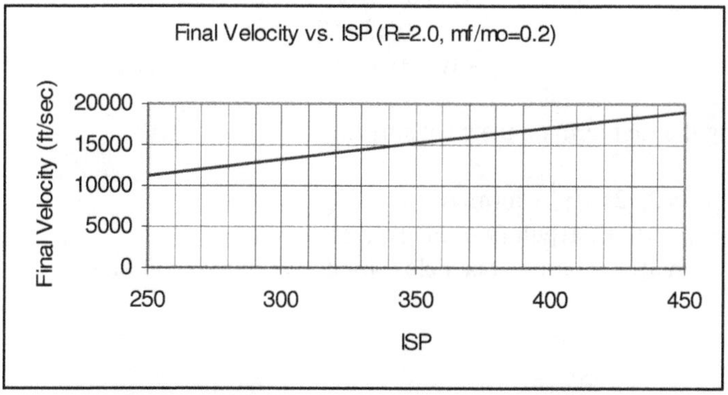

Fig. 1.2.2-6 Final Velocity vs. ISP (R=2, mf/mo=0.2)

If a 10,000 lbm satellite is to be placed in an orbit using a 100,000-lbm launch vehicle, assuming 80% of the total mass is fuel, then,

$$mf/mo = 20{,}000\ [=10{,}000\ (\text{satellite}) + 10{,}000\ (\text{structure})] / 100{,}000 = 0.2$$

Usually the F/W ratio is about 1.2. Now, if Isp is 450 sec., then the final velocity will be about 15,000 ft/sec. (Fig. 1.2.2-4), which is only 60% of the desired velocity of 25,000 ft/sec. Therefore, we need the second stage to achieve the velocity of 25,000 ft/sec.

1.2.3 Range Computation

In this derivation, it is assumed that the fuel consumption rate remains constant.

$$m' = (mo - mf)/tf = \text{constant}$$

From Eqn. (1.2-1),

$$V = v \ln(mo/mf) - gt + Vo, \quad to=0.0,\ t=tf,\ \text{and}\ V=Vf$$

Integrating the equation above from t = 0,

$$[t1 = 0, t2=tf] \int V\,dt = [t1 = 0, t2= tf \int (v \ln(mo/mf) - gt + Vo)\,dt = [t1 = 0, t2= tf] \int v \ln(mo/mf)dt - [t1 = 0, t2= tf] \int gt\,dt + [t1 = 0, t2=tf] \int Vo\,dt \qquad (1.2.3\text{-}1)$$

Integrating the first term on the right side first,

$$[t1 = 0, t2= tf] \int v \ln(mo/mf)dt = [t1 = 0, t2=tf] v \int \ln(mo/mf) \{mf^{(2)}/[m' (mo - mf)]\} d(mo/mf) \quad \text{(See below for derivation.)} \qquad (1.2.3\text{-}2)$$

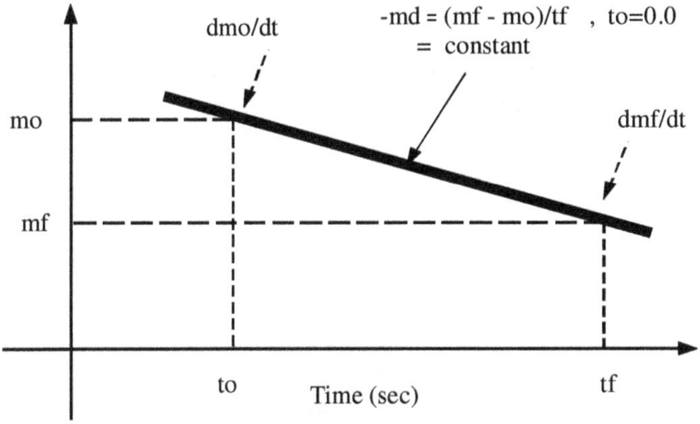

Fig. 1.2.3-1 Mass Reduction as a Function of Time

$$
\begin{aligned}
d(mo/mf)/dt &= d(mo\ mf^{\wedge}(-1))/dt \\
&= (dmo/dt)mf^{\wedge}(-1)\ +(-1)mo\ mf^{\wedge}(-2)dmf/dt
\end{aligned}
$$

From Fig. 1.2.3-1 above,

$$dm/dt = dmo/dt = dmf/dt = -m'$$

Thus, we can write,

$$
\begin{aligned}
d(mo/mf)/dt &= (dm/dt)mf^{\wedge}(-1) - mo{*}mf^{\wedge}(-2)dm/dt \\
&= (dm/dt)[\ mf^{\wedge}(-1) - mo{*}mf^{\wedge}(-2)] \\
&= m'\ \{[mf^{\wedge}(-2)](mf - mo)\}
\end{aligned}
$$

Rearranging the above equation for dt,

$$
\begin{aligned}
dt &= d(mo/mf)/\{m'\ \{[mf^{\wedge}(-2)](mf - mo)\}]\} \\
&= d(mo/mf)/\{[mf^{\wedge}(-2)]\ m'\ (mf - mo)\} \\
&= d(mo/mf)[mf^{\wedge}(2)]/[m'\ (mf - mo)] \\
&= \{[mf^{\wedge}(2)]/[m'\ (mo - mf)]\}\ d(mo/mf)
\end{aligned}
$$

Now substitute "dt" into Eqn. (1.2.3-2)

= v [t1=1, t2=r] ∫[(mf^2/(m'(mo- mf))]ln(mo/mf) d(mo/mf)

=v [t1=1, t2=r] ∫[(mf^2/(m'(mo- mf))] ln(mo/mf) d(mo/mf)

Substituting m'= (mf-mo)/tf

=v [t1=1, t2=r] ∫[(mf^2 tf/((mf-mo)(mo- mf))] ln(mo/mf) d(mo/mf)

Substituting tf = (Isp)(1/ℜ)[1-(mf/mo)] [see Eqn. (1.2.2-1)]

= v (Isp)(1/ℜ) [t1=1, t2=r] ∫[(mf^2 [1-(mf/mo)] /((mf-mo)(mo- mf))] ln(mo/mf)d(mo/mf)

= v (Isp)(1/ℜ) [t1=1, t2=r] ∫[(mf^2 [(mo-mf)/mo)] /((mf-mo)(mo- mf))] ln(mo/mf)d(mo/mf)

Canceling "(mo-mf)" from the numerator and denominator,

= v (Isp)(1/ℜ) [t1=1, t2=r] ∫{mf^2 /[mo(mf-mo)]} ln(mo/mf)d(mo/mf)

= v (Isp)(1/ℜ) [t1=1, t2=r] ∫{mf^2 /[mo mf(1-mo/mf)]} ln(mo/mf)d(mo/mf)

Canceling "mf" from the numerator and denominator,

= v (Isp)(1/ℜ) [t1=1, t2=r] ∫{(mf/mo) /(1-mo/mf)]} ln(mo/mf)d(mo/mf)

Now let X=(mf/mo), then

= v (Isp)(1/ℜ) [t1=1, t2=r] ∫{(X) /(1-1/X)]} ln(1/X)d(1/X)

= v (Isp)(1/ℜ) [t1=1, t2=r] ∫{(X) /(1-1/X)]} ln(1/X)[-X^(-2)]dX

= v (Isp)(1/ℜ) [t1=1, t2=r] ∫[-1/(X-1)] ln(1/X)dX

= v (Isp)(1/ℜ) [t1=1, t2=r] ∫[-1/(X-1)][- ln(X)]dX

= v (Isp)(1/ℜ) [t1=1, t2=r] ∫[1/(X-1)]ln(X)dX

(1.2.3-3)

Expanding ln(X),

ln(X) = 2{(X-1)/(X+1) + (X-1)^(3)/[3(X+1)^(3)] + (X-1)^(5)/[5(X+1)^(5)] + . . .}

Since 1.0 > X > 0.0, ln(X) can be approximated as,

ln(X) ≈ 2{(X-1)/(X+1) + (X-1)^(3)/[3(X+1)^(3)]}

Then, Eqn. (1.2.3-3) can be written as,

[t1 = 0, t2= tf] ∫ v ln(mo/mf)dt
=v (Isp)(1/ℜ) [t1=1, t2=r] ∫{[1/(X-1)] 2{(X- 1)/(X+1) + (X-1)^(3)/[3(X+1)^(3)]}dX

Canceling "(X- 1)" from the numerator and denominator,
= 2v (Isp)(1/ℜ) [t1=1, t2=r] ∫ {1/(X+1) + (X-1)^(2)/[3(X+1)^(3)]}dX

Integrating the first term,

[t1=1, t2=r] ∫ 1/(X+1)dX = [t1=1, t2=r] ln |X+1|
= ln |r+1| - ln |1+1|
= ln |r+1|-0.693

the second term,

[t1=1, t2=r] ∫ {(X- 1)^(2)/[3(X+1)^(3)]}dX
= (1/3){ [t1=1, t2=r] ∫ {(X^(2)/[(X+1)^(3)]}dX
- [t1=1, t2=r] ∫ {2X/[(X+1)^(3)]}dX
+[t1=1, t2=r] ∫ {1/[(X+1)^(3)]}}dX

=(1/3){[t1=1, t2=r] { ln| X+1| + 2/(X+1) – 1/[2(X+1)^(2)]}
- [t1=1, t2=r] 2{-1/(X+1) + 1/[2(X+1)^(2)]}
+ [t1=1, t2=r] {-1/[2(X+1)^(2)]}

= (1/3){{ln| r+1| + 2/(r+1) – 1/[2(r+1)^(2)]}- { ln| 1+1| + 2/(1+1) – 1/[2(1+1)^(2)]} – 2 {-1/(r+1) + 1/[2(r+1)^(2)]}+2{-

1/(1+1) + 1/[2(1+1)^(2)]}+ {-1/[2(r+1)^(2)]}- {-1/[2(1+1)^(2)]}}

= (1/3){{ln| r+1| + 2/(r+1) – 1/[2(r+1)^(2)]} – 2 {-1/(r+1) + 1/[2(r+1)^(2)]}+{-1/[2(r+1)^(2)]} -2.191}

= 2v (Isp)(1/ℜ) {ln |r+1|-0.693 + (1/3){{ln| r+1| + 2/(r+1) - 1/[2(r+1)^(2)]} - 2 {-1/(r+1) + 1/[2(r+1)^(2)]}+{-1/[2(r+1)^(2)]} -2.191}

=(8/3)v (Isp)(1/ℜ){ ln |r+1| + [1/(r+1)] – 1/[2(r+1)^(2)] – 1.06725}

(1.2.3-4)

Now integrating the second term of Eqn. (1.2.3-1),

[t1 =0, t2= tf] ∫ gt dt = (1/2)g tf^(2)

Substituting tf = (Isp)(1/ℜ)[1-(mf/mo)] [see Eqn. (1.2.2-2)],

=(1/2)g{(Isp)(1/ℜ)[1-(mf/mo)] }^(2)
=(1/2)g{(Isp)(1/ℜ)[1-r] }^(2)

(1.2.3-5)

Integrating the third term of Eqn. (1.2.3-1),

[t1 =0, t2= tf] ∫ Vo dt = Vo tf
= Vo (Isp)(1/ℜ)[1-(mf/mo)]
= Vo (Isp)(1/ℜ)[1-r] (1.2.3-6)

The integration of Eqn. (1.2.3-1) is then the sum of Eqn. (1.2.3-4), Eqn. (1.2.3-5), and Eqn. (1.2.3-6).

hf (traveled distance)= [t1 =0, t2=tf] ∫ V dt
=(Isp)(1/ℜ){(8/3)v { ln |r+1| + [1/(r+1)] - 1/[2(r+1)^(2)] -1.06725} + (1/2)g(Isp)(1/ℜ)[1-r]^(2) + Vo(1-r)}

(1.2.3-7)

where

Isp = specific impulse
$\Re$ = T/gmo
v = - (-u + V)
r = ratio of final mass over initial mass [(mf/mo)≤ 1.0] at
t = tf

Now the rocket velocity and distance traveled can be computed by using Eqn. (1.2.2-3) and Eqn. (1.2.3-7), respectively.

At the end of the powered flight, the rocket will coast in free flight mode. In this mode, gravity will be the only force affecting the flight. In order to compute the distance over which the rocket will continue to fly, the work done by gravity is set equal to the kinetic energy dissipation from the end of the powered flight.

1) Kinetic energy dissipation (force times distance) from tf to tend,

[r=r1, r=r2] ∫ F dr = [t=t1, t=t2] ∫ m(dV/dt)Vdt
= [t=t1, t=t2] ∫ m((dV/dt)V)dt
= [t=t1, t=t2](1/2)∫ m(d(V*V)/dt)dt

----

d(V*V)/dt = V(dV/dt) + (dV/dt)V
= 2.0 (dV/dt)V

(dV/dt)V = (1/2)* d(V*V)/dt

= [V=V1, V=V2](1/2)∫ m(d(V^(2))
= (1/2)m [V=V1, V=V2] (V^(2))
= (1/2)m[V2^(2) – V1^(2)] (1.2.3-8)

2) Work done by gravity

Here the altitude is high enough such that gravity variation cannot be ignored,

gr = g (R/r)^2

a simplified form, where R is the earth's radius.

[r =r1, r=r2]∫m gr dr = [r =r1, r=r2]∫m g(R/r)^2 dr
= mgR^(2) [r =r1, r=r2]∫ r^(-2)dr
= mgR^(2) [r =r1, r=r2](-r^(-1))

$$= mgR^{2}[-r2^{-1} + r1^{-1}]$$

Since r2=r1+rco, (rco: coasting distance)

$$= mgR^{2}[\ 1/r1 - 1/(r1+rco)]$$
$$= mgR^{2}\{(r1+rco-r1)/[r1(r1+rco)]\}$$
$$= mgR^{2}\{rco/[\ r1(r1+rco)]\} \quad (1.2.3\text{-}9)$$

Now equating Eqns. (1.2.3-8) and (1.2.3-9),

$$mgR^{2}\{rco/[\ r1(r1+rco)]\} = (1/2)m[V2^{2} - V1^{2}]$$

Solving for rco,

$$rco= Qr1^{2}/(gR^{2} - Q\ r1), \qquad (1.2.3\text{-}10)$$

where

$$Q=(1/2)V2^{2} - V1^{2}]$$

Notice that rco is independent of mass.

Therefore, total distance can be obtained by adding hf from Eqn. (1.2.3-7) and rco from Eqn. (1.2.3-10).

$$ht = hf\ + rco \qquad (1.2.3\text{-}11)$$

To study the sensitivity of the final velocity or distance with respect to Isp, $\mathfrak{R}$, or (mf/mo), partial differentiations with respect to these variables are derived. From Eqn. (1.2.2-3),

$$Vf\ = fv(Isp, \mathfrak{R}, (mo/mf)\ )$$
$$= g\ Isp\{\ \ln(mo/mf) - (1/\mathfrak{R})[1-(mf/mo)]\}$$
$$= g\ Isp\ [\ \ln(1/rfo) - (1/\mathfrak{R})(1-rfo)], \qquad rfo=mf/mo$$

Final velocity sensitivity,

$$dVf= (\partial Vf/\partial Isp)Isp + (\partial Vf\ /\partial \mathfrak{R})\mathfrak{R} + \ (\partial Vf\ /\partial rfo)rfo$$

Final distance sensitivity,

$$dhf = (\partial hf/\partial Isp)Isp + (\partial hf/\partial \mathfrak{R})\mathfrak{R} + (\partial hf/\partial rfo)rfo$$

The sensitivity coefficients are:

1) $(\partial Vf/\partial Isp)$

$$(\partial Vf/\partial Isp) = g\,[\ln(1/rfo) - (1/\mathfrak{R})(1-rfo)] \quad (1.2.3\text{-}12)$$

2) $(\partial Vf/\partial \mathfrak{R})$

$$(\partial Vf/\partial \mathfrak{R}) = g\,Isp\,(1/\mathfrak{R}^{(2)})(1-rfo) \quad (1.2.3\text{-}13)$$

3) $(\partial Vf/\partial rfo)$

$$(\partial Vf/\partial rfo) = g\,Isp\,[rfo + (1/\mathfrak{R})] \quad (1.2.3\text{-}14)$$

4) $$(\partial hf/\partial Isp) = (1/\mathfrak{R})\{(8/3)v\,\{\ln|rfo+1| + [1/(rfo+1)] - 1/[2(rfo+1)^{(2)}] - 1.06725\} + Vo(1-rfo)\} + g(Isp)(1/\mathfrak{R})^{(2)}[1-rfo]^{(2)}$$

(1.2.3-15)

5) $$(\partial hf/\partial \mathfrak{R}) = (-Isp/\mathfrak{R}^{(2)})\{(8/3)v\,\{\ln|rfo+1| + [1/(rfo+1)] - 1/[2(rfo+1)^{(2)}] - 1.06725\} + Vo(1-rfo)\} + (1/2)g(Isp)^{(2)}(1/\mathfrak{R})^{(2)}[1-rfo]^{(2)}$$

(1.2.3-16)

6) $$[\partial hf/\partial(rfo)] = (Isp)(1/\mathfrak{R})\{(8/3)v\,[(rfo+1)^{(-1)} - (rfo+1)^{(-2)} + (rfo+1)^{(-3)}] - g(Isp)(1/\mathfrak{R})[1-rfo] - Vo)\}$$

(1.2.3-17)

1.2.4 Optimal Rocket Sizing for Multi-Stage Rockets

From Eqn. (1.2-1),

$$Vf - Vo = v\ln(mo/mf) - g\,tf,\; to=0.0$$
$$Vf = v\ln(mo/mf) - g\,tf + Vo$$

For stage 1,

$$Vf1 = v1 \ln(mo1/mo2) - g\, tf1 + Vo1 \qquad (1.2.4\text{-}1)$$

For stage 2,

$$Vf2 = v2 \ln(mo2/mo3) - g\, tf2 + Vf1 \qquad (1.2.4\text{-}2)$$

Substituting "Vf1" from Eqn. (1.2.4-1),

$$Vf2 = v2 \ln(mo2/mo3) + v1 \ln(mo1/mo2) - g\, tf1 - g\, tf2 + Vo1$$

For multi-stage rockets, an equation shown below can be used.

$$Vfn = [i=1, i=n]\Sigma vi \ln[moi/mo(i+1)] - g[i=1,in]\ \Sigma\ tfi + Vo1$$

$$(1.2.4\text{-}3)$$

The term "Vfn" is the final velocity of a rocket, and in order to place a satellite on an orbit, we design the rocket system such that Vfn meets the velocity requirement for that particular orbit. Once we specify the velocity requirement, then we determine the optimal number of stages and the optimal ratio of [moi/mo(i+1)] for each i that minimizes (mo1/P), where P is the weight of a payload (satellite).

Consider the first term on the right-hand side of Eqn. (1.2.4-3),

$$Vfn = [i=1,i=n]\Sigma vi \ln[moi/mo(i+1)] + Vo1 \qquad (1.2.4\text{-}4)$$

Let's define mi = moi/mo(i+1),

$$Vfn = [i=1,i=n]\Sigma vi \ln(mi) + Vo1$$

Now we write,

$$\frac{mo1}{P} = \frac{mo1}{mo2} * \frac{mo2}{mo3} * \frac{mo3}{mo4} * \text{----} * \frac{mo(n)}{P}$$

where P is payload mass.

$$= \frac{mo1}{(mo1-mf1-ms1)} \frac{mo2}{(mo2-mf2-ms2)} \frac{mo3}{(mo3-mf3-ms3)}$$

$$\ldots * \frac{mo(n-1)}{[mo(n-1) - mf(n-1) - ms(n-1)]} \frac{mo(n)}{[mo(n)-mf(n)-ms(n)]}$$

(1.2.4-5)

where

mo2 = (mo1-mf1-ms1)
mo3 = (mo2-mf2-ms2)
. . . .
p = [mo(n)-mf(n)-ms(n)], e.g. mo(1)= mo1

mo (i) : total mass at the beginning of stage i
mf (i) : fuel mass in stage i
ms (i) : structural mass separated after stage i

e.g. At the launch pad, mo1 = 100
mf1 = 50
ms1 = 7
m02 = mo1-mf1-ms1
= 43

Expanding the first term,

$$\frac{mo1}{mo1-mf1-ms1}. = \frac{mo1}{(mo1-mf1-ms1)}. \frac{(mo1-mf1)(mf1+ms1)(mf1)}{(mo1-mf1)(mf1+ms1)(mf1)}$$

$$= \frac{mo1}{(mo1-mf1)} \frac{mf1}{(mf1+ms1)} \frac{1}{(mf1)} \frac{(mo1-mf1)(mf1+ms1)}{(mo1-mf1-ms1)}$$

****

Expanding the denominators of the last two terms,

(mf1)(mo1-mf1-ms1)
= mf1* mo1- mf1*mf1- mf1*ms1

Adding "(mo1*ms1 – mo1*ms1)=0.0"
= mf1* mo1- mf1*mf1- mf1*ms1 + mo1*ms1 –
mo1*ms1

Rearranging the equation above

$$= mf1*mo1 + mo1*ms1 - mf1\hat{}(2) - mf1*ms1 - mo1*ms1$$
$$= mo1\,(mf1 + ms1) - mf1(ms1 + mf1) - mo1*ms1$$
$$= (mf1 + ms1)(mo1 - mf1) - mo1*ms1$$

****

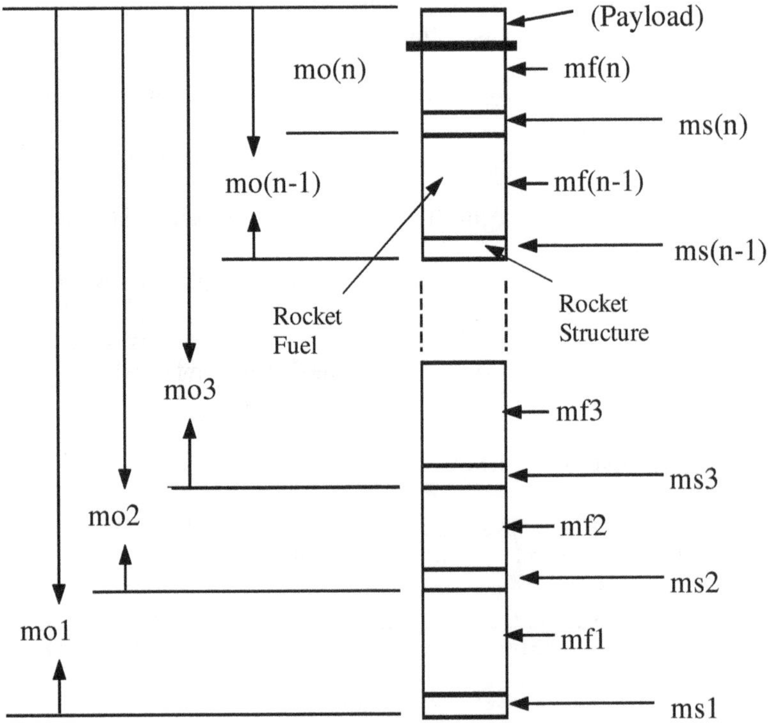

Fig. 1.2 .4-1 Multi Stage Rocket Mass Distribution

Substituting $(mf1)(mo1-mf1-ms1) = (mf1 + ms1)(mo1 - mf1) - mo1*ms1$ and dividing numerator and denominator by "[(mo1-mf1)(mf1+ms1) (mf1+ms1)],"

$$= \frac{[\,(mo1)\ (mf1)\ (mo1-mf1)\ (mf1+ms1)\,]/[\,(mo1-mf1)\ (mf1+ms1)(mf1+ms1)]}{\{(mo1-mf1)(mf1+ms1)[(mf1+ms1)(mo1-mf1)-mo1*ms1]\}/[\,(mo1-mf1)(mf1+ms1)(mf1+ms1)]}$$

Canceling "(mo1-mf1) (mf1+ms1)" in the numerator and "(mf1+ ms1)" in the denominator,

$$= \frac{(mo1)\ (mf1)/(mf1+ms1)}{\{(mo1-mf1)[(mf1+ms1)(mo1-mf1)-mo1*ms1]\}/[(mo1-mf1)(mf1+ms1)]}$$

A part of the numerator = (mf1)/(mf1+ms1)
= (mf1 –ms1 +ms1)/(mf1+ms1)
= [(mf1 + ms1) - ms1]/(mf1+ms1)
= 1 - ms1/(mf1+ms1)

A part of the denominator =[(mf1+ ms1)(mo1- mf1)-mo1*ms1]}/[(mo1-mf1)(mf1+ms1)]
= 1 – (mo1)(ms1)/[(mo1-mf1)(mf1+ms1)]
=1 – [mo1/(mo1-mf1)][ms1/(mf1+ms1)]

$$= \frac{(mo1)\quad [1 - ms1/(mf1+ms1)]}{\{(mo1-mf1)\quad \{ 1 - [mo1/(mo1-mf1)][ms1/(mf1+ms1)]\}}$$

Now defining,

$$\alpha(i) = moi/(moi-mfi) \qquad (1.2.4\text{-}6)$$
$$\beta(i) = msi/(mfi+msi) \qquad (1.2.4\text{-}7)$$

Substituting them,

$$\frac{mo1}{(mo1-mf1-ms1)} = \alpha(1)\ [1 - \beta(1)]/[1 - \alpha(1)\ \beta(1)]$$

Now Eqn. (1.2.4-5) can be rewritten as,

$$mo1/P = \{\alpha(1)\ [1 - \beta(1)]/[1 - \alpha(1)\ \beta(1)]\}\{ \alpha(2)\ [1 - \beta(2)]/[1 - \alpha(2)\ \beta(2)]\}\{\alpha(3)\ [1 - \beta(3)]/[1 - \alpha(3)\ \beta(13)]\}$$
$$\ldots$$
$$\{\alpha(n)\ [1 - \beta(n)]/[1 - \alpha(n)\ \beta(n)]\}$$

$$= [i=1, i=n]\ \Pi\ \{\alpha(i)\ [1 - \beta(i)]/[1 - \alpha(i)\ \beta(i)]\} \qquad (1.2.4\text{-}8)$$

The desire is to launch a satellite into an orbit with minimum mass, which is cost effective. To meet this goal, "mo1/P" must be minimized. To derive an equation that minimizes the ratio, the natural logarithm (ln) is taken on both sides of the Eqn. (1.2.4-8).

$$\begin{aligned} \ln(mo1/P) &= \ln\ [i=1, i=n]\ \Pi\ \{\alpha(i)\ [1 - \beta(i)]/[1 - \alpha(i)\ \beta(i)]\} \\ &= [i=1, i=n]\ \Sigma\ \ln\{\alpha(i)\ [1 - \beta(i)]/[1 - \alpha(i)\ \beta(i)]\} \\ &= [i=1, i=n]\ \Sigma\ \{\ln\alpha(i) + \ln[1 - \beta(i)] - \ln[1 - \alpha(i)\ \beta(i)]\} \end{aligned}$$

Now a Lagrange multiplier "λ" is multiplied to Eqn. (1.2.4-3), and the product is added to the equation above, ignoring the gravity term. The Eqn. (1.2.4-3) is used as a constraint and characterizes the rocket velocity profile.

$$\begin{aligned} \ln(mo1/P) &= [i=1, i=n]\ \Sigma\{\ln\alpha(i) + \ln[1 - \beta(i)] - \ln[1 - \alpha(i)\ \beta(i)]\} + \\ &\quad \lambda\{[i=1,i=n]\Sigma vi\ \ln[moi/mo(i+1)] + Vo1 - Vfn\} \\ &= [i=1, i=n]\ \Sigma\{\ln\alpha(i) + \ln[1 - \beta(i)] - \ln[1 - \alpha(i)\ \beta(i)] + \\ &\quad \lambda\ vi\ \ln[moi/(moi-mfi)]\ \} + \lambda(Vo1 - Vfn) \\ &= [i=1, i=n]\ \Sigma\{\ln\alpha(i) + \ln[1 - \beta(i)] - \ln[1 - \alpha(i)\ \beta(i)] + \\ &\quad \lambda\ vi\ \ln\alpha(i)\ \} + \lambda(Vo1 - Vfn),\ \ \alpha(i) = [moi/(moi-mfi)] \end{aligned}$$

Differentiating the equation above with respect to $\alpha(i)$ and set it equal to zero for minimization,

$$\begin{aligned} [d\ln(mo1/P)/d(\ln\alpha(i))] &= [i=1, i=m]\ \Sigma\{1/\alpha(i) + \beta(i)/[1 - \alpha(i)\beta(i)] + \\ &\quad \lambda\ vi/\alpha(i)\ \} \\ &= 0.0 \end{aligned} \qquad (1.2.4\text{-}9)$$

From Eqn. (1.2.4-9),

$$\{1/\alpha(i) + \beta(i)/[1 - \alpha(i)\beta(i)] + \lambda\ vi/\alpha(i)\} = 0 \text{ for all i.}$$

Solving for $\alpha(i)$,

$$\alpha(i) = (1 + \lambda\ vi)/\ [\lambda\ vi\ \beta(i)] \qquad (1.2.4\text{-}10)$$

Rewriting Eqn. (1.2.4-3) and ignoring the gravity term,

$$Vfn = [i=1, i=n]\Sigma vi \ln[moi/mo(i+1)] + Vo1 \qquad (1.2.4\text{-}3)$$

and Eqn. (1.2.4-6),

$$\alpha(i) = moi/(moi-mfi) \qquad (1.2.4\text{-}6)$$

Here, the term, mo(i+1) in Eqn. (1.2.4-3) means the rocket mass after the fuel burned out and the (moi-mfi) in Eqn. (1.2.4-6) also means the rocket mass after the fuel burned out. Thus, from Eqn. (1.2.4-6) and Eqn. (1.2.4-10),

$$moi/mo(i+1) = (1 + \lambda\ vi)/ [\lambda\ vi\ \beta(i)]$$

and,

$$Vfn = [i=1,i=n]\Sigma vi \ln((1 + \lambda\ vi)/ [\lambda\ vi\ \beta(i)]) + Vo1 \qquad (1.2.4\text{-}11)$$

Note that moi where i=1- n in Eqn. (1.2.4-5) is different from the moi in Eqn. (1.2.4-3). The moi in Eqn. (1.2.4-3) is the mass after the fuel burned out and the moi in Eqn. (1.2.4-5) is the mass after the fuel burned out and structural jettison.

An example is shown.

Let's assume that we have a two-stage rocket. The following data are given for the rocket.

Vo1 = 1,500 ft/sec

v1 = gIsp1
= (32.174 ft/sec^2)(400 sec)
= 12,870 ft/sec
$\beta(1)$ = ms1/(mf1+ms1) = 0.1

v2 = gIsp2
= (32.174 ft/sec^2)(380 sec)
= 12,226 ft/sec
$\beta(2)$ = ms1/(mf1+ms1) = 0.1

Now from Eqn. (1.2.4-11),

$$Vf2 = [i=1,i=2]\Sigma vi \ln((1 + \lambda\, vi)/ [\lambda\, vi\, \beta(i)]) + Vo1$$
$$=\{v1\ln((1 + \lambda\, v1)/ [\lambda\, v1\beta(1)])\} + \{v2 \ln((1 + \lambda\, v2)/ [\lambda\, v2\beta(2)]) + Vo1$$

$$Vf2 = 12{,}870 \ln\{(1+12{,}870\lambda)/[(12{,}870)(0.1)\lambda]\}+ 12{,}226 \ln\{(1+12{,}226\lambda)/[(12{,}226)(0.1)\lambda]\} + Vo1$$

Fig. 1.2.4-2 shows a plot of the above equation. If the required final velocity is 25,000 ft/sec, which is the approximate velocity required to place a satellite on a low, earth-bound orbit, then from the plot we can find that λ is equal to -0.000107. Now, α (1) can be computed.

$$\alpha (1) = (1 + \lambda\, v1)/[\lambda\, v1\beta (1)]$$
$$= (1 + (- 0.000107* 12{,}870)/(-0.000107*12870*0.1)$$
$$= 2.738$$

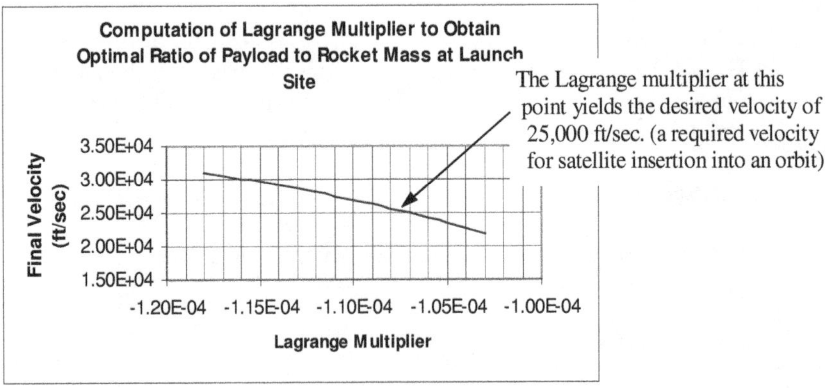

Fig. 1.2.4-2. Computation of Lagrange Multiplier

From Eqn. (1.2.4-6)

$$2.738 = mo1/(mo1 - mf1)$$

Defining mf1 = mo1*x1

$$2.738 = mo1/(mo1 - mo1*x1)$$

$= 1/(1 - x1)$

Solving the above equation for x1,

$$x1 = 0.635$$
$$= mf1/mo1 \quad \text{by the definition of x1 defined above.} \qquad (1.2.4\text{-}12)$$

This ratio is the ratio of fuel in the 1st stage alone to the total mass of the rocket (1st stage fuel and structure, and 2nd stage fuel and structure). That is, the mass of the fuel in the 1st stage is 63.5 % of the total mass at the launch site for an optimal flight.

From Eqn. (1.2.4-7),

$$\beta(1) = ms1/(mf1+ms1)$$

Defining mf1=ms1*y1,

$$\beta(1) = ms1/(mf1+ms1)$$
$$= ms1/(ms1*y1 + ms1)$$
$$= 1 / (y1 + 1)$$

It is given that $\beta(1) = 0.1$,

$$0.1 = 1 / (y1 + 1)$$

Solving for y1, we find that y1 = 9. By the definition of y1,

$$mf1/ms1 = 9.0.$$

The first stage fuel mass is 9.0 times larger than the structural mass. From Eqn. (1.2.4-12),

$$0.635 = mf1/mo1$$
$$= 9ms1/mo1$$

And

$$mo1/ms1 = 14.173.$$

The mass of the total rocket at the launch site is 14.173 times of the mass of the 1st stage structure.

In the 2nd stage,

$$\alpha\,(2) = [1 + (-0.000107*12{,}226)]/[(-0.000107*12{,}226*0.1)] = 2.356$$

From Eqn. (1.2.4-6)

$$2.356 = mo2/(mo2 - mf2)$$

Defining mf2 = mo2*x2

$$2.356 = mo2/(mo2 - mo2*x2) = 1/(1 - x2)$$

Solving the above equation for x2,

$$x2 = 0.575 = mf2/mo2 \quad \text{by the definition of x2 defined above.} \qquad (1.2.4\text{-}13)$$

This ratio is the ratio of mass of fuel in the second stage to the mass of the rocket's second stage. The mass of the fuel in the 2nd stage is 57.5 % of the mass of the total 2nd stage rocket.

From Eqn. (1.2.4-7),

$$\beta\,(2) = ms2/(mf2+ms2)$$

Defining mf2=ms2*y2,

$$\beta\,(2) = ms2/(mf2+ms2) = ms2/(ms2*y2 + ms2) = 1\,/\,(y2 + 1)$$

It is given that $\beta\,(2) = 0.1$,

$$0.1 = 1\,/\,(y2 + 1)$$

Solving for y2, we find that y2 = 9.0. By the definition of y2, mf2/ms2 = 9.0

The 2nd stage fuel mass is 9.0 times of the structural mass. From Eqn. (1.2.4-13),

0.575 = mf2/mo2
= 9ms2/mo2

And

mo2/ms2 = 15.652

The mass of the 2nd stage rocket is 15.652 times of the mass of the 2nd stage structure. Fig. 1.2.4-3 shows the mass ratios graphically. It is seen here that for optimal performance of satellite insertion into a low, earth-bound orbit, the rocket mass must be approximately 10 times the payload (satellite) mass. The ratio of the 1st stage to 2nd stage is approximately 6.36 to 3.64.

Lamda	Final vel.	Alpha1	mo1	mf1	y1	ms1	mf1/mo1
-0.000103	21895.24	2.46	10.00	5.93	9.00	0.66	0.59
-0.000104	22714.99	2.53	10.00	6.05	9.00	0.67	0.60
-0.000105	23493.63	2.60	10.00	6.15	9.00	0.68	0.62
-0.000106	24234.54	2.67	10.00	6.25	9.00	0.69	0.63
-0.000107	24940.72	2.74	10.00	6.35	9.00	0.71	0.63
-0.000108	25614.84	2.81	10.00	6.44	9.00	0.72	0.64
-0.000109	26259.26	2.87	10.00	6.52	9.00	0.72	0.65
-0.000110	26876.13	2.94	10.00	6.59	9.00	0.73	0.66
-0.000111	27467.36	3.00	10.00	6.67	9.00	0.74	0.67
-0.000112	28034.67	3.06	10.00	6.73	9.00	0.75	0.67

Lamda	Final vel.	Alpha 2	mo2	mf2	y2	ms2	mf2/mo2	Payload	mo2/mo1
-0.000103	21895.24	2.06	3.41	1.76	9.00	0.20	0.51	1.46	0.34
-0.000104	22714.99	2.14	3.28	1.75	9.00	0.19	0.53	1.34	0.33
-0.000105	23493.63	2.21	3.16	1.73	9.00	0.19	0.55	1.24	0.32
-0.000106	24234.54	2.28	3.05	1.71	9.00	0.19	0.56	1.15	0.31

-0.000107	24940.72	2.36	2.95	1.70	9.00	0.19	0.58	1.06	0.29
-0.000108	25614.84	2.43	2.85	1.68	9.00	0.19	0.59	0.99	0.28
-0.000109	26259.26	2.50	2.76	1.65	9.00	0.18	0.60	0.92	0.28
-0.000110	26876.13	2.56	2.67	1.63	9.00	0.18	0.61	0.86	0.27
-0.000111	27467.36	2.63	2.59	1.61	9.00	0.18	0.62	0.81	0.26
-0.000112	28034.67	2.70	2.52	1.58	9.00	0.18	0.63	0.76	0.25

Table 1.2.4-1 Optimal Values for Given Conditions

Table 1.2.4-1 provides optimal values for the given condition. It should be noticed that for the given rocket mass at the launch site, the payload mass reduces as the desired final velocity increases. The mo2/mo1 ratio also decreases as the final velocity increases. This implies that we need larger 1st stage to achieve higher final velocity. It should be also noticed that the fuel mass increases in both stages. The 1st stage structure mass increases but the 2nd stage structure mass decreases slightly.

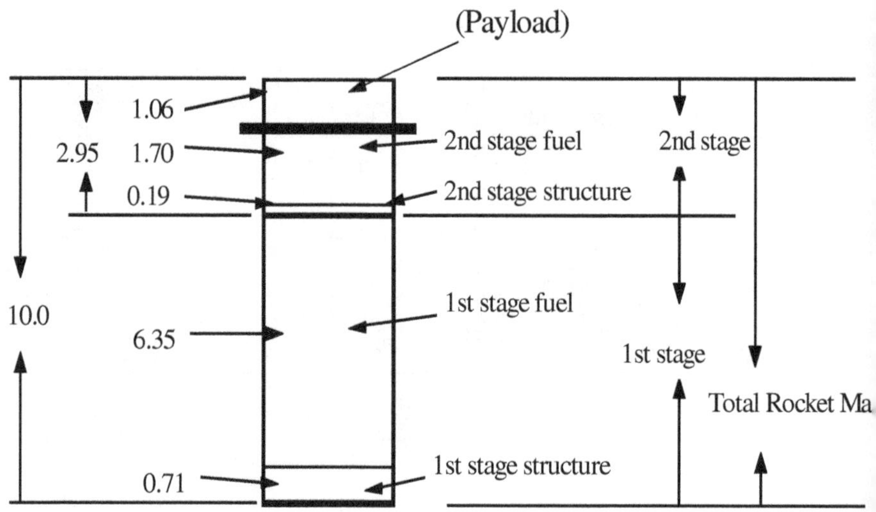

Fig. 1.2.4-3 Two Stage Rocket Mass Ratio

To check that the parameters obtained after optimization truly yields the desired final velocity, the following optimal values are substituted into Eqn. (1.2.4-4).

Vo = 1,500 ft/sec

Isp1 = 400 sec
Isp2 = 380 sec
g = 32.174 ft/sec^2
mo1 = 10.00
mf1 = 6.35
mo2 = 2.95
mf2 = 1.70

Solving for P in Eqn. (1.2.4-8)

$$P = mo1/[\{\alpha(1)\ [1 - \beta(1)]/[1 - \alpha(1)\,\beta(1)]\}\{\alpha(2)\ [1 - \beta(2)]/[1 - \alpha(2)\,\beta(2)]\}]$$
$$=10/\{[2.738(1-0.1)/(1-2.738*0.1)]*[2.356(1-0.1)/(1-2.356*0.1)]\}$$
$$=1.06 \text{ (Payload mass)}$$

$$Vf = (400)*(32.174)*\ln(10.0/(10.0-6.35)) + (380)*(32.174)*\ln(2.95/(2.95-1.70)) + 1{,}500$$
$$= 24{,}968.83 \text{ ft/sec.}$$

The final velocity computed is 0.125 % slower from 25,000 ft/sec. Now, to show that these are truly the optimal values, we vary these parameters while maintaining the desired velocity to determine whether the ratio of the payload mass to the rocket mass at the launch site is maximized for these optimal values. Here again, we use Eqn. (1.2.4-4) and vary α (1) and α (2) as shown in the table below.

The following equation is obtained from Eqn. (1.2.4-4). We use this equation to construct the table.

$$Vf = Isp1*g*\ln(mo1/(mo1-mf1)) + Isp2*g*\ln(mo2/(mo2-mf2))$$

Final vel.	Alpha1	Alpha 2	mo1	mf1	ms1	mf1/mo1	mo2	mf2	ms2	mf2/mo2	Payload	(mo2/mo1) (%)
24981.95	6.20	1.00	10	8.39	0.93	0.84	0.68	0.00	0.00	0.00	0.68	6.81
24984.32	3.21	2.00	10	6.88	0.76	0.69	2.35	1.18	0.13	0.50	1.04	23.50
24970.42	2.74	2.36	10	6.35	0.71	0.64	2.94	1.70	0.19	0.58	1.06	29.44
24971.58	1.66	4.00	10	3.98	0.44	0.40	5.58	4.19	0.47	0.75	0.93	55.82
24943.65	1.34	5.00	10	2.54	0.28	0.25	7.18	5.74	0.64	0.80	0.80	71.81
24978.99	1.13	6.00	10	1.15	0.13	0.12	8.72	7.27	0.81	0.83	0.65	87.22

Table 1.2.4-2 Maximum Payload Conditions

It is shown that the ratio of the payload to the rocket at the launch site is maximum when the parameters are optimal (underlined in the table). From the table, it is seen that

Alpha 1(α (1))	, Alpha 2(α (2))	I	
mf1	, mf2	I	are inversely related.
ms1	, ms2	I	

1.3 Flight Optimization

Optimizations of trajectory and fuel consumption are considered in this section. In section 1.3.1, a flight trajectory optimization algorithm is developed that maximizes the horizontal velocity. Two examples are presented to demonstrate how it works. In section 1.3.2, fuel optimization is briefly discussed.

1.3.1 Trajectory Optimization

It is assumed here that there is no aerodynamic force, and the flight occurs in a plane parallel to the gravity force field. The variables in Fig. 1.3.1-1 are:

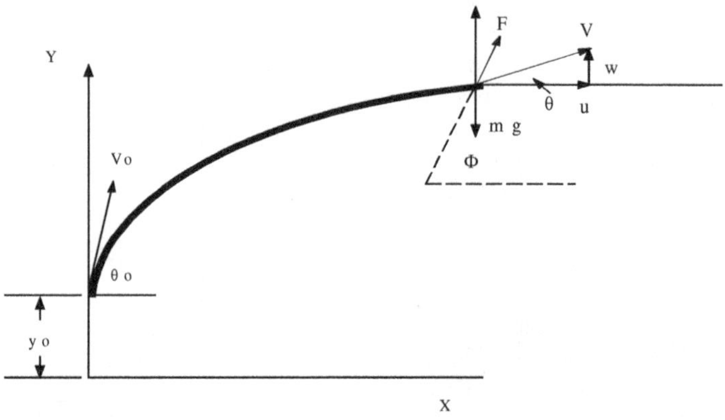

Fig. 1.3.1-1. Definition of Flight Trajectory Variables
Vo = initial velocity

θo = angle of the initial velocity from horizontal axis
yo = initial altitude from the center of the earth
F = thrust vector of a rocket
m = rocket mass
g = gravity
V = rocket velocity vector
ϕ = thrust attitude angle
θ = flight path angle (velocity angle from the horizontal axis)
u = horizontal component of the velocity vector, V
w = vertical component of the velocity vector, V

In this figure, the vehicle acceleration in the X (horizontal) and Y (vertical) directions can be written as,

1) Horizontal direction (X)

$$u' = F(t)*\cos\phi(t)/ m(t) \;,\; u' = du/dt \qquad (1.3.1\text{-}1)$$

2) Vertical direction (Y)

$$w' = F(t)*\sin\phi(t)/ m(t) - g \;,\; w'=dw/dt \qquad (1.3.1\text{-}2)$$

To maximize " u " at a given altitude, yD , Eqn. (1.3.1-1) is integrated.

$$u = \int u' \, dt$$
$$= [t=0, t=\tau] \int \{[F(t)*\cos\phi(t)]/ m(t)\}dt + uo \qquad (1.3.1\text{-}3)$$

Now maximize " u" while meeting the following constraints.

1) $y = yD$, at $t = \tau$
2) $w = 0$, at $t = \tau$

"w = 0" implies that $y' = dy/dt = 0.0$.

By the definition of the variables,

$$y' = w \qquad (1.3.1\text{-}4)$$

Now from Eqn. (1.3.1-2) and Eqn. (1.3.1-4),

$$w' - F(t)*\sin\phi(t)/ m(t) + g = 0 \qquad (1.3.1\text{-}5)$$

$$y' - w = 0 \qquad (1.3.1\text{-}6)$$

Add Eqn. (1.3.1-5) and Eqn.(1.3.1-6) to Eqn. (1.3.1-3) with Lagrange multipliers, λ_1 and λ_2.

$$u = \int u'\,dt$$
$$= [t=0, t=\tau] \int \{[F(t)*\cos\phi(t)]/ m(t) + \lambda_1[w' - F(t)*\sin\phi(t) / m(t) + g] +\lambda_2(y' - w)\}\,dt + u_o \qquad (1.3.1\text{-}7)$$

Boundary conditions are:

1) At $t = 0$

$$X(0) = 0 \qquad (1.3.1\text{-}8)$$
$$Y(0) = y_o \qquad (1.3.1\text{-}9)$$
$$u(0) = v_o \cos(\theta_o) \qquad (1.3.1\text{-}10)$$
$$w(0) = v_o \sin(\theta_o) \qquad (1.3.1\text{-}11)$$

2) At $t = \tau$

$$X(\tau) = \text{open} \qquad (1.3.1\text{-}12)$$
$$Y(\tau) = y_D \qquad (1.3.1\text{-}13)$$
$$u(\tau) = \text{max.} \qquad (1.3.1\text{-}14)$$
$$w(\tau) = 0 \qquad (1.3.1\text{-}15)$$

Taking variation of integration of Eqn. (1.3.1-7),

$$\delta u = \delta u_o + \delta\{ [t=0, t=\tau] \int \{[F(t)*\cos\phi(t)]/ m(t) + \lambda_1[w' - F(t)*\sin\phi(t)/ m(t) + g] +\lambda_2(y' - w)\}\,dt \} \qquad (1.3.1\text{-}16)$$

$$= - v_o \sin(\theta_o)\delta\theta_o \text{ [see Fig. (1.3.1-1)]} + \delta \{[t=0, t=T] \int f(\phi, w, w', y', F/m)\,dt \} \qquad (1.3.1\text{-}17)$$

At this point, we review the variation of calculus briefly and then apply it to our problem.

****

The Variation of Calculus

Consider an integral of the following equation.

$$I = \int f(t,z,dz/dt)dt \qquad (1.3.1\text{-}18)$$

The value of the integral, " I " will depend on the path of integration it takes. The integral taken along the path zo will be different from that taken along z1 (see Fig. 1.3.1-2). An equation relating zo and z1 is written as,

$$z1 = zo + \delta z \qquad (1.3.1\text{-}19)$$

It is to be noted that δz (variation) and dz (differentiation) are different. The difference is clearly shown in Fig. 1.3.1-2 and Fig. 1.3.1-3. "δz" is the difference (variation) in Z between two paths (curves) at any given point in time, while "dz" is the increment in Z along the same path (curve) as the time increases incrementally (dt). Now by taking the Taylor series expansion of the integrand of Eqn. (1.3.1-18),

$$f(t, z+\delta z, z'+\delta z') = f(t, z, z') + (\partial f/\partial z)\delta z + (\partial f/\partial z')\delta z'$$

Note that the time is fixed and the variation occurs along the independent variables.

$$f(t, z+\delta z, z'+\delta z') - f(t, z, z') = (\partial f/\partial z)\delta z + (\partial f/\partial z')\delta z'$$

$$\delta f(t, z, z') = (\partial f/\partial z)\delta z + (\partial f/\partial z')\delta z' \qquad (1.3.1\text{-}20)$$

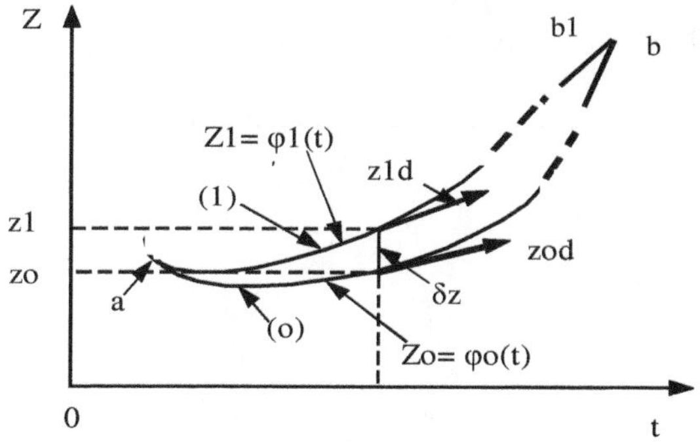

Fig. 1.3.1-2. Definition of Variables

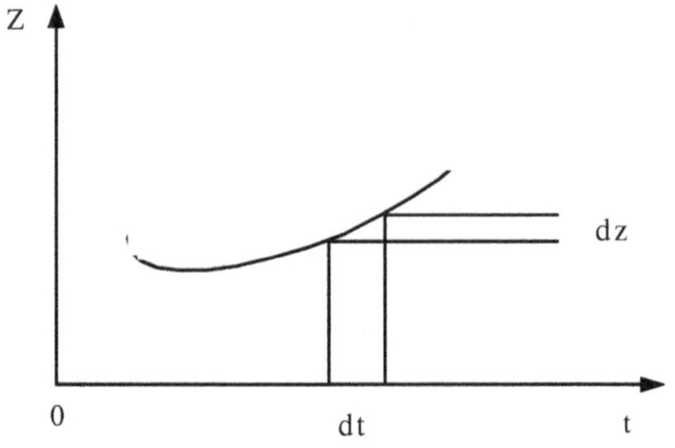

Fig. 1.3.1-3. Definition of dz

Now, the variation of Eqn. (1.3.1-18) is,

$$\delta I = \int \delta\, f(t,z,dz/dt)dt \qquad (1.3.1\text{-}21)$$

and using Eqn. (1.3.1-20),

$$= \int [(\partial f/\partial z)\delta z + (\partial f/\partial z')\delta z']dt$$
$$= \int (\partial f/\partial z)\delta z\, dt + \int (\partial f/\partial z')\delta z' dt$$

Integrating the second term by using integrating by part method,

$u = \partial f/\partial z'$ $\quad du = d(\partial f/\partial z')$
$v = \partial z$ $\quad dv = d(\partial z)$
$\qquad = \delta z' dt$
(See below for derivation.)

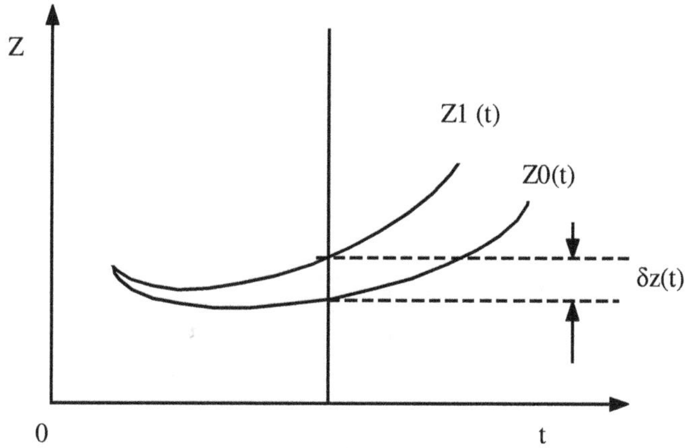

Fig. 1.3.1-4. Definition of δz

From Fig. 1.3.1-4,

$$\delta z(t) = z1(t) - zo(t)$$

Taking the time derivative,

$$\begin{aligned} d[\delta z(t)]/dt &= d\,[z1(t) - zo(t)]/dt \\ &= d\,[z1(t)]/dt - d[zo(t)]/dt \\ &= z1'(t) - zo'(t) \end{aligned} \qquad (1.3.1\text{-}22)$$

From Fig. 1.3.1-5,

$$\delta z'(t) = z1'(t) - zo'(t) \qquad (1.3.1\text{-}23)$$

Equating Eqn. (1.3.1-22) and Eqn. (1.3.1-23),

$$d[\delta z(t)]/dt = \delta z'(t)$$

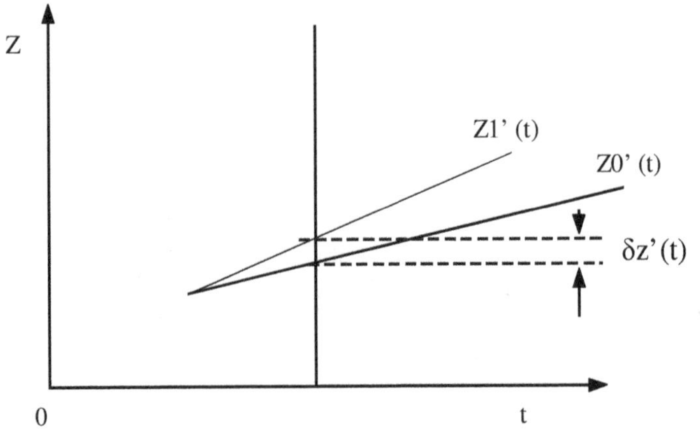

Fig. 1.3.1-5. Definition of δz'

$$
\begin{aligned}
\int u\,dv &= \int (\partial f/\partial z')\ \delta z' \delta t \\
&= uv - \int v\,du \\
&= (\partial f/\partial z')\partial z \quad - \int \partial z\ d(\partial f/\partial z') \\
&= (\partial f/\partial z')\partial z \quad - \int \partial z\ [d(\partial f/\partial z')]\ (dt/dt) \\
&= (\partial f/\partial z')\partial z \quad - \int [d(\partial f/\partial z')/dt]\ \delta z\ dt
\end{aligned}
$$

$$
\delta I = \int (\partial f/\partial z)\delta z dt \ + \ (\partial f/\partial z')\delta z \quad - \int [d(\partial f/\partial z')/dt]\ \delta z\ dt \tag{1.3.1-24}
$$

For optimal solution, δI is set equal to zero.

$$
0 = \int (\partial f/\partial z)\delta z dt \ + \ (\partial f/\partial z')\delta z \quad - \int [d(\partial f/\partial z')/dt]\ \delta z\ dt
$$

Rearranging the equation above and introducing integration boundary [a,b]

$$
\begin{aligned}
0 &= (\partial f/\partial z')\delta z \ + \int (\partial f/\partial z)\delta z dt \ - \int [d(\partial f/\partial z')/dt]\ \delta z\ dt \\
&= (\partial f/\partial z')\delta z \ + \int [(\partial f/\partial z) - d(\partial f/\partial z')/dt]\ \delta z\ dt \\
&= [t{=}a, t{=}b]\ (\partial f/\partial z')\delta z + \int [(\partial f/\partial z) - d(\partial f/\partial z')/dt]\ \delta z\ dt
\end{aligned} \tag{1.3.1-25}
$$

The path to be found is the optimal path that leads to the destination. There can be many different paths; however, the starting point and the destination point are common to all different paths. It is shown in Fig, 1.3.1-2 that b and b1 are at the same point and $\delta z = 0$ at that point. And at the starting point, δz is also equal to zero. Thus, the first term in the equation above is equal to zero.

$$[t=a,\ t=b]\ (\partial f/\partial z')\delta z = 0$$

And the integrand of the second term is set equal to zero to satisfy the optimal condition.

$$[(\partial f/\partial z) - d(\partial f/\partial z')/dt] = 0 \qquad (1.3.1\text{-}26)$$

Eqn. (1.3.1-26) is the well-known Euler's Equation.

Now the optimal path is the path [$z = \varphi\ (t)$] that satisfies the Euler's Equation.

****

The variables in Eqn. (1.3.1-17) are: ϕ, w, y. F and m are not treated as variables because it is assumed that the ratio of F and m is the same for all the paths. Now the Calculus of Variation is applied to the second term with respect to the three variables. Using Eqn. (1.3.1-25),

$$\begin{aligned}\delta u = &- vo \sin(\theta o)\delta\theta o \\ &+[t=0,\ t=\tau]\ (\partial f/\partial\phi')\delta\ \phi + \int [(\partial f/\partial\phi - d\ (\partial f/\partial\phi')/dt]\ \delta\phi\ dt \\ &+[t=0,\ t=\tau]\ (\partial f/\partial w')\delta w + \int [(\partial f/\partial w - d\ (\partial f/\partial w')/dt]\ \delta w\ dt \\ &+[t=0,\ t=\tau]\ (\partial f/\partial y')\delta y + \int [(\partial f/\partial y - d\ (\partial f/\partial y')/dt]\ \delta y\ dt \\ = &\ 0.0 \qquad (1.3.1\text{-}28)\end{aligned}$$

Here, the following terms need to be computed.

$$(\partial f/\partial\phi)\ ,\ (\partial f/\partial\phi')$$
$$(\partial f/\partial w)\ ,\ (\partial f/\partial w')$$
$$(\partial f/\partial y)\ ,\ (\partial f/\partial y')$$

where

$$f = \{(F(t)*\cos(\phi(t))/ m(t) + \lambda_1[(w' - F(t)*\sin(\phi(t))/ m(t) + g] + \lambda_2 (y' - w)\}$$

[See Eqn. (1.3.1-7).]

Computing each term,

$(\partial f/\partial\phi) = - [F(t)/m(t)] [\sin\phi(t) + \lambda1*\cos\phi(t)]$
$(\partial f/\partial\phi') = 0$

$(\partial f/\partial w) = - \lambda2$
$(\partial f/\partial w') = \lambda1$

$(\partial f/\partial y) = 0$
$(\partial f/\partial y') = \lambda2$

Substituting these terms into Eqn. (1.3.1-28),

$$\begin{aligned} 0 = &- vo \sin(\theta o)\delta\theta o \\ &+[t=0, t=\tau] (0.0* \partial\varphi)-\int[F(t)/m(t)] [\sin\phi(t) + \lambda1*\cos\phi(t)]\delta\phi\, dt \\ &+[t=0, t=\tau] \lambda1\, \delta w + \int [- \lambda2 - d(\lambda1)/dt]\, \delta w\, dt \\ &+[t=0, t=\tau] \lambda2\, \delta y + \int [0 - d(\lambda2)/dt]\, \delta y\, dt \end{aligned}$$

$$\begin{aligned} = &- vo \sin(\theta o)\delta\theta o + [t=0, t=\tau] \lambda1\, \delta w + [t=0, t=\tau] \lambda2\, \delta y \\ &-\int\{[F(t)/m(t)] [\sin\phi(t) + \lambda1*\cos\phi(t)]\delta\phi + [(\lambda2 + d(\lambda1)/dt]\, \delta w \\ &+ [d(\lambda2)/dt]\, \delta y\}\, dt \end{aligned}$$

$$\begin{aligned} = &- vo \sin(\theta o)\delta\theta o + \lambda1 [\delta w(\tau) - \delta w(0)] + \lambda2 [\delta y(\tau) - \delta y(0)] \\ &-\int\{[F(t)/m(t)] [\sin\phi(t) + \lambda1*\cos\phi(t)]\delta\phi + [(\lambda2 + d(\lambda1)/dt]\, \delta w \\ &+ [d(\lambda2)/dt]\, \delta y\}\, dt \end{aligned} \qquad (1.3.1\text{-}29)$$

****

In the following, it is shown how $\delta w(\tau)$, $\delta w(0)$, $\delta y(\tau)$, and $\delta y(0)$ are obtained, which are needed for Eqn. (1.3.1-29).

1) $\delta w(\tau)$

From Eqn. (1.3.1-2),

$$w' = F(t)*\sin(\phi(t))/ m(t) - g$$

Taking time integration,

$$
\begin{aligned}
w(\tau) &= \int w'(t)\,dt \\
&= \int [(F(t)*\sin(\phi(t))/ m(t) \;-\; g\,]dt \\
&= \int [(F(t)*\sin(\phi(t))/ m(t)]dt - \int g\,dt \\
&= [t=0, t=\tau]\int [(F(t)*\sin(\phi(t))/ m(t)]dt + vo \sin \theta o \\
&\quad - [t=0, t=\tau]gt
\end{aligned}
$$

Consider a variation of $w(\tau)$ due to the variation of τ,

$$
\begin{aligned}
\delta\, w(\tau) &= [t=\tau, t=\tau+\delta\tau]\int [(F(t)*\sin(\phi(t))/ m(t)]dt \\
&\quad - [t=\tau, t=\tau+\delta\tau]gt
\end{aligned}
$$

Note that "vo sin θo" is deleted because it is not a function of $\delta\tau$, and at the highest point of a flight, there will be no thrust, that is, $F(\tau) = 0.0$ and $F(\tau + \delta\tau) = 0.0$.

$$
\begin{aligned}
&= 0 + (\tau+\delta\tau)g - \tau g \\
&= \delta\tau \text{ (at } t=\tau) \; g \qquad (1.3.1\text{-}30)
\end{aligned}
$$

2) $\delta w\,(0)$

From Eqn. (1.3.1-11),

$w(0)= vo \sin(\theta o)$

Its variation at $t = 0$ due to the variation of the initial angle, θo, is,

$$\delta\, w(0) = vo \cos(\theta o)\, \delta\theta o \qquad (1.3.1\text{-}31)$$

3) $\delta y(\tau)$

From Eqn. (1.3.1-13)

$y(\tau) = yD$

Variation of $y(\tau)$ at $t = \tau$ is zero because it is a fixed value, that is, no variable is involved.

$$\delta\, y(\tau) = 0 \qquad (1.3.1\text{-}32)$$

4) $\delta y(\tau)$

From Eqn. (1.3.1-9)

$Y(0) = yo$

Variation of y(0) at t = 0 is zero because it is a fixed value, that is, no variable is involved.

$$\delta\, y(0) = 0 \qquad (1.3.1\text{-}33)$$

****

Substituting Eqn. (1.3.1-30), Eqn. (1.3.1-31), Eqn. (1.3.1-32), and Eqn. (1.3.1-33) into Eqn. (1.3.1-29),

$$\begin{aligned} 0 = &- vo \sin(\theta o)\delta\theta o - \lambda 1\, vo \cos(\theta o)\, \delta\theta o - \lambda 1(\text{at } t=\tau)\ \delta\tau\, g \\ &+ \lambda 2\,[0 - 0] - \textstyle\int\{[F(t)/m(t)]\,[\,\sin\phi(t) + \lambda 1^{*}\cos\phi(t)]\delta\phi + [\,\lambda 2 + d \\ &(\lambda 1)/dt]\ \delta w + [d(\lambda 2)/dt]\ \delta y\}\ dt \end{aligned}$$

$$\begin{aligned} 0 = &- vo\,[\sin(\theta o) + \lambda 1 \cos(\theta o)\,]\ \delta\theta o \\ &- \lambda 1(\text{at } t=\tau)\ g\ \delta\tau \\ &- \textstyle\int\{[F(t)/m(t)]\,[\,\sin\phi(t) + \lambda 1^{*}\cos\phi(t)]\ \delta\varphi \\ &+ [\lambda 2 + d\,(\lambda 1)/dt]\ \delta w \\ &+ [d(\lambda 2)/dt]\ \delta y\}\ dt \end{aligned}$$

It is seen in the equation above that since the variables $\delta\theta o$, $\delta\varphi$, δw, δy, and $\delta\tau$ are not necessarily zero at all times, we need to set the coefficients equal to zero to satisfy the equation. That is,

$$[\sin(\theta o) + \lambda 1 \cos(\theta o)\,] = 0 \qquad (1.3.1\text{-}34)$$
$$\lambda 1\ g\ \ (\text{at } t = \tau) = 0 \qquad (1.3.1\text{-}35)$$
$$[\,\sin\phi(t) + \lambda 1 \cos\phi(t)] = 0 \qquad (1.3.1\text{-}36)$$
$$[(\,\lambda 2 + d\,(\lambda 1)/dt] = 0 \qquad (1.3.1\text{-}37)$$
$$[d(\lambda 2)/dt] = 0 \qquad (1.3.1\text{-}38)$$

These are the conditions that must be satisfied for optimal trajectory. Now, solving for the Lagrange multipliers, λ_1 and λ_2,

From Eqn. (1.3.1-34),

$$\lambda_1 \text{ (at } t = 0\text{)} = -\tan\theta_o \qquad (1.3.1\text{-}39)$$

from Eqn. (1.3.1-35),

$$\lambda_1 \text{ (at } t = \tau\text{)} = 0 \qquad (1.3.1\text{-}40)$$

from Eqn. (1.3.1-36),

$$\lambda_1 = -\tan\phi(t) \qquad (1.3.1\text{-}41)$$

from Eqn. (1.3.1-37),

$$[d(\lambda_1)/dt] = -\lambda_2 \qquad (1.3.1\text{-}42)$$

and from Eqn. (1.3.1-38),

$$[d(\lambda_2)/dt] = 0. \qquad (1.3.1\text{-}43)$$

From Eqn. (1.3.1-43),

$$\lambda_2 = C2,$$

and Eqn. (1.3.1-42) can be written as,

$$[d(\lambda_1)/dt] = -C2$$
$$d(\lambda_1) = -C2\, dt$$
$$\lambda_1 = -C2\, t + C1 \qquad (1.3.1\text{-}44)$$
$$\lambda_1(\text{at } t = \tau) = -C2\, \tau + C1 \qquad (1.3.1\text{-}45)$$

From Eqn. (1.3.1-39) and Eqn. (1.3.1-45), we obtain,

$$C1 = -\tan\theta_o \qquad (1.3.1\text{-}46)$$

From Eqn. (1.3.1-45),

$$C2 = [C1 - \lambda_1(\text{at } t=\tau)]/\tau$$
$$= C1/\tau, \text{ for } \lambda_1(\text{at } t=\tau) = 0, \text{ Eqn.(1.3.1-40)}$$
$$= -(1/\tau)\tan\theta_o \qquad (1.3.1\text{-}47)$$

Substituting Eqn. (1.3.1-46) and Eqn. (1.3.1-47) into Eqn.(1.3.1-45),

$$\lambda_1 = [\ (t/\tau) - 1]\tan\theta o \qquad (1.3.1\text{-}48)$$

Now equating Eqn. (1.3.1-41) and Eqn. (1.3.1-48),

$$\tan\phi(t) = [1 - (t/\tau)\ ]\tan\theta o \qquad (1.3.1\text{-}49)$$

Check the boundary conditions,

when $t = 0$,
$\tan\phi(0) = [1 - (0/\tau)\ ]\tan\theta o$
$= \tan\theta o$

This implies that $\phi(0)$ is equal to θo. That is, the thrust attitude angle at the time of rocket launch (t=0) is equal to the angle of the initial velocity for optimal trajectory. The angle is measured from the horizontal axis.

When $t = \tau$,
$\tan\phi(\tau) = [1 - (\tau/\tau)\ ]\tan\theta o$
$= 0$

The thrust attitude angle at $t=\tau$ is equal to zero, which is the horizontal direction. Thus, the results verify the accuracy of the equation.

A procedure to generate an optimal trajectory is presented in sequence.

1) Determine,

   a. final time (τ)
   b. initial velocity ($|vo|$, θo) → uo and wo
   c. vertical position (yo).

2) Using Eqn. (1.3.1-49),

$$\tan\phi(t) = [1 - (t/\tau)\ ]\tan\theta o$$

3) Now using Eqn. (1.3.1-1) and Eqn. (1.3.1-2), compute

u'= F(t)*cos(φ(t))/ m(t)
w' = F(t)*sin(φ(t))/ m(t) - g

4) Compute u and w.

$$u = u' t + uo$$
$$w = w' t + wo$$

5) Compute the flight path angle

$$\theta = \arctan(w/u) \qquad (1.3.1\text{-}50)$$

6) Compute x (range) and y (altitude: altitude computation).

$$x = (0.5)u't^2 + uot + xo \qquad (1.3.1\text{-}51)$$
$$y = (0.5)w't^2 + wot + yo \qquad (1.3.1\text{-}52)$$

7) Go back to 2)

Example 1. Ground to ground missile :

Final Time of the Flight -------------	118 s
Theta (initial angle) ------------------	45 deg
Initial Velocity -----------------------	0 ft/sec
Initial Position ------------------------	0 ft
Altitude where the horizontal velocity is to be maximized -----------	10.266 miles above ground

The objective of this optimization is to maximize the horizontal velocity at a specific altitude. The altitude specified here is 10.266 miles above the ground.

The time history of (F/m), an axial acceleration, is assumed for this example as shown in Fig. 1.3.1-6. It is seen that the maximum acceleration, 950 ft/sec^2, occurs at t = 1.0 sec. and the thrust stops at t =5.0 sec. As suggested above, first, the optimal thrust vector angle [$\phi(t)$] is computed using Eqn. (1.3.1-49). Then, the horizontal and vertical accelerations are computed using Eqn. (1. 3-1) and Eqn. (1.3.1-2) in which the thrust vector angle and the axial acceleration histories are used. From the acceleration, the horizontal and vertical velocities are computed, from which the optimal flight path angle (θ) can be obtained by using Eqn. (1.3.1-50). Fig. 1.3.1-7 shows the time history of the optimal thrust vector angle and the flight path angle. Notice that at the middle of the flight, near 60 seconds after launch, the flight path angle is zero (horizontal angle) and thereafter the angle becomes less than zero as the rocket comes down toward the target on the ground. The rocket attitude, i.e., the thrust vector attitude, remains positive until the last phase of the flight.

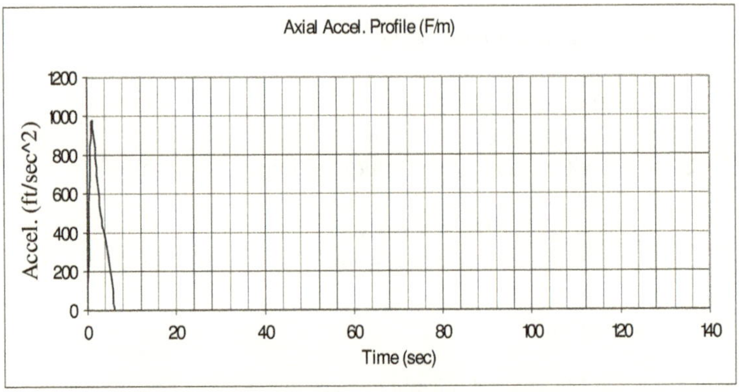

Fig. 1.3.1-6. Axial Acceleration Profile

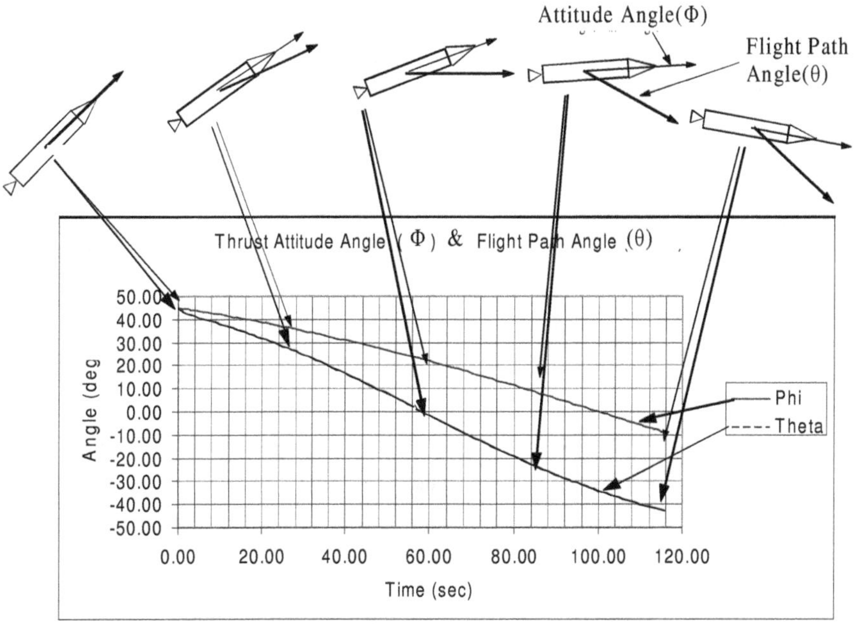

Fig. 1.3.1-7. Thrust Attitude Angle and Flight Path Angle for an Optimal Trajectory

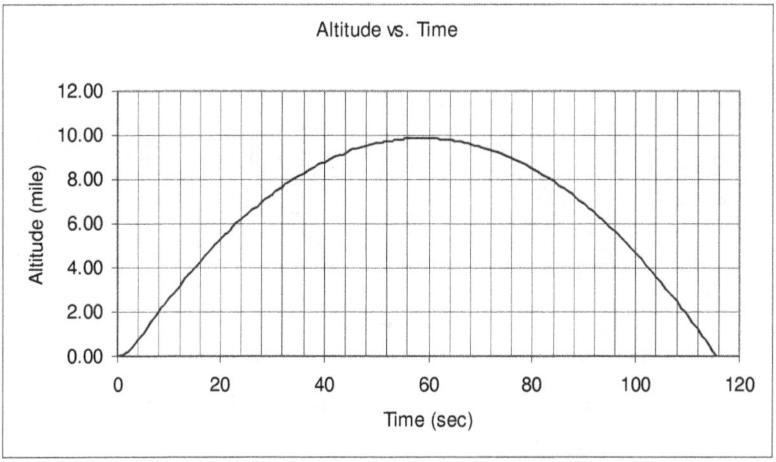

Fig. 1.3.1-8. Altitude vs. Time for an Optimal Trajectory

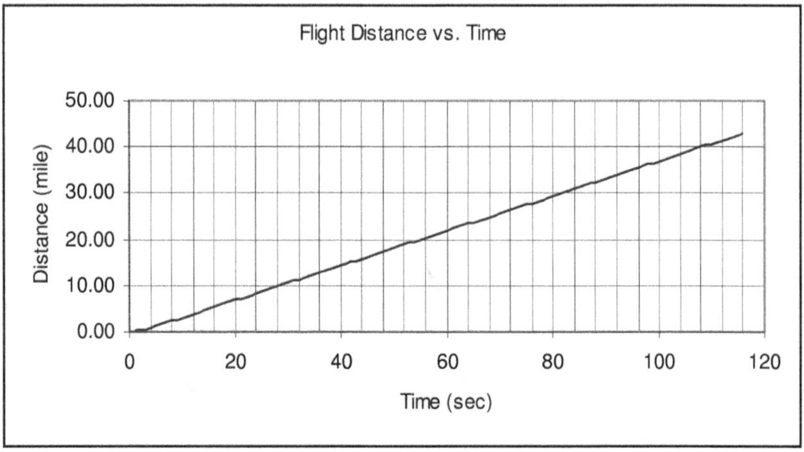

Fig. 1.3.1-9. Flight Distance Projected on Flat Ground

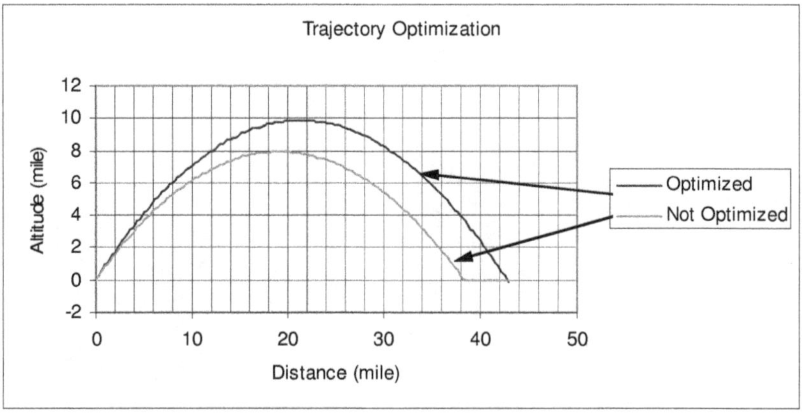

Fig. 1.3.1-10. Comparison of Optimal and Non-Optimal Trajectories

Fig. 1.3.1-8 shows that the maximum altitude is 9.86 miles from the ground, and it occurs at t = 59 sec. It is 3.95 % less than the desired altitude. Fig. 1.3.1-9 is a plot of distance traveled. It is to be noticed that once the horizontal velocity reaches its maximum (986.32 ft/sec) at t = 6.0 sec, the velocity remains constant throughout the rest of the flight. Fig. 1.3.1-10 compares optimized and non-optimized trajectories. The "Non-Optimized" trajectory is constructed by perturbing the time history of thrust vector angle (Φ) slightly.

Fig. 1.3.1-11 demonstrates the optimality by comparing three trajectories: the first one (Alt. var1 in Fig. 1.3.1-11) is an altitude trajectory when the Φ is 0.01 rad greater than optimal value, the second one (Alt. var2 in Fig. 1.3.1-11) is an altitude trajectory when the Φ is 0.01 rad less than the optimal value, and the third one (Alt. opt in Fig. 1.3.1-11) is an altitude trajectory when the Φ is optimal. It is seen here that the Alt. var1 case has higher altitude, but the velocity at the maximum altitude is 1937.364 ft/sec, which is less than the optimal velocity [1949.987 ft/sec (A)]. On the other hand, when the Φ is decreased by 0.01 rad (Alt. var2), the trajectory does not even reach the desired altitude. This demonstrates that the trajectory obtained by using Eqn. (1.3.1.1-49) is optimal (Alt. opt).

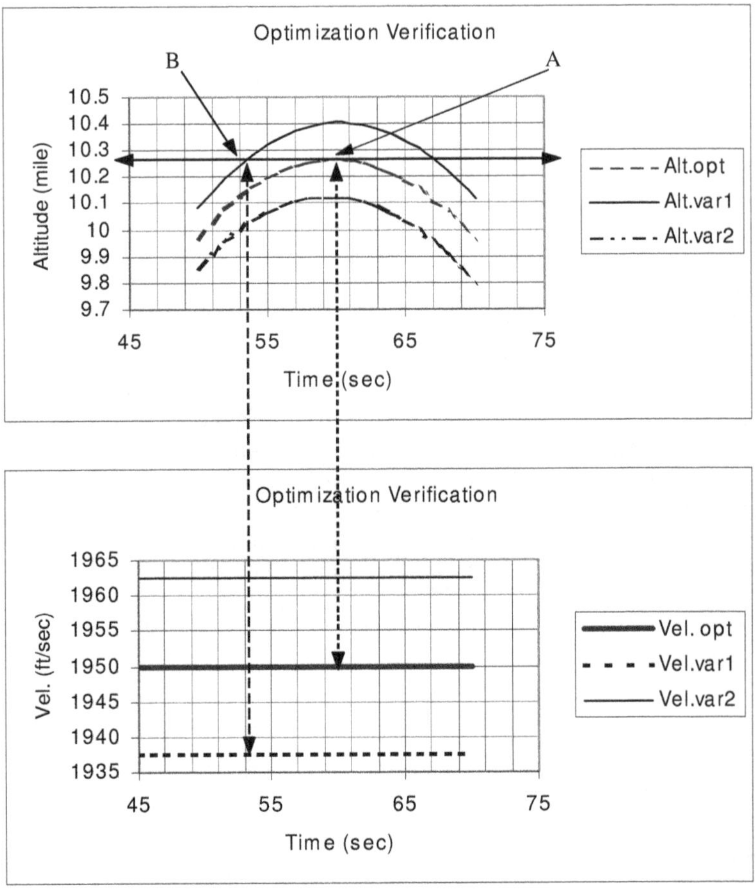

Fig. 1.3.1-11 Verification of Optimality

Example 2. Satellite Launch:

The objective of this mission is to achieve a horizontal velocity of 19,396.26 ft/sec when the launch vehicle reaches an altitude of 6,251.65 miles above the surface of the earth. The axial acceleration time history (Fig. 1.3.1.1-12) is designed to achieve the mission goal. From the figure, it is seen that the launch vehicle consists of two stages and the second stage starts at t=280.0 sec. It also should be noted that the optimal trajectory starts 60.0 sec after the lift off, and the initial thrust vector attitude angle is 78.0 deg.

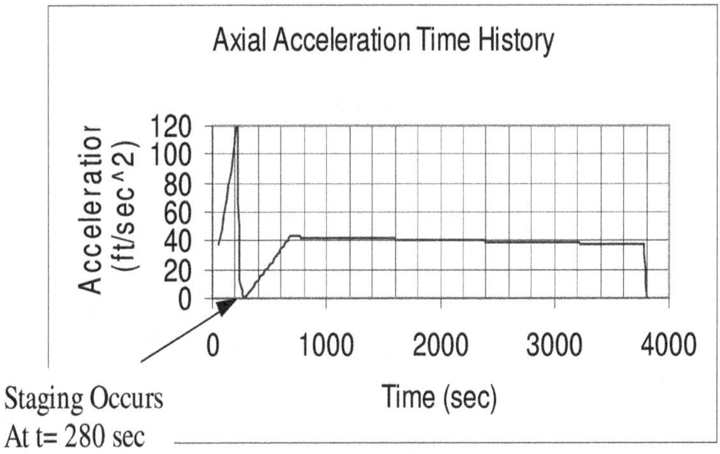

Fig. 1.3.1-12 Axial Acceleration Time History

Summarizing the boundary conditions,

Initial Values:

Time = 60 sec.
Horizontal Acceleration = 7.69 ft/sec^2
Horizontal Velocity = 1521.76 ft/sec (at equator)
Horizontal distance = 0.0 ft
Vertical Acceleration = 4.02 ft/sec^2
Vertical Velocity = 580.0 ft/sec
Vertical Position (Altitude) = 60.0 ft

Final Values:

Time = 3,800 sec.
Horizontal Acceleration = 0 ft/sec^2
Horizontal Velocity = 71,810.12 ft/sec
Horizontal distance = 19,396.26 miles
Vertical Acceleration = -32.174 ft/sec^2
Vertical Velocity = 2.2 ft/sec
Vertical Position (Altitude) = 6251.65 miles

Using the trajectory optimization algorithm (see page 46) developed earlier, thrust attitude and flight path angle time histories are computed as shown in Fig. 1.3.1-13.

The optimization algorithm maximizes the horizontal velocity for the given acceleration time history. It is seen in Fig. 1.3.1-14 that the desired velocity, 71,810.12 ft/sec, is obtained at the specified time, t=3,800 sec. Fig. 1.3.1-15 shows that the required altitude, 6,252.65 miles, is also reached at the specified time.

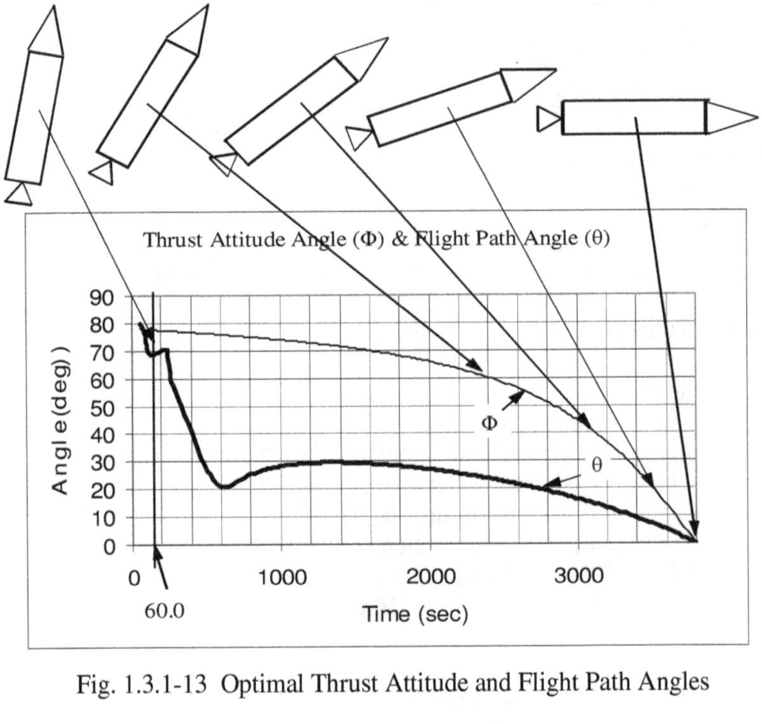

Fig. 1.3.1-13 Optimal Thrust Attitude and Flight Path Angles

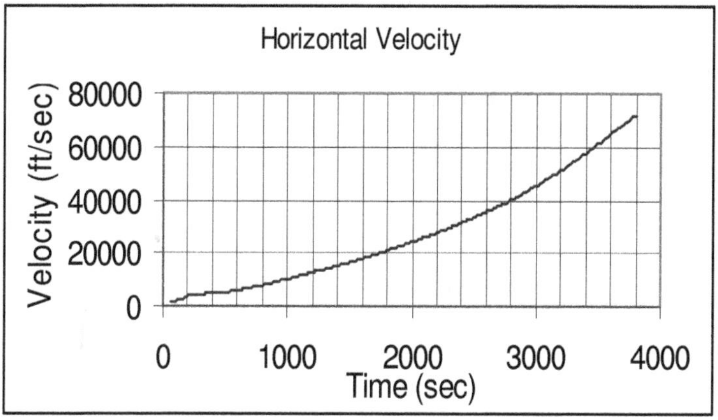

Fig. 1.3.1-14 Maximizing the Horizontal Velocity.

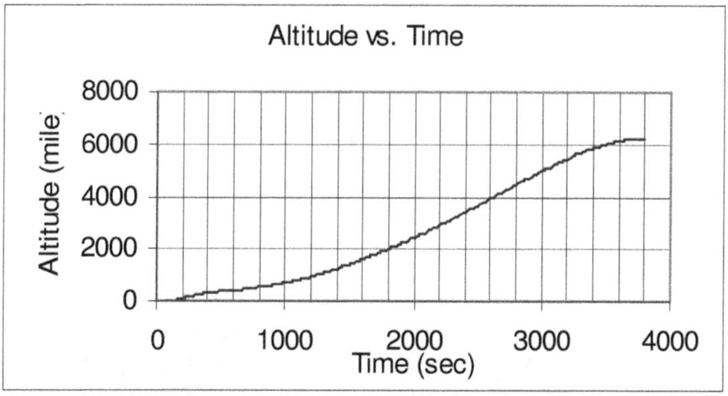

Fig. 1.3.1-15 Launch Vehicle Altitude vs. Time

1.3.2 Optimal Fuel Consumption

The optimal fuel consumption trajectory can be attained by use of fuel optimization. From Eqn. (1.3.1-7),

$$u = uo + \int \{F(t)*\cos(\phi(t))/ m(t) + \lambda 1[(w' - F(t)*\sin(\phi(t))/ m(t) + g] + \lambda 2(y' - w)\}\, dt$$
$$= uo + \int \{[F(t) / m(t)]*[\cos\phi(t) - \lambda 1\sin\phi(t)] + \lambda 1[(w' + g] + \lambda 2(y' - w)\}dt \quad (1.3.2\text{-}1)$$

Since,

$$dV/dt = F(t) / m(t)$$
$$= V'$$
$$uo = 0.0$$

Eqn. (1.3.2-1) can be written as,

$$u = \int \{V'*[\cos\phi(t) - \lambda 1\sin\phi(t)] + \lambda 1[(w'+ g] +\lambda 2(y' - w)\} dt$$
$$= \int f\, dt$$

where

$$f = \{V'*[\cos\phi(t) - \lambda 1\sin\phi(t)] + \lambda 1[V'\sin \theta(t)+ g] , \text{ for } (y' - w)=0$$

[See Eqn. (1.3.1-4).]

Now using calculus of variation V and V',

$$(\partial f/\partial V) = 0$$
$$(\partial f/\partial V') = [\cos\phi(t) - \lambda 1\sin\phi(t)] + \lambda 1 \sin \theta(t)$$

For optimality with respect to V, using Eqn. (1.3.1-25),

$$\delta u = 0$$
$$= [t=0, t=\tau] (\partial f/\partial V')\delta V + \int [(\partial f/\partial V) - d(\partial f/\partial V')/dt] \delta V\, dt$$
$$= \{[\cos\phi(\tau) - \lambda 1\sin\phi(\tau)+ \lambda 1 \sin \theta - [\cos\phi(0) - \lambda 1\sin\phi(0) + \lambda 1 \sin \theta o]\} \delta V -\int d([\cos\phi(t) - \lambda 1\sin\phi(t) + \lambda 1 \sin \theta(t)])/dt] \delta V\, dt$$

$$= \{[\cos\phi(\tau) - \lambda 1\sin\phi(\tau)+ \lambda 1 \sin \theta(\tau)]\} \delta V -\int d([\cos\phi(t) - \lambda 1\sin\phi(t) + \lambda 1 \sin \theta(t)])/dt] \delta V\, dt,$$

for $\cos\phi(0) =0$, and $\lambda 1\sin\phi(0) = \lambda 1 \sin \theta o$ where $\phi(0)= \theta o=$ 90.0 deg.

$$= -\{\lambda 1\sin\phi(\tau) - \cos\phi(\tau) - \lambda 1 \sin \theta(\tau) + \int d([\cos\phi(t) - \lambda 1\sin\phi(t) + \lambda 1 \sin\theta(t)])/dt]\, dt \} \delta V$$
$$= Q\, \delta V$$

where

$$Q = -\{\lambda 1 \sin\phi(\tau) - \cos\phi(\tau) - \lambda 1 \sin\theta(\tau) + \int d([\cos\phi(t) - \lambda 1 \sin\phi(t) + \lambda 1 \sin\theta(t)])/dt]\, dt\}$$

Since Q cannot be zero consistently, δV has to be zero for optimality, which implies that velocity must be constant to achieve the fuel consumption optimality. Constant velocity can be achieved in two ways. One way is to remain at the initial velocity all the way to the end and then at the end consume the fuel instantly to achieve the desired velocity. The other way is to consume the fuel instantly in the beginning and obtain the maximum velocity. In the real world, the fuel is consumed as fast as possible in the beginning to reach the maximum velocity, and then that velocity is maintained until the end of the flight. Fig. 1.3.2-1 shows time history of the optimal fuel utilization. It is seen here that the maximum velocity must be achieved at the beginning of the flight for the optimal fuel utilization.

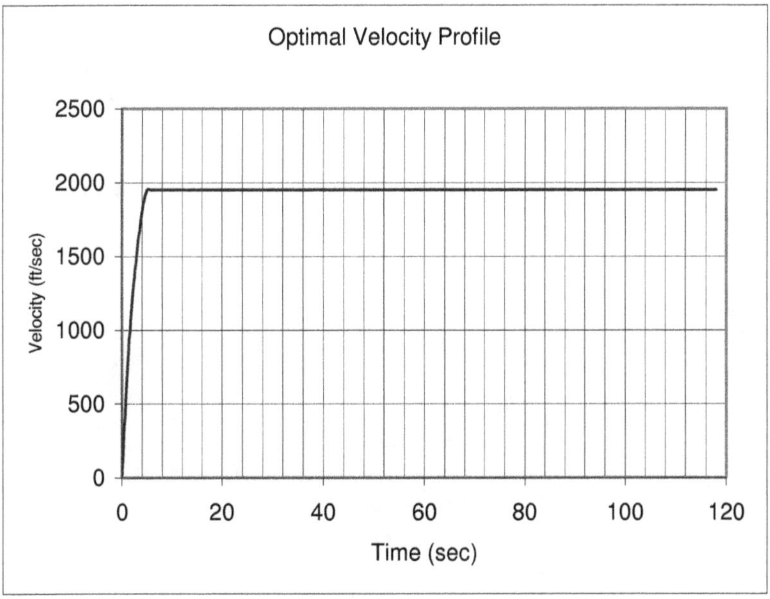

Fig. 1.3.2-1. Optimal Fuel Utilization Velocity Profile

1.4 Gravity Turn Trajectory Generation

The angle of attack must be minimized during the atmospheric (high air density) flight to reduce the rocket's aerodynamic load on the rocket structure. The large angle of attack may cause unacceptably high rocket structural bending loads, and the rocket could be disintegrated. It also causes the attitude control load to be unacceptably heavier. The angle of attack [= thrust attitude angle (Φ) – vehicle velocity angle (θ), Fig. 1.4-1] must be kept close to zero during this phase of flight.

The gravity turn trajectory enables the rocket to maintain the angle of attack as small as possible. In this trajectory, the acceleration vector angle is close to the thrust vector angle and thus it minimizes the lifting force.

We define a notation, ψ, as the angle between the velocity vector and the vertical line as shown in Fig. 1.4-1. Writing tangential and normal force equations, we have

Tangential force:

$$F - mg\cos\psi = m\,(dV/dt) \qquad (1.4\text{-}1)$$

Normal force:

$$mg\sin\psi = mv(d\psi/dt) \qquad (1.4\text{-}2)$$

[See below for derivation of $v(d\psi/dt)$.]

****

$V = vi$

$a = (dV/dt)$

$= (dv/dt)\,i + v\,(di/dt)$

| $di/dt = [\Delta t \to 0] \lim [i(t+\Delta t) - i(t)]/\Delta t$

| $= [\Delta t \to 0] \lim (1\Delta\psi j / \Delta t)$ (see Fig. 1.4-2)

| $= (d\psi/dt)\,j$

| $= \omega\,j$

$= (dv/dt)\,i + \underline{\mathbf{v\,(d\psi/dt)}}\,j$

Normal acceleration

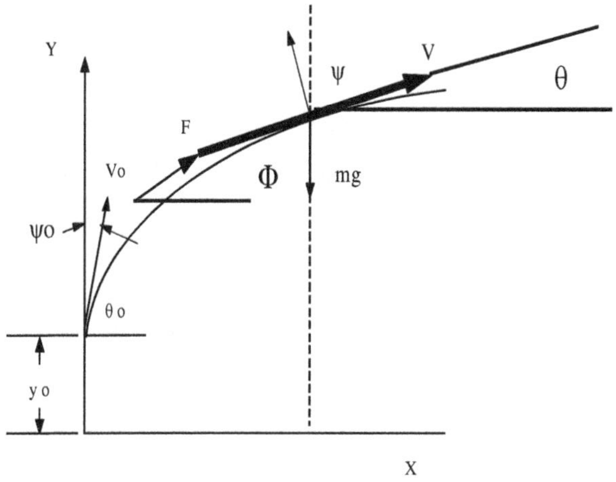

Fig. 1.4-1. Gravity Turn Trajectory

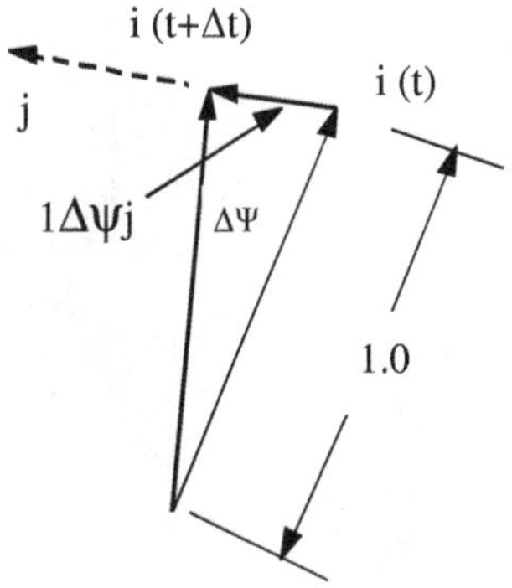

Fig. 1.4-2. Vector Analysis

****

Rearranging Eqn. (1.4-1) and Eqn. (1.4-2)

$$(1/g)(dV/dt) = F/mg - \cos\psi \qquad (1.4\text{-}3)$$

$$(V/g)(d\psi/dt) = \sin\psi \qquad (1.4\text{-}4)$$

To understand the meaning of the gravity turn, we rearrange Eqn. (1.4-3).

$$(dV/dt) = F/m - g\cos\psi \qquad (1.4\text{-}5)$$

In this equation, V is the velocity obtained by integrating an acceleration that is computed by subtracting the gravity acceleration from the rocket thrust acceleration. [See Eqn. (1.4-5).] Now, let the rocket turn gradually in the pitch plane, following the dynamics specified in Eqn. (1.4-3) and Eqn. (1.4-4), in the clockwise direction, which means that ψ approaches 90.0 deg. (See Fig. 1.4-1.) This turn reduces the angle of attack because the second term, "$(g\cos\psi)$," in Eqn. (1.4-5) approaches zero, and, consequently, the rocket acceleration (dV/dt) approaches the thrust vector (F/m). The dynamics is forced to cause the angle of attack equal to zero as it starts with $\sin\psi > 0.0$ in Eqn. (1.4-4). If a rocket follows a trajectory generated by these equations, it satisfies the gravity turn trajectory requirement that the angle of attack stays close to zero.

Here, we need to solve for V in terms of ψ as we increase it by $\Delta\psi$ (clockwise rotation) and compute the time increase, Δt, required to change ψ to $\psi + \Delta\psi$. The V, Δt, and $\psi + \Delta\psi$ are the terms we need to compute to generate the gravity turn trajectory.

To solve for V, a variable, z, is defined as follows.

$$\begin{aligned} z &= \tan(\psi/2) \\ &= \mathrm{sqrt}((1-\cos\psi)/(1+\cos\psi)) \qquad \text{From math table} \\ &= \sin\psi/(1+\cos\psi) \qquad \text{From math table} \end{aligned} \qquad (1.4\text{-}6)$$

Squaring both sides of the equation above,

$$z^2 = (1-\cos\psi)/(1+\cos\psi))$$

Rearranging the equation above for $\cos\psi$,

$$\cos\psi = (1 - z^2)/(1 + z^2) \qquad (1.4\text{-}7)$$

Now differentiating Eqn. (1.4-6),

$$(dz/dt) = d\,[\tan(\psi/2)]$$
$$= \{[\sec(\psi/2)]^2\}(1/2)(d\psi/dt) \qquad (\text{From math table})$$

Rearranging the equation above again for $(d\psi/dt)$, and using $(\cos A)^2 = (1/2)(\cos 2A + 1)$,

$$(d\psi/dt) = (1 + \cos\psi)\,(dz/dt) \qquad (1.4\text{-}8)$$

Substituting Eqn. (1.4-7) into Eqn. (1.4-3),

$$(1/g)(dV/dt) = F/mg - (1 - z^2)/(1 + z^2) \qquad (1.4\text{-}9)$$

and substituting Eqn. (1.4-8) into Eqn. (1.4-4),

$$(V/g)(1 + \cos\psi)\,(dz/dt)) = \sin\psi$$
$$(V/g)(dz/dt) = \sin\psi/(1 + \cos\psi)$$
$$= z \quad (\text{see Eqn. } (1.4\text{-}6) \qquad (1.4\text{-}10)$$

From Eqn. (1.4-10),

$$dt = (V/g)(dz/z) \qquad (1.4\text{-}11)$$

Substituting Eqn. (1.4-11) into Eqn. (1.4-9),

$$(1/g)[gz\,dV/(V\,dz)] = F/mg - (1 - z^2)/(1 + z^2)$$
$$[z\,dV/(V\,dz)] = F/mg - (1 - z^2)/(1 + z^2)$$

Multiply both sides by (dz/z),

$$(dV/V) = (F/mg)\,(dz/z) - [(1 - z^2)/(1 + z^2)]\,(dz/z)$$

Integrating both sides,

$$\int (dV/V) = \int (F/mg)(dz/z) - \int [(1 - z^2)/(1 + z^2)](dz/z) \tag{1.4-12}$$

The first term on the right side,

$$\int (F/mg)(dz/z) = (F/mg)\ln z + C1 \quad \text{for } z > 0 \text{ because } \psi > 0$$

The second term on the right side,

$$\begin{aligned}\int [(1 - z^2)/(1 + z^2)](dz/z) &= \int [1/(1 + z^2)](dz/z) - \int [z^2/(1 + z^2)](dz/z)\\ &= (1/2)\ln [z^2/(1+z^2)] - (1/2)\ln (1+z^2) + C2\\ &\quad \text{(From math table)}\\ &= (1/2)\ln \{[z^2/(1+z^2)]/(1+z^2)\} + C2\\ &= (1/2)\ln [z^2/(1+z^2)^2] + C2\end{aligned}$$

Substituting these terms into Eqn. (1.4-12),

$$\begin{aligned}\int (dV/V) &= \ln V\\ &= (F/mg)\ln z - (1/2)\ln [z^2/(1+z^2)^2] + \ln C\\ &\quad (\ln C = C1 + C2)\\ &= (F/mg)\ln z + (1/2)\ln [(1+z^2)^2/z^2] + \ln C\\ &= (F/mg)\ln z + \ln [(1+z^2)/z] + \ln C\\ &= \ln [z^{(F/mg)}]\, C\, [(1+z^2)/z]\\ &= \ln [z^{(F/mg)}] z^{(-1)}\, C\, [(1+z^2)]\\ &= \ln \{z^{[(F/mg)-1]}\} C\, [(1+z^2)]\end{aligned}$$

From the equation above, solving for V,

$$V = [z^{(\eta-1)}]C[(1+z^2)] \quad , \quad \eta = F/mg \tag{1.4-13}$$

Substituting Eqn. (1.4-13) into Eqn. (1.4-10),

$$z = [z^{(\eta-1)}]\, C\, [(1+z^2)]\,/g)(dz/dt)$$
$$1 = [z^{(\eta-2)}]\, C\, [(1+z^2)]\,/g)(dz/dt)$$

Rearranging for dt,

$$\begin{aligned} dt &= [z^{(\eta-2)}]\, C\, [(1+z^2)] /g)(dz) \\ &= C[z^{(\eta-2)}]\ [(1+z^2)] /g)(dz) \\ &= C[z^{(\eta-2)} + z^{(\eta-2)} z^2)] /g)(dz) \\ &= C[z^{(\eta-2)} + z^{(\eta)}] /g)(dz) \\ &= (C/g)\, [z^{(\eta-2)} (dz) + z^{(\eta)}(dz)] \end{aligned}$$

Integrating both sides and assuming piece-wise constant value for η,

$$\begin{aligned} t &= (C/g)\, [\int z^{(\eta-2)} (dz) + \int z^{(\eta)}(dz)] \\ &= (C/g)\, [z=zo, z=Z]\, [z^{(\eta-1)}/(\eta-1) + z^{(\eta+1)}/(\eta+1)] \\ &= (C/g)\, [z=zo, z=Z]\, z^{(\eta-1)}[(1/(\eta-1) + z^2/(\eta+1)] \end{aligned} \qquad (1.4\text{-}14)$$

A procedure that generates the Gravity Turn Trajectory is summarized.

1. Set up a set of initial values

   Vo = Initial velocity
   ψo = Initial angle
   yo = Initial attitude

2. Using Eqn. (1.4-13) and the initial values given above, compute the "C." Here the assumption that $\eta(=F/mg)$ is a constant is valid because the integration time in this process is short.

   $$C = Vo / \{(1+zo^2)\, [zo^{(\eta-1)}]\, \}$$

   $zo = \tan(\psi o/2)$, [See Eqn. (1.4-6).]

3. Increase ψ by $\Delta\psi$

   $$\psi = \psi o + \Delta\psi$$

4. Using Eqn. (1.4-14), the time elapsed is computed.

   $$t = (C/g)\, \{z^{(\eta-1)}[(1/(\eta-1) + z^2/(\eta+1)] - zo^{(\eta-1)}[(1/(\eta-1) + zo^2/(\eta+1)]\}$$

   $\Delta t = t$, a short time

where

$$zo = \tan(\psi o/2)$$
$$z = \tan(\psi/2)$$
$$\eta = F/(mg)$$

5. Using Eqn. (1.4-13), compute V.

$$V = [z^{\wedge}(\eta\text{-}1)]C\ [(1+z^{\wedge}2)]$$

6. Compute Δx and Δy (see Fig. 1.4-1)

$$\Delta x = (1/2)\ [Vo \sin \psi o + V \sin \psi]\ \Delta t$$
$$\Delta y = (1/2)\ [Vo \cos \psi o + V \cos \psi]\ \Delta t$$

7. Generate a trajectory

$$x\ (i+1) = x(i) + \Delta x$$
$$y\ (i+1) = y(i) + \Delta y$$

8. Go back to 3) and repeat the process until the aerodynamic load is no longer significant.

Example

1. Set up a set of initial values

$$Vo = 2500 \text{ ft/sec}$$
$$\psi o = 15.0 \text{ deg}$$
$$yo = 7.0 \text{ nm}$$

2. $zo = \tan(15/2)$
$= 0.132$

$$C = Vo\ /\ \{(1+zo^{\wedge}2)\ [zo^{\wedge}(\eta\text{-}1)]\ \}$$
$$2500/\ \{(1 + 0.132^{\wedge}2)(\ 0.132^{\wedge}(1.38\text{-}1)]$$
$$= 5310$$

3. $\psi = \psi o + \Delta\psi$

$= 15 + 0.5$

4. z = tan (15.5/2) = 0.14
   zo = tan (15/2) = 0.13

   Δt = (5310/32.174) {0.14^(1.38-1)[(1/(1.38-1) + 0.14^2/(1.38+1)] - 0.13^(1.38-1)[(1/(1.38-1) + 0.3^2/(1.38+1)]}
   = 2.60

5. V = [0.14^(1.38-1)]*5310*[(1+0.14^2)]
   = 2534.68

6. Δx = (1/2) [2500 sin 15 + 2534.68 sin 15.5]*2.60
   = 867.22
   Δy = (1/2) [2500 cos15 + 2534.68 cos15.5]* 2.60
   = 6477.22

6. x (i+1) = 0.0 + 867.22/5280, 1.0 nm = 5280 ft
   = 0.16
   y (i+1) = 7.0 + 6477.22/5280
   = 8.23

The trajectory computed is tabulated in Table 1.4-1 and plotted in Fig. 1.4-3.

time (sec)	Psi (deg)	z(i)	z(i +1)	C	delta t (sec)	V(ft/sec)	del x (ft)	del y (ft)	x(mile)	y(mile)
0.00	15.00					2500.00			0.00	7.00
2.60	15.50	0.13	0.14	5310.00	2.60	2534.68	867.22	6477.22	0.16	8.23
5.15	16.00	0.14	0.14	5310.00	2.55	2568.94	891.65	6445.75	0.17	1.22
7.65	16.50	0.14	0.14	5310.00	2.51	2602.80	916.16	6416.60	0.17	1.22
10.12	17.00	0.14	0.15	5310.00	2.47	2636.31	940.78	6389.63	0.18	1.21
12.54	17.50	0.15	0.15	5310.00	2.43	2669.49	965.50	6364.73	0.18	1.21
14.93	18.00	0.15	0.16	5310.00	2.39	2702.37	990.35	6341.81	0.19	1.20
17.29	18.50	0.16	0.16	5310.00	2.35	2734.97	1015.33	6320.77	0.19	1.20
19.61	19.00	0.16	0.17	5310.00	2.32	2767.32	1040.47	6301.53	0.20	1.19
.										
.										
.										
122.78	46.50	0.42	0.43	5310.00	1.71	4562.86	3044.84	7129.76	0.58	1.35

124.49	47.00	0.43	0.43	5310.00	1.71	4601.00	3101.74	7176.15	0.59	1.36
126.20	47.50	0.43	0.44	5310.00	1.71	4639.50	3159.83	7223.83	0.60	1.37
127.90	48.00	0.44	0.45	5310.00	1.71	4678.37	3219.14	7272.82	0.61	1.38
129.61		0.45	0.45	5310.00	1.71	4717.61	3279.70	7323.14	0.62	1.39
131.32	49.00	0.45	0.46	5310.00	1.71	4757.24	3341.55	7374.81	0.63	1.40
133.03	49.50	0.46	0.46	5310.00	1.71	4797.26	3404.73	7427.86	0.64	1.41
134.74	50.00	0.46	0.47	5310.00	1.71	4837.68	3469.27	7482.30	0.66	1.42
136.46	50.50	0.47	0.47	5310.00	1.71	4878.52	3535.22	7538.17	0.67	1.43
138.17	51.00	0.47	0.48	5310.00	1.72	4919.78	3602.61	7595.49	0.68	1.44
139.89	51.50	0.48	0.48	5310.00	1.72	4961.46	3671.50	7654.29	0.70	1.45
141.61	52.00	0.48	0.49	5310.00	1.72	5003.59	3741.91	7714.59	0.71	1.46
143.34	52.50	0.49	0.49	5310.00	1.72	5046.17	3813.91	7776.42	0.72	1.47

Table 1.4-1. Gravity Turn Trajectory Computation

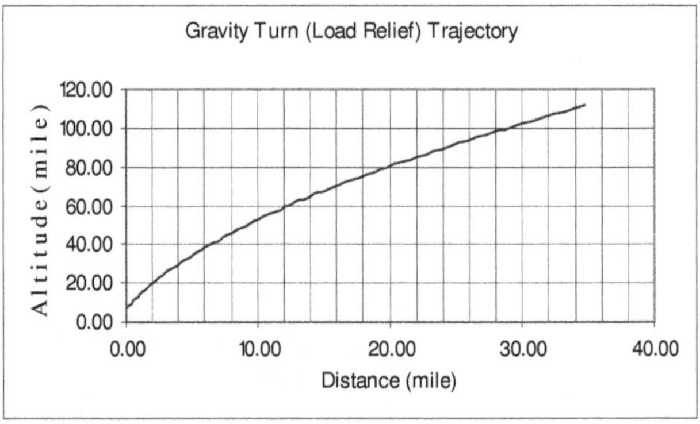

Fig. 1.4-3. Gravity Turn Trajectory

1.5 Rocket Propulsion Systems

The rocket propulsion system is not directly related to the attitude control system, however understanding the basic concept will broaden the knowledge-base that systems engineers need. In this section, propulsion thermodynamics is briefly introduced. We start from the most general form of the principle of energy conservation applied to a "control volume."

The control volume in thermodynamics means a closed volume that encloses a system such as rocket engine. (See Fig. 1.5-1.)

The energy conservation equation is,

$$(dEcv/dt) = Qcv' - Wcv' + \Sigma\, mi'[hi + (Vi^2/2) + gzi] - \Sigma\, mj'[hj + (Vj^2/2) + gzj] \quad (1.5\text{-}1)$$

where

Ecv = energy in closed volume
Qcv' = rate of heat transfer
Wcv' = work done in thermodynamics sense
(It is said that work is done by a system on its surroundings if the sole effect on everything external to the system could have been used to exert energy. One example of the sole effect is expansion of volume due to pressure increase.)

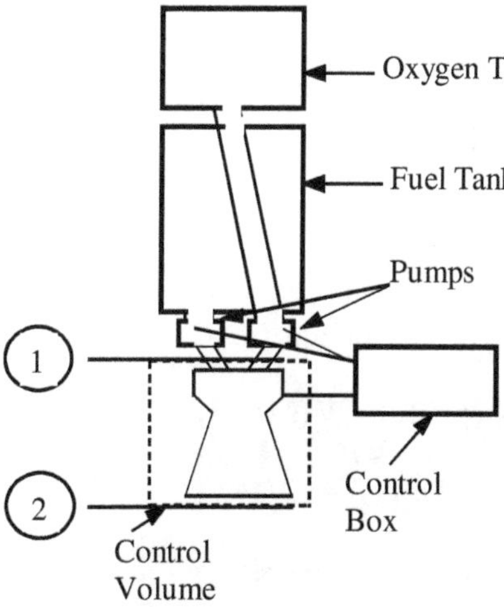

Fig. 1.5-1. Rocket Propulsion System

m' = mass flow rate

h = enthalpy per unit mass
(= u + pv)

(The word, "enthalpy" came from the Greek word, "enthalpein,"
meaning "to warm within.")

u	= internal energy
p	= pressure
V	= velocity
g	= gravity
z	= altitude

In steady state, the energy level in the control volume remains constant, therefore (dEcv/dt)=0. Now since the mass flow process can be assumed to be adiabatic (meaning no significant heat transfer between the control volume and its surroundings), Qcv' = 0, and there is no expansion of the control volume, i.e., Wcv' = 0. Mass-flow into and mass-flow out of the control volume are equal, i.e., m'1= m'2 and (gz1 – gz2) ≈ 0. All these conditions and system states listed above are imposed on Eqn. (1.5-1) and reflected in Eqn. (1.5-2),

0= [h1 +(V1^2/2)] – [h2+(V2^2/2)] (1.5-2)

Rearranging Eqn. (1.5-2) for V2,

V2= sqrt [2(h1-h2) + V1^2] (1.5-3)

For ideal gas, the enthalpy can be written as a product of specific heat cp and the absolute temperature T. Thus, in Eqn. (1.5-3),

V2 = sqrt [2(cp T1 – cp T2) + V1^2]
= sqrt [2cp (T1 - T2) + V1^2]
= sqrt { 2 cp T1[1 –(T2/T1)] + V1^2 } (1.5-4)

For an isentropic (constant entropy) flow process,

T2/T1 = (p2/p1)^[(k-1)/k]

And Eqn. (1.5-4) can be written as,

V2= sqrt { 2 cp T1{ 1 - (p2/p1)^[(k-1)/k]}+ V1^2 }
(1.5-5)

For an ideal gas, the ratio of specific heat at constant pressure, cp , to the specific heat at constant volume, cv , is constant over wide range of temperatures, and the ratio, k, is,

$$k = cp / cv$$

and

$$cp - cv = R \text{ (gas constant)} \qquad (1.5\text{-}6)$$

From these two equations,

$$Cp = kR/(k-1) \qquad (1.5\text{-}7)$$

Substituting the equation above into Eqn. (1.5-5).

$$V2 = \text{sqrt}\{ [2 kR/(k-1)] T1\{1 - (p2/p1)^\wedge[(k-1)/k]\} + V1^\wedge 2 \} \qquad (1.5\text{-}8)$$

Using Eqn. (1.5-8), the velocity at '2' (Fig. 1.5-2) can be computed.

For an isentropic flow process,

$$T1/T2 = (p1/p2)^\wedge[(k-1)/k] = (V2/V1)^\wedge(k-1) \qquad (1.5\text{-}9)$$

From Eqn. (1.5-2),

$$(h1 - h2) = (V2^\wedge 2 - V1^\wedge 2) / 2 \qquad (1.5\text{-}10)$$

For an ideal gas, the enthalpy can be expressed as the product of the specific heat, cp, and the absolute temperature, T.

$$h1 = cp\ T1$$
$$h2 = cp\ T2$$

Now Eqn. (1.5-10) can be written as,

$$(h1 - h2) = (V2^\wedge 2 - V1^\wedge 2) / 2$$
$$= cp\ (T1 - T2) \qquad (1.5\text{-}11)$$

If location 2 (Fig. 1.5-2) is the stagnation point, then

$$V2=0,$$

And from Eqn. (1.5-11),

$$T2 = T1 + V1^2 / (2\ cp) \qquad (1.5\text{-}12)$$

Dividing the equation above by T1,.

$$(T2/T1) = 1 + V1^2 / (2\ cp\ T1)$$

Using Eqn. (1.5-9),

$$\begin{aligned}(p1/p2) &= [1 + V1^2 / (2\ cp\ T1)]^{[k/(k-1)]} \\ &= (V2/V1)^k \qquad (1.5\text{-}13)\end{aligned}$$

The velocity of sound, a, in an ideal gas is,

$$a = \mathrm{sqrt}(kRT)$$

And the Mach Number, M, is defined as,

$$\begin{aligned}M &= V/a \\ &= V/\mathrm{sqrt}(kRT) \qquad (1.5\text{-}14)\end{aligned}$$

Rewriting Eqn, (1.5-12),

$$T2 = T1[1+ V1^2/ (2\ cp\ T1)]$$

Substituting Eqn. (1.5-7) for cp,

$$\begin{aligned}T2 &= T1[1+(1/2)(k-1)\ V1^2 /(kRT1)] \\ &= T1\{1+(1/2)(k-1)[V1 /\mathrm{sqrt}(kRT1)]^2\}\end{aligned}$$

Using Eqn. (1.5-14),

$$T2 = T1[1+(1/2)(k-1)\ M1^2] \qquad (1.5\text{-}15)$$

Solving for M1,

$$M1 = \mathrm{sqrt}\{[2/(k-1)][(T2/T1)-1]\} \qquad (1.5\text{-}16)$$

An equation to compute the stagnation pressure, p2, is derived below. Using Eqn. (1.5-15) and Eqn. (1.5-9),

$$[1+(1/2)(k-1)M1^2] = T2/T1 = (p2/p1)^{[(k-1)/k]} \qquad (1.5\text{-}17)$$

Solving for p2,

$$p2 = p1\,[1+(1/2)(k-1)M1^2]^{[k/(k-1)]} \qquad (1.5\text{-}18)$$

Solving for M1 from Eqn. (1.5-16) by substituting (T2/T1) in Eqn. (1.5-17) into Eqn.(1.5-16)

$$M1 = \text{sqrt}\{[2/(k-1)]\{(p2/p1)^{[(k-1)/k]} - 1\}\} \qquad (1.5\text{-}19)$$

Now, an equation relating nozzle cross section area ratio to Mach number ratio is derived below. From the mass conservation requirement,

$$\rho 1\, A1\, V1 = \rho 2\, A2\, V2$$

$\rho 1$, $\rho 2$	= density at section 1 and 2, respectively
A1, A2	= cross sectional area at section 1 and 2, respectively
V1, V2	= velocity at section 1 and 2, respectively

From the equation above,

$$(A2/A1) = (\rho 1\, V1/ \rho 2 V2)$$

Using Eqn. (1.5-14),

$$= [\rho 1\, M1\, \text{sqrt}(kRT1)]/[\rho 2 M2\, \text{sqrt}(kRT2)]$$
$$= (\rho 1\, M1/ \rho 2\, M2)\, \text{sqrt}(T1/T2)$$
$$= (M1/M2)(\rho 1 / \rho 2)\, \text{sqrt}(T1/T2) \qquad (1.5\text{-}20)$$

For an ideal gas,

$$pv = RT,$$

p = pressure
v = volume
R = gas constant
T = absolute temperature, ^{o}K

$$p(m/\rho) = RT,$$

m = mass

ρ = density

and solving for ρ,

$$\rho = pm/(RT).$$

At the "1" (see Fig. 1.5-2),

$$\rho 1 = p1m/(RT1)$$

and at "2" (see Fig. 1.5-2),

$$\rho 2 = p2m/(RT2)$$

Thus, the ratio of the density between "1" and "2",

$$(\rho 1/ \rho 2) = (p1T2/p2T1)$$

Substituting the ratio into Eqn. (1.5-20).

$$\begin{aligned}(A2/A1) &=(M1/M2)\ (p1T2/p2T1)\mathrm{sqrt}\ (T1/T2)\\ &=(M1/M2)\ (p1/p2)\mathrm{sqrt}\ (T2/T1) \qquad (1.5\text{-}21)\end{aligned}$$

Introducing po and To,

$$(A2/A1)=(M1/M2)[(p1/po)/(p2/po)]\ \mathrm{sqrt}[(T2/To)/(T1/To)]$$

where using Eqn. (1.5-15) and Eqn. (1.5-18), (Fig. 1.5-2),

$$\begin{aligned}(po/p1) &= [1+(1/2)\ (k-1)\ M1\hat{}2]\hat{}[k/(k-1)]\\ (po/p2) &= [1+(1/2)\ (k-1)\ M2\hat{}2]\hat{}[k/(k-1)]\\ (To/T1) &= [1+(1/2)\ (k-1)\ M1\hat{}2]\\ (To/T2) &= [1+(1/2)\ (k-1)\ M2\hat{}2]\end{aligned}$$

Expanding the equation above,

$$\begin{aligned}(A2/A1)=&(M1/M2)\{\ [1+(1/2)\ (k-1)\ M1\hat{}2]\hat{}[-k/(k-1)]\}/\ \{[1+(1/2)\\ &(k-1)M2\hat{}2]\hat{}[-k/(k-1)]\ \}\ \mathrm{sqrt}\{[1+(1/2)\ (k-1)\ M1\hat{}2]/\\ &[1+(1/2)(k-1)\ M2\hat{}2]\}\end{aligned}$$

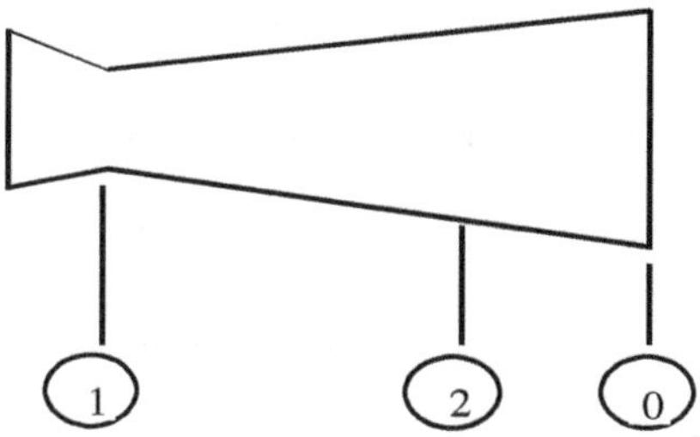

Fig. 1.5-2. Rocket Nozzle Configuration

$$(A_2/A_1)=(M_1/M_2)\left\{[1+(1/2)(k-1)M_2^2]^{[k/(k-1)]}\right\}/\left\{[1+(1/2)(k-1)M_1^2]^{[k/(k-1)]}\right\}\left\{[1+(1/2)(k-1)M_1^2]^{(1/2)}[1+(1/2)(k-1)M_2^2]^{(-1/2)}\right\}$$

$$=(M_1/M_2)\left\{[1+(1/2)(k-1)M_2^2]^{[k/(k-1)]}[1+(1/2)(k-1)M_2^2]^{(-1/2)}\right\}/\left\{[1+(1/2)(k-1)M_1^2]^{[k/(k-1)]}[1+(1/2)(k-1)M_1^2]^{(-1/2)}\right\}$$

$$=(M_1/M_2)\left\{[1+(1/2)(k-1)M_2^2]^{[k/(k-1)-(1/2)]}\right\}/\left\{[1+(1/2)(k-1)M_1^2]^{[k/(k-1)-(1/2)]}\right\}$$

$$=(M_1/M_2)\left\{[1+(1/2)(k-1)M_2^2]^{\{(k+1)/[2(k-1)]\}}\right\}/\left\{[1+(1/2)(k-1)M_1^2]^{\{(k+1)/[2(k-1)]\}}\right\}$$

$$=(M_1/M_2)\,\mathrm{sqrt}\left\{[1+(1/2)(k-1)M_2^2]^{\{(k+1)/[(k-1)]\}}\right\}/\left\{[1+(1/2)(k-1)M_1^2]^{\{(k+1)/[(k-1)]\}}\right\} \qquad (1.5\text{-}22)$$

Fig. 1.5-3 shows plots of temperature ratio [Eqn. (1.5-15)], pressure ratio [Eqn. (1.5-18)], and area ratio [Eqn. (1.5-22)] between the "1" and the "2" when $M_1=1.0$ and $k=1.3$.

From the figure, area ratio, pressure ratio, and temperature ratio can be found for a given Mach number. For example, if Mach number 1.6 ($M_2 = 1.6$) is desired at 2, then the following ratios must be provided. Here $M_1=1.0$.

$A_2/A_1 = 1.57$
$p_2/p_1 = 0.244$

$$T2/T1 = 0.722$$

Higher Mach numbers demand higher area ratios, lower pressure ratios, and lower temperature ratios.

Derivation of a simplified rocket propulsion equation, which conceptually explains how rocket propulsion takes place, is provided below. In the derivation, several parameters such as the properties of compressible gas flow and the pressure force produced are ignored. Fig. 1.5-4 conceptually shows a linear momentum change in rocket propulsion.

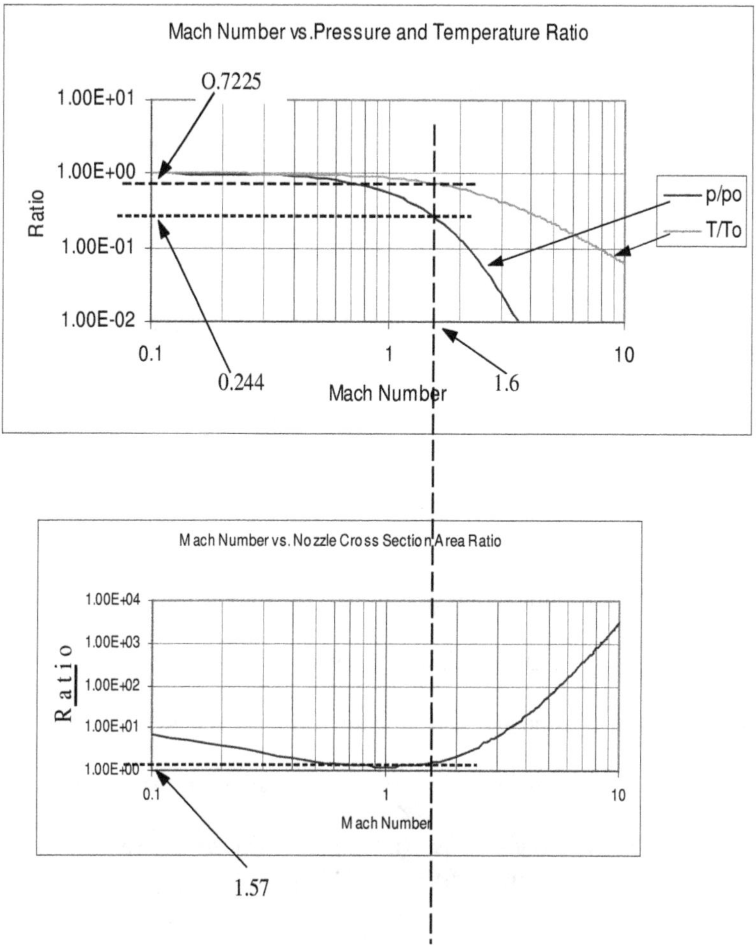

Fig. 1.5-3. Nozzle Parameters vs. Mach Number

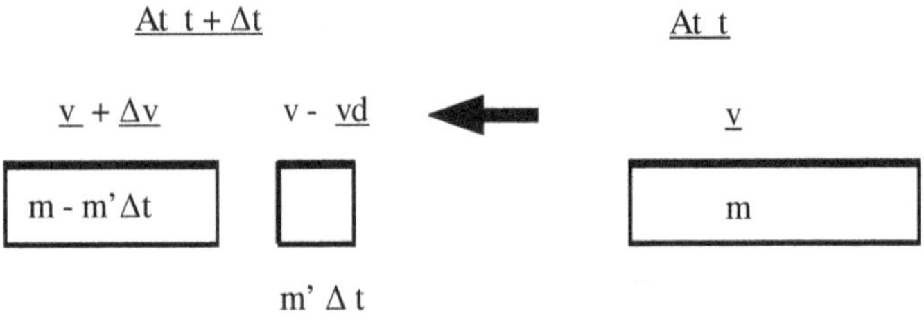

$$m' =: dm/dt$$

Fig. 1.5-4. Linear Momentum Change in a Rocket System

From the Fig. 1.5-4, momentums before and after propulsion can be written as:

Momentum before propulsion = mv
Momentum after propulsion = (m- m' Δt)(v + Δv) + (m' Δt)(v - vd)

Momentum conservation property allows us to write the following equation.

$$mv = (m - m'\,\Delta t)(v + \Delta v) + (m'\,\Delta t)(v - vd)$$

Expanding the equation and canceling mv from both sides,

$$mv = m\,v - (m'\,\Delta t)\,v + m\,\Delta v - (m'\,\Delta t)\,\Delta v + (m'\,\Delta t)\,\underline{v} - (m'\,\Delta t)\,vd$$

$$\begin{aligned} 0 &= -(m'\,\Delta t)\,v + m\,\Delta v - (m'\,\Delta t)\,\Delta v + (m'\,\Delta t)\,v - (m'\,\Delta t)\,vd \\ &= m\,\Delta v - (m'\,\Delta t)\,\Delta v - (m'\,\Delta t)\,vd \\ &= [m - (m'\,\Delta t)]\,\Delta v - (m'\,\Delta t)\,vd \end{aligned}$$

Solving for Δv

$$\begin{aligned} \Delta v &= (m'\,\Delta t)\,vd/[m - (m'\,\Delta t)] \\ &= vd/[m/(m'\,\Delta t)-1] \end{aligned}$$

From the equation above, following observations can be made:

1. Δv becomes large if vd is large, i.e., the velocity of the rocket depends on the rocket exhaust velocity.

2. Δv increases drastically, as the rate of mass exhaustion (m') is such that the ratio of $[m/(m'\Delta t)]$ comes close to 1.

3. Δv decreases, as m' becomes smaller.

Chapter 2 H-infinity, Loop Shaping, and Coprime Factorization

NOTE: Presentation in this chapter is not as rigorous as in references [9], [10], [13],[14],[15],[41],[42],[51],[52], and [53]. A layman approach was taken to explain only a part of the theory for easy understanding for engineers who have never been exposed to this field of study.

The objective of this chapter is to present a mathematical explanation of the concept of H-infinity, loop shaping, and coprime factorization based on which optimal control system design algorithm can be developed. In Section 2.1, the Hamiltonian matrix and the Riccati equation are derived. The Riccati equation is important because it plays a key roll in H-infinity optimization. In Section 2.2, the H-infinity preliminary is introduced and its effectiveness is demonstrated. Section 2.3 deals with the mathematical explanation of the H-infinity design concept. Section 2.4 discusses briefly the concept of loop shaping and the requirement of the coprime factorization. H-infinity control theory is complex, and to reduce its complexity, only a brief description of the portion that is applicable to LV attitude control system design is presented here.

2.1 **Optimization by use of the Riccati Equation**

Consider a linear system as shown in Fig. 2.1-1.

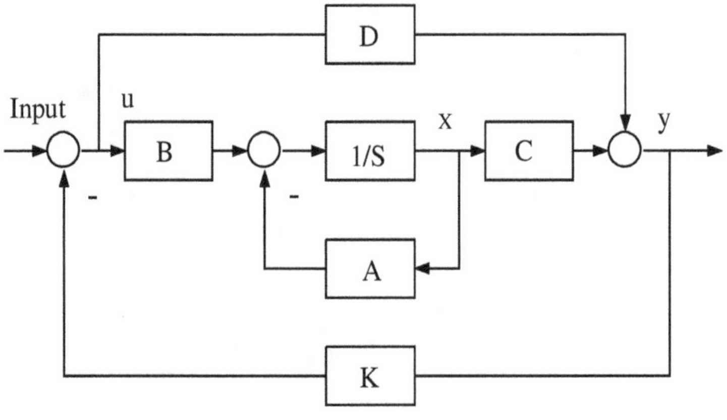

Fig. 2.1-1. Linear System Block Diagram

From Fig. 2.1-1, a system equation can be written as,

$$xd = Ax + Bu \qquad (2.1\text{-}1)$$
$$y = Cx + Du$$

where

xd (x dot) ≡ dx/dt (a new notation used in this book)

For this study, we set D equal to zero. Then, from the figure we can write,

$$u = -Ky \qquad (2.1\text{-}2)$$

Note that the input also is set equal to zero because linear system stability is independent of input, which implies that the zero input assumption will not affect control system design. In other words, once a stable system is designed, then the system is stable no matter any type of input is applied.

Now, we construct an optimization (minimization) cost function,

$$J = \int [y^*(Qr)y + u^*Ru]dt \qquad (2.1\text{-}3)$$

where

* : transpose
Qr : positive definite or semi-definite
R : positive definite.

Using Eqn. (2.1-1) and Eqn. (2.1-2) for D=0,

$$u^*Ru = x^*C^*K^*RKCx \qquad (2.1\text{-}4)$$

Substituting Eqn. (2.1-4) into Eqn. (2.1-3),

$$J = \int x^*(C^*QrC + C^*K^*RKC)\, x\, dt$$

Minimization of J implies bounding the state vector, x. Now we set the integrant equal to the derivative of -(x*Px). The reason for this set up will be explained later.

$$x^*(Q + C^*K^*RKC)x = -(d/dt)(x^*Px)$$
$$= -[(xd)^*Px + x^*(Pd)x + x^*P(xd)]$$

(2.1-5)

where

Q = C*Qr C
P: positive definite real symmetric matrix
Pd (P dot) ≡ dP/dt (a new notation used in this book)

From Eqn. (2.1-1) and Eqn.(2.1-2),

xd = Ax – BKCx
= (A – BKC)x (2.1-6)

Substituting Eqn. (2.1-6) into Eqn. (2.1-5)

x*(Q + C*K*RKC)x = -x *(A – BKC)*Px-x*(Pd)x - x*P(A – BKC)x
= -x*((A – BKC)*P + (Pd) + P(A – BKC))x
(2.1-7)

From Eqn. (2.1-7),

Q + C*K*RKC = - [(A – BKC)*P + (Pd) + P(A – BKC)]

Let R = T*T (R is a symmetric matrix), and rearranging it,

(A* – C*K*B*)P + P(A – BKC)) + Q + C*K*T*TKC = -(Pd)

A*P – C*K*B*P + PA – PBKC + Q + C*K*T*TKC = -(Pd)

(2.1-8)

Rewriting C*K*B*P,

C*K*B*P = C*K*(T*)Inv(T*)*B*P (2.1-9)

And

PBKC = PB(T*)Inv(T*)KC (2.1-10)

Substituting Eqn. (2.1-9) and (2.1-10) into Eqn. (2.1-8),

A*P – C*K*(T*)Inv(T*)*B*P + PA - PB(T*)Inv(T*)KC + Q +
C*K*T*TKC=-(Pd) (2.1-11)

Adding and subtracting "PB[Inv(T*)]*[Inv(T*)]B*P" from Eqn. (2.1-11),

we have,

A*P – C*K*(T*)Inv(T*)*B*P + PA - PB(T*)Inv(T*)KC + Q +
C*K*T*TKC + S –S = -(Pd)

A*P – C*K*(T*)Inv(T*)*B*P + PA - PB(T*)Inv(T*)KC + Q +
C*K*T*TKC + PB[Inv(T*)]*[Inv(T*)]B*P –
PB[Inv(T*)]*[Inv(T*)]B*P = -(Pd)

Rearranging the equation above to construct a square term,

A*P + PA+ C*K*T*TKC – C*K*(T*)Inv(T*)*B*P -
PB(T*)Inv(T*)KC + PB[Inv(T*)]*[Inv(T*)]B*P –
PB[Inv(T*)]*[Inv(T*)]B*P + Q =-(Pd) (2.1-12)

The underlined part of Eqn. (2.1-12) can be written as a square term as shown below.

A*P + PA+{TKC-[Inv(T*)]B*P)}*{TKC-[Inv(T*)]B*P}*-
PB[Inv(T*)]*[Inv(T*)]B*P + Q =-(Pd)

The term second to the last in Eqn. (2.1-12),

PB[Inv(T*)]*[Inv(T*)]B*P = PB Inv(R)B*P

| [Inv(T*)]*[Inv(T*)] = Inv(T)*Inv(T*)
| = Inv(T*T)
| = Inv (R)

Rewriting the equation above,

A*P + PA+[TKC-Inv(T*)B*P]*[TKC-Inv(T*)B*P]* - PB
Inv(R)B*P + Q = -(Pd)

One of the solutions of the equation above can be obtained by setting

$$TKC - Inv(T^*)B^*P = 0 \qquad (2.1\text{-}13)$$

and

$$A^*P + PA - PB\, Inv(R)B^*P + Q = -(Pd) \qquad (2.1\text{-}14)$$

Eqn. (2.1-14) is known as the Riccati equation.

From Eqn. (2.1-13), we can compute the feedback gain, K.

$$KC = Inv(T)Inv(T^*)B^*P$$

Multiplying the transpose of C on both sides,

$$KC C^* = Inv(T)Inv(T^*)B^*PC^*$$

Since,

$C \in R(rxn)$, $r \leq n$ and rank (C) = $r \leq$, the inverse of (CC^*) exists,

$$\begin{aligned} K &= Inv(T^*T)B^*P\,C^*\, Inv\,(CC^*) \\ &= Inv\,(R)\,B^*\,P\,C^*\, Inv\,(CC^*) \qquad (2.1\text{-}15) \end{aligned}$$

The Riccati equation, Eqn. (2.1-14), is very important in control system design. From this equation, we solve for P, and the P is used to construct the feedback gain, K, in Eqn. (2.1-15).

Now we go back to the cost function J [Eqn. (2.1-3)], and using Eqn. (2.1-4) and Eqn. (2.1-5), we obtain,

$$\begin{aligned} J &= \int (x^*Qx + u^*Ru)dt \\ &= -\int (d/dt)(x^*Px)dt \\ &= -\int d(x^*Px) \end{aligned}$$

Integrating from $t=0$ to $t=\infty$

$$= -\ [x(\infty)^*Px(\infty)] + [x(0)^*Px(0)]$$

It is seen here that if the system is stable [i.e., $x(\infty)$ is bounded], then J will be bounded, and it will be a minimum if the P matrix is a minimum. The solution of the Riccati equation provides the P matrix as stated earlier, and as J converges, we achieve minimization.

In order to solve the Riccati equation, we introduce a variable, λ, where

$$\lambda = Ph \qquad (2.1\text{-}16)$$

$$\lambda d = (Pd)h + P(hd) \qquad (2.1\text{-}17)$$

where

λd (λ dot) $\equiv d\lambda/dt$
Pd (P dot) $\equiv dP/dt$
hd (h dot) $\equiv dh/dt$

From Eqn. (2.1-12),

{A*P + PA+ C*K*T*TKC – C*K*(T*)Inv(T*)*B*P - PB(T*)Inv(T*)KC + PB[Inv(T*)]*[Inv(T*)]B*P – PB[Inv(T*)]*[Inv(T*)]B*P + Q} h + (Pd)h =0

Now substituting K =Inv (R) B* P Inv (C) [Eqn. (2.1-15)],

{A*P + PA+ C* Inv (R) B* P Inv (C)*T*T Inv (R) B* P Inv (C)C – C* Inv (R) B* PInv(C)*(T*)Inv(T*)*B* P-PB(T*)Inv(T*) Inv (R) B* P Inv (C)C+PB[Inv(T*)]* [Inv(T*)]B*P – PB[Inv(T*)]*[Inv(T*)]B*P + Q} h + (Pd)h = 0

{A*P + PA+ C* Inv (R) B* P Inv (C)RInv (R) B* P – C* Inv (R) B* P Inv (C)*(T*)Inv(T*)*B*P - PB(T*)Inv(T*) Inv (R) B* P Inv (C)C + PB[Inv(T*)]*[Inv(T*)]B*P – PB[Inv(T*)]*[Inv(T*)]B*P + Q} h + (Pd)h = 0

(underlined terms are cancelled out)

{A*P + PA- PB(T*)Inv(T*) Inv (R) B* P + PB[Inv(T*)]*[Inv(T*)]B*P – PB[Inv(T*)]*[Inv(T*)]B*P + Q} h+ (Pd)h = 0

(underlined terms are cancelled out)

$$[A*P + PA- PB\underline{(T*)Inv(T*)} \, Inv \, (\, R \,) \, B* \, P \, + Q] \, h + (Pd)h = 0$$

$$[A*P + PA- PB \, Inv \, (\, R \,) \, B* \, P \, + Q] \, h + (Pd)h = 0$$

$$A*Ph + PAh- PB \, Inv \, (\, R \,) \, B* \, Ph \, + Qh + (Pd)h = 0$$

$$A*Ph + P[Ah- B \, Inv \, (\, R \,) \, B* \, Ph \,] \, + Qh + (Pd)h= 0$$

Solving for (Pd)h from the equation above,

$$(Pd)h = -A*Ph - P[Ah- B \, Inv \, (\, R \,) \, B* \, Ph \,] \, - Qh$$

$$(2.1-18)$$

Set the term, "[Ah- B Inv (R) B* Ph]", in Eqn. (2.1-18) equal to derivative of h.

$$(hd) = Ah - \, BInv(R)B*Ph \qquad (2.1-19)$$

Substituting Eqn. (2.1-18) and Eqn. (2.1-19) into Eqn. (2.1-17)

$$\begin{aligned}
\lambda d &= (Pd)h + P(hd) \\
&= -A*Ph - PAh + PB \, Inv \, (\, R \,) \, B* \, Ph \, - Qh + P(hd) \\
&= -A*Ph - PAh + PB \, Inv \, (\, R \,) \, B* \, Ph \, - Qh + P(Ah - \\
&\quad BInv(R)B*Ph) \\
&= -A*Ph - \underline{PAh} + \underline{PB \, Inv \, (\, R \,) \, B* \, Ph} \, - Qh + \underline{PAh} \\
&\quad -\underline{PBInv(R)B*Ph}
\end{aligned}$$

(underlined terms are cancelled out)

$$= - \, A*\lambda \, - Qh \qquad (2.1-20)$$

$$| \, \lambda = Ph$$
$$| \, \lambda d = (Pd)h + P(hd), \quad \text{Eqn. (2.1-17)}$$

From Eqn. (2.1-19) and Eqn. (2.1-16)

$$(hd) = Ah - BInv(R)B^*\lambda \qquad (2.1\text{-}21)$$

From Eqn. (2.1-20) and (2.1-21),

$$\begin{vmatrix} (hd) \\ \lambda d \end{vmatrix} = \begin{vmatrix} A & -BInv(R)B^* \\ -Q & -A^* \end{vmatrix} \begin{vmatrix} h \\ \lambda \end{vmatrix} \qquad (2.1\text{-}22)$$

$$= H \begin{vmatrix} h \\ \lambda \end{vmatrix}$$

The matrix, H, is known as the Hamiltonian matrix. Solution of Eqn. (2.1-22) yields "h" and "λ," and, using Eqn. (2.1-16), we solve the Riccati equation [Eqn. (2.1-14)]. The solution is,

$$P=\lambda\ Inv\ (h) \qquad (2.1\text{-}23)$$

A better method of solving for P is presented below. It is known in linear algebra that the modal matrix of a matrix diagonalizes the matrix. The modal matrix consists of eigenvectors of the matrix, and the diagonal elements of the diagonalized matrix are the eigenvalues of the matrix. Applying this linear algebra theorem to our problem, we let M be a modal matrix of H. Then, we write,

$$\begin{aligned}\Lambda &= Inv\,|M|\,H\,|M| \\ &= Inv\,|M| \begin{vmatrix} A & -BInv(R)B^* \\ -Q & -A^* \end{vmatrix} |M|\end{aligned}$$

where

Λ : diagonal matrix

Rearranging the equation above,

$$|M|\,\Lambda = \begin{vmatrix} A & -BInv(R)B^* \\ -Q & -A^* \end{vmatrix} |M| \qquad (2.1\text{-}24)$$

Where M (2nx2n) is a modal matrix of the Hamiltonian matrix, H (2nx2n), and Λ(2nx2n) is a diagonal matrix. The diagonal elements of "Λ " are the eigenvalues of H. Notice that there are 2n eigenvectors in M. Now, we select n eigenvectors that are the bases of an invariant subspace of H.

****

The meaning of invariant subspace is that S is invariant of A if the image of S under A is contained in S. Let a set of three vectors, x1, x2, and x3, be eigenvectors of a matrix A and span a space S. Then, Ax1, Ax2, and Ax3 can be expressed as a linear combination of x1, x2, and x3, i.e., these vectors belong to the same space, S. For example,

$$A = \begin{vmatrix} 2 & 5 & 9 \\ 3 & 7 & 4 \\ 6 & 8 & 1 \end{vmatrix}$$

Eigenvalues and corresponding eigenvectors of A are:

$$\lambda 1 = 14.7859 \qquad x1 = \begin{vmatrix} 1.0000 \\ 0.8676 \\ 0.9387 \end{vmatrix}$$

$$\lambda 2 = -5.9933 \qquad x2 = \begin{vmatrix} 1.0000 \\ 0.0513 \\ -0.9166 \end{vmatrix}$$

$$\lambda 3 = 1.2074 \qquad x3 = \begin{vmatrix} 1.0000 \\ -0.7416 \\ 0.3239 \end{vmatrix}$$

Now we can show that the vector generated by transforming x1 by the transformation matrix, A, remains in the same space spanned by the set of the eigenvectors. That is,

$$Ax1 = \begin{vmatrix} 14.7863 \\ 12.8280 \\ 13.8795 \end{vmatrix} = 14.7863 * x1 + 0.0 * x2 + 0.0 * x3$$

The right side of the equation above is,

$$14.7863 * x1 + 0.0 * x2 + 0.0 * x3 = 14.7863 * \begin{vmatrix} 1.000 \\ 0.8676 \\ 0.9387 \end{vmatrix} + 0.0 * \begin{vmatrix} 1.000 \\ 0.0513 \\ -0.9166 \end{vmatrix} + 0.0 * \begin{vmatrix} 1.000 \\ -0.7416 \\ 0.3239 \end{vmatrix}$$

$$= \begin{vmatrix} 14.7863 \\ 12.8280 \\ 13.8795 \end{vmatrix}$$

It is shown that the Ax1 matrix belongs to the space that is spanned by x1, x2, and x3. Thus, the space is an invariant subspace.

****

As we discussed earlier, any vector that belongs to this invariant subspace will remain in the same subspace after being transformed by the H matrix.

The selected n eigenvectors, x , are,

$$|x| = \begin{vmatrix} x(1,1) & \cdots & x(1,n) \\ \vdots & & \vdots \\ \vdots & & \vdots \\ x(2n,1) & \cdots & x(2n,n) \end{vmatrix}$$

From Eqn. (2.1-24), we construct the following equation.

$$\begin{vmatrix} x(1,1) & \cdots & x(1,n) \\ \vdots & & \vdots \\ \vdots & & \vdots \\ x(2n,1) & \cdots & x(2n,n) \end{vmatrix} \begin{vmatrix} \lambda(1,1) & 0 & \cdots & 0 \\ 0 & \cdots & & 0 \\ 0 & \cdots & & \lambda(n,n) \end{vmatrix} = H \begin{vmatrix} x(1,1) & \cdots & x(1,n) \\ \vdots & & \vdots \\ \vdots & & \vdots \\ x(2n,1) & \cdots & x(2n,n) \end{vmatrix} \qquad (2.1\text{-}25)$$

Now define Λ, P1, and P2,

$$\Lambda = \begin{vmatrix} \lambda(1,1) & 0 & \cdots & 0 \\ 0 & \cdots & & 0 \\ 0 & \cdots & & \lambda(n,n) \end{vmatrix},$$

$$P1 = \begin{vmatrix} x(1,1) & \cdots & x(1,n) \\ \vdots & & \vdots \\ x(n,1) & \cdots & x(n,n) \end{vmatrix}$$

and

$$P2 = \begin{vmatrix} x(n+1,1) & \cdots & x(n+1,n) \\ \vdots & & \vdots \\ x(2n,1) & \cdots & x(2n,n) \end{vmatrix}$$

Then, Eqn. (2.1-25) can be written as,

$$\begin{vmatrix} P1 \\ P2 \end{vmatrix} \Lambda = \begin{vmatrix} A & -BInv(R)B^* \\ -Q & -A^* \end{vmatrix} \begin{vmatrix} P1 \\ P2 \end{vmatrix}$$

Post-multiplying both sides of the equation above by Inv(P1),

$$\begin{vmatrix} P1 \\ P2 \end{vmatrix} \Lambda \; Inv(P1) = \begin{vmatrix} A & -BInv(R)B^* \\ -Q & -A^* \end{vmatrix} \begin{vmatrix} P1 \\ P2 \end{vmatrix} Inv(P1) \qquad (2.1\text{-}26)$$

the left side of Eqn. (2.1-26) can be rewritten as

$$\begin{vmatrix} P1 \\ P2 \end{vmatrix} \Lambda \; Inv(P1) = \begin{vmatrix} P1 \\ (P2)Inv(P1)(P1) \end{vmatrix} \Lambda \; Inv(P1)$$

$$= \begin{vmatrix} I \\ (P2)Inv(P1) \end{vmatrix} P1 \; \Lambda \; Inv(P1)$$

Now defining P as

$$P =: (P2)[Inv(P1)] \qquad (2.1\text{-}27)$$

From Eqn. (2.1-26) and Eqn. (2.1-27),

$$\begin{vmatrix} I \\ P \end{vmatrix} P1 \; \Lambda \; Inv(P1) = \begin{vmatrix} A & -BInv(R)B^* \\ -Q & -A^* \end{vmatrix} \begin{vmatrix} I \\ P \end{vmatrix} P1 \; Inv(P1)$$

Pre-multiplying both sides by $\begin{vmatrix} -P & I \end{vmatrix}$,

$$\begin{vmatrix} -P & I \end{vmatrix} \begin{vmatrix} I \\ P \end{vmatrix} P1 \; \Lambda \; Inv(P1) = \begin{vmatrix} -P & I \end{vmatrix} \begin{vmatrix} A & -BInv(R)B^* \\ -Q & -A^* \end{vmatrix} \begin{vmatrix} I \\ P \end{vmatrix}$$

$$(-P + P) \; P1 \; \Lambda \; Inv(P1) = \begin{vmatrix} -PA - Q & P\,BInv(R)B^* - A^* \end{vmatrix} \begin{vmatrix} I \\ P \end{vmatrix}$$

$$0 = [-PA - Q + PBInv(R)B^*P - A^*P]$$

$$0 = A^*P + PA - PBInv(R)B^*P + Q$$

(2.1-28)

From Eqn. (2.1-14),

A*P + PA- PB Inv(R)B*P + Q = -(Pd)

Since the derivative of the steady state of P for a stable control system is equal to zero, i.e.,

Pd ≡ dP/dt =0.0,

Eqn. (2.1-14) can be written as,

A*P + PA- PB Inv(R)B*P + Q = 0.0 (2.1-29)

Now, note that Eqn. (2.1-28) is derived from Eqn. (2.1-24) and in the process we set P as,

P = P2[Inv(P1)].

The P satisfies the equation, Eqn. (2.1-28), and that equation is identical to the Riccati equation, Eqn. (2.1-29). Thus, the P satisfies the Riccati equation, and the P is indeed a solution of the Riccati equation.

Note that |P1 | are sets of eigenvectors of the Hamiltonian matrix, and they
|P2 |
are the bases of an invariant subspace.

Therefore, to solve the Riccati equation, instead of solving the matrix differential equation, Eqn. (2.1-22), we construct the Hamiltonian matrix from the system equation and then compute eigenvectors. From this set of eigenvectors, we construct the solution of the Riccati equation according to Eqn. (2.1-27). Note that not all the eigenvectors are part of the solution. Only those sets of eigenvectors that are the bases for an invariant subspace of the Hamiltonian matrix are the ones that satisfy the Riccati equation.

Following is an example, demonstrating how we can solve Riccati equation by use of the Hamiltonian matrix.

Example:

A plant has the following system matrices

$$A=\begin{vmatrix} -18.0 & -4.7 & 0 \\ 6.8 & 2.0 & 0 \\ 0 & 4.6 & 0 \end{vmatrix} \quad C=\begin{vmatrix} 0 & 5.6 & 4.9 \end{vmatrix}$$

$$B=\begin{vmatrix} 4.2 \\ 0.0 \\ 0.0 \end{vmatrix} \quad D=\begin{vmatrix} 0 \end{vmatrix}$$

$$R=\begin{vmatrix} 2 \end{vmatrix}$$

$$Q=\begin{vmatrix} 1 & 0 & 0 \\ 0 & 1 & 0 \\ 0 & 0 & 1 \end{vmatrix}$$

Then, the Hamiltonian matrix can be constructed by using the following matrices.

$$A=\begin{vmatrix} -18.0 & -4.7 & 0 \\ 6.8 & 2.0 & 0 \\ 0 & 4.6 & 0 \end{vmatrix}$$

$$-B\mathrm{Inv}(R)B^* = -\begin{vmatrix} 4.2 \\ 0.0 \\ 0.0 \end{vmatrix} \begin{vmatrix} 0.5 \end{vmatrix} \begin{vmatrix} 4.2 & 0.0 & 0.0 \end{vmatrix}$$

$$= \begin{vmatrix} -8.82 & 0 & 0 \\ 0 & 0 & 0 \\ 0 & 0 & 0 \end{vmatrix}$$

$$-A^* = \begin{vmatrix} 18.0 & -6.8 & 0 \\ 4.7 & -2.0 & -4.6 \\ 0 & 0 & 0 \end{vmatrix}$$

$$-Q = \begin{vmatrix} -1 & 0 & 0 \\ 0 & -1 & 0 \\ 0 & 0 & -1 \end{vmatrix}$$

Using Eqn. (2.1-22),

$$H = \begin{vmatrix} -18.0 & -4.7 & 0 & -8.82 & 0 & 0 \\ 6.8 & 2.0 & 0 & 0 & 0 & 0 \\ 0 & 4.6 & 0 & 0 & 0 & 0 \\ -1 & 0 & 0 & 18.0 & -6.8 & 0 \\ 0 & -1 & 0 & 4.7 & -2.0 & -4.6 \\ 0 & 0 & -1 & 0 & 0 & 0 \end{vmatrix}$$

Using "[V D]=eig(H).m" MATLAB program, we compute the eigenvectors and eigenvalues of the H as shown below.

v1	v2	v3	v4
0.9335	-0.2253	0.0865 + 0.1721i	0.0865 - 0.1721i
-0.3436	-0.1059	-0.2412 - 0.2096i	-0.2412 + 0.2096i
0.0960	-0.0296	0.6190	0.6190
0.0214	0.9369	-0.0609 - 0.1894i	-0.0609 + 0.1894i
-0.0288	0.2437	-0.1466 - 0.5905i	-0.1466 + 0.5905i
0.0058	0.0018	0.1968 - 0.1709i	0.1968 + 0.1709i

v5	v6
0.0595 + 0.0525i	0.0595 - 0.0525i
-0.2591 + 0.2251i	-0.2591 - 0.2251i
-0.6648	-0.6648
-0.0046 - 0.2272i	-0.0046 + 0.2272i
0.0323 - 0.5503i	0.0323 + 0.5503i
0.2113 + 0.1836i	0.2113 - 0.1836i

and eigenvalues are,

Val(1) = 16.4719
Val(2) = -16.4719
Val(3) = 1.7927 + 1.5575i
Val(4) = 1.7927 - 1.5575i
Val(5) = -1.7927 + 1.5575i
Val(6) = -1.7927 - 1.5575i

Constructing P1.

$$P1 = \begin{vmatrix} v1(1) & v2(1) & v3(1) \\ v1(2) & v2(2) & v3(2) \\ v1(3) & v2(3) & v3(3) \end{vmatrix}$$

```
= | 0.9335     -0.2253     0.0865 + 0.1721i  |
  |-0.3436     -0.1059    -0.2412 - 0.2096i  |
  | 0.0960     -0.0296     0.6190            |

P2 = | v1(4)   v2(4)   v3(4) |
     | v1(5)   v2(5)   v3(5) |
     | v1(6)   v2(6)   v3(6) |

   = |0.0214      0.9369     -0.0609 - 0.1894i  |
     |-0.0288     0.2437     -0.1466 - 0.5905i  |
     |0.0058      0.0018      0.1968 - 0.1709i  |
```

and

```
P = P2[Inv(P1)]

  = | -1.6108 + 0.1766i   -4.9303 + 0.0519i   -1.7631 - 1.5317i  |
    | -0.4195 + 0.1450i   -1.2324 + 0.0427i   -0.6325 - 1.2582i  |
    | -0.0366 + 0.0318i   -0.0277 + 0.0094i    0.3179 - 0.2762i  |
```

To check the answer, we substitute the P into Eqn.(2.1-29).

A*P + PA- PB Inv(R)B*P + Q = 0

```
| -18.0  6.8    0 || -1.6108 + 0.1766i   -4.9303 + 0.0519i   -1.7631 - 1.5317i |
|  -4.7  2.0  4.6 || -0.4195 + 0.1450i   -1.2324 + 0.0427i   -0.6325 - 1.2582i |
|   0    0      0 || -0.0366 + 0.0318i   -0.0277 + 0.0094i    0.3179 - 0.2762i |

  | -1.6108 + 0.1766i   -4.9303 + 0.0519i   -1.7631 - 1.5317i || -18.0  -4.7   0 |
+ | -0.4195 + 0.1450i   -1.2324 + 0.0427i   -0.6325 - 1.2582i ||   6.8   2.0   0 |
  | -0.0366 + 0.0318i   -0.0277 + 0.0094i    0.3179 - 0.2762i ||   0     4.6   0 |

  | -1.6108 + 0.1766i   -4.9303 + 0.0519i   -1.7631 - 1.5317i || 4.2 || ½ ||4.2 0 0|
- | -0.4195 + 0.1450i   -1.2324 + 0.0427i   -0.6325 - 1.2582i || 0   |
  | -0.0366 + 0.0318i   -0.0277 + 0.0094i    0.3179 - 0.2762i || 0   |

| -1.6108 + 0.1766i   -4.9303 + 0.0519i   -1.7631 - 1.5317i |     | 1 0 0 |
| -0.4195 + 0.1450i   -1.2324 + 0.0427i   -0.6325 - 1.2582i |  +  | 0 1 0 | =
| -0.0366 + 0.0318i   -0.0277 + 0.0094i    0.3179 - 0.2762i |     | 0 0 1 |
```

(1.0e-013)*

-0.3197 + 0.1066i	-0.7105 + 0.2309i	-0.4263 - 0.1066i
-0.0178 + 0.0622i	0 + 0.1865i	-0.0711 + 0.1021i
-0.0244 + 0.0155i	-0.0711 + 0.0400i	-0.0444 + 0.0039i

Multiplying each element by 1.0e(-13), we can see that the matrix is practically a null matrix, and this proves that the P satisfies the Riccati equation.

Now we summarize what we have learned so far in this section.

1) Construct a linearized plant to be controlled (see Fig. 2.1-2)

2) Set up an optimization equation to optimize the control system [Eqn. (2.1-3)]

   $J = \int (y^*Qy + u^*Ru)dt$

   where Q and R are diagonal weighting matrices.

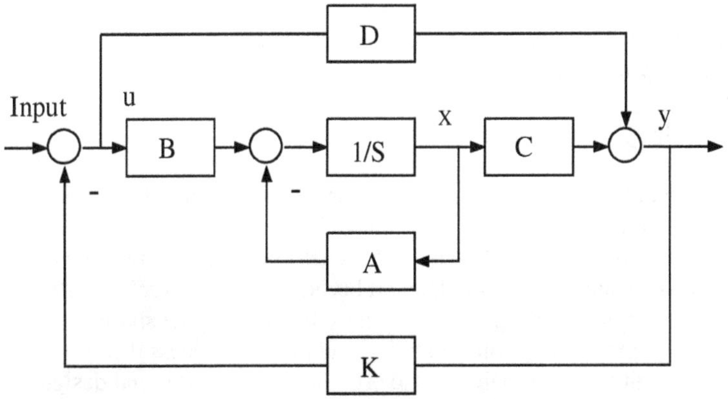

Fig. 2.1-2 Typical Control Systems Block Diagram

3) Construct a Riccati equation [Eqn. (2.1-29), Pd =0]

   $A^*P + PA - PBInv(R)B^*P + Q = 0$

where A and B are from the linearized plant model.

4) Construct the Hamiltonian matrix and solve for P [Eqn. (2.1-22)].

$$H = \begin{vmatrix} A & -BInv(R)B^* \\ -Q & -A^* \end{vmatrix}$$

5) Compute a set of eigenvectors that is the bases of an invariant subspace of H

and set it to $\begin{vmatrix} P1 \\ P2 \end{vmatrix}$.

6) Construct the solution, P [Eqn. (2.1-27)]

$$P = P2^*Inv(P1)$$

7) Construct the feedback loop block, K [Eqn.(2.1- 15)]

$$K = Inv(R)B^* P\, Inv(C)$$

A typical block diagram representing a control system in classical control theory is shown in Fig. 2.1-2.

2.2 H-infinity Preliminary

The history of the control system theory can be divided into three time periods. It started in 1930, and since then, classical control theories, such as Bode, Nyquist, and Root Locus, were developed. Beginning in 1960, modern control theories emerged, and it became possible to solve multi-input multi-output control problems by introducing the state space approach. Controllability, observability, and random process theories developed in this period enabled us to estimate system states and design optimal feedback loops. In 1980, Zames introduced the H-infinity method, and, in 1987, Francis published his now- famous book entitled "A Course in H-infinity Control Theory." Since then, Glover and Doyle came up with techniques of obtaining practical solutions, and numerous other researchers published valuable papers that contributed to the refinement of H-infinity control theory (ref. 42) during this time period.

The H-infinity method, rather than being a theory, is basically a technique by which an optimization-based control system can be designed. Major advantages of the H-infinity method are that it guarantees robustness and stability by appropriate design and that it can be applied to multi-input, multi-output control system design. It was named after an eminent mathematician, G. H. Hardy, who investigated the behaviors of many stable transfer functions. The name actually refers to a mathematical framework (ref. 37).

The mathematical complexity of this method often discourages many engineers from even attempting to use it. Here, a basic concept is introduced. Consider the block diagram in Fig. 2.2-1a.

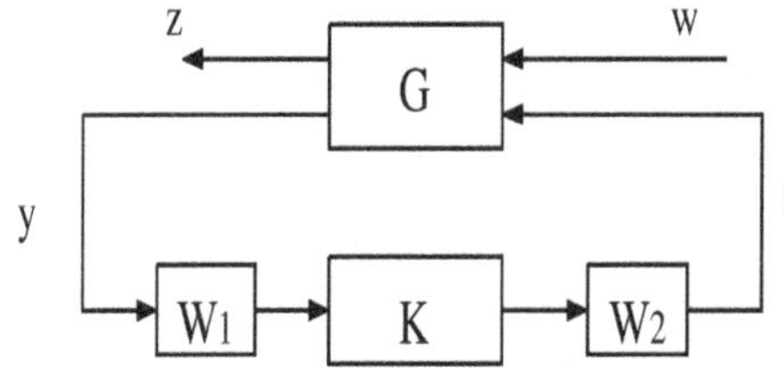

Fig. 2.2-1a. H-infinity Block Diagram

In the figure,

G	= plant
K	= controller
W1	= weighting function 1
W2	= weighting function 2
z	= error signal
y	= measurable output
u	= controller output and input to plant, G.
w	= input (external disturbance)

The H-infinity design constructs K for given G, W1, and W2 in such a way that K minimizes the largest singular value (see page 101 for definition) of the transfer function from w to z for all frequencies of interest to the designer.

In other words, minimization of the largest singular value means minimization of "z" (error), which implies that the effect of "w" (disturbance) on the system response is minimized (robustness). In a sense, the H-infinity control objective is similar to minimization of the error "e" in classical control system (Fig. 2.2-1b) but under some disturbances and with math modeling errors without requiring state estimation such as Kalman filtering, which is a very powerful control system design tool.

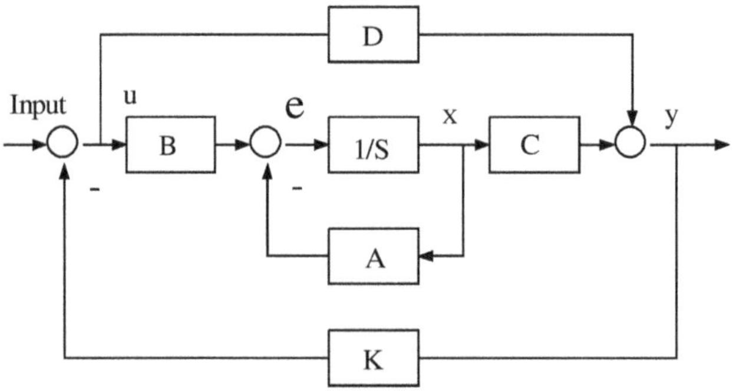

Fig. 2.2-1b Classical Control System Block Diagram

If the largest singular value is less than or equal to some value, γ, then the system can tolerate error less than $(1/\gamma)$. This means that if the minimized largest singular value is small, then the system can tolerate large modeling errors. The mathematical expression of the statement above is as follows:

$$\| Tzw \|_\infty = \sup \sigma max \, [Tzw\,(j\omega)] \text{ for all } \omega$$

and

$$\text{if } \| Tzw \|_\infty \leq \gamma, \text{ then } \| \Delta \|_\infty \leq 1/\gamma$$

where $\| \Delta \|_\infty$ means the largest singular value of the modeling error.

H-infinity control system design is basically a control system synthesis. The mathematical manipulation of the synthesis technique is a major issue in H-infinity control theory. Numerous approaches to solve this mathematical problem were reported in the past, but these approaches were not applicable to the state-space method until in 1984 when for the first time Doyle presented the solution to a general state space multi-input multi-output problem in H-infinity frame. Since then many researchers published papers on H-infinity control synthesis problems (ref. 42). The following is an introduction of one of the solutions.

In this approach two important functions are used. These are the Riccati equations and weighting functions. It requires the use of the Riccati equation twice, and also requires two weighting functions, W1 and W2. The Riccati equation is used for optimization, and the two weighting functions are used for loop shaping. The loop shaping is an important design technique that virtually determines the success of design. The weighting functions are used to weight the singular values in such a way that the signals in the noisy frequency range are to be attenuated and those in the desired signal frequency range are amplified. Fig. 2.2-2 shows a comparison of singular values of two control systems. The first control system has two weighting functions, W1 and W2. The plot shows amplification of the desired signal in the frequency range less than 10 rad/sec, while the disturbing signals of frequencies above 10 rad/sec undergo considerable attenuation. The second control system is not equipped with the weighting functions. It displays flat singular values throughout the whole frequency and fails to attenuate the noisy signals.

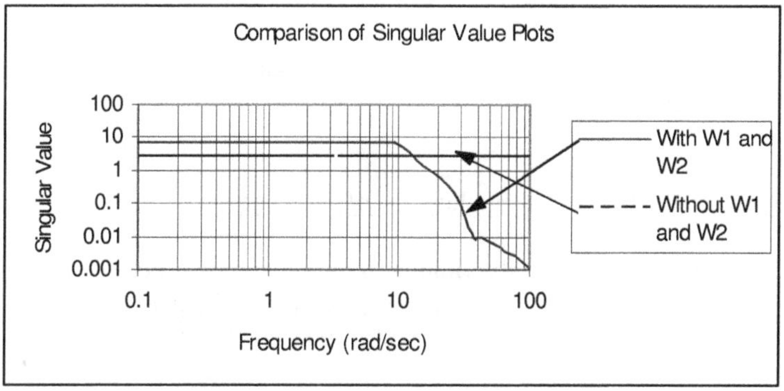

Fig. 2.2-2. Singular Plots with and without Weighting Functions

Fig. 2.2-3 shows another comparison of the singular value plots of poor and better design of the weighting functions. Both plots show attenuation of the noisy signal, but the poor design has a hump from 9.0 rad/sec to 25.0 rad/sec.

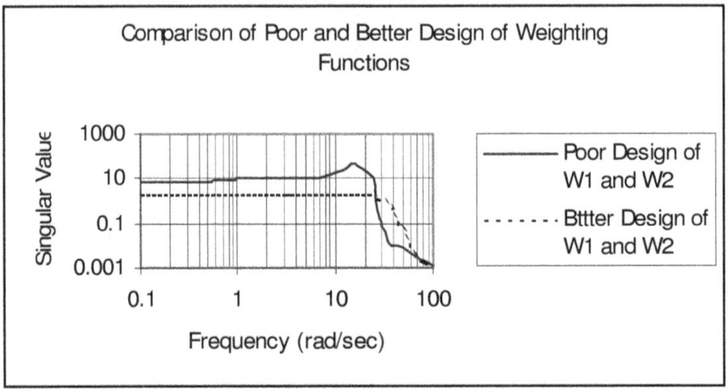

Fig. 2.2-3. Comparison of Poor and Better Design of the Weighting Functions

A Preliminary Mathematical Explanation of The Theory

The H-infinity optimization algorithm designs a feedback block (K) that provides a desired response from a plant. (See Fig.2.2-4.) The desirable system requirements are stability and fast response without excessive over- or under-shoot.

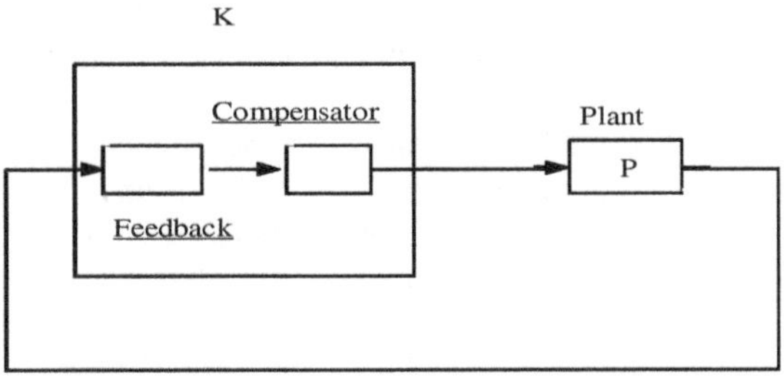

Fig. 2.2-4. Reformulated Linear Control System Block Diagram

In classical control system design, the modeling errors, parameter variations, and disturbances were apparently excluded from the design process. However, in modern control theory, all these uncertainties are included in system design. Fig. 2.2-5 shows a standard feedback configuration that includes them.

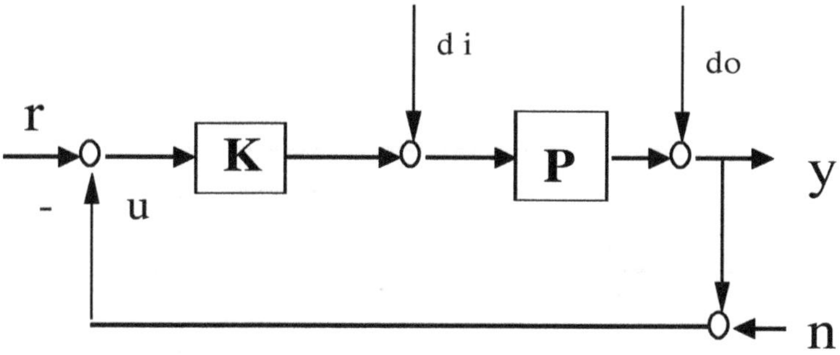

Fig. 2.2-5. Standard H-infinity Feedback Configuration

In Fig. 2.2-5,

P	= plant (system to be controlled)
K	= controller (block to be designed, weighting functions included)
r	= system input or command
di	= plant input disturbance
do	= plant output disturbance
y	= system output
n	= sensor noise

In control system design, internal stability must be assured. In order to investigate the internal stability condition, the block diagram in Fig. 2.2-5 is reformulated as shown below. Here, the internal stability condition means that KP and PK do not have eigenvalues in the Right Half Plane.

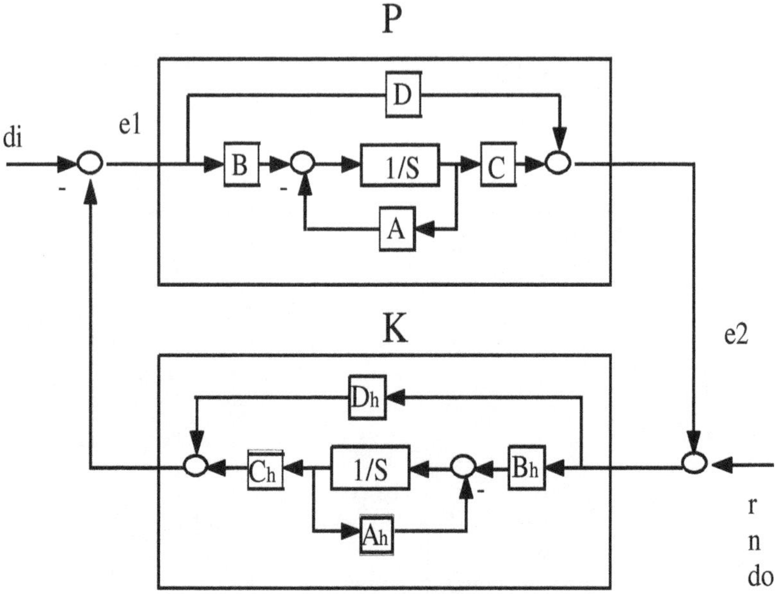

Fig. 2.2-6. Internal Stability Block Diagram

From Fig. 2.2-6,

$$xd = Ax + B\,e1 \qquad (2.2\text{-}1)$$
$$(xd \equiv dx/dt)$$
$$e2 = Cx + D\,e1 \qquad (2.2\text{-}2)$$

$$xhd = Ahx\,h + Bh\,e2 \qquad (2..2\text{-}3)$$
$$(xhd \equiv dxh/dt)$$
$$e1 = Ch\,xh + Dh\,e2 \qquad (2.2\text{-}4)$$

From Eqn. (2.2-2) and Eqn. (2.2-4),

$$\begin{vmatrix} e1 \\ e2 \end{vmatrix} = \begin{vmatrix} Ch & 0 \\ 0 & C \end{vmatrix} \begin{vmatrix} xh \\ x \end{vmatrix} + \begin{vmatrix} Dh & 0 \\ 0 & D \end{vmatrix} \begin{vmatrix} e2 \\ e1 \end{vmatrix}$$

$$\begin{vmatrix} e1 \\ e2 \end{vmatrix} - \begin{vmatrix} Dh & 0 \\ 0 & D \end{vmatrix} \begin{vmatrix} e2 \\ e1 \end{vmatrix} = \begin{vmatrix} Ch & 0 \\ 0 & C \end{vmatrix} \begin{vmatrix} xh \\ x \end{vmatrix}$$

$$\begin{vmatrix} 1 & 0 \\ 0 & 1 \end{vmatrix} \begin{vmatrix} e1 \\ e2 \end{vmatrix} + \begin{vmatrix} 0 & -Dh \\ -D & 0 \end{vmatrix} \begin{vmatrix} e1 \\ e2 \end{vmatrix} = \begin{vmatrix} Ch & 0 \\ 0 & C \end{vmatrix} \begin{vmatrix} xh \\ x \end{vmatrix}$$

$$\begin{vmatrix} 1 & -Dh \\ -D & 1 \end{vmatrix} \begin{vmatrix} e1 \\ e2 \end{vmatrix} = \begin{vmatrix} 0 & Ch \\ C & 0 \end{vmatrix} \begin{vmatrix} x \\ xh \end{vmatrix}$$

$$\begin{vmatrix} e1 \\ e2 \end{vmatrix} = \begin{vmatrix} 1 & -Dh \\ -D & 1 \end{vmatrix}^{(-1)} \begin{vmatrix} 0 & Ch \\ C & 0 \end{vmatrix} \begin{vmatrix} x \\ xh \end{vmatrix} \qquad (2.2\text{-}5)$$

From Eqn. (2.2-1) and Eqn. (2.2-3)

$$\begin{vmatrix} xd \\ xh \end{vmatrix} = \begin{vmatrix} A & 0 \\ 0 & Ah \end{vmatrix} \begin{vmatrix} x \\ xh \end{vmatrix} + \begin{vmatrix} B & 0 \\ 0 & Bh \end{vmatrix} \begin{vmatrix} e1 \\ e2 \end{vmatrix} \qquad (2.2\text{-}6)$$

Substituting Eqn. (2.2-5) into Eqn. (2.2-6),

$$\begin{vmatrix} xd \\ xh \end{vmatrix} = \begin{vmatrix} Ac(1,1) & Ac(1,2) \\ Ac(2,1) & Ac(2,2) \end{vmatrix} \begin{vmatrix} x \\ xh \end{vmatrix}$$

where

$$|Ac| = \begin{vmatrix} Ac(1,1) & Ac(1,2) \\ Ac(2,1) & Ac(2,2) \end{vmatrix} = \begin{vmatrix} A & 0 \\ 0 & Ah \end{vmatrix} + \begin{vmatrix} B & 0 \\ 0 & Bh \end{vmatrix} \begin{vmatrix} 1 & -Dh \\ -D & 1 \end{vmatrix}^{(-1)} \begin{vmatrix} 0 & Ch \\ C & 0 \end{vmatrix}$$

The system (Fig. 2.2-6) is said to be internally stable if $|Ac|$ has all its eigenvalues in the open Left-Half Plane. The internal stability is required for H-infinity control system design. It “seems” that the internal stability comes about in the process of H-infinity design because the minimization of the max singular value (H-infinity design) apparently forces the system design to be stable.

H-infinity is a mathematical technique applied to the set of all infinite numbers of stable transfer functions. It provides:

1. optimal control system design method when there are;

    a. modeling errors
    b. disturbances

c. system parameter variations

2. robustness
3. application to MIMO (Multi-Input Multi-Output)
4. easy tradeoffs technique between performance and robustness

The H-infinity design mainly concerns system response to disturbances and system uncertainties. Therefore, terms such as gain margin, phase margin, or eigenvalues are not of primary concern. Here we use the term, maximum singular value, which measures directly the performance of the closed loop system. In modern control system design, eigenvalues are computed from the system matrix, "A" in "dx/dt = Ax + Bu." But in H-infinity control system design, the singular values are computed from the transfer function that relates disturbances to system performance error. Theorem 16.4 on page 124 suggests how to design the control system satisfying the H-infinity requirement on Page 95. In other words, satisfying the conditions listed in the theorem, automatically we design a control system that minimizes the largest singular value ($\|Tzw\|_\infty$ on page 95), thereby we minimize the influence of disturbance and modeling errors on the system performance. Note that we do not need to compute the maximum singular to minimize it if we follow the Theorem 16.4.

The maximum singular value is defined below. A system matrix, P, can be decomposed as,

$$P = U\,S\,V^* \qquad (2.2\text{-}7)$$

where

U = unitary matrix
V* = complex conjugate transpose of a unitary matrix, V
S = diagonal matrix

The unitary matrix is a matrix whose columns are orthonormal (meaning that the cross product of the column vectors is a unit vector normal to the columns). The diagonal elements of the diagonal matrix, S, are the singular values and the first one, S(1,1), is the maximum singular value and the last one, S(n,n) is the minimum singular value. Singular values are positioned in the matrix, S, in the order of maximum to minimum values.

Example:

A system matrix, P, is given as,

$$P = \begin{vmatrix} 2 & 3 & 4 \\ 5 & 6 & 7 \\ 8 & 9 & 10 \end{vmatrix}$$

$$= U\, S\, V^*$$

Now using a software program, svd, in MATLAB, U, S, and V* are computed, where

$$U = \begin{vmatrix} -0.2721 & 0.8714 & 0.4082 \\ -0.5357 & 0.2153 & -0.8165 \\ -0.7994 & -0.4408 & 0.4082 \end{vmatrix}$$

$$S = \begin{vmatrix} 19.5743 & 0 & 0 \\ 0 & 0.9196 & 0 \\ 0 & 0 & 0 \end{vmatrix}$$

and

$$V^* = \begin{vmatrix} -0.4913 & -0.7694 & 0.4082 \\ -0.5734 & -0.0670 & -0.8165 \\ -0.6555 & 0.6353 & 0.4082 \end{vmatrix}$$

Checking the orthonormality of V matrix,

$$C1 = \begin{vmatrix} -0.4913 \\ -0.5734 \\ -0.6555 \end{vmatrix}$$

$$C2 = \begin{vmatrix} -0.7694 \\ -0.0670 \\ 0.6353 \end{vmatrix}$$

$$C3 = C1 \times C2$$

= | i j k |
|-0.4913 -0.5734 -0.6555 |
| -0.7694 -0.0670 0.6353 |
= (-0.4082) i + 0.8165j + (-0.4083)k

|C3|= |C1 X C2|
= sqrt((-0.4082)^2+0.8165^2+(-0.4083)^2)
= 1.00 (unit vector)

The C3 is a cross product of C1 and C2, therefore it is orthogonal to the plane that contains C1 and C2. And it is shown that the magnitude of the cross product is one. As stated earlier, the first element of the matrix S, S(1,1), is the maximum singular value that we found, and its value is 19.57.

The singular values can also be computed from the eigenvalues as shown below.
For a given matrix, P, we can write that

P = U S V*
P = M^(-1)DM

Now by equating the two equations above,

U S V* = M^(-1)DM

Solving for S,

S = U^(-1)M^(-1) D M V*^(-1)

Where

M: modal matrix
U: unitary matrix
V*: unitary matrix.

Eigenvalues and singular values are identical when U^(-1)M^(-1)=MV^(-1)= I

Example:

$$\text{Let } P = \begin{vmatrix} 2 & 3 & 4 \\ 5 & 6 & 7 \\ 8 & 9 & 10 \end{vmatrix}$$

Then, P=U S V*,
Where

$$U = \begin{vmatrix} -0.2721 & 0.8714 & 0.4082 \\ -0.5357 & 0.2153 & -0.8165 \\ -0.7994 & -0.4408 & 0.4082 \end{vmatrix}$$

$$S = \begin{vmatrix} 19.5743 & 0 & 0 \\ 0 & 0.9196 & 0 \\ 0 & 0 & 0.0000 \end{vmatrix}$$

$$V^* = \begin{vmatrix} -0.4913 & -0.7694 & 0.4082 \\ -0.5734 & -0.0670 & -0.8165 \\ -0.6555 & 0.6353 & 0.4082 \end{vmatrix}$$

And P= M^(-1) D M,
Where

$$M^{(-1)} = \begin{vmatrix} -0.4963 & -0.5897 & -0.6830 \\ -0.8939 & -0.2150 & 0.4640 \\ 0.4082 & -0.8165 & 0.4082 \end{vmatrix}$$

$$D = \begin{vmatrix} 18.9499 & 0 & 0 \\ 0 & -0.9499 & 0 \\ 0 & 0 & 0.0000 \end{vmatrix}$$

$$M = \begin{vmatrix} -0.2826 & -0.7753 & 0.4082 \\ -0.5383 & -0.0740 & -0.8165 \\ -0.7939 & 0.6273 & 0.4082 \end{vmatrix}$$

The matrix equation that relates eigenvalues to singular values is,

$$S = U^{(-1)} M^{(-1)}\, D\, M\, V^{*(-1)}$$

$$= \begin{vmatrix} 0.2875 & 0.9283 & -0.3891 \\ -0.8049 & -0.2002 & -0.6753 \\ 0.6940 & -0.3985 & -0.4911 \end{vmatrix} \; D \; \begin{vmatrix} 0.9020 & -0.1193 & -0.1405 \\ -0.0119 & 0.9803 & -0.0275 \\ 0.0742 & 0.0799 & 1.0857 \end{vmatrix}$$

Thus, the singular values can be computed from the eigenvalues.

One other way of computing the singular value is shown below. From Eqn. (2.2-7), post-multiplying V,

$$PV = U\,S\,V^*V$$

Since $V^*V = I$ where V is a unitary matrix,

$$PV = U\,S \qquad (2.2\text{-}8)$$

Rewriting Eqn. (2.2-8),

$$P\,|\,v1\ v2\ \ldots vn\,| = U \begin{vmatrix} s1 & 0 & 0 & \ldots & 0 \\ 0 & s2 & 0 & \ldots & 0 \\ 0 & 0 & s3 & \ldots & 0 \\ \ldots & & & & \\ 0 & 0 & 0 & \ldots & sn \end{vmatrix}$$

$$= |\,u1s1\ \ u2s2\ \ldots\ unsn\,|$$

From the equation above,

$$P\,vi = ui\,si,\quad i = 1, 2, 3,\ \ldots\ n \qquad (2.2\text{-}9)$$

Now pre-multiplying U^* both sides of Eqn. (2.2-7),

$$U^*P = U^*U\,S\,V^*$$
$$= S\,V^*, \qquad \text{for } U^*U = I$$

Taking conjugate transposition of both side of the equation above,

$$(U^*P)^* = (S\,V^*)^*$$
$$P^*\,U = V\,S^*$$

Rewriting the equation above,

$$P^*\,|\,u1\ u2\ u3\ \ldots\ un\,| = V \begin{vmatrix} s1 & 0 & 0 & \ldots & 0 \\ 0 & s2 & 0 & \ldots & 0 \\ 0 & 0 & s3 & \ldots & 0 \\ \ldots & & & & \\ 0 & 0 & 0 & \ldots & sn \end{vmatrix}$$

= | v1s1 v2s2 ... vnsn |

From the equation above,

P*ui = vi si , i=1,2,3, ... n (2.2-10)

Pre-multiplying, "ui" both sides of Eqn. (2.2-9).

ui* P vi = ui*ui si,
= si (2.2-11)

Post-ultiplying each side of Eqn. (2.2-11) to each side of Eqn. (2.2-10),

P* ui ui* P vi = vi si si

P*P vi = vi si^2
P̲ vi = Si vi (2.2-12)

Here, P̲ = P*P and Si := si^2 and vi si^2=Sivi, because si is an element. The structure of Eqn. (2.2-12) reveals that vi is an eigenvector of P̲, and Si an eigenvalue of P̲. Thus, the singular value (si) can be computed by using the following equation.

si = square root of (eigenvalues of P*P) (2.2-13)

The eigenvalues are the terms often used in the classical control system stability analysis, and the singular values are the terms used to determine system performance, robustness, and sensitivity. The performance, robustness and sensitivity are the three basic parameters that are of concern in H-infinity control system design.

Performance

Performance can be managed by shaping a sensitivity function. The sensitivity function is defined as a transfer function from the output disturbance (d) to the system output (y). From Fig. 2.2-7, we derive the sensitivity function as shown below.

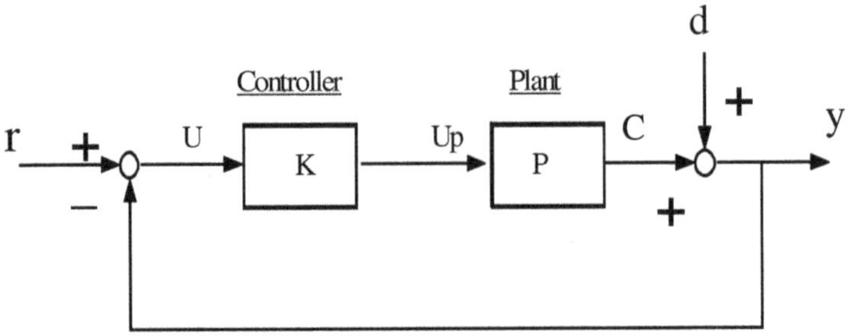

Fig. 2.2-7. Sensitivity Function Block Diagram

C = -P(s)K(s)*y
y = C + d
y = -P(s)K(s)*y + d
[I + P(s)K(s)] y = d
y = {Inv. of [I + P(s)K(s)]}*d
= [I + P(s)K(s)]^(-1)*d

Since,

P(s) = Cp(SI – Ap)^(-1)Bp + Dp
K(s) = Ck(SI-Ak)^(-1)Bk + Dk

y ={I + [Cp(SI – Ap)^(-1)Bp + Dp] [Ck(SI-Ak)^(-1)Bk + Dk]}^(-1)*d

Note that Fig. 2.2-7 does not show the sub-blocks such as Ap, Bp, … The inverse of [I +P(s)K(s)] is the sensitivity function. It is basically a system response to the disturbance (d). Now we compute the singular value of the sensitivity transfer function for each frequency and pick the maximum singular value. If that value is less than 1.0, then we know that the response to the disturbance will diminish eventually.

[I + P(jω)*K(jω)]^(-1) = USV* (2.2-14)

where

U, V* = unitary matrix
S = diagonal matrix of singular values
ω = all frequencies

For an example, we assume that the inverse of [I + P(jω)*K(jω)] at ω=3.5 rad/sec is,

```
|2  5  8|
|1  6  3|
|9  2  8|
```

Then,

```
U =|-0.5830  -0.4483  -0.6776|
   |-0.3364  -0.6259   0.7036|
   |-0.7395   0.6381   0.2142|

V* = |-0.5244   0.6654   0.5312|
     |-0.4122  -0.7443   0.5255|
     |-0.7450  -0.0566  -0.6646|

S =  |15.5564    0         0       |
     | 0         6.3427    0       |
     | 0         0         2.4020  |
```

Here, U, S, and V* are computed by using a MATLAB program, "svd." The maximum singular value at a frequency of 3.5 rad/sec is 15.56.

In order to construct a plot of maximum singular value versus frequency, we compute the maximum singular value at each frequency of interest and plot them like a Bode plot. If the plot shows a low value in the high frequency range, then the system is able to reject high-frequency disturbances, which is desired in system design.

Example:

Let the P be a stable system,

P = [C(SI-A)^(-1)B + D]

where

C : Identity Matrix

$$A = \begin{vmatrix} 0 & 1 & 0 \\ 0 & 0 & 1 \\ -5 & -6 & -7 \end{vmatrix}$$

B : Identity Matrix

D : Null Matrix

Then,

$$P = (SI - A)^{\wedge}(-1)$$
$$= \text{Inverse of } \begin{vmatrix} S & -1 & 0 \\ 0 & S & -1 \\ 5 & 6 & S+7 \end{vmatrix}$$

and let the K be an identity matrix for simplicity. Then, the inverse of the sensitivity function is,

$$I + PK = I + P$$

$$= I + \text{Inverse of } \begin{vmatrix} S & -1 & 0 \\ 0 & S & -1 \\ 5 & 6 & S+7 \end{vmatrix}$$

Now compute the maximum singular values of Inverse of (I+P) as the frequency varies over the frequency range of interest. For this example, let the frequency, S=ωj, vary from ω=0.01 rad/sec to ω=100 rad/sec with an increment of 0.01 rad/sec. The results are plotted in Fig. 2.2-8. A simple Matlab m-file is code below to compute the max singular values.

```
a=[0 1 0;0 0 1;-5 -6 -7];
b=[1 0 0;0 1 0;0 0 1];
c=[1 0 0;0 1 0;0 0 1];
d=[0 0 0;0 0 0;0 0 0];
p=pck(a, b, c, d);
om=logspace(-2,2,50);
pg=frsp(p,om);
s=vsvd(pg)
vplot('liv, lm',vsvd(pg),1);
title('Sensitivity Before H-inf Controller Design')
ylabel('Max. Singular Values');
xlabel('Freq. (rad/sec)');
```

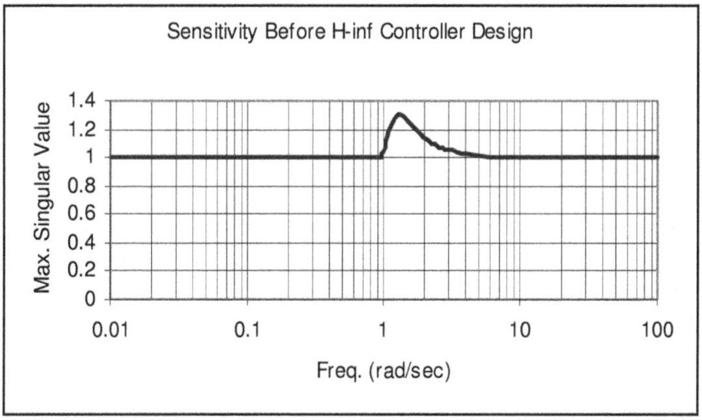

Fig. 2.2-8. Sensitivity Before H-infinity Controller Design

It is seen here that the system is more sensitive to disturbances of frequencies near 1.5 rad/sec.

Robustness

The robustness is measured by the inverse of Allowable Maximum Modeling Error. When a control system can tolerate a large modeling error, the robustness index decreases. It is usually the case that a math model that exactly represents the real plant is not obtainable. Therefore, we need to accommodate a controller that minimizes the error that occurs due to the discrepancy between the plant and the math model. Our objective here is to design a controller that can tolerate large uncertainty (modeling error). Here again, the robust design can be achieved by minimizing the maximum singular value of a system distorted by modeling errors. The following example simulates a case in which a modeling error as well as an impulse disturbance at the output exists.

Example:

Consider a plant (P), the true math model of which is,

$$A= \begin{vmatrix} -18.0 & -4.70 & 0 \\ 6.8 & 2 & 0 \\ 0 & 4.6 & 0 \end{vmatrix}$$

$$B = \begin{vmatrix} 4.2 \\ 0 \\ 0 \end{vmatrix}$$

$$C = \begin{vmatrix} 0 & 5.6 & 4.9 \end{vmatrix}$$

$$D = \begin{vmatrix} 0 \end{vmatrix}$$

Using the H-infinity technique, we design a controller that suppresses output disturbances and at the same time enables the control system to tolerate modeling error. In the following simulation study, we show plant's response to a step input and demonstrate improvement when an H-infinity controller is implemented. First, we show two cases of system performance before implementing the H-infinity controller. The first case is when there is no disturbance (Fig. 2.2-9), and the second case, when there is a disturbance at output (Fig. 2.2-10).

In Fig 2.2-9, it is seen that the system's response to a step input has a large overshoots during the transient period, but it displays stability after it reaches steady state. Notice that there is a unity controller (K=I) implemented in this simulation. In Fig. 2.2-10, a disturbance occurs at the output from t = 3.0s to t = 4.0s (top of the Figure). Here again, no designed controller is implemented. The output response plotted at the bottom of the Figure shows clearly that it is affected by the disturbance. Now, it will be shown that the H-infinity optimal controller deletes the overshoot and the hump generated by the disturbance at output.

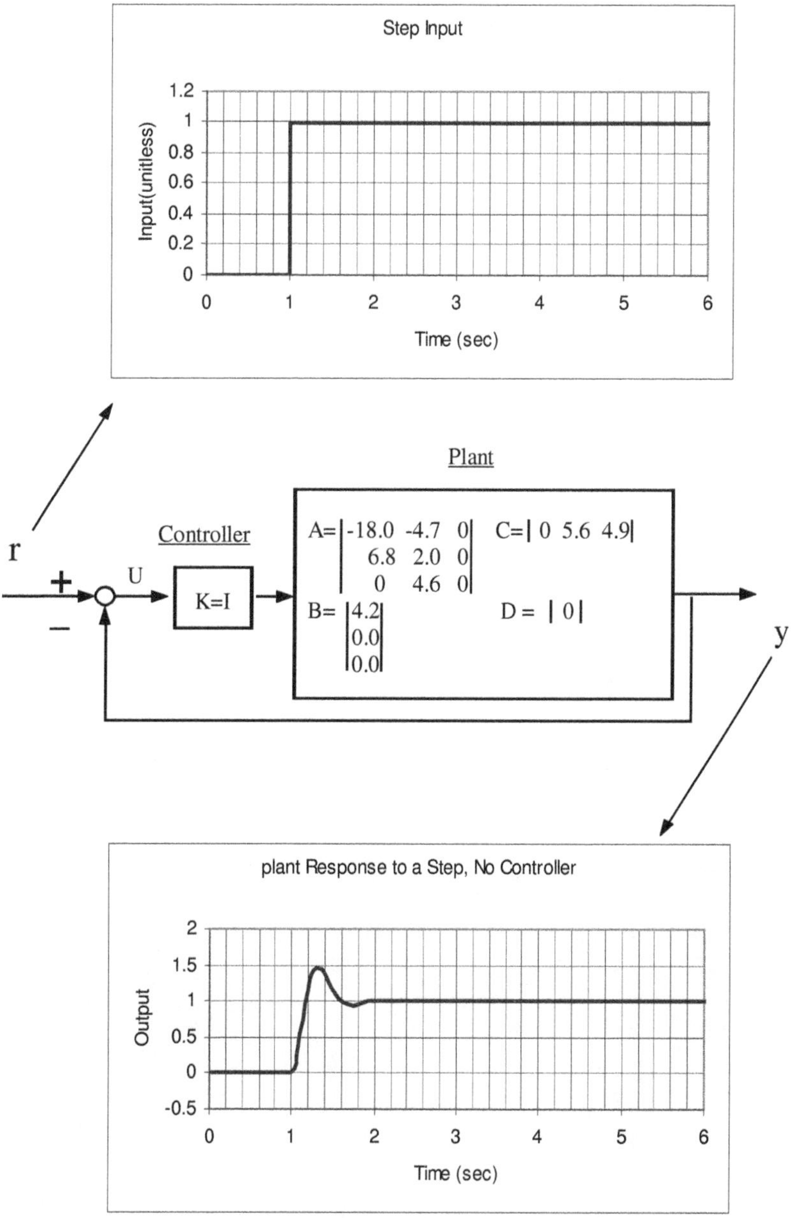

Fig. 2.2-9. Step Response Before Disturbance

(Blank Page)

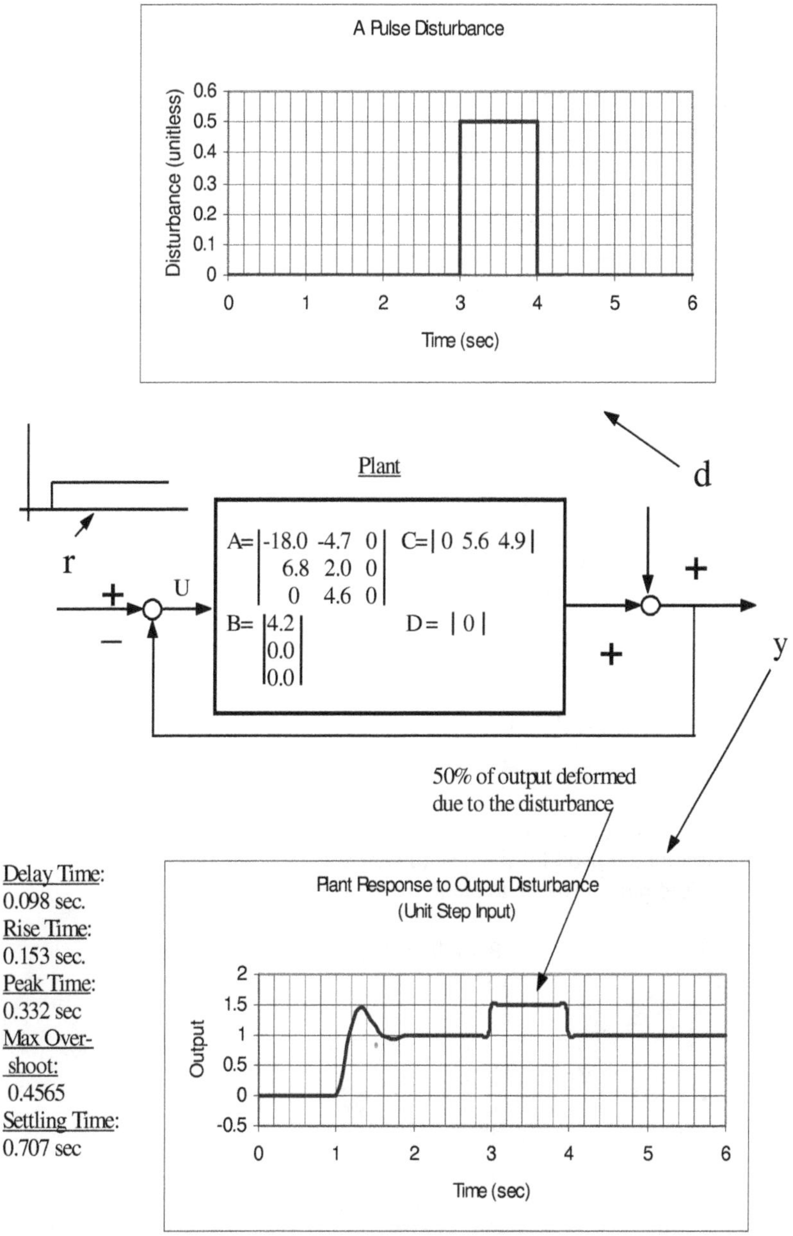

Fig. 2.2-10 Plant Response to a Step Input with Disturbance at Output

Fig. 2.2-11 shows the effectiveness of the H-infinity design. An H-infinity feedback loop block, KH, is designed by using a Matlab program, "hinfsyn.m," and it is inserted in the feedback loop of the same system shown in Fig. 2.2-10. Two weighting functions are also designed and added to the system to allocate proper weights on the signal. The weighting functions, W1 and W2, and the KH are:

W1 = (50 S + 8,000)/(S + 10,000)
W2 = (0.015 S + 0.615)/(S + 0.003)
KH = C (SI – A)^(-1)B

where

```
A= [ -19.4184    -11.9640    -4.3081     -1.0385     58.2961;
       6.8000     -6.0882    -7.0771      0.0000      0;
       0           0.6595    -3.4480      0           0;
       0           0          0          -0.0030      0;
    -237.6090  -778.0908  -337.7719   -173.9587   -234.5338 ]

B=[ 3.3071;
   10.2128;
    4.9757;
    4.7410;
    0      ]

C= [-0.0478  -0.1564  -0.0679  -0.0350  1.9629]

D = [ 0.0e+000]
```

It should be restated here that when the A, B, C, and D in KH are designed, the plant model used for the design was not the same as the true plant model (see page 110). The plant model used is:

```
A= | -18.0  -4.70  0 |
   |   6.8   2     0 |
   |   0     4.6   0 |

B= | 4.2 |
   | 0   |
   | 0   |

C= | 0  9.6  0 |

D = | 0 |
```

Notice that the math modeling error occurs in the C matrix. More comprehensive modeling error analysis is presented in Chapter 4.4.

At the top of the Fig. (2.2-11), a disturbing impulse of strength 0.5 and width of 1.0 s is displayed. That impulse is applied at the output as shown in the middle of the figure. The bottom plot is the response of the system to the disturbing impulse. Note that the input is applied at t = 1.0 sec., not t = 0.0 sec. It is shown here that the over-shoot and the hump due to the disturbing impulse (shown in Fig. 2.2-10) have almost disappeared.

A similar simulation is shown in Fig. 2.2-12. The basic structure is the same except that the feedback loop block is designed by using the H-infinity, Loop Shaping, and Coprime Factorization. The weighting functions and KH designed are:

W1 = (2 S + 0.01)/(S + 1)
W2 = (0.043 S + 1.783)/(S + 3)
KH = C (SI – A)^(-1)B

where

```
A=[ -25.5103    11.4715   -1.8332 ;
     11.2358    -9.1830    2.2812;
     -1.7943     2.2795   -1.8431 ]

B=[ -5.0110 ;  1.5365;  -0.2191 ]

C=[  -4.9045   1.5354  -0.2191  ]

D=[ -1.6776  ]
```

As shown in the figure, this design technique improves the system response more than the H-infinity controller presented earlier does. The feedback block system matrices, A, B, C, and D are obtained by using a Matlab program, "ncfsyn.m."

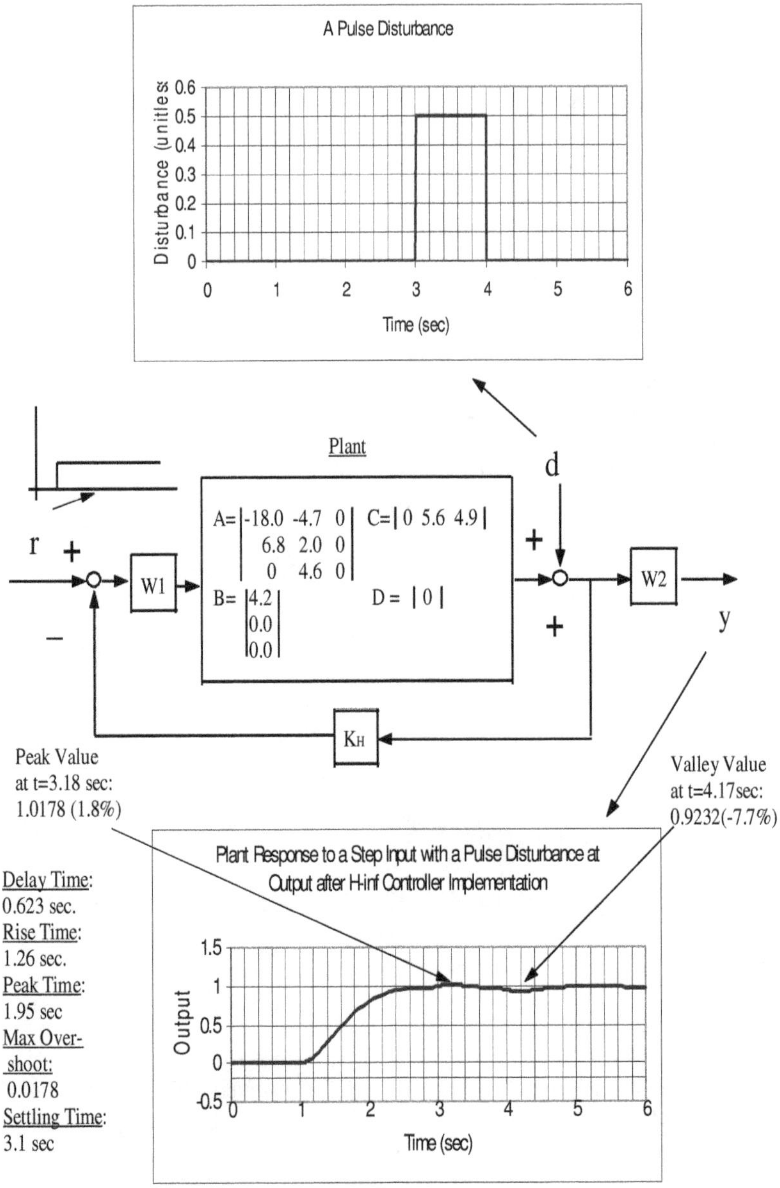

Fig. 2.2-11. After Implementation of H-infinity Controller

(Blank Page)

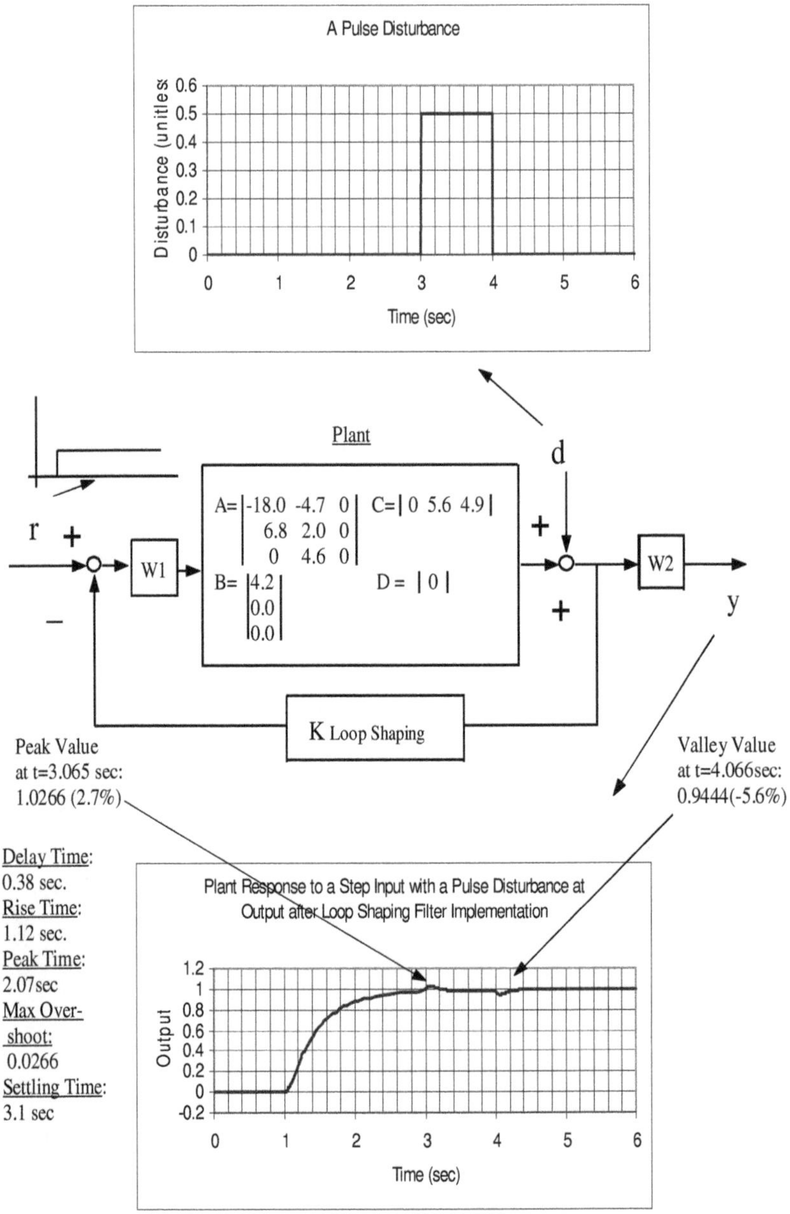

Fig. 2.2-12. After Implementation of Loop Shaping Controller

Fig. 2.2-13 shows the differences in sensitivity in three different cases: 1) Fig. 2.2-10; 2) Fig. 2.2-11; and 3) Fig. 2.2-12. The data label "Before Controller Implementation" in Fig. (2.2-13) means the case 1) plot (Fig. 2.2-10), and it is seen here that the sensitivity near 10 rad/sec is very high and remains high all the way.. The sensitivities of the other two cases are only about 15.0 % of the first case, which are verified in Fig. 2.2-11 and Fig. 2.2-12. A MATLAB m-file coded to plot the figure is shown below.

Feedback Loop Block Designed

```
ak=[-25.5103  11.4715  -1.8332 ;
     11.2358  -9.1830   2.2812;
     -1.7943   2.2795  -1.8431 ];
bk=[ -5.0110 ;
      1.5365;
     -0.2191  ];
ck=[ -4.9045   1.5354  -0.2191  ];
dk=[-1.6776 ];
```

Plant System Matrices

\

```
ap=[-18.0    -4.7    0;
      6.8     2.0    0;
      0       4.6    0];
bp=[4.2 ;
     0 ;
     0];
cp=[0 5.6 4.9];
dp=[0];
```

Initial Values

```
id=1;
w=0.01;
```

Iteration Loop Begins

```
for i=1:1000
 w=w+0.01;
```

Weighting Function 1

```
        w1=cw1*inv(w*j-aw1)*bw1 + dw1;

        where
                aw1=-1.0;
                bw1= 1.3784;
                cw1= -1.3784;
                dw1= 2.0 ;

Weighting Function 2

        w2=cw2*inv(w*j-aw2)*bw2 + dw2;

        where

                aw2 = -3.0 ;
                bw2 = 1.2861;
                cw2 = 1.2861 ;
                dw2 = 0.0430;

Feedback Loop Block

        k=ck*inv(jw-ak)*bk;

Plant Equation

        p=cp*inv(jw-ap)*bp;
        pw1=p*w1;

        sen1=inv(id + k*pw1);
        sen=w2*sen1;

        x(i)=w;
        y(i)=svd(sen);
end
```

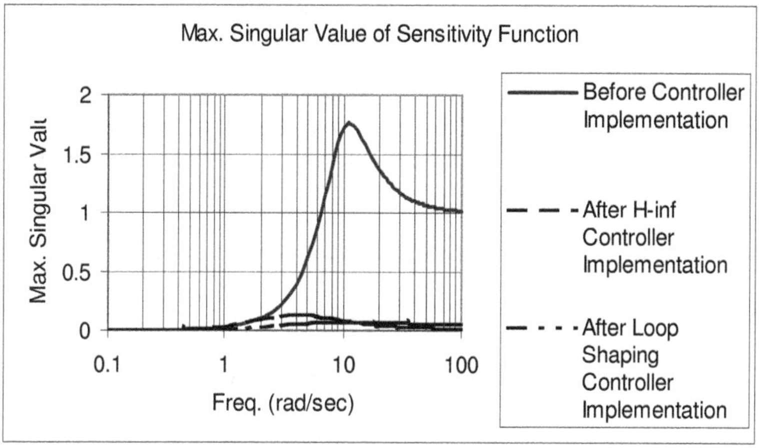

Fig. 2.2-13. Comparison of Maximum Singular Values

2.3 H-infinity Design

The H-infinity design is basically a computation procedure by which multi-input multi-output (MIMO) optimal design is accomplished. It enables us to design a controller that minimizes the maximum singular value of a system exposed to disturbances and subject to modeling errors. Most of the current satellite launch vehicle attitude control systems are designed based on classical control theory. One critical drawback of the classical control theory is that it cannot be applied to MIMO systems. The advantage of MIMO system design technique is that the wind disturbance can be treated as a part of input. Therefore, once a design is completed, then the system remains stable regardless how strong the wind is. This statement is valid because, in linear system theory, once a system is designed stable, then it remains stable independent of the inputs, that is, no input can make the system unstable.

The wind has been one of the major threats in vehicle launch operations. Wind measurements must be made days ahead and continue until shortly before the launch. If the wind profile prediction proves to be unfavorable, then the launch must be canceled. Cancellation of a launch is costly, because the fuel must be drained to maintain the rocket's structural integrity, and the launch-ready process must be repeated. In an effort to remove this vulnerability, the H-infinity, loop shaping, and coprime factorization are chosen and applied to the design in this book.

Since the H-infinity control theory is mathematically complex, only one aspect of its design concept is discussed here. Ref. 42 (Zhou) is recommended for extensive study. We start from a theorem on page 419 of the reference. The theorem is quoted below:

Theorem 16.4: There exists an admissible controller, K, such that

$$\|Tzw\|_\infty < \gamma$$

if the following three conditions hold:

i) $H\infty \in \text{dom(Ric)}$ and $X\infty := \text{Ric}(H\infty) \geq 0$
ii) $J\infty \in \text{dom(Ric)}$ and $Y\infty := \text{Ric}(J\infty) \geq 0$
iii) $\rho(X\infty\, Y\infty) < \gamma^{\wedge}(2)$

A controller is said to be admissible if it can internally stabilize a system. Several symbols are used in this theorem, and they are briefly explained below.

$\|Tzw\|_\infty$: largest singular value of a transfer function from w to z (see Fig. 2.2-1a) covering all frequencies of interest
$\|Tzw\|_\infty < \gamma$: essence of H-infinity concept and γ is a boundary.
$H\infty \in \text{dom(Ric)}$: $H\infty$ belongs to "Ric" domain where "Ric" means stabilizing solution of an ARE (Algebraic Riccati Equation), where the Hamiltonian matrix, $H\infty$ is:

$$H\infty = \begin{vmatrix} A & \gamma^{\wedge}(-2)\, B1B1^* - B2B2^* \\ -C1^*C1 & -A^* \end{vmatrix}$$

A, B1, B2, and C1 are defined in Fig. 2.3-1.
" * " means complex conjugate transpose.

The ARE (Algebraic Riccati Equation) for $H\infty$ is,

$$A^*X\infty + X\infty A + \gamma^{\wedge}(-2)\, X\infty\, B1B1^*\, X\infty - X\infty\, B2B2^*\, X\infty = 0 \tag{2.3-1}$$

$X\infty := \text{Ric}(H\infty) \geq 0$: $X\infty$ is a solution of the ARE, Eqn. (2.3-1), and semi-positive definite.

$J\infty \in \text{dom(Ric)}$: $J\infty$ belongs to Ric domain where Ric means stabilizing solution of an ARE (Algebraic Riccati Equation), where the Hamiltonian matrix, $J\infty$ is:

$$J\infty = \begin{vmatrix} A^* & \gamma^{(-2)}\,C1C1^* - C2C2^* \\ -B1^*B1 & -A \end{vmatrix}$$

The ARE for $J\infty$ is,

$$A^*Y\infty + Y\infty A + B1^*B1 + \gamma^{(-2)}\,Y\infty\,C1C1^*\,Y\infty - Y\infty\,C2C2^*\,Y\infty = 0$$

$$(2.3-2)$$

$Y\infty := \text{Ric}(J\infty) \geq 0$: $Y\infty$ is a solution of the ARE, Eqn. (2.3-2) and semi-positive definite.

$\rho(X\infty\,Y\infty)$: spectral radius of product of $X\infty$ and $Y\infty$ (Spectral radius is the largest eigenvalue of a matrix in the absolute sense.)

The proof of Theorem 16.4 is not presented here. It can be found on page 436 of ref. 42. In this book, a concept based on which H-infinity control system design algorithm can be developed is presented. The presentation procedure is briefly summarized below.

First, a system equation is constructed from Fig.2.3-1. Then, a quadratic cost function is defined and a series mathematical manipulations is applied to generate a Riccati equation and a set of functions that brings about the conversion of the quadratic cost function. In the process, the system equation constructed (Eqn. 2.3-3) is used. The mathematical manipulations yield a higher order system equation, a more complex mathematical expression. The simple system shown in Fig. 2.3-1 evolves into the system shown in Fig. 2.3-2 (excluding the "K" block). Next, a feedback loop block, "K," is devised to improve system performance and guarantee convergence. All these processes are needed to comply with Theorem 16.4. Detail follows.

G(S)

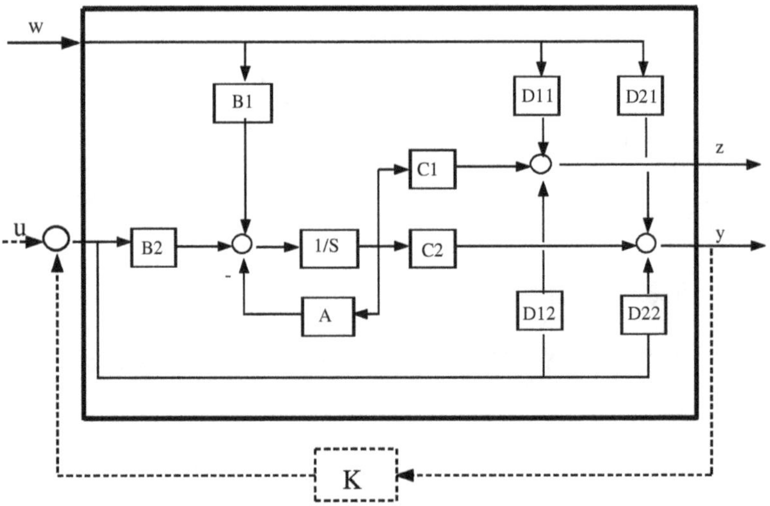

Fig. 2.3-1. Linearized Plant

A math model of the plant in Fig. 2.3-1 is,

$$dx/dt = Ax + (B1)w + (B2)u$$

$$z = (C1)x + (0)w + (D12)u$$

$$y = (C2)x + (D21)w + (0)u \qquad (2.3\text{-}3)$$

Rewriting the set of equations above in matrix form,

$$\begin{vmatrix} xd \\ z \\ y \end{vmatrix} = |G| \begin{vmatrix} x \\ w \\ u \end{vmatrix} \qquad (2.3\text{-}4)$$

$$xd := dx/dt$$

$$G = \begin{vmatrix} A & B1 & B2 \\ C1 & 0 & D12 \\ C2 & D21 & 0 \end{vmatrix}$$

Now consider a quadratic function as shown below.

$$C(t) = x(t)^* (Px)\, x(t) \qquad (2.3\text{-}5)$$

In this equation, we find (Px) such that C(t) converges as t increases. The convergence can be obtained by assigning an optimal or sub-optimal value to (Px). Since (Px) is a constant matrix (a solution of the Riccati equation), the convergence of C(t) implies the convergence of x(t), which means the system is stable. The C(t) convergence can be achieved if (dC(t)/dt) approaches 0.0 as t increases. Taking a time derivative of Eqn. 2.3-5,

$$dC(t)/dt = (dx/dt)^* (Px)\, x(t) + x(t)^* (Px)\, (dx/dt) \qquad (2.3\text{-}6)$$

Substituting Eqn. (2.3-3) into Eqn. (2.3-6),

$$\begin{aligned} dC(t)/dt &= (Ax + (B1)w + (B2)\,u)^* (Px)\; x + x^* (Px)\,(Ax + (B1)w + (B2)\,u) \\ &= x^*A^* (Px)\; x + w^*(B1)^* (Px)\; x + u^*(B2)^* (Px)\; x + x^* (Px)\, Ax \\ &\quad + x^* (Px)\; (B1)w + x^* (Px)\,(B2)\,u \\ &= x^*[A^*(Px) + (Px)A]x + 2\, w^*(B1)^* (Px)\, x + 2u^*(B2)^* (Px)\, x \end{aligned}$$

(2.3-7)

Here, we add the following two terms to Eqn. (2.3-7) and subtract the same two terms from the same equation, which should not alter the equation. The two terms are:

$$\gamma\text{^}(-2)x\, (Px)\; (B1)(B1)^* (Px)\, x$$

and

$$x\, (Px)\, (B2)B2^*(Px)\, x$$

Then, Eqn. (2.3-7) becomes,

$$\begin{aligned} &= x^* [A^*(Px) + (Px)\, A]x + \gamma\text{^}(-2)x\, (Px)\; (B1)(B1)^* (Px)\, x - x(Px) \\ &\quad (B2)B2^*(Px)x - \gamma\text{^}(-2)x(Px)(B1)(B1)^* (Px)x + x(Px)(B2)B2^*(Px)x \\ &\quad + 2w^*(B1)^* (Px)x + 2u^*(B2)^*(Px)\, x \end{aligned}$$

$$\begin{aligned} &= x^* [A^*(Px) + (Px)A + \gamma\text{^}(-2)\, (Px)\, (B1)(B1)^* (Px) - (Px)\, (B2)B2^*(Px)]\, x \\ &\quad -\gamma\text{^}(-2)\, \|(B1)^* (Px)\, x\|\text{^}(2) + \|B2^*(Px)\, x\|\text{^}(2) + 2\, w^*(B1)^* (Px)\, x \\ &\quad + 2u^*(B2)^* (Px)\, x \end{aligned}$$

(2.3-8)

If we set [see Eqn. (2.3-1)],

$$[A^*(Px) + (Px)A + \gamma^{-2}(Px)(B1)(B1)^*(Px) - (Px)(B2)B2^*(Px)] = 0, \qquad (2.3\text{-}9)$$

then, Eqn. (2.3-8) becomes,

$$\begin{aligned} dC(t)/dt = {} & -\gamma^{-2}\,\|(B1)^*(Px)\,x\|^{2} + 2\,w^*(B1)^*(Px)x + \|B2^*(Px)\,x\|^{2} \\ & + 2u^*(B2)^*(Px)\,x \end{aligned}$$

Here again, adding and subtracting the two terms, $-\gamma^{2}w^{2}$ and u^{2},

$$\begin{aligned} dC(t)/dt &= -\gamma^{-2}(\|B1)^*(Px)x\|^{2} + 2w^*(B1)^*(Px)x - \gamma^{2}w^{2} \\ &\quad + \gamma^{-2}w^{2} + [\|B2^*(Px)x\|^{2} + 2u^*(B2)^*(Px)x + u^{2}] - u^{2} \\ &= -\gamma^{2}[\gamma^{-4}\|B1^*(Px)x\|^{2} - 2\gamma^{-2}(w^*(B1)^*(Px)x \\ &\quad + w^{2})] + [\|B2^*(Px)x\|^{2} + 2u^*(B2)^*(Px)\,x + u^{2}] \\ &\quad + \gamma^{2}w^{2} - u^{2} \\ &= -\gamma^{2}[\gamma^{-2}\|B1^*(Px)\,x\| - w]^{2} + [\|B2^*(Px)\,x\| + u]^{2} \\ &\quad + \gamma^{2}w^{2} - u^{2} \end{aligned}$$

And let,

$$r = w - \gamma^{-2}\|(B1)^*(Px)\,x\| \qquad (2.3\text{-}10)$$

and

$$v = u + \|(B2)^*(Px)\,x\| \qquad (2.3\text{-}11)$$

Now we design a control system (i.e. "K" in Fig. 2.3-2) that drives "r" [Eqn. (2.3-10)] and "v" [Eqn. (2.3-11)] to zero. Then,

$$dC(t)/dt = \gamma^{2}w^{2} - u^{2}$$

Now by setting,

$$\gamma^{2} = u^{2} / w^{2} \qquad (2.3\text{-}12)$$

we have

$$dC(t)/dt = 0$$

which means system's convergence. Eqn. (2.3-12) implies that for a given u, if "w" gets larger, then "γ" becomes smaller.

Now, we find (Px) that satisfies the Riccati equation, Eqn. (2.3-9) and then design a control system that drives "r" [Eqn. (2.3-10)] and "v" [Eqn. (2.3-11)] to zero. The convergence of "r" and "v" and Eqn. (2.3-12) implies that the time derivative of C(t), dC(t)/dt, converges to zero. All this process is one of the schemes that can be used to improve system performance in compliance with Theorem 16.4. Note the difference between stability and performance. Good performance includes stability, but performance here is not directly related to stability, i.e., the performance is not a measure of stability. The word, "performance," is used to emphasize the capability of handling modeling errors and outside disturbances.

Now we show how the structure of control system computation software is formulated to accomplish the optimal design. From Eqn. (2.3-3),

$$dx/dt = Ax + (B1)w + (B2)u \qquad (2.3\text{-}13)$$
$$z = (C1)x + (0)w + (D12)u \qquad (2.3\text{-}14)$$

Substituting Eqn. (2.3-11) into "u" in Eqn. (2.3-13),

$$\begin{aligned} dx/dt &= Ax + (B1)w + (B2)[v - \|(B2)^* (Px)\, x\|] \\ &= [A - (B2)(B2)^*(Px)]x + \gamma^{\wedge}(-1)(B1)\gamma^{\wedge}(1)w + (B2)\, v \end{aligned}$$

(2.3-15)

Substituting Eqn. (2.3-11) into "u" in Eqn. (2.3-14),

$$\begin{aligned} z &= (C1)x + (0)w + (D12)[v - \|(B2)^* (Px)\, x\|] \\ &= [C1 - (D12)(B2)^* (Px)]x + (0)w + (D12)\, v \end{aligned}$$

(2.3-16)

Pre-multiplying "γ" both sides of Eqn. (2.3-10),

$$\gamma r = -\gamma^{\wedge}(-1)(B1)^* (Px)\, x + \gamma w \qquad (2.3\text{-}17)$$

From Eqn. (2.3-15), Eqn.(2.3-16), and Eqn. (2.3-17), we construct a system matrix equation that becomes a part of G(S) in Fig. 2.3-2.

$$\begin{vmatrix} dx/dt \\ z \\ \gamma r \end{vmatrix} = \begin{vmatrix} [A - (B2)(B2)^* (Px)] & \gamma^{(-1)}(B1) & B2 \\ [C1 - (D12)(B2)^* (Px)] & 0 & D12 \\ -\gamma^{(-1)}(B1)^* (Px) & I & 0 \end{vmatrix} \begin{vmatrix} x \\ \gamma w \\ v \end{vmatrix} \qquad (2.3\text{-}18)$$

Now from Eqn. (2.3-3) again,

$$dx/dt = Ax + (B1)w + (B2)u \qquad (2.3\text{-}19)$$
$$y = (C2)x + (D21)w + (0)u \qquad (2.3\text{-}20)$$

Substituting Eqn. (2.3-10) into "w" in Eqn. (2.3-19),

$$\begin{aligned} dx/dt &= Ax + (B1)(r + \gamma^{(-2)}(B1)^*(Px)x + (B2)u \\ &= [A + \gamma^{(-2)} (B1)(B1)^* (Px)]x + (B1) r + (B2)u \end{aligned} \qquad (2.3\text{-}21)$$

Substituting Eqn. (2.3-10) into "w" in Eqn. (2.3-20),

$$\begin{aligned} y &= (C2)x + (D21) [r + \gamma^{(-2)}(B1)^* (Px)x] \\ &= [C2 + \gamma^{(-2)} (D21) (B1)^* (Px)] x + (D21) r \end{aligned} \qquad (2.3\text{-}22)$$

From Eqn. (2.3-11),

$$v = u + \|(B2)^* (Px)x\| \qquad (2.3\text{-}23)$$

From Eqn. (2.3-21), Eqn. (2.3-22), and Eqn. (2.3-23), we construct another system matrix equation that becomes a part of G(S) in Fig. 2.3-2.

$$\begin{vmatrix} dx/dt \\ v \\ y \end{vmatrix} = \begin{vmatrix} [A + \gamma^{(-2)} (B1)(B1)^* (Px)] & B1 & B2 \\ (B2)^*(Px) & 0 & I \\ [C2 + \gamma^{(-2)} (D21) (B1)^* (Px)] & D21 & 0 \end{vmatrix} \begin{vmatrix} x \\ r \\ u \end{vmatrix} \qquad (2.3\text{-}24)$$

Fig. 2.3-2 shows a new configuration constructed from Eqn.(2.3-18) and Eqn.(2.3-24). Note that it does not alter the plant dynamics. In this figure, a state estimation feedback dynamic block, K(S), is connected.

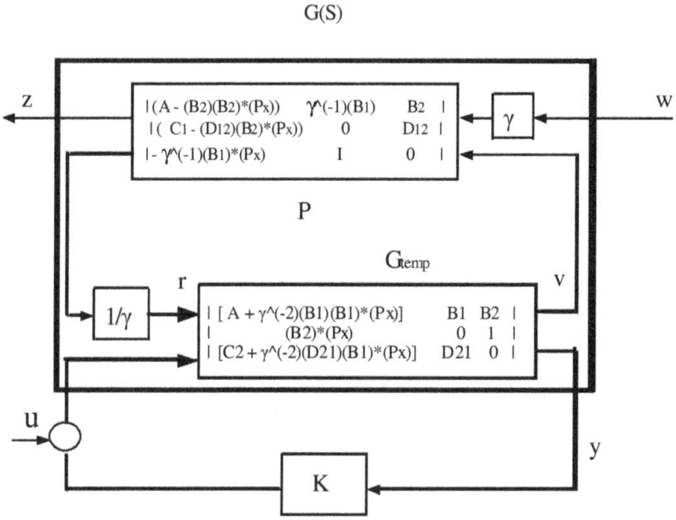

Fig. 2.3-2. New Configuration Constructed from Eqn. (2.3-18) and Eqn. (2.3-24)

Now we design "K" such that the two variables, "r" and "v," converge to zero while minimizing "γ". As we can see in the block diagram in Fig. 2.3-2, a smaller "γ" implies less effect of "w" (disturbance) on the control system. However, excessively small values of "γ" may cause convergence problem. It is also interesting to note that we do not have to design more than one feedback controller in spite of the fact that there are four separate channels that we need to be concerned about ensuring good performance. The four channels are:

1. From w to z
2. From w to y
3. From u to z
4. From u to y

To design the controller, K, (read Chapters 3 and 12 of ref. 42 for detail), we introduce a reconfigured block diagram that represents the plant shown in Fig. 2.3-2 but in a different structure (Fig. 2.3-3).

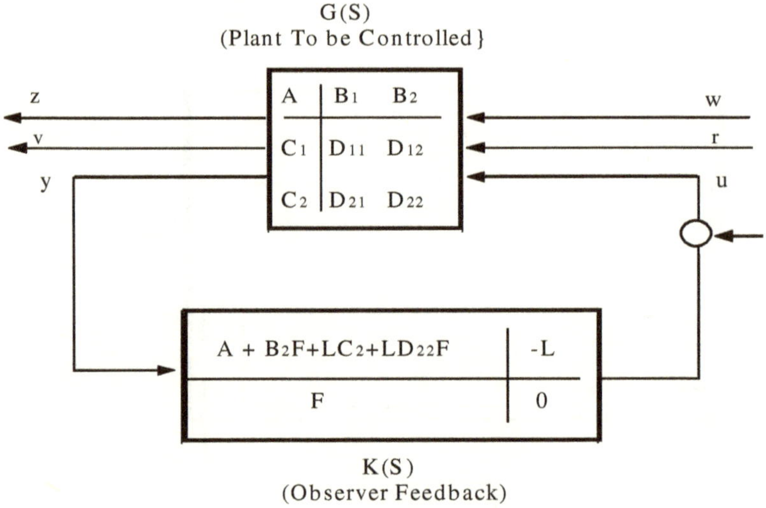

Fig. 2.3-3. System with Observer Feedback

The Theorem 16.4 suggests that there can be many different types of structures for K(S). The K(S) chosen in Fig. 2.3-3 is a Luenberger type observer that estimates the plant's states and at the same time provides state feedback to achieve the desired system performance. Derivation of K(S) may be found in ref. 42. Here, it is shown that if we can find F and L such that (A + LC2) and (A + B2F) are stable, then the closed loop system in Fig. 2.3-3 is stable. From G(S) in Fig. 2.3-3,

$$xd = A\,x + B1\,w + B2\,u \quad (2.3\text{-}25)$$
$$z = C1\,x + D11\,w + D12\,u \quad (2.3\text{-}26)$$
$$y = C2\,x + D21\,w + D22\,u \quad (2.3\text{-}27)$$

and from K(S) in Fig. 2.3-3,

$$xhd = (A + B2F + LC2 + LD22F)\,xh - Ly \quad (2.3\text{-}28)$$
$$u = F\,xh \quad (2.3\text{-}29)$$

where

xh : K(S) state vector

From Eqn. (2.3-25) and Eqn. (2.3-29),

$$xd = A\,x + B1\,w + B2\,F\,xh \tag{2.3-30}$$

From Eqn. (2.3-27), Eqn. (2.3-28), and Eqn.(2.3-29)

$$\begin{aligned} xhd &= (A + B2F + LC2 + LD22F)\,xh - L(C2\,x + D21\,w + D22\,u) \\ &= (A + B2F + LC2 + LD22F)xh - L(C2\,x + D21\,w + D22\,F\,xh) \\ &= (A + B2F + LC2 + LD22F - L\,D22\,F)\,xh - LC2\,x - LD21\,w \end{aligned} \tag{2.3-31}$$

From Eqn. (2.3-30) and Eqn. (2.3-31), we construct the following matrix system equation.

$$\begin{vmatrix} xd \\ xhd \end{vmatrix} = \begin{vmatrix} A & B2\,F \\ -LC2 & A + B2F + LC2 \end{vmatrix} \begin{vmatrix} x \\ xh \end{vmatrix} + \begin{vmatrix} B1 \\ -LD21 \end{vmatrix} w \tag{2.3-32}$$

Now we need to show that the eigenvalues of the following matrix are in LHP (Left Half Plane).

$$\begin{vmatrix} A & B2\,F \\ -LC2 & A + B2F + LC2 \end{vmatrix} \tag{2.3-33}$$

This matrix can be transformed into the following matrix.

$$\begin{vmatrix} A + LC2 & 0 \\ -LC2 & A + B2F \end{vmatrix} \tag{2.3-34}$$

The transformation will be justified by showing that the eigenvalue equation (or characteristic equation) of Eqn. (2.3-33) and that of Eqn. (2.3-34) are identical.

First, we derive a characteristic equation from Eqn. (2.3-33).

$$\left|\begin{matrix}\lambda & 0\\ 0 & \lambda\end{matrix}\right| - \left|\begin{matrix}A & B2\,F\\ -LC2 & A + B2F + LC2\end{matrix}\right|$$

$$= \left|\begin{matrix}\lambda - A & B2\,F\\ LC2 & \lambda - (A + B2F + LC2)\end{matrix}\right|$$

$$= \lambda^{(2)} - (2A + B2F + LC2)\,\lambda + [A(A + B2F + LC2) + LC2\; B2F]$$

(2.3-35)

now from Eqn. (2.3-34).

$$\left|\begin{matrix}\lambda & 0\\ 0 & \lambda\end{matrix}\right| - \left|\begin{matrix}A + LC2 & 0\\ -LC2 & A + B2F\end{matrix}\right|$$

$$= \left|\begin{matrix}\lambda - (A + LC2) & 0\\ LC2 & \lambda - (A + B2F)\end{matrix}\right|$$

$$= \lambda^{(2)} - (2A + B2F + LC2)\,\lambda + [A(A + B2F + LC2) + LC2\; B2F]$$

(2.3-36)

The two characteristic equations, Eqn. (2.3-35) and Eqn. (2.3-36), are identical. Thus, the transformation is justified.

Since Eqn. (2.3-34) is a lower diagonal matrix, if we can find L such that A + LC2 is stable and F such that A + B2F is stable for a given set of matrices, A, B2, and C2 , then Eqn. (2.3-32) is stable. This implies that the closed loop system with the feedback block, K(S), is stable. The convergence rate can be controlled by assigning appropriate values to each element in the L and F matrices.

Using the results obtained above, we design a controller, K(S), for the plant in Fig. 2.3-2. From the K(s) structure in Fig. 2.3-3 and sub-blocks of Eqn (2.3-24), we find that

$$K = \left|\begin{array}{c|c} A + \gamma^{(-2)} (B1)(B1)^{*}(Px) + B2\,F + L\,[C2 + \gamma^{(-2)} (D21)\,(B1)^{*}(Px)] & -L\\ \hline F & 0\end{array}\right|$$

by identifying the following terms ,

$$
\begin{array}{ll}
A \text{ in } K = (A + \gamma^{\wedge}(-2)\,(B1)(B1)^*(Px)) & \text{in Eqn (2.3-24)} \\
B2 \text{ in } K = B2 & \text{in Eqn (2.3-24)} \\
C2 \text{ in } K = (C2 + \gamma^{\wedge}(-2)\,(D21)\,(B1)^*(Px)) & \text{in Eqn (2.3-24)}
\end{array}
$$

The remaining task is to determine “F” and “L,” which influence the speed of convergence.

Summarizing,

1) Determine the system matrices, A, B1, B2, C1, and C2.
2) Solve the Riccati equation for Px,

$$A^*(Px) + (Px)\,A + 2\,w^*(B1)^*\,(Px) + 2u^*(B2)^*\,(Px) = 0$$

3) Construct the K(S),

$$K = \left|\begin{array}{c|c} A + \gamma^{\wedge}(-2)\,(B1)(B1)^*(Px) + B2\,F + L\,[C2 + \gamma^{\wedge}(-2)\,(D21)\,(B1)^*(Px)] & -L \\ \hline F & 0 \end{array}\right|$$

4) Determine “F” and “L” that minimize the maximum singular value of the transfer function from w to z . Here we need to use an optimization algorithm to achieve these objectives. As noted earlier, excessively small “γ” may cause a convergence problem. A set of solutions for “F” and “L” is shown below ((see page 419 and 437 of ref. 42)

$$
\begin{array}{l}
F = -\,(B2)^*\,(Px) \\
L = -\,(I - \gamma^{\wedge}(-2)\,(Py)(Px)\,)^{\wedge}(-1)\,(Py)\,(C2)^*
\end{array}
$$

where “Py” is a solution of the following Riccati equation.

$$A(Py) + (Py)\,A^* + (B1)^*\,(B1) + \gamma^{\wedge}(-2)\,(Py)\,(C1)(C1)^*(Py) - (Py)\,(C2)(C2)^*(Py) = 0 \qquad (2.3\text{-}25)$$

Final block diagram is shown in Fig. 2.3-4, which is designed according to Theorem 16.4. Notice that Eqn. (2.3-9) and Eqn. (2.3-25) are corresponding to Eqn. (2.3-1) and Eqn. (2.3-2) in the Theorem 16.4 respectively.

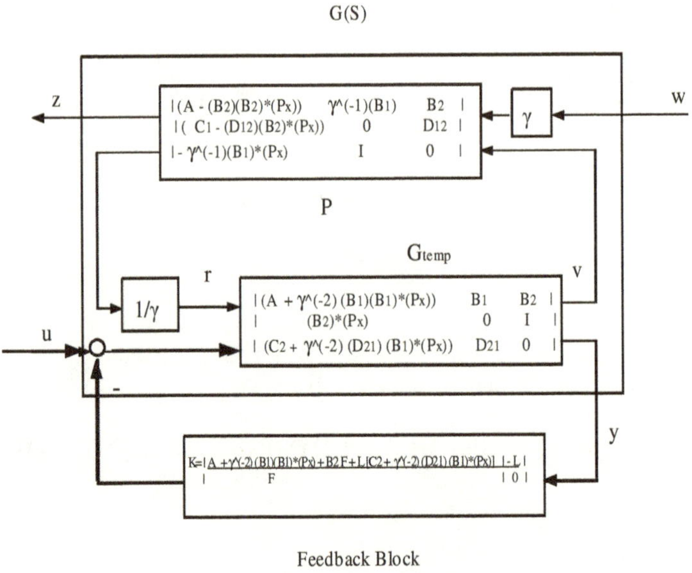

Fig. 2.3-4. Implementation of H-infinity Feedback Block Design

2.4 Loop Shaping and Coprime Factorization

The loop shaping and coprime factorization improve the control system performance extensively. The loop shaping technique means that the weighting functions (or compensators) are added before and after a plant to tailor the slope and shape of the singular value vs. frequency curve to the specifications.

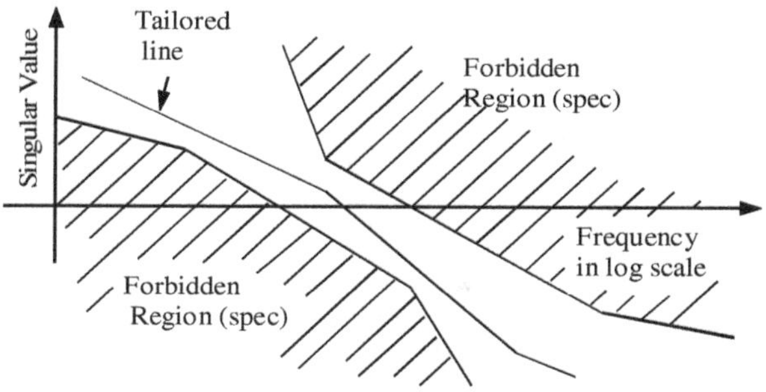

Fig. 2.4-1. Concept of Loop Shaping

As seen in Fig. 2.4-1, the line is tailored to pass signals in the low frequency range and filter the signals (noise or interference) in the high frequency range. This is a typical loop shaping.

The coprime factorization basically means that it ensures no pole zero cancellation possible in RHP in the system equation.

In this section, we show that coprime factorization P and coprime factorization K in Fig. 2.4-2 are required for the system shown in the figure to be stable. To start with, it will be shown that the control system is stable, iff the following matrices hold, that is, the inverses stand.

$$Sys = \left| \begin{matrix} I & K \\ P & I \end{matrix} \right|^{(-1)} \qquad (2.4\text{-}1a)$$

$$Sys = \left| \begin{matrix} I & P \\ K & I \end{matrix} \right|^{(-1)} \qquad (2.4\text{-}1b)$$

where the "P" includes the pre and post weighting functions.

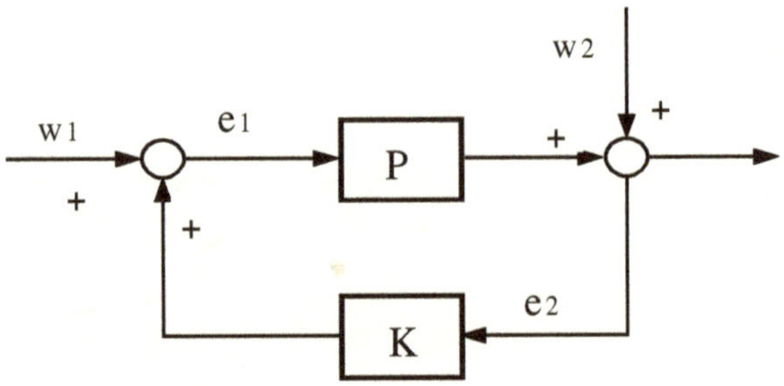

Fig. 2.4-2. Feedback Control System Block Diagram

Eqn. (2.4-1) is derived from Fig. 2.4-2 as shown below. From the figure, we write a set of equations,

$$e1 = w1 + Ke2 \qquad (2.4\text{-}2)$$
$$e2 = w2 + Pe1 \qquad (2.4\text{-}3)$$

Substituting Eqn. (2.4-3) into (2.4-2),

$$\begin{aligned} e1 &= w1 + K(w2 + Pe1) \\ &= w1 + Kw2 + KPe1 \\ &= (I - KP)^{\wedge}(-1)\,(w1 + Kw2) \qquad (2.4\text{-}4) \end{aligned}$$

Substituting Eqn. (2.4-2) into (2.4-3),

$$\begin{aligned} e2 &= w2 + P\,(w1 + Ke2) \\ &= w2 + P\,w1 + PKe2 \\ e2 &= (I - PK)^{\wedge}(-1)\,(w2 + Pw1) \qquad (2.4\text{-}5) \end{aligned}$$

From Eqn. (2.4-4) and Eqn. (2.4-5),

$$\begin{vmatrix} e1 \\ e2 \end{vmatrix} = \begin{vmatrix} (I - KP)^{\wedge}(-1) & (I - KP)^{\wedge}(-1)K \\ (I - PK)^{\wedge}(-1)P & (I - PK)^{\wedge}(-1) \end{vmatrix} \begin{vmatrix} w1 \\ w2 \end{vmatrix}$$

(2.4-6)

The Eqn. (2.4-6) holds and the system is stable if (I – KP) and (I – PK) are invertible. The invertibility is shown below.

1) Show that (I-PK) is invertible.

$$(I - PK) = \begin{vmatrix} I & K \\ P & I \end{vmatrix} \qquad (2.4\text{-}7)$$

We let the P and K be written as follows

$$P = NM^{(-1)} \qquad (2.4\text{-}8)$$

And

$$K = UV^{(-1)} \qquad (2.4\text{-}9)$$

where

1) N(S), M(S), U(S), and V(S) are polynomials
2) N(S) and M(S) are coprime
3) U(S) and V(S) are coprime

Substituting Eqn. (2.4-8) and Eqn. (2.4-9) into Eqn. (2.4-7).

$$\begin{vmatrix} I & K \\ P & I \end{vmatrix} = \begin{vmatrix} I & UV^{(-1)} \\ NM^{(-1)} & I \end{vmatrix}$$

$$= \begin{vmatrix} M & U \\ N & V \end{vmatrix} \begin{vmatrix} M^{(-1)} & 0 \\ 0 & V^{(-1)} \end{vmatrix}$$

Taking inverse on both sides,

$$\begin{vmatrix} I & K \\ P & I \end{vmatrix}^{(-1)} = \begin{vmatrix} M & 0 \\ 0 & V \end{vmatrix} \begin{vmatrix} M & U \\ N & V \end{vmatrix}^{(-1)} \qquad (2.4\text{-}10)$$

If $\begin{vmatrix} M & 0 \\ 0 & V \end{vmatrix}$ and $\begin{vmatrix} M & U \\ N & V \end{vmatrix}$ are coprime, then (I-PK) is invertible.

It will be shown below that the two matrices are coprime.

Before showing the coprime, we briefly explain the meaning of coprime factorization. It means that we factor P into N and M such that the greatest common divisor for them is 1. For example,

$$P(s) = \frac{(s+a)(s+b)}{(s+b)(s+c)} = N(s)M(s)^{\wedge}(-1)$$

$N(s) = (s + a)(s + b)$
$M(s) = (s + b)(s + c)$

Here, they are not coprime, because the greatest common divisor is not 1.0 but (s+b). However, N(s) = (s + a) and M(s)=(s + c) are coprime because their common divisor is 1.0. The coprime property of a transfer function actually implies that there is no pole zero cancellation possible. This property is particularly important for system stability, because it assures that there is no common divisor in the RHP (Right Half Plane).

Coprime can be checked out by the following equation. Let m(s) and n(s) are two transfer functions or polynomials. These two functions are coprime if we can find x and y such that

$$xm + yn = I$$

Example:

Check whether (s+a) and (s+c) are coprime. Now let

$x= 1/(a-c)$
$y= 1/(c-a)$

Then,

$$x (s + a) + y(s+c) = 1$$

Now for $\begin{vmatrix} M & 0 \\ 0 & V \end{vmatrix}$ and $\begin{vmatrix} M & U \\ N & V \end{vmatrix}$,

we find,

$$x = \begin{vmatrix} a & (aM-1)/N \\ (dV-1)/U & d \end{vmatrix}$$

$$y = \begin{vmatrix} 0 & (1-aM)/N \\ (1-dV)/U & 0 \end{vmatrix}$$

that

$$x \left|\begin{matrix} M & 0 \\ 0 & V \end{matrix}\right| + y \left|\begin{matrix} M & U \\ N & V \end{matrix}\right| = I$$

Thus,

$$\left|\begin{matrix} M & 0 \\ 0 & V \end{matrix}\right| \text{ and } \left|\begin{matrix} M & U \\ N & V \end{matrix}\right| \text{ are coprime.}$$

2) Show that (I - KP) is invertible.

$$(I - KP) = \left|\begin{matrix} 1 & P \\ K & 1 \end{matrix}\right|$$

$$= \left|\begin{matrix} I & NM^{-1} \\ UV^{-1} & I \end{matrix}\right|$$

$$= \text{Transpose of } \left| \left|\begin{matrix} M & U \\ N & V \end{matrix}\right| \left|\begin{matrix} M^{-1} & 0 \\ 0 & N^{-1} \end{matrix}\right| \right|$$

$$\left|\begin{matrix} 1 & P \\ K & 1 \end{matrix}\right|^{-1} = \left|\begin{matrix} M & 0 \\ 0 & N \end{matrix}\right| \left|\begin{matrix} M & N \\ U & V \end{matrix}\right|^{-1}$$

If $\left|\begin{matrix} M & 0 \\ 0 & N \end{matrix}\right|$ and $\left|\begin{matrix} M & N \\ U & V \end{matrix}\right|$ are coprime, then (I-KP) is invertible.

It will be shown below that the two matrices are coprime.

For $\left|\begin{matrix} M & 0 \\ 0 & N \end{matrix}\right|$ and $\left|\begin{matrix} M & N \\ U & V \end{matrix}\right|$,

we find,

$$x = \left|\begin{matrix} a & V(aM-1)/(NU) \\ (dN-1)/N & d \end{matrix}\right|$$

$$y = \left|\begin{matrix} 0 & (1-aM)/U \\ (1-dN)/N & 0 \end{matrix}\right|$$

that

$$x \begin{vmatrix} M & 0 \\ 0 & V \end{vmatrix} + y \begin{vmatrix} M & U \\ N & V \end{vmatrix} = I$$

Thus,

$\begin{vmatrix} M & 0 \\ 0 & N \end{vmatrix}$ and $\begin{vmatrix} M & N \\ U & V \end{vmatrix}$ are coprime.

Otherwise, the vector of

$$\begin{vmatrix} e1 \\ e2 \end{vmatrix}$$

diverges.

2.5 H-infinity Requirement on Feedback Block, "K"

We derive an equation that must be satisfied as we design the feedback block, "K," that assures convergence when

$$\| \Delta \|_\infty = \| \Delta M \;\; \Delta N \|_\infty < \delta.$$

Now consider a small perturbation ΔM in M and ΔN in N in Eqn. 2.4-10. The perturbed system is,

$$\begin{vmatrix} (M + \Delta M) & U \\ (N + \Delta N) & V \end{vmatrix}^{\wedge}(-1) = [(M + \Delta M)V - (N + \Delta N)U]^{\wedge}(-1)$$
$$= [(MV - NU) + (\Delta MV - \Delta NU)]^{\wedge}(-1)$$
$$= [I + (\Delta MV - \Delta NU)(MV-NU)^{\wedge}(-1)]^{\wedge}(-1)\;(MV-NU)^{\wedge}(-1)$$

(2.5-1)

Since (MV-NU) is invertible (see Lemma 5.10 on page 128, ref. 42), if we can show that "[I +(ΔMV- ΔNU) (MV-NU)^(-1)] " in Eqn. (2.5-1) is invertible, then the system with the small perturbation is stable. The invertibility of "[I +(ΔMV- ΔNU) (MV-NU)^(-1)]" is shown below.

$$I + (\Delta MV - \Delta NU)(MV-NU)^{\wedge}(-1) = I - \begin{vmatrix} \Delta N & -\Delta M \end{vmatrix} \begin{vmatrix} U \\ V \end{vmatrix} (MV-NU)^{\wedge}(-1)$$
$$= I - |\Delta| |R|$$

(2.5-2)

where
$$|\Delta| = |\Delta N \quad \Delta M|$$

$$|R| = \begin{vmatrix} U \\ V \end{vmatrix} (MV-NU)^{\wedge}(-1)$$

Now, set boundaries on the singular values of $\| \Delta \|$ and $\| R \|$ as shown below,

$$\| \Delta \|\infty < 1/\gamma, \tag{2.5-3a}$$

and

$$\| R \| \infty = \left\| \begin{vmatrix} U \\ V \end{vmatrix} (MV-NU)^{\wedge}(-1) \right\|_\infty < \gamma \tag{2.5-3b}$$

where γ is less than 1.0

Then, the right-hand side of Eqn. (2.5-2) is greater than zero but less than 1.0, because the singular values are greater than zero (singular value property), that is,

$$\|\Delta\|\infty \, \| R\|\infty < (1/\gamma)\,(\gamma)$$
$$< 1.0$$

Thus, the "$[I +(\Delta MV- \Delta NU)\,(MV-NU)^{\wedge}(-1)]$" is invertible for,

$$(I - \|\Delta\|\infty \, \| R\|\infty\,) \neq 0.0$$

and the system is stable iff Eqn. (2.5-3a) and Eqn. (2.5-3b) hold.

Rearranging Eqn. (2.5-3a),

$$\left\| \begin{vmatrix} U \\ V \end{vmatrix} (MV-NU)^{\wedge}(-1) \right\|_\infty = \left\| \begin{vmatrix} UV^{\wedge}(-1) \\ I \end{vmatrix} V(MV-NU)^{\wedge}(-1) \right\|_\infty$$

From Eqn. (2.4-9),

$$K = UV^{\wedge}(-1)$$

And since,

$$V(MV-NU)^{-1} = [(MV-NU)V^{-1}]^{-1}$$
$$= [(MVV^{-1} - NUV^{-1}]^{-1},$$

the matrix equation can be rewritten as,

$$\left\| \begin{bmatrix} U \\ V \end{bmatrix} (MV-NU)^{-1} \right\|_\infty = \left\| \begin{bmatrix} K \\ I \end{bmatrix} [MVV^{-1}-NU\,V^{-1}]^{-1} \right\|_\infty$$

$$= \left\| \begin{bmatrix} K \\ I \end{bmatrix} (M-NK)^{-1} \right\|_\infty$$

Since,

$$(M-NK)^{-1} = \{M\,[I - M^{-1} N K]\}^{-1}$$
$$= [I - M^{-1} N K]^{-1} M^{-1},$$

$$\left\| \begin{bmatrix} U \\ V \end{bmatrix} (MV-NU)^{-1} \right\|_\infty = \left\| \begin{bmatrix} K \\ I \end{bmatrix} [I - M^{-1}NK]^{-1} M^{-1} \right\|_\infty < \gamma$$

(2.5-4)

Here, we design a controller (K) that satisfies the Eqn. (2.5-4) by using the H-infinity minimization algorithm. Here, γ is less than 1. The perturbation must meet the following limitation. That is, if we design K satisfying the Eqn. (2.5-4), then the control system is stable as long as the math modeling errors are within the boundary shown below.

$$\| \Delta \|_\infty < 1/\gamma$$

where

$$\| \Delta \| = \| \Delta N \quad \Delta M \|$$

2.6 A Summary of an H-infinity Control System Design Algorithm

Concept of the design algorithm is summarized below (see Fig. 2.6-1).

1) Construct a math model of the plant to be controlled

a. Derive differential equations describing the plant dynamics
b. Linearize the differential equations if they are non-linear.
c. Construct a state space linear system equation from the linearized differential equations, thus define, A, B1, B2, C1, C2, D11, D12, D21, and D22 (Fig. 2.6-1). Note that D11 = D22 = 0.0 for simplicity.

2) Find pre-weighting function, W1(S) and post-weighting function, W(S) such that the gain of W2(S) P(S) W1(S) meets the loop shaping requirements.

3) Construct $P_w(S) = W1(S)\, P(S)\, W2(S)$

$$= \left| \begin{array}{ccc} A_w & B_{w1} & B_{w2} \\ C_{w1} & 0 & D_{12w} \\ C_{w2} & D_{21w} & 0 \end{array} \right|$$

4) Construct Riccati equations and solve for X_∞ and Y_∞ (Theorem 16.4)

$$A_w^* X_\infty + X_\infty A_w + C_{w1}{}^{*} C_{w1} + \gamma^{\wedge}(-2)\, X_\infty B_{w1} B_{w1}^* X_\infty - X_\infty B_{w2} B_{w2}^* X_\infty = 0$$

$$A_w^* Y_\infty + Y_\infty A_w + B_{w1}^* B_{w1} + \gamma^{\wedge}(-2)\, Y_\infty C_{w1} C_{w1}^* Y_\infty - Y_\infty C_{w2} C_{w2}^* Y_\infty = 0$$

Solving the Riccati equation requires iteration as $\gamma^{\wedge}$ changes from the initial guess until $\rho\,(X_\infty Y_\infty) < \gamma^{\wedge}(2)$.

5) Construct the K(S)

$$K = \left| \begin{array}{c|c} A + \gamma^{\wedge}(-2)\,(B_{w1})(B_{w1})^*(X_\infty) + B_{w2} F + L\, C_{w2} & -L \\ \hline F & 0 \end{array} \right|$$

where

$$F = -\,(B_{w2})^*\,(X_\infty)$$

$$L = -\,(I - \gamma^{\wedge}(-2)\,(Y_\infty)(X_\infty))^{\wedge}(-1)\,(Y_\infty)\,(C_{w2})^*$$

Note: 1. " * " means complex conjugate.
2. "K " can be of different structure.

6) Check the following singular value inequality [Eqn. (2.5-4)].

$$\left\| \begin{bmatrix} K(j\omega) \\ I \end{bmatrix} [I - M(j\omega)^{-1}N(j\omega)K(j\omega)]^{-1} M(j\omega)^{-1} \right\|_{\infty} < \gamma$$

where

$$P_w(j\omega) = N(j\omega)\, M(j\omega)^{-1}$$
$$\gamma < 1.0$$

We need γ less than 0.3
If this condition is not met, then go back to 2.

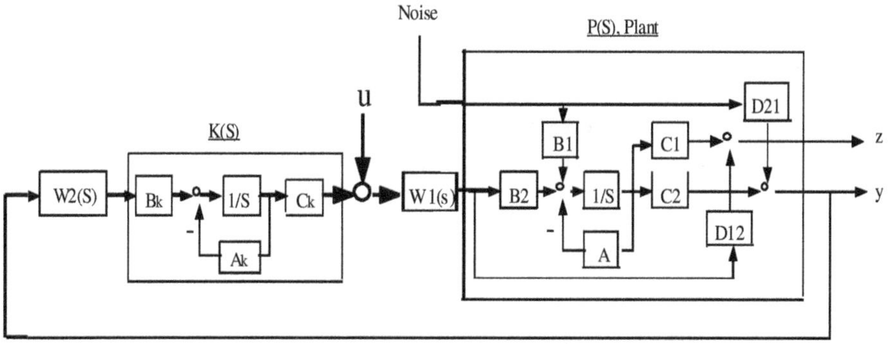

K= | A + γ^(-2) (Bw1)(Bw1)*(Px) + Bw2 F + L Cw2 | - L |
| F | 0 |

where,

F = - (Bw2)* (Px)
L = - [I - γ^(-2) (Py) (Px)]^(-1) (Py) (Cw2)*

Note that * means complex conjugate.
See text for Bw1, Bw2,Cw2,Px,and Py

u : Input (Command)
w : Noise (Input)
y : Feedback Output
(Response to Noise)
z : System Output

$$W1(S) = \frac{\Pi (s + a1(i))}{\Pi (s + b1(i))}$$

$$W2(S) = \frac{\Pi (s + a2(i))}{\Pi (s + b2(i))}$$

Fig. 2.6-1. H-infinity Optimized Control System Block Diagram

****

Symbols:

P(S) = | A | B |
| C | D |

= C(SI-A)^(-1) + D

= | c1 c2 | | s + a 0 | ^(-1) | b1 |
| c3 c4 | | 0 s+b | | b2 |

= | c1 (s+a)^(-1)b1 c2(s+b)^(-1)b2 |
| c3(s+a)^(-1)b1 c4(s+b)^(-1)b2|

= | c1b1/(s+a) c2b2/(s+a) |

| c3b1/(s+a)   c4b2/(s+b) |

If c2 = 0 and c3 = 0, then

= | c1b1/(s+a)   0 |
  | 0   c4b2/(s+b) |

= M^(-1) N

where

M = | (s+a)   0 |
    | 0   (s+b) |

N = |c1b1   0 |
    | 0   c4b2 |

****

It is to be noted that as we apply the Theorem 16.4 in our design process, we do not need to compute the maximum singular value of the system and minimize the singular value.

Paraphrasing the Theorem,

if we satisfy the following:

1) $H_\infty \in$ dom(Ric) and $X_\infty :=$ Ric $(H_\infty) \geq 0$
2) $J_\infty \in$ dom(Ric) and $Y_\infty :=$ Ric $(J_\infty) \geq 0$
3) $\rho\ (X_\infty\ Y_\infty) < \gamma$^(2)

Then, we meet the following H-infinity requirement:

$$\|Tzw\|_\infty < \gamma$$

and there exists an admissible controller, K.

The "ncfsyn.m." in Matlab synthesizes an H-infinity loop shaping controller to robustly stabilize a family of systems by a ball of uncertainty in the normalized coprime factors of the system. It takes basically the following steps (see Matlab Toolbox for detail).

1. Realization (Construction of A, B, C, D plant system matrices)

2. Reduced the order of the system optimally by using Hankel Norm Approximation (ref. 42).
3. Balance (Eliminate uncontrollable and unobservable states)
4. Coprime factorization (making it sure that greatest common divisor of numerator and denominator polynomials is 1)
5. Find pre-weighting function, W1(S) and post-weighting function, W(S) such that the gain of W2(S) P(S) W1(S) is sufficiently high at frequencies where desired signals are associated and sufficiently low where noise signals are associated (loop shaping).
6. Unpack [W2(S) P(S) W1(S)]
   [A, B, C, D] =unpack [W2(S) P(S) W1(S)]
7. Normalization

   If A, B, C, and D are such that

   W2(S) P(S) W1(S) =N*Inv(M)

   and

   $$M^*(-S)M(S) + N^*(-S)N(S) = I$$

   where

   A, B, C, and D : system matrices

8. Check the K(jω) whether it meets the following requirement.

   $$\left\| \begin{bmatrix} K(j\omega) \\ I \end{bmatrix} [I - M(j\omega)^{-1}N(j\omega)K(j\omega)]^{-1} M(j\omega)^{-1} \right\|_\infty < \gamma$$

   where

   $\gamma < 0.3$

   If not satisfied, redesign W1(S) and W2(S) (back to step 5).

Our objective here is to find K such that

$$\left\| \begin{bmatrix} K \\ I \end{bmatrix} [I - M^{-1}NK]^{-1} M^{-1} \right\|_\infty < \gamma$$

where

{[W2] P [W1] } (jw) = N M^(-1)

W1: front weighting function
W2: rear weighting function
W1 and W2 are designed to meet the loop shaping requirements.

P : plant
N, M : coprime of W1PW2 (jw)
$\gamma < 1.0$, which implies $\| \Delta \|_\infty < 1/\gamma$.

The smaller γ is, the larger the modeling errors are allowed while maintaining stability. [See Eqn. (2.5-4).]

Chapter 3 Launch Vehicle Attitude Systems Dynamics

The launch vehicle attitude control system design is an extremely complex process requiring many years of research and development, and a large amount of funding and human resources controlled and managed by experienced personnel under a well-organized operational structure.

The first step for a successful launch operation is to determine a set of specific mission requirements such as 1) functional requirements, 2) flight systems engineering and integration requirements, 3) operational requirements, and 4) constraints. This book addresses a critical part of the flight systems engineering and integration requirements, i.e., the design of launch vehicle (LV) attitude control systems. Development and design of the control systems require supports from many disciplines such as:

1. Propulsion Systems

   - Thrust time history, fuel mix ratio, and mass flow rate, etc.
   - Fuel tank configuration and dimensions
   - Slosh damping mechanism
   - Engine gimbal dynamic model and nozzle load
   - Actuator design and analysis

2. Mass properties and structures

   - Weight management and total weight integration
   - Bending and load analysis
   - Mass property time history requirements
   - Vehicle layouts
   - Instrument locations analysis
   - Load strength requirements
   - Stiffness requirement

3. Aerodynamics

   - Aerodynamic coefficients vs. trajectory
   - Fairing design
   - Nozzle aerodynamic loads
   - Wind tunnel testing

4. Performance

- Baseline trajectory
- Multi-mission trajectory
- Trajectory dispersion
- Payload capabilities
- Vehicle steering requirements
- Launch azimuth

The main goal of all these functions is to support successful design of attitude and flight control systems. With these supports, the flight systems group designs:

Guidance, navigation, and control
Avionics
Inertial measurement units for navigation
Computer and flight software
Telemetry and communication
Range and safety

Systems analysis is required for a successful mission, and we need to perform numerous analyses such as:

Flight mission analysis
Propulsion performance analysis
Control system analysis
Wind analysis
Load analysis
Structural analysis
Aerodynamic analysis
Thermodynamic analysis
Environmental analysis
Vehicle/Spacecraft integration analysis

The attitude control systems group designs a control system that satisfies the following requirements:

Controllability
Stability
Transient response
Load relief stability
Control system and guidance feedback loop schedule
Engine actuator angle limit

Flight software constant

In addition to these, we need to design and analyze the staging process to ensure that there will be no contact between stage I and stage II as they separate. We also need to analyze the post-separation orbit profile to meet the requirements concerning the residual disposal in space. We will focus our attention on the design of the attitude control systems. First, we develop LV system dynamic equations that are compatible with H-infinity control theory so that we can utilize the optimal design techniques, namely H-infinity, Loop Shaping, and Coprime Factorization.

The coordinate systems that will be used in this book are:

Body Frame (BF: Xb,Yb,Zb) :
a. right-hand coordinate system
b. origin at vehicle geometric center
c. rotates with the vehicle body with respect to inertial frame

Inertial Frame (IF: Xi, Yi, Zi):
a. right-hand coordinate system
b. origin at any point determined by the analyst
c. fixed in inertial space

Earth Center Inertial Frame: (ECI: Zeci, Yeci, Zeci)
a. right-hand coordinate system
b. origin at the center of the earth
c. fixed in inertial space

The objective of this chapter is to derive a state space equation that describes launch vehicle atttitude system dynamics. We will start with a derivation of three-dimensional system equations, and later we will reduce the math model to a pitch plane motion only, which will be used for our design example in Chapter 4. The reason for deriving the state space equation is that in that form, we can easily apply H-infinty control theory.

$$\underline{\mathbf{xd}} = A*\underline{\mathbf{x}} + B*\underline{\mathbf{u}}$$

$$\underline{\mathbf{y}} = C*\underline{\mathbf{x}} + D*\underline{\mathbf{u}}$$

where

$$\underline{\mathbf{xd}} = \underline{\mathbf{dx/dt}}$$

x = [**θ θd q1 q1d q2 q2d Γ1 Γ1d Γ2 Γ2d θPG θRG (θRG)d**]

θ = attitude angle in pitch (rad)
θd = attitude angular rate in pitch (rad/sec)
q1 = generalized coordinate of the 1st bending mode in pitch (ft)
q1d = rate of generalized coordinate of the 1st bending mode in pitch (ft/sec)
q2 = generalized coordinate of the 2nd bending mode in pitch (ft)
q2d = rate of generalized coordinate of the 2nd bending mode in pitch (ft/sec)
Γ1 = pendulum angle of fuel tank 1 sloshing (rad)
Γ1d = pendulum angular rate of fuel tank 1 sloshing (rad/sec)
Γ2 = pendulum angle of fuel tank 2 sloshing (rad)
Γ2d = pendulum angular rate of fuel tank 2 sloshing (rad/sec)
θPG = position gyroscope output (rad)
θRG = rate gyroscope output (rad/sec)
θRG)d = time derivative rate gyroscope output (rad/sec^(2))

u = [**δ** ; **α** ; **Ww**], (transpose of [**δ α Ww**])

δ = engine deflection angle (rad)
α = angle of attack (rad)
Ww = lateral velocity (due to wind) (ft/sec)
[Note: In this study, we consider "Ww" alone in α = (w-Ww)/U. See page 182. "w" is lateral velocity due to some other sources such as bending and sloshing.]

y = [**θ** ; **θd**]

where

θd = dθ/dt

A =

0	1	0	0	0	0	0	0	0	0	0	0	0
C1	C2	C3	C4	C6	C7	C9	0	C10	0	0	0	0
0	0	0	1	0	0	0	0	0	0	0	0	0
E1	E2	E4	E5	E6	E7	E9	0	E10	0	0	0	0
0	0	0	0	0	1	0	0	0	0	0	0	0

G1	G2	G4	G5	G7	G8	G9	0	G10	0	0	0	0
0	0	0	0	0	0	0	1	0	0	0	0	0
0	0	0	0	0	0	H1	0	H2	0	0	0	0
0	0	0	0	0	0	0	0	0	1	0	0	0
0	0	0	0	0	0	J1	0	J2	0	0	0	0
K1	0	K2	0	K3	0	0	0	0	0	K4	0	0
0	0	0	0	0	0	0	0	0	0	0	0	1
0	L1	0	L2	0	L3	0	0	0	0	0	L4	L5

B =

0	0	0
C11	C13	C14
0	0	0
E11	E13	E14
0	0	0
G11	G13	G14
0	0	0
H2	0	0
0	0	0
J2	0	0
0	0	0
0	0	0
0	0	0

C =

1	0	0	0	0	0	0	0	0	0	0	0	0
0	1	0	0	0	0	0	0	0	0	0	0	0

D =

0	0	0
0	0	0

Equations that compute the elements of the matrices, A, B, C, and D are in Section 3.8.

3.1 Launch Vehicle Rigid Body

Fig. 3.1-1 depicts a launch vehicle's flight in ECI frame. To start with, we consider an element (E) in the vehicle located $|\underline{\rho}|$ away from the origin. Dynamic equations will be developed for this element, and then these equations will be integrated over the LV to complete the derivation of the system equations. The position vector of the element is written in BF.

$$\underline{\rho i} = xi\,\underline{ib} + yi\,\underline{jb} + zi\,\underline{kb} \qquad (3.1\text{-}1)$$

Linear and angular velocity vectors of the vehicle in BF are

$$\underline{\mu} = u\,\underline{ib} + v\,\underline{jb} + w\,\underline{kb} \qquad (3.1\text{-}2)$$
$$\underline{\omega} = P\,\underline{ib} + Q\,\underline{jb} + R\,\underline{kb} \qquad (3.1\text{-}3)$$

where

$\underline{\mu}$ = linear velocity vector
$\underline{\omega}$ = angular velocity vector
$\underline{ib}$, $\underline{jb}$, $\underline{kb}$ are unit vectors along Xb, Yb, and Zb in BF, respectively.

Note that the underline symbol means vector quantity. The velocity vector of the element can be written as a vector sum of the linear velocity vector of the vehicle and the angular velocity vector. Note that the subscript, "i," indicates that the term is related to the element (E).

$$vi = \underline{\mu i} + d\,\underline{\rho i}/dt$$
$$= \underline{\mu i} + \underline{\omega i} \times \underline{\rho i} \qquad (3.1\text{-}4)$$

The acceleration is,

$$ai = dvi / dt$$
$$= d\underline{\mu i}/dt + d(\underline{\omega i} \times \underline{\mu i})/dt \qquad (3.1\text{-}5)$$

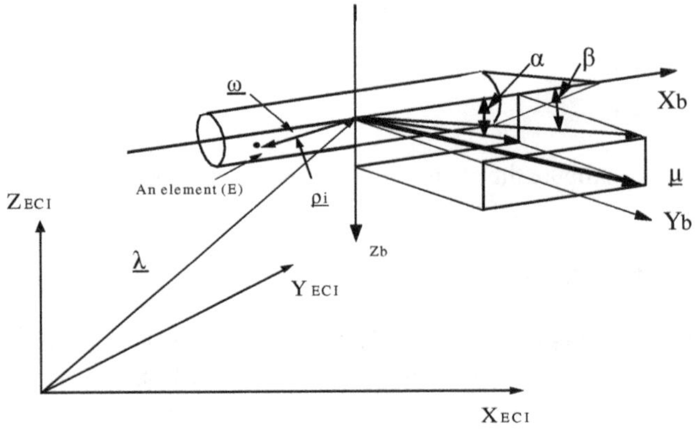

Fig. 3.1-1. Launch Vehicle Flight in ECI Frame

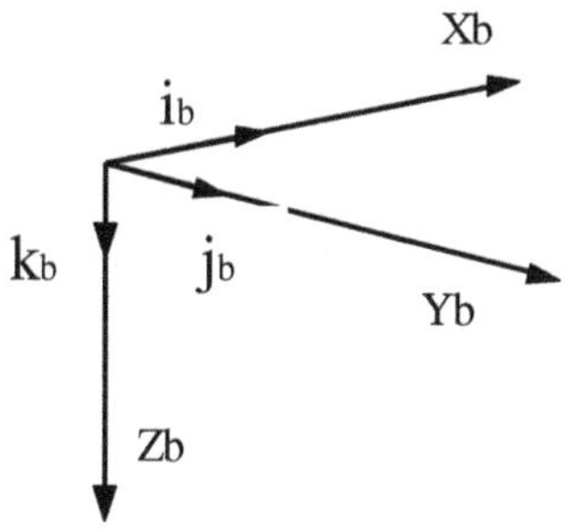

Fig. 3.1-2. Body Frame Orientation

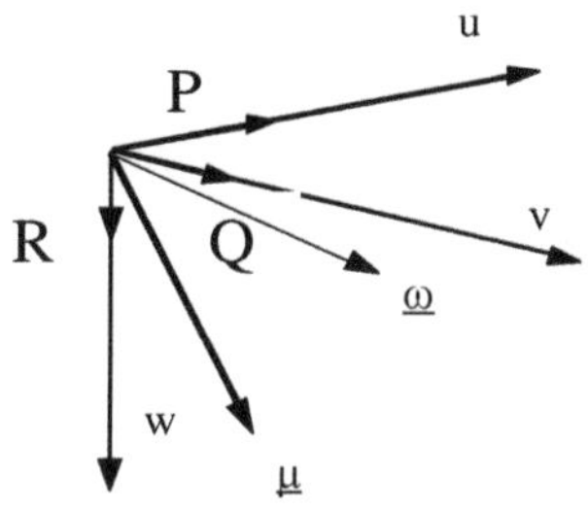

Velocity Vector
$\underline{\mu} = [\, u \;\; v \;\; w \,]$
Angular Rate Vector
$\underline{\omega} = [P \;\; Q \;\; R]$

Fig. 3.1-3 Vector Notation

$$= d\,\underline{\mu}/dt + (\partial\,\underline{\omega}/\partial t) \times \underline{\rho i} + \underline{\omega} \times (d\,\underline{\rho i}/dt)$$
$$= (\partial\underline{\mu}/\partial t + \underline{\omega} \times \underline{\mu}) + (\partial\,\underline{\omega}/\partial t) \times \underline{\rho i} + \underline{\omega} \times (\underline{\omega} \times \underline{\rho i}) \qquad (3.1\text{-}6)$$

Note that $(\partial / \partial t)$ means partial time derivative in BF frame.

Using Newton's second law,

Linear motion

$$
\begin{aligned}
\underline{F} &= \int \underline{a}\, dm \\
&= \int [(\partial\underline{\mu} / \partial t + \underline{\omega} \times \underline{\mu}) + (\partial\, \underline{\omega} / \partial t) \times \underline{\rho i} + \underline{\omega} \times (\underline{\omega} \times \underline{\rho i})]\, dm \\
&= \int [(\partial\underline{\mu} / \partial t + \underline{\omega} \times \underline{\mu})\, dm + \int (\partial\, \underline{\omega} / \partial t) \times \underline{\rho i}\, dm + \int \underline{\omega} \times (\underline{\omega} \times \underline{\rho i})\, dm \\
&= [(\partial\underline{\mu} / \partial t + \underline{\omega} \times \underline{\mu}) \int dm + (\partial\, \underline{\omega} / \partial t) \times \int \underline{\rho i}\, dm + \underline{\omega} \times (\underline{\omega} \times \int \underline{\rho i}\, dm) \\
&= [(\partial\underline{\mu} / \partial t + \underline{\omega} \times \underline{\mu})\, mo + (\partial\, \underline{\omega} / \partial t) \times (\underline{\rho c}\, mo) + \underline{\omega} \times [\underline{\omega} \times (\underline{\rho c}\, mo)]
\end{aligned}
$$

(3.1-7)

where

$$
\begin{aligned}
&\int dm = mo \\
&\int \underline{\rho i}\, dm = \underline{\rho c}\, mo \\
&\underline{\rho c}\,(cg) = xcg\, \underline{ib} + ycg\, \underline{jb} + zcg\, \underline{kb}
\end{aligned}
$$

Sunstituting Eqn. (3.1-1), Eqn. (3.1-2), and Eqn. (3.1-3) into Eqn. (3.1-7),

$$
\begin{aligned}
= &(u'\underline{ib} + v'\underline{jb} + w'\underline{kb})\, mo \\
&+ (P\, \underline{ib} + Q\, \underline{jb} + R\, \underline{kb}) \times (u\, \underline{ib} + v\, \underline{jb} + w\, \underline{kb})\, mo \\
&+ (P'\, \underline{ib} + Q'\, \underline{jb} + R'\, \underline{kb}) \times (xcg\, \underline{ib} + ycg\, \underline{jb} + zcg\, \underline{kb})mo \\
&+ (P\, \underline{ib} + Q\, \underline{jb} + R\, \underline{kb}) \times [(P\, \underline{ib} + Q\, \underline{jb} + R\, \underline{kb}) \times (xcg\, \underline{ib} + \\
&ycg\, \underline{jb} + zcg\, \underline{kb})mo
\end{aligned}
$$

(3.1-8)

The second term is expanded by cross product.

$$
\begin{vmatrix} \underline{ib} & \underline{jb} & \underline{kb} \\ P & Q & R \\ u & v & w \end{vmatrix} = (Q\,w - v\,R)\underline{ib} + (uQ - Pv)\underline{jb} + (Pv - uQ)\underline{kb}
$$

Expanding other terms in similar manner, we have

$$
\begin{aligned}
\underline{F} = &\underline{ib}[mo(u'+QW-RV) - mo\, xcg(Q^2 + R^2) - mo\, ycg(R'-PQ) + mo\, zcg(Q'+ \\
&PR)] + \underline{jb}[mo(v'+RU-PW) + mo\, xcg(R'+PQ) - mo\, ycg(P^2+R^2) - mo \\
&zcg(P' - QR)] + \underline{kb}[mo(w'+PV-QU) - mo\, xcg(Q' - PR) + mo\, ycg(P'+QR) - \\
&mo\, zcg(P^2 + Q^2)]
\end{aligned}
$$

$$
= Fx\, \underline{ib} + Fy\, \underline{jb} + Fz\, \underline{kb}
$$

(3.1-9)

where

Fx = [mo(u'+QW–RV)-mo xcg(Q^2 + R^2)–mo ycg(R'–PQ) + mo zcg(Q'+ PR)]
Fy = [mo(v'+RU-PW)+mo xcg(R'+PQ)–mo ycg(P^2+R^2) - mo zcg(P'- QR)]
Fz = [mo(w'+PV–QU)-mo xcg(Q'- PR) + mo ycg(P'+QR) - mo zcg(P^2+ Q^2)]

Angular Motion

a) Angular Momentum Equation

Angular momentum is a cross product of a momentum and its position in ECI. That is,

$$H = r \times mv \tag{3.1-10}$$

where

m : mass
v : velocity
r : position vector in ECI.

Thus, the angular momentum of an element, E, in ECI can be written as,

$$\underline{HECI}\ \underline{i} = (\underline{\lambda} + \underline{\rho i}) \times mi\ (\underline{\lambda'} + \underline{\rho i'}) \tag{3.1-11}$$

where

$\underline{\lambda'} \equiv d\underline{\lambda}/dt$
$\underline{\rho i'} \equiv d\underline{\rho i}/dt$
mi ≡ i_{th} mass element (not modal mass here in this definition. This symbol is used as a modal mass later in this chapter)

To obtain the launch vehicle angular momentum, we integrate $\underline{HECI}\ \underline{i}$ throughout the total vehicle.

$$\begin{aligned} \underline{HECI} &= \Sigma\ \underline{HECI}\ \underline{i} \\ &= \Sigma\ (\underline{\lambda} \times \underline{\lambda'} + \underline{\lambda} \times \underline{\rho i'} + \underline{\rho i} \times \underline{\lambda'} + \underline{\rho i} \times \underline{\rho i}')\ mi \\ &= \Sigma\ (\underline{\lambda} \times \underline{\lambda'} + \underline{\lambda} \times \underline{\rho i'} + \underline{\rho i} \times \underline{\lambda'})mi + \Sigma\ (\underline{\rho i} \times \underline{\rho i}')\ mi \end{aligned} \tag{3.1-12 a}$$

$$= \Sigma\ (\underline{\lambda} \times \underline{\lambda'} + \underline{\lambda} \times \underline{\rho i'} + \underline{\rho i} \times \underline{\lambda'})mi + \underline{H}LVECI$$

(3.1-12 b)

The "$\underline{H}$LVECI (LVECI : LV body in ECI)" in Eqn. (3.1-12 b) is the angular momentum of the vehicle as a whole expressed in a frame that is inertial and all the axes are parallel to ECI, but the origin is at the geometric center of the vehicle's body. The rate of change of the angular momentum, d $\underline{HECI}$ /dt, in ECI is

$$d\,\underline{HECI}/dt = \Sigma\,(\underline{\lambda'} \times \underline{\lambda'} + \underline{\lambda} \times \underline{\lambda''} + \underline{\lambda'} \times \underline{\rho i'} + \underline{\lambda} \times \underline{\rho i''} + \underline{\rho} i' \times \underline{\lambda'} + \underline{\rho} i \times \underline{\lambda}'')mi + d\underline{H}LVECI/dt$$

$$= \Sigma\,(\cancel{\underline{\lambda'} \times \underline{\lambda'}} + \underline{\lambda} \times \underline{\lambda''} + \cancel{\underline{\lambda'} \times \underline{\rho i'}} + \underline{\lambda} \times \underline{\rho} i'' + \cancel{\underline{\rho I'} \times \underline{\lambda'}}\ \underline{\rho} i \times \underline{\lambda}'')mi + d\underline{H}LVECI/dt$$

where

$$\underline{\lambda'} \times \underline{\lambda}' = 0$$

$$\underline{\lambda'} \times \underline{\rho i'} + \underline{\rho} i' \times \underline{\lambda'} = \underline{\lambda'} \times \underline{\rho i'} - \underline{\lambda'} \times \underline{\rho i'} = 0$$

$$= \Sigma\,(\underline{\lambda} \times mi\,(\underline{\lambda}'' + \underline{\rho i}'') + \Sigma\,(mi\,\underline{\rho} i) \times \underline{\lambda}'' + d\underline{H}LVECI/dt$$
$$= \Sigma\,(\underline{\lambda} \times mi\,(\underline{\lambda}'' + \underline{\rho i}'') + (mo\,\underline{\rho} c) \times \underline{\lambda}'' + d\underline{H}LVECI/dt$$
$$= \underline{\lambda} \times \Sigma\, mi\,(\underline{\lambda}'' + \underline{\rho i}'') + (mo\,\underline{\rho} c) \times \underline{\lambda}'' + d\underline{H}LVECI/dt$$
$$= \underline{\lambda} \times \underline{F}o + (mo\,\underline{\rho} c) \times \underline{\lambda}'' + d\underline{H}LVECI/dt$$

(3.1-13)

b) Angular Momentum Applied

$$MECI = \Sigma\,(\underline{\lambda} + \underline{\rho} i) \times \underline{F}i$$
$$= \Sigma\,\underline{\lambda} \times \underline{F}i + \Sigma\,\underline{\rho} i \times \underline{F}i$$
$$= \underline{\lambda} \times (\Sigma\,\underline{F}i) + \Sigma\,\underline{\rho} i \times \underline{F}i$$
$$= \underline{\lambda} \times \underline{F}o + \Sigma\,\underline{\rho} i \times \underline{F}i$$

where

$\underline{\lambda} \times \underline{F}o$: torque applied to LV at the origin of the body frame in inertial frame

$\Sigma\,\underline{\rho} i \times \underline{F}i$: sum of torque at each element due to external force in body frame

$$= \underline{\lambda} \times \underline{F}o + MLVECI \qquad (3.1\text{-}14)$$

Since

$$MECI = d\,\underline{HECI}\,/dt,$$

and substituting Eqn. (3.1-13) and Eqn. (3.1-14) into the equation above,

$$\underline{\lambda} \times \underline{F}o + (mo\,\underline{\rho}c) \times \underline{\lambda}'' + d\underline{H}LVECI/dt = \underline{\lambda} \times \underline{F}o + MLVECI$$

Thus, by canceling "$\underline{\lambda} \times \underline{F}o$ " from both sides of the equation above, we have

$$MLVECI = d\underline{H}LVECI/dt + (mo\,\underline{\rho}c) \times \underline{\lambda}'' \qquad (3.1\text{-}15)$$

The moment of the external forces (MLVECI) is equal to the sum of the change of angular momentum in LV body in ECI and the product of the LV mass acceleration and the moment arm, i.e., the distance from the geometric body frame origin to the center of gravity of the inertial frame. If the center of gravity is chosen as the origin of the geometric body frame, then the second term will be deleted.

And,

$$d\underline{H}LVECI/dt = d\underline{H}LV/dt + \underline{\omega} \text{ cross } \underline{H}LV$$
(Derivation is shown below.)

(3.1- 16)

where

$$\underline{H}LV = (HLVx\,\underline{i}LV + HLVy\underline{j}LV + HLVz\underline{k}LV)$$

$$\begin{vmatrix} HLVx \\ HLVy \\ HLVz \end{vmatrix} = \begin{vmatrix} Ixx & -Ixy & -Ixz \\ -Iyx & Iyy & -Iyz \\ Ixx & -Ixy & -Ixz \end{vmatrix} \begin{vmatrix} P \\ Q \\ R \end{vmatrix}$$

$\underline{\omega} = [\,P \;\; Q \;\; R\,]$
= Angular rate of LV with respect to ECI

****

Derivation of $d\underline{H}A/dt = d\underline{H}B/dt + \underline{\omega}B/A$ cross $\underline{H}B$

We assume that we have two coordinate frames, A and B. There will be a rotation of frame B with respect to frame A, and the rotation rate vector is $\underline{\omega}B/A$. Without losing generality, we can assume that both frames share the same origin. Then, we can write that for a vector, $\underline{H}$,

$$\underline{H} = HAx\ iA + HAy\ jA + HAz\ kA \quad \text{in Frame A}$$
$$= HBx\ iB + HBy\ jB + HBz\ kB \quad \text{in Frame B}$$

The same vector is expressed in frame A and frame B. (See Fig. 3.1-4 for the two-dimensional case.)

Now, computing a time derivative of $\underline{H}$,

$$d\,\underline{H}/dt = \lim\,[\underline{H}(t + \Delta t) - \underline{H}(t)] / \Delta t, \quad \Delta t \to 0$$
$$= \lim\,[(HAx(t + \Delta t)\ \underline{i}A\ (t + \Delta t) + HAy(t + \Delta t)\ \underline{j}A(t + \Delta t) + HAz\,(t + \Delta t)\ \underline{k}A\ (t + \Delta t)) - (HAx\,(t)\underline{i}A\,(t) + HAy(t)\ \underline{j}A\,(t) + HAz(t)\ \underline{k}A\,(t))] / \Delta t \quad \underline{\text{in Frame A}}$$

(3.1-16)

$$= \lim\,[(HBx(t + \Delta t)\ \underline{i}B\ (t + \Delta t) + HBy(t + \Delta t)\ \underline{j}B(t + \Delta t) + HBz\,(t + \Delta t)\ \underline{k}B(t + \Delta t)) - (HBx\,(t)\underline{i}B(t) + HBy(t)\ \underline{j}B\,(t) + HBz(t)\ \underline{k}B(t))]/ \Delta t \quad \underline{\text{in Frame B}}$$

(3.1-17)

The length of the unit vector, $\underline{i}A$, remains a unit, and the angle is the only parameter that changes as time increases. Since the small angle can be treated as a vector, the cross product of a vector, $\underline{\Delta\theta}$, and the unit vector, $\underline{i}A$ (t), is in the plain of $\underline{i}A$ (t) and $\underline{i}A$ (t +Δt). Now, we can write,

$$\underline{i}A\,(t + \Delta t) = \underline{i}A\,(t) + \underline{\Delta\theta}jk \text{ cross } \underline{i}A\,(t) \qquad (3.1\text{-}18)$$

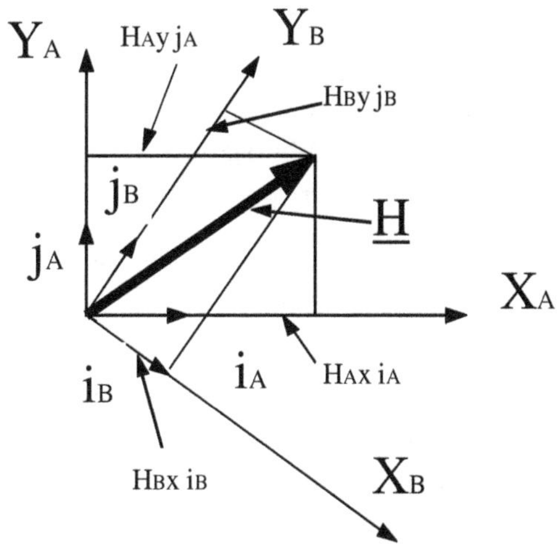

Fig. 3.1-4. Coordinate of Angular Momentum, H

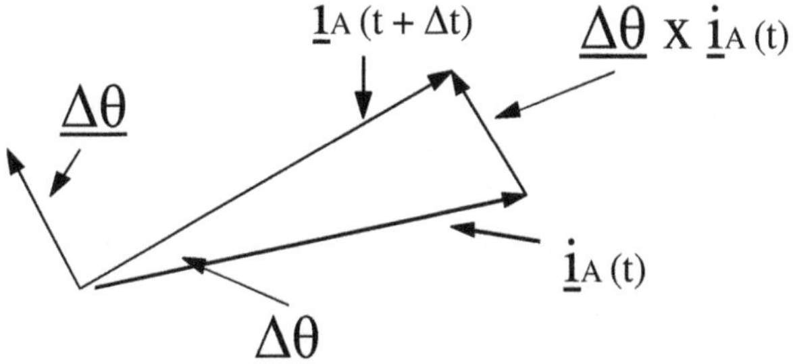

Fig. 3.1-5. $\underline{i}A$ $(t + \Delta t)$ = $\underline{i}A$ $(t + \Delta t) = \underline{i}A$ $(t) + \Delta\theta \times \underline{i}A$ (t)

If Δt is equal to zero, then $\underline{\Delta\theta}$ is also equal to zero, and the equation above verifies its validity. Applying Eqn. (3.1-18) to Eqn. (3.1 – 16) and Eqn. (3.1 – 17),

lim {(HAx(t + Δt) [$\underline{i}$A (t) + $\underline{\Delta\theta}$ Ajk cross $\underline{i}$A (t)] + HAy(t + Δt) [$\underline{j}$A (t) + $\underline{\Delta\theta}$ Aki cross $\underline{j}$A (t)] + HAz (t +

Δt)[kA (t) + Δθ Aij cross kA (t)] – [HAx (t)iA (t) + HAy(t) jA (t) + HAz(t) kA (t)]} / Δt
= lim {(HBx(t + Δt) [iB (t) + Δθ Bjk cross iB (t)]+ HBy(t + Δt) [jB (t) + Δθ Bki cross jB (t)] + HBz (t + Δt)[kB (t) + Δθ Bij cross kB (t)] – [HBx (t)iB(t) + HBy(t) jB (t) + HBz(t) kB(t)]}/ Δt

Rearranging and collecting the terms that have the angular rotation angle,

lim {HAx(t + Δt) iA (t) + HAy(t + Δt)jA (t) + HAz (t + Δt)kA (t)] + HAx(t + Δt) Δθ Ajk cross iA (t) + HAy(t + Δt) Δθ Aki cross jA (t) + HAz (t + Δt)Δθ Aij cross kA (t)] - [HAx (t)iA (t) + HAy(t) jA (t) + HAz(t) kA (t)]} / Δt
= lim{ HB[x(t + Δt) iB (t) + HBy(t + Δt)jB (t) + HBz (t + Δt)kB (t)] + HBx(t + Δt) Δθ Bjk cross iB (t) + HBy(t + Δt) Δθ Bki cross jB(t) + HBz (t + Δt)Δθ Bij cross kB(t)] - [Hax (t)iB (t) + HBy(t) jB(t) + HBz(t) kB (t)]} / Δt

Using the dot and cross product format,

lim {[HA(t + Δt) dot UA] - [HA(t) dot UA]}/ Δt + lim {HAx(t + Δt) Δθ Ajk cross iA (t) + HAy(t + Δt) Δθ Aki cross jA (t) + HAz (t + Δt)Δθ Aij cross kA (t)] }/ Δt
= lim {[HB(t + Δt) dot UB] - [HB(t) dot UB]}/ Δt + lim {HBx(t + Δt) Δθ Bjk cross iB (t) + HBy(t + Δt) Δθ Bki cross jB (t) + HBz (t + Δt)Δθ Bij cross kB (t)] }/ Δt

(3.1-19)

Before proceeding further, we need to show that,

[HAx(t + Δt) Δθ Ajk cross iA (t) + HAy(t + Δt) Δθ Aki cross jA (t) + HAz (t + Δt)Δθ Aij cross kA (t)] = (ΔθA) cross [HA(t + Δt)]

(3.1-20)

Interpretation of Δθ Ajkx, Δθ Aki y, and Δθ Aijz,

Δθ Ajk = ΔθAy j + ΔθAz k

$\underline{\Delta\theta}$ Aki = ΔθAz $\underline{k}$ + ΔθAx $\underline{i}$
$\underline{\Delta\theta}$ Aij = ΔθAx $\underline{i}$ + ΔθAy $\underline{j}$

Substituting the equations above into Eqn. (3.1-20),

LHS = { [HAx(t + Δt)] [(ΔθAy $\underline{j}$ + ΔθAz $\underline{k}$) cross $\underline{i}$A (t)] + [HAy(t + Δt)][(ΔθAz $\underline{k}$ + ΔθAx $\underline{i}$) cross $\underline{j}$A (t)] + [HAz (t + Δt)] [(ΔθAx $\underline{i}$ + ΔθAy $\underline{j}$) cross $\underline{k}$A (t)]}/ Δt

Carrying out the cross products,

= {[HAx(t + Δt)] (ΔθAz $\underline{j}$ - ΔθAy $\underline{k}$)+ [HAy(t + Δt)] (ΔθAx $\underline{k}$ -ΔθAz $\underline{i}$) + [HAz (t + Δt)] [(ΔθAy $\underline{i}$ - ΔθAx $\underline{j}$) cross $\underline{k}$A(t)]}/ Δt

Rearranging the equation above,

= i {[HAz (t + Δt)] ΔθAy - [HAy(t + Δt)] ΔθAz } +
j {[HAx(t + Δt)] ΔθAz - [HAz (t + Δt)] ΔθAx } +
k {HAy(t + Δt)] ΔθAx - [HAx(t + Δt)] ΔθAy

= $\underline{\Delta\theta}$A cross $\underline{H}$A(t + Δt)

where

$\underline{H}$A(t + Δt) = [HAx(t + Δt)] $\underline{i}$ + HAy(t + Δt)] j + [HAz (t + Δt)]$\underline{k}$

$\underline{\Delta\theta}$A = ΔθAx i + ΔθAy j + ΔθAz k

Then, Eqn. (3.1 – 19) can be written as,

lim {[HA(t + Δt) dot $\underline{U}$A] - [HA(t) dot $\underline{U}$A]}/ Δt + lim {$\underline{\Delta\theta}$A cross $\underline{H}$A(t + Δt) }/ Δt = lim {[HB(t + Δt) dot $\underline{U}$B] - [HB(t) dot $\underline{U}$B]}/ Δt + lim {$\underline{\Delta\theta}$B cross $\underline{H}$B(t + Δt) }/ Δt

Rearranging the equation above,

lim {[HA(t + Δt) dot $\underline{U}$A] - [HA(t) dot $\underline{U}$A]}/ Δt = lim {[HB(t + Δt) dot $\underline{U}$B] - [HB(t) dot $\underline{U}$B]}/ Δt + lim

{(ΔθB/Δt) cross HB(t + Δt) } - lim {(ΔθA/Δt) cross HA(t + Δt) }

Now, taking the limit,

d [HA(t)]/dt = d [HB(t)]/dt + ωB cross HB(t) - ωA cross HA(t)
= d [HB(t)]/dt + ωB/A cross HB(t)

****

The CG location in LV frame is,

ρc = xcg iLV + ycg jLV + zcg kLV (3.1- 21)

Acceleration of the origin of the LV is,

λ'' = d μ /dt (ECI frame)
= d μ /dt (LV frame) + ω cross μ
= (u' i LV + v' j LV + w' k LV) + (P i LV + Q j LV + R k LV) cross (u i LV + v j LV + w k LV)

(3.1-22)

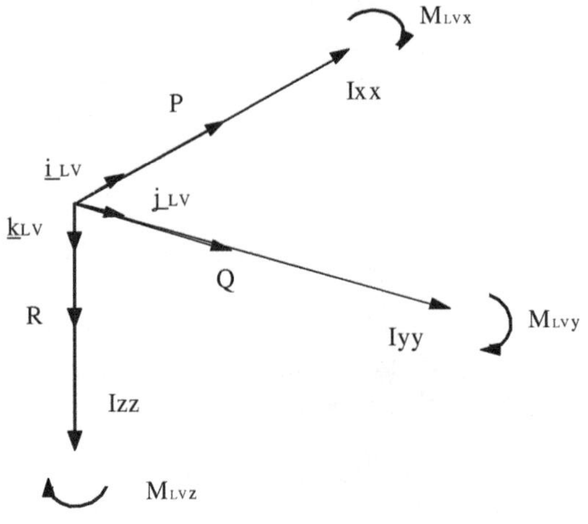

Fig. 3.1-6. MOI and Moment

Substituting Eqn. (3.1-16), Eqn. (3.1-21), and Eqn. (3.1-22) into Eqn. (3.1-15) and expanding terms and rearranging the equation, we have,

$$\begin{aligned} MLVECI = {} & \underline{i}LV\ [Ixx\,P' - Ixy(Q' - PR) - Ixz(R' + PQ) + Iyz(R^2 - Q^2) + \\ & (Izz - Iyy)QR + mo\ ycg\,(w' + Pv - Qu) - mo\ zcg\,(v' + Ru - \\ & - R^2) + (IxxIzz)PR - mo\ xcg\,(w' + Pv - Qu) + mo\ zcg\,(u' + \\ & QW - RV)] + \underline{k}LV\ [-Ixz\,(P' - QR) - Iyz(Q' + PR) + Izz\,R' + \\ & Ixy(Q^2 - P^2) + (Iyy - Ixx)PQ + mo\ xcg(v' + Ru - Pw) - mo \\ & ycg\,(u' + QW - RV)] \end{aligned} \quad (3.1-23)$$

To study the stability, we disturb the system from stability and find out whether we can bring it back to stability by applying some control rule. In the analysis, we disturb both linear velocity (U, V, W) and angular velocity (P, Q, R) as shown below:

$$\begin{aligned} u &= uo + u & \qquad P &= Po + p \\ v &= vo + v & \qquad Q &= Qo + q \\ w &= wo + w & \qquad R &= Ro + r \end{aligned}$$

First, we study the case of perturbation in linear motion and then the case of perturbation in angular motion.

Perturbation in Linear Motion

Here, we show the derivation of the perturbation equation in the x-axis alone. Equations for the other axes can be obtained by following the same procedure shown for x-axis. From Eqn. (3.1-9),

$$\begin{aligned} Fx = {} & [mo(u'+Qw-Rv)-mo\ xcg(Q^2 + R^2)-mo\ ycg(R'-PQ) + \\ & mo\ zcg(Q'+ PR)] \end{aligned} \quad (3.1-24)$$

Adding disturbances,

$$\begin{aligned} Fx + \Delta Fx = {} & \{mo[(uo'+u')+(Qo+q)(wo+w)-(Ro+r)(Vo + v)]\ -mo \\ & xcg[(Qo+q)^2 + (Ro+r)^2] - mo\ ycg[(Ro'+r') - (Po + \\ & p)(Qo + q)] + mo\ zcg[(Qo' + q') + (Po + p)(Ro + r)] \end{aligned}$$

$$= mo\,(uo' + u' + Qo\,wo + q\,wo + Qo\,w + q\,w - Ro\,Vo - r\,Vo - Ro\,v - rv) - mo\,xcg\,(Qo^2 + q^2 + 2\,Qo\,q + Ro^2 + r^2 + 2\,Ro\,r) - mo\,ycg\,(Ro' + r' - Po\,Qo + pQo + Poq + pq) + mo\,zcg(Qo' + q' + Po\,Ro + p\,Ro + Por + pr)$$

(3.1-25)

Note that the Eqn. (3.1-25) without disturbances is the same as Eqn. (3.1-24) except that t u, v, w, P, Q, and R are replaced by uo, vo, wo, Po, Qo, and Ro, respectively. Thus, the equation before applying the disturbances will be,

$$Fx = [mo(uo' + Qowo - Rovo) - mo\,xcg(Qo^2 + Ro^2) - mo\,ycg(Ro' - PoQo) + mo\,zcg(Qo' + PoRo)]$$

(3.1-26)

Subtracting Eqn. (3.1-26) from Eqn. (3.1 -25),

$$\begin{aligned}\Delta Fx = {} & mo\,(u' + qwo + q\,w - r\,Vo - Ro\,v - rv) \\ & - mo\,xcg\,(q^2 + 2\,Qo\,q + r^2 + 2\,Ro\,r) \\ & - mo\,ycg\,(r' + pQo + Poq + pq) \\ & + mo\,zcg(q' + p\,Ro + Por + pr)\end{aligned}$$

Assuming the disturbances are so small in magnitude that the higher order terms can be ignored, then

$$\begin{aligned}\Delta Fx = {} & mo\,(u' - r\,vo - Ro\,v) \\ & - mo\,xcg\,(2\,Qo\,q + 2\,Ro\,r) \\ & - mo\,ycg\,(r' + pQo + Poq) \\ & + mo\,zcg(q' + p\,Ro + Por)\end{aligned} \qquad (3.1\text{-}27)$$

And since the order of magnitudes of these terms (vo, Po, Qo, and Ro) is the same as the disturbance, we can ignore the terms that are products of these terms as well. Now, we have,

$$\Delta Fx = mo\,(u' - ycg\;r' + zcg\,q') \qquad (3.1\text{-}28)$$

Following the derivation just shown, we can obtain the disturbance equations for the Y and Z axes.

$$\Delta Fy = mo\,(v' + uo\,r + xcg\;r' - zcg\,p') \qquad (3.1\text{-}29)$$
$$\Delta Fz = mo\,(w' - uo\,q - xcg\;q' + ycg\,p') \qquad (3.1\text{-}30)$$

where

ΔFx	= external disturbance force in X axis
ΔFy	= external disturbance force in Y axis
ΔFz	= external disturbance force in Z axis
mo	= total mass
uo	= axial velocity
u'	= acceleration perturbation in X axis
v'	= acceleration perturbation in Y axis
w'	= acceleration perturbation in Z axis
xcg	= LV CG (Center of Gravity) location in X axis
ycg	= LV CG (Center of Gravity) location in Y axis
zcg	= LV CG (Center of Gravity) location in Z axis
p'	= angular acceleration in X axis
q'	= angular acceleration in Y axis
r'	= angular acceleration in Z axis

The external forces are:

Gravity
Thrust
Aerodynamic load
Propellant sloshing
Engine inertia

Perturbation in Angular Motion

Here again we show a derivation of angular perturbation equations in the x-axis alone. Equations for the other axes can be obtained following the procedure shown for the x-axis. From Eqn. (3.1- 23), we have for the x-axis,

$$MECIX = [Ixx\,P' - Ixy(Q' - PR) - Ixz(R' + PQ) + Iyz(R^2 - Q^2) + (Izz - Iyy)QR + mo\,ycg\,(w' + Pv - Qu) - mo\,zcg\,(v' + Ru - Pw)]$$

(3.1-31)

Adding perturbation,

$$MECIX + \Delta MECIX = [Ixx (Po'+p') - Ixy[(Qo'+q') - (Po + p) (Ro + r) - Ixz((Ro' + r') + (Po + p) (Qo + q)] + Iyz[(Ro + r)^2 - (Qo + q)^2] + (Izz - Iyy) (Qo + q)(Ro + r) + mo\ ycg [(wo' + w') + (Po + p)(vo + v) - (Qo + q)(uo + u)] - mo\ zcg [(vo' + v') + (Ro + r)(uo + u) - (Po + p)(wo + w)]$$

$$= Ixx (Po'+p') - Ixy(Qo'+q' - PoRo - pRo - Por - pr) - Ixz(Ro' + r' + PoQo + pQo + Poq + qq) + Iyz(Ro^2 +2Ro\ r + r^2 - Qo^2 - 2Qoq - q^2)+ (Izz - Iyy)(Qo\ Ro+ q\ Ro +Qor+ qr) + mo\ ycg [wo' + w' + Povo + pvo + Pov + pv - Qouo - quo - Qou - qu) - mo\ zcg (vo' + v' + Rouo + ruo+Rou+ ru - Powo - pwo - Powo - pw)$$

(3.1-32)

Subtracting Eqn. (3.1-31) from Eqn. (3.1-32),

$$\Delta MECIX = Ixx (p') - Ixy(q' - pRo - Por - pr) - Ixz (r'+ pQo + Poq + qq) + Iyz(+2Ro\ r + r^2 - 2Qoq - q^2)+ (Izz - Iyy)(q\ Ro +Qor+ qr) + mo\ ycg [w' + pvo + Pov + pv - quo - Qou - qu) - mo\ zcg (v' + ruo+Rou+ ru - pwo - pw)$$

Since in LV, the axial velocity (u) is very large compared to other terms, i.e.,

$$uo >> vo,\ Po,\ Qo, \text{ and } Ro$$

we can reduce the equation above further.

$$\Delta MECIX = Ixx (p') - Ixy(q') - Ixz (r') + mo\ ycg (w'- q\ uo) - mo\ zcg (v' +r\ uo)$$

(3.1 – 33)

Following the same procedure, we derive angular momentum perturbation equations for Y and Z axes as shown below:

$$\Delta MECIY = Iyy\,(q') - Ixy(p') - Iyz\,(r') - mo\,xcg\,(w' - q\,uo) + mo\,zcg\,(u') \qquad (3.1-34)$$

$$\Delta MECIXZ = Izz\,(r') - Ixz(p') - Iyz\,(q') + mo\,xcg\,(v' + r\,uo) - mo\,ycg\,(u') \qquad (3.1-35)$$

where

ΔMECIX	= External moment in X axis
ΔMECIY	= External moment in Y axis
ΔMECIZ	= External moment in Z axis
Ixx	= Moment of inertia in X axis
Iyy	= Moment of inertia in Y axis
Izz	= Moment of inertia in Z axis
Ixy	= Product of inertia in XY
Ixz	= Product of inertia in XZ
Iyz	= Product of inertia in YZ
Mo	= Total LV mass
Uo	= LV axial velocity(X axis)
xcg	= CG location in X axis
ycg	= CG location in Y axis
zcg	= CG location in Z axis
p'	= Angular acceleration in X axis
q'	= Angular acceleration in Y axis
r'	= Angular acceleration in Z axis

We have derived a set of simplified perturbation equations in all axes for linear and angular motions. Now, we consider external forces that cause the perturbations. The external forces are:

Gravity
Thrust
Aerodynamic load
Propellant sloshing
Engine inertia

3.2 **Computation of Forces and Moments**

In this section we derive equations to compute the forces and moments, which we need to solve the linear motion perturbation equations, Eqn. (3.1-28), Eqn. (3.1-29), and Eqn. (3.1-30), and the angular motion perturbation equations, Eqn. (3.1-33), Eqn. (3.1-34), and Eqn. (3.1-35). Specifically,

$$\Delta FX = Fxg + FxT + FxA + FxS + FxE \quad (3.2\text{-}1a)$$
$$\Delta FY = Fyg + FyT + FyA + FyS + FyE \quad (3.2\text{-}1b)$$
$$\Delta FZ = Fzg + FzT + FzA + FzS + FzE \quad (3.2\text{-}1c)$$

And

$$\Delta MECIX = Mxg + MxT + MxA + MxS + MxE \quad (3.2\text{-}2a)$$
$$\Delta MECIY = Myg + MyT + MyA + MyS + MyE \quad (3.2\text{-}2b)$$
$$\Delta MECIZ = Mzg + MzT + MzA + MzS + MzE \quad (3.2\text{-}2c)$$

where the suffixes, g , T, A, S, and E stand for gravity, thrust, aerodynamics, sloshing, and engine inertia, respectively.

Gravity

For simplicity in equation derivation and without losing generality, we set up the coordinate systems, ECI and BF frames, such that the plane of XECI and ZECI and the plane of Xbody and Zbody are on the same plane. (See Fig. 3.2-1.) Then, from the figure, we write

$$Fxg = -\,mg\cos\theta \quad (3.2\text{-}3a)$$
$$Fyg = 0.0 \quad (3.2\text{-}3b)$$
$$Fzg = -\,mg\sin\theta \quad (3.2\text{-}3c)$$

The Ybody component of gravity (Fyg) is zero, because that was the way we set it up. However, when a disturbance occurs, this component will no longer be zero. Any disturbance will cause the rotation of the LV body axis with respect to the original LV body orientation, and, assuming the disturbance is small, we can write a transformation matrix as follows:

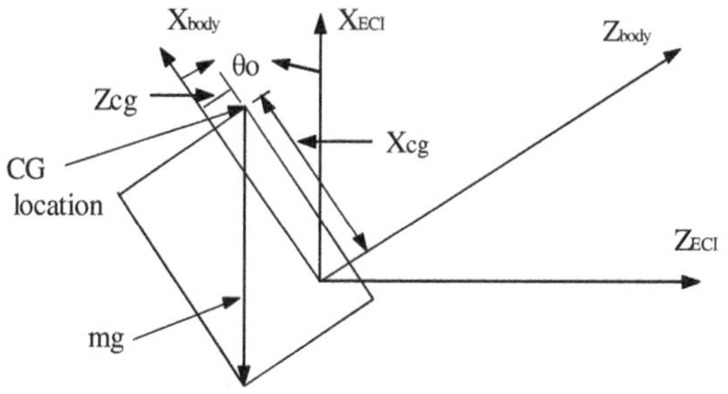

Fig. 3.2-1. ECI and Body Frame

$$\begin{vmatrix} \Delta Fxg \\ \Delta Fyg \\ \Delta Fzg \end{vmatrix} = (A - I) \begin{vmatrix} Fxg(0) \\ Fyg(0) \\ Fzg(0) \end{vmatrix} \qquad (3.2\text{-}4)$$

where

$$A = \begin{vmatrix} 1 & \psi & -\theta \\ -\psi & 1 & \varphi \\ \theta & -\varphi & 1 \end{vmatrix}$$

φ, θ, and ψ are small angles of rotation around the X, Y, and Z axes, respectively.

From Eqn. (3.2-3a,b,c) and Eqn.(3.2-4),

$$\Delta Fxg = mg\ \theta\ \sin\theta o \qquad (3.2\text{-}5a)$$
$$\Delta Fyg = mg\ (\psi \cos\theta o\ - \varphi \sin\theta o\ ) \qquad (3.2\text{-}5b)$$
$$\Delta Fzg = -mg\ \theta \cos\theta o \qquad (3.2\text{-}5c)$$

Fig. 3.2-2 is presented here to understand how the moment disturbance equations can be written as;

$$\Delta Mxg = ycg\ \Delta Fzg\ -\ zcg\ \Delta Fyg \qquad (3.2\text{-}6a)$$
$$\Delta Myg = -xcg\ \Delta Fzg\ +\ zcg\ \Delta Fxg \qquad (3.2\text{-}6b)$$
$$\Delta Mzg = xcg\ \Delta Fyg\ -\ ycg\ \Delta Fxg \qquad (3.2\text{-}6c)$$

From Eqn. (3.2 –5a,b,c) and Eqn. (3.2-6a,b,c), we obtain the moment disturbance equations.

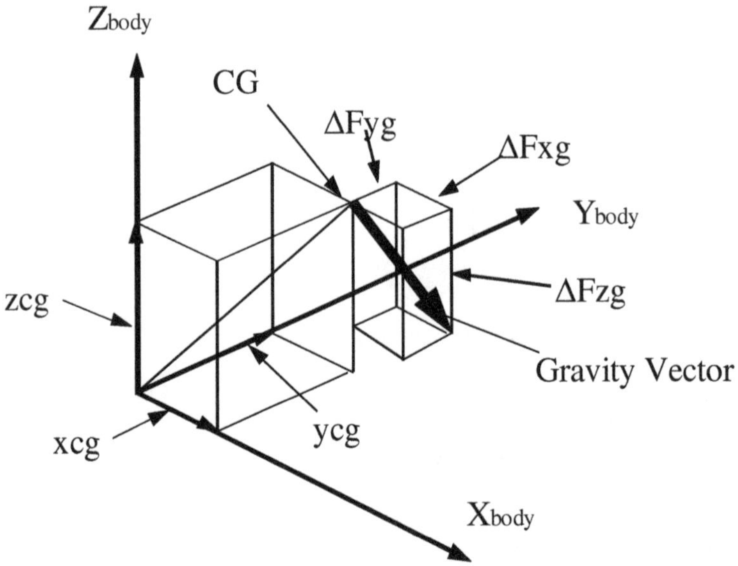

Fig. 3.2-2. Moment Disturbance

$$\Delta Mxg = mg\,(zcg\,\varphi \sin\theta o - ycg\;\theta\cos\theta o - zcg\,\psi\cos\theta o) \tag{3.2-7a}$$

$$\Delta Myg = mg\,(xcg\,\theta\cos\theta o + zcg\,\theta\sin\theta o) \tag{3.2-7b}$$

$$\Delta Mzg = mg\,(xcg\,\psi\cos\theta o - xcg\,\varphi\sin\theta o - ycg\;\theta\sin\theta o) \tag{3.2-7c}$$

Thrust Disturbance

Derivation of force and moment disturbance equations due to the thrust involves consideration of TVC activation and LV structural bending. Before the derivation, we show that the axial (Xbody axis) thrust variation can be ignored. In Fig. 3.2-3, Tu stands for ungimballed thrust, and Td stands for disturbance thrust including control thrust by TVC [Thrust Vector Control (gimbaled engine)].

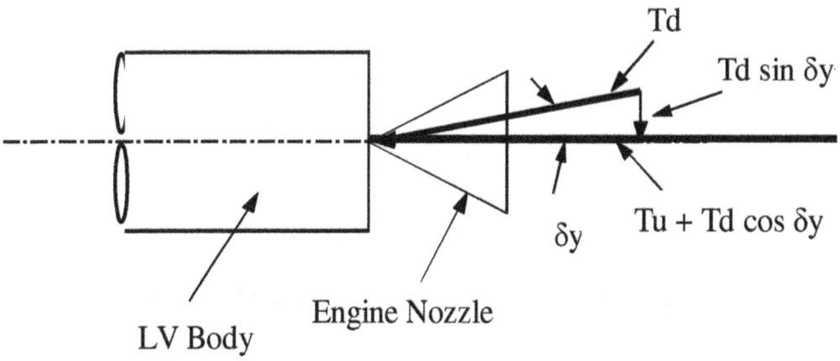

Fig. 3.2-3 Axial Thrust Variation

The axial direction thrust disturbance can be computed by subtracting the thrust without the disturbance from the thrust with disturbance. From Fig. 3.2-3,

$$\begin{aligned} \Delta FxT &= (Tu + Td \cos \delta y) - (Tu + Td) \\ &= Td (\cos \delta y - 1) \\ &\approx 0 \text{ for small } \delta y \end{aligned}$$

Thus, we showed that the axial thrust variation could be ignored.

a. Linear Disturbance

i. Disturbance in Y Axis

a) Due to TVC

From Fig. 3.2-3,

$$\Delta FYTVC = Td \sin \delta y \approx Td\ \delta y \qquad (3.2\text{-}8)$$

b) Due to Bending

1) <u>Bending Mode 1</u>

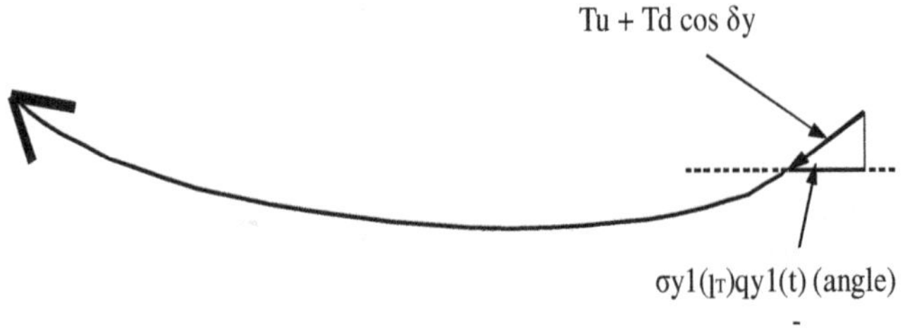

Fig. 3.2-4. LV First Mode Bending

From Fig. 3.2-4,

$$\begin{aligned} \Delta FYB1 &= (Tu + Td \cos \delta) \sin (\sigma y1\,(t) qy1(t)) \\ &\approx (Tu + Td)(\sigma y1(t) qy1(t)) \end{aligned} \qquad (3.2\text{-}9)$$

2) <u>Bending Mode 2</u>

From Fig. 3.2-5,

$$\begin{aligned} \Delta FYB2 &= (Tu + Td \cos \delta) \sin (\sigma y2\,(t) qy2(t)) \\ &\approx (Tu + Td)(\sigma y2\,(t) qy2(t)) \end{aligned} \qquad (3.2\text{-}10)$$

.

.

.

n) <u>Bending Mode n</u>

$$\begin{aligned} \Delta FYBn &= (Tu + Td \cos \delta) \sin (\sigma yn\,(t) qyn(t)) \\ &\approx (Tu + Td)(\sigma yn\,(t) qyn(t)) \end{aligned} \qquad (3.2\text{-}11)$$

Now, adding all the thrust disturbances from Eqn. (3.2-8), Eqn. (3.2-9), Eqn. (3.2-10), and Eqn. (3.2-11),

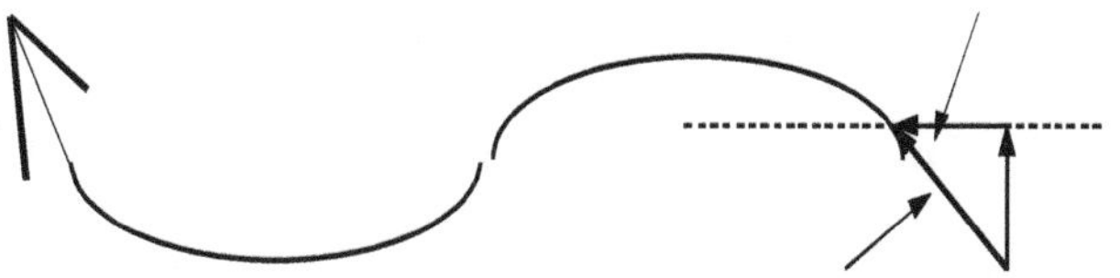

Fig. 3.2-5. LV Second Mode Bending

$$\Delta FYT = \Delta FYTVC + \Delta FYB1 + \Delta FYB2 + \ldots + \Delta FYBn$$
$$= \Delta FYTVC + \Sigma\, \Delta FYBi$$
$$= Td\ \delta y - (Tu + Td)(i=1,n)\ \Sigma\, (\sigma yi\ (\ell T) qyi(t))$$

(3.2- 12)

ii. Disturbance in Z Axis

In a similar manner, we obtain,

$$\Delta FZT = Td\ \delta z - (Tu + Td)(i=1,n)\ \Sigma\, [\sigma zi\ (t) qzi(t)]$$

(3.2-13)

b. Moments

Fig. 3.2-6 shows the case of bending mode 1 in Xbody-Zbody plane. From the Figure, we can write the moment disturbance equation for mode 1.

$$MYT = \ell c\ FzT - \Phi y1(lT)qy1(t)\ \{(Td \cos \delta + Tu) \cos [\sigma y1(lT)qy1(t)]\}$$

Now, adding higher mode of bending,

$$MYT \;=\; \ell c\, FzT \;-\; (i=1,n)\Sigma\, \Phi zi(lT)qzi(t)\, \{(Td \cos \delta z \;+\; Tu) \cos[\sigma zi(lT)qzi(t)]\}$$

where

$$FzT = Td\;\; \delta z \;\;-\;\; (Tu + Td)\,(i =:1,\, i =:n)\; \Sigma\; [\sigma zi\; (lT)qzi(t)]$$

Note that Fig. 3.2-6 is exaggerated in angle and the angles δz and ($\sigma zi\ (lT)qzi(t)$) are very small.

Substituting FzT,

$$= \ell c\, [\, Td\;\; \delta z \;\;-\;\; (Tu + Td)\,(i =:1,\, i =:n)\; \Sigma\; (\sigma zi\; (lT)qzi(t)\,)] \;-\; (i =1,\, n)\Sigma\Phi zi(lT)qzi(t)\; \{(Td \cos \delta z \;+ Tu) \cos [\sigma zi(lT)qzi(t)]\}$$

$$= \ell c\, [\, Td\;\; \delta z \;\;-\;\; (Tu + Td)\,(i =1,n)\; \Sigma\; (\sigma zi\; (lT)qzi(t)\,)\,] \;-\; (Td \;+ Tu)\,(i =1,\, n)\Sigma\; \Phi zi(lT)qzi(t),$$

$$\text{for } \cos\, [\sigma zi(t)qzi(t)] =1 \text{ and } \cos \delta z \;= 1$$

(3.2-14)

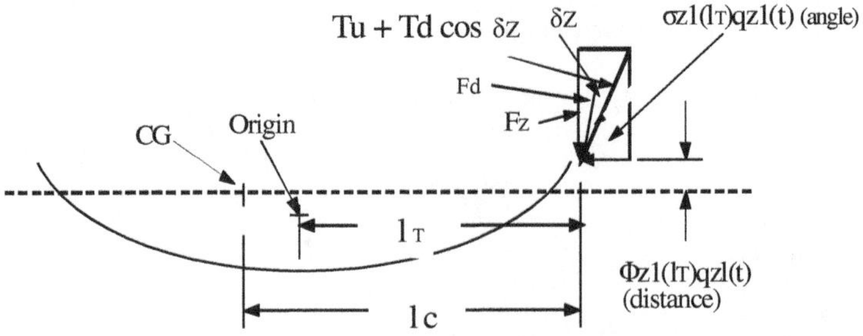

Fig. 3.2-6. Pitch (Y-axis) Moment

In a similar manner, we find that

MZT = - ℓc [Td δy - (Tu + Td) (i =:1, i =:n) Σ (σyi (lT)qyi(t))] + (Td + Tu) (i =:1, i=:n)Σ Φyi(lT)qyi(t)

(3.2- 15)

For roll moment disturbance, we have,

MXT = Tr δr

where

Tr = Thrust from the roll control engine
δr = Roll control signal

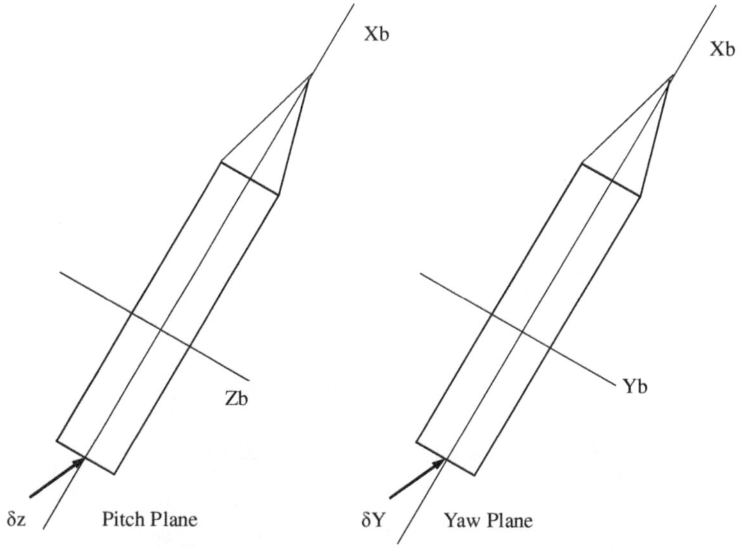

Fig. 3.2-7. Pitch Plane and Yaw Plane

3.3 **Aerodynamics**

The forces and moments exerted on LV by the aerodynamic loads are:

FxA (axial force) = (1/2) ρ U^(2) A1 ∫ CA dℓ (3.3-1)

FyA (side force) = $(1/2)\,\rho\,U^{2}\,A2 \int CS\, d\ell$ (3.3-2)
FzA (normal force) = $-\,(1/2)\,\rho\,U^{2}\,A3 \int CN\, d\ell$ (3.3-3)

MR (roll moment) = $(1/2)\,\rho\,U^{2}\,A4\,\ell 1 \int CR\, d\ell$ (3.3-4)
MP (pitch moment) = $(1/2)\,\rho\,U^{2}\,A5\,\ell 2 \int CM\, d\ell$ (3.3-5)
NY (yaw moment) = $(1/2)\,\rho\,U^{2}\,A6\,\ell 3 \int CY\, d\ell$ (3.3-6)

where

Ai = reference area
ℓi = reference length
ρ = air density
U = axial velocity
Ci = aero coefficient

The negative sign in FzA comes from the assumption that FzA acts in the negative Zb direction, which is in accordance with common practice (pp 29, Ref. 18).

The aerodynamic coefficients (Ci) are functions of many variables. These variables are:

α = pitch angle of attack
α' = pitch angle of attack rate
β = yaw angle of attack
β' = yaw angle of attack rate
p = angular rate perturbation in X axis
q = angular rate perturbation in Y axis
r = angular rate perturbation in Z axis

Thus, we can write,

$$Ci = Ci\,o + (\partial Ci/\partial\alpha)\alpha + (\partial Ci/\partial\alpha')\alpha' + (\partial Ci/\partial\beta)\beta + (\partial Ci/\partial\beta')\beta' + (\partial Ci/\partial p)p + (\partial Ci/\partial q)q + (\partial Ci/\partial r)r \qquad (3.3\text{-}7)$$

Since FyA (side force), FzA (normal force), MP (pitch moment), and NY (yaw moment) are the variables that we are interested in for attitude control, we will expand Eqn. (3.3-2), Eqn. (3.3-3), Eqn. (3.3-5), and Eqn. (3.3-6).

From Eqn. (3.3-3),

$$FzA \text{ (normal force)} = -(1/2) \, \rho \, U^{(2)} \, A3 \int CN \, d\ell \quad (3.3-8)$$

In this equation, we expand the aerodynamic coefficient, CN. Using Eqn. (3.3-7),

$$CN = CN_0 + (\partial CN / \partial\alpha)\alpha + (\partial CN / \partial\alpha')\alpha' + (\partial CN / \partial\beta)\beta + (\partial CN$$
$$/\partial\beta')\beta' + (\partial CN / \partial p)p + (\partial CN / \partial q)q + (\partial CN / \partial r)r$$
$$\approx (\partial CN / \partial\alpha)\alpha \quad \text{for all the other terms are relatively small.}$$

Then, Eqn. (3.3-8) can be written as,

$$FzA \text{ (normal force)} = -(1/2) \, \rho \, U^{(2)} \, A3 \int (\partial CN / \partial\alpha)\alpha \, d\ell$$
$$(3.3-9)$$

The angle of attack consists of α and, in addition, angles created by 1) pitch rate, 2) slope due to bending, 3) the ratio of bending speed and LV axial velocity, and 4) the ratio of wind velocity and LV axial velocity.

1) Pitch Rate

From Fig. 3.3-1,

$$\Delta\alpha\theta = \Delta\alpha 1 + \Delta\alpha 2 + \Delta\alpha 3$$
$$= (i = 1,3) \sum \Delta\alpha i$$

From the equation above, we can induce,

$$\Delta\alpha\theta = (i = 1, n) \sum \Delta\alpha i \quad , n = \infty$$

where

$$\Delta\alpha i = [(\ell\alpha - \ell i) \, (d\theta/dt)] / U$$

$$= (i = 1, n) \sum [(\ell\alpha - \ell i) \, (d\theta/dt)] / U, \quad n = \infty$$

If the pitch rate is zero, then there will be no variation in the angle of attack. In Fig. 3.3-1, $L\alpha = \ell\alpha$, $L = \ell$, $L1 = \ell1$, $L2 = \ell2$, and $L3 = \ell3$.

2) Slope Due to Bending

In a similar manner, we obtain,

$$\Delta\xi i = \partial\xi(\ell p,t)/\partial\ell$$

3) The Ratio of Bending Speed and LV Axial Velocity,

In a similar manner, we obtain,

$$\Delta\xi R = [\partial\xi(\ell p,t)/\partial t]/U$$

1) The Ratio of the Wind Velocity (Ww) and LV Axial Velocity.

$$\alpha w = (w - Ww) / U$$
$$= - Ww / U, \quad w \approx 0.0$$

We consider only the wind impact (Ww) on the angle of attack. In flight, $w \neq 0$ because of the bending and sloshing which cause the center of gravity shift, generating the angle of attack. The angle of attack variation due to the bending and sloshing can be differentiated from that due to the wind by analyzing gyro data from gyroscopes located at the front end and stern end. The case when $w \neq 0$ needs to be studied.

These terms must be taken into account to represent true angle of attack, α .

$$\alpha = \alpha o - [(\ell a - \ell i)(d\theta/dt)] / U - \partial\xi(\ell p,t)/\partial\ell - [\partial\xi(\ell p,t)/\partial t]/U + \alpha w \qquad (3.3\text{-}10)$$

Inserting Eqn. (3.3-10) into Eqn. (3.3-9),

$$FzA = - (1/2)\rho U^{(2)} A3 \int (\partial CN/\partial\alpha)\{\alpha o - [(\ell a - \ell i)(d\theta/dt)] / U - \partial\xi(\ell p,t)/\partial\ell - [\partial\xi(\ell p,t)/\partial t]/U + \alpha w\}\, d\ell$$

$$= - (1/2)\rho U^{(2)} A3 \{\alpha o \int (\partial CN/\partial\alpha)\, d\ell - \int (\partial CN/\partial\alpha)\, [[(\ell a - \ell i)\theta'] / U]\, d\ell - \int [(\partial CN/\partial\alpha)\, \partial\xi(\ell p,t)/\partial\ell]\, d\ell - \int (\partial CN/\partial\alpha)\, [[\partial\xi(\ell p,t)/\partial t]/U]\, d\ell + \int \alpha w\, (\partial CN/\partial\alpha)\, d\ell\}$$

$$= -(1/2)\,\rho\,U^{2}\,A3\,\{\alpha o \int (\partial CN/\partial\alpha)\,d\ell - (\theta'/U)\int(\ell a - \ell)(\partial CN/\partial\alpha)\,d\ell + \Sigma\, qp(t)\int \sigma p(\ell)(\partial CN/\partial\alpha)\,d\ell - \Sigma\,[qp'(t)/U]\int \varphi p(\ell)(\partial CN/\partial\alpha)\,d\ell - (Ww/U)\int(\partial CN/\partial\alpha)\,d\ell\} \quad (3.3\text{-}11)$$

where

Ww = Wind Velocity in Z direction

Following similar procedure taken for FzA, we can obtain the following equation for FYA:

$$FYA = (1/2)\,\rho\,U^{2}\,A2\,\{\beta o \int (\partial CN/\partial\beta)\,d\ell - (\varphi'/U)\int(\ell a - \ell)(\partial CN/\partial\beta)\,d\ell + \Sigma\, qy(t)\int \sigma y(\ell)(\partial CN/\partial\beta)\,d\ell - \Sigma\,[qy'(t)/U]\int \varphi y(\ell)(\partial CN/\partial\beta)\,d\ell - (Vw/U)\int(\partial CN/\partial\beta)\,d\ell\} \quad (3.3\text{-}12)$$

where

Vw = Wind Velocity in Y direction

Moment equations are:

$$MYA = (\ell a - \ell)(-FzA)$$
$$= (1/2)\,\rho\,U^{2}\,A3\,\{\alpha o \int (\ell a - \ell)(\partial CN/\partial\alpha)\,d\ell - (\theta'/U)\int(\ell a - \ell)^{2}(\partial CN/\partial\alpha)\,d\ell + \Sigma\, qp(t)\int(\ell a - \ell)\,\sigma p(\ell)(\partial CN/\partial\alpha)\,d\ell - \Sigma\,[qp'(t)/U]\int(\ell a - \ell)\varphi p(\ell)(\partial CN/\partial\alpha)\,d\ell - (Ww/U)\int(\ell a - \ell)(\partial CN/\partial\alpha)\,d\ell\} \quad (3.3\text{-}13)$$

$$MZA = (\ell a - \ell)\,FYA$$
$$= (1/2)\,\rho\,U^{2}\,A2\,\{\beta o \int(\ell a - \ell)(\partial CN/\partial\beta)\,d\ell - (\varphi'/U)\int(\ell a - \ell)^{2}(\partial CN/\partial\beta)\,d\ell + \Sigma\, qy(t)\int(\ell a - \ell)\,\sigma y(\ell)(\partial CN/\partial\beta)\,d\ell - \Sigma\,[qy'(t)/U]\int(\ell a - \ell)\,\varphi y(\ell)(\partial CN/\partial\beta)\,d\ell - (Vw/U)\int(\ell a - \ell)(\partial CN/\partial\beta)\,d\ell\} \quad (3.3\text{-}14)$$

Derivations of force and moment equations for propellant sloshing and engine inertia will be presented after the structural bending analysis.

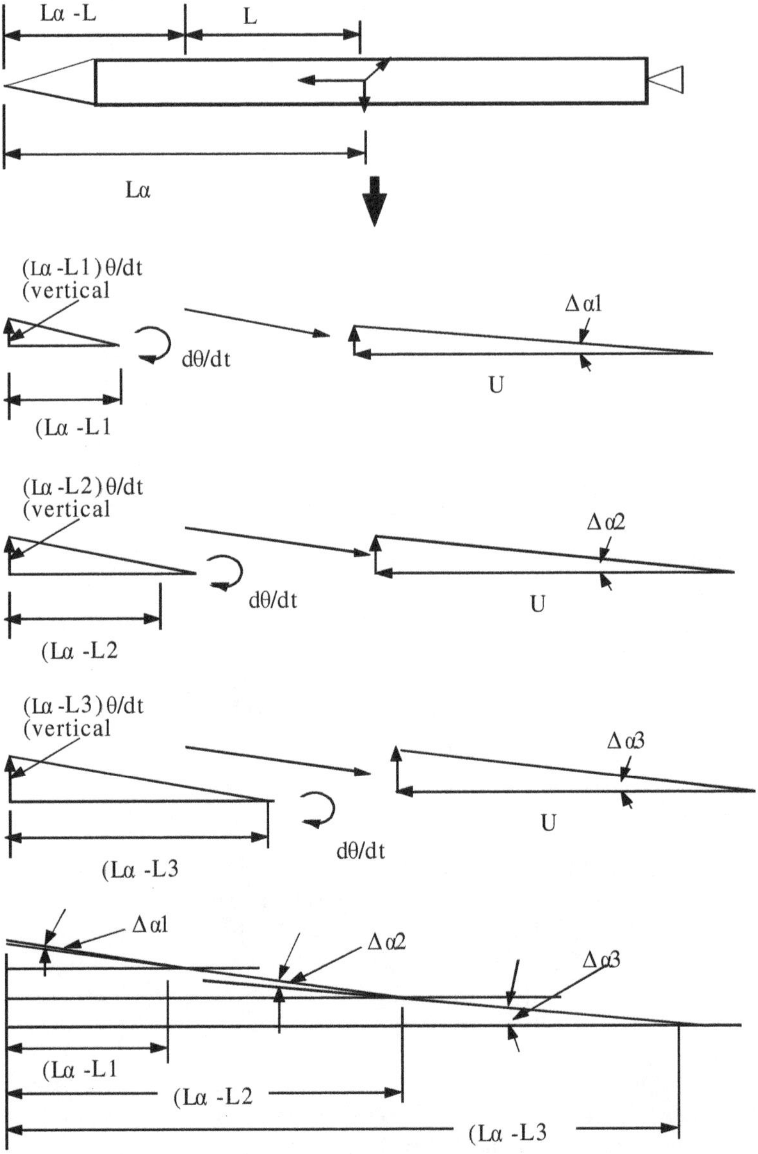

Fig. 3.3-1. Variation of Angle of Attack (L=Length)

3.4 Launch Vehicle Structural Bending

Launch vehicle bending is one of the major issues we have to deal with for successful flight. Usually, the structure that supports the weight of the vehicle is approximately 10% of the total weight; therefore, bending is inevitable. We would like to minimize the total structural weight, that minimizes fuel consumption. As launch vehicle weight increases, the amount of fuel required for a successful launch also increases. Since the impact of vehicle bending on the attitude control system design is so critical, we must allocate a large amount of our design effort for bending analysis. Fig. 3.4-1 shows the first bending mode of an LV.

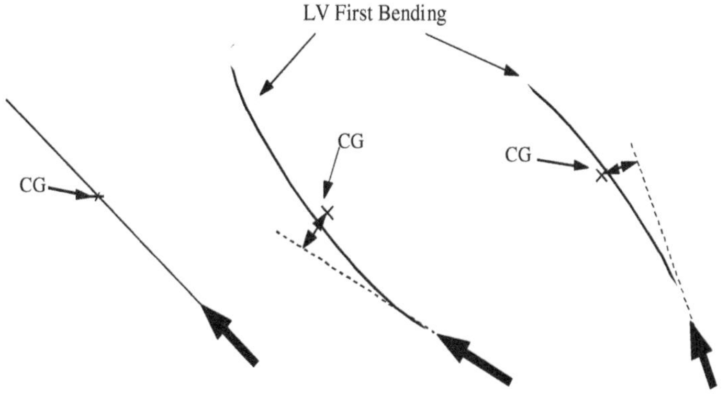

Fig. 3.4-1 First Bending Mode of an LV

Before bending, the vehicle does not experience angular rotation. However, as bending progresses, the CG shifts and the vehicle starts to rotate because the moment arm is no longer zero. A sensor detects the rotation and sends a message to the control system. The control system sends out a command to rotate the thrust vector to align with the CG, but the CG keeps changing because of the bending. In general, we need to track deflection of several important points in the LV, particularly for higher bending modes. It is seen here that bending is a phenomenon we cannot ignore if we expect to have a successful flight.

Structural vibration occurs in several directions: longitudinal, torsional, and lateral. The longitudinal vibration may not be significant in general, but it may be significantly important in fuel tank control systems.

Torsional vibration is not of a major concern in attitude control, since the mode of vibration is orthogonal to lateral vibration, which does not affect attitude control significantly. However, in practice, we should analyze the impact of the torsional vibration on the control system. The lateral vibration is a very important phenomenon that we must filter out from the sensor signals, because this mode of vibration can lead to control system instability or structural failure. Ideally, the output of sensor should contain the vehicle attitude/attitude rate information only, that is, filtering out any signal coming from the structural bending or sloshing.

Here we introduce the basic concept of bending analysis. An equation that describes a motion of a point, "A," as bending occurs in the LV, is,

$$\underline{\rho} = \underline{\rho}R + \underline{\xi} \qquad (3.4\text{-}1)$$

where

$\underline{\rho}$ = position of a point, A, in LV in LV frame after bending

$\underline{\rho}R$ = position of the point, A, in LV in LV frame before bending

$\underline{\xi}$ = displacement of the point, A, due to the bending.

We are interested in computing the displacement, $\underline{\xi}$, which can be computed using the equation shown below. (The derivation of this equation is not presented here.)

$$\underline{\xi}(\ell,t) = (i =:1, i = \infty)\, \Sigma\, \underline{\Phi}i\,(\ell)\, qi(t) \qquad (3.4\text{-}2)$$

$$d\,\underline{\xi}(\ell,t)/d\ell = (i =:1, i = \infty)\, \Sigma\, (\underline{\Phi}i\,(\ell)/d\ell)\, qi(t)$$
$$= -\,(i =:1, i = \infty)\, \Sigma\, \sigma i(\ell)\, qi(t) \qquad (3.4\text{-}3)$$

where

i = mode shape number

$\underline{\Phi}i(\ell)$ = i th normalized mode shape which is a function of spatial variable, ℓ

qi(t) = generalized coordinate of bending mode i [qi(t) computation is shown later]

$d\,\underline{\xi}(\ell,t)/d\ell$ = modal slope

$\sigma i(\ell)$ = normalized mode slop of the ith bending mode

****

Note:

- Mode shape number and normalized mode shape for uniform beam.

  LV bending consists of many bending modes. The most significant mode is the first bending mode (i = 1). Fig. 3.4-2 shows the shape of normalized bending modes 1, 2, and 3. Each mode has it own unique spatial frequency (note: not time frequency), and these can be computed using the following equations.

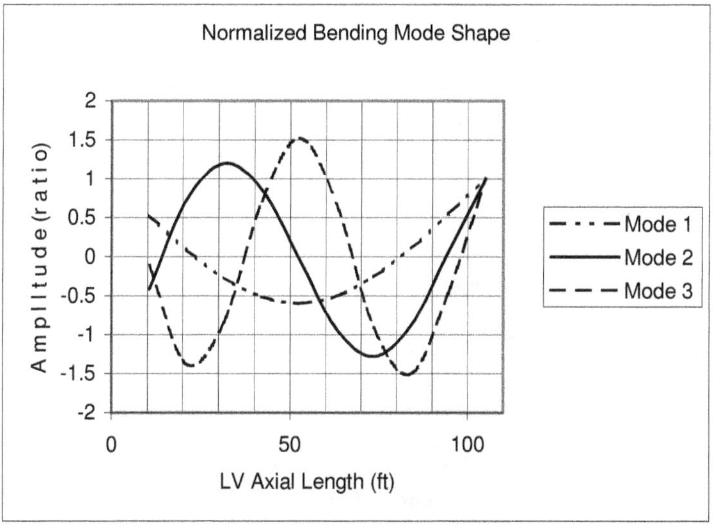

Fig. 3.4-2. Normalized Bending Mode Shape

Bending Mode Shape (Phi)			
LV Length (ft)	Mode 1	Mode 2	Mode 3
10.50	0.52	-0.41	-0.11
21.00	0.09	0.72	-1.38
31.50	-0.27	1.20	-0.84

42.00	-0.51	0.86	0.71
52.50	-0.60	-0.02	1.52
63.00	-0.50	-0.92	0.67
73.50	-0.26	-1.28	-0.91
84.00	0.11	-0.88	-1.50
94.50	0.54	0.05	-0.47
105.00	1.00	1.00	1.00

Table 3.4-1. Bending Mode Shape

Mode 1 : ω = 22.504 * sqrt(EI/m)/L^2
Mode 2 : ω = 61.685 * sqrt(EI/m)/L^2
Mode 3 : ω = 120.903 * sqrt(EI/m)/L^2

Where

E = Young's modulus
I = moment of inertia
m = mass per unit length
L = total length

$\underline{\Phi}i(\ell) = \phi i(\ell) / \phi i(L)$

$\phi i(\ell)$ = C1(sinh w) ℓ + C2(cosh w) ℓ + C3(sin w) ℓ + C4(cos w) ℓ

(3.4-4)

where

w = sqrt(ωi / a)

a = EI/m

C1, C2, C3, and C4 are obtained satisfying following boundary conditions.

$\phi i''(\ell) = 0$, $\phi i'''(\ell) = 0$, $\phi i''(L) = 0$, and $\phi i'''(L) = 0$

- Generalized coordinate of the bending mode [qi(t)]

  The generalized coordinate is a solution of the following dynamic equation.

$$qi(t)'' + 2\,\zeta i\,\omega i\,qi(t)' + \omega i\text{^}2\,qi(t) = Qi / Mi \quad (3.4\text{-}5)$$

where

ζi = damping ratio of mode i
ωi = frequency of mode i
Qi = generalized force

$$Qi = (\ell=0, L) \int f(\ell,t)\,\phi i\,(\ell)\,d\ell$$

$f(\ell,t)$ =force applied

Mi = generalized mass (modal mass)
$= (\ell=0, L) \int m(\ell)\,[\phi i\,(\ell)]\text{^}2\,d\ell$

where

$m(\ell)$ = mass per unit length

Generalized coordinate means an independent variable used with others to describe the state of a physical or chemical system, a different concept from X-Y coordinate. Here, the coordinates of modes 1, 2, 3, … n are independent. That is, the solutions of Eqn. (3.4-5) for mode i and for mode (i + j), where j is not zero, are independent. For better understanding, we show an example. Fig. 3.4-3 depicts a situation in which a double pendulum swings and generates two angles, θ1 and θ2. It can be seen here that the two variables vary independently.

Fig. 3.4-4 shows generalized coordinates when the forcing function is sinusoidal. Table 3.4-2 is constructed to tabulate the bending deflection ξ (ℓ,t) in Eqn. (3.4 –2). Let's assume that we are interested in the amount of the deflection of a point in LV located at ℓ = 52.5 ft from the

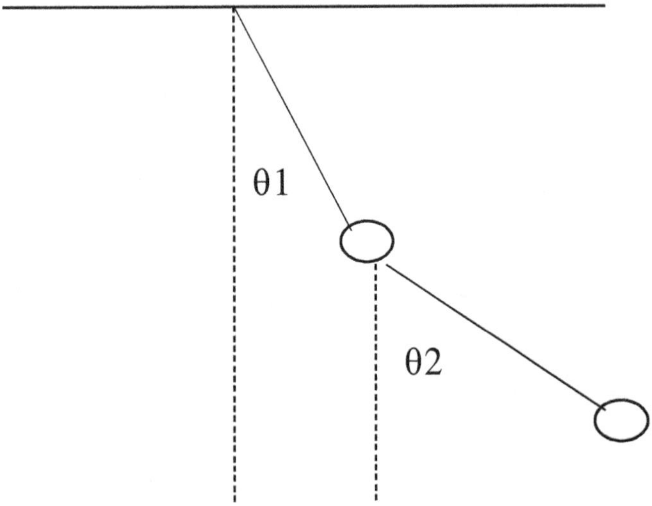

Fig. 3.4-3. Double Pendulum Swing

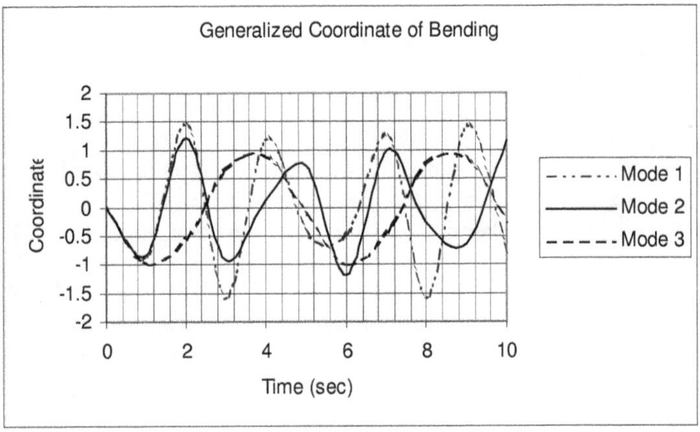

Fig. 3.4-4. Generalized Coordinate of Bending

tip of the LV nose. From Table 3.4-1, at ℓ = 52.5, we have,

For mode 1, $\Phi 1$ = -0.60
For mode 2, $\Phi 2$ = -0.02
For mode 3, $\Phi 3$ = 1.52

We ignored the higher modes to simplify our example. Then, the deflection time history (Eqn. 3.4-2) for 10 seconds is,

Generalized Coordinate			
Time (sec)	q1	q2	q3
0	0	0	0
1	-0.87	-0.83	-0.96
2	1.46	1.20	-0.55
3	-1.58	-0.91	0.65
4	1.19	0.13	0.92
5	-0.42	0.73	-0.12
6	-0.49	-1.19	-0.99
7	1.24	0.99	-0.44
8	-1.59	-0.25	0.74
9	1.43	-0.63	0.86
10	-0.81	1.16	-0.25

Table 3.4-2. Generalized Coordinate

ξ (52.5,t) = (i =:1, i = 3) $\Sigma\ \Phi$i (52.5) qi(t), for t=0-10
= Φ1(52.5) q1(t) + Φ2(52.5) q2(t) + Φ3(52.5) q3(t) for t=0 –10
Note that q1(t), q2(t), and q3(t) are from Table 3.4-2

ξ (52.5, 0) = 0

ξ (52.5, 1) = (-0.60)*(-0.87) +(-0.02)*(-0.83) + (1.52)*(-0.96)
= -0.32

ξ (52.5, 2) = (-0.60)*(1.46) +(-0.02)*(1.20) + (1.52)*(-0.55)
= -2.52

$$\ddot{\xi}(52.5, 10) = (-0.60)*(-0.81) + (-0.02)*(1.16) + (1.52)*(-0.25) = 0.67$$

These values are plotted in Fig. 3.4-5. The maximum deflection during the 10-second period is 3 ft in the positive direction, which is the distance from the X-axis in the pitch plane. We can compute deflection as a function of time at any point in the LV, which will be used in force and moment computations.

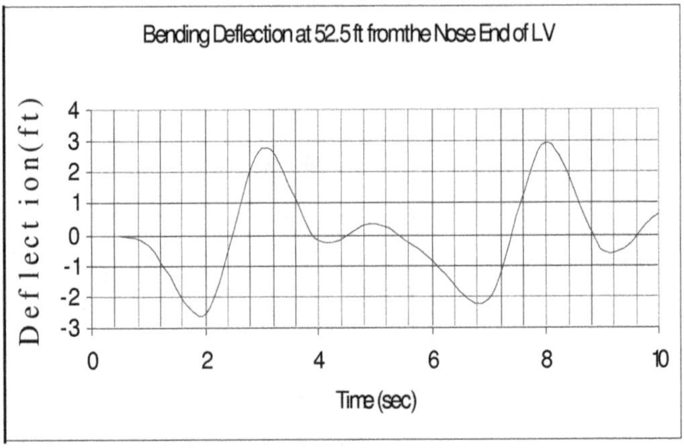

Fig. 3.4-5. Bending Deflection at 52.5 ft from the Nose End of the LV

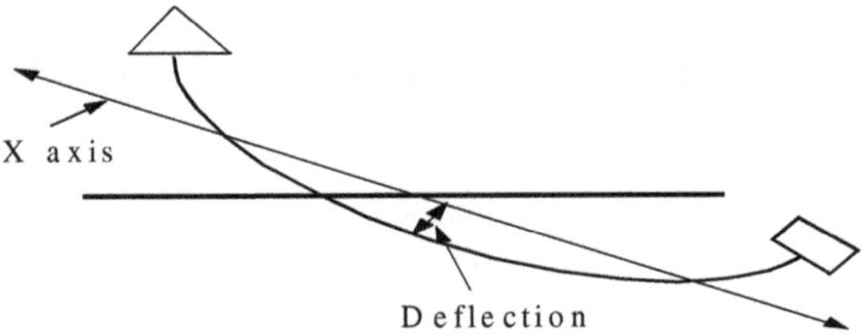

Fig. 3.4-6. Definition of Deflection

****

Derivation of Generalized Coordinate Differential Equations

In general, the main frame of the launch vehicle is made of aluminum alloy, and because of the elongated nature of the LV structure, bending cannot be avoided. Bending of the LV in flight is a phenomenon that cannot be overlooked in attitude control system design. Ignoring bending will cause instability and eventually cause the launch operation to fail. For example, bending at the attitude sensor will cause erroneous attitude information and mislead the flight computer to issue an attitude command that directs the engine nozzle in the wrong direction.

In this section, we derive a generalized coordinate of bending modes, which will be used to compute the structural deflection. The generalized coordinate means that the coordinates are independent. In bending analysis, we desire to have a coordinate per each bending mode that is independent of that of the other bending mode, so that we can simply add structural deflections after we compute deflections in each mode. In3.4-7, we describe a trace of a point "A" after some bending has occurred.

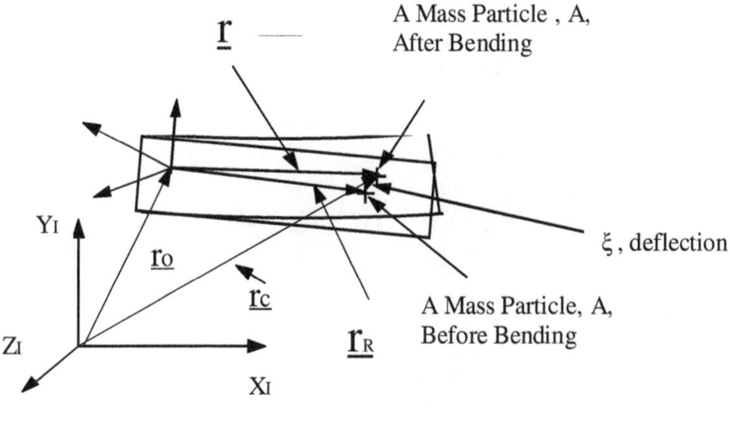

Fig. 3.4-7. Trace of Point "A" During Bending

From Fig. 3.4-7,

$$\underline{r} = \underline{rR} + \underline{\xi} \qquad (3.4\text{-}6)$$

The deflection, ξ, can be computed by using Hooke's Law.

$$\mu \left[\partial\partial\underline{\xi} / \partial^{(2)}t\right] + \pounds\,(\underline{\xi}) = \underline{F} \qquad (3.4\text{-}7)$$

Here, μ and £ are linear differential operators that are unique to each system, e.g., string, bar, membrane, and plate.

From Fig. 3.4-7 again,

$$\underline{rc} = \underline{ro} + \underline{r} \qquad (3.4\text{-}8)$$
$$\underline{rc} = \underline{ro} + \underline{rR} + \underline{\xi} \qquad (3.4\text{-}9)$$

The dynamic equation of motion of particle, A, can be written as,

$$\begin{aligned} d\underline{F} &= (\rho dV)\left[d^{(2)}\underline{rc}/dt^{(2)}\right] \\ &= (\rho dV)\left[d^{(2)}(\underline{ro} + \underline{rR} + \underline{\xi})/dt^{(2)}\right] \\ &= (\rho dV)\left[d^{(2)}(\underline{\xi})/dt^{(2)}\right] \end{aligned} \qquad (3.4\text{-}10)$$

where

(ρdV) = mass element

$[d^{(2)}\underline{rc}/dt^{(2)}]$ = acceleration

And

$$\begin{aligned} d\underline{m} &= \underline{r} \times d\underline{F} \\ &= (\underline{rR} + \underline{\xi}) \times (\rho dV)\left[d^{(2)}(\underline{\xi})/dt^{(2)}\right] \end{aligned} \qquad (3.4\text{-}11)$$

where

dm = moment

r = moment arm

dF = force

The " $\underline{\xi}$ " is a function of time and a deflection in the LV. The deflection varies depending on where the point of interest (ℓ) is located and the time (t) when the deflection occurs. Thus, we write,

$$\underline{\xi}(\ell, t) = \Phi(\ell)\, q(t) \qquad (3.4\text{-}12)$$

Substituting Eqn. (3.4-12) into Eqn. (3.4-7)

$$\underline{F} = \mu\, \{\partial\partial[\Phi(\ell)\, q(t)] / \partial^{(2)}t\} + £\, [\Phi(\ell)\, q(t)]$$
$$= \mu\, \Phi(\ell)\, q(t)'' + £\, [\Phi(\ell)]q(t) \qquad (3.4\text{-}13)$$

For free vibration in a vacuum, we set $\underline{F} = 0.0$. Then, Eqn. (3.4-13) becomes,

$$0 = q(t)'' + \{£\, [\Phi(\ell)]/[\mu\, \Phi(\ell)]\}q(t)$$
$$= q(t)'' + \omega^{(2)}q(t)$$

where

$$\omega^{(2)} = \{£\, [\underline{\Phi}(\ell)]/[\mu\, \underline{\Phi}(\ell)]\}$$

From the equation above,

$$£\, [\underline{\Phi}(\ell)] - \mu\, \omega^{(2)}\, \underline{\Phi}(\ell) = 0 \qquad (3.4\text{-}14)$$

We define $\underline{\Phi}(\ell)$ such that its fundamental property is orthogonality. This is the property that allows us to express the motion in terms of natural modes. We derive the orthogonality condition as follows:

If $\underline{\Phi i}(\ell)$ and $\underline{\Phi j}(\ell)$ are two distinct modes, then from Eqn. (3.4-14),

$$£\, [\underline{\Phi i}(\ell)] - \mu\, \omega i^{(2)}\, \underline{\Phi i}(\ell) = 0 \qquad (3.4\text{-}15)$$
$$£\, [\underline{\Phi j}(\ell)] - \mu\, \omega j^{(2)}\, \underline{\Phi j}(\ell) = 0 \qquad (3.4\text{-}16)$$

Taking the dot product of Eqn. (3.4-15) and $\underline{\Phi j}(\ell)$, and of Eqn. (3.4-16) and $\underline{\Phi i}(\ell)$, and then integrating them over the whole volume,

$$\int £\, [\underline{\Phi i}(\ell)] \text{dot}\, \underline{\Phi j}(\ell) dv - \mu\, \omega i^{(2)} \int \underline{\Phi i}(\ell)\, \text{dot}\, \underline{\Phi j}(\ell)\, dv = 0$$
$$(3.4\text{-}17)$$

$$\int \pounds \, [\underline{\Phi j}(\ell)]\text{dot } \underline{\Phi i}(\ell)dv \; - \mu \, \omega j^{\wedge}(2) \int \underline{\Phi j}(\ell) \text{ dot } \underline{\Phi i}(\ell) \, dv = 0$$

$$(3.4\text{-}18)$$

Subtracting Eqn. (3.4-17) from Eqn. (3.4-18),

$$[\omega i^{\wedge}(2) - \omega j^{\wedge}(2)] \, \mu \int \underline{\Phi j}(\ell) \text{ dot } \underline{\Phi i}(\ell) \, dv + \int \{\pounds \, [\underline{\Phi j}(\ell)]\text{dot } \underline{\Phi i}(\ell) - \pounds [\underline{\Phi i}(\ell)]\text{dot } \underline{\Phi j}(\ell)\}dv \; = 0$$

In the equation above, if

$$\{\pounds \, [\underline{\Phi j}(\ell)]\text{dot } \underline{\Phi i}(\ell) - \pounds \, [\underline{\Phi i}(\ell)]\text{dot } \underline{\Phi j}(\ell)\} = 0.0,$$

$$(3.4\text{-}19)$$

then

$$\text{``}\underline{\Phi j}(\ell) \text{ dot } \underline{\Phi i}(\ell) = 0.0 \text{''}$$

which implies that $\underline{\Phi j}(\ell)$ and $\underline{\Phi i}(\ell)$ are orthogonal. Thus, Eqn. (3.4-19) is the orthogonality condition.

Back to Hooke's law for lateral vibration of a bar, we learned from the Hooke's law that

$$\mu = \rho \, A \, (\ell)$$
$$= m(\ell) \qquad\qquad (3.4\text{-}20)$$
$$\pounds = [\partial^{\wedge}(2)/ \, (\partial \ell)^{\wedge}(2)]EI \, [\partial^{\wedge}(2)/ \, (\partial \ell)^{\wedge}(2)] \qquad (3.4\text{-}21)$$

where

$$\rho \; = \text{density}$$
$$A = \text{cross section area}$$
$$E = \text{Young's modulus}$$
$$I \; = \text{centroidal moment of inertia of } A$$

For a system of forced vibration of a non-uniform beam, we can write the Hooke's law as,

$$m(\ell)[\partial^{\wedge}(2)\xi / \, \partial^{\wedge}(2)t] \; + \; [\partial^{\wedge}(2)/ \, (\partial \ell)^{\wedge}(2)]EI \, [\partial^{\wedge}(2) \, \xi / \, (\partial \ell)^{\wedge}(2)]$$
$$= \sum Fz$$

$$(3.4\text{-}22)$$

196

Adding many modes of the vibration, Eqn. (3.4-12) can be written as the sum of these modes for that location and at that specific time.

$$\underline{\xi}(\ell, t) = \sum \Phi i(\ell)\, qi(t) \qquad (3.4\text{-}23)$$

Differentiating Eqn. (3.4-23) twice with respect to time and substituting into Eqn. (3.4-22),

$$m(\ell)\sum \underline{\Phi}i(\ell)\, qi(t)'' + [\partial^{(2)}/ (\partial\ell)^{(2)}]EI\, [\partial^{(2)} [\sum \Phi i(\ell)\, qi(t)]/ (\partial\ell)^{(2)}] = \sum Fz$$

Taking the dot product of both sides of the equation above by $\underline{\Phi}j(\ell)$ and integrating over the length,

$$\sum qi(t)'' \int m(\ell)\, \underline{\Phi}i(\ell) \cdot \underline{\Phi}j(\ell)\, d\ell + \sum qi(t) \int [\partial^{(2)}/ (\partial\ell)^{(2)}]EI\, [\partial^{(2)} [\Phi i(\ell)]/ (\partial\ell)^{(2)}] \cdot \underline{\Phi}j(\ell)\, d\ell = \sum Fz \cdot \int \underline{\Phi}j(\ell)\, d\ell \qquad (3.4\text{-}24)$$

From Eqn. (3.4-21),

$$£(\xi) = [\partial^{(2)}/ (\partial\ell)^{(2)}]EI\, [\partial^{(2)} \xi / (\partial\ell)^{(2)}] \qquad (3.4\text{-}25)$$

Substituting Eqn. (3.4-23) into Eqn. (3.4-25) and rearranging it,

$$qi(t) \sum £\, [\Phi i(\ell)] = \sum [\partial^{(2)}/ (\partial\ell)^{(2)}]EI\, [\partial^{(2)} \Phi i(\ell)/ (\partial\ell)^{(2)}]\, qi(t) \qquad (3.4\text{-}26)$$

From Eqn. (3.4-26),

$$£\, [\Phi(\ell)] = [\partial^{(2)}/ (\partial\ell)^{(2)}]EI\, [\partial^{(2)} \Phi(\ell)/ (\partial\ell)^{(2)}] \qquad (3.4\text{-}27)$$

From Eqn. (3.4-14), Eqn. (3.4-20), and Eqn. (3.4-27),

$$[\partial^{(2)}/ (\partial\ell)^{(2)}]EI\, [\partial^{(2)} \Phi i(\ell)/ (\partial\ell)^{(2)}] = m(\ell)\, \omega^{(2)}\, \underline{\Phi i}(\ell) \qquad (3.4\text{-}28)$$

Substituting Eqn. (3.4-28) into Eqn. (3.4-24),

$$\sum q_i(t)'' \int m(\ell)\, \underline{\Phi}_i(\ell) \cdot \underline{\Phi}_j(\ell)\, d\ell + \sum q_i(t) \int m(\ell)\, \omega^{2}\, \underline{\Phi}_i(\ell) \cdot \underline{\Phi}_j(\ell)\, d\ell = \sum F_z \cdot \int \underline{\Phi}_j(\ell)\, d\ell$$

Since $\underline{\Phi}_i(\ell) \cdot \underline{\Phi}_j(\ell) = 0$ when $i \neq j$, ($\underline{\Phi}_i(\ell)$ and $\underline{\Phi}_j(\ell)$ are orthogonal).

$$q_i(t)'' \int m(\ell)\, \underline{\Phi}_i(\ell)^{2}\, d\ell + q_i(t) \int m(\ell)\, \omega^{2}\, \underline{\Phi}_i(\ell)^{2}\, d\ell = F_z \cdot \int \underline{\Phi}_i(\ell)\, d\ell$$

Rewriting the equation above,

$$m_i \left[q_i(t)'' + \omega^{2}\, q_i(t) \right] = Q_i \qquad (3.4\text{-}29)$$

where

$m_i = \int m(\ell)\, \underline{\Phi}_i(\ell)^{2}\, d\ell$: ith mode modal mass
$Q_i = \sum F_z \cdot \int \underline{\Phi}_i(\ell)\, d\ell$: ith mode modal force

The orthogonal property ensures that all modes are uncoupled. Therefore, theoretically we can compute deflection independently and add the results to obtain the total deflection. However, we will see that in case of a launch vehicle, coupling arises between the elastic, slosh, and rigid-body modes, because the applied forces in each case are dependent on the motion of all of the these modes. To simplify the numerical computation, the elastic modes are computed with the sloshing mass removed and the engine fixed. Slosh modes are computed assuming a rigid vehicle. In this sense, the modes are often referred to as artificially uncoupled. The coupling is reintroduced via the forcing terms and inertial forces (from ref. 17).

Taking into account structural damping, we rewrite Eqn. (3.4-29),

$$\left[q_i(t)'' + 2\, \omega_i\, \xi_i\, q_i(t)' + \omega_i^{2}\, q_i(t) \right] = Q_i / m_i$$

(3.4-30)

where

$m_i = \int m(\ell)\, \underline{\Phi}_i(\ell)^{2}\, d\ell$
$Q_i = \sum F_z \cdot \int \underline{\Phi}(\ell)\, d\ell$

$$\sum Fz \text{ dot } \underline{\Phi}(\ell) = Fzg \text{ dot} \int \underline{\Phi}(\ell)d\ell + FzT \text{ dot} \int \underline{\Phi}(\ell)d\ell + FzA \text{ dot} \int \underline{\Phi}(\ell)d\ell + Fzs \text{ dot} \int \underline{\Phi}(\ell)d\ell + Fze \text{ dot} \int \underline{\Phi}(\ell)d\ell$$

where Fzg, FzT, FzA, Fzs, and Fze are from Section 3.2, Section 3.3, Section 3.5, and Section 3.6, and these equations are collected below.

a. Gravity [Eqn. (3.2-5)]

$$Fzg = -mT\, g\, \theta \cos\theta o = -\left[\int m(\ell)d\ell\right] g (\cos\theta o)\, \theta$$

b. Thrust [Eqn. (3.2-12)]

$$FzT = Td\, \delta - (Tu + Td)\Sigma\sigma i\, qi$$

c. Aerodynamics [Eqn. (3.3-11)]

$$FzA = -(1/2)\rho U^{(2)} A3 \{\alpha o \int (\partial CN/\partial\alpha)\, d\ell - (\theta'/U)\int (\ell a - \ell)(\partial CN/\partial\alpha)d\ell + \Sigma\, qi(t)\int \sigma i(\ell)(\partial CN/\partial\alpha)\, d\ell - \Sigma\, [qi'(t)/U] \int \varphi i(\ell)(\partial CN/\partial\alpha)\, d\ell - (Ww/U)\int (\partial CN/\partial\alpha)\, d\ell\}$$

d. Sloshing [Eqn. (3.5-41)]

$$Fzs = Uo' \Sigma mpi\, \Gamma pi$$

e. Engine Inertia [Eqn. (3.6-24)]

$$Fze = -mR\{w' - u'\theta + \ell C\, \theta'' - \ell R\, (\delta'' - \theta'') + \Sigma\, qTi''(t)\, [\ell R\, \sigma i(\ell T) + \varphi i(\ell T)]\}$$

Substituting Fzg, FzT, FzA, Fzs, and Fze back to $Qi = \sum Fz \text{ dot} \int \underline{\Phi}(\ell)d\ell$,

$$Qi = \{-[\int \underline{\Phi i}(\ell)m(\ell)d\ell]\, g (\cos\theta o) + mR\, \underline{\Phi i}(\ell T)\, u'\}\theta + qD\, A3\, (1/U)\int \underline{\Phi i}(\ell)\, (\ell a - \ell)(\partial CN/\partial\alpha)d\ell\, \theta' - mR\, \underline{\Phi i}(\ell T)(\ell C + \ell R)\theta' - \{(Tu + Td)\underline{\Phi i}(\ell T)\, \Sigma\sigma j + qD\, A3\, \Sigma \int \underline{\Phi i}(\ell)\sigma j(\ell)(\partial CN/\partial\alpha)\, d\ell\}qj + qD\, A3\, \Sigma\, [1/U] \int \underline{\Phi i}(\ell)\, \underline{\Phi j}(\ell)(\partial CN/\partial\alpha)\, d\ell\, qi' - mR\, \underline{\Phi i}(\ell T)\, \Sigma\, [\ell R\, \sigma j(\ell T) + \underline{\Phi j}(\ell T)]\, qTi'' + Uo'\, \underline{\Phi i}(\ell pk)\, \Sigma mpk\, \Gamma pi + Td\, \underline{\Phi i}(\ell T)\, \delta$$

$$+ mR\, \underline{\Phi i}(\ell T) \quad \ell R\, \delta'' - qD\, A3 \int \underline{\Phi i}(\ell)\ (\partial CN / \partial \alpha)\, d\ell\ \ \alpha o$$
$$+ qD\, A3\, (1/U) \int \underline{\Phi i}(\ell)(\partial CN / \partial \alpha)\ d\ell \}\ Ww - mR\, \underline{\Phi i}(\ell T) w'$$

(3.4-31)

The lateral acceleration, " w', " in the last term is a function of many variables. Now, we substitute " w' " [Eqn. (3.7-5) in Section 3.7] into the equation above and then write two equations, one for the first bending mode and the other for the second bending mode.

First bending mode,

$$Q1 = \{- mR\, \Phi 1(\ell T) \{(-mT\, g \cos \theta o + mR\, u' - mT\, Uo) / (mT + mR)\} + \{- [\int \Phi 1 (\ell)\ m(\ell) d\ell]\ g (\cos \theta o) + mR\, \Phi 1(\ell T)\, u'\}\} \theta\ +$$

$$\{- mR\, \Phi 1(\ell T)\{ [qD\, A3\ (1/U) \int (\ell a - \ell)\, (\partial CN / \partial \alpha) d\ell]/(mT + mR) \} + qD\, A3\, (1/U) \int \Phi(\ell)\, (\ell a - \ell)\, (\partial CN / \partial \alpha) d\ell \}\ \theta'\ +$$

$$\{- mR\ \Phi 1(\ell T)\, [-mR\, (\ell R + \ell C) / (mT + mR)] - mR\, \Phi 1(\ell T)\, (\ell C + \ell R)\} \theta''\ +$$

$$\{ mR\, \Phi 1(\ell T)\, \{[(Tu + Td) \sigma 1(\ell T) + qD\, A3 \int \sigma 1(\ell)(\partial CN / \partial \alpha)\, d\ell\ ]/(mT + mR)\} - \{(Tu + Td)\, \Phi 1(\ell T)\ \sigma 1(\ell T) + qD\, A3 \int \Phi 1(\ell)\, \sigma 1(\ell)(\partial CN / \partial \alpha)\, d\ell\}\} q1\ +$$

$$\{- mR\ \Phi 1(\ell T)\{[\, qD\, A3\ (1/U) \int \Phi 1(\ell)\, (\partial CN / \partial \alpha)\, d\ell\ ] / (mT + mR)\} + qD\, A3\, [1/U] \int \Phi 1(\ell)\, \Phi 1(\ell)\, (\partial CN / \partial \alpha)\, d\ell\ \}\ q1'\ +$$

$$\{ mR\, \Phi 1(\ell T)\{mR\, [\ell R\ \sigma 1(\ell T) + \Phi 1(\ell T)] / (mT + mR)\} - mR\, \Phi 1(\ell T) \Sigma\, [\ell R\, \sigma 1(\ell T) + \Phi 1(\ell T)]\}\ q1''\ +$$

$$\{ mR\, \Phi 1(\ell T)\, \{[(Tu + Td) \sigma 2(\ell T) + qD\, A3 \int \sigma 1(\ell)(\partial CN / \partial \alpha)\, d\ell\ ]/(mT + mR)\} - \{(Tu + Td)\, \Phi 1(\ell T)\ \sigma 2(\ell T) + qD\, A3 \int \Phi 1(\ell)\, \sigma 2(\ell)(\partial CN / \partial \alpha)\, d\ell\}\} q2\ +$$

$$\{- mR\ \Phi 1(\ell T)\{[\, qD\, A3\ (1/U) \int \Phi 1(\ell)\, (\partial CN / \partial \alpha)\, d\ell\ ] / (mT + mR)\} + qD\, A3\, \Sigma\, [1/U] \int \Phi 1(\ell)\, \Phi 2(\ell)\, (\partial CN / \partial \alpha)\, d\ell\ \}\ q2'\ +$$

$$\{ mR\, \Phi 1(\ell T)\{mR\, [\ell R\ \sigma 2(\ell T) + \Phi 2(\ell T)] / (mT + mR)\} - mR\, \Phi 1(\ell T)\, [\ell R\, \sigma 2(\ell T) + \Phi 2(\ell T)]\}\ q2''\ +$$

$$\{ - mR\, \Phi 1(\ell T)\, [Uo' mp1 / (mT + mR)] + Uo'\ \Phi 1(\ell p1)\ \}\ mp1\}\ \Gamma p1\ +$$

$$\{ - mR\, \Phi 1(\ell T)\, [Uo' mp2 / (mT + mR)] + Uo'\ \Phi 1(\ell p2)\ \}\ mp2\}\ \Gamma p2\ +$$

$$\{- mR\, \Phi 1(\ell T)\, [\, Td / (mT + mR)] + Td\, \Phi 1(\ell T)\ \}\ \delta\ +$$

$$\{- mR\, \Phi 1(\ell T)\, [-mR\, \ell R / (mT + mR)] + mR\, \Phi 1(\ell T)\, \ell R\}\ \delta''\ +$$

$$\{ - mR\, \Phi 1(\ell T)\{-qD\, A3 \int (\partial CN / \partial \alpha)\, d\ell] / (mT + mR)\} - qD\, A3 \int \Phi 1(\ell)\, (\partial CN / \partial \alpha)\, d\ell\ \}\ \alpha o$$

$$\{ - mR\, \Phi 1(\ell T)\{-qD\, A3\ (1/U) \int (\partial CN / \partial \alpha)\, d\ell / (mT + mR)\} + qD\, A3\, (1/U) \int \underline{\Phi}(\ell)(\partial CN / \partial \alpha)\ d\ell\}\}\ Ww$$

(3.4-32)

Now, the first bending mode generalized coordinate, q1, can be written from Eqn. (3.4-30) and Eqn (3.4-32) after moving the q1'' term to the left, and rearranging it,

$$q1'' = E1\ \theta + E2\ \theta' + E3\ \theta'' + E4\ q1 + E5\ q1' + E6\ q2 + E7\ q2' + E8\ q2'' + E9\ \Gamma p1 + E10\ \Gamma p2 + E11\ \delta + E12\ \delta'' + E13\ \alpha o + E14 Ww \qquad (3.4\text{-}33)$$

where

$$E1 = (1/m1\ \Psi q1)\ \{-\ mR\ \Phi1(\ell T)\ \{(-mT\ g \cos\theta o + mR\ u' - mT\ Uo)/(mT + mR)\} + \{-[\int\Phi1(\ell)\ m(\ell)d\ell]\ g(\cos\theta o) + mR\ \Phi1(\ell T)\ u'\}\}$$

$$E2 = (1/m1\ \Psi q1)\ \{-\ mR\ \Phi1(\ell T)\{[qD\ A3\ (1/U)\int(\ell a - \ell)(\partial CN/\partial\alpha)d\ell]/(mT + mR)\} + qD\ A3\ (1/U)\ \int\Phi(\ell)(\ell a - \ell)(\partial CN/\partial\alpha)d\ell\}$$

$$E3 = (1/m1\ \Psi q1)\ \{-\ mR\ \Phi1(\ell T)\ [-mR\ (\ell R + \ell C)/(mT + mR)] - mR\ \Phi1(\ell T)\ (\ell C + \ell R)\}$$

$$E4 = (1/\ \Psi q1)\ \{\{-\ \omega1\text{\textasciicircum}(2) + (1/m1)\{\ mR\ \Phi1(\ell T)\ \{[(Tu + Td)\sigma1(\ell T) + qD\ A3\int\sigma1(\ell)(\partial CN/\partial\alpha)\ d\ell]/(mT + mR)\} - \{(Tu + Td)\ \Phi1(\ell T)\ \sigma1(\ell T) + qD\ A3\int\Phi1(\ell)\sigma1(\ell)(\partial CN/\partial\alpha)\ d\ell\}\}\}$$

$$E5 = (1/\ \Psi\ q1)\ \{\{-2\ \omega1\ \xi1 + (1/m1)\ \{-\ mR\ \ \Phi1(\ell T)\{[\ qD\ A3\ (1/U)\int\Phi1(\ell)(\partial CN/\partial\alpha)\ d\ell\ ]/(mT + mR)\} + qDA3\ [1/U]\int\Phi1(\ell)\ \Phi1(\ell)\ (\partial CN/\partial\alpha)\ d\ell\ \}\}$$

$$E6 = (1/m1\ \Psi q1)\ \{\ mR\ \Phi1(\ell T)\ \{[(Tu + Td)\sigma2(\ell T) + qD\ A3\int\sigma1(\ell)(\partial CN/\partial\alpha)\ d\ell]/(mT + mR)\} - \{(Tu + Td)\ \Phi1(\ell T)\ \sigma2(\ell T) + qD\ A3\int\Phi1(\ell)\ \sigma2(\ell)(\partial CN/\partial\alpha)\ d\ell\}\}$$

$$E7 = (1/m1\ \Psi q1)\ \{-\ mR\ \ \Phi1(\ell T)\{[\ qD\ A3\ (1/U)\int\Phi1(\ell)(\partial CN/\partial\alpha)\ d\ell\ ]/(mT + mR)\} + qDA3\ \Sigma\ [1/U]\int\Phi1(\ell)\ \Phi2(\ell)\ (\partial CN/\partial\alpha)\ d\ell\ \}$$

$$E8 = (1/m1\ \Psi q1)\ \{\ mR\ \Phi1(\ell T)\{mR\ [\ell R\ \ \sigma2(\ell T) + \Phi2(\ell T)]/(mT + mR)\} - mR\ \Phi1(\ell T)\ [\ell R\ \sigma2(\ell T) + \Phi2(\ell T)]\}$$

$$E9 = (1/m1\ \Psi q1) -\ mR\ \Phi1(\ell T)\ [Uo'mp1/(mT + mR)] + Uo'\ \ \Phi1(\ell p1)\ \}\ mp1\}$$

$$E10 = (1/m1\ \Psi q1) -\ mR\ \Phi1(\ell T)\ [Uo'mp2/(mT + mR)] + Uo'\ \ \Phi1(\ell p2)\ \}\ mp2\}$$

$$E11 = (1/m1\ \Psi q1)\ \{-\ mR\ \Phi1(\ell T)\ [\ Td/(mT + mR)] + Td\ \Phi1(\ell T)\ \}$$

$$E12 = (1/m1\ \Psi q1)\ \{-\ mR\ \Phi1(\ell T)\ [-mR\ \ell R/(mT + mR)] + mR\ \Phi1(\ell T)\ \ell R\}$$

$$E13 = (1/m1\ \Psi q1)\ \{\ -\ mR\ \Phi1(\ell T)\{-qD\ A3\int(\partial CN/\partial\alpha)\ d\ell]/(mT + mR)\} - qD\ A3\int\Phi1(\ell)\ (\partial CN/\partial\alpha)\ d\ell\ \}$$

$$E14 = (1/m1\ \Psi q1)\ \{\ -\ mR\ \Phi1(\ell T)\{-qD\ A3\ \ (1/U)\int(\partial CN/\partial\alpha)\ d\ell/(mT + mR)\} + qD\ A3\ (1/U)\underline{\int\Phi}(\ell)(\partial CN/\partial\alpha)\ \ d\ell\}\}\ Ww$$

where

$$\Psi q1=\{1 - (1/m1) \{ mR\, \Phi1(\ell T)\{mR\, [\ell R\, \sigma1(\ell T) + \Phi1(\ell T)] / (mT + mR)\} - mR\, \Phi1(\ell T)\Sigma\, [\ell R\, \sigma1(\ell T) + \Phi1(\ell T)]\}\}$$

The second bending mode,

$$Q(2) = \{- mR\, \Phi2(\ell T)\, \{(-mT\, g \cos \theta o + mR\, u' - mT\, Uo) / (mT + mR)\} + \{- [\int \Phi2(\ell)\, m(\ell) d\ell]\, g\, (\cos \theta o) + mR\, \Phi2(\ell T)\, u'\}\}\theta +$$

$$\{- mR\, \Phi2(\ell T)\{ [qD\, A3\, (1/ U) \int (\ell a - \ell)\, (\partial CN /\partial\alpha) d\ell]/(mT + mR) \} + qD\, A3\, (1/ U) \int \Phi(\ell)\, (\ell a - \ell)\, (\partial CN /\partial\alpha) d\ell \}\, \theta' +$$

$$\{- mR\, \Phi2(\ell T)\, [-mR\, (\ell R + \ell C) / (mT + mR)] - mR\, \Phi2(\ell T)\, (\ell C + \ell R)\}\theta'' +$$

$$\{ mR\, \Phi2(\ell T)\, \{[(Tu + Td)\sigma1(\ell T) + qD\, A3 \int \sigma1(\ell)(\partial CN /\partial\alpha)\, d\ell]/(mT + mR)\} -\{(Tu + Td)\, \Phi2(\ell T)\, \sigma1(\ell T) + qD\, A3 \int \Phi2(\ell)\, \sigma1(\ell)(\partial CN /\partial\alpha)\, d\ell\}\}q1 +$$

$$\{- mR\, \Phi2(\ell T)\{[qD\, A3\, (1/ U) \int \Phi2(\ell)\, (\partial CN /\partial\alpha)\, d\ell] / (mT + mR)\} + qD\, A3\, [1/U] \int \Phi2(\ell)\, \Phi1(\ell)\, (\partial CN /\partial\alpha)\, d\ell \}\, q1' +$$

$$\{ mR\, \Phi2(\ell T)\{mR\, [\ell R\, \sigma1(\ell T) + \Phi1(\ell T)] / (mT + mR)\} - mR\, \Phi2(\ell T)\Sigma\, [\ell R\, \sigma1(\ell T) + \Phi1(\ell T)]\}\, q1'' +$$

$$\{ mR\, \Phi2(\ell T)\, \{[(Tu + Td)\sigma2(\ell T) + qD\, A3 \int \sigma2(\ell)(\partial CN /\partial\alpha)\, d\ell]/(mT + mR)\} -\{(Tu + Td)\, \Phi2(\ell T)\, \sigma2(\ell T) + qD\, A3 \int \Phi2(\ell)\, \sigma2(\ell)(\partial CN /\partial\alpha)\, d\ell\}\}q2 +$$

$$\{- mR\, \Phi2(\ell T)\{[qD\, A3\, (1/ U) \int \Phi2(\ell)\, (\partial CN /\partial\alpha)\, d\ell] / (mT + mR)\} + qD\, A3\, [1/U] \int \Phi2(\ell)\, \Phi2(\ell)\, (\partial CN /\partial\alpha)\, d\ell \}\, q2' +$$

$$\{ mR\, \Phi2(\ell T)\{mR\, [\ell R\, \sigma2(\ell T) + \Phi2(\ell T)] / (mT + mR)\} - mR\, \Phi2(\ell T)\, [\ell R\, \sigma2(\ell T) + \Phi2(\ell T)]\}\, q2'' +$$

$$\{ - mR\, \Phi2(\ell T)\, [Uo' mp1 / (mT + mR)] + Uo'\, \Phi2(\ell p1) \}\, mp1\}\, \Gamma p1 +$$

$$\{ - mR\, \Phi2(\ell T)\, [Uo' mp2 / (mT + mR)] + Uo'\, \Phi2(\ell p2) \}\, mp2\}\, \Gamma p2 +$$

$$\{- mR\, \Phi2(\ell T)\, [Td / (mT + mR)] + Td\, \Phi2(\ell T) \}\, \delta +$$

$$\{- mR\, \Phi2(\ell T)\, [-mR\, \ell R / (mT + mR)] + mR\, \Phi2(\ell T)\, \ell R\}\, \delta'' +$$

$$\{ mR\, \Phi2(\ell T)\{-qD\, A3 \int (\partial CN /\partial\alpha)\, d\ell]/ (mT + mR)\} - qD\, A3 \int \Phi2(\ell)\, (\partial CN /\partial\alpha)\, d\ell \}\, \alpha o$$

$$\{ - mR\, \Phi2(\ell T)\{-qD\, A3\, (1/U) \int (\partial CN /\partial\alpha)\, d\ell/ (mT + mR)\} + qD\, A3\, (1/U) \int \Phi(\ell)(\partial CN /\partial\alpha)\, d\ell\}\}\, Ww \qquad (3.4\text{-}34)$$

Similarly, the second bending mode generalized coordinate, q2, can be written from Eqn. (3.4-30) and Eqn (3.4-34) after moving the q2'' term to the left, and rearranging it,

$$q2'' = G1\,\theta + G2\,\theta' + G3\,\theta'' + G4\,q1 + G5\,q1' + G6\,q1'' + G7\,q2 + G8\,q2' + G9\,\Gamma p1 + G10\,\Gamma p2 + G11\,\delta + G12\,\delta'' + G13\,\alpha o + G14\,Ww \quad (3.4\text{-}35)$$

where

$$G1 = (1/m2\,\Psi q2)\{-\,mR\,\Phi2(\ell T)\,\{(-mT\,g\cos\theta o + mR\,u' - mT\,Uo)/(mT + mR)\} + \{-[\int\Phi2(\ell)\,m(\ell)d\ell]\,g(\cos\theta o) + mR\,\Phi2(\ell T)\,u'\}\}$$

$$G2 = (1/m2\,\Psi q2)\{-\,mR\,\Phi2(\ell T)\{[qD\,A3\,(1/U)\int(\ell a - \ell)(\partial CN/\partial\alpha)d\ell]/(mT + mR)\} + qDA3\,(1/U)\int\Phi(\ell)(\ell a - \ell)(\partial CN/\partial\alpha)d\ell\}$$

$$G3 = (1/m2\,\Psi q2)\{-\,mR\,\Phi2(\ell T)[-mR(\ell R + \ell C)/(mT + mR)] - mR\,\Phi2(\ell T)(\ell C + \ell R)\}$$

$$G4 = (1/m2\,\Psi q2)\{mR\,\Phi2(\ell T)\{[(Tu + Td)\sigma1(\ell T) + qD\,A3\int\sigma1(\ell)(\partial CN/\partial\alpha)\,d\ell]/(mT + mR)\} - \{(Tu + Td)\,\Phi2(\ell T)\,\sigma1(\ell T) + qD\,A3\int\Phi2(\ell)\,\sigma1(\ell)(\partial CN/\partial\alpha)\,d\ell\}\}$$

$$G5 = (1/m2\,\Psi q2)\{-\,mR\,\Phi2(\ell T)\{[qD\,A3\,(1/U)\int\Phi2(\ell)(\partial CN/\partial\alpha)\,d\ell]/(mT + mR)\} + qD\,A3\,[1/U]\int\Phi2(\ell)\,\Phi1(\ell)(\partial CN/\partial\alpha)\,d\ell\}$$

$$G6 = (1/m2\,\Psi q2)\{mR\,\Phi2(\ell T)\{mR[\ell R\,\sigma1(\ell T) + \Phi1(\ell T)]/(mT + mR)\} - mR\,\Phi2(\ell T)\Sigma[\ell R\,\sigma1(\ell T) + \Phi1(\ell T)]\}$$

$$G7 = (1/\Psi q2)\{-\,\omega1^{(2)} + (1/m2)\{mR\,\Phi2(\ell T)\{[(Tu + Td)\sigma2(\ell T) + qD\,A3\int\sigma2(\ell)(\partial CN/\partial\alpha)\,d\ell]/(mT + mR)\} - \{(Tu + Td)\,\Phi2(\ell T)\,\sigma2(\ell T) + qD\,A3\int\Phi2(\ell)\,\sigma2(\ell)(\partial CN/\partial\alpha)\,d\ell\}\}\}$$

$$G8 = (1/\Psi q2)\{-\,2\,\omega2\xi2 + (1/m2)\{-\,mR\,\Phi2(\ell T)\{[qD\,A3\,(1/U)\int\Phi2(\ell)(\partial CN/\partial\alpha)\,d\ell]/(mT + mR)\} + qD\,A3\,[1/U]\int\Phi2(\ell)\,\Phi2(\ell)(\partial CN/\partial\alpha)\,d\ell\}\}$$

$$G9 = (1/m2\,\Psi q2)\{-\,mR\,\Phi2(\ell T)[Uo'mp1/(mT + mR)] + Uo'\,\Phi2(\ell p1)\}\,mp1\}$$

$$G10 = (1/m2\,\Psi q2)\{-\,mR\,\Phi2(\ell T)[Uo'mp2/(mT + mR)] + Uo'\,\Phi2(\ell p2)\}\,mp2\}$$

$$G11 = (1/m2\,\Psi q2)\{-\,mR\,\Phi2(\ell T)[Td/(mT + mR)] + Td\,\Phi2(\ell T)\}$$

$$G12 = (1/m2\,\Psi q2)\{-\,mR\,\Phi2(\ell T)[-mR\,\ell R/(mT + mR)] + mR\,\Phi2(\ell T)\,\ell R\}$$

$$G13 = (1/m2\,\Psi q2)\{mR\,\Phi2(\ell T)\{-qD\,A3\int(\partial CN/\partial\alpha)\,d\ell]/(mT + mR)\} - qD\,A3\int\Phi2(\ell)(\partial CN/\partial\alpha)\,d\ell\}$$

$$G14 = (1/m2\,\Psi q2)\{-\,mR\,\Phi2(\ell T)\{-qD\,A3\,(1/U)\int(\partial CN/\partial\alpha)\,d\ell/(mT + mR)\} + qD\,A3\,(1/U)\int\Phi(\ell)(\partial CN/\partial\alpha)\,d\ell\}\}$$

where

$$\Psi q2 = \{ 1 - (1/m2) \{ mR\ \Phi 2(\ell T)\{ mR\ [\ell R\ \sigma 2(\ell T) + \Phi 2(\ell T)] / (mT + mR)\} - mR\ \Phi 2(\ell T)\ [\ell R\ \sigma 2(\ell T) + \Phi 2(\ell T)]\}\}$$

3.5 Sloshing

Liquid hydrogen in fuel tank and liquid oxygen in oxidizer tank are the sources of sloshing. Modern rockets use these liquid fuels because of their high Isp values. In order to ensure stable flight, we must deal with sloshing, a complex phenomenon that depends on the shape of the tank, the liquid level, and the density of the liquid.

Because of its complexity, we simplify the sloshing math model by using either a pendulum or spring mass system. Parameters associated with the pendulum model are 1) slosh mass, 2) pendulum length, and 3) hinge-point location. These parameters vary with the shape of the tank and change as the liquid level increases or decreases. There is a program called CA31 that computes these parameters as a function of liquid level when the tank shape and the density of its liquid contents are given. The CFD (Computational Fluid Dynamics) technique is also available for computation of these parameters. If the spring mass model is used, the CA31 program calculates spring mass, spring constant, and distance. In this book, we model sloshing as a pendulum. The dynamic equation of the pendulum model is derived and its impacts on force and moment exerting on LV are discussed.

To start with, we write the velocity of the pendulum mass, mpi (Fig. 3.5-1) in ECI (inertial coordinate frame),

$$\begin{aligned} \underline{\mu} pi &= \underline{\mu} + d\ \underline{\rho} pi / dt, \\ &= \underline{\mu} + \partial\ \underline{\rho} pi / \partial t + \underline{\omega} \times \underline{\rho} pi \end{aligned} \qquad (3.5\text{-}1)$$

where

$\underline{\mu} pi$ = d Rpi /dt , the subscript "p" stands for pitch plane
$\underline{\mu}$ = d RLV /dt , RLV : position of LV coordinate origin
= u $\underline{i}$b + w $\underline{k}$b
$\underline{\omega}$ = ωyb $\underline{j}$ (ωyb: rotation rate of LV about Yb axis)

Now, from Fig. 3.5-1,

$$\underline{\rho}pi = (\ell pi - Lpi \cos \Gamma pi)\, \underline{i}b + (Lpi \sin \Gamma pi + \xi p(\ell pi, t))\, \underline{k}b \quad (3.5\text{-}2)$$

Taking a partial differentiation of $\underline{\rho}pi$ with respect to time,

$$(\partial \underline{\rho}pi / \partial t) = (Lpi\ \Gamma pi' \sin \Gamma pi)\, \underline{i}b + (Lpi\ \Gamma pi' \cos \Gamma pi + \xi p'(\ell pi, t))\, \underline{k}b \quad (3.5\text{-}3)$$

The cross product of $\underline{\omega}$ and $\underline{\rho}pi$,

$$\underline{\omega} \times \underline{\rho}pi = \begin{vmatrix} ib & jb & \underline{k}b \\ 0 & \omega yb & 0 \\ (\ell pi - Lpi \cos \Gamma pi) & 0 & (Lpi \sin \Gamma pi + \xi p(\ell pi, t)) \end{vmatrix}$$

$$= \omega yb\, [Lpi \sin \Gamma pi + \xi p(\ell pi, t)]\, \underline{i}b - \omega yb\, (\ell pi - Lpi \cos \Gamma pi)\, \underline{k}b \quad (3.5\text{-}4)$$

Substituting Eqn. (3.5-3) and Eqn. (3.5-4) into Eqn. (3.5-1),

$$\begin{aligned} \underline{\mu}pi &= \underline{\mu} + \partial \underline{\rho}pi / \partial t + \underline{\omega} \times \underline{\rho}pi \\ &= (u\, \underline{i}b + w\, \underline{k}b) + [(Lpi\ \Gamma pi' \sin \Gamma pi)\, \underline{i}b + [Lpi\ \Gamma pi' \cos \Gamma pi + \xi p'(\ell pi, t)]\, \underline{k}b] + \omega yb\, (Lpi \sin \Gamma pi + \xi p(\ell pi, t))\, \underline{i}\, b - \omega yb\, (\ell pi - Lpi * \cos \Gamma pi)\, \underline{k}b \\ &= \{u + Lpi\ \Gamma pi' \sin \Gamma pi + \omega yb\, [Lpi \sin \Gamma pi + \xi p(\ell pi, t)]\}\, \underline{i}\, b + [w + Lpi\ \Gamma pi' \cos \Gamma pi + \xi p'(\ell pi, t) - \omega yb\, (\ell pi - Lpi \cos \Gamma pi)]\underline{k}b \end{aligned} \quad (3.5\text{-}5)$$

Now, we use the following Lagrange equation to derive the pendulum dynamic equation,

$$d(\partial L / \partial \Gamma pi')/dt - (\partial L / \partial \Gamma pi) = 0 \quad (3.5\text{-}6)$$

where

$$L = K - P$$

T= kinetic energy
P= potential energy

For the pendulum, the potential energy can be written as,

$$P = m_{pi}\, g\, L_{pi}\, (1- \cos \Gamma_{pi}) \qquad (3.5\text{-}7)$$
≈ 0 for $(1- \cos \Gamma_{pi}) \approx 0$ assuming Γ_{pi} is less than 10 deg.

The kinetic energy, K, is

$$K = (0.5)^* m_{pi}\, (\underline{\mu}_{pi} \text{ dot } \underline{\mu}_{pi}) \qquad (3.5\text{-}8)$$

Substituting Eqn. (3.5-7) and Eqn. (3.5-8) into Eqn. (3.5-6),

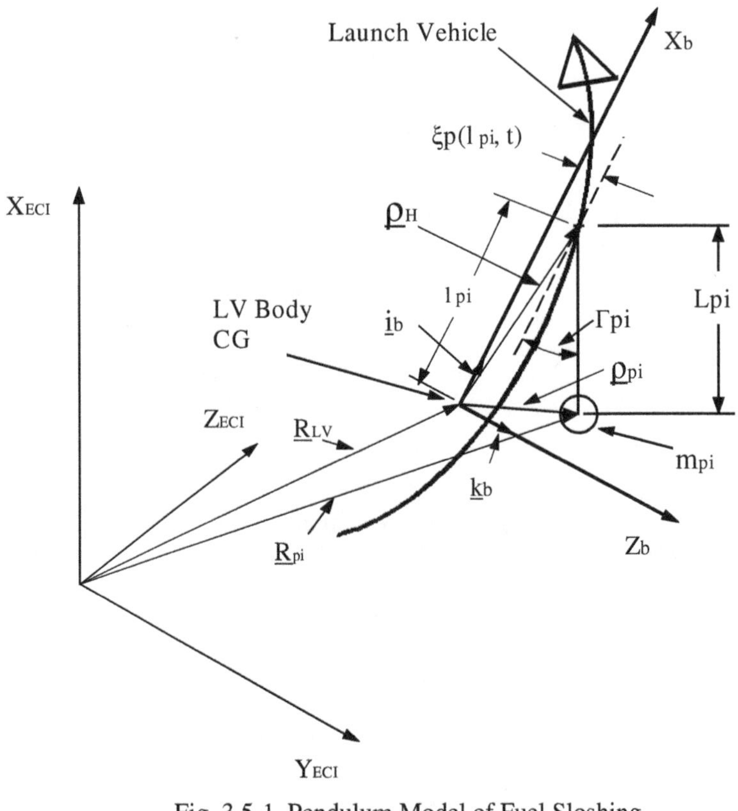

Fig. 3.5-1. Pendulum Model of Fuel Sloshing

$$d\{[\partial(\underline{\mu}pi \text{ dot } \underline{\mu}pi)] / \partial \Gamma pi'\}/dt - \partial[(\underline{\mu}pi \text{ dot } \underline{\mu}pi)] / \partial \Gamma pi = 0$$

$$(0.5)d\{ [(\partial \underline{\mu}pi / \partial \Gamma pi') \text{ dot } \underline{\mu}pi]]\}/dt - (0.5)[(\partial \underline{\mu}pi / \partial \Gamma pi) \text{ dot } \underline{\mu}pi] =0$$

$$d [(\partial \underline{\mu}pi / \partial \Gamma pi') \text{ dot } \underline{\mu}pi] /dt - [(\partial \underline{\mu}pi / \partial \Gamma pi) \text{ dot } \underline{\mu}pi] =0 \quad (3.5\text{-}9)$$

By the small angle approximation of Γpi, we can rewrite Eqn. (3.5-5) as,

$$\underline{\mu}pi = \{u + Lpi\ \Gamma pi'\ \Gamma pi + \omega yb\,[Lpi\,\Gamma pi + \xi p(\ell pi, t)]\}\ \underline{i}\,b + [w + Lpi\ \Gamma pi' + \xi p'(\ell pi, t) - \omega yb\,(\ell pi - Lpi)]\underline{k}b \quad (3.5\text{-}10)$$

And taking partial differentiation of Eqn. (3.5-10) with respect to "$\partial \Gamma pi'$,"

$$\partial \underline{\mu}pi / \partial \Gamma pi' = Lpi\,\Gamma pi\ \underline{i}\,b + Lpi\ \underline{k}b$$

The dot product of $\partial \underline{\mu}pi / \partial \Gamma pi'$ and $\underline{\mu}pi$,

$$\begin{aligned}(\partial \underline{\mu}pi/\partial \Gamma pi') \text{ dot } \underline{\mu}pi &= (Lpi\,\Gamma pi\ \underline{i}\,b + Lpi\ \underline{k}b) \text{ dot } (\{u + Lpi\ \Gamma pi'\ \Gamma pi + \omega yb\,[Lpi\,\Gamma pi \\ &\quad + \xi p(\ell pi, t)]\}\ \underline{i}\,b + [w + Lpi\ \Gamma pi' + \xi p'(\ell pi, t) - \omega yb\,(\ell pi - Lpi)]\underline{k}b \\ &= (Lpi\,\Gamma pi)\,\{u + Lpi\ \Gamma pi'\ \Gamma pi + \omega yb\,[Lpi\,\Gamma pi + \xi p(\ell pi, t)]\} + Lpi\,[w + Lpi\,\Gamma pi' \\ &\quad + \xi p'(\ell pi, t) - \omega yb\,(\ell pi - Lpi)] \\ &= (Lpi\,\Gamma pi)\,u + (Lpi\,\Gamma pi)\,Lpi\ \Gamma pi'\ \Gamma pi + (Lpi\,\Gamma pi)\,\omega yb\,[Lpi\,\Gamma pi + \xi p(\ell pi, t)]\} + \\ &\quad Lpi\ w + Lpi\,Lpi\ \Gamma pi' + Lpi\ \xi p'(\ell pi, t) \\ &\quad - Lpi\,\omega yb\,(\ell pi - Lpi)\end{aligned}$$

Since $\Gamma pi\ \Gamma pi <= 0.03046$ in general,

$$Lpi\,\Gamma pi\,Lpi\ \Gamma pi'\ \Gamma pi \approx 0.0$$
$$Lpi\,\Gamma pi\ \omega yb\,Lpi\,\Gamma pi \approx 0.0$$

Then,

(∂ µpi / ∂ Γpi’) dot µpi = (Lpi Γpi) u + (Lpi Γpi) ωyb ξp(ℓpi , t)
+ Lpi w + Lpi Lpi Γpi’ + Lpi
ξp’(ℓpi , t) - Lpi ωyb ℓpi + Lpi ωyb
Lpi

Now, taking the time derivative,

d [(∂ µpi / ∂ Γpi’) dot µpi]/dt = Lpi’ Γpi u + Lpi Γpi ‘ u + Lpi
Γpi u’ + Lpi Γpi’ ωyb ξp(ℓpi , t)
+ Lpi Γpi ωyb’ξp(ℓpi , t)
+ Lpi
Γpi ωyb ξp’(ℓpi , t) + Lpi w’ +
Lpi Lpi Γpi’’ + Lpi
ξp’’(ℓpi , t) - Lpi ωyb’ ℓpi -
Lpi Qℓpi’ + Lpi ωyb’ Lpi

(3.5-11)

The second term of Eqn. (3.5-9),

[(∂ µpi / ∂ Γpi) dot µpi] = [(Lpi Γpi’ + ωyb Lpi) i b] dot {u +
Lpi Γpi’ Γpi + ωyb
[Lpi Γpi + ξp(ℓpi , t)]} i b + [w + Lpi
Γpi’ + ξp’(ℓpi , t) - ωyb (ℓpi - Lpi)]kb

= (Lpi Γpi’ + ωyb Lpi) {u + Lpi Γpi’
Γpi + ωyb [Lpi Γpi + ξp(ℓpi , t)]}

= Lpi Γpi’ u + (Lpi Γpi’)^(2) Γpi +
(Lpi)^(2) Γpi’ ωyb Γpi + Lpi Γpi’ ωyb
ξp(ℓpi , t) + ωyb Lpi u + ωyb
(Lpi)^(2)Γpi’ Γpi + (ωyb Lpi)^(2)Γpi
+ ωyb Lpi ωyb ξp(ℓpi , t)

(3.5-12)

Now, substituting Eqn. (3.5-11) and Eqn. (3.5-12) into Eqn. (3.5-9),

d [(∂ µpi / ∂ Γpi’) dot µpi] /dt - [(∂ µpi / ∂ Γpi) dot µpi] =0

[Lpi’ Γpi u + Lpi Γpi ‘ u + Lpi Γpi u’ + Lpi Γpi’ ωyb ξp(ℓpi , t) + Lpi Γpi ωyb’ξp(ℓpi , t) + Lpi Γpi ωyb ξp’(ℓpi , t) + Lpi w’ + Lpi Lpi Γpi’’ + Lpi ξp’’(ℓpi , t) - Lpi ωyb’ ℓpi - Lpi ωyb ℓpi’ + Lpi ωyb’ Lpi] – [Lpi Γpi’ u + (Lpi Γpi’)^(2) Γpi + (Lpi)^(2) Γpi’ ωyb Γpi + Lpi Γpi’ ωyb ξp(ℓpi , t) + ωyb Lpi u + ωyb (Lpi)^(2)Γpi’ Γpi + (ωyb Lpi)^(2)Γpi + ωyb Lpi ωyb ξp(ℓpi , t)] = 0

Canceling the terms,

[~~Lpi Γpi’ u~~ + Lpi Γpi u’ + ~~Lpi Γpi’ωyb ξp(lpi , t)~~ + Lpi Γpi ωyb’ξp(lpi , t) + Lpi Γpi ωyb ξp’(lpi , t) + Lpi w’ + Lpi Lpi Γpi’’ + Lpi ξp’’(lpi , t) - Lpi ωyb’ lpi + Lpi ωyb ’ Lpi] – [~~ Lpi Γpi’ u~~ + (Lpi Γpi’)^(2) Γpi + ~~(Lpi)^(2) Γpi’ ωyb~~Γpi + Lpi Γpi’ ωyb ξp(lpi , t) + ωybLpi u + ωyb (Lpi)^(2)Γpi’ Γpi + (ωybLpi)^(2)Γpi + ωybLpi ωyb

Lpi Γpi u’ + Lpi Γpi ωyb’ξp(ℓpi , t) + Lpi Γpi ωyb ξp’(ℓpi , t) + Lpi w’ + Lpi Lpi Γpi’’ + Lpi ξp’’(ℓpi , t) - Lpi ωyb’ ℓpi + Lpi ωyb’ Lpi] – [(Lpi Γpi’)^(2) Γpi + (Lpi)^(2) Γpi’ ωyb Γpi + ωyb Lpi u + ωyb (Lpi)^(2)Γpi’ Γpi + (ωyb Lpi)^(2)Γpi + ωyb Lpi ωyb ξp(ℓpi , t) = 0

The product of Γpi and Γpi’ is small because Γpi is usually less than 10 deg (0.1745 rad) and Γpi’ is less than 10.0 deg/sec (0.1745 rad/sec). The boundary of the product is about 0.03. Thus, we may neglect the terms that have “Γpi Γpi’.” Then,

Lpi Γpi u’ + Lpi Γpi ωyb’ξp(ℓpi , t) + Lpi Γpi ωyb ξp’(ℓpi , t) + Lpi w’ + Lpi Lpi Γpi’’ + Lpi ξp’’(ℓpi , t) - Lpi ωyb’ ℓpi + Lpi ωyb’ Lpi] – [ωyb Lpi u + (ωyb Lpi)^(2)Γpi + ωyb Lpi ωyb ξp(ℓpi , t) = 0

Rearranging the equation above,

Γpi’’ + ωpi ^(2) Γpi = - (1/ Lpi) [w’ + ξp’’(ℓpi , t) - θ’’ (ℓpi - Lpi)– θ’u]

(3.5-13)

where

$$
\begin{aligned}
\omega pi\ \hat{}(2) &= (1/\,Lpi\,)[\,u' + \theta''\xi p(\ell pi\,,t) + \theta'\xi p'(\ell pi\,,t) - \\
&\quad (\theta')\hat{}(2)\,Lpi\,] \\
\theta' &= \omega yb \\
\theta'' &= \omega yb' \\
\xi p(\ell pi\,,t)] &= \Sigma\, qpj(t)\,\varphi pj(\ell pi) \\
\xi p(\ell pi\,,t)] &= \Sigma\, qpj'(t)\,\varphi pj(\ell pi) \\
\xi p''(\ell pi\,,t)] &= \Sigma\, qpj''(t)\,\varphi pj(\ell pi)
\end{aligned}
$$

We can obtain the equation of motion of the ith pendulum in the yaw plane in an analogous manner.

From the equation above, we notice that the pendulum frequency is a function of axial acceleration (u'), pendulum length (Lpi), LV angular rate (θ'), LV angular acceleration (θ''), elastic deflection [ξp(ℓpi , t)], and elastic deflection rate [ξp'(ℓpi , t)]. The pendulum driving force is a function of Coriolis force (θ'u), lateral acceleration (w'), tangential acceleration [θ'' (ℓpi - Lpi)], and elastic deflection acceleration [ξp''(ℓpi , t)].

The next step is to find the acceleration of the pendulum mass. Fig. 3.5-1 is modified in Fig. 3.5-2 to define the position and angular rate vectors we need for derivation of the acceleration equation.

From Fig. 3.5-2,

$$
\begin{aligned}
\underline{r}p &= \underline{R} + \underline{\rho}H + \underline{\rho}P \qquad (3.5\text{-}14)\\
\underline{r}p' &= \underline{R}' + (\underline{\rho}H)' + (\underline{\rho}P)'
\end{aligned}
$$

****

$$
\begin{aligned}
(\underline{\rho}H)' &= (\rho HX\,\underline{i}b + \rho HY\underline{j}b + \rho HZ\underline{k}b\,)' \\
&= (\rho HX'\,\underline{i}b + \rho HY'\underline{j}b + \rho HZ'\underline{k}b\,) + (\rho HX\,\underline{i}b' + \\
&\quad \rho HY\underline{j}b' + \rho HZ\underline{k}b'\,) \\
&= (\underline{\rho}H'\,) + (\rho HX\ \omega x\,\underline{i}b + \rho HY\,\omega x\,\underline{j}b + \rho HZ\,\omega x\,\underline{k}b\,) \\
&= (\underline{\rho}H\,') + \underline{\omega}\ x\ \underline{\rho}H
\end{aligned}
$$

****

$$= \underline{R}' + [(\underline{\rho}H') + \underline{\omega} \times \underline{\rho}H] + [(\underline{\rho}P') + \underline{\omega} \times \underline{\rho}p] \qquad (3.5\text{-}15)$$

We define two new notations, $(\underline{\rho}H)'$ and $(\underline{\rho}H')$.

$$\begin{aligned}(\underline{\rho}H)' &= d(\underline{\rho}H)/dt \\ (\underline{\rho}H') &= d(\rho HX)/dt\,\underline{i}b + d(\rho HX)/dt\,\underline{j}b + d(\rho HX)/dt\,\underline{k}b \\ &= \rho HX'\,\underline{i}b + \rho HY'\,\underline{j}b + \rho HZ'\,\underline{k}b\end{aligned} \qquad (3.5\text{-}16)$$

It should be noted that the angular rate, "$\underline{\omega}$," in $[(\underline{\rho}P') + \underline{\omega} \times \underline{\rho}p]$ is not the pendulum frame angular rate. In LV, the LV body frame and the sloshing pendulum frame are different because of the LV's structural bending; therefore, these two frames have different angular frame rates. The "$\underline{\omega}$" is an angular rate vector of pendulum frame with respect to the LV body frame. We differentiate Eqn. (3.5-15) one more time to get acceleration,

$$\begin{aligned}(\underline{\rho}H)'' &= \underline{R}'' + (\underline{\rho}H')' + \underline{\omega}' \times \underline{\rho}H + \underline{\omega} \times (\underline{\rho}H)' + (\underline{\rho}P')' + \underline{\omega}' \times \underline{\rho}p + \underline{\omega} \times (\underline{\rho}p)' \\ &= \underline{R}'' + [(\underline{\rho}H'') + \underline{\omega} \times (\underline{\rho}H')] + \underline{\omega}' \times \underline{\rho}H + \underline{\omega} \times [(\underline{\rho}H') + \underline{\omega} \times \underline{\rho}H] + [(\underline{\rho}P'') + \underline{\omega} \times (\underline{\rho}P')] + \underline{\omega}' \times \underline{\rho}p + \underline{\omega} \times [(\underline{\rho}P') + \underline{\omega} \times \underline{\rho}P] \\ &= \underline{R}'' + (\underline{\rho}H'') + 2\,\underline{\omega} \times (\underline{\rho}H') + \underline{\omega}' \times \underline{\rho}H + \underline{\omega} \times (\underline{\omega} \times \underline{\rho}H) + (\underline{\rho}P'') + 2\,\underline{\omega} \times (\underline{\rho}P') + \underline{\omega}' \times \underline{\rho}p + \underline{\omega} \times (\underline{\omega} \times \underline{\rho}p)\end{aligned} \qquad (3.5\text{-}17)$$

Rewriting each term,

$$\underline{R}'' = ax\,\underline{i}b + ay\,\underline{j}b + az\,\underline{k}b \qquad (3.5\text{-}18)$$

$$(\underline{\rho}H'') = \rho HX''\,\underline{i}b + \rho HY''\,\underline{j}b + \rho HZ''\,\underline{k}b \qquad (3.5\text{-}19)$$

$$2\,\underline{\omega} \times (\underline{\rho}H') = 2(\omega Y\,\rho HZ' - \omega Z\,\rho HY')\,\underline{i}b + 2(\omega Z\,\rho HX' - \omega X\,\rho HZ')\,\underline{j}b + 2(\omega X\,\rho HY' - \omega Y\,\rho HX')\,\underline{k}b \qquad (3.5\text{-}20)$$

$$\begin{aligned}\underline{\omega}' \times \underline{\rho}H &= (\omega X'\,\underline{i}b + \omega Y'\,\underline{j}b + \omega Z'\,\underline{k}b) \times (\rho HX\,\underline{i}b + \rho HY\,\underline{j}b + \rho HZ\,\underline{k}b) \\ &= (\omega Y'\,\rho HZ - \omega Z'\,\rho HY)\,\underline{i}b + (\omega Z'\,\rho HX - \omega X'\,\rho HZ)\,\underline{j}b + (\omega X'\,\rho HY - \omega Y'\,\rho HX)\,\underline{k}b\end{aligned} \qquad (3.5\text{-}21)$$

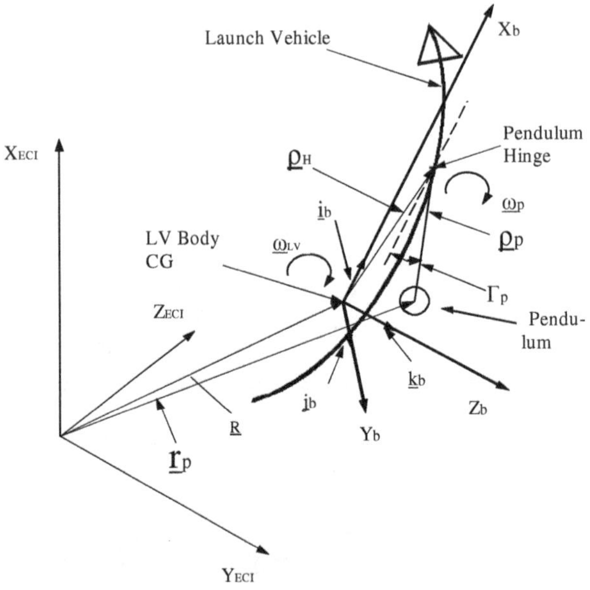

Fig. 3.5-2. Pendulum Position Vector in ECI

$$\underline{\omega} \times (\underline{\omega} \times \underline{\rho}H) = [\omega Y(\omega X\,\rho HY - \omega Y\,\rho HX) - \omega Z(\omega Z\,\rho HX - \omega X\,\rho HZ)]\,\underline{i}b + [\omega Z(\omega Y\,\rho HZ - \omega Z\,\rho HY) - \omega X(\omega X\,\rho HY - \omega Y\,\rho HX)]\,\underline{j}b + [\omega X(\omega Z\,\rho HX - \omega X\,\rho HZ) - \omega Y(\omega Y\,\rho HZ - \omega Z\,\rho HY)]\,\underline{k}b \quad (3.5\text{-}22)$$

$$(\underline{\rho}P'') = (\rho PX''\,\underline{i}b + \rho PY''\,\underline{j}b + \rho PZ''\,\underline{k}b) \quad (3.5\text{-}23)$$

$$2\,\underline{\omega} \times (\underline{\rho}P') = 2(\omega Y\,\rho PZ' - \omega Z\,\rho PY')\,\underline{i}b + 2(\omega Z\,\rho PX' - \omega X\,\rho PZ')\,\underline{j}b + 2(\omega X\,\rho PY' - \omega Y\,\rho PX')\,\underline{k}b \quad (3.5\text{-}24)$$

$$\underline{\omega}' \times \underline{\rho}p = (\omega Y'\,\rho PZ - \omega Z'\,\rho PY)\,\underline{i}b + (\omega Z'\,\rho PX - \omega X'\,\rho PZ)\,\underline{j}b + (\omega X'\,\rho PY - \omega Y'\,\rho PX)\,\underline{k}b \quad (3.5\text{-}25)$$

$$\underline{\omega} \times (\underline{\omega} \times \underline{\rho}p) = [\omega Y(\omega X\,\rho PY - \omega Y\,\rho PX) - \omega Z(\omega Z\,\rho PX - \omega X\,\rho PZ)]\,\underline{i}b + [\omega Z(\omega Y\,\rho PZ - \omega Z\,\rho PY) - \omega X(\omega X\,\rho PY - \omega Y\,\rho PX)]\,\underline{j}b + [\omega X(\omega Z\,\rho PX - \omega X\,\rho PZ) - \omega Y(\omega Y\,\rho PZ - \omega Z\,\rho PY)]\,\underline{k}b \quad (3.5\text{-}26)$$

Without losing generality, we set a coordinate system such that the plane that contains the pendulum motion is in the Xb-Zb plane, and then we write,

$$\underline{R}'' = ax\,\underline{i}b + 0\,\underline{j}b + az\,\underline{k}b$$

$$\underline{\rho}H = \rho HX\,\underline{i}b + 0\,\underline{j}b + \rho HZ\,\underline{k}b$$
$$(\underline{\rho}H') = (\rho HX'\,\underline{i}b + 0\,\underline{j}b + \rho HZ'\,\underline{k}b)$$

$$\underline{\rho}P = \rho px\,\underline{i}b + 0\,\underline{j}b + \rho PZ\,\underline{k}b$$
$$(\underline{\rho}P') = (\rho PX'\,\underline{i}b + 0\,\underline{j}b + \rho PZ'\,\underline{k}b)$$
$$(\underline{\rho}P'') = (\rho PX''\,\underline{i}b + 0\,\underline{j}b + \rho PZ''\,\underline{k}b)$$

$$\underline{\omega} = (0\,\underline{i}b + \omega Y\,\underline{j}b + 0\,\underline{k}b)$$
$$\underline{\omega}' = (0\,\underline{i}b + \omega Y'\,\underline{j}b + 0\,\underline{k}b)$$

Substituting these vectors into Eqn. (3.5-19) through Eqn. (3.5-26),

$$\underline{R}'' = ax\,\underline{i}b + 0\,\underline{j}b + az\,\underline{k}b \quad (3.5\text{-}27)$$
$$(\underline{\rho}H'') = \rho HX''\,\underline{i}b + \rho HZ''\,\underline{k}b \quad (3.5\text{-}28)$$
$$2\,\underline{\omega} \times (\underline{\rho}H') = 2(\omega Y\,\rho HZ')\,\underline{i}b + 2(-\omega Y\,\rho HX')\,\underline{k}b \quad (3.5\text{-}29)$$
$$\underline{\omega}' \times \underline{\rho}H = \omega Y'\,\rho HZ\,\underline{i}b - \omega Y'\,\rho HX\,\underline{k}b \quad (3.5\text{-}30)$$
$$\underline{\omega} \times (\underline{\omega} \times \underline{\rho}H) = -\omega Y^{\wedge}(2)\,\rho HX\,\underline{i}b - \omega Y^{\wedge}(2)\,\rho HZ\,\underline{k}b \quad (3.5\text{-}31)$$
$$(\underline{\rho}P)'' = \rho PX''\,\underline{i}b + \rho PZ''\,\underline{k}b \quad (3.5\text{-}32)$$
$$2\underline{\omega} \times (\underline{\rho}P') = 2(\omega Y\,\rho PZ')\,\underline{i}b + 2(-\omega Y\,\rho PX')\,\underline{k}b \quad (3.5\text{-}33)$$
$$\underline{\omega}' \times \underline{\rho}p = \omega Y'\,\rho PZ\,\underline{i}b - \omega Y'\,\rho PX\,\underline{k}b \quad (3.5\text{-}34)$$
$$\underline{\omega} \times (\underline{\omega} \times \underline{\rho}p) = -\omega Y^{\wedge}(2)\,\rho PX\,\underline{i}b - \omega Y^{\wedge}(2)\,\rho PZ\,\underline{k}b \quad (3.5\text{-}35)$$

Substituting Eqn. (3.5-18) through Eqn. (3.5-26), into Eqn. (3.5-17), and rearranging after the substitution,

$$\begin{aligned}(\underline{\rho}H)'' = {} & [ax + \rho HX'' + 2(\omega Y\,\rho HZ') + \omega Y'\,\rho HZ - \omega Y^{\wedge}(2)\,\rho HX \\ & + \rho PX'' + 2(\omega Y\,\rho PZ') + \omega Y'\,\rho PZ - \omega Y^{\wedge}(2)\,\rho PX]\,\underline{i}b + \\ & [az + \rho HZ'' + 2(-\omega Y\,\rho HX') - \omega Y'\,\rho HX - \omega Y^{\wedge}(2)\rho HZ + \\ & \rho PZ'' + 2(-\omega Y\,\rho PX') - \omega Y'\,\rho PX - \omega Y^{\wedge}(2)\,\rho PZ\,]\,\underline{k}b \\ = {} & axp\,\underline{i}b + azp\,\underline{k}b\end{aligned} \quad (3.5\text{-}36)$$

where

$$axp = [ax + \rho HX'' + 2(\omega Y\ \rho HZ') + \omega Y'\ \rho HZ - \omega Y^{(2)}\ \rho HX + \rho PX'' + 2(\omega Y\ \rho PZ') + \omega Y'\ \rho PZ - \omega Y^{(2)}\ \rho PX] \tag{3.5-37}$$

$$azp = [az + \rho HZ'' + 2(-\omega Y\ \rho HX') - \omega Y'\ \rho HX - \omega Y^{(2)}\ \rho HZ + \rho PZ'' + 2(-\omega Y\ \rho PX') - \omega Y'\ \rho PX - \omega Y^{(2)}\ \rho PZ\ ] \tag{3.5-38}$$

Redefining the variable names and rewriting the vector components,

$$\omega Y = \theta'$$
$$\omega Y^{(2)} = \theta'^{(2)}$$
$$\omega Y' = \theta''$$
$$\rho PX = -|\underline{\rho}P| \cos \Gamma P$$
$$(\rho PX') = |\underline{\rho}P|\,\Gamma P' \sin \Gamma P$$
$$(\rho PX'') = [(|\underline{\rho}P|\,\Gamma P' \sin \Gamma P)'\ ] = (|\underline{\rho}P|'\ \Gamma P' \sin \Gamma P + |\underline{\rho}P|\,\Gamma P'' \sin \Gamma P + |\underline{\rho}P|\,\Gamma P'\Gamma P' \cos \Gamma P\ ) = |\underline{\rho}P|\,(\Gamma P'' \sin \Gamma P + \Gamma P'^{(2)} \cos \Gamma P\ )$$

$$\rho PZ = |\underline{\rho}P| \sin \Gamma P$$
$$(\rho PZ') = |\underline{\rho}P|\ \Gamma P' \cos \Gamma P$$
$$(\rho PZ'') = |\underline{\rho}P|\,(\Gamma P'' \cos \Gamma P - \Gamma P'^{(2)} \sin \Gamma P\ )$$

$$\rho HX = |\underline{\rho}H| \cos \theta B$$
$$\theta B =: \sin^{(-1)} (\xi i / |\underline{\rho}H|\,)$$
$$\xi i = \Sigma qi(t)\varphi i(\ell)$$
$$(\rho HX') = |\underline{\rho}H|\,\theta B' \sin \theta B$$
$$(\rho HX'') = |\underline{\rho}H|\,(\theta B'' \sin \theta B + \theta B'^{(2)} \cos \theta B\ )$$

$$\rho HZ = |\underline{\rho}H| \sin \theta B$$
$$(\rho HZ') = |\underline{\rho}H|\,\theta B' \cos \theta B$$
$$(\rho HZ'') = |\underline{\rho}H|\,(\theta B'' \cos \theta B - \theta B'^{(2)} \sin \theta B\ )$$

Substituting these terms into Eqn. (3.5-37) and Eqn. (3.5-38)

$$axp = ax + [|\underline{\rho}H|\ \ \Gamma H'^{(2)} + 2\theta'\,|\underline{\rho}H|\,\theta B' - \theta'^{(2)}\,|\underline{\rho}H|\,]\cos \theta B + [|\underline{\rho}H|\,\Gamma H'' + \theta''\,|\underline{\rho}H|\,]\sin \theta B +$$

$$[|\underline{\rho}P|\,\Gamma P'^{(2)} + 2\theta'\,|\underline{\rho}P|\,\Gamma P' + \theta'^{(2)}|\underline{\rho}P|]\cos\Gamma P +$$
$$[|\underline{\rho}P|\,\Gamma P'' + \theta''\,|\underline{\rho}P|]\sin\Gamma P \qquad (3.5\text{-}39)$$

$$azp = az +$$
$$[|\underline{\rho}H|\,\theta B'' - \theta''\,|\underline{\rho}H|]\cos\theta B -$$
$$[|\underline{\rho}H|\,\theta B'^{(2)} + 2\,\theta'\,|\underline{\rho}H|\,\theta B' + \theta'^{(2)}\,|\underline{\rho}H|]\sin\theta B$$
$$[|\underline{\rho}P|\,\Gamma P'' + \theta''\,|\underline{\rho}P|]\cos\Gamma P -$$
$$[|\underline{\rho}P|\,\Gamma P'^{(2)} + 2\theta'\,|\underline{\rho}P|\,\Gamma P' + \theta'^{(2)}\,|\underline{\rho}P|]\sin\Gamma P \qquad (3.5\text{-}40)$$

where

axp	= axial acceleration due to fuel sloshing
ax	= LV axial acceleration
azp	= lateral acceleration due to fuel sloshing
az	= LV lateral acceleration
$|\underline{\rho}H|$	= distance from LV CG to pendulum hinge point
$|\underline{\rho}P|$	= pendulum length
θ	= LV angular displacement (clockwise: positive)
θB	= angle between axial axis and position vector to hinge point (Clockwise: positive)
ΓP	= pendulum angle [Eqn. (3.5-13), counter clockwise: positive]

Now, we write the sloshing force and moment equations for one pendulum. Here again, we define the LV body frame in such a way that the pendulum motion is in the plane of Xbody and Zbody. In general, there are four tanks in a satellite launch vehicle, and thus we need to write four sets of these equations to complete.

Forces

$$Fxs = mp\,[axp^{(2)} + azp^{(2)}]\cos\Gamma p$$
$$= mp\,ap\cos\Gamma p \qquad (3.5\text{-}41a)$$
$$Fys = 0 \qquad (3.5\text{-}41b)$$
$$Fzs = mp\,ap\sin\Gamma p \qquad (3.5\text{-}41c)$$

Where

mp = Pendulum mass

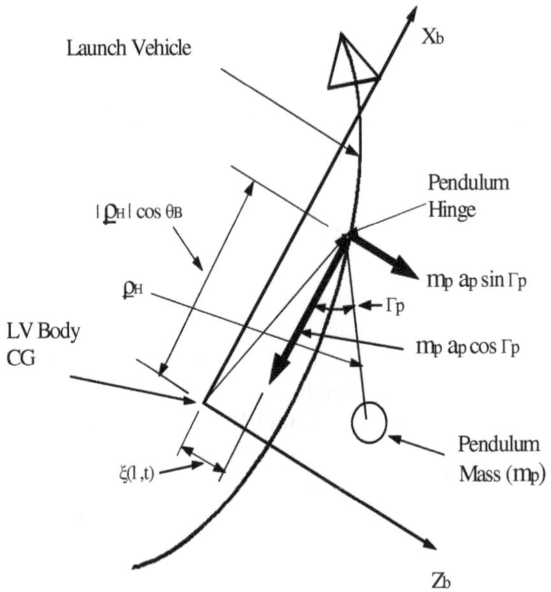

Fig. 3.5-3. Pendulum Force and Moment

axp = Axial acceleration due to fuel sloshing [Eqn. (3.5-39)]

azp = Lateral acceleration due to fuel sloshing [Eqn. (3.5-40)]

Moments

$$\text{Moment} = \underline{F}p \times \underline{\rho}H$$
$$= Mxp\, \underline{i}b + Mxp\, \underline{j}b + Mxp\, \underline{k}b$$
$$= \begin{vmatrix} \underline{i}b & \underline{j}b & \underline{k}b \\ mp\, ap \cos \Gamma p & 0 & mp\, ap \sin \Gamma p \\ -|\underline{\rho}H| \cos \theta B & 0 & |\underline{\rho}H| \sin \theta B \end{vmatrix}$$

$$Mxs = 0.0 \qquad (3.5\text{-}42a)$$

$$MyS = -[(mp\ ap \cos \Gamma p)(\ |\rho H| \sin \theta B\ ) + (\ mp\ ap \sin \Gamma p)(\ |\rho H| \cos \theta B)\ ] \qquad (3.5\text{-}42b)$$

$$Mzs = 0.0 \qquad (3.5\text{-}42c)$$

Where

$$\theta B = \sin^{(-1)} (\xi i \,/\, |\ \rho H\ |\ )$$

$$\xi i = \Sigma qi(t)\varphi i(\ell)$$

For small θB, we can simplify further,

$$MxS = 0.0 \qquad (3.5\text{-}43a)$$

$$MyS = -\ (\ mp\ ap \sin \Gamma p)(\ |\rho H| \qquad (3.5\text{-}43b)$$

$$MzS = 0.0 \qquad (3.5\text{-}43c)$$

In Eqn. (3.5-13),

$$w'\ ,\ \xi p''(\ell pi\ ,\ t),\quad \theta''\ (\ell pi - Lpi\ ) << \theta' u$$

and

$$\theta''\xi p(\ell pi\ ,\ t),\quad \theta'\xi p'(\ell pi\ ,\ t),\quad (\theta')^{(2)}\ Lpi << u''$$

we may simplify the Eqn. (3.5-13),

$$\Gamma pi'' + (u'/\ Lpi\ )\Gamma pi = (1/\ Lpi\ )\ \theta' u \qquad (3.5\text{-}44)$$

Summarizing,

$$\Gamma pi'' + (u'/\ Lpi\ )\Gamma pi = (1/\ Lpi\ )\ \theta' u \qquad (3.5\text{-}45)$$

$$FxS = \sum mpi\ ap \cos \Gamma pi \qquad (3.5\text{-}46a)$$

$$FyS = 0 \qquad (3.5\text{-}46b)$$

$$FzS \approx 0 \qquad (3.5\text{-}46c)$$

$$MxS = 0 \qquad (3.5\text{-}47a)$$

$$MyS = -\sum (\ mpi\ ap \sin \Gamma pi)(\ |\rho Hi| \qquad (3.5\text{-}47b)$$

$$MzS = 0 \qquad (3.5\text{-}47c)$$

In practice, we install ring-shaped baffles in the tanks to increase the damping ratio to minimize sloshing impact on LV flight. Eqn. (3.5-45) explains that if the velocity (u) is very large, it is essential to avoid large pitch rate command (θ') to minimize "sin Γpi " in Eqn.(3.5-47b) unless the sloshing mass, "mpi" in Eqn. (3.5-47b), is small. Minimizing the sloshing angle (Γpi) will reduce the unwanted "MyS" in Eqn. (3.5-47b).

State Space Equation of Sloshing

In Eqn. (3.7-4) in Section 3.7, we assume 1) no bending, 2) $\theta o = 90$ deg., 3) $\theta' \approx 0.0$, 4) Ww =0.0, 5) no engine inertia impact, 6) small angle θ, and 7) small mR. Then the equation can be written as,

$$\Sigma Fz = Td\,\delta - L\alpha\,\alpha o + Uo'mp1\Gamma p1 + Uo'mp2\Gamma p2 \tag{3.5-48}$$

where

$$L\alpha = qD\,A3 \int (\partial CN / \partial\alpha)\, d\ell$$

Ignoring terms that are insignificant in Eqn. (3.7-2) in Section 3.7,

$$\begin{aligned}\Sigma my &= \ell c\,Td\;\delta + L\alpha\,(\ell a - \ell)\,\alpha o - (\,mp1ap)\,|\underline{\rho}H1|\;\Gamma p1 - (\,mp2\\ &\quad ap)\,|\underline{\rho}H2|\;\Gamma p2\\ &= Iyy\;\theta''\end{aligned} \tag{3.5-49}$$

From Eqn. (3.5-49),

$$\begin{aligned}\theta'' &= (\ell c\,Td / Iyy)\;\delta + L\alpha\,(\ell a - \ell)\,\alpha o/Iyy - [(\,mp1ap)\,|\underline{\rho}H1|/Iyy]\\ &\quad \Gamma p1 - [(\,mp2ap)\,|\underline{\rho}H2|\,/Iyy]\;\Gamma p2\\ &= \mu c\;\delta + \mu\alpha\,(\ell a - \ell)\,\alpha o - [(\,mp1ap)\,|\underline{\rho}H1|/Iyy]\;\Gamma p1 - [(\,mp2\\ &\quad ap)\,|\underline{\rho}H2|\,/Iyy]\;\Gamma p2\end{aligned} \tag{3.5-50}$$

where

$$\mu c = (\ell c\,Td / Iyy)$$
$$\mu\alpha = L\alpha / Iyy$$

From Eqn. (3.7-4) in Section 3.7, the force equation along the Z-axis,

$$mT \ (w' - Uo\,\theta') = \Sigma\, Fz \qquad (3.5\text{-}51)$$

Rewriting Eqn. (3.5-51),

$$(w' - Uo\,\theta') = \Sigma\, Fz \ / \ mT \qquad (3.5\text{-}52)$$

From Eqn. (3.5-13), if we assume a rigid body, i.e., we ignore bending, then Eqn. (3.5-13) can be written as,

$$\Gamma pi'' + \omega pi\,^\wedge(2)\, \Gamma pi = -\,(1/\,Lpi\,)\,[w' - \theta''\,(\ell pi - Lpi\,) - \theta' u]$$

$$(3.5\text{-}53)$$

where

$$\omega pi\,^\wedge(2) = (1/\,Lpi\,)[\,u' - (\theta')^\wedge(2)\,Lpi\,]$$

Rearranging Eqn. (3.5-53)

$$\begin{aligned}\Gamma pi'' &= -\,(1/\,Lpi\,)[\,u' - (\theta')^\wedge(2)\,Lpi\,]\,\Gamma pi - (1/\,Lpi\,)\,[w' - \theta''\,(\ell pi - Lpi\,) - \theta' u] \\ &= (1/\,Lpi\,)\,\{[\,-u'\,\Gamma pi + (\theta')^\wedge(2)\,Lpi\,\Gamma pi - [(w' - \theta' u) - \theta''\,(\ell pi - Lpi)]\,\}\end{aligned}$$

$$(3.5\text{-}54)$$

Substituting Eqn. (3.5-50) and Eqn. (3.5-51) into Eqn. (3.5-54),

$$\begin{aligned}\Gamma pi'' = {} & (1/\,Lpi\,)\,\{[\,-u'\,\Gamma pi + (\theta')^\wedge(2)\,Lpi\,\Gamma pi - [(\Sigma\, Fz \ / \ mT\,) - \{\mu c\ \delta + \ \mu\alpha\ (\ell a - \ell)\alpha o - [(\,mp1ap)\,|\underline{\rho}H1|/Iyy]\ \Gamma p1 - [(\,mp2\ ap)\,|\underline{\rho}H2|\,/Iyy]\ \Gamma p2\ \}(\ell pi - Lpi)]\ \} \\ = {} & (1/\,Lpi\,)\,\{\,-u'\,\Gamma pi + (\theta')^\wedge(2)\,Lpi\,\Gamma pi - (\Sigma\, Fz \ / \ mT\,) + \mu c\ \delta(\ell pi - Lpi) + \mu\alpha\ (\ell a - \ell)\ \alpha o\,(\ell pi - Lpi) - [(\,mp1ap)\,|\underline{\rho}H1|/Iyy]\ \Gamma p1\ (\ell pi - Lpi) - [(\,mp2\ ap)\,|\underline{\rho}H2|\,/Iyy]\ \Gamma p2\ (\ell pi - Lpi)\}\end{aligned}$$

$$(3.5\text{-}55)$$

Substituting Eqn. (3.5-48) into Eqn. (3.5-55).

$$= (1/\,Lpi\,)\,\{\,-u'\,\Gamma pi + (\theta')^\wedge(2)\,Lpi\,\Gamma pi - [(Td\,\delta\ - L\alpha\,\alpha o + Uo' mp1\Gamma p1 + Uo' mp2\Gamma p2\,)/\ mT\,] + \mu c\ \delta(\ell pi - Lpi) +$$

$$\mu\alpha \ (\ell a - \ell)\,\alpha o\,(\ell pi - Lpi) - [(mp1ap)\,|\underline{\rho}H1|/Iyy]\ \Gamma p1$$
$$(\ell pi - Lpi) - [(\ mp2\ ap)\,|\underline{\rho}H2|\,/Iyy]\ \Gamma p2\,(\ell pi - Lpi)\}$$

$$= (1/\,Lpi\,)\,\{\ -u'\ \Gamma pi\ + (\theta')^{\wedge}(2)\,Lpi\,\Gamma pi - Td\ \delta/mT\ + L\alpha$$
$$\alpha o/mT - Uo'mp1\Gamma p1/mT\ -\ Uo'mp2\Gamma p2\,/mT\ +\ \mu c$$
$$\delta(\ell pi - Lpi) + \mu\alpha\ (\ell a - \ell)\,\alpha o\,(\ell pi - Lpi) - [(mp1ap)\,|$$
$$\underline{\rho}H1|/Iyy]\ \Gamma p1\,(\ell pi - Lpi) - [(\ mp2\ ap)\,|\underline{\rho}H2|\,/Iyy]\ \Gamma p2$$
$$(\ell pi - Lpi)\}$$

$$= (1/\,Lpi\,)\,\{-\,Td\ \delta/mT\ + L\alpha\ \alpha o/mT$$
$$-Uo'mp1\Gamma p1\,/mT\ -\ Uo'mp2\Gamma p2\,/mT$$
$$+ (\ell c\ Td\,/\,Iyy)\ \delta(\ell pi - Lpi)$$
$$+ L\alpha\,/Iyy\ (\ell a - \ell)\,\alpha o\,(\ell pi - Lpi)$$
$$-u'\ \Gamma pi\ -[(\ mp1ap)\,|\underline{\rho}H1|/Iyy]\ \Gamma p1\,(\ell pi - Lpi)$$
$$-[(\ mp2\ ap)\,|\underline{\rho}H2|\,/Iyy]\ \Gamma p2\,(\ell pi - Lpi)\}$$

where

$$L\alpha = (1/2)\ \rho\ U^{\wedge}(2)\ A3 \int\ [\partial CN(\ell)/\partial\alpha]\ d\ell$$

Here, we can assume that the $\rho \approx 0.0$ in high altitude then,

$$= (1/\,Lpi\,)\,\{-\,Td\ \delta/mT\ -Uo'mp1\Gamma p1\,/mT\ -\ Uo'mp2\Gamma p2$$
$$/mT + (\ell c\ Td\,/\,Iyy)\ \delta(\ell pi - Lpi)\ -u'\ \Gamma pi\ -[(\ mp1ap)\,|$$
$$\underline{\rho}H1|/Iyy]\ \Gamma p1\,(\ell pi - Lpi)\ -[(\ mp2\ ap)\,|\underline{\rho}H2|\,/Iyy]\ \Gamma p2$$
$$(\ell pi - Lpi)\}$$

$$\Gamma pi'' = -\,(1/\,Lpi\,)\,u'\ \Gamma pi\,\{-\,(1/\,Lpi\,)\,Td\,/mT\ + (1/\,Lpi\,)\,(\ell c\ Td$$
$$/\,Iyy)\ (\ell pi - Lpi)\}\ \delta + \{-\,(1/\,Lpi\,)\,Uo'mp1/mT\ - (1/$$
$$Lpi\,)\,[(\ mp1ap)\,|\underline{\rho}H1|/Iyy]\ (\ell pi - Lpi)\,\}\Gamma p1 + \{-\,(1/\,Lpi\,)$$
$$Uo'mp2/mT\ - (1/\,Lpi\,)\,[(\ mp2\ ap)\,|\underline{\rho}H2|\,/Iyy]\ (\ell pi -$$
$$Lpi)\}\ \Gamma p2$$

Since we have two sloshings, $i = 1, 2$.

When $i = 1$,

$$\Gamma p1'' = H1\ \Gamma p1\ +\ H2\ \Gamma p2\ +\ H3\ \delta \qquad\qquad (3.5\text{-}56)$$

where

$$H1 = (-1/Lp1)\{ u' + Uo'mp1/mT + [(mp1ap) |_\rho H1|/Iyy] (\ell p1 - Lp1) \}$$
$$H2 = (-1/Lp1)\{Uo'mp2/mT + [(mp2\ ap) |_\rho H2| /Iyy] (\ell p1 - Lp1)\}$$
$$H3 = (-1/Lp1)\{Td/mT - (\ell c\ Td/Iyy)(\ell p1 - Lp1)\}$$

When i = 2,

$$\Gamma p2'' = J1\ \Gamma p1 + J2\ \Gamma p2 + J3\ \delta \quad (3.5\text{-}57)$$

where

$$J1 = -(1/Lp2)\{Uo'mp1/mT + [(mp1ap) |_\rho H1|/Iyy] (\ell p2 - Lp2) \}$$
$$J2 = -(1/Lp2)\{ u' + Uo'mp2/mT + [(mp2\ ap) |_\rho H2| /Iyy] (\ell p2 - Lp2)\}$$
$$J3 = -(1/Lp2)\{Td/mT - (\ell c\ Td/Iyy)(\ell p2 - Lp2)\}$$

3.6 Engine Inertia

The engine inertia "Tail Wags Dog" (TWD) is a vibrational phenomenon that we observe in launch vehicle in flight as the engine body swivels around the hinge point at the tail section of the LV. This term must have come from a physical phenomenon occurs when a dog swivels his tail, a motion that arises in the dog's body as a reaction to the swiveling of the tail. This phenomenon cannot be ignored in LV attitude control systems because it could cause instability. Theoretically, it creates a pair of complex zeroes in a root-locus plot, and because of this pair, the operating Root Locus trace may enter into the right half plane. Fig. 3.6-1 shows the basic mechanism that results in swiveling.

Notice that while the engine rotates in the counterclockwise, the rocket rotates in the clockwise in reaction to the engine's swiveling. In the following, we derive the engine inertia force and moment equations by taking the Lagrangian approach.

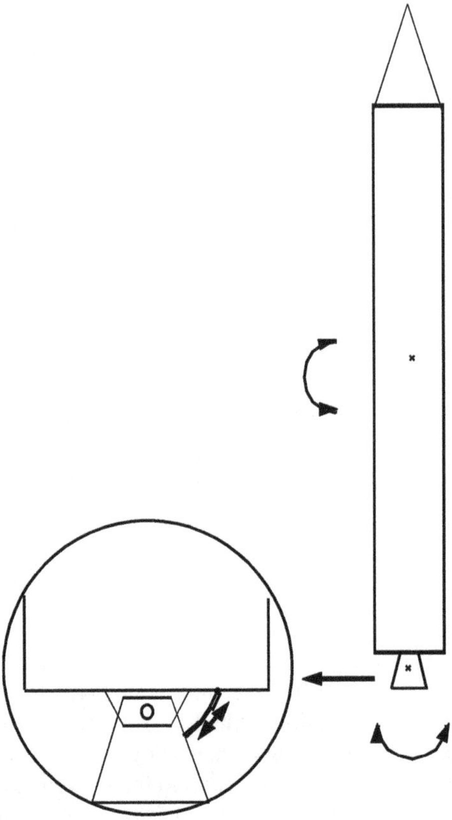

Fig. 3.6-1. Basic Swiveling Mechanism

First, we need to compute the engine CG location in ECI.

$$
\begin{aligned}
Xe &= XG - \ell RX \\
&= XG - \ell R \cos \zeta \\
&= XG - \ell R \cos (\delta po + \delta p + \delta RB) \\
&= XG - \ell R \cos \{\delta po + \delta p + [(\delta \xi p(\ell T, t)/ \delta \ell) - (\theta o + \theta)]\}
\end{aligned}
\qquad (3.6\text{-}1)
$$

where

Xe = engine CG location in X axis
XG = engine gimbal location in X axis

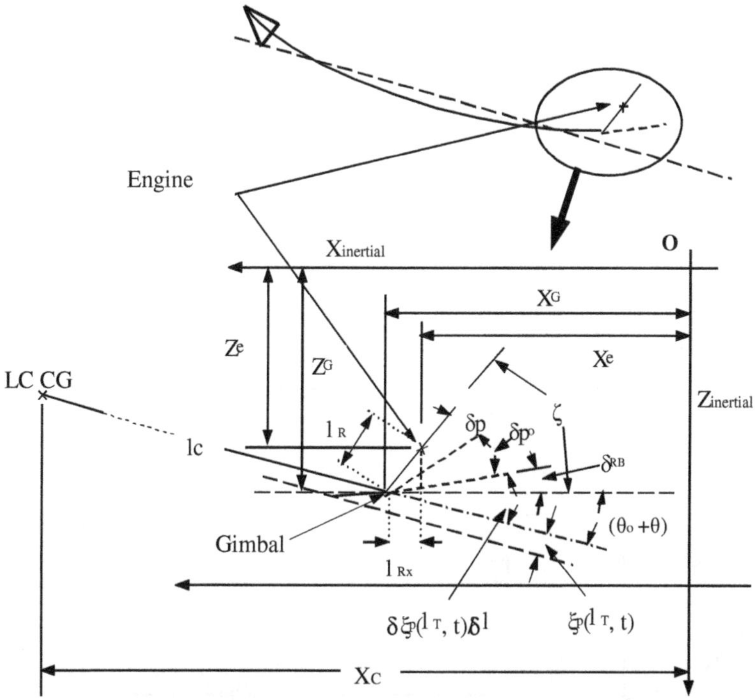

Fig. 3.6-2. Inertial Coordinate of the Engine CG

In similar manner,

$$Ze = ZG - \ell R \sin\{\delta po + \delta p + [(\delta\xi p(\ell T, t)/\delta\ell) - (\theta o + \theta)] \quad (3.6\text{-}2)$$

where

Ze = engine CG location in Z axis

ZG = engine gimbal location in Z axis

Now, we compute the inertial coordinate of the LV CG. From Fig. 3.6-3,

$$Xc = XG + \xi p (\ell T ,t) \sin (\theta o + \theta) + \ell C \cos (\theta o + \theta) \tag{3.6-3}$$

$$Zc = ZG + \xi p (\ell T ,t) \cos (\theta o + \theta) - \ell C \sin (\theta o + \theta) \tag{3.6-4}$$

From Eqn. (3.6-1) through Eqn. (3.6-4), we derive equations that relate (Xe, Ze) to (Xc, Zc).

$$Xe = Xc - \xi p (\ell T ,t) \sin (\theta o + \theta) - \ell C \cos (\theta o + \theta) - \ell R \cos [\delta po + \delta p + \delta \xi p(\ell T ,t)/ \delta \ell - (\theta o + \theta)] \tag{3.6-5}$$

$$Ze = Zc - \xi p (\ell T ,t) \cos (\theta o + \theta) + \ell C \sin (\theta o + \theta) - \ell R \sin \{\delta po + \delta p + [(\delta \xi p(\ell T ,t)/ \delta \ell) - (\theta o + \theta)]\} \tag{3.6-6}$$

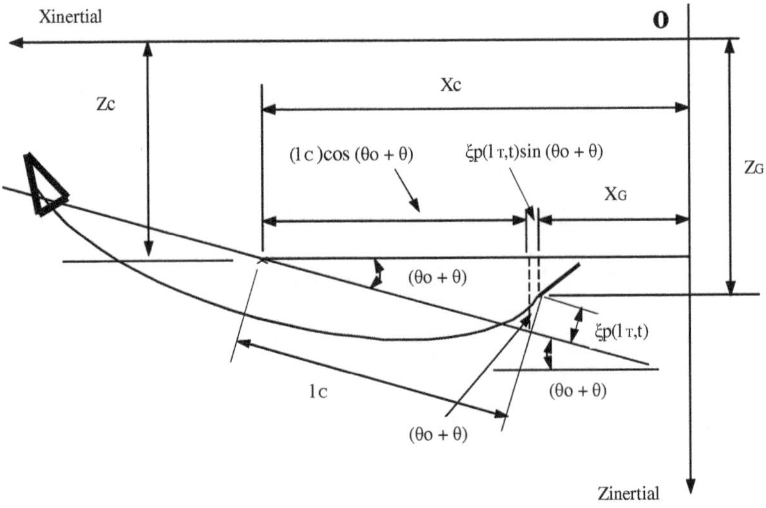

Fig. 3.6-3. LV CG Inertial Coordinate

Differentiating Eqn. (3.6-5),

$$Xe' = Xc' - \xi p'(\ell T,t)\sin(\theta o + \theta) - \xi p(\ell T,t)\,\theta'\cos(\theta o + \theta) + \ell C\,\theta'\sin(\theta o + \theta) + \ell R\,\{\delta p' + d[\delta\xi p(\ell T,t)/\delta\ell]/dt - \theta'\}\sin[\delta po + \delta p + \delta\xi p(\ell T,t)/\delta\ell - (\theta o + \theta)]$$

We set $\theta o = 0$ and $\delta po = 0$ for our analysis. In fact, we can do this because we design our control system at the operating point as a reference, and the reference is zero. Then,

$$= Xc' - \xi p'(\ell T,t)\sin(\theta) - \xi p(\ell T,t)\,\theta'\cos(\theta) + \ell C\,\theta'\sin(\theta) + \ell R\,\{\delta p' + d[\delta\xi p(\ell T,t)/\delta\ell]/dt - \theta'\}\sin[\delta p + \delta\xi p(\ell T,t)/\delta\ell - (\theta)]$$

Now, we assume that the perturbation angle, θ, and the rate of change of θ, θ', are small,

$$= Xc' - \xi p'(\ell T,t)(\theta) - \xi p(\ell T,t)\,\theta' + \ell C\,\theta'(\theta) + \ell R\,\{\delta p' + d[\delta\xi p(\ell T,t)/\delta\ell]/dt - \theta'\}\,[\delta p + \delta\xi p(\ell T,t)/\delta\ell - (\theta)]$$

$$= Xc' - \xi p'(\ell T,t)(\theta) - \xi p(\ell T,t)\,\theta' + \ell C\,\theta'(\theta) + \ell R\;A'\,A \qquad (3.6\text{-}7)$$

where

$$A = \delta p + \delta\xi p(\ell T,t)/\delta\ell - (\theta)]$$
$$A' = \delta p' + d[\delta\xi p(\ell T,t)/\delta\ell]/dt - \theta'$$

Canceling the higher order terms,

$$= Xc' - \xi p'(l_T,t)\,\theta - \xi p(l_T,t)\,\theta' + l C\,\theta'\theta + l R\,\{\delta p' + d[\delta\xi p(l_T,t)/\delta l]/dt - \theta'\}\sin[\delta p + \delta\xi p(l_T,t)/\delta l - (\theta)]$$

$$= Xc' - \xi p'(\ell T,t)\,\theta - \xi p(\ell T,t)\,\theta' + \ell R\,(\delta p' - \theta')(\delta p - \theta)$$

$$Xe'' = Xc'' - \xi p''(\ell T,t)\,\theta - \xi p'(\ell T,t)\,\theta' - \xi p'(\ell T,t)\,\theta' - \xi p(\ell T,t)\,\theta''$$, for small angle of $(\delta po + \delta p - \theta)]$

$$\approx Xc'' \tag{3.6-8}$$

In a similar manner, we start with differentiation of Eqn. (3.6-6),

$$Ze' = Zc' - \xi p'(\ell T,t)\cos(\theta o+\theta) + \xi p(\ell T,t)\,\theta'\sin(\theta o+\theta) + \ell C\,\theta'\cos(\theta o+\theta) - \ell R\,\{\delta p' + d[(\delta\xi p(\ell T,t)/\delta\ell)]/dt - \theta')]\}\cos\{\delta po + \delta p + [(\delta\xi p(\ell T,t)/\delta\ell) - (\theta o+\theta)]\} \tag{3.6-9}$$

For $\theta o = \delta po = 0$, and θ is small

$$= Zc' - \xi p'(\ell T,t) + \xi p(\ell T,t)\,\theta'\theta + \ell C\,\theta' - \ell R\,A'\cos A \tag{3.6-10}$$

We retain cos A here because it has δp in it. We need to take a derivative of Ze' with respect to δp later.

$$= Zc' - \xi p'(\ell T,t) + \xi p(\ell T,t)\,\theta'\theta + \ell C\,\theta' - \ell R\,A' \tag{3.6-11}$$

$$\begin{aligned} Ze'' &= Zc'' - \xi p''(\ell T,t) + \xi p(\ell T,t)\,\theta''\theta + \xi p(\ell T,t)\,\theta'^{(2)} + \ell C\,\theta'' - \ell R\,A'' \\ &= Zc'' - \xi p''(\ell T,t) + \ell C\,\theta'' - \ell R\,A'' \end{aligned} \tag{3.6-12}$$

The kinetic energy equation for the engine motion is,

$$\begin{aligned} KE &= (1/2)\,[mR\,Ve^{(2)} + IR\,\omega e^{(2)}] \\ &= (1/2)\,mR\,[Xe'^{(2)} + Ze'^{(2)}] + (1/2)\,IR\,\{\delta p' + d[(\delta\xi p(\ell T,t)/\delta\ell)]/dt - \theta'\}^{(2)} \end{aligned} \tag{3.6-13}$$

where

IR = engine moment of inertia

The potential energy can be ignored as we set the energy reference such that it can be ignored. Now, we write the Lagrange equations along Xc and Zc axes.

$$d(\partial KE/\partial Xe')/dt - \partial KE/\partial Xe = Fxe \quad (3.6\text{-}14)$$

$$d(\partial KE/\partial Ze')/dt - \partial KE/\partial Ze = Fze \quad (3.6\text{-}15)$$

From Eqn. (3.6-13),

$$\partial KE/\partial Xe' = mR\ Xe'$$

$$d(\partial KE/\partial Xe')/dt = mR\ Xe''$$

$$\partial KE/\partial Xe = 0$$

Thus, Eqn. (3.6-14) can be written as,

$$Fxe = mR\ Xe'' \quad (3.6\text{-}16)$$

In Eqn. (3.6-15),

$$(\partial KE/\partial Ze') = \partial\{(1/2)\, mR\ [Xe'^{(2)} + Ze'^{(2)}] + (1/2)\, IR\ \{\delta p' + d[(\delta\xi p(\ell T\ ,t)/\ \delta\ell)]/dt - \theta'\}^{(2)}\}\ /\partial Ze'$$

$$= mR\ Ze'$$

$$d(\partial KE/\partial Ze')/dt = mR\ Ze''$$

$$KE/\partial Ze = 0$$

Substituting these terms into Eqn. (3.6-15),

$$Fze = mR\ Ze'' \quad (3.6\text{-}17)$$

From Fig. 3.6-4, the forces acting on the vehicle in Xb- and Zb-axes are,

$$Fxe = Fze \sin(\theta o + \theta) + Fxe \cos(\theta o + \theta)$$

$$= Fxe, \text{ for small angle of } ((\theta o + \theta)$$

$$= mR\ Xe'' \quad (3.6\text{-}18)$$

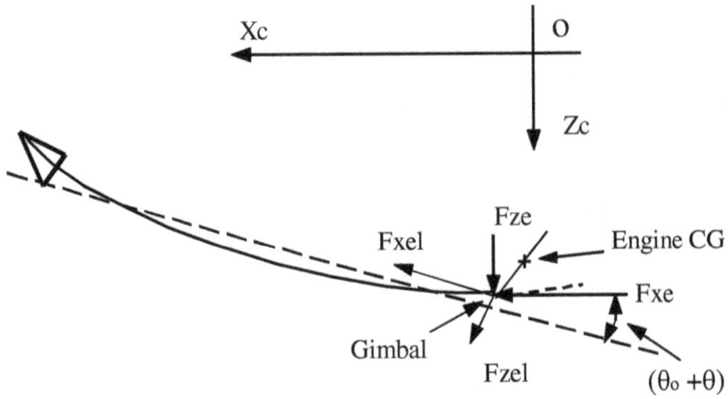

Fig. 3.6-4. Forces Acting on the Vehicle in Xb- and Zb-Axes

$$
\begin{aligned}
Fze &= Fze\cos(\theta o + \theta) + Fxe\sin(\theta o + \theta) \\
&= Fze\,, \quad \text{for small angle of } ((\theta o + \theta) \\
&= mR\ Ze''
\end{aligned}
$$

Substituting Eqn. (3.6-12),

$$
= mR\,\{Zc'' - \xi p''(\ell T, t) + \ell C\,\theta'' - \ell R\,[\delta p'' + d[d[(\delta\xi p(\ell T, t)/\delta\ell)]/dt]/dt - \theta'']\} \qquad (3.6\text{-}19)
$$

From Fig. 3.6-4, the lateral acceleration, w', is,

$$
w' = Xc''\sin(\theta o + \theta) + Zc''\cos(\theta o + \theta)
$$

We set $\theta o = 0.0$ and for small angle, θ,

$$
w' = Xc''\,\theta + Zc''
$$

Solving for Zc'',

$$
\begin{aligned}
Zc'' &= w' - Xc''\,\theta \\
&= w' - u'\,\theta
\end{aligned}
$$

Now, Eqn. (3.6-19) can be written,

$$
Fze = mR\,\{w' - u'\,\theta - \xi p''(\ell T, t) + \ell C\,\theta'' - \ell R\,[\delta p'' + d[d[(\delta\xi p(\ell T, t)/\delta\ell)]/dt]/dt - \theta'']\} \qquad (3.6\text{-}20)
$$

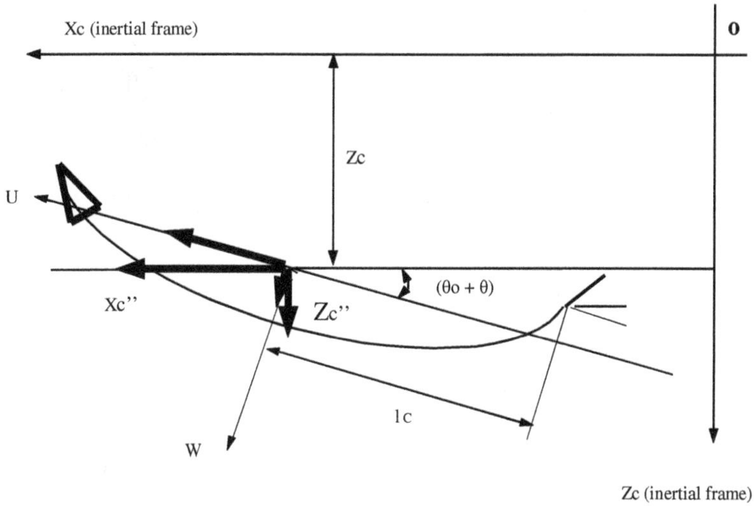

Fig. 3.6-5. Acceleration Vectors at LV CG

****

Expanding a term, "$\delta\xi p(\ell T,t)/\delta\ell$" in Eqn. (3.6-20)

$$\delta\xi p(\ell T,t)/\delta\ell = \Sigma\, qPT(t)[\partial\varphi PT(\ell)/\partial\ell]$$
$$= -\Sigma\, qPT(t)\,\sigma PT(\ell)$$
$$d[(\delta\xi p(\ell T,t)/\delta\ell)]/dt = -\Sigma\, qPT'(t)\,\sigma PT(\ell)$$
$$d[d[(\delta\xi p(\ell T,t)/\delta\ell)]/dt]/dt = -\Sigma\, qPT''(t)\,\sigma PT(\ell)$$

(3.6-21)

$$\xi p(\ell T,t) = \Sigma\, qPT(t)\varphi PT(\ell)$$
$$\xi p''(\ell T,t) = \Sigma\, qPT''(t)\varphi PT(\ell)$$
(3.6-22)

Adding Eqn. (3.6-21) and Eqn. (3.6-22),

$$d[d[(\delta\xi p(\ell T,t)/\delta\ell)]/dt]/dt - \xi p''(\ell T,t)$$
$$= -\Sigma\, qPT''(t)\,\sigma PT(\ell) - \Sigma\, qPT''(t)\varphi PT(\ell)$$
$$= \Sigma\, qPT''(t)\,[-\sigma PT(\ell) - \varphi PT(\ell)]$$
(3.6-23)

****

Substituting Eqn. (3.6-23) into Eqn. (3.6-20),

$$Fze = mR \{w' - u'\theta + \ell C\,\theta'' - \ell R\,(\delta p'' - \theta'') + \Sigma\, qPT''(t)\,[\ell R\,\sigma PT(\ell) + \varphi PT(\ell)]\,\} \quad (3.6\text{-}24)$$

Now, we compute the moment on the LV caused by engine inertia. First, we compute the moment applied to the engine.

The Lagrange equation for moment is,

$$d(\partial KE/\partial\,\delta p')/dt - \partial KE/\partial\,\delta p = Myee \quad (3.6\text{-}25)$$

Myee is an external moment applied to the engine body by the electrical or hydraulic actuator.

For easy reference, we repeat Eqn. (3.6-13)

$$KE = (1/2)\,mR\,[Xe'^{(2)} + Ze'^{(2)}] + (1/2)\,IR\,\{\delta p' + d[(\delta\xi p(\ell T, t)/\delta\ell)]/dt - \theta'\}^{(2)}$$

Taking a derivative with respect to δp',

$$\partial KE/\partial\,\delta p' = mR\,Xe'(\partial Xe'/\partial\,\delta p') + mR\,Ze'(\partial Ze'/\partial\,\delta p') + IR\,A'\,(1) \quad (3.6\text{-}26)$$

where

$$A = \{\delta p + [\delta\xi p(\ell T, t)/\delta\ell] - \theta'\}$$
$$A' = \{\delta p' + d[(\delta\xi p(\ell T, t)/\delta\ell)]/dt - \theta'\}$$

Expanding the first term in Eqn. (3.6-26) using Eqn. (3.6-7)

$$\begin{aligned} mR\,Xe'(\partial Xe'/\partial\,\delta p') &= mR\,\{[Xc' - \xi p'(\ell T, t)\,\theta - \xi p(\ell T, t)\,\theta' + \ell C\,\theta'\theta + \ell R\,A'\,A]\,\{\partial[Xc' - \xi p'(\ell T, t)\,\theta - \xi p(\ell T, t)\,\theta' + \ell C\,\theta'\theta + \ell R\,A'\,A]/\partial p'\,\} \\ &= mR\,[Xc' - \xi p'(\ell T, t)\,\theta - \xi p(\ell T, t)\,\theta' + \ell C\,\theta'\theta + \ell R\,A'\,A]\,\ell R\,A \end{aligned} \quad (3.6\text{-}27)$$

where

$$A = \delta p + \delta\xi p(\ell T,t)/\delta\ell - \theta$$
$$A' = \delta p' + d[\delta\xi p(\ell T,t)/\delta\ell]/dt - \theta'$$

Expanding the second term using Eqn. (3.6-11),

$$mR\ Ze'(\partial Ze'/\partial\delta p') = mR\,[Zc' - \xi p'(\ell T,t) + \xi p(\ell T,t)\theta'\theta + \ell C\,\theta' - \ell R\,A']\{\partial[Zc' - \xi p'(\ell T,t) + \xi p(\ell T,t)\theta'\theta + \ell C\,\theta' - \ell R\,A']/\partial\delta p'\}$$

$$= mR\,[Zc' - \xi p'(\ell T,t) + \xi p(\ell T,t)\theta'\theta + \ell C\,\theta' - \ell R\,A'](-\ell R)$$

(3.6-28)

Substituting Eqn. (3.6-27) and Eqn. (3.6-28) into Eqn. (3.6-26),

$$\partial KE/\partial\delta p' = mR\ \{[Xc' - \xi p'(\ell T,t)\,\theta - \xi p(\ell T,t)\,\theta' + \ell C\,\theta'\theta + \ell R\,A'A]\,\ell R\,A + [Zc' - \xi p'(\ell T,t) + \xi p(\ell T,t)\,\theta'\theta + \ell C\,\theta' - \ell R\,A'](-\ell R) + IR\,A'$$
$$= mR\ \ell R\,[Xc'A - \xi p'(\ell T,t)\,\theta A - \xi p(\ell T,t)\,\theta' A + \ell C\,\theta'\theta A + \ell R\,A'A^{(2)} - Zc' + \xi p'(\ell T,t) - \xi p(\ell T,t)\,\theta'\theta - \ell C\,\theta' + \ell R\,A'] + IR\,A'$$

(3.6-29)

Taking the time derivative of Eqn. (3.6-29),

$$d(\partial KE/\partial\delta p')/dt = (\partial KE/\partial\delta p')'$$
$$= mR\ \ell R\,[Xc''A + Xc'A' - \xi p''(\ell T,t)\,\theta A - \xi p'(\ell T,t)\,\theta' A - \xi p'(\ell T,t)\,\theta A' - \xi p'(\ell T,t)\,\theta' A - \xi p(\ell T,t)\,\theta'' A - \xi p(\ell T,t)\,\theta' A' + \ell C\,\theta''\theta A + \ell C\,\theta'\theta' A + \ell C\,\theta'\theta A' + \ell R\,A''A^{(2)} + 2\ell R\,A'AA' - Zc'' + \xi p''(\ell T,t) - \xi p'(\ell T,t)\,\theta'\theta - \xi p(\ell T,t)\,\theta''\theta - \xi p(\ell T,t)\,\theta'\theta' - \ell C\,\theta'' + \ell R\,A''] + IR\,A''$$

(3.6-30)

Expanding the second term of Eqn. (3.6-25),

$$\partial KE/\partial \delta p = \partial \{(1/2)\, mR\ [Xe'^{(2)} + Ze'^{(2)}] + (1/2)\, IR\ \{\delta p' + d[(\delta\xi p(\ell T,t)/\delta\ell)]/dt - \theta'\}^{(2)}\}/\partial \delta p$$
$$= mR\ Xe'(\partial Xe'/\partial \delta p) + mR\ Ze'(\partial Ze'/\partial \delta p) + 0$$

(3.6-31)

Expanding the first term using Eqn. (3.6-7),

$$mR\ Xe'(\partial Xe'/\partial \delta p) = mR\ [Xc' - \xi p'(\ell T,t)\ \theta - \xi p(\ell T,t)\ \theta' + \ell C\ \theta'\theta + \ell R\ A'\ A]\{\partial [Xc' - \xi p'(\ell T,t)\ \theta - \xi p(\ell T,t)\ \theta' + \ell C\ \theta'\theta + \ell R\ A'\ A]/\partial \delta p\}$$
$$= mR\ [Xc' - \xi p'(\ell T,t)\ \theta - \xi p(\ell T,t)\ \theta' + \ell C\ \theta'\theta + \ell R\ A'\ A]\ \ell R\ A'$$

Expanding the second term using Eqn. (3.6-10),

$$mR\ Ze'(\partial Ze'/\partial \delta p) = mR\ [Zc' - \xi p'(\ell T,t) + \xi p(\ell T,t)\ \theta'\theta + \ell C\ \theta' - \ell R\ A' \cos A]\{\partial [Zc' - \xi p'(\ell T,t) + \xi p(\ell T,t)\ \theta'\theta + \ell C\ \theta' - \ell RA' \cos A]/\partial \delta p\}$$
$$= mR\ [Zc' - \xi p'(\ell T,t) + \xi p(\ell T,t)\ \theta'\theta + \ell C\ \theta' - \ell R\ A' \cos A]\ (\ell R\ A' \sin A)$$
$$= mR\ [Zc' - \xi p'(\ell T,t) + \xi p(\ell T,t)\ \theta'\theta + \ell C\ \theta' - \ell R\ A']\ (\ell R\ A'A)\text{, small A}$$
$$= mR\ [Zc'\ (\ell R\ A'A) - \xi p'(\ell T,t)\ (\ell R\ A'A) + \xi p(\ell T,t)\ \theta'\theta(\ell R\ A'A) + \ell C\ \theta'(\ell R\ A'A) - \ell R\ A'(\ell R\ A'A)]$$

Substituting "mR Xe'(∂ Xe'/∂ δp)" and "mR Ze'(∂Ze'/∂ δp)" into Eqn. (3.6-31),

$$\partial KE/\partial \delta p = mR\ [Xc'\ \ell R\ A' - \xi p'(\ell T,t)\ \theta\ \ell R\ A' - \xi p(\ell T,t)\ \theta'\ell R\ A' + \ell C\ \theta'\theta\ \ell R\ A' + (\ell R\ A')^{(2)} A + Zc'\ (\ell R\ A'A) - \xi p'(\ell T,t)\ (\ell R\ A'A) + \xi p(\ell T,t)\ \theta'\theta(\ell RA'A) + \ell C\ \theta'(\ell R\ A'A) - \ell R\ A'(\ell R\ A'A)]$$

(3.6-32)

Now, subtracting Eqn. (3.6-32) from Eqn. (3.6-30), Eqn. (3.6-25) becomes,

$$Myee = mR\ [\ell R\ Xc''\ A + \ell R\ Xc'\ A' - \ell R\ \xi p''(\ell T,t)\ \theta\ A - \ell R\ \xi p'(\ell T,t)\ \theta' A - \ell R\ \xi p'(\ell T,t)\ \theta\ A' - \ell R\ \xi p'(\ell T,t)\ \theta'\ A$$

- ℓR ξp (ℓT ,t) θ'' A - ℓR ξp (ℓT ,t)θ' A' + ℓR ℓC θ''(θ) A + ℓR ℓC θ'θ' A + ℓR ℓC θ'θ A' + ℓR ℓR A''A^(2) + 2ℓR ℓR A'AA' - ℓR Zc'' + ℓR ξp''(ℓT ,t) - ℓR ξp'(ℓT ,t) θ'θ - ℓR ξp (ℓT ,t) θ''θ - ℓR ξp (ℓT ,t) θ'θ'- ℓR ℓC θ''+ ℓR ℓR A'' - Xc' ℓR A' + ξp' (ℓT ,t) θ ℓR A' + ξp (ℓT ,t) θ'ℓR A' - ℓC θ'θ ℓR A' - (ℓR A')^(2) A – Zc' (ℓR A'A) + ξp' (ℓT ,t) (ℓR A'A) - ξp (ℓT ,t) θ'θ(ℓR A'A)- ℓC θ'(ℓR A'A) + ℓR A'(ℓR A'A)] + IR A''

= mR [lR Xc'' A + lR Xc' A'- lR ξp'' (lT ,t) θ A - lR ξp' (lT ,t) θ'A- lR ξp' (lT ,t) θ A' - lR ξp' (lT ,t) θ' A - lR ξp (lT ,t) θ'' A - lR ξp (lT ,t)θ' A' + lR lc θ'' θ A + lR lc θ' θ' A + lR lc θ' θ A' + lR lR A'' A^(2) + 2lR lR A'AA' -lR Zc'' + lR ξp''(lT ,t) - lR ξp'(lT ,t) θ'θ - lR ξp (lT ,t) θ''θ - lR ξp (lT ,t) θ'θ'- lR lc θ''+ lR lR A'' - Xc' lR A' + ξp' (lT ,t) θ lR A' + ξp (lT ,t) θ' lR A' - lc θ'θ lR A' - (lR A')^(2) A – Zc' (lR A'A) + ξp' (lT ,t) (lR A'A) - ξp (lT ,t) θ'θ(lR A'A)- lc θ'(lR A'A) + lR A'(lR A'A)] + IR A''

= mR ℓR [Xc'' A- Zc'' + ξp''(ℓT ,t) - ℓC θ''] + IR A''

From Fig. 3.6-6, it is seen that the total moment applied to the LV by the engine inertia is,

Mye = Myee - ℓc Fze

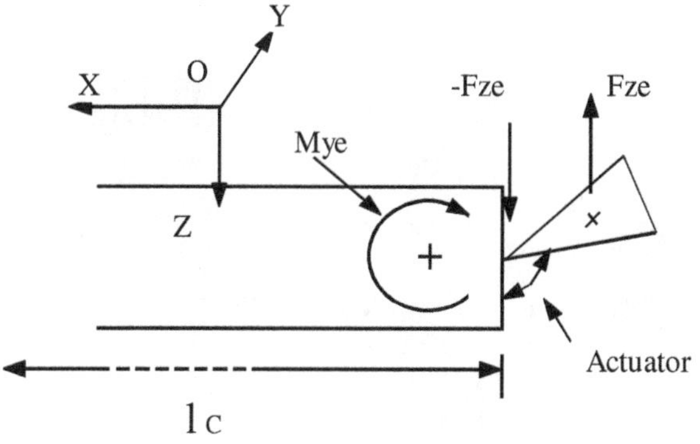

Fig. 3.6-6. Moment Applied to LV

$= \{mR\ \ell R\ [\ Xc''\ A-\ Zc'' + \xi p''(\ell T,t) - \ell C\ \theta''] +\ \ IR\ \{\ \delta p'' + d[d[\delta\ \xi p\ (\ell T,t)/\ \delta\ell]/dt\ ]/dt-\ \theta''\} - \ell c\ mR\ \{w' - u'\ \theta\ +\ \ell C\ \theta'' - \ell R\ \delta p'' + \ell R\ \ \theta''\ -\ \ell R\ \ d[d[\delta\ \xi p\ (\ell T,t)/\ \delta\ell]/dt\ ]/dt\ -\ \xi p''(\ell T,t)\}$

[See Eqn. (3.6-23).]

$= -mR\ \ell R\ Xc''\ A-\ mR\ \ell R\ Zc'' + mR\ \ \ell R\ \xi p'\ '(\ell T,t) - mR\ \ \ell R\ \ \ell C\ \theta''\ +\ IR\ \delta p''\ -\ IR\ \Sigma\ qPT''\ (t)\ \sigma PT\ (\ell)\ -\ IR\ \theta'' - \ell c\ mR\ w' + \ell c\ mR\ u'\ \theta\ -\ \ell c\ mR\ \ell C\ \theta'' - \ell c\ mR\ \ell R\ \ \delta p'' - \ell c\ mR\ \ell R\ \ \theta''\ -\ \ell c\ mR\ \ell R\ \Sigma\ qPT''\ (t)\ \sigma PT\ (\ell)\ +\ \ell c\ mR\ \xi p''(\ell T,t)$

[See Eqn. (3.6-26).]

$= mR\ \ \ell R\ Xc''\ \delta p - mR\ \ \ell R\ Xc''\Sigma\ qPT\ (t)\ \sigma PT\ (\ell) - mR\ \ \ell R\ Xc''\ \theta - mR\ \ \ell R\ Zc'' + mR\ \ell R\ \Sigma\ qPT\ ''(t)\varphi PT\ (\ell) -\ mR\ \ \ell R\ \ell C\ \theta''\ +\ IR\ \delta p''\ -\ IR\ \Sigma\ qPT''\ (t)\ \sigma PT\ (\ell)\ -\ IR\ \theta'' + \ell c\ mR\ w' + \ell c\ mR\ u'\ \theta\ -\ \ell c\ mR\ \ell C\ \theta'' + \ell c\ mR\ \ell R\ \ \delta p'' - \ell c\ mR\ \ell R\ \ \theta''\ -\ \ell c\ mR\ \ell R\ \Sigma\ qPT''\ (t)\ \sigma PT\ (\ell)\ +\ \ell c\ mR\ \Sigma\ qPT\ ''(t)\varphi PT\ (\ell)$

$= (\ IR\ +\ \ell c\ mR\ \ell R\ )\ \delta p''\ +\ mR\ \ \ell R\ Xc''\ \delta p - (mR\ \ \ell R\ \ell C + mR\ \ \ell c\ \ell C\ +mR\ \ \ell c\ \ell R + IR\ )\theta''\ -\ mR\ \ (\ell R\ Xc''\ -\ell c\ u'\ )\theta\ \ -\ mR\ (\ell c\ \ w' +\ \ell R\ Zc'')\ -\ mR\ \ \ell R\ Xc''\ \Sigma\ qPT\ (t)\ \sigma PT\ (\ell)\ -\ \Sigma\ \{[IR\ +\ \ell c\ mR\ \ell R]\ \sigma PT\ (\ell)\ -\ mR\ \ [\ell R\ \ \ +\ \ell c]\varphi PT\ (\ell)]qPT''\ (t)$

where the third term of the equation above can be rewritten:

$[IR + mR\ \ell c\ (\ \ell C + 2\ell R\ )\ ]\ \theta'' = \{IR + mR\ [\ell c^2 + 2\ell C\ \ell R + \ell R^2 - \ \ \ell R^2\ ]\}\theta''$

$= [IR + mR\ (\ell c + \ell R\ )^2 - mR\ \ell R^2\ ]\ \theta''$

$= \{[IR - mR\ \ell R^2] + mR\ (\ell c + \ell R\ )^2\}\theta''$

$= [\ Io + mR\ (\ell c + \ell R\ )^2]\theta''$

Replacing it,

$Mye\ =\ (\ IR\ +\ \ell c\ mR\ \ell R\ )\ \delta p''$
$+\ mR\ \ \ell R\ u'\ \delta p$
$-\ [\ Io + mR\ (\ell c + \ell R\ )^2]\theta''$
$-\ mR\ \ (\ell R - \ell c\ )\ u'\theta$
$-\ mR\ (\ell c\ \ +\ \ell R\ )\ w'$
$-\ mR\ \ \ell R\ Xc''\ \Sigma\ qPT\ (t)\ \sigma PT\ (\ell)$
$-\ \Sigma\ \{[IR + \ell c\ mR\ \ell R]\ \sigma PT\ (\ell)\ -\ mR\ \ [\ell R\ \ \ +\ \ell c]\varphi PT\ (\ell)\}qPT''\ (t)$ \hfill (3.6-36)

Proceeding in a completely analogous manner, we can obtain equations for Fye, Mzee, and Mze.

3.7 Derivation of Launch Vehicle System Dynamic Equations

Here, we narrow our scope down to pitch plane motion alone.

Attitude Differential Equation

We have discussed the issues of bending, gravity, thrust, sloshing, engine inertia, and aerodynamics. These entities directly influence the vehicle attitude, and an attitude differential equation can be written as,

$$I_{yy}\theta'' = M_{yT} + M_{yS} + M_{yE} + M_{yA}$$

where

MyT from Eqn. (3.2-14) -- Thrust
MyS from Eqn. (3.5-47b) – Sloshing
MyE from Eqn. (3.6-36) – Engine Inertia
MyA from Eqn. (3.3-13) – Aerodynamics

Substituting these equations,

$$= \ell_c [T_d \delta - (T_u + T_d) \Sigma (\sigma_i (\ell_T) q_i(t))] - (T_d + T_u) \Sigma \Phi_i(\ell_T) q_i(t) - \sum (m_{pi}\, a_p \sin \Gamma_{pi}) |\rho H_i| + (I_R + \ell_c m_R \ell_R) \delta'' + m_R \ell_R u' \delta - [I_o + m_R(\ell_c + \ell_R)^{2}]\theta'' - m_R (\ell_R - \ell_c) u'\theta - m_R (\ell_c + \ell_R) w' - m_R \ell_R X_c'' \Sigma q_i (t) \sigma_i (\ell_T) - \Sigma \{[I_R + \ell_c m_R \ell_R] \sigma_i(\ell_T) - m_R [\ell_R + \ell_c]\varphi(\ell_T)\}q''(t) + (1/2) \rho U^{2} A_3 \{\alpha_o \int (\ell_a - \ell) (\partial C_N /\partial \alpha) d\ell - (\theta'/ U) \int (\ell_a - \ell)^{2} (\partial C_N /\partial \alpha) d\ell + \Sigma q_i(t) \int(\ell_a - \ell) \sigma_i(\ell)(\partial C_N /\partial \alpha) d\ell - \Sigma [q_i'(t)/U] \int(\ell_a - \ell) \varphi_i(\ell)(\partial C_N /\partial \alpha) d\ell - (W_w/U) \int(\ell_a - \ell) (\partial C_N /\partial \alpha) d\ell\} \quad (3.7\text{-}1)$$

We simplify the equation above so that we can understand the attitude control problem at least conceptually. There are two areas we can work on to simplify the problem. These are bending and sloshing. The number of bending modes can go as high as 200. Usually higher modes can be ignored because their magnitudes are small enough to be insignificant. However, it has been observed that some higher-mode magnitudes grow larger for some reason as the LV flight proceeds. In that case, control system instability could result. Therefore, caution should be exercised when the number of bending modes to be included in bending analysis is determined. Here, we consider only two modes that simplify our analysis.

Sloshing has been an issue since we started using more efficient liquid fuels and oxidizer. Each stage has two tanks to support the propulsion system. One is for fuel, liquid Hydrogen and the other is for the oxidizer, liquid Oxygen. As pointed out earlier, a satellite launch vehicle must have two stages, and consequently in our analysis we need to consider four sloshings at the same time. However, analysis shows that when the tank is filled, the sloshing pendulum mass is small, and therefore its impacts on the control systems are not significant. When we design rocket systems, we put in a tremendous effort in minimizing the structural weight. Because of the structural minimization, the size of the tanks also must be minimized, and in fact they are just large enough to contain just about the right amount of fuel and oxidizer to place a satellite on an orbit. This implies that there will be only two significant sloshing motions in each stage of flight needed to be considered because the tanks in the second stage will remain filled during the first-stage flight, and the tanks in the first stage will be removed during the second-stage flight. Therefore, in our work here, we consider only two sloshings.

Reflecting the statement above, Eqn. (3.7-1) is rewritten as shown below.

$$I_{yy}\theta'' = \Sigma\, my$$
$$= - mR\ (\ell R - \ell c\ )\, u'\theta\ -$$

$$[(1/2)\ \rho\ U^{\wedge}(2)\ A3\ (1/\ U) \int (\ell a - \ell)^{\wedge}(2)\ (\partial CN /\partial\alpha) d\ell]\ \theta'\ -$$

$$[\ Io + mR\ (\ell c + \ell R\ )^{\wedge}(2)]\theta''\ +$$

$$[-\ \ell c\ (Tu + Td)\sigma 1(\ell T) -\ (Td\ + Tu)\ \Phi 1(\ell T) - mR\ \ell R$$
$$Xc''\ \sigma 1\ (\ell T) + (1/2)\ \rho\ U^{\wedge}(2)\ A3\int(\ell a - \ell)\ \sigma 1(\ell)(\partial CN /\partial\alpha)$$
$$d\ell]\ q1(t)\ +$$

$$- (1/2)\, \rho\, U^{2} A_3 \,[1/U] \int(\ell_a - \ell)\, \varphi_1(\ell)(\partial C_N /\partial\alpha)\, d\ell\; q_1'(t) \;-$$

$$\{[I_R + \ell_c m_R \ell_R]\, \sigma_1(\ell_T) - m_R[\ell_R + \ell_c]\varphi_1(\ell_T)\}\, q_1''(t) +$$

$$[- \ell_c (T_u + T_d)\sigma_2(\ell_T) - (T_d + T_u)\, \Phi_2(\ell_T) - m_R \ell_R X_c''\, \sigma_2(\ell_T) + (1/2)\, \rho\, U^{2} A_3 \int(\ell_a - \ell)\, \sigma_2(\ell)(\partial C_N /\partial\alpha)\, d\ell]\; q_2(t) \;-$$

$$-(1/2)\, \rho\, U^{2} A_3 \,[1/U] \int(\ell_a - \ell)\, \varphi_2(\ell)(\partial C_N /\partial\alpha)\, d\ell]\; q_2'(t) -$$

$$\{[I_R + \ell_c m_R \ell_R]\, \sigma_2(\ell_T) - m_R[\ell_R + \ell_c]\varphi_2(\ell_T)\}\, q_2''(t) \;-$$

$$(m_{p1} a_p)\, |\underline{\rho}_{H1}|\; \Gamma_{p1} \;-$$

$$(m_{p2}\, a_p)\, |\underline{\rho}_{H2}|\; \Gamma_{p2} \;+$$

$$(\ell_c T_d + m_R\, \ell_R\, u')\, \delta \;+$$

$$(I_R + \ell_c m_R \ell_R)\, \delta'' \;+$$

$$\{(1/2)\, \rho\, U^{2} A_3 \int (\ell_a - \ell)\, (\partial C_N /\partial\alpha)\, d\ell\}\, \alpha_o \;-$$

$$m_R (\ell_c + \ell_R)\, w' -$$

$$(1/2)\, \rho\, U^{2} A_3\, (W_w/U) \int(\ell_a - \ell)\, (\partial C_N /\partial\alpha)\, d\ell\} \qquad (3.7\text{-}2)$$

Eqn. (3.7-2) has 11 states and 4 inputs. The 11 states are: 1) θ, 2) θ', 3) θ'', 4) q_1, 5) q_1', 6) q_1'', 7) q_2, 8) q_2', 9) q_2'', 10) Γ_{p1}, and 11) Γ_{p2}. The 4 inputs are: 1) δ, 2) δ'', 3) w' (lateral acceleration of LV), and 4) W_w (lateral wind) . Note that the positive rotation of δ is opposite to that of α and θ (see Fig. 3.7-1).

The input, w'(LV acceleration in the Z-axis), is a function of all the states. Therefore, in order to complete Eqn. (3.7-2), we need to derive an equation for w' as a function of all these states, and then substitute it back into Eqn. (3.7-2). And then, we rearrange Eqn. (3.7-2) to complete the derivation of the state space equation. In the following we will go through all of these processes.

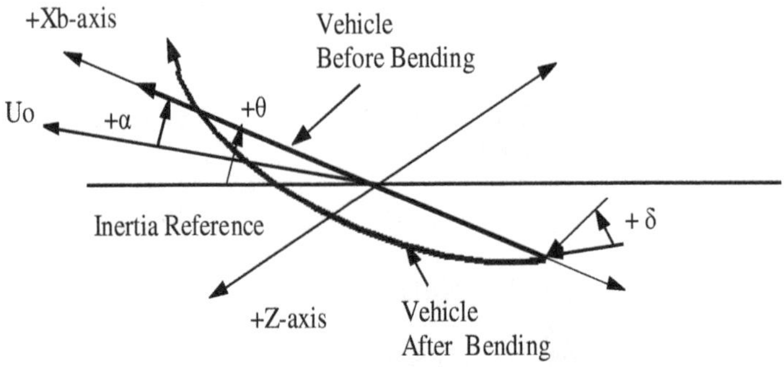

Fig. 3.7-1 Rotation Direction definition

First, we collect all the forces that contribute LV acceleration in the Z-axis.

1. Gravity [Eqn. (3.2-5)]

   $Fzg = -mT\ g\ \theta \cos \theta o$

2. Thrust [Eqn. (3.2-12)]

   $FzT = Td\ \delta - (Tu + Td)\Sigma\sigma i\ qi$

3. Aerodynamics[Eqn. (3.3-11)]

   $FzA = -(1/2)\ \rho\ U^{(2)}\ A3\ \{\alpha o \int (\partial CN/\partial\alpha)\ d\ell - (\theta'/U) \int (\ell a - \ell)(\partial CN/\partial\alpha) d\ell + \Sigma\ qi(t) \int \sigma i(\ell)(\partial CN/\partial\alpha)\ d\ell - \Sigma\ [qi'(t)/U] \int \varphi i(\ell)(\partial CN/\partial\alpha)\ d\ell - (Ww/U) \int (\partial CN/\partial\alpha)\ d\ell\}$

4. Sloshing [Eqn. (3.4-41)]

   $Fzs = Uo'\ \Sigma mpi\ \Gamma pi$

5. Engine Inertia [Eqn. (3.6-24)]

Fze-on-LV = - Fze
$$= - mR \{w' - u' \theta + \ell C \theta'' - \ell R (\delta'' - \theta'') - \Sigma qTi'' (t) [-\ell R \sigma i(\ell T) - \varphi i (\ell T)] \}$$

Now, we can write an equation of motion in the Z-axis.

mT (w'- Uo θ') = Σ Fz
$$= Fzg + FzT + FzA + Fzs + Fze\text{-on-LV}$$
$$= -mT g \theta \cos \theta o + Td \delta -(Tu + Td)\Sigma\sigma i qi - (1/2) \rho U^{\wedge}(2) A3 \{\alpha o\!\int (\partial CN /\partial\alpha) d\ell - (\theta'/ U) \int (\ell a - \ell) (\partial CN /\partial\alpha)d\ell+ \Sigma qi(t) \int \sigma i(\ell)(\partial CN /\partial\alpha) d\ell - \Sigma [qi'(t)/U] \int\varphi i(\ell)(\partial CN /\partial\alpha) d\ell - (Ww/U) \int (\partial CN /\partial\alpha) d\ell\} - Uo' \Sigma mpi \Gamma pi - mR \{w' - u' \theta + \ell C \theta'' - \ell R (\delta'' - \theta'') - \Sigma qTi'' (t) [-\ell R \sigma i(\ell T) - \varphi i (\ell T)] \}$$

(3.7-3)

Since in our study, we consider only two bending modes, Eqn. (3.7-3) can be rewritten as,

mT (w'- Uo θ') = Σ Fz
$$= (-mT g \cos \theta o) \theta + Td \delta -(Tu + Td)\Sigma\sigma yi qyi - (1/2) \rho U^{\wedge}(2) A3\{\alpha o\!\int (\partial CN /\partial\alpha) d\ell - (\theta'/ U) \int (\ell a - \ell) (\partial CN /\partial\alpha)d\ell+ q1(t) \int \sigma 1(\ell)(\partial CN /\partial\alpha) d\ell + q2(t) \int \sigma 2(\ell)(\partial CN /\partial\alpha) d\ell -[q1'(t)/U] \int\varphi 1(\ell)(\partial CN /\partial\alpha) d\ell - [q2'(t)/U] \int\varphi 2(\ell)(\partial CN /\partial\alpha) d\ell-(Ww/U) \int (\partial CN /\partial\alpha) d\ell\} - Uo'mp1\Gamma p1 - Uo'mp2 \Gamma p- mR \{w'- u' \theta + \ell C \theta'' - \ell R (\delta'' - \theta'') - qT1'' (t) [-\ell R \sigma 1(\ell T) + \varphi 1(\ell T)] - qT2'' (t) [-\ell R \sigma 2(\ell T) - \varphi 2 (\ell T)] \}$$

$$= (-mT g \cos \theta o) \theta + Td \delta -(Tu + Td)\Sigma\sigma yi qyi - qD A3 \{\alpha o\!\int (\partial CN /\partial\alpha) d\ell - (\theta'/ U) \int (\ell a - \ell) (\partial CN /\partial\alpha)d\ell+ q1(t) \int \sigma 1(\ell)(\partial CN /\partial\alpha) d\ell + q2(t) \int \sigma 2(\ell)(\partial CN /\partial\alpha) d\ell-[q1'(t)/U] \int\varphi 1(\ell)(\partial CN /\partial\alpha) d\ell - [q2'(t)/U] \int\varphi 2(\ell)(\partial CN /\partial\alpha) d\ell -(Ww/U) \int (\partial CN /\partial\alpha) d\ell\} - Uo'mp1\Gamma p1 - Uo'mp2 \Gamma p -mR w' + mR u' \theta - mR \ell C \theta'' + mR \ell R (\delta'' - \theta'') + mR q T1'' (t) [-\ell R \sigma 1(\ell T) + \varphi 1(\ell T)] + mR q2'' (t) [-\ell R \sigma 2(\ell T) - \varphi 2 (\ell T)] \}$$

$$= (-mT g \cos \theta o + mR u') \theta +$$

$qD\ A3\ (1/U) \int (\ell a - \ell)(\partial CN/\partial\alpha) d\ell]\ \theta' -$
$mR\ (\ell R + \ell C)\theta'' + [-(Tu + Td)\sigma 1 + (1/2)\ \rho$
$U^{(2)}\ A3 \int \sigma 1(\ell)(\partial CN/\partial\alpha)\ d\ell\ ]\ q1(t) +$
$[qD\ A3\ (1/U) \int \varphi 1(\ell)(\partial CN/\partial\alpha)\ d\ell\ ]\ q1' +$
$mR\ [-\ell R\ \sigma 1(\ell T) - \varphi 1(\ell T)]\ q1'' +$
$[-(Tu + Td)\sigma 2 - qDA3 \int \sigma 2(\ell)(\partial CN/\partial\alpha)\ d\ell]\ q2(t)$
$[qD\ A3\ (1/U) \int \varphi 2(\ell)(\partial CN/\partial\alpha)\ d\ell\ ]\ q2' +$
$mR\ [-\ell R\ \sigma 2(\ell T) - \varphi 2\ (\ell T)]\ q2'' -$
$Uo' mp1 \Gamma p1 -$
$Uo' mp2\ \Gamma p2 +$
$Td\ \delta +$
$mR\ \ell R\ \delta'' -$
$qD\ A3\ \alpha o \int (\partial CN/\partial\alpha)\ d\ell -$
$qD\ A3\ (Ww/U) \int (\partial CN/\partial\alpha)\ d\ell -$
$mR\ w'$

(3.7-4)

where

$qD = (1/2)\ \rho\ U^{(2)}$
= dynamic pressure

Rearranging Eqn. (3.7-4) to obtain an equation for w',

$w' = \{(-mT\ g \cos\theta o + mR\ u')/(mT + mR)\}\ \theta +$
$\{ [qD\ A3\ (1/U) \int (\ell a - \ell)(\partial CN/\partial\alpha) d\ell + mT\ Uo]/(mT +$
$mR)\ \}\ \theta' +$
$[-mR\ (\ell R + \ell C)/(mT + mR)]\theta'' +$
$\{[-(Tu + Td)\sigma 1 - qD\ A3 \int \sigma 1(\ell)(\partial CN/\partial\alpha)\ d\ell\ ]/(mT +$
$mR)\} q1(t) +$
$\{[\ qD\ A3\ (1/U) \int \varphi 1(\ell)(\partial CN/\partial\alpha)\ d\ell\ ]/(mT + mR)\}\ q1' -$
$\{-mR\ [-\ell R\ \sigma 1(\ell T) - \varphi 1(\ell T)]/(mT + mR)\}\ q1'' +$
$\{[\ -(Tu + Td)\sigma 2 - qD\ A3 \int \sigma 2(\ell)(\partial CN/\partial\alpha)\ d\ell]/(mT + mR)\}$
$q2(t) +$
$\{[\ qD\ A3\ (1/U) \int \varphi 2(\ell)(\partial CN/\partial\alpha)\ d\ell]/(mT + mR)\}\ q2' -$
$\{-mR\ [-\ell R\ \sigma 2(\ell T) - \varphi 2\ (\ell T)]/(mT + mR)\}\ q2'' -$
$[Uo' mp1/(mT + mR)]\ \Gamma p1 -$
$[Uo' mp2/(mT + mR)]\ \Gamma p2 +$
$[\ Td/(mT + mR)]\delta -$
$[-mR\ \ell R/(mT + mR)]\ \delta'' +$
$\{-qD\ A3 \int (\partial CN/\partial\alpha)\ d\ell]/(mT + mR)\}\ \alpha o -$

$$\{-qD\ A3\ (1/U) \int (\partial CN/\partial\alpha)\, d\ell / (mT + mR)\}\ Ww \tag{3.7-5}$$

Substituting Eqn. (3.7-5) into Eqn. (3.7-2) and rearranging it,

$$\theta'' = C1\ \theta + C2\ \theta' + C3\, q1 + C4\, q1' + C5\, q1'' + C6\, q2 + C7\, q2' + C8\, q2'' + C9\, \Gamma p1 + C10\ \Gamma p2 + C11\ \delta + C12\ \delta'' + C13\ \alpha o + C14\ W \tag{3.7-6}$$

Where

$$C1 = \{\{-mR\,(\ell c + \ell R)\,\{(-mT\, g \cos\theta o + mR\, u')/(mT + mR)\} - mR\,(\ell R - \ell c)\, u'\}/\Psi\}$$

$$C2 = \{\{-mR\,(\ell c + \ell R)\,\{[qD\ A3\ (1/U)\int(\ell a - \ell)(\partial CN/\partial\alpha)d\ell + mT\ Uo]/(mT + mR)\} - [(1/2)\,\rho\, U^{(2)}\, A3\ (1/U)\int(\ell a - \ell)^{(2)}(\partial CN/\partial\alpha)d\ell]\}/\Psi\}$$

$$C3 = \{\{-mR\,(\ell c + \ell R)\,\{[-(Tu + Td)\sigma 1 - qD\ A3 \int \sigma 1(\ell)(\partial CN/\partial\alpha)\, d\ell]/(mT + mR)\} + [-\ \ell c\,(Tu + Td)\sigma 1(\ell T) - (Td + Tu)\,\Phi 1(\ell T) - mR\ \ell R\, Xc''\, \sigma 1(\ell T) + qD\ A3\int(\ell a - \ell)\,\sigma 1(\ell)(\partial CN/\partial\alpha)\, d\ell]\}/\Psi\}$$

$$C4 = \{\{-mR\,(\ell c + \ell R)\ \{[qD\ A3\ (1/U)\int \Phi 1(\ell)(\partial CN/\partial\alpha)\, d\ell]/(mT + mR)\} - [qD\ A3\ (1/U)\int(\ell a - \ell)\,\Phi 1(\ell)(\partial CN/\partial\alpha)\, d\ell]\}/\Psi\}$$

$$C5 = \{-\{mR\,(\ell c + \ell R)\,\{-mR\,[-\ell R\ \sigma 1(\ell T) - \Phi 1(\ell T)]/(mT + mR)\} - \{[IR + \ell c\, mR\ \ell R]$$

$$C6 = \sigma 1(\ell T) - mR\ [\ell R + \ell c]\,\Phi 1(\ell T)\}\}/\Psi\}\, q1''(t) + \{-\{mR\,(\ell c + \ell R)\,\{[-(Tu + Td)\sigma 2 - qD\ A3\int \sigma 2(\ell)(\partial CN/\partial\alpha)\, d\ell]/(mT + mR)\} + [-\ \ell c\,(Tu + Td)\sigma 2(\ell T) - (Td + Tu)\,\Phi 2(\ell T) - mR\ \ell R\, Xc''\, \sigma 2\ (\ell T) + qD\ A3\int(\ell a - \ell)\,\sigma 2(\ell)(\partial CN/\partial\alpha)\, d\ell]\}/\Psi\}$$

$$C7 = \{-\{mR\,(\ell c + \ell R)\,\{[qD\ A3\ (1/U)\int \Phi 2(\ell)(\partial CN/\partial\alpha)\, d\ell]/(mT + mR)\} - [qD\ A3\ [1/U]\int(\ell a - \ell)\,\Phi 2(\ell)(\partial CN/\partial\alpha)\, d\ell]\}/\Psi\}$$

$$C8 = \{-\{mR\,(\ell c + \ell R)\,\{-mR\,[-\ell R\ \sigma 2(\ell T) - \Phi 2\,(\ell T)]/(mT + mR)\} - \{[IR + \ell c\ mR\ \ell R]\,\sigma 2(\ell T) - mR\ [\ell R + \ell c]\,\Phi 2(\ell T)\}\}/\Psi\}$$

$$C9 = \{\{mR\,(\ell c + \ell R)\,[Uo' mp1/(mT + mR)] - (mp1ap)\,|\underline{\rho}H1|\}/\Psi\}$$

$$C10 = \{\{mR\,(\ell c + \ell R)\,[Uo' mp2/(mT + mR)] - (mp2\ ap)\,|\underline{\rho}H2|\}/\Psi\}$$

$$C11 = \{\{-mR\,(\ell c + \ell R)\ [Td/(mT + mR)] + (\ell c\, Td + mR\ \ell R\, u')\}/\Psi\}$$

$$C12 = \{\{-mR\,(\ell c + \ell R)\,[-mR\,\ell R/(mT + mR)] + (IR + \ell c\, mR\,\ell R)\}/\Psi\}$$

$$C13 = \{\{-mR\,(\ell c + \ell R)\,\{-qD\ A3\int(\partial CN/\partial\alpha)\, d\ell]/(mT + mR)\} + qD\ A3\int(\ell a - \ell)(\partial CN/\partial\alpha)\, d\ell\}/\Psi\}$$

$$C14 = \{\{-mR\,(\ell c + \ell R)\,\{-qD\ A3\ (1/U)\int(\partial CN/\partial\alpha)\, d\ell/(mT + mR)\} - (1/2)\,\rho\, U^{(2)}\, A3(1/U)\int(\ell a - \ell)(\partial CN/\partial\alpha)\, d\ell\}/\Psi\}$$

where

$$\Psi=\{ I_{yy} - \{ - m_R (\ell_c + \ell_R) [-m_R (\ell_R + \ell_C) / (m_T + m_R)] - [I_o + m_R (\ell_c + \ell_R)^{2}]\}\}$$

Position Gyro Math Model

$$\theta_{PG}'' = K_1 \theta + K_2 q_1 + K_3 q_2 + K_4 \theta_{PG} \qquad (3.7\text{-}7)$$

where

$K_1 = 1/\tau$, τ : time constant
$K_2 = \sigma_1(\ell_{PG}) / \tau$
$K_3 = \sigma_2(\ell_{PG}) / \tau$
$K_4 = -1/\tau$

Rate Gyro Math Model

$$\theta_{RG}'' = L_1 \theta' + L_2 q_1' + L_3 q_2' + L_4 \theta_{RG} + L_5 \theta_{RG}' \qquad (3.7\text{-}8)$$

where

$L_1 = \omega_R^{2} K_R$, K_R : Rate Gyro Gain
$L_2 = \omega_R^{2} K_R \sigma_1(\ell_{RG})$
$L_3 = \omega_R^{2} K_R \sigma_2(\ell_{RG})$
$L_4 = - \omega_R^{2}$
$L_5 = - 2 \xi_R \omega_R$

Now, collecting all the equations that constitute the attitude control system dynamic equations, i.e., Eqn. (3.7-6), Eqn. (3.4-33), Eqn. (3.4-35), Eqn. (3.5-56), Eqn. (3.5-57), Eqn. (3.7-7), and Eqn. (3.7-8),

a. Moment Equation [Eqn. (3.7-6)]

$$\theta'' = C1\ \theta + C2\ \theta' + C3\, q1 + C4\, q1' + C5\, q1'' + C6\, q2 + C7\, q2' + C8\, q2'' + C9\, \Gamma p1 + C10\ \Gamma p2 + C11\ \delta + C12\ \delta'' + C13\ \alpha o + C14\ W$$

where

$$C1 = \{\{-mR\,(\ell c + \ell R)\,\{(-mT\,g\cos\theta o + mR\,u')\,/\,(mT - mR)\} - mR\,(\ell R - \ell c)\,u'\}\,/\,\Psi\}$$

$$\Psi = \{\,Iyy - \{-mR\,(\ell c + \ell R)\,[-mR\,(\ell R + \ell C)\,/\,(mT + mR)] - [\,Io + mR\,(\ell c + \ell R)^{\wedge}(2)]\}\}$$

$$C2 = \{\{-mR\,(\ell c + \ell R)\,\{\,[qD\,A3\,(1/U)\int(\ell a - \ell)\,(\partial CN/\partial\alpha)d\ell + mT\,Uo]/(mT - mR)\,\} - [(1/2)\,\rho\,U^{\wedge}(2)\,A3\,(1/U)\int(\ell a - \ell)^{\wedge}(2)\,(\partial CN/\partial\alpha)d\ell]\,\}/\,\Psi\}$$

$$C3 = \{\{-mR\,(\ell c + \ell R)\,\{[-(Tu + Td)\sigma 1 - qD\,A3\int\sigma 1(\ell)(\partial CN/\partial\alpha)\,d\ell\,]/(mT - mR)\} + [\,-\,\ell c\,(Tu + Td)\sigma 1(\ell T) - (Td + Tu)\,\Phi 1(\ell T) - mR\,\ell R\,Xc''\,\sigma 1\,(\ell T) + qD\,A3\int(\ell a - \ell)\,\sigma 1(\ell)(\partial CN/\partial\alpha)\,d\ell]\}/\,\Psi\}$$

$$C4 = \{\{-mR\,(\ell c + \ell R)\,\{[\,qD\,A3\,(1/U)\int\Phi 1\,(\ell)(\partial CN/\partial\alpha)\,d\ell\,]\,/\,(mT - mR)\} - [qD\,A3\,(1/U)\int(\ell a - \ell)\,\Phi 1(\ell)(\partial CN/\partial\alpha)\,d\ell\,]\}\,/\,\Psi\}$$

$$C5 = \{-\{mR\,(\ell c + \ell R)\,\{-mR\,[-\ell R\,\sigma 1(\ell T) - \Phi 1(\ell T)]\,/\,(mT - mR)\} - \{[IR + \ell c\,mR\,\ell R]$$

$$C6 = \sigma 1(\ell T) - mR\,[\ell R + \ell c]\,\Phi 1\,(\ell T)\}\}\,/\,\Psi\}\,q1''\,(t) + \{-\{\,mR\,(\ell c + \ell R)\,\{[-(Tu + Td)\sigma 2 - qD\,A3\int\sigma 2(\ell)(\partial CN/\partial\alpha)\,d\ell]/(mT - mR)\} + [-\,\ell c\,(Tu + Td)\sigma 2(\ell T) - (Td + Tu)\,\Phi 2(\ell T) - mR\,\ell R\,Xc''\,\sigma 2\,(\ell T) + qD\,A3\int(\ell a - \ell)\,\sigma 2(\ell)(\partial CN/\partial\alpha)\,d\ell]\}/\,\Psi\}$$

$$C7 = \{-\{\,mR\,(\ell c + \ell R)\,\{[qD\,A3\,(1/U)\int\Phi 2(\ell)(\partial CN/\partial\alpha)\,d\ell]/\,(mT - mR)\} - [qD\,A3\,[1/U]\int(\ell a - \ell)\,\Phi 2(\ell)(\partial CN/\partial\alpha)\,d\ell]\}/\,\Psi\}$$

$$C8 = \{-\{mR\,(\ell c + \ell R)\,\{-mR\,[-\ell R\,\sigma 2(\ell T) - \Phi 2\,(\ell T)]\,/\,(mT - mR)\} - \{[IR + \ell c\,mR\,\ell R]\,\sigma 2(\ell T) - mR\,[\ell R + \ell c]\,\Phi 2(\ell T)\}\}\,/\,\Psi\}$$

$$C9 = \{\{mR\,(\ell c + \ell R)\,[Uo'mp1\,/\,(mT - mR)] - (\,mp1ap)\,|\underline{\rho}H1|\,\}/\,\Psi\}$$

$$C10 = \{\{mR\,(\ell c + \ell R)\,[Uo'mp2\,/\,(mT - mR)] - (\,mp2\,ap)\,|\underline{\rho}H2|\,\}/\,\Psi\}$$

$$C11 = \{\{-\,mR\,(\ell c + \ell R)\,[\,Td\,/\,(mT - mR)] + (\ell c\,Td + mR\,\ell R\,u')\}\,/\,\Psi\}$$

$$C12 = \{\{\,-mR\,(\ell c + \ell R)\,[-mR\,\ell R\,/\,(mT - mR)] + (\,IR + \ell c\,mR\,\ell R\,)\}\,/\,\Psi\}$$

$$C13 = \{\{-mR\,(\ell c + \ell R)\,\{\,-\,qD\,A3\int(\partial CN/\partial\alpha)\,d\ell]/\,(mT - mR)\} + qD\,A3\int(\ell a - \ell)\,(\partial CN/\partial\alpha)\,d\ell\,\}/\,\Psi\}$$

$$C14 = \{\{-mR\,(\ell c + \ell R)\,\{-qD\,A3\,(1/U)\int(\partial CN/\partial\alpha)\,d\ell/\,(mT - mR)\} - (1/2)\,\rho$$

U^(2) A3(1/U) ∫(ℓa - ℓ) (∂CN /∂α) dℓ }/ Ψ }

where

Ψ={ Iyy -{ - mR (ℓc + ℓR) [-mR (ℓR + ℓC) / (mT - mR)] - [Io + mR (ℓc+ ℓR)^(2)]}}

b. First Bending Mode Generalized Coordinate [Eqn. (3.4-33)]

q1'' = E1 θ + E2 θ' + E3 θ'' + E4 q1 + E5 q1' + E6 q2 + E7 q2' + E8 q2'' + E9 Γp1 + E10 Γp2 + E11 δ + E12 δ'' + E13 αo + E14 Ww

where

E1 = (1/m1 Ψq1) {- mR Φ1(ℓT) {(-mT g cos θo + mR u' - mT Uo) / (mT + mR)} + {- [∫Φ1(ℓ) m(ℓ)dℓ] g (cos θo) + mR Φ1(ℓT) u'}}

E2 = (1/m1 Ψq1) {- mR Φ1(ℓT){ [qD A3 (1/ U) ∫ (ℓa - ℓ) (∂CN /∂α)dℓ]/(mT + mR) } + qD A3 (1/ U) ∫ Φ(ℓ) (ℓa - ℓ) (∂CN /∂α)dℓ }

E3 = (1/m1 Ψq1) {- mR Φ1(ℓT) [-mR (ℓR + ℓC) / (mT + mR)]-mR Φ1(ℓT) (ℓC + ℓR)}

E4 = (1/ Ψq1) {{- ω1^(2) + (1/m1) { mR Φ1(ℓT) {[(Tu + Td)σ1(ℓT) + qD A3 ∫ σ1(ℓ)(∂CN /∂α) dℓ]/ (mT +mR)} –{(Tu + Td) Φ1(ℓT) σ1(ℓT) +qD A3 ∫ Φ1(ℓ)σ1(ℓ)(∂CN /∂α) dℓ}}}

E5 = (1/ Ψ q1) {{- 2 ω1 ξ1 + (1/m1) {- mR Φ1(ℓT){[qD A3 (1/ U) ∫ Φ1(ℓ) (∂CN /∂α) dℓ] / (mT + mR)} + qDA3 [1/U] ∫ Φ1(ℓ) Φ1(ℓ) (∂CN /∂α) dℓ }}

E6 = (1/m1 Ψq1) { mR Φ1(ℓT) {[(Tu + Td)σ2(ℓT) + qD A3 ∫ σ1(ℓ)(∂CN /∂α) dℓ]/(mT +mR)} –{(Tu + Td) Φ1(ℓT) σ2(ℓT) +qD A3 ∫ Φ1(ℓ) σ2(ℓ)(∂CN /∂α) dℓ}}

E7 = (1/m1 Ψq1) {- mR Φ1(ℓT){[qD A3 (1/ U) ∫ Φ1(ℓ) (∂CN /∂α) dℓ] / (mT + mR)} + qDA3 Σ [1/U] ∫ Φ1(ℓ) Φ2(ℓ) (∂CN /∂α) dℓ }

E8 = (1/m1 Ψq1) { mR Φ1(ℓT){mR [ℓR σ2(ℓT) + Φ2(ℓT)] / (mT + mR)} - mR Φ1(ℓT) [ℓR σ2(ℓT) + Φ2(ℓT)]}

E9 = (1/m1 Ψq1) - mR Φ1(ℓT) [Uo'mp1 / (mT + mR)] +Uo' Φ1(ℓp1) } mp1}

E10 = (1/m1 Ψq1) - mR Φ1(ℓT) [Uo'mp2 / (mT + mR)] +Uo' Φ1(ℓp2) } mp2}

E11 = (1/m1 Ψq1) {- mR Φ1(ℓT) [Td / (mT + mR)]+ Td Φ1(ℓT) }

E12 = (1/m1 Ψq1) {- mR Φ1(ℓT) [-mR ℓR / (mT + mR)] + mR Φ1(ℓT) ℓR}

$E13 = (1/m1\ \Psi q1)\ \{ -\ mR\ \Phi 1(\ell T)\{-qD\ A3\ \int (\partial CN/\partial\alpha)\ d\ell]/ (mT + mR)\}\ -qD\ A3\ \int \Phi 1(\ell)\ (\partial CN/\partial\alpha)\ d\ell\ \}$

$E14 = (1/m1\ \Psi q1)\ \{ -\ mR\ \Phi 1(\ell T)\{-qD\ A3\ (1/U) \int (\partial CN/\partial\alpha)\ d\ell/ (mT + mR)\} + qD\ A3\ (1/U) \int \underline{\Phi}(\ell)(\partial CN/\partial\alpha)\ d\ell\}\}$

where

$\Psi q1=\{1 - (1/m1)\ \{ mR\ \Phi 1(\ell T)\{mR\ [\ell R\ \sigma 1(\ell T) + \Phi 1(\ell T)] / (mT + mR)\} - mR\ \Phi 1(\ell T)\Sigma\ [\ell R\ \sigma 1(\ell T) + \Phi 1(\ell T)]\}\}$

c. Second Bending Mode Generalized Coordinate [Eqn. (3.4-35)]

$$q2'' = G1\ \theta + G2\ \theta' + G3\ \theta'' + G4\ q1 + G5\ q1' + G6\ q1'' + G7\ q2 + G8\ q2' + G9\ \Gamma p1 + G10\ \Gamma p2 + G11\ \delta + G12\ \delta'' + G13\ \alpha o + G14\ Ww$$

where

$G1 = (1/m2\ \Psi q2)\ \{-\ mR\ \Phi 2(\ell T)\ \{(-mT\ g \cos\theta o + mR\ u' - mT\ Uo) / (mT + mR)\} + \{- [\int \Phi 2(\ell)\ m(\ell) d\ell]\ g\ (\cos\theta o\ )\_+\ mR\ \Phi 2(\ell T)\ u'\}\}$

$G2 = (1/m2\ \Psi q2)\ \{-\ mR\ \Phi 2(\ell T)\{\ [qD\ A3\ (1/U) \int (\ell a - \ell)\ (\partial CN/\partial\alpha) d\ell]/(mT + mR)\ \} + qDA3\ (1/U)\ \int \Phi(\ell)\ (\ell a - \ell)\ (\partial CN/\partial\alpha) d\ell\ \}$

$G3 = (1/m2\ \Psi q2)\ \{-\ mR\ \Phi 2(\ell T)\ [-mR\ (\ell R + \ell C\ ) / (mT + mR)]-mR\ \Phi 2(\ell T)\ (\ \ell C + \ell R\ )\}$

$G4 = (1/m2\ \Psi q2)\ \{\ mR\ \Phi 2(\ell T)\ \{[(Tu + Td)\sigma 1(\ell T) + qD\ A3\ \int \sigma 1(\ell)(\partial CN/\partial\alpha)\ d\ell\ ]/(mT + mR)\} - \{(Tu + Td)\ \Phi 2(\ell T)\ \sigma 1(\ell T) + qD\ A3\ \int \Phi 2(\ell)\ \sigma 1(\ell)(\partial CN/\partial\alpha)\ d\ell\}\}$

$G5 = (1/m2\ \Psi q2)\ \{-\ mR\ \Phi 2(\ell T)\{[\ qD\ A3\ (1/U) \int \Phi 2(\ell)\ (\partial CN/\partial\alpha)\ d\ell\ ] / (mT + mR)\} + qD\ A3\ [1/U] \int \Phi 2(\ell)\ \Phi 1(\ell)\ (\partial CN/\partial\alpha)\ d\ell\ \}$

$G6 = (1/m2\ \Psi q2)\ \{\ mR\ \Phi 2(\ell T)\{mR\ [\ell R\ \sigma 1(\ell T) + \Phi 1(\ell T)] / (mT + mR)\} - mR\ \Phi 2(\ell T)\Sigma\ [\ell R\ \sigma 1(\ell T) + \Phi 1(\ell T)]\}$

$G7 = (1/\Psi q2)\ \{-\ \omega 1\wedge(2) + (1/m2)\ \{\ mR\ \Phi 2(\ell T)\ \{[(Tu + Td)\sigma 2(\ell T) + qD\ A3\ \int \sigma 2(\ell)(\partial CN/\partial\alpha)\ d\ell\ ]/(mT + mR)\} - \{(Tu + Td)\ \Phi 2(\ell T)\ \sigma 2(\ell T) + qD\ A3\ \int \Phi 2(\ell)\ \sigma 2(\ell)(\partial CN/\partial\alpha)\ d\ell\}\}\}$

$G8 = (1/\Psi q2)\ \{-\ 2\ \omega 2\xi 2 + (1/m2)\ \{-\ mR\ \Phi 2(\ell T)\{[\ qD\ A3\ (1/U) \int \Phi 2(\ell)\ (\partial CN/\partial\alpha)\ d\ell\ ] / (mT + mR)\} + qD\ A3\ [1/U] \int \Phi 2(\ell)\ \Phi 2(\ell)\ (\partial CN/\partial\alpha)\ d\ell\ \}\}$

$G9 = (1/m2\ \Psi q2)\ \{ -\ mR\ \Phi 2(\ell T)\ [Uo' mp1 / (mT + mR)] + Uo'\ \Phi 2(\ell p1)\ \}\ mp1\}$

$G10 = (1/m2\ \Psi q2)\ \{ -\ mR\ \Phi 2(\ell T)\ [Uo' mp2 / (mT + mR)] + Uo'\ \Phi 2(\ell p2)$

$\} \, mp2\}$

$G11 = (1/m2 \, \Psi q2) \{- mR \, \Phi 2(\ell T) \, [\, Td / (mT + mR)] + Td \, \Phi 2(\ell T) \, \}$

$G12 = (1/m2 \, \Psi q2) \{- mR \, \Phi 2(\ell T) \, [-mR \, \ell R / (mT + mR)] + mR \, \Phi 2(\ell T) \, \ell R\}$

$G13 = (1/m2 \, \Psi q2) \{ \, mR \, \Phi 2(\ell T)\{-qD \, A3 \int (\partial CN / \partial \alpha) \, d\ell] / (mT + mR)\} -$
$qD \, A3 \int \Phi 2(\ell) \, (\partial CN / \partial \alpha) \, d\ell \, \}$

$G14 = (1/m2 \, \Psi q2) \{ - mR \, \Phi 2(\ell T)\{-qD \, A3 \, (1/U) \int (\partial CN / \partial \alpha) \, d\ell / (mT +$
$mR)\} + qD \, A3 \, (1/U) \int \underline{\Phi}(\ell)(\partial CN / \partial \alpha) \, d\ell\}\}$

where

$\Psi q2 = \{ \, 1 - (1/m2) \, \{ \, mR \, \Phi 2(\ell T)\{mR \, [\ell R \, \sigma 2(\ell T) + \Phi 2(\ell T)] / (mT + mR)\}$
$- mR \, \Phi 2(\ell T) \, [\ell R \, \sigma 2(\ell T) + \Phi 2(\ell T)]\}\}$

d. Sloshing 1 [Eqn. (3.5-56)]

$\Gamma p1'' = H1 \, \Gamma p1 + H2 \, \Gamma p2 + H3 \, \delta$

where

$H1 = (-1/ Lp1)\{ \, (1 + mp1/mT) \, Uo' + \mu P1 \, (\ell p1 - Lp1) \, \}$
$H2 = (-1/ Lp1)\{Uo' mp2/mT + \mu P2 \, (\ell p1 - Lp1)\}$
$H3 = (-1/ Lp1)\{Td /mT - \mu c \, (\ell p1 - Lp1)\}$

$\mu P1 = [(\, mp1ap) \, |\underline{\rho} H1| / Iyy]$
$\mu P2 = [(\, mp2ap) \, |\underline{\rho} H1| / Iyy]$
$\mu c = (\ell c \, Td / Iyy)$

Note: $Uo \approx u$ since Uo is steady state value of u.

e. Sloshing 2 [Eqn. (3.5-57)]

$\Gamma p2'' = J1 \, \Gamma p1 + J2 \, \Gamma p2 + J3 \, \delta$

where

$J1 = \{- (1/ Lp2)\{Uo' mp1/mT + \mu P1 \, (\ell p2 - Lp2) \, \}$
$J2 = \{- (1/ Lp2)\{ \, (1 + mp2/mT) \, Uo' + \mu P2 \, (\ell p2 - Lp2)\}$
$J3 = \{-1/ Lp2)\{Td /mT - \mu c \, (\ell p2 - Lp2)\}$

f. Position Gyros Math Model [Eqn. (3.7-7)]

$\theta PG'' = K1 \, \theta + K2 \, q1 + K3 \, q2 + K4 \, \theta PG$

where

$K1 = 1/\tau$, τ : time constant
$K2 = \sigma 1(\ell PG)/\tau$
$K3 = \sigma 2(\ell PG)/\tau$
$K4 = -1/\tau$

g. Rate Gyro Math Model [Eqn. (3.7-8)]

$\theta RG'' = L1\ \theta' + L2\ q1' + L3\ q2' + L4\ \theta RG + L5\ \theta RG'$

where

$L1 = \omega R^{\wedge}(2)\ KR$, KR : Rate Gyro Gain
$L2 = \omega R^{\wedge}(2)\ KR\ \sigma 1(\ell RG)$
$L3 = \omega R^{\wedge}(2)\ KR\ \sigma 2(\ell RG)$
$L4 = -\omega R^{\wedge}(2)$
$L5 = -2\ \xi R\ \omega R$

3.8 Construction of LV State Space Equation

A simplified math model of the state space equation is constructed for desktop computer simulation. This model includes only two bending modes, two sloshings, a simplified thrust engine actuator, and simple position and rate gyros. Equations computing the elements of the system matrices A and B are derived as shown below:

$\underline{xd} = A*\underline{x} + B*\underline{u}$

$\underline{y} = C*\underline{x} + D*\underline{u}$

where

$\underline{x}$ = [**θ θd q1 q1d q2 q2d Γ1 Γ1d Γ2 Γ2d θPG θRG θRGd**]

θ = attitude angle in pitch (rad)
θd = attitude angular rate in pitch (rad/sec)
q1 = generalized coordinate of the 1st bending mode in pitch (ft)

q1d= rate of generalized coordinate of the 1st bending mode in pitch (ft/sec)
q2 = generalized coordinate of the 2nd bending mode in pitch (ft)
q2d= rate of generalized coordinate of the 2nd bending mode in pitch (ft/sec)
Γ1 = pendulum angle of fuel tank 1 sloshing (rad)
Γ1d= pendulum angular rate of fuel tank 1 sloshing (rad/sec)
Γ2 = pendulum angle of fuel tank 2 sloshing (rad)
Γ2d = pendulum angular rate of fuel tank 2 sloshing (rad/sec)
θPG = position gyro output (rad)
θRG = rate gyro output (rad/sec)
θRGd= time derivative rate gyro output [rad/sec^(2)]

u = [**δ** ; **α** ; **Ww**], (transpose of [**δ α Ww**])

δ = engine deflection angle (rad)
α = perturbation angle of attack (rad)
Ww = wind velocity (ft/sec)

y = [**θ θd**]

A =

0	1	0	0	0	0	0	0	0	0	0	0	0
C1	C2	C3	C4	C6	C7	C9	0	C10	0	0	0	0
0	0	0	1	0	0	0	0	0	0	0	0	0
E1	E2	E4	E5	E6	E7	E9	0	E10	0	0	0	0
0	0	0	0	0	1	0	0	0	0	0	0	0
G1	G2	G4	G5	G7	G8	G9	0	G10	0	0	0	0
0	0	0	0	0	0	0	1	0	0	0	0	0
0	0	0	0	0	0	H1	0	H2	0	0	0	0
0	0	0	0	0	0	0	0	0	1	0	0	0
0	0	0	0	0	0	J1	0	J2	0	0	0	0
K1	0	K2	0	K3	0	0	0	0	0	K4	0	0
0	0	0	0	0	0	0	0	0	0	0	0	1
0	L1	0	L2	0	L3	0	0	0	0	0	L4	L5

B =

0	0	0
C11	C13	C14
0	0	0
E11	E13	E14
0	0	0

G11	G13	G14
0	0	0
H3	0	0
0	0	0
J3	0	0
0	0	0
0	0	0
0	0	0

C =

1	0	0	0	0	0	0	0	0	0	0	0	0
0	1	0	0	0	0	0	0	0	0	0	0	0

D =

0	0	0
0	0	0

Equations that compute the elements of the matrices A and B are derived in the following.

1) <u>Derivation of "C" Row</u>

Starting from Eqn. (3.7-1),

$$Iyy\,\theta dd = \ell c\,[TC\,\boldsymbol{\delta} - (TC + Ts)\,\Sigma j\,\mathbf{qj}\,\sigma j\,(\ell T)] - (TC + Ts)\,\Sigma j\,\mathbf{qj}\,\varphi j\,(\ell T) - \Sigma k\; mpk\;\ell pk\; Uodot\;\boldsymbol{\Gamma}\,\mathbf{pk} + (IR + mR\,\ell R\,\ell c)\,\boldsymbol{\delta}\mathbf{dd} + mR\,\ell R\; Uodot\,\boldsymbol{\delta} - [IR + mR\,(\ell c+\ell R)^{(2)}]\,\boldsymbol{\theta}\mathbf{dd} - mR\,(\ell R + \ell c)\mathbf{wd} + mR\,\ell c\; Uodot\,\boldsymbol{\theta} - mR\,\ell R\; Uodot\,\Sigma j\,\sigma j(\ell T)\mathbf{qj} - \Sigma j\,[\,mR\,(\ell R + \ell c)\varphi j\,(\ell T) + (IR + mR\,\ell R\,\ell c)\,\sigma j(\ell T)]\,\mathbf{qdd} + qD\,A[\boldsymbol{\alpha}\int(\delta CN/\delta\alpha)(\ell a - \ell)d\ell - (\boldsymbol{\theta}\mathbf{d}/Uo)\int(\delta CN/\delta\alpha)(\ell a - \ell)^{(2)}d\ell + \Sigma j\,\mathbf{qi}\int(\delta CN/\delta\alpha)(\ell a - \ell)\,\sigma j\,(\ell)d\ell - \Sigma j\,(\mathbf{qdi}/Uo)\int(\delta CN/\delta\alpha)(\ell a - \ell)\varphi j\,(\ell)d\ell - \mathbf{Ww}\,(1/U)\int(\ell a - \ell)\,(\partial CN/\partial\alpha)\,d\ell]$$

where

$$\theta dd = d(d\theta/dt)/dt$$

Deleting a term, $\Sigma j\,[\,mR\,(\ell R + \ell c)\,\varphi j\,(\ell T) + (IR + mR\,\ell R\,\ell c)\,\sigma j(\ell T)]\,\mathbf{qdd}$

$$Iyy\,\theta dd = \ell c\,TC\,\boldsymbol{\delta} - \ell c\,(TC + Ts)\,(\mathbf{q1}\,\sigma 1\,(\ell T) + \mathbf{q2}\,\sigma 2\,(\ell T)) - (TC + Ts)\,(\mathbf{q1}\,\varphi 1\,(\ell T) + \mathbf{q2}\,\varphi 2\,(\ell T)) - mp1\,\ell p1\; Uodot\;\boldsymbol{\Gamma}\,\mathbf{p1} - mp2\,\ell p2\; Uodot\;\boldsymbol{\Gamma}\,\mathbf{p2} + (IR + mR\,\ell R\,\ell c)\,\boldsymbol{\delta}\mathbf{dd} + mR\,\ell R\; Uodot\,\boldsymbol{\delta} - (IR - mR\,\ell c^{(2)})\,\boldsymbol{\theta}\mathbf{dd} - mR\,(\ell R + \ell c)\,\mathbf{wd} + mR\,\ell c\; Uodot\,\boldsymbol{\theta} - mR\,\ell R\; Uodot\,(\sigma 1(\ell T)\mathbf{q1} + \sigma 2(\ell T)\mathbf{q2}) + qD\,A\,\boldsymbol{\alpha}\int(\delta CN/)(\ell a - \ell)d\ell - qD\,A$$

(**θd**/Uo) ∫ (δCN/ δα)(ℓa - ℓ)^(2)dℓ + qD A(**q1** ∫ (δCN/ δα)(ℓa - ℓ) σ1(ℓ)dℓ + **q2** ∫ (δCN/ δα)(ℓa - ℓ) σ2 (ℓ)dℓ)- qD A ((**qd1**/Uo) ∫ (δCN/ δα)(ℓa - ℓ)φ1 (ℓ)dℓ - (**q2d**/Uo) ∫ (δCN/ δα)(ℓa - ℓ)φ2(ℓ)dℓ) - **Ww** qD A (1/U) ∫(ℓa - ℓ) (∂CN /∂α) dℓ

Rearranging the above equation,

θdd =

\+ {mR ℓc Uodot /[Iyy + (IR + mR (ℓc+ℓR)^(2))] }**θ**

+{- (qD A (1/Uo) ∫ (δCN/ δα)(ℓa - ℓ)^(2)dℓ)/ [Iyy + (IR + mR (ℓc+ℓR)^(2))] }**θd**

\+ {[–ℓc (TC + Ts) σ1 (ℓT) - (TC + Ts) φ1 (ℓT) - mR ℓR Uodot σ1(ℓT) + qD A ∫ (δCN/ δα)(ℓa – ℓ) σ1 (ℓ)dℓ]/ [Iyy + (IR + mR (ℓc+ℓR)^(2))] }**q1**

\+ {[- qD A (1/Uo) ∫ (δCN/ δα)(ℓa - ℓ)φ1 (ℓ)dℓ]/[Iyy + (IR + mR (ℓc+ℓR)^(2))] }**q1d**

\+ {[- ℓc (TC + Ts) σ2 (ℓT)) - (TC + Ts) φ2 (ℓT) - mR ℓR Uodot σ2(ℓT) +qD A ∫ (δCN/ δα)(ℓa –ℓ) σ2 (ℓ)dℓ)]/ [Iyy + (IR + mR (ℓc+ℓR)^(2))] }**q2**

+{ [- qD A (1/Uo) ∫ (δCN/ δα)(ℓa - ℓ)φ2 (ℓ)dℓ)]/[Iyy + (IR + mR (ℓc+ℓR)^(2))] }**q2d**

\+ {[-mp1 ℓp1 Uodot]/ [Iyy + (IR + mR (ℓc+ℓR)^(2))] }**Γ p1**

\+ { [- mp2 ℓp2 Uodot]/ [Iyy + (IR + mR (ℓc+ℓR)^(2))] }**Γ p2**

\+ {[ℓc TC + mR ℓR Uodot]/ [Iyy + (IR + mR (ℓc+ℓR)^(2))] }**δ**

\+ {(IR + mR ℓR ℓc) /[Iyy + (IR + mR (ℓc+ℓR)^(2))] }**δdd**

\+ {qD A ∫ (δCN/ δα)(ℓa - ℓ)dℓ /[Iyy + (IR + mR (ℓc+ℓR)^(2))] }**α**

\+ {[- mR (ℓR + ℓc)]/ [Iyy + (IR + mR (ℓc+ℓR)^(2))] }**wd**

+{[- qD A (1/U) ∫(ℓa - ℓ) (∂CN /∂α) dℓ]/ [Iyy + (IR + mR (ℓc+ℓR)^(2))] } **Ww**

= A1* **θ** + A2 * **θd** + A3***q1** +A4* **q1d** + A5* **q2** + A6***q2d** + A7* **Γ p1** + A8* **Γ p2** + A9* **δ** + A10* **δdd** + A11* **α** + A12* **wd** + A13* **Ww**

where

A1 = {mR ℓc Uodot /[Iyy + (IR + mR (ℓc+ℓR)^(2))] }

A2 = {- (qD A (1/Uo) ∫ (δCN/ δα)(ℓa - ℓ)^(2)dℓ)/ [Iyy + (IR + mR (ℓc+ℓR)^(2))] }

A3 = {[–ℓc (TC + Ts) σ1 (ℓT) - (TC + Ts) φ1 (ℓT) - mR ℓR Uodot σ1(ℓT) + qD A ∫ (δCN/ δα)

(ℓa – ℓ) σ1 (ℓ)dℓ]/ [Iyy + (IR + mR (ℓc+ℓR)^(2))] }

A4 = {[- qD A (1/Uo) ∫ (δCN/ δα)(ℓa - ℓ)φ1 (ℓ)dℓ]/[Iyy + (IR + mR (ℓc+ℓR)^(2))] }

A5 = {[- ℓc (TC + Ts) σ2 (ℓT)) - (TC + Ts) φ2 (ℓT) - mR ℓR Uodot σ2(ℓT) +qD A ∫ (δCN/ δα)
(ℓa – ℓ) σ2 (ℓ)dℓ)]/ [Iyy + (IR + mR (ℓc+ℓR)^(2))] }

A6 = { [- qD A (1/Uo) ∫ (δCN/ δα)(ℓa - ℓ)φ2 (ℓ)dℓ)]/[Iyy + (IR + mR (ℓc+ℓR)^(2))] }

A7 = {[-mp1 ℓp1 Uodot]/ [Iyy + (IR + mR (ℓc+ℓR)^(2))] }

A8 = { [- mp2 ℓp2 Uodot]/ [Iyy + (IR + mR (ℓc+ℓR)^(2))] }

A9 = {[ℓc TC + mR ℓR Uodot]/ [Iyy + (IR + mR (ℓc+ℓR)^(2))] }

A10 = {(IR + mR ℓR ℓc) /[Iyy + (IR + mR (ℓc+ℓR)^(2))] }

A11 = {qD A ∫ (δCN/ δα)(ℓa - ℓ)dℓ /[Iyy + (IR + mR (ℓc+ℓR)^(2))] }

A12 = {[- mR (ℓR + ℓc)]/ [Iyy + (IR + mR (ℓc+ℓR)^(2))] }

A13 = {[- qD A (1/U) ∫(ℓa - ℓ) (∂CN /∂α) dℓ]/ [Iyy + (IR + mR (ℓc+ℓR)^(2))] }

Normal Force Equation

mo(**wd** - Uo **θd**) = -mT gcos(θo) **θ** + TC **δ** – (TC + Ts) Σj **qj** σj (ℓT)] - Σk mpkUodot
Γ pk + mR {ℓR **δdd** - (ℓc + ℓR) **θdd** – **wd** + Uodot **θ** - Σj [φj (ℓT) + ℓR
σj(ℓT)] **qdd** - qD A[**α** ∫ (δCN/ δα)dℓ - (**θd**/Uo) ∫ (δCN/ δα)(ℓa -
ℓ)dℓ + Σj **qi** ∫ (δCN/ δα) σj (ℓ)dℓ - Σj (**qdi**/Uo) ∫ (δCN/ δα)φj(ℓ)dℓ]
+ qD A (1/U) ∫(ℓa - ℓ) (∂CN /∂α) dℓ **Ww**

Rearranging the above equation,

mo **wd** = -mT gcos(θo) **θ** + mo Uo **θd** + TC **δ** – (TC + Ts)**q1** σ1(ℓT) – (TC + Ts)
q2 σ2 (ℓT) + mp1Uodot **Γ p1** + mp2Uodot **Γ p2** + mR ℓR **δdd** - mR
(ℓc + ℓR) **θdd** – mR **wd** + mR Uodot **θ** - mR [φ1 (ℓT) + ℓR σ1(ℓT)] **q1dd** - mR [φ2
(ℓT) + ℓR σ2(ℓT)] **q2dd** - qD A**α** ∫ (δCN/ δα)dℓ + qD A (**θd**/Uo) ∫ (δCN/ δα)
(ℓa - ℓ)dℓ -qD A **q1** ∫ (δCN/ δα) σ1(ℓ)dℓ -qD A **q2** ∫ (δCN/ δα) σ2 (ℓ)dℓ
+ qD A (**q1d**/Uo) ∫ (δCN/ δα)φ1(ℓ)dℓ + qD A (**q2d**/Uo) ∫ (δCN/ δα)φ2(ℓ)dℓ
+ qD A (1/U) ∫(ℓa - ℓ) (∂CN /∂α) dℓ **Ww**

wd =

{[mR Uodot -mT gcos(θo)]/ (mo + mR) } **θ**

+ {[mo Uo+ qD A (1/Uo) ∫ (δCN/ δα) (ℓa - ℓ)dℓ]/ (mo + mR) } **θd**

+[-mR (ℓc + ℓR)/ (mo + mR)] **θdd**

+ {[– (TC + Ts) σ1(ℓT) -qD A ∫ (δCN/ δα) σ1(ℓ)dℓ]/ (mo + mR)} **q1**

+ {[qD A (1/Uo) ∫ (δCN/ δα)φ1(ℓ)dℓ] / (mo + mR) } **q1d**

+ {[- mR [φ1 (ℓT) + ℓR σ1(ℓT)] /(mo + mR)} **q1dd**

+ {[– (TC + Ts) σ2 (ℓT) -qD A ∫ (δCN/ δα) σ2 (ℓ)dℓ]/(mo + mR)} **q2**

+ {[qD A (1/Uo) ∫ (δCN/ δα)φ2(ℓ)dℓ]/(mo + mR)} **q2d**

+ {[- mR [φ2 (ℓT) + ℓR σ2(ℓT)]/ (mo + mR) } **q2dd**

+ [mp1Uodot / (mo + mR)] **Γ p1**

+ [mp2Uodot /(mo + mR)] **Γ p2**

+ [TC/ (mo + mR)]**δ**

+ [mR ℓR / (mo + mR)]**δdd**

+ {[- qD A ∫ (δCN/ δα)dℓ]/(mo + mR) } **α**

+ qD A (1/U) ∫(ℓa - ℓ) (∂CN /∂α) dℓ **Ww**

wd = B1* **θ** + B2 * **θd** +B3* **θdd** + B4***q1** +B5* **q1d** + B6* **q1dd** + B7* **q2** + B8***q2d**

+B9* **q2dd** + B10* **Γ p1** + B11* **Γ p2** + B12* **δ** + B13* **δdd** + B14* **α**

+ B15* **Ww**

where

B1 = {[mR Uodot -mT gcos(θo)]/ (mo + mR) }

B2 = {[mo Uo+ qD A (1/Uo) ∫ (δCN/ δα) (ℓa - ℓ)dℓ]/ (mo + mR) }

B3 = [- mR (ℓc + ℓR)/ (mo + mR)]

B4 = {[– (TC + Ts) σ1(ℓT) -qD A ∫ (δCN/ δα) σ1(ℓ)dℓ]/ (mo + mR)}

B5 = {[qD A (1/Uo) ∫ (δCN/ δα)φ1(ℓ)dℓ] / (mo + mR) }

B6 = {[- mR [φ1 (ℓT) + ℓR σ1(ℓT)] /(mo + mR)}

B7 = {[– (TC + Ts) σ2 (ℓT) -qD A ∫ (δCN/ δα) σ2 (ℓ)dℓ]/(mo + mR)}

$B8 = \{[qD\,A\,(1/Uo) \int (\delta CN/\,\delta\alpha)\varphi 2(\ell)d\ell]/(mo + mR\,)\}$

$B9 = \{[\,-\,mR\,[\,\varphi 2\,(\ell T) + \ell R\;\;\sigma 2(\ell T)]/\,(mo + mR\,)\,\}$

$B10 = [mp1Uodot\,/\,(mo + mR\,)]$

$B11 = [mp2Uodot\,/(mo + mR\,)]$

$B12 = [TC/\,(mo + mR\,)]$

$B13 = \;[\,mR\,\ell R\,/\,(mo + mR\,)\,]$

$B14 = \{[-\,qD\,A \int (\delta CN/\,\delta\alpha)d\ell]/(mo + mR\,)\,\}$

$B15 = \{[qD\,A\,(1/U) \int(\ell a - \ell)\,(\partial CN\,/\partial\alpha)\,d\ell]/(mo + mR\,)\,\}$

Substituting **wd** into **θdd** equation, and rearranging it,

$$\mathbf{\theta dd} = A1*\mathbf{\theta} + A2*\mathbf{\theta d} + A3*\mathbf{q1} + A4*\mathbf{q1d} + A5*\mathbf{q2} + A6*\mathbf{q2d} + A7*\mathbf{\Gamma\,p1} + A8*\mathbf{\Gamma\,p2}$$

$$A12*B1*\mathbf{\theta} + A12*B2*\mathbf{\theta d} + A12*B3*\mathbf{\theta dd} + A12*B4*\mathbf{q1} + A12*B5*\mathbf{q1d} +$$

$$A12*B6*\mathbf{q1dd} + A12*B7*\mathbf{q2} + A12*B8*\mathbf{q2d} + A12*B9*\mathbf{q2dd} + A12*B10*\mathbf{\Gamma\,p1}$$

$$+ A12*B11*\mathbf{\Gamma\,p2} + A12*B12*\mathbf{\delta} + A12*B13*\mathbf{\delta dd} + A12*B14*\mathbf{\alpha} + A9*\mathbf{\delta} +$$

$$A10*\mathbf{\delta dd} + A11*\mathbf{\alpha} + A13*\mathbf{Ww}$$

$\mathbf{\theta dd} =$

$[(A1 + A12*B1)/(1 - A12*B3)]*\mathbf{\theta}$

$+ [(A2 + A12*B2)/(1 - A12*B3)]*\mathbf{\theta d}$

$+ [\,(A3 + A12*B4)/(1 - A12*B3)]*\mathbf{q1}$

$+ [(A4 + A12*B5)/(1 - A12*B3)]*\mathbf{q1d}$

$+ [(\,A12*B6)/(1 - A12*B3)]*\mathbf{q1dd}$

$+ [(A5 + A12*B7)/(1 - A12*B3)]*\mathbf{q2}$

$+ [\,(A6 + A12*B8)/(1 - A12*B3)]*\mathbf{q2d}$

$+ [(A12*B9)/(1 - A12*B3)]*\mathbf{q2dd}$

$+ [(A7 + A12*B10)/(1 - A12*B3)]*\mathbf{\Gamma\,p1}$

$+ [(A8 + A12*B11)/(1 - A12*B3)]*\mathbf{\Gamma\,p2}$

$+ [\,A9 + A12*B12)/(1 - A12*B3)]*\mathbf{\delta}$

$+ [(A10 + A12*B13)/(1 - A12*B3)]*\mathbf{\delta dd}$

$+ [(A11 + A12*B14)/(1- A12*B3)] * \boldsymbol{\alpha}$

$+ [(A13 + A12*B15)/(1- A12*B3)]*\mathbf{Ww}$

$= C1* \boldsymbol{\theta} + C2 * \boldsymbol{\theta}\mathbf{d} + C3*\mathbf{q1} + C4* \mathbf{q1d} + C5* \mathbf{q1dd} + C6* \mathbf{q2} +$

$C7*\mathbf{q2d} + C8 * \mathbf{q2dd} + C9* \boldsymbol{\Gamma}\,\mathbf{p1} + C10* \boldsymbol{\Gamma}\,\mathbf{p2} + C11* \boldsymbol{\delta} + C12* \boldsymbol{\delta}\mathbf{dd} +$

$C13* \boldsymbol{\alpha} + C14* \mathbf{Ww}$

where

$C1 = [(A1+ A12*B1)/(1- A12*B3)]$

$C2 = [(A2 + A12*B2)/(1- A12*B3)]$

$C3 = [(A3 + A12*B4)/(1- A12*B3)]$

$C4 = [(A4+ A12*B5)/(1- A12*B3)]$

$C5 = [(A12*B6)/(1- A12*B3)]$

$C6 = [(A5 + A12*B7)/(1- A12*B3)]$

$C7 = [(A6+ A12*B8)/(1- A12*B3)]$

$C8 = [(A12*B9)/(1- A12*B3)]$

$C9 = [(A7 + A12*B10)/(1- A12*B3)]$

$C10 = [(A8 + A12*B11)/(1- A12*B3)]$

$C11 = [A9+ A12*B12)/(1- A12*B3)]$

$C12 = [(A10 + A12*B13)/(1- A12*B3)]$

$C13 = [(A11 + A12*B14)/(1- A12*B3)]$

$C14 = [(A13 + A12*B15)/(1- A12*B3)]$

2) Derivation of "E" and "G" Rows

Generalized Force (Moment) of ith Bending Mode [from Eqn. (3.4-31)]

$Qi = [mR*Uodot * \varphi i(\ell T) – g*\cos(\theta o) \int \varphi i(\ell)m(\ell)d\ell]\ \boldsymbol{\theta}$

$+[(qDA/Uo) \int \varphi i(\ell)(\delta CN/ \delta\alpha)(\ell a - \ell)d\ell]\ \boldsymbol{\theta}\mathbf{d}$

$- [mR *\varphi i(\ell T)*(\ell c + \ell R)]\ \boldsymbol{\theta}\mathbf{dd} + [TC\ \varphi i(\ell T)]\ \boldsymbol{\delta}$

$+ [mR*\ell R *\varphi i(\ell T)]\ \boldsymbol{\delta}\mathbf{dd} - [qDA\int \varphi i(\ell)(\delta CN/ \delta\alpha)\ d\ell]\ \boldsymbol{\alpha}$

- [mR φi(ℓT)] **wd** + Σk [Uodot * mpk φi(ℓpk)] **Γ pk**

- Σj [(TC + Ts) φi(ℓT) σj (ℓT) + qDA∫ φi(ℓ)(δCN/ δα) σj (ℓ)dℓ] **qj**

+ Σj [(qDA/Uo) ∫ φi(ℓ)(δCN/ δα) φj(ℓ)dℓ]**qjd**

- Σj [mR φi(ℓT) (φj(ℓT) + ℓR σj (ℓT))] **qjdd**

+ qD A3 (1/U) ∫ Φi(ℓ)(∂CN /∂α) dℓ} **Ww**

<u>(E Row)</u>

When i=1,

Q1/M1 = {[mR*Uodot * φ1(ℓT) – g*cos(θo) ∫φ1(ℓ)m(ℓ)dℓ]/M1} **θ**

+{[(qDA/Uo) ∫φ1(ℓ)(δCN/ δα)(ℓa - ℓ)dℓ]/M1} **θd**

+{- [mR *φ1(ℓT)*(ℓc + ℓR)] /M1 } **θdd**

+ {-[(TC + Ts) φ1(ℓT) σ1 (ℓT) + qDA∫ φ1(ℓ)(δCN/ δα) σ1 (ℓ)dℓ] /M1 }**q1**

+{ [(qDA/Uo) ∫ φ1(ℓ)(δCN/ δα) φ1(ℓ)dℓ]/M1}**q1d**

- { [mR φ1(ℓT) (φ1(ℓT) + ℓR σ1(ℓT))] /M1 } **q1dd**

+ {-[(TC + Ts) φ1(ℓT) σ2 (ℓT) + qDA∫ φ1(ℓ)(δCN/ δα) σ2 (ℓ)dℓ] /M1 }**q2**

+ { [(qDA/Uo) ∫ φ1(ℓ)(δCN/ δα) φ2(ℓ)dℓ] /M1}**q2d**

- {[mR φ1(ℓT) (φ2(ℓT) + ℓR σ2(ℓT))] /M1 } **q2dd**

+ [Uodot * mp1 φ1(ℓp1) /M1] **Γ p1** + [Uodot * mp2 φ1(ℓp2) /M1] **Γ p2**

+{[TC φ1(ℓT)] /M1 } **δ**

+ {[mR*ℓR *φ1(ℓT)] /M1 } **δdd**

+ {- [qDA∫ φ1(ℓ)(δCN/ δα) dℓ] /M1}**α**

+[- mR φ1(ℓT) /M1] **wd**

+ qD A3 (1/U) ∫ Φ1(ℓ)(∂CN /∂α) dℓ} **Ww**

Q1/M1 = D1* **θ** + D2* **θd** + D3* **θdd** + D4* **q1** + D5* **q1d** + D6***q1dd**

D7* **q2** + D8* **q2d** + D9* **q2dd** + D10* **Γ p1** + D11* **Γ p2**

D12***δ** + D13* **δdd** + D14* **α** + D15* **wd** + D16* **Ww**

Where

wd = B1* **θ** + B2 * **θd** +B3* **θdd** + B4***q1** +B5* **q1d** + B6* **q1dd**

+ B7* **q2** + B8***q2d** +B9* **q2dd** + B10* **Γ p1** + B11* **Γ p2**

+ B12* **δ** + B13* **δdd** + B14* **α** + B15* **Ww**

and,

D1 = {[mR*Uodot * φ1(ℓT) – g*cos(θo) ∫φ1(ℓ)m(ℓ)dℓ]/M1}

D2 = {[(qDA/Uo) ∫φ1(ℓ)(δCN/ δα)(ℓa - ℓ)dℓ]/M1}

D3 = {- [mR *φ1(ℓT)*(ℓc + ℓR)] /M1 }

D4 = {-[(TC + Ts) φ1(ℓT) σ1 (ℓT) + qDA∫ φ1(ℓ)(δCN/ δα) σ1 (ℓ)dℓ] /M1 }

D5 = { [(qDA/Uo) ∫ φ1(ℓ)(δCN/ δα) φ1(ℓ)dℓ]/M1}

D6 = - { [mR φ1(ℓT) (φ1(ℓT) + ℓR σ1(ℓT))] /M1 }

D7 = {-[(TC + Ts) φ1(ℓT) σ2 (ℓT) + qDA∫ φ1(ℓ)(δCN/ δα) σ2 (ℓ)dℓ] /M1 }

D8 = { [(qDA/Uo) ∫ φ1(ℓ)(δCN/ δα) φ2(ℓ)dℓ] /M1}

D9 = - {[mR φ1(ℓT) (φ2(ℓT) + ℓR σ2(ℓT))] /M1 }

D10 = [Uodot * mp1 φ1(ℓp1) /M1]

D11 = [Uodot * mp2 φ1(ℓp2) /M1]

D12 = {[TC φ1(ℓT)] /M1 }

D13 = {[mR*ℓR *φ1(ℓT)] /M1 }

D14 = {- [qDA∫ φ1(ℓ)(δCN/ δα) dℓ] /M1}

D15 = [- mR φ1(ℓT) /M1]

D16 = [qD A3 (1/U) ∫ Φ1(ℓ)(∂CN /∂α) dℓ]/M1

Substituting **wd,**

Q1/M1 = (D1+ D15*B1)* **θ** + (D2 + D15* B2) * **θd** + (D3 + D15*B3)* **θdd** +

(D4 + D15*B4)* **q1** + (D5 + D15*B5)* **q1d** + (D6 + D15*B6)* **q1dd** +

(D7 + D15*B7)* **q2** + (D8 + D15*B8)* **q2d** + (D9 + D15*B9)* **q2dd** +

(D10 + D15*B10)* **Γ p1** + (D11 + D15*B11)* **Γ p2** + (D12 + D15*B12)* **δ** +

(D13 + D15*B13)* **δdd** + (D14 + D15*B14)* **α** + (D16 + D15*B15) **Ww**

q1dd = -2.0ζ1ω1* **q1d** - ω1^(2)* **q1** + Q1/M1

= -2.0ζ1ω1* **q1d** - ω1^(2)* **q1** + (D1+ D15*B1)* **θ** + (D2 + D15* B2) * **θd** + (D3 + D15*B3)* **θdd** + (D4 + D15*B4)* **q1** + (D5 + D15*B5)* **q1d** + (D6 + D15*B6)* **q1dd** + (D7 + D15*B7)* **q2** + (D8 + D15*B8)* **q2d** + (D9 + D15*B9)* **q2dd** + (D10 + D15*B10)* **Γ p1** + (D11 + D15*B11)* **Γ p2** + (D12 + D15*B12)* **δ** + (D13 + D15*B13)* **δdd** + (D14 + D15*B14)* **α** + (D16 + D15*B15) **Ww**

[1 - (D6 + D15*B6)]* **q1dd** = -2.0ζ1ω1* **q1d** - ω1^(2)* **q1** + (D1+ D15*B1)* **θ** + (D2 + D15* B2) * **θd** + (D3 + D15*B3)* **θdd** + (D4 + D15*B4) * **q1** + (D5 + D15*B5)* **q1d** + (D7 + D15*B7)* **q2** + (D8 + D15*B8)* **q2d** + (D9 + D15*B9)* **2dd** + (D10 + D15*B10)* **Γ p1** + (D11 + D15*B11)* **Γ p2** + (D12 + D15*B12)* **δ** + (D13 + D15*B13)* **δdd** + (D14 + D15*B14)* **α** + (D16 + D15*B15) **Ww**

q1dd = E1* **θ** + E2* **θd** + E3* **θdd** + E4* **q1** + E5* **q1d** + E6* **q2** + E7* **q2d** + E8* **q2dd** + E9* **Γ p1** + E10* **Γ p2** + E11* **δ** + E12* **δdd** + E13* **α** + E14***Ww**

where

E1 = (D1+ D15*B1) / [1 - (D6 + D15*B6)]

E2 = (D2 + D15* B2) / [1 - (D6 + D15*B6)]

E3 = (D3 + D15*B3) / [1 - (D6 + D15*B6)]

E4 = (- ω1^(2) + D4 + D15*B4) / [1 - (D6 + D15*B6)]

E5 = (- 2.0 ζ 1 * ω1 + D5 + D15*B5) / [1 - (D6 + D15*B6)]

E6 = (D7 + D15*B7) / [1 - (D6 + D15*B6)]

E7 = (D8 + D15*B8) / [1 - (D6 + D15*B6)]

E8 = (D9 + D15*B9) / [1 - (D6 + D15*B6)]

$$E9 = (D10 + D15*B10) / [1 - (D6 + D15*B6)]$$

$$E10 = (D11 + D15*B11) / [1 - (D6 + D15*B6)]$$

$$E11 = (D12 + D15*B12) / [1 - (D6 + D15*B6)]$$

$$E12 = (D13 + D15*B13) / [1 - (D6 + D15*B6)]$$

$$E13 = (D14 + D15*B14) / [1 - (D6 + D15*B6)]$$

$$E14 = (D16 + D15*B15)/ [1 - (D6 + D15*B6)]$$

<u>(G Row)</u>

When i = 2,

$$Q2/M2 = \{[mR*Uodot * \varphi2(\ell T) - g*\cos(\theta o) \int \varphi2(\ell)m(\ell)d\ell]/M2\} \boldsymbol{\theta}$$

$$+\{[(qDA/Uo) \int \varphi2(\ell)(\delta CN/ \delta\alpha)(\ell a - \ell)d\ell]/M2\} \mathbf{\theta d}$$

$$+\{- [mR *\varphi2(\ell T)*(\ell c + \ell R)] /M2 \} \mathbf{\theta dd}$$

$$+ \{-[(TC + Ts) \varphi2(\ell T) \sigma1 (\ell T) + qDA\int \varphi2(\ell)(\delta CN/ \delta\alpha) \sigma1 (\ell)d\ell] /M2 \}\mathbf{q1}$$

$$+\{ [(qDA/Uo) \int \varphi2(\ell)(\delta CN/ \delta\alpha) \varphi1(\ell)d\ell]/M2\}\mathbf{q1d}$$

$$- \{ [mR \varphi2(\ell T) (\varphi1(\ell T) + \ell R \ \sigma1(\ell T))] /M1 \} \mathbf{q1dd}$$

$$+ \{-[(TC + Ts) \varphi2(\ell T) \sigma2 (\ell T) + qDA\int \varphi2(\ell)(\delta CN/ \delta\alpha) \sigma2 (\ell)d\ell] /M2 \}\mathbf{q2}$$

$$+ \{ [(qDA/Uo) \int \varphi2(\ell)(\delta CN/ \delta\alpha) \varphi2(\ell)d\ell] /M1\}\mathbf{q2d}$$

$$- \{[mR \varphi2(\ell T) (\varphi2(\ell T) + \ell R \ \sigma2(\ell T))] /M1 \} \mathbf{q2dd}$$

$$+ [Uodot * mp1 \ \varphi2(\ell p1) /M2] \mathbf{\Gamma \ p1}$$

$$+ [Uodot * mp2 \ \varphi2(\ell p2) /M2] \mathbf{\Gamma \ p2}$$

$$+\{[TC \ \varphi2(\ell T)] /M2 \} \boldsymbol{\delta}$$

$$+ \{[mR*\ell R *\varphi2(\ell T)] /M2 \} \mathbf{\delta dd}$$

$$+ \{- [qDA\int \varphi2(\ell)(\delta CN/ \delta\alpha) d\ell] /M2\}\boldsymbol{\alpha}$$

$$+[- mR \ \varphi2(\ell T) /M2] \mathbf{wd}$$

$$+ [qD \ A3 \ (1/U) \int \Phi2(\ell)(\partial CN /\partial\alpha) \ d\ell/M2] \mathbf{Ww}$$

$$Q2/M2 = F1* \boldsymbol{\theta} + F2* \mathbf{\theta d} + F3* \mathbf{\theta dd} + F4* \mathbf{q1} + F5* \mathbf{q1d} + F6*\mathbf{q1dd} + F7* \mathbf{q2} +$$

$$F8* \mathbf{q2d} + F9* \mathbf{q2dd} + F10* \mathbf{\Gamma \ p1} + F11* \mathbf{\Gamma \ p2} + F12*\boldsymbol{\delta} + F13* \mathbf{\delta dd} +$$

$F14^* \boldsymbol{\alpha} + F15^* \mathbf{wd} + F16^* \mathbf{Ww}$

where

$$\mathbf{wd} = B1^* \boldsymbol{\theta} + B2 ^* \boldsymbol{\theta d} + B3^* \boldsymbol{\theta dd} + B4^* \mathbf{q1} + B5^* \mathbf{q1d} + B6^* \mathbf{q1dd} + B7^* \mathbf{q2} + B8^* \mathbf{q2d} + B9^* \mathbf{q2dd} + B10^* \boldsymbol{\Gamma} \mathbf{p1} + B11^* \boldsymbol{\Gamma} \mathbf{p2} + B12^* \boldsymbol{\delta} + B13^* \boldsymbol{\delta dd} + B14^* \boldsymbol{\alpha} + B15^* \mathbf{Ww}$$

and,

$F1 = \{[mR^*Uodot * \varphi2(\ell T) - g^*\cos(\theta o) \int\varphi2(\ell)m(\ell)d\ell]/M2\}$

$F2 = \{[(qDA/Uo) \int\varphi2(\ell)(\delta CN/ \delta\alpha)(\ell a - \ell)d\ell]/M2\}$

$F3 = \{- [mR *\varphi2(\ell T)*(\ell c + \ell R)] /M2 \}$

$F4 = \{-[(TC + Ts) \varphi2(\ell T) \sigma1 (\ell T) + qDA\int \varphi2(\ell)(\delta CN/ \delta\alpha) \sigma1 (\ell)d\ell] /M2 \}$

$F5 = \{ [(qDA/Uo) \int \varphi2(\ell)(\delta CN/ \delta\alpha) \varphi1(\ell)d\ell]/M2\}$

$F6 = - \{ [mR \varphi2(\ell T) (\varphi1(\ell T) + \ell R \ \sigma1(\ell T))] /M2 \}$

$F7 = \{-[(TC + Ts) \varphi2(\ell T) \sigma2 (\ell T) + qDA\int \varphi2(\ell)(\delta CN/ \delta\alpha) \sigma2 (\ell)d\ell] /M2 \}$

$F8 = \{ [(qDA/Uo) \int \varphi2(\ell)(\delta CN/ \delta\alpha) \varphi2(\ell)d\ell] /M2\}$

$F9 = - \{[mR \varphi2(\ell T) (\varphi2(\ell T) + \ell R \ \sigma2(\ell T))] /M2 \}$

$F10 = [Uodot * mp1 \ \varphi2(\ell p1) /M2]$

$F11 = [Uodot * mp2 \ \varphi2(\ell p2) /M2]$

$F12 = \{[TC \ \varphi2(\ell T)] /M2 \}$

$F13 = \{[mR^*\ell R *\varphi2(\ell T)] /M2 \}$

$F14 = \{- [qDA\int \varphi2(\ell)(\delta CN/ \delta\alpha) d\ell] /M2\}$

$F15 = [- mR \ \varphi2(\ell T) /M2]$

$F16 = [qD \ A3 \ (1/U) \int \Phi2(\ell)(\partial CN /\partial\alpha) \ d\ell/M2]$

Substituting **wd,**

$$Q2/M2 = (F1+ F15^*B1)^* \boldsymbol{\theta} + (F2 + F15^* B2) ^* \boldsymbol{\theta d} + (F3 + F15^*B3)^* \boldsymbol{\theta dd} + (F4 + F15^*B4)^* \mathbf{q1} + (F5 + F15^*B5)^* \mathbf{q1d} + (F6 + F15^*B6)^* \mathbf{q1dd} + (F7 + F15^*B7)^* \mathbf{q2} + (F8 + F15^*B8)^* \mathbf{q2d} + (F9 + F15^*B9)^* \mathbf{q2dd} + (F10 + F15^*B10)^* \boldsymbol{\Gamma} \mathbf{p1} + (F11 + F15^*B11)^* \boldsymbol{\Gamma} \mathbf{p2} + (F12 + F15^*B12)^* \boldsymbol{\delta} +$$

(F13 + F15*B13)* **δdd** + (F14 + F15*B14)* **α** + (F16 + F15*B15)**Ww**

q2dd = -2.0ζ2ω2* **q2d** - ω1^(2)* **q2** + Q2/M2

= -2.0ζ2ω2* **q2d** - ω1^(2)* **q2** + (F1+ F15*B1)* **θ** + (F2 + F15* B2) * **θd** + (F3 + F15*B3)* **θdd** + (F4 + F15*B4)* **q1** + (F5 + F15*B5)* **q1d** + (F6 + F15*B6)* **q1dd** + (F7 + F15*B7)* **q2** + (F8 + F15*B8)* **q2d** + (F9 + F15*B9)* **q2dd** + (F10 + F15*B10)* **Γ p1** + (F11 + F15*B11)* **Γ p2** + (F12 + F15*B12)* **δ** + (F13 + F15*B13)* **δdd** + (F14 + F15*B14)* **α** + (F16 + F15*B15) **Ww**

[1 - (F9 + F15*B9)]* **q2dd** = -2.0ζ2ω2* **q2d** - ω1^(2)* **q2** + (F1+ F15*B1)* **θ** + (F2 + F15* B2) * **θd** + (F3 + F15*B3)* **θdd** + (F4 + F15*B4)* **q1** + (F5 + F15*B5)* **q1d** + (F6 + F15*B6)* **q1dd** + (F7 + F15*B7)* **q2** + (F8 + F15*B8)* **q2d** + (F10 + F15*B10)* **Γ p1** + (F11 + F15*B11)* **Γ p2** + (F12 + F15*B12)* **δ** + (F13 + F15*B13)* **δdd** + (F14 + F15*B14)* **α** + (F16 + F15*B15) **Ww**

q2dd = G1* **θ** + G2* **θd** + G3* **θdd** + G4* **q1** + G5* **q1d** + G6* **q1dd** + G7* **q2** + G8* **q2d** + G9* **Γ p1** + G10* **Γ p2** + G11* **δ** + G12* **δdd** + G13* **α** + G14***Ww**

where

G1 = (F1+ F15*B1) /[1 - (F9 + F15*B9)]

G2 = (F2 + F15* B2) /[1 - (F9 + F15*B9)]

G3 = (F3 + F15*B3) /[1 - (F9 + F15*B9)]

G4 = (- ω2^(2) + F4 + F15*B4) /[1 - (F9 + F15*B9)]

G5 = (- 2.0 ζ 2 * ω2 + F5 + F15*B5) /[1 - (F9 + F15*B9)]

G6 = (F6 + F15*B6) /[1 - (F9 + F15*B9)]

G7 = (F7 + F15*B7) /[1 - (F9 + F15*B9)]

G8 = (F8 + F15*B8) /[1 - (F9 + F15*B9)]

G9 = (F10 + F15*B10) /[1 - (F9 + F15*B9)]

G10 = (F11 + F15*B11) /[1 - (F9 + F15*B9)]

G11 = (F12 + F15*B12) /[1 - (F9 + F15*B9)]

G12 = (F13 + F15*B13) /[1 - (F9 + F15*B9)]

G13 = (F14 + F15*B14) /[1 - (F9 + F15*B9)]

G14 = (F16 + F15*B15) /[1 - (F9 + F15*B9)]

3) <u>Derivation of "H" Row</u>

Γ1dd = H1* **Γ1** + H2 **Γp2** + H3 **δ**

where

H1 = - (1/ Lp1) {Uo'mp1 /mT + u' + μp1 (ℓp1– Lp1)}
H2 = - (1/ Lp1) { μp2 (ℓp1 – Lp1)}
H3 = -(1/ Lp1) { Td /mT - μc (ℓp1– Lp1)}

4) <u>Derivation of "J" Row</u>

<u>Γ2dd</u> = J1 Γp1 + J2 Γp2 + J3 δ

Where

J1 = - (1/ Lp2) {Uo'mp1 /mT + u' + μp1 (ℓp2– Lp2)}
J2 = - (1/ Lp2) μp2 (ℓp2 – Lp2)}
J3 = -(1/ Lp2) { Td /mT - μc (ℓp2 – Lp2)}

5) <u>Derivation of "K" Row</u>

<u>Position Gyro Math Model</u>

(θPG)d = K1 θ + K2 q1 + K3 q2 + K4 θPG

where

K1 = 1/ τ , τ : time constant
K2 = σ1(ℓPG) / τ
K3 = σ2(ℓPG) / τ
K4 = -1/ τ

6) <u>Derivation of "L" Row</u>

Rate Gyro Math Model (Eqn. 3.7-8)

$(\theta_{RG})_{dd} = L1\ \theta d + L2\ q1d + L3\ q2d + L4\ \theta_{RG} + L5\ (\theta_{RG})_d$

where

$L1 = \omega R^{\wedge}(2)\ KR$, KR : Rate Gyro Gain
$L2 = \omega R^{\wedge}(2)\ KR\ \sigma 1(\ell RG)$
$L3 = \omega R^{\wedge}(2)\ KR\ \sigma 2(\ell RG)$
$L4 = -\ \omega R^{\wedge}(2)$
$L5 = -\ 2\ \xi R\ \omega R^{\wedge}(2)$

Chapter 4 Satellite Launch Vehicle Attitude Control System Design Example and Analysis

Theories have been introduced, expanded, and explained in the previous chapters. In this chapter, these theories are applied, and it is demonstrated that a robust attitude control system can be designed by use of these new control theories, more specifically and prominently the H-infinity control theory.

For our design example, we use a set of system parameters and other related data found in ref. 18. From these, a 13-state state space equation is constructed. Using this math model, an optimized dynamic feedback loop is designed that stabilizes the attitude control system. A step input is applied to show its step response, by which the system performance and the stability can be evaluated. This step response will be used as a reference when we study math modeling error analysis in Section 4.4.

We start with a simple rigid body, and then later we add more system parameters, including:

Sloshing
Bending
Engine dynamics

The H-infinity, Loop Shaping, and Coprime Factorization approaches are chosen for the system design. In this approach, unlike classical control theory, singular values, rather than eigenvalues, are chosen for the design and analysis parameter. These singular values are computed as frequency varies over the range of our interest, i.e., 0.01 rad/sec to 100 rad/sec.

4.1 Rigid Body

Design parameters, from ref . 18, for the rigid body are listed below:

mo (lv mass without slosh mass)	: 5,058 slugs
mT (total mass, excluding slosh mass)	: 5,058 slugs
Tc (control gimballed thrust)	: 410,400 lbf
Iyy (pitch axis product of inertia)	: 1,694,000 slugs ft^2
uo (nominal axial velocity)	: 3,836 ft/sec
udot (axial acceleration)	: 66.3 ft/sec^2
μc (control moment coefficient)	: 4.56 sec^(-2)

Lα (aerodynamic load)	: 198,000 lb/rad
lα (distance to center of pressure)	: 34.1 ft
lc (distance to engine swivel point)	: -32.3 ft
θo (flight path angle)	: 60.0 deg
IR (engine moi about swivel point)	: 377 slugs ft^2
lR (dist. from engine cg to swivel point)	: 2.52 ft
g(gravity)	: 32.174 ft/sec^2

Using these parameters, a state space system equation is constructed as shown below. Equations derived in Chapter 3 are used to compute the elements of the system matrices A and B. The CD-ROM attached to this book (see back cover) that contains an Excel program can be used to computes these elements.

xd = Ax + Bu
y = Cx + Du

where

$$A = \begin{vmatrix} 0 & 1 & 0 & 0 & 0 \\ 3.9848 & -0.00794 & 0 & 0 & 0 \\ 25 & 0 & -25 & 0 & 0 \\ 0 & 0 & 0 & 0 & 1 \\ 0 & 8479.7440 & 0 & -21199.36 & -116.4800 \end{vmatrix}$$

$$B = \begin{vmatrix} 0 & 0 & 0 \\ -7.8235 & -9.413 & 0.00245 \\ 0 & 0 & 0 \\ 0 & 0 & 0 \\ 0 & 0 & 0 \end{vmatrix}$$

$$C = \begin{vmatrix} 1 & 0 & 0 & 0 & 0 \\ 0 & 1 & 0 & 0 & 0 \end{vmatrix}$$

$$D = \begin{vmatrix} 0 & 0 & 0 \\ 0 & 0 & 0 \end{vmatrix}$$

where

xd = dx/dy
x = Transpose of [x1, x2, x3, x4, x5]
x1 = attitude angle
x2 = attitude angular rate

x3 = position gyro output
x4 = rate gyro output
x5 = time derivative of the rate gyro output

u1 = attitude angle command
u2 = angle of attack
u3 = wind

y1 = attitude output
y2 = attitude angular rate output

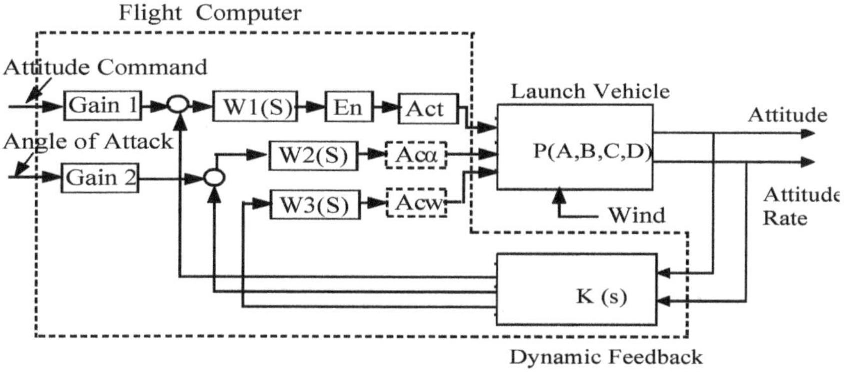

Fig. 4.1-1a. Control System Configuration

Fig. 4.1-1a shows control system block diagram. W1(s), W2(s), and W3(s) are the weighting functions that are used to suppress noisy signals and enhance the desired signals (Loop Shaping). In our design, these are very important design elements, and finding the right functions is not a simple task and requires a trial-and-error approach with some experience. Output of "En" block is the engine command signal and the "Act" block is an engine nozzle actuator (simplified).

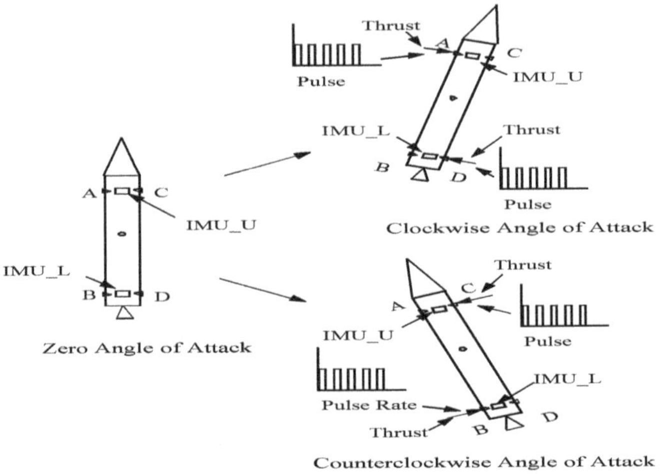

Fig. 4.1-1b Angle of Attack Reaction Jets

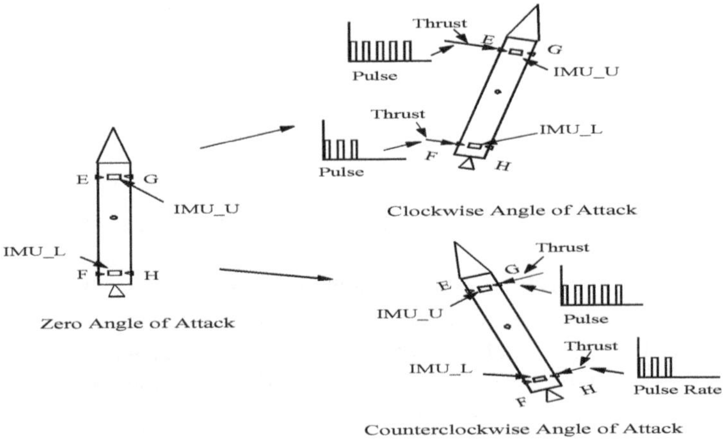

Fig. 4.1-1c Wind Reaction Jets

P(A,B,C,D) block is a linearized math model of the launch vehicle and K(S) is the feedback block that is to be designed to ensure the desired system performance, that is, an optimized robust performance

The launch vehicle configuration to which the H-infinity MIMO attitude control system technique can be applied has a configuration that is different from the conventional launch vehicle configuration. In the H-infinity MIMO attitude control system, the flight computer computes not only attitude feedback but also angle of attack and the lateral velocity (due to the wind alone in our study) feedbacks.

Therefore, this new breed launch vehicle requires installation of unconventional subsystems to accommodate the last two feedback variables, that is, the angle of attack and wind effect. The attitude feedback is treated as it has been done in the conventional rockets.

To accommodate the angle of attack feedback, "Acα" (see Fig. 4.1-1a and Fig. 4.1-1d) subsystem is added. All it does is to create angle attack to respond to the feedback. How it works is shown in Fig. 4.1-1b. In pitch plane, two pairs of reaction jets, (A,D) and (C,B) are installed. When a clockwise angle of attack is to be generated, then the reaction jet pair (A,D) is on. Note that the reaction jet pulse rate of "A" and that of "D" are identical and the reactions are in opposite direction so that this pair causes clockwise rotation of the angle of attack but does not generate lateral displacement. When a counter-clockwise angle of attack is to be generated, then the reaction jet pair (C,B) is on. The reaction jet pulse rate of "C" and that of "B" are also identical and the reactions are in opposite direction so that this pair causes counter-clockwise rotation of the angle of attack but not lateral displacement unlike the wind effect.

To accommodate the wind feedback, "Acw" (see Fig. 4.1-1a and Fig. 4.1-1d) subsystem is added. All it does is to create the wind effect to respond to the feedback. How it works is shown in Fig. 4.1-1c. In pitch plane, two pairs of reaction jets, (E,F) and (G,H) are installed. When a clockwise wind reaction is to be generated, then the reaction jet pair (E,F) is on. Note that the reaction jet pulse rates of "E" and that of "F" are different and the reactions are in the same direction, thus this pair creates the wind effect. When a counter-clockwise wind reaction is to be generated, then the reaction jet pair (G,H) are on. The reaction jet pulse rates are also different and the reactions are in the same direction, thus this pair also creates the wind effect.

When SISO design technique is used as has been done in the past, the flight computer software cannot compute the angle of attack and wind effect feedbacks which we need to improve LV flight performance. Therefore, it is impossible for SISO control system to operate subsystems like "Acα" and "Acw" which are the kind of subsystem that enables the launch vehicle to withstand the strong wind.

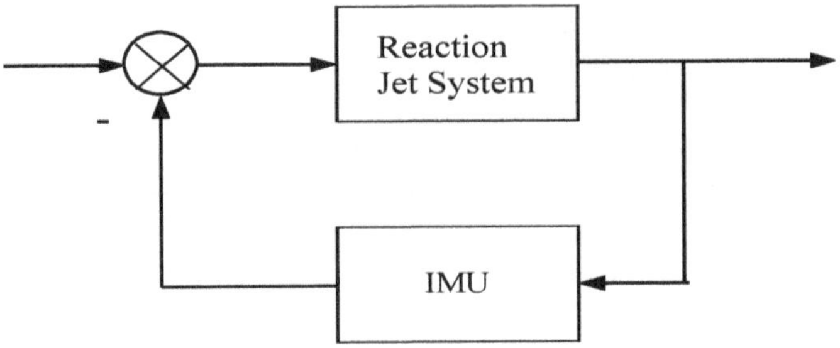

Fig. 4.1-1d Block Diagram for "Acα" or "Acw"

The "Acα" and "Acw" are subsystems independent of P(A,B,C,D) in Fig 4.1-1a. The inputs to these subsystems are from K(S) in Fig. 4.1-1a. Strictly speaking, the "Acα" and "Acw are not computing systems, but they support the flight computer. The IMU block consists of a gyro and an accelerometer. It measure rotation and lateral motion. For "Acα," the amount of rotation at IMU_U and IMU_L will be the same and in the same direction. The accelerometer measures the same amount of lateral motion also but in opposite direction. For "Acw, " the amount of rotation at IMU_U and IMU_L will be the same and in the same direction. But the amount of lateral motion that the accelerometer at IMU_U measures is different from that at IMU_L but in same direction.

The function of "Acα" and "Acw" blocks is to execute the commands coming from the feedback block, K(S). If there is no wind, then the input to "Acw" block is null, but the input to the "Acα" is not zero because the sloshing and bending. The sloshing and bending generate the angle of attack. If there is a high velocity wind, then the inputs to these two blocks will be optimally generated by the K(S) block, and activate the angle of attack and wind effect reaction jets to execute the commands (that is, responding to the inputs).

As stated in the Introduction, unlike classical control system theory, here we construct a multi-input multi-output system. In this example, there are three inputs and two outputs. The inputs are: 1) attitude command, 2) angle of attack, and 3) wind velocity effect. The outputs are: 1) the attitude angle and 2) attitude angular rate. Because the wind and the angle of attack are treated as inputs in this approach, the control system is able to maintain its stability in piece wise linearized sense regardless how strong the wind and angle of attack disturbances are once a stable control system is designed. One other interesting observation is that the wind effect feedback and the wind input are summed up by the vehicle maneuver.

To observe how the eigenvalues and the singular values versus frequencies vary as the attitude control system design improves, first we compute them when there is no feedback (i.e. open-loop system) and second, when a non-optimal feedback loop is added (i.e. closed-loop system before optimization), and third, when an optimal feedback loop is designed (i.e. H-infinity optimization).

i. <u>Open-Loop System</u>

The eigenvalues of the open-loop system (no feedback loop) are:

-25.00
1.99
-2.00
-58.24 + 133.44i
-58.24 - 133.44i

The positive eigenvalue, 1.99, indicates that the system is unstable. The

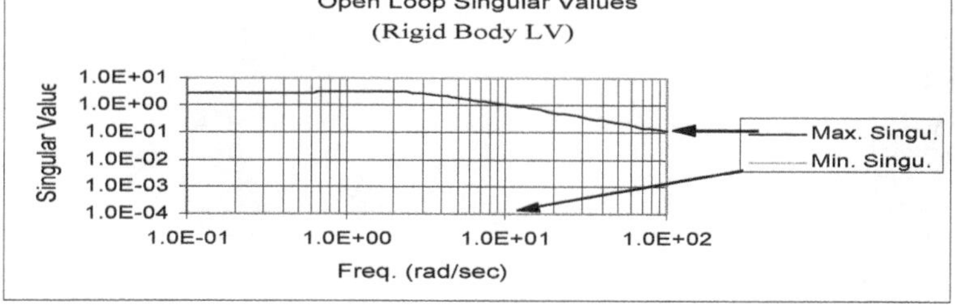

Fig. 4.1-2. Open-Loop Singular Value

singular value vs. frequency plot is shown in Fig. 4.1-2. The maximum singular value is 3.5395 at 1.3257 rad/sec.

Fig. 4.1-3 shows a 1.0 deg step input response of the open-loop system. It is seen here that the attitude increases more than 30.0 deg in less than 2.0 sec.

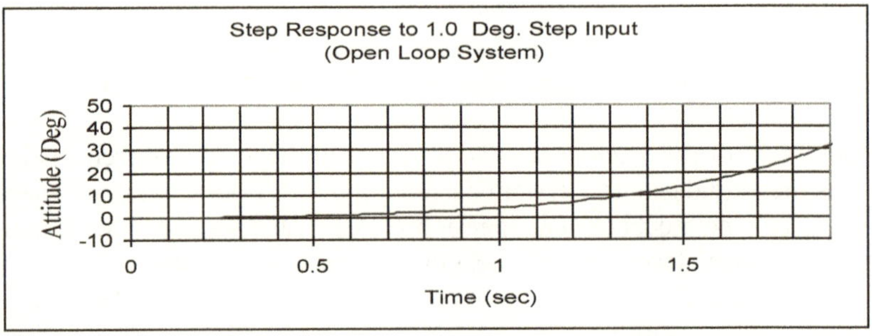

Fig. 4.1-3. Open-Loop Response to a 1.0 Deg Step Input

ii. Closed-Loop System before Optimization

The eigenvalues of the non-optimal system are:

-25.00
-58.24 + 133.44i
-58.24 - 133.44i
-63.45
-20.68
- 3.08
-1.98 + 1.16i
-1.98 - 1.16i
-0.28
-0.80 + 0.33i
-0.80 - 0.33i
-0.87

The system is stable, but it is not optimized. The plot of singular value vs. frequency is shown in Fig. 4.1-4.

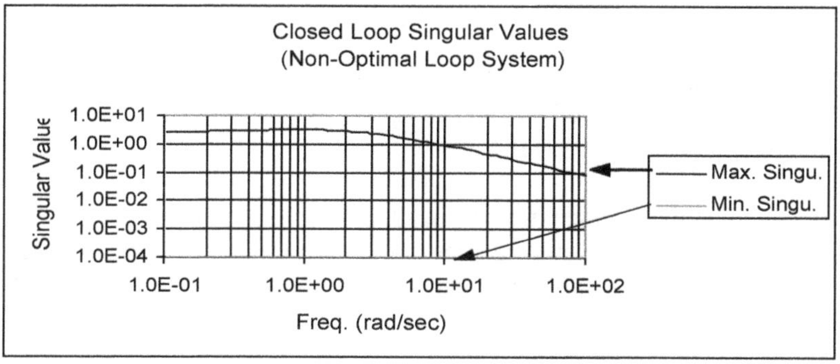

Fig. 4.1-4. Closed-Loop System Singular Values before Optimization

The maximum singular value is 3.2088 at 0.9103 rad/sec.

The 1.0 deg step input response of the non-optimal loop system is shown in Fig. 4.1-5. A 50% overshoot is displayed.

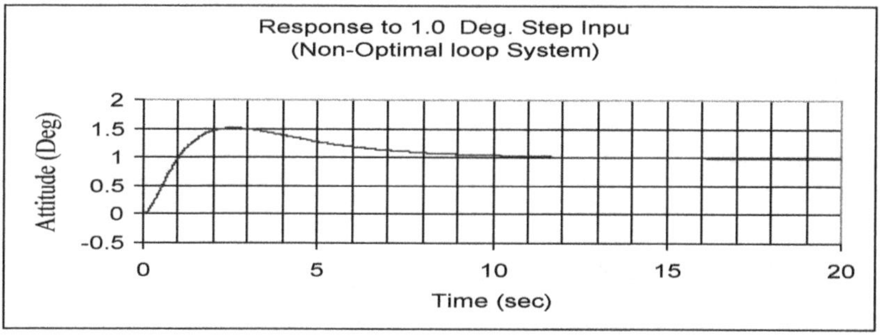

Fig. 4.1-5. Non-Optimal Response to a 1.0 Deg Step Input

Because of the log scale used in vertical axis, it looks that there is not much difference in singular values between the open loop and the closed loop.

iii. <u>H-infinity optimization (Optimization of the Closed- Loop System)</u>

In this section, the H-infinity, Loop Shaping, and Coprime Factorization are used. A Matlab program, "ncfsyn.m," used extensively in this book, is programmed based on papers published by D. C. McFarlane, Keith Glover in 1984 (ref.13) , 1989 (ref. 51) and 1992 (ref. 52), and a PhD dissertation by Vinnicombe G. in 1993 (ref. 53).

Before running the "ncfsyn.m," we need to assume a set of weighting functions. Usually, we start with some weighting functions of lower order and then increase the order as we approach an optimal design. That will be the beginning of the trial and error process. The objective here is to minimize "emax," which is an optimization cost computed by "ncfsyn.m." The "emax" cannot be greater than 1.0. An "emax" within the range of 0.3 to 0.4 is usually acceptable . Finding the right sets of weighting functions that minimize "emax" is not straightforward at the present time. However, it is very important to find the best weighting functions because the performance of the control system depends heavily on the selected weighting functions.

The Ap, Bp, C, and D matrices of the rigid body system are repeated below for convenience for our optimal design:

Ap=

$$\begin{vmatrix} 0 & 1 & 0 & 0 & 0 \\ 3.9848 & -0.00794 & 0 & 0 & 0 \\ 25 & 0 & -25 & 0 & 0 \\ 0 & 0 & 0 & 0 & 1 \\ 0 & 8479.7440 & 0 & -21199.36 & -116.4800 \end{vmatrix}$$

Bp =

$$\begin{vmatrix} 0 & 0 & 0 \\ -7.8235 & -9.413 & 0.00245 \\ 0 & 0 & 0 \\ 0 & 0 & 0 \\ 0 & 0 & 0 \end{vmatrix}$$

C =

$$\begin{vmatrix} 1 & 0 & 0 & 0 & 0 \\ 0 & 1 & 0 & 0 & 0 \end{vmatrix}$$

D =
| 0   0   0 |
| 0   0   0 |

Engine command signal, δc is:

$$\delta c = Ka(Ki/S + 1)(\theta c - \theta f)$$

In Fig. 4.1-1a, the "En" block generates the engine command signal.

Here, we adopt a simplest model.

$$\delta c = (\theta c - \theta f)$$

where Ka=1, Ki = 0

A case when Ka=1.0 and Ki =1.5 is considered later (see page 339)

The actuator math model is:

$$Act(S) = 3/(S + 3)$$

The weighting functions are:

$$W1(S) = (20 S + 11.6667)/(S + 2)$$

$$W2(S) = [(20 S + 7)/(S^{2} + 3.7 S + 5)]$$

$$W3(S) = 1.0$$

Fig. 4.1-6 and Fig. 4.1-7 show the magnitude plots of these two functions.

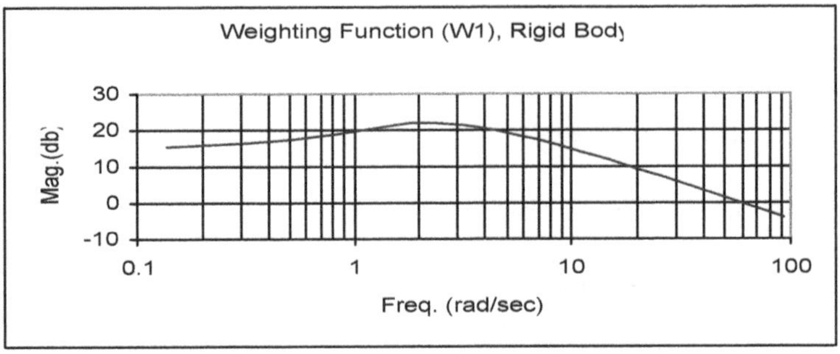

Fig. 4.1-6. Weighting Function, W1

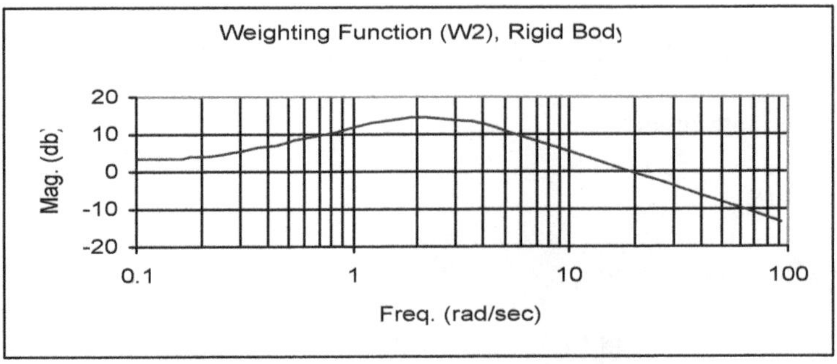

Fig. 4.1-7. Weighting Function, W2

The system matrices of the optimal dynamic feedback block, K(s), obtained by using "ncfsyn.m" are:

Afeedback =

-1.7604	6.9965	-0.4878	1.6256	-0.2634
5.7712	-51.8648	3.7254	-11.6048	1.8893
-0.1754	1.9169	-0.2577	1.9559	-0.2047
0.1839	-1.8693	-1.3827	-3.9817	1.3972
-0.0443	0.6071	-0.1796	-1.1314	-0.8159

Bfeedback =

1.5506	2.8568
-0.4551	-9.7387

$$\begin{vmatrix} 0.0157 & 0.3201 \\ -0.0202 & -0.3727 \\ 0.0045 & 0.0773 \end{vmatrix}$$

Cfeedback =

$$\begin{vmatrix} -0.4824 & 8.4756 & -0.4439 & 1.3257 & -0.2126 \\ -0.2106 & 3.5187 & -0.2006 & 0.5154 & -0.0879 \\ -0.0000 & -0.0009 & 0.0001 & -0.0002 & 0.0000 \end{vmatrix}$$

Dfeedback =

$$\begin{vmatrix} 0.1221 & 2.0751 \\ 0.0502 & 0.8528 \\ -0.0000 & -0.0003 \end{vmatrix}$$

The gains are:

Gain1 = -0.78215
Gain2 = 1.0

The closed loop eigenvalues are:

-25.00
-58.24 + 133.44i
-58.24 - 133.44i
-23.78
-16.05 + 15.76i
-16.05 - 15.76i
-1.95 + 1.15i
-1.95 - 1.15i
-1.95 + 1.15i
-1.95 - 1.15i
-1.58
-0.55
-0.56
-1.00

In classical control theory, the major concern is the system stability. The stability is measured by phase margin and gain margin in Nyquist plot, Bode plot, or Polar plot, or location of operating point in Root Locus. In H-infinity control, the major concerns are Performance and Robustness.

They are measured by closed loop system singular value. Mathematically (see Chapter 2.2) ,

$$\text{If} \parallel Tzw \parallel \infty \leq \gamma, \text{ then } \parallel \Delta \parallel \infty \leq 1/\gamma$$

> where
> $\parallel Tzw \parallel \infty$ = largest closed loop system singular value over all frequencies
> γ = some scalar value
> $\parallel \Delta \parallel \infty$ = largest singular value of the modeling error.

Basically, the mathematical expression above means that if the largest singular value is less than γ, then the control system can tolerate the modeling less than $(1/\gamma)$. Therefore, the smaller the largest singular value of a closed loop control system is, the larger the modeling error is allowed.

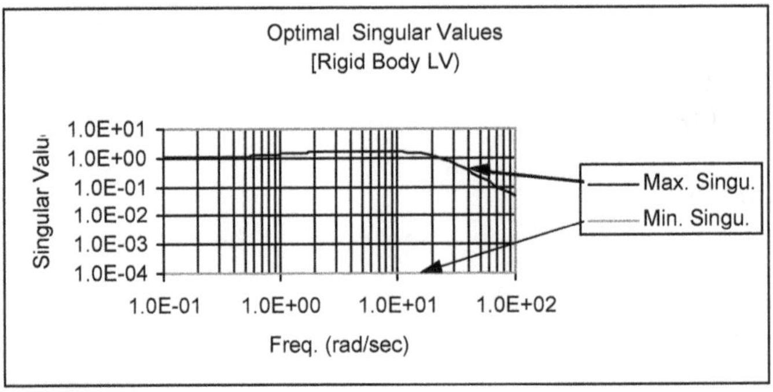

Fig. 4.1-8. Rigid-Body LV Optimal System Singular Values

The singular value vs. frequency plot is shown in Fig. 4.1-8. The singular value plot also shows that the undesired signals are attenuated, while the desired signals are relatively enhanced. The maximum singular value is 1.7343 at 4.900 rad/sec. Notice that the maximum singular value is reduced down to 49.0 % of the open loop case and to 54.0 % of the closed loop case (not optimized).

The unit step input response and state vector time histories are plotted below:

a) Step Response

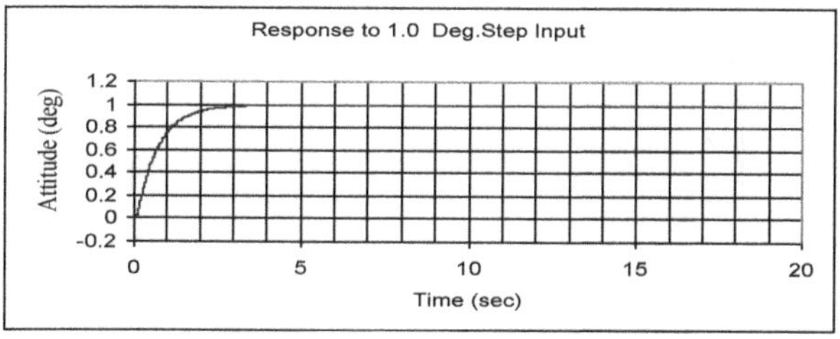

Fig. 4.1-9. Optimal Response to 1.0 Deg Step Input

b) Attitude Rate

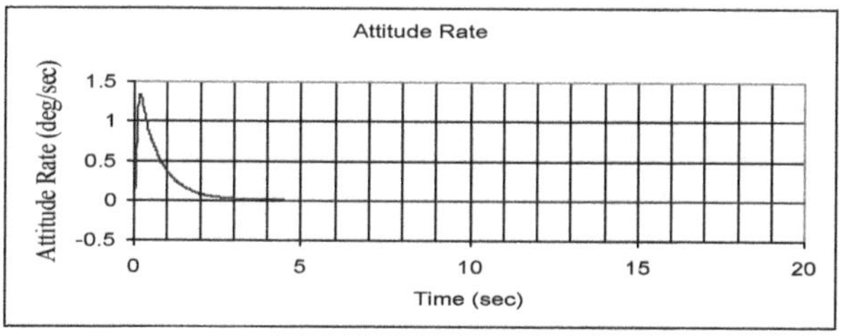

Fig. 4.1-10. Attitude Rate [x(2)]

c) Position Gyro Read

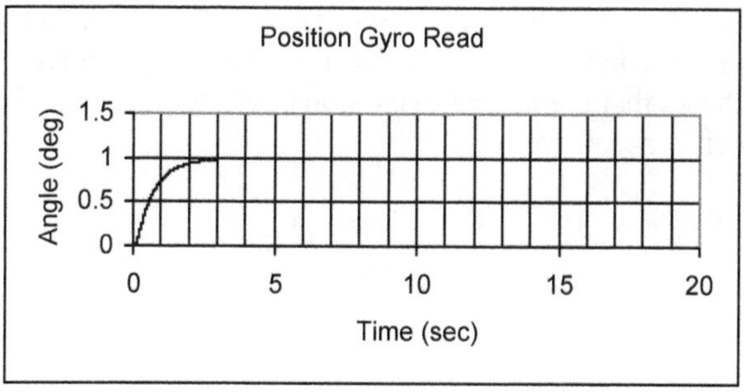

Fig. 4.1-11. Position Gyro Read

d) Rate Gyro

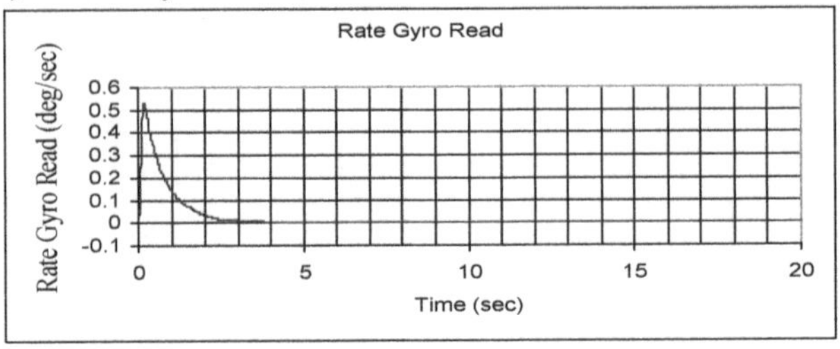

Fig. 4.1-12. Rate Gyro Read

e) Rate Gyro Rate Read

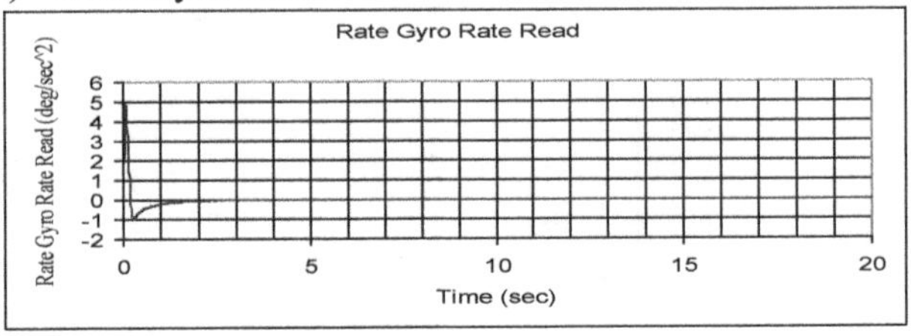

Fig. 4.1-13. Rate Gyro Rate Read

Since the swiveling angle span of the rocket actuator is usually limited to a small angle (e.g., 8.0 deg in one direction), the maximum engine deflection angle required for any flight must be estimated before flight. The estimated angle is greater than the swiveling angle allowed, then some new adjustments are needed to meet the swiveling angle requirement. Simulation shows that the largest excursion angle is -1.72 deg (See Fig. 4.1-14.), which is acceptable.

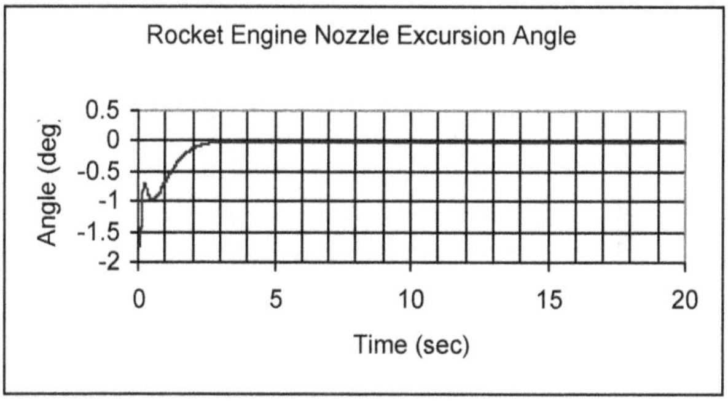

Fig. 4.1-14. Engine Nozzle Excursion Angle

4.2 Flex Body with Two Bending Modes

Before adding the bending to the math model, bending behavior is simulated in order to understand how the bending of the two modes configures the final shape of the LV's structure. Structural deflection is computed every 4.0 sec for 20.0 sec starting from t = 0.0 sec. It is assumed that there is no bending at t = 0.0 as shown in Fig. 4.2-1, Fig. 4.2-2, and Fig. 4.2-3. This is an analytical assumption. Actually, there is substantial bending at the launch pad even before the launch because of the heavy fuel weight.

- When t = 0.0 sec.

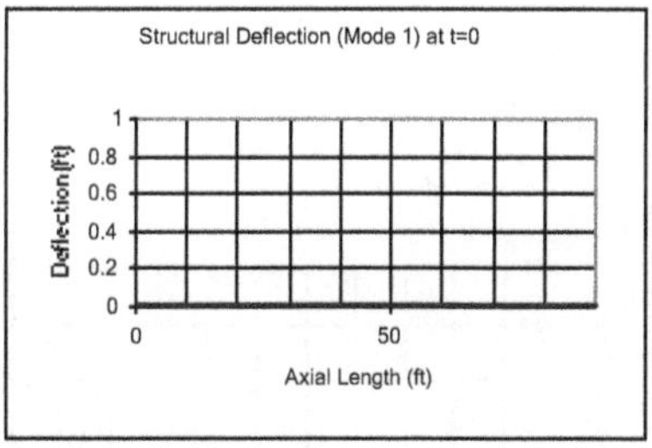

Fig. 4.2-1 Mode 1 Structural Deflection at t=0.0 sec.

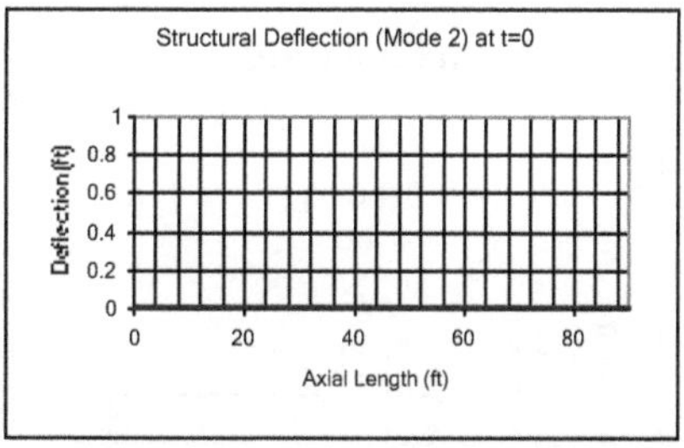

Fig. 4.2-2 Mode 2 Structural Deflection at t=0.0 sec.

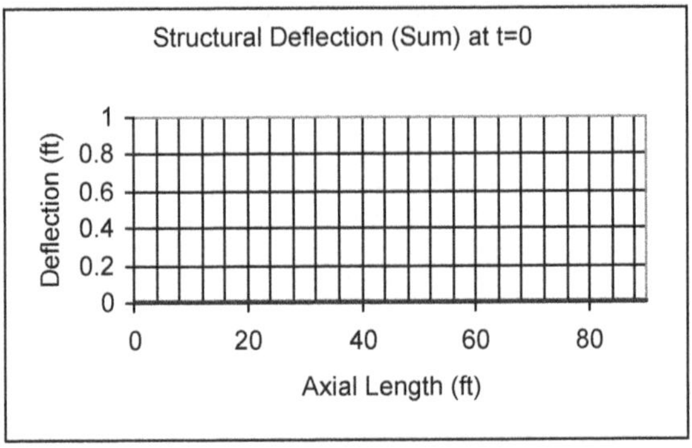

Fig. 4.2-3. Sum of the Structural Deflections (Mode 1 and Mode 2) at t = 0.0 sec.

Fig. 4.2-4 shows the LV's mode 1 structural deflection at t = 4.0 sec, and Fig. 4.2-5 shows the LV's mode 2 structural deflection at t = 4.0 sec. It is seen here that mode 1 bending is very close to being a second order function, and its peak occurs at 40 feet from the left. The total length of the LV is 90.0 feet. Mode 2 bending, shown in Fig. 4.2.5, is similar to a third order function. These two modes are added to show the LV's total structural deflection in Fig. 4.2-6. The maximum deflection is about 0.45 ft. The ratio of the maximum deflection to the total length is about 0.005. Notice that the tail end is deflected significantly.

- When t = 4.0 sec.

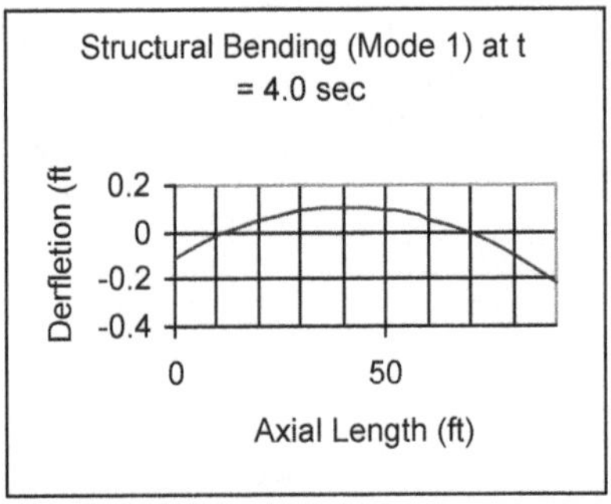

Fig. 4.2-4 Mode 1 Structural Deflection at t=4.0 sec.

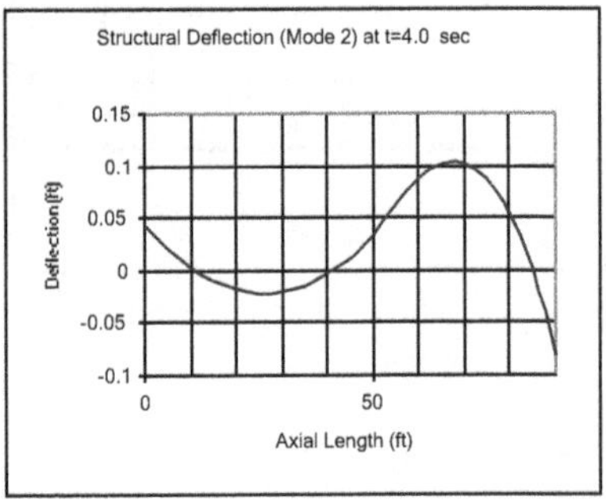

Fig. 4.2-5 Mode 2 Structural Deflection at t=4.0 sec.

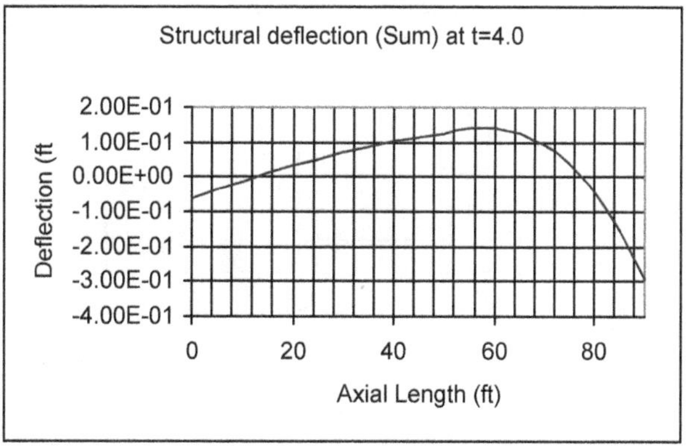

Fig. 4.2-6. Sum of Mode 1 and Mode 2 Structural Deflections at t = 4.0 sec.

Fig. 4.2-7 through Fig. 4.2-18 show the bending shape at t = 8.0 sec., t = 12.0 sec., t = 16.0 sec., and t = 20.0 sec.

- When t = 8.0 sec.

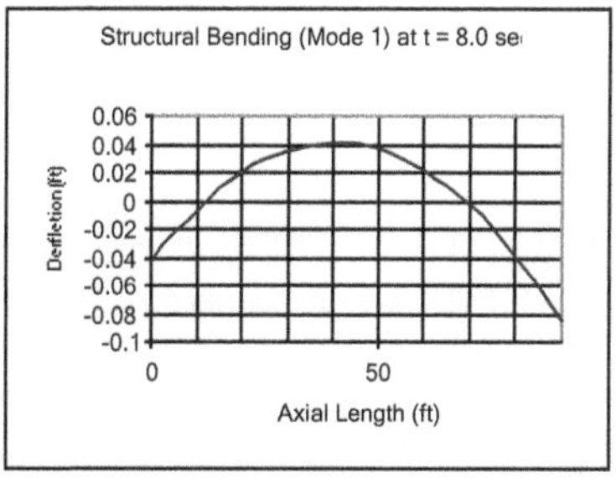

Fig. 4.2-7 Mode 1 Structural Deflection at t=8.0 sec.

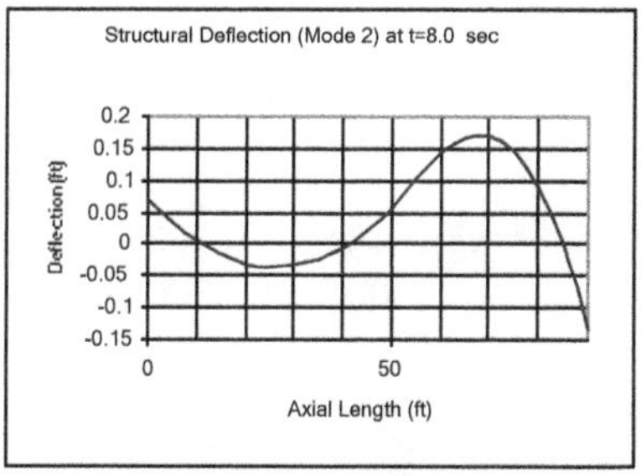

Fig. 4.2-8 Mode 2 Structural Deflection at t=8.0 sec.

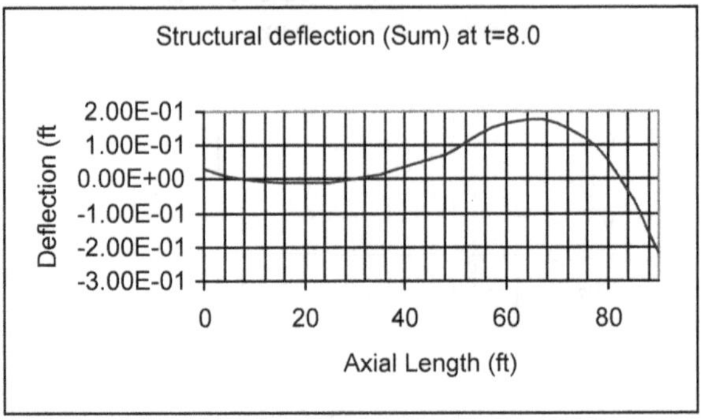

Fig. 4.2-9. Sum of Mode 1 and Mode 2 Structural Deflections at t = 8.0 sec.

- When t = 12.0 sec.

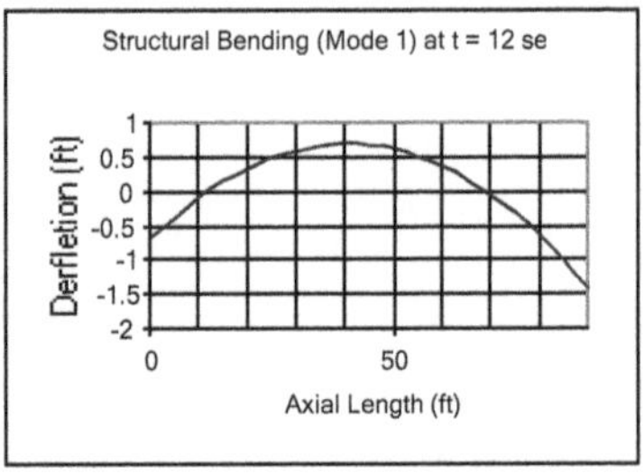

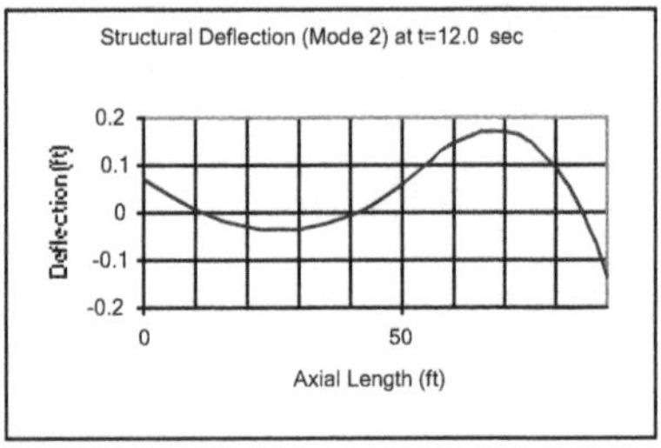

Fig. 4.2-10 Mode 1 Structural Deflection at t= 12.0 sec.

Fig. 4.2-11 Mode 2 Structural Deflection at t=12.0 sec.

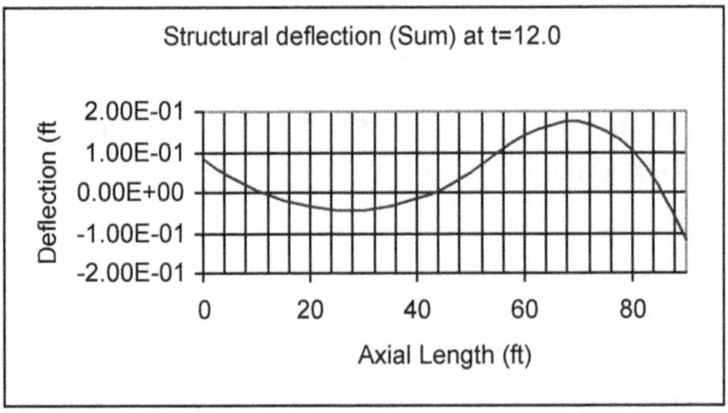

Fig. 4.2-12. Sum of Mode 1 and Mode 2 Structural Deflections at t = 12.0 sec.

- When t = 16.0 sec.

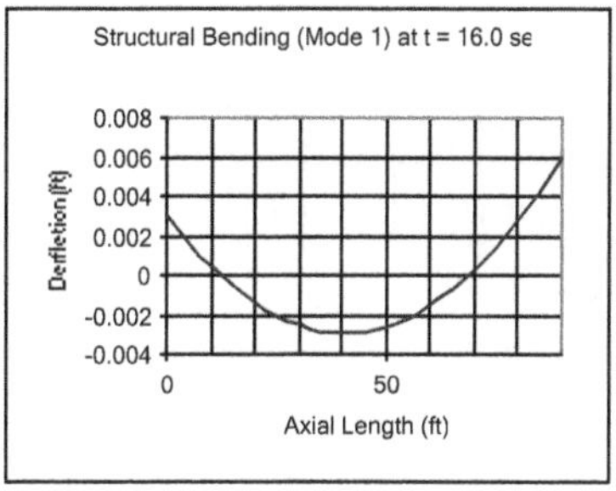

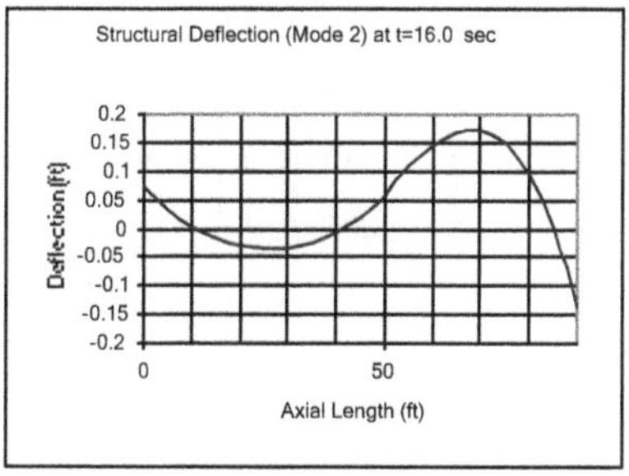

Fig. 4.2-13 Mode 1 Structural Deflection at t=16.0 sec

Fig. 4.2-14 Mode 2 Structural Deflection at at t=16.0 sec.

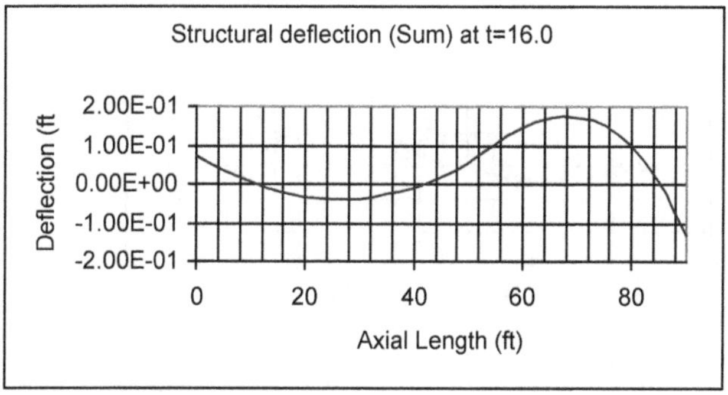

Fig. 4.2-15. Sum of Mode 1 and Mode 2 Structural Deflections at t = 16.0 sec.

- When t = 20.0 sec.

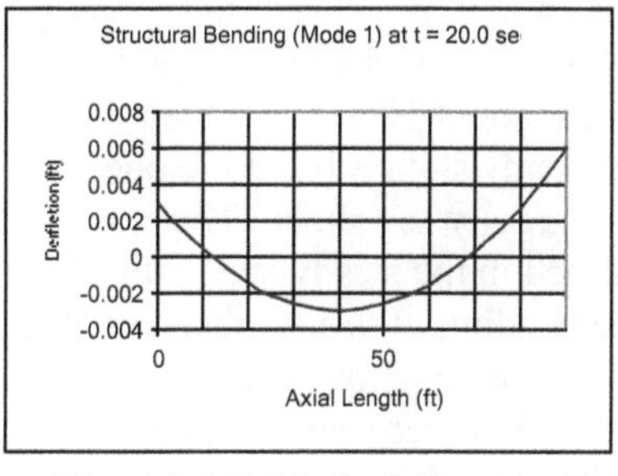

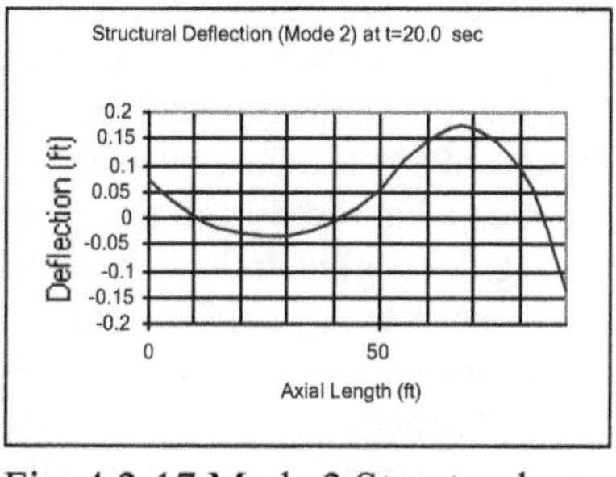

Fig. 4.2-16 Mode 1 Structural Deflection at at t=20.0 sec

Fig. 4.2-17 Mode 2 Structural Deflection at t=20.0 sec.

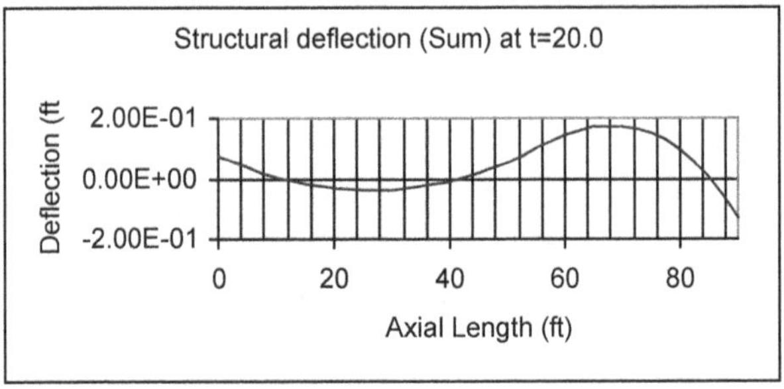

Fig. 4.2-18. Sum of Mode 1 and Mode 2 Structural Deflections at t=20.0 sec.

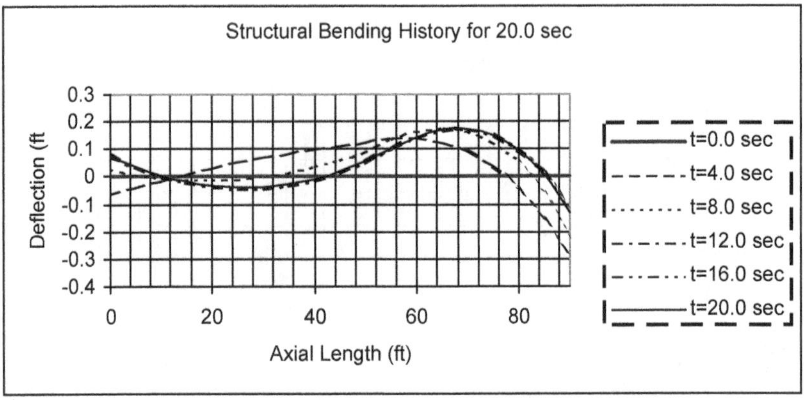

Fig. 4.2-19. Structural Bending History for 20 Seconds

The 6 sets of the plots are collected in Fig. 4.2-19 to observe the bending time history. It can be seen here that as a step input is applied, bending starts from zero deflection and increases for about 10 sec, and then it reaches steady state. The structure in the steady state is biased to one direction causing the shift of the center of gravity (cg shift). The reason for having the engine deflection angle biased in the steady state attitude is partially due to the cg shift caused by the bending. It is also observed that, since the mode 1 generalized coordinate reaches a zero steady state while the mode 2 generalized coordinate reaches a non-zero steady state, the final structural configuration looks like the mode 2 bending.

Here in our study, only two bending modes are added to the rigid body math model. System data used here are from ref. 18, and they are listed in Appendix I: Launch Vehicle System Parameter Table. The system matrices, A and B are computed by using an Excel program stored in the CD attached at the back of this book.

The system matrices, A, B, C, and D are:

A =

0.000E+00	1.000E+00	0.000E+00	0.000E+00	0.000E+00	0.000E+00	0.000E+00	0.000E+00	0.000E+00
3.985E+00	-7.944E-03	-2.504E-01	-1.590E-04	0.000E+00	-7.348E-01	0.000E+00	0.000E+00	0.000E+00
0.000E+00	0.000E+00	0.000E+00	1.000E+00	0.000E+00	0.000E+00	0.000E+00	0.000E+00	0.000E+00
5.718E+02	-2.743E-01	-3.749E+02	-2.646E+01	-4.826E+01	2.897E-02	0.000E+00	0.000E+00	0.000E+00
0.000E+00	0.000E+00	0.000E+00	0.000E+00	0.000E+00	1.000E+00	0.000E+00	0.000E+00	0.000E+00
2.512E+02	-1.205E-01	-3.692E+03	-8.494E+01	-2.120E+01	1.562E-02	0.000E+00	0.000E+00	0.000E+00
2.500E+01	0.000E+00	1.100 E-01	0.000E+00	-6.900E-01	0.000E+00	-2.5000E+01	0.000E+00	0.000E+00
0.000E+00	0.000E+00	0.000E+00	0.000E+00	0.000E+00	0.000E+00	0.000E+00	0.000E+00	1.000E+00
0.000E+00	8.479E+03	0.000E+00	3.7300E+01	0.000E+00	-2.340E+00	0.000E+00	-2.120E+04	-1.165E+02

B=

0.000E+00	0.000E+00	0.000E+00
-7.823E+00	-9.413E+00	2.454 E-03
0.000E+00	0.000E+00	0.000E+00
-2.701E+02	4.025E+01	1.049E-02
0.000E+00	0.000E+00	0.000E+00
-1.187E+02	1.768E+01	8.931E-05
0.000E+00	0.000E+00	0.000E+00
0.000E+00	0.000E+00	0.000E+00
0.000E+00	0.000E+00	0.000E+00

C=

1.000E+00	0.000E+00	0.000E+00	0.000E+00	0.000E+00	0.000E+00	0.000E+00	0.000E+00	0.000E+00
0.000E+00	1.000E+00	0.000E+00	0.000E+00	0.000E+00	0.000E+00	0.000E+00	0.000E+00	0.000E+00

D =

0.000E+00	0.000E+00	0.000E+00
0.000E+00	0.000E+00	0.000E+00

As was done in Section 4.1, the following 3 cases are evaluated to understand how the eigenvalues and the singular values vs. frequency change as the attitude control system design improves.

- Open Loop (no feedback)
- Closed Loop System after Optimization

i. <u>Open Loop System</u>

The eigenvalues of the open loop system (no feedback) are:

-25.00
-58.24 + 133.44i
-58.24 - 133.44i
17.26
-8.87 + 21.22i
-8.87 - 21.22i
0.44
-15.16
-11.26

Notice that there are three eigenvalues that are identical to eigenvalues found in the rigid body open loop case. These are:

-25.00
-58.24 + 133.44i
-58.24 - 133.44i

One positive eigenvalue, 17.26, indicates that the system is unstable.

The singular values vs. frequency plot is shown in Fig. 4.2-20.

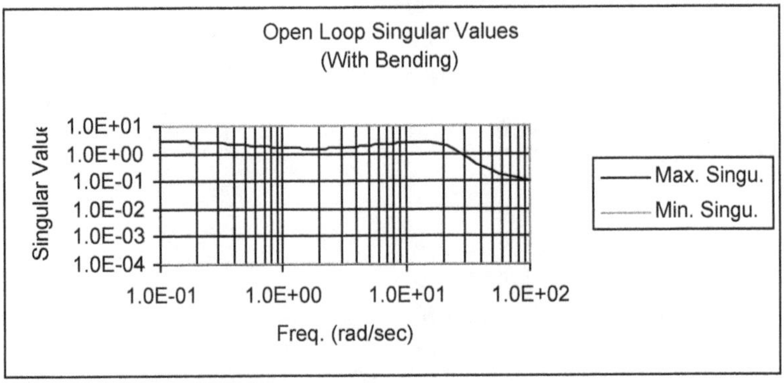

Fig. 4.2-20. Open Loop Singular Values

The maximum singular value is 2.8025 at 0.01 rad/sec.

The open loop response to a 1.0 deg step input is shown in Fig. 4.2-21. This figure shows rapid divergence. The attitude grows to over 6.0 deg in 0.3 sec.

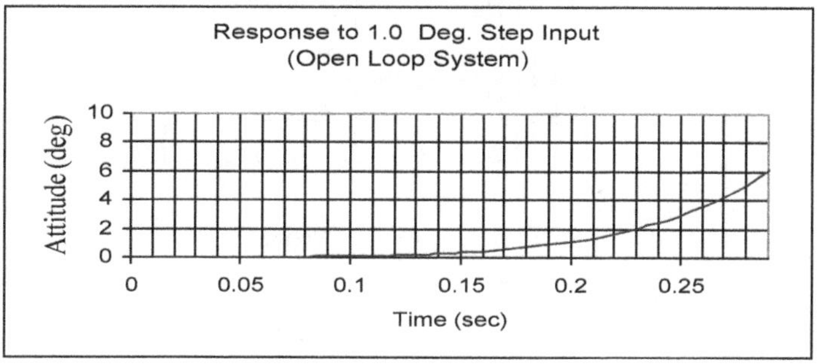

Fig. 4.2-21. Response to a 1.0 Deg Step Input

ii. <u>Closed Loop System after Optimization</u>

A Matlab program, ncfsyn.m, is used to design the dynamic feedback. The control system block diagram is shown in Fig. 4.1-1.
Actuator:

Act (S) = 16.602/(s + 16.602)

The weighting functions are:

W1(S) = (0.0060233 S^2 + 0.12046745 S + 0.602332)/(S^2 + 1.3979 S + 1.807)

W2(S) = (S^2 + 28 S + 115)/(S^3 + 36 S^2 + 10 S + 100.0)

W3(S) = 1.0

Fig. 4.2-22 and Fig. 4.2-23 show the magnitude plots of these two weighting functions.

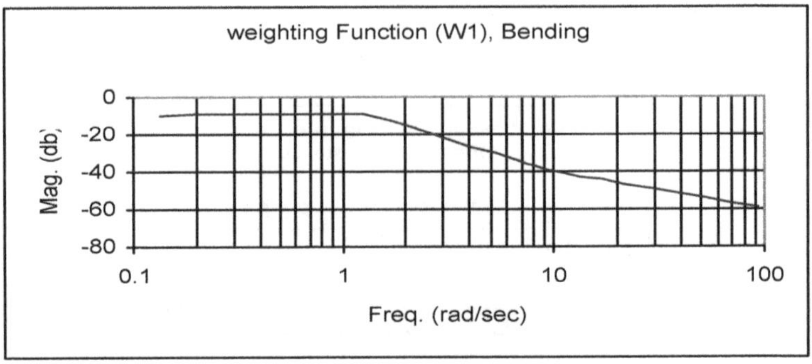

Fig. 4.2-22. Weighting Function, W1

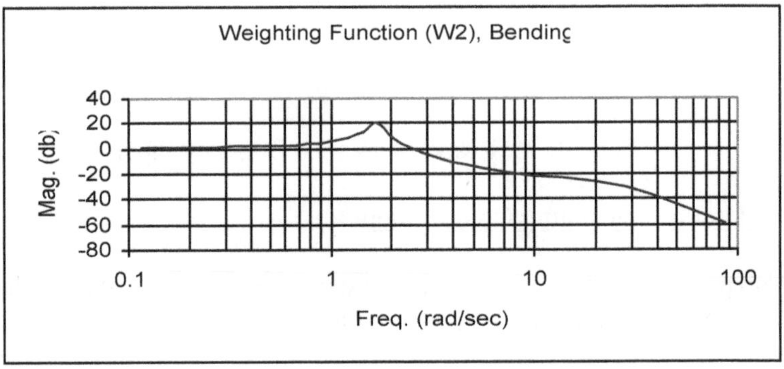

Fig. 4.2-23. Weighting Function, W2

The system matrices of the optimal dynamic feedback block, K(s), obtained by using "ncfsyn.m" are:

Afeedback =

Columns 1 through 4

-121.8023	-53.6592	-159.6336	-40.0596
-69.2649	-32.1314	-91.0709	-23.0310
28.5652	14.1685	35.0613	8.6811
39.2803	18.1319	51.7501	12.8161
40.9893	18.7223	55.0209	12.6390
86.4312	40.0368	112.3801	27.4987

15.0706	7.0351	19.1981	4.6912
-17.3904	-7.9286	-23.5017	-5.6608
-13.7263	-6.3454	-17.9319	-4.3654
1.8275	0.8259	2.5215	0.6043
0.3932	0.1848	0.4917	0.1212

Columns 5 through 8

83.4274	-145.5388	42.6994	-41.5524
47.1497	-82.0968	24.0676	-23.4438
-18.1013	31.5319	-9.2400	8.9971
-25.7847	46.2013	-13.5110	13.1505
-29.8839	54.6011	-15.9863	15.4695
-60.6395	84.8867	-10.1477	18.7857
-9.6384	-1.6401	-8.5285	13.2554
13.3931	-19.9344	0.7153	-15.9066
9.6149	-11.4400	7.4622	-7.2563
-1.4946	2.7348	0.0554	3.8602
-0.2407	0.1237	-0.3784	-0.2878

Columns 9 through 11

20.5321	7.1402	-2.1351
11.5777	4.0275	-1.2038
-4.4452	-1.5459	0.4622
-6.5020	-2.2606	0.6762
-7.6607	-2.6645	0.7965
-9.4496	-3.2621	0.9754
-5.7643	-1.9802	0.5914
12.9560	4.3886	-1.3214
-23.4888	-17.2763	5.0641
0.7364	-25.8743	14.6697
2.8616	4.3614	-13.2564

Bfeedback =

5.6470 222.7428

```
   1.5539  125.5509
  -0.9683  -50.1543
  -1.3257  -72.4449
  -1.3797  -76.4630
  -2.8648 -158.5132
  -0.4947  -27.3714
   0.5877   32.5208
   0.4561   25.2371
  -0.0624   -3.4534
  -0.0128   -0.7075

Cfeedback =

  Columns 1 through 4

-144.5588  -66.3538 -191.8293  -48.4297
  79.3661   36.3769  103.1164   26.1187
   1.2222    0.5614    1.6218    0.4098

  Columns 5 through 8

 100.2787 -174.7788   51.2672  -49.9078
 -54.0695   94.2645  -27.6528   26.9182
  -0.8481    1.4785   -0.4337    0.4222

  Columns 9 through 11

  24.6553    8.5751   -2.5637
 -13.2982   -4.6251    1.3828
  -0.2086   -0.0725    0.0217

Dfeedback =

   4.8466  268.1885
  -2.6141 -144.6516
  -0.0410   -2.2687
```

Gain1 = 0.0321734
Gain2 = 1.0

The feedback dynamics has 11 states in this design. It can be reduced to a lower degree, which is needed for actual flight computation. However, some degradation in performance may occur. The effect of the feedback system order reduction on the control system performance is presented in Section 4.5. The closed loop eigenvalues are:

1.0e+002 *

-0.2500
-0.5824 + 1.3344i
-0.5824 - 1.3344i
-0.0887 + 0.2122i
-0.0887 - 0.2122i
-0.0887 + 0.2122i
-0.0887 - 0.2122i
-0.3580
-0.3580
-0.1792 + 0.0127i
-0.1792 - 0.0127i
-0.1486 + 0.0125i
-0.1486 - 0.0125i
-0.1515
-0.1131
-0.1126
-0.0090 + 0.0196i
-0.0090 - 0.0196i
-0.0096 + 0.0187i
-0.0096 - 0.0187i
-0.0053 + 0.0120i
-0.0053 - 0.0120i
-0.0096 + 0.0108i
-0.0096 - 0.0108i
-0.0074
-0.0079

The singular values vs. frequency plot is shown in Fig. 4.2-24.

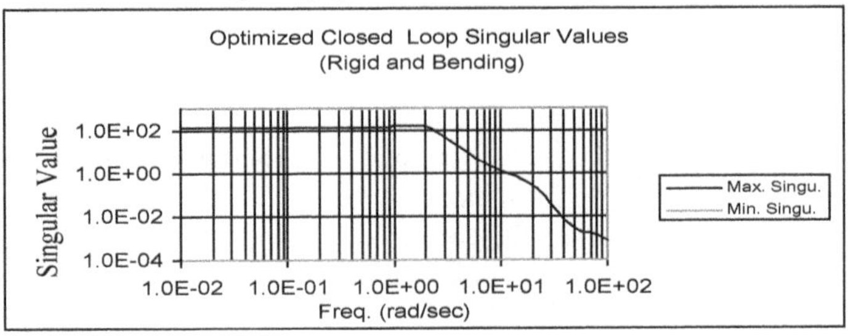

Fig. 4.2-24. Optimal Closed Loop System Singular Values

The singular value plot displays that the noise signals are attenuated considerably while the desired signals are extensively amplified and weighted uniformly, which improves performance significantly. The maximum singular value is 155.09 at 1.5999 rad/sec.

Unit step input response and state vector time histories are plotted below.

a) Step Response

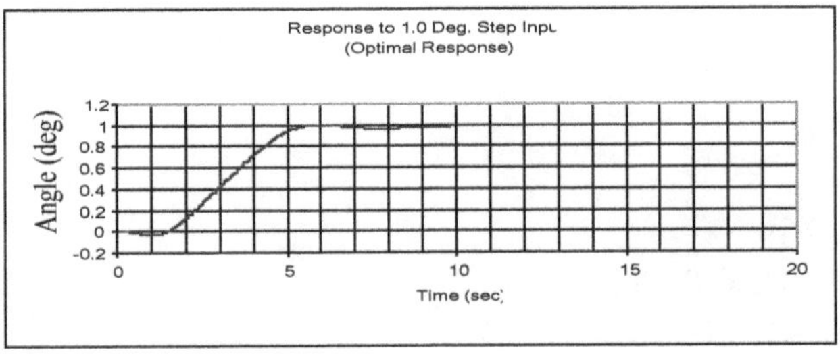

Fig. 4.2-25. Response to a 1.0 Deg Step Input

It takes about 5.5 seconds to reach the steady state attitude.

b) Engine Deflection Time History

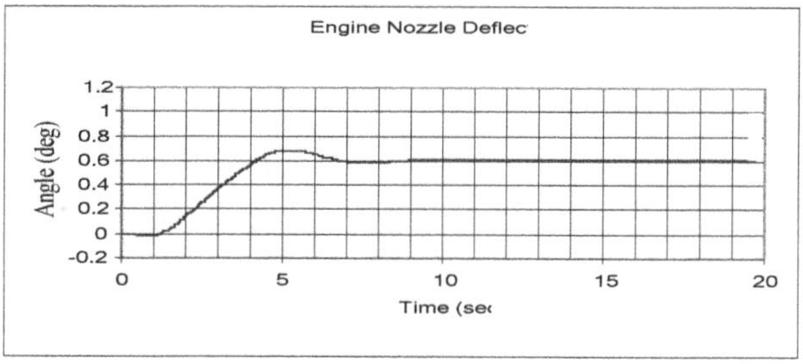

Fig. 4.2-26. Engine Nozzle Deflection Time History

It is seen here that the nozzle angle did not return to zero even after the attitude reached steady state. The reason is that the LV's cg is no longer on the axial axis due to bending of the LV. In the rigid body case, the engine nozzle angle returns to zero at steady state.

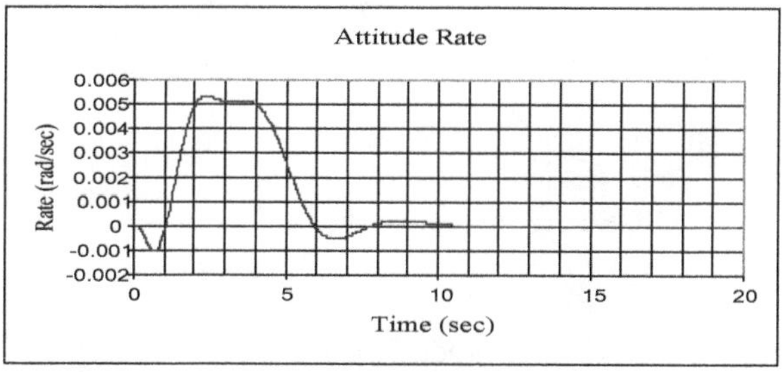

Fig. 4.2-27 Attitude Rate [x(2)]

c) Generalized Coordinate, Mode 1(q1)

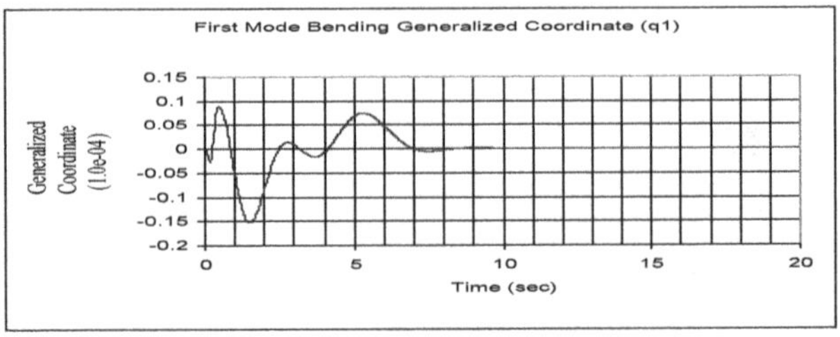

Fig. 4.2-28 Mode 1 Bending Generalized Coordinate (q1)

d) Generalized Coordinate Rate, Mode 1 (q1d)

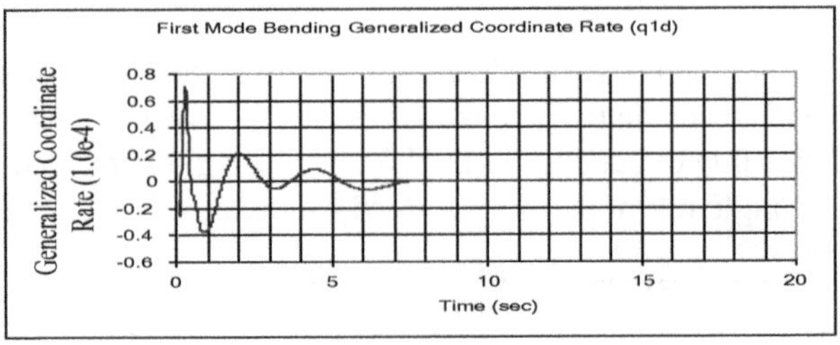

Fig. 4.2-29 Mode 1 Bending Generalized Coordinate Rate (q1d)

f) Generalized Coordinate, Mode 2(q2)

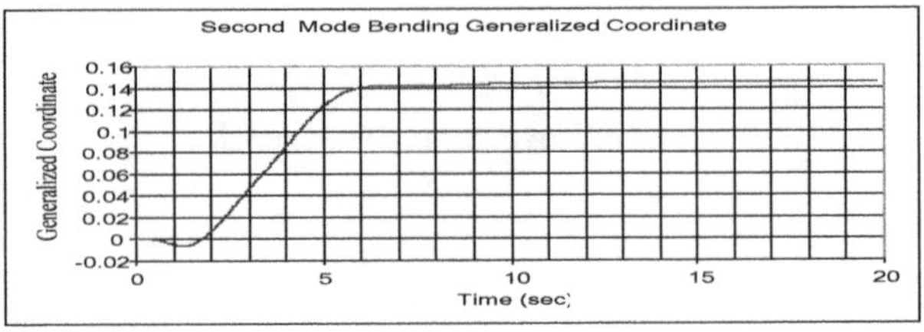

Fig. 4.2-30 Mode 2 Bending Generalized Coordinate (q2)

e) Generalized Coordinate Rate, Mode 2(q2d)

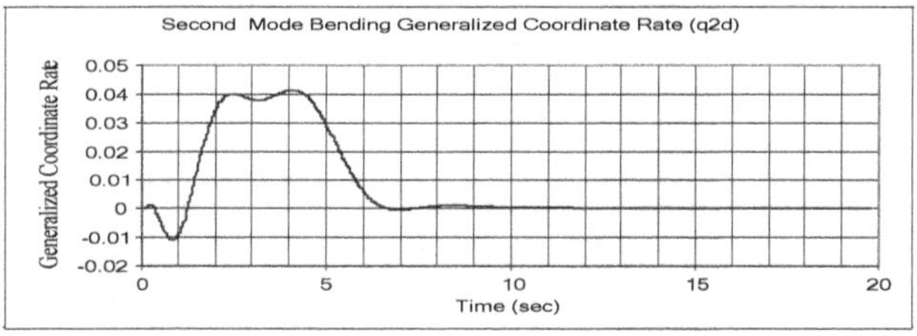

Fig. 4.2-31 Mode 2 Bending Generalized Coordinate Rate (q2d)

h) Position Gyro Read

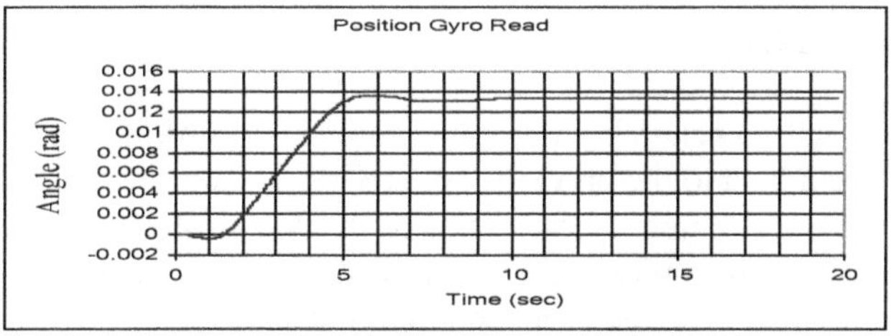

Fig. 4.2-32 Position Gyro Read

i) Rate Gyro

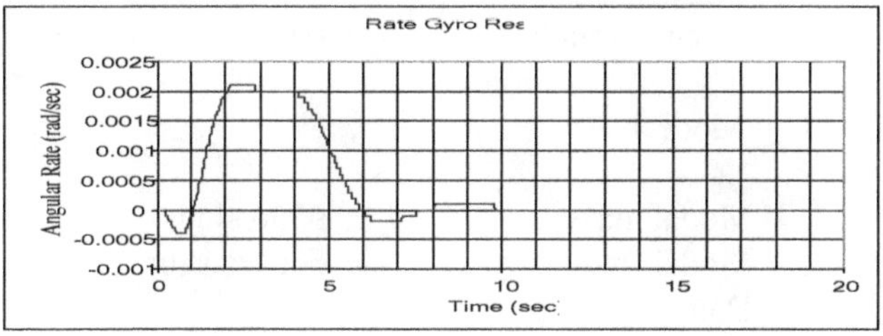

Fig. 4.2-33 Rate Gyro Read

j) Rate Gyro Rate

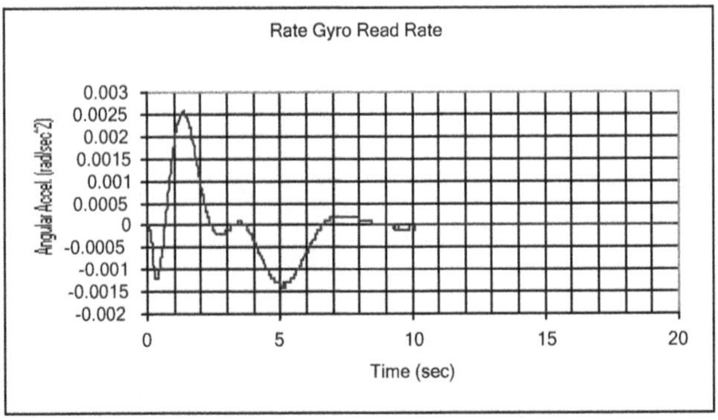

Fig. 4.2-34 Rate Gyro Rate Readout

4.3 Simplified Satellite Launch Vehicle Math Model and Analysis

In this section, two bending modes, two fuel tanks (sloshing), and engine dynamics are added to the previous system, which completes a simplified launch vehicle math model. Properties of the two sloshings and rocket engine swiveling dynamics added to the system are listed below. The sloshing motion is modeled as a pendulum motion.

Bending Mode 1

Modal mass	: 1590 slugs
Modal frequency	: 18.9 rad/sec
Damping ratio	: 0.7
Slope at tail	: -0.0645 rad/ft
Slope at middle	: 0.0044 rad/ft

Bending Mode 1

Modal mass	: 9550 slugs
Modal frequency	: 60.7 rad/sec
Damping ratio	: 0.7
Slope at tail	: -0.171 rad/ft
Slope at middle	: -0.0276 rad/ft

Sloshing 1

Mass	: 388 slugs
Frequency	: 5.98 rad/sec
Pendulum length	: 2.0 ft
Hinge location	: 5.56 ft

Sloshing 2

Mass	: 280 slugs
Frequency	: 5.17 rad/sec
Pendulum length	: 2.75 ft
Hinge location	: -20.62 ft

Rocket engine physical specifications are:

Mass	: 30.8 slugs
ℓR(Engine cg location from the hinge point)	: 2.52 ft
IR (Engine MOI)	: 377 slugs ft^2

The system block diagram is shown in Fig. 4.1-1a.

Elements of the system matrices, Ap and Bp, are computed by using an Excel program stored in the CD attached to this book at back cover.

The system matrices, Ap, Bp, Cp, and Dp, are:

Ap=
[0.0e+000 1.0e+000 0.0e+000 0.0e+000 0.0e+000 0.0e+000 0.0e+000 0.0e+000 0.0e+000 0.0e+000 0.0e+000 0.0e+000 0.0e+000;
3.9e+000 2.0e+000 -2.4e-001 0.0e+000 0.0e+000 -7.1e-001 0.0e+000 -8.0e-002 2.2e-001 0.0e+000 0.0e+000 0.0e+000 0.0e+000;
0.0e+000 0.0e+000 0.0e+000 1.0e+000 0.0e+000 0.0e+000 0.0e+000 0.0e+000 0.0e+000 0.0e+000 0.0e+000 0.0e+000 0.0e+000;
2.3e+002 7.5e+001 -3.7e+002 -2.6e+001 -4.7e+001 2.8e-002 -3.9e-001 0.0e+000 -4.8e+000 0.0e+000 0.0e+000 0.0e+000 0.0e+000;
0.0e+000 0.0e+000 0.0e+000 0.0e+000 0.0e+000 1.0e+000 0.0e+000 0.0e+000 0.0e+000 0.0e+000 0.0e+000 0.0e+000 0.0e+000;
9.9e+001 3.3e+001 -3.6e+003 -8.3e+001 -2.0e+001 1.5e-002 -1.7e-001 0.0e+000 2.2e+000 0.0e+000 0.0e+000 0.0e+000 0.0e+000;
0.0e+000 0.0e+000 0.0e+000 0.0e+000 0.0e+000 0.0e+000 0.0e+000 1.0e+000 0.0e+000 0.0e+000 0.0e+000 0.0e+000 0.0e+000;
0.0e+000 0.0e+000 0.0e+000 0.0e+000 0.0e+000 0.0e+000 -3.6e+001 0.0e+000 -1.3e+000 0.0e+000 0.0e+000 0.0e+000 0.0e+000;
0.0e+000 0.0e+000 0.0e+000 0.0e+000 0.0e+000 0.0e+000 0.0e+000 0.0e+000 0.0e+000 1.0e+000 0.0e+000 0.0e+000 0.0e+000;

0.0e+000 0.0e+000 0.0e+000 0.0e+000 0.0e+000 0.0e+000 -1.1e+000 0.0e+000 -2.7e+001 0.0e+000 0.0e+000 0.0e+000 0.0e+000;
2.5e+001 0.0e+000 1.1e-001 0.0e+000 -6.9e-001 0.0e+000 0.0e+000 0.0e+000 0.0e+000 0.0e+000 -2.5e+001 0.0e+000 0.0e+000;
0.0e+000 0.0e+000 0.0e+000 0.0e+000 0.0e+000 0.0e+000 0.0e+000 0.0e+000 0.0e+000 0.0e+000 0.0e+000 0.0e+000 1.0e+000;
0.0e+000 8.5e+003 0.0e+000 3.7e+001 0.0e+000 -2.3e+002 0.0e+000 0.0e+000 0.0e+000 0.0e+000 0.0e+000 -2.1e+004 -1.2e+002]

Bp=

0.0e+000	0.0e+000	0.0e+000
-7.7e+000	-9.3e+000	2.0e-003
0.0e+000	0.0e+000	0.0e+000
-2.6e+002	3.9e+001	1.0e-002
0.0e+000	0.0e+000	0.0e+000
-1.1e+002	1.7e+001	0.0e+000
0.0e+000	0.0e+000	0.0e+000
-2.8e+001	0.0e+000	0.0e+000
0.0e+000	0.0e+000	0.0e+000
-6.5e+001	0.0e+000	0.0e+000
0.0e+000	0.0e+000	0.0e+000
0.0e+000	0.0e+000	0.0e+000
0.0e+000	0.0e+000	0.0e+000

Cp =

| 1 0 0 0 0 0 0 0 0 0 0 0 0 |
| 0 1 0 0 0 0 0 0 0 0 0 0 0 |

Dp=

| 0 0 0 |
| 0 0 0 |

i. Open Loop

The eigenvalues of the open loop system (no feedback loop) are:

1.0e+002 *

-0.2500
-0.5824 + 1.3344i
-0.5824 - 1.3344i
0.2037
-0.1052 + 0.2539i

-0.1052 - 0.2539i
-0.2244
0.0105
-0.0172
-0.0000 + 0.0597i
-0.0000 - 0.0597i
-0.0000 + 0.0514i
-0.0000 - 0.0514i

There are two positive eigenvalues (20.37 and 1.05), indicating that the system is unstable. The singular value vs. frequency plot is shown in Fig. 4.3-1.

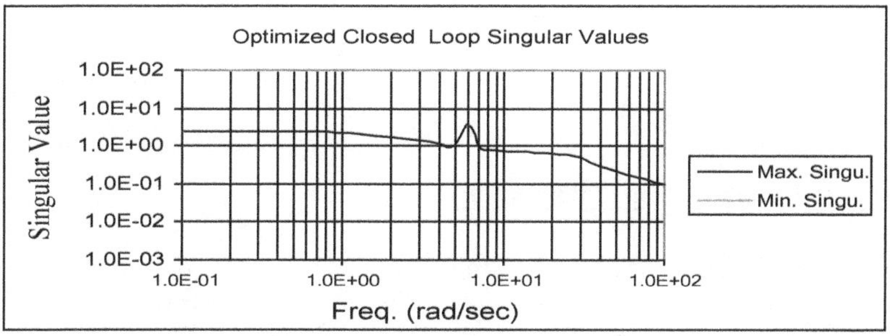

Fig. 4.3-1. Open Loop System Singular Values

The maximum singular value is 2.4752 at 0.01 rad/sec.

The response of the open loop system to a 1.0 deg step input is shown in Fig. 4.3-2.

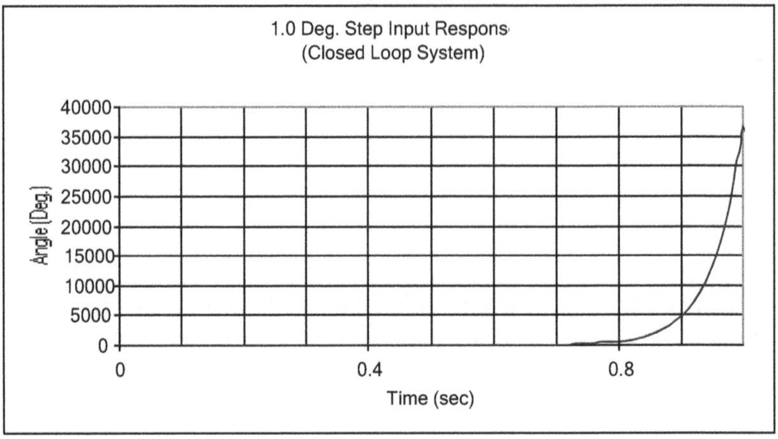

Fig. 4.3-2. Response to a1.0 Deg Step Input

ii. Closed Loop Systems after Optimization

Here, three different sets of weighting functions are selected. For each set, an optimal control system is designed. This study demonstrates that control system performance depends on the weighting functions selected.

- First Set

Actuator:

Act(S)= 16.60465 /(S + 16.60465)

The weighting functions are:

W1(S) = (0.0060224 S^2 + 0.120448 S + 0.1806722)/(S^2 + 1.39535 S + 1.8308125)

W2(S) = (S^2 + 28 S + 115)/(S^3 + 36 S^2 + 10 S + 95)

W3(S) = 1.0

Fig. 4.3-5 and Fig. 4.3-6 show the magnitude plots of these two functions.

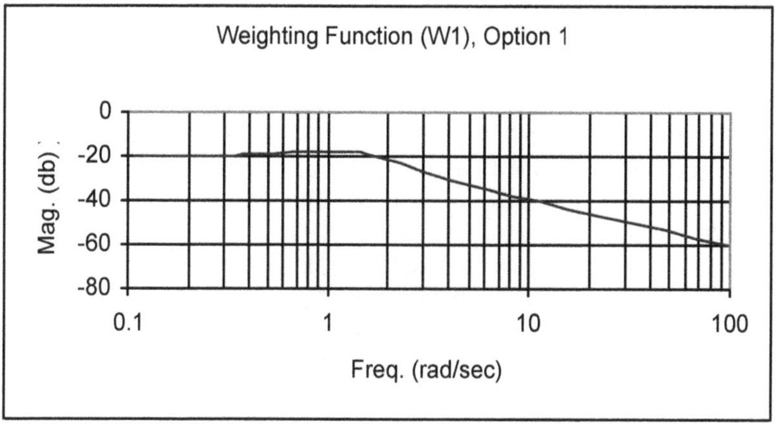

Fig. 4.3-5. Weighting Function, W1

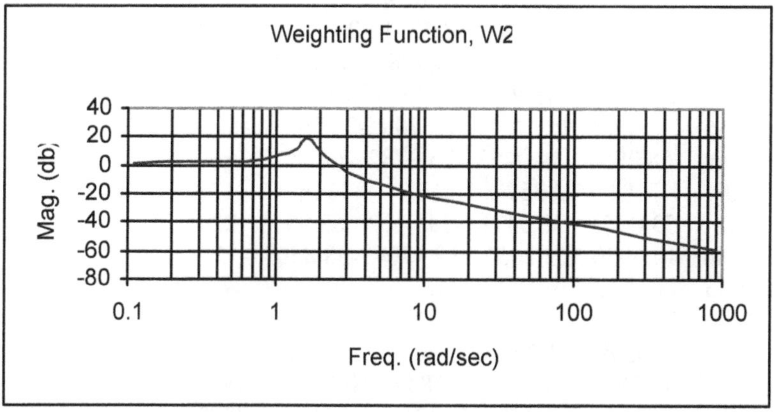

Fig. 4.3-6. Weighting Function, W2

The A, B, C, and D matrices of the optimal dynamic feedback block, K(s), obtained by using "ncfsyn.m," are:

Afeedback =

Columns 1 through 6

71.1392	23.0135	-2.9746	-0.4014	0.9478	-10.2110
-53.8266	-19.8974	4.5308	-0.3307	-1.8346	7.3692

-21.8304	-9.4368	0.9188	5.4973	-0.2381	2.9694
-31.9926	-11.6605	-4.1461	-0.2491	-1.0675	4.1008
2.9349	2.0466	-0.1934	0.6044	0.0393	-5.4396
25.7874	8.2522	-1.0909	0.1913	5.3832	-3.4750
68.0972	25.4255	-2.9827	0.9187	0.8946	-8.9057
1.2545	-0.3091	-0.0090	-0.1653	0.0357	-0.1815
-65.5059	-22.0915	2.7544	-0.2098	-0.9278	8.8229
49.6233	16.8656	-2.0944	0.1891	0.6998	-6.6815
-56.1547	-18.3374	2.3255	-0.0398	-0.8100	7.5758
45.0449	15.1911	-1.8941	0.1441	0.6381	-6.0673
-22.6192	-7.6073	0.9499	-0.0675	-0.3209	3.0471
-6.2451	-2.2405	0.2706	-0.0513	-0.0852	0.8385
-1.4441	-0.4925	0.0610	-0.0059	-0.0203	0.1944

Columns 7 through 12

-126.5575	-59.9853	34.9840	24.6600	-67.8688	-24.1449
94.9539	45.5110	-26.7671	-18.8248	51.8593	18.4462
37.5210	17.7916	-10.4154	-7.3351	20.1962	7.1843
55.6650	26.6143	-15.6327	-10.9987	30.2949	10.7760
-5.2169	-2.4885	1.4506	1.0225	-2.8139	-1.0011
-44.2156	-20.8068	12.1650	8.5709	-23.5951	-8.3935
-122.2004	-58.2959	34.0453	23.9782	-66.0039	-23.4821
0.1109	-1.7084	1.8352	1.2125	-3.4644	-1.1438
112.4143	53.0653	-32.0471	-23.5372	63.9676	22.5635
-85.5646	-40.6950	25.2055	15.4704	-38.8503	-13.3004
94.5956	48.1132	-29.1028	-19.3132	30.2389	-7.8940
-77.3124	-37.0669	22.5013	12.2359	-18.3426	-20.4422
38.7604	18.6899	-11.3536	-6.5075	14.8662	14.1408
11.1197	4.7781	-3.0117	-1.7847	10.7361	1.5291
2.4949	1.1746	-0.7187	-0.4136	1.2548	0.7570

Columns 13 through 15

-12.8890	10.5247	0.7388
9.8493	-8.0416	-0.5646
3.8355	-3.1318	-0.2199
5.7536	-4.6977	-0.3298
-0.5344	0.4364	0.0306
-4.4809	3.6589	0.2569
-12.5360	10.2361	0.7186
-0.6176	0.5043	0.0353
12.0852	-9.8644	-0.6925

-7.4607	6.0677	0.4261
2.0530	-1.4835	-0.0991
-14.6622	11.3749	0.7972
-10.2521	20.4527	1.4057
-0.5567	-34.9769	-4.8274
-0.9468	-2.5865	-20.9125

Bfeedback =

-2.3612	155.6045
-0.7578	-118.3005
0.1014	-46.1861
-0.2661	-69.2361
-0.0538	6.4410
-0.2413	53.8345
-0.3123	149.1294
-0.0066	1.0065
0.1034	-138.5882
-0.0787	105.2740
0.0879	-117.4180
-0.0713	95.2925
0.0358	-47.8029
0.0101	-13.5191
0.23	-3.0674

Cfn =

Columns 1 through 6

309.8635	104.9641	-13.1099	1.0649	4.3791	-41.6462
39.5921	14.6512	-1.7676	0.3561	0.5506	-5.4253
-4.1756	-1.4142	0.1774	-0.0137	-0.0591	0.5603

Columns 7 through 12

-533.1994	-252.9435	148.2120	104.3427	-287.3261	-102.2086
-71.8111	-33.8255	19.8312	13.9634	-38.4526	-13.6777
7.1865	3.4095	-1.9977	-1.4065	3.8730	1.3777

Columns 13 through 15

-54.5678	44.5553	3.1279

-7.3024 5.9625 0.4186
0.7355 -0.6006 -0.0422

Dfeedback =

-0.4914 656.6797
-0.0658 87.8786
0.66 -8.8517

The gains are:

Gain1 = -0.2925
Gain2 = 1.0

The feedback dynamics that optimizes the control system has 15 states. The number of states will be reduced later, and the functional degradation of the system due to the reduction will be studied.

Closed-loop eigenvalues, the singular vs. frequency plot, and step input response are presented below. They demonstrate its optimality.

The closed loop eigenvalues are:

1.0e+002 *

-0.2500
-0.5824 + 1.3344i
-0.5824 - 1.3344i
-0.1052 + 0.2539i
-0.1052 - 0.2539i
-0.1052 + 0.2539i
-0.1052 - 0.2539i
-0.3579
-0.3579
-0.2244 + 0.0000i
-0.2244 - 0.0000i
-0.2037
-0.1660
-0.1661
-0.0000 + 0.0597i
-0.0000 - 0.0597i

-0.0000 + 0.0597i
-0.0000 - 0.0597i
-0.0001 + 0.0514i
-0.0001 - 0.0514i
-0.0001 + 0.0514i
-0.0001 - 0.0514i
-0.0114 + 0.0224i
-0.0114 - 0.0224i
-0.0114 + 0.0224i
-0.0114 - 0.0224i
-0.0231
-0.0230
-0.0101
-0.0101
-0.0070 + 0.0116i
-0.0070 - 0.0116i
-0.0070 + 0.0116i
-0.0070 - 0.0116i

All the eigenvalues are in LHP (Left Half Plane) indicating stability.

The singular value vs. frequency plot is shown in Fig. 4.3-7.

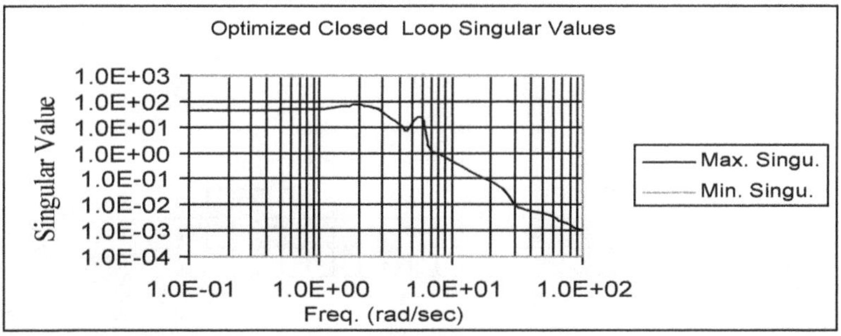

Fig. 4.3-7. Optimized Closed-Loop System Singular Values

The singular-value plot displays that the noise signals are attenuated considerably, while the desired signals are amplified substantially, which improves performance significantly. The maximum singular value is 73.803 at 1.9307 rad/sec.

a) Step Response

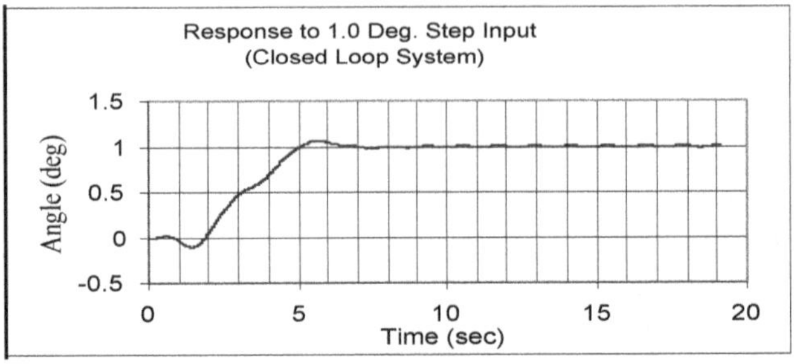

Fig. 4.3-8. Response to a 1.0 Deg Step Input

Unit step input response is plotted above. The response reaches steady state at 6.00 sec, but it reaches 1.0 deg in 5.0 sec and overshoots by 0.06 deg. State vector time histories of the optimal system are plotted below.

b) Engine Deflection Angle

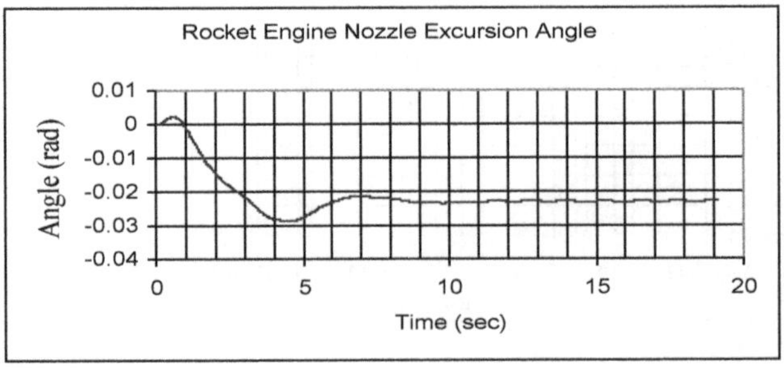

Fig. 4.3-9. Engine Nozzle Deflection Time History

The nozzle angle does not return to zero even after the attitude reaches steady state. The reason is that the LV's cg is no longer on the axial axis due to bending, liquid fuel sloshing, and, in addition, the non-zero angle of attack (to be shown later).

c) Attitude Rate

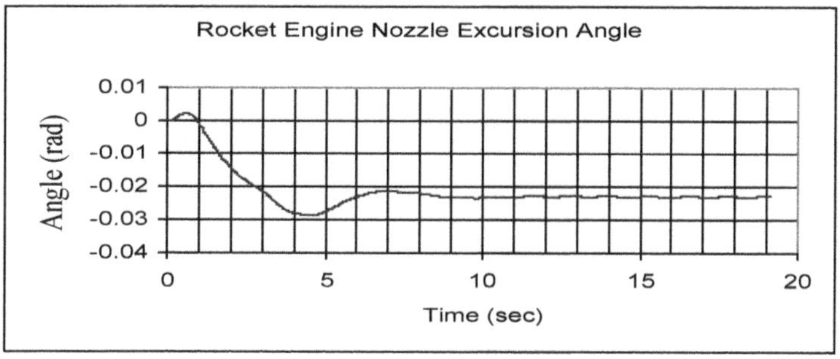

Fig. 4.3-10 Attitude Rate [x(2)]

d) Generalized Coordinate, Mode 1 (q1)

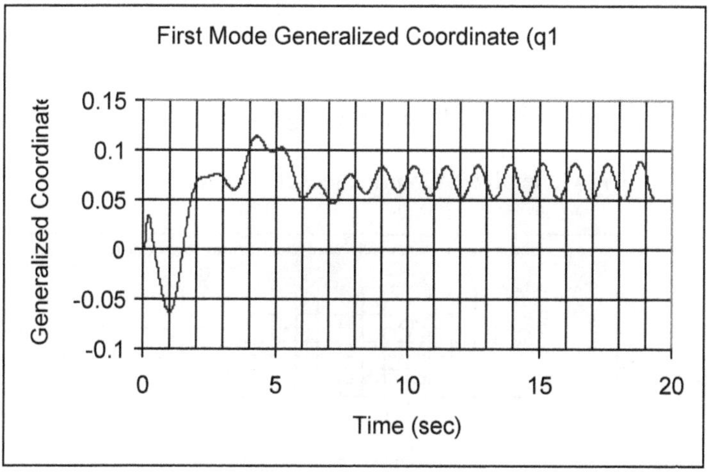

Fig. 4.3-11 Mode 1 Bending Generalized Coordinate

e) Generalized Coordinate Rate, Mode 1 (q1d)

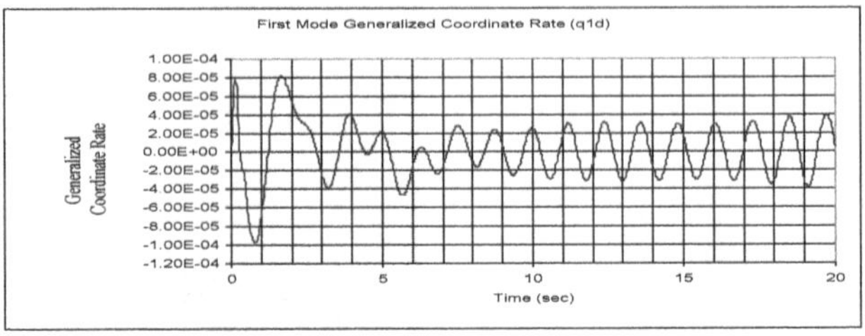

Fig. 4.3-12 Mode 1 Bending Generalized Coordinate Rate (q1d)

f) Generalized Coordinate , Mode 2 (q2)

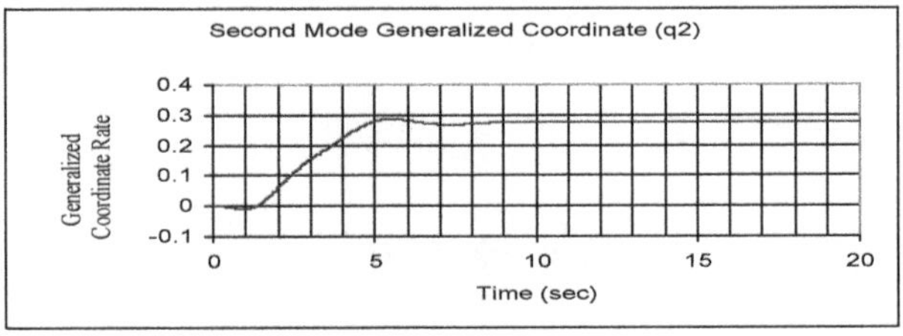

Fig. 4.3-13 Mode 2 Bending Generalized Coordinate (q2)

g) Generalized Coordinate Rate, Mode 2 (q2d)

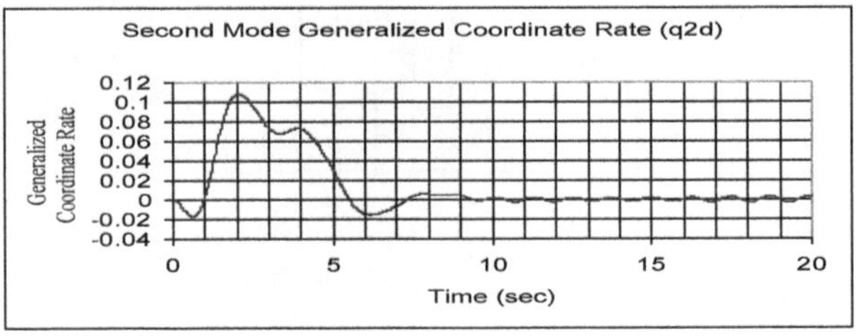

Fig. 4.3-14 Mode 2 Bending Generalized Coordinate Rate (q2d)

h) Fuel Sloshing 1

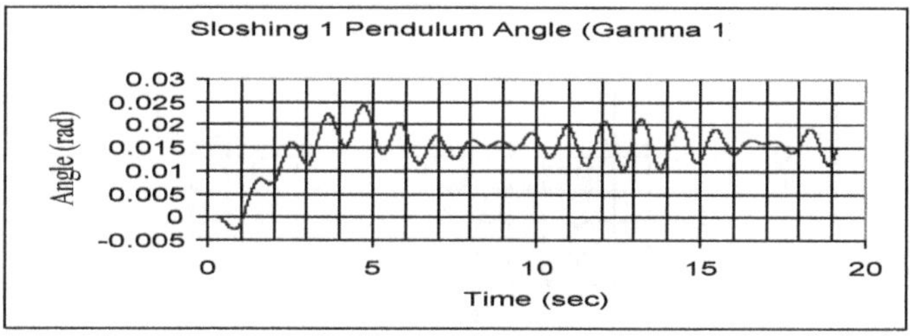

Fig. 4.3-15 Sloshing 1 Pendulum Angle (Γ1)

i) Fuel Sloshing Rate 1

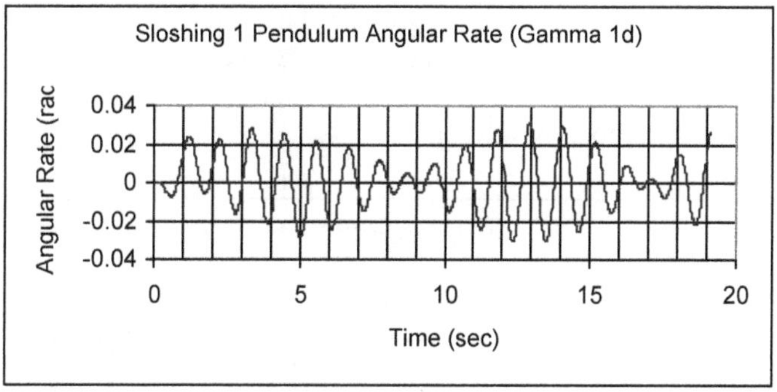

Fig. 4.3-16 Sloshing 1 Pendulum Angular Rate (Γ1d)

j) Fuel Sloshing 2

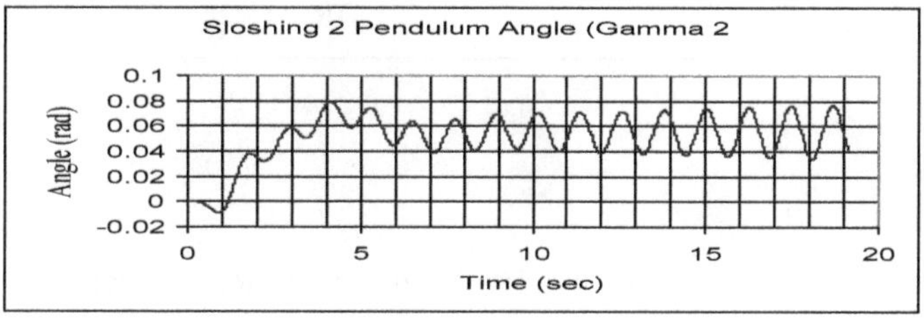

Fig. 4.3-17 Sloshing 2 Pendulum Angle (Γ2)

k) Fuel Sloshing Rate 2

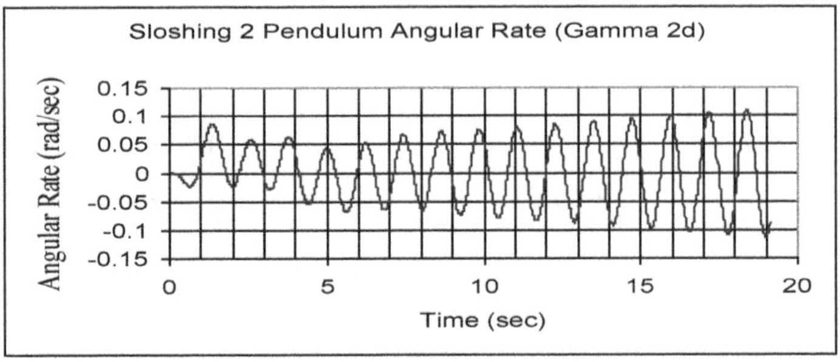

Fig. 4.3-18 Sloshing 2 Pendulum Angular Rate (Γ2d)

The amplitude seems to grow in the beginning. However, it reaches a finite steady state near 200 sec.

l) Position Gyro Read

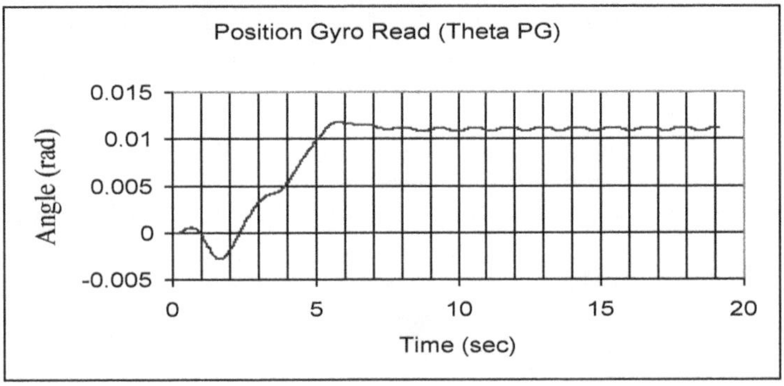

Fig. 4.3-19 Position Gyro Read (θPG)

m) Rate Gyro Read

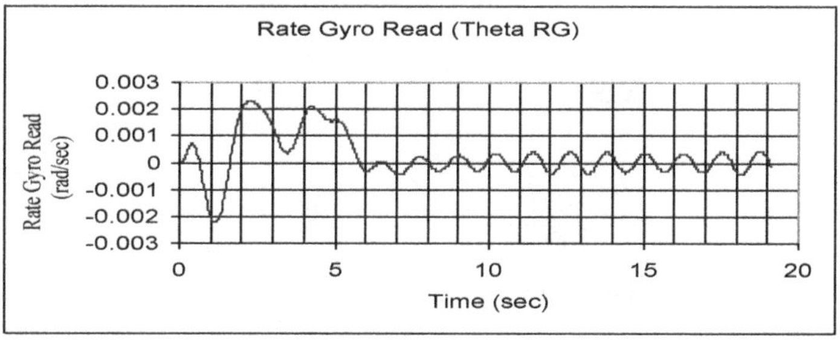

Fig. 4.3-20 Rate Gyro Read (θRG)

n) Rate Gyro Rate Read

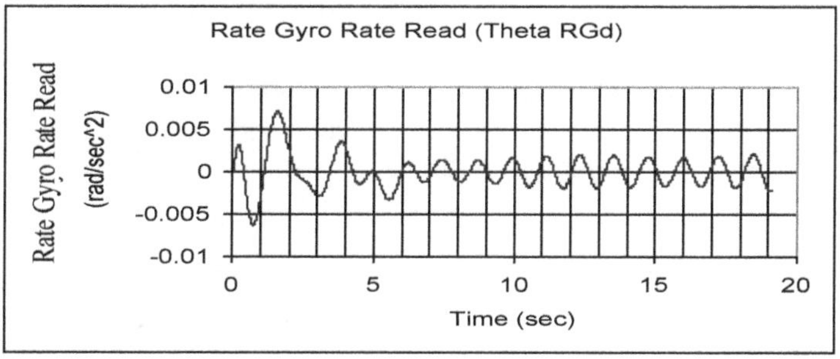

Fig. 4.3-21 Rate Gyro Rate Read (θRGd)

o) Plot of Attitude Angle, Nozzle Angle, and the Angle of Attack Angle

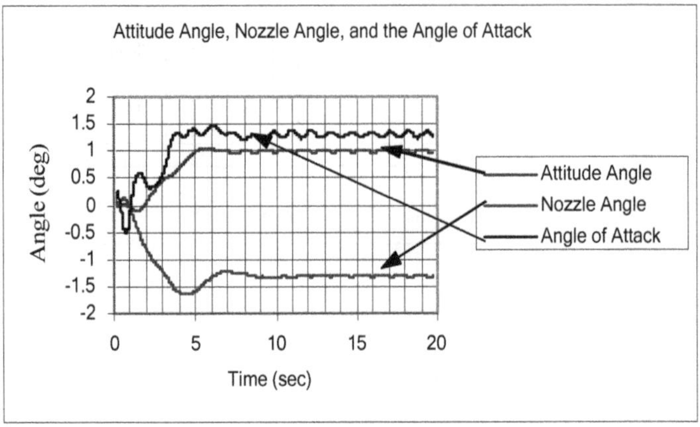

Fig. 4.3-22 Attitude Angle and Nozzle Angle

In Fig. 4.3-22, it is shown that the attitude control system achieves the desired attitude of 1.0 deg. The nozzle angle and the angle of attack are optimally settled while the bending and sloshing are acting.

- <u>Second Set</u>

Actuator:

Act(S)= 16.60465 /(S + 16.60465)

The weighting functions are:

W1(S) = (0.0060233 S^2 + 0.120488 S + 3.3123252)/(S^2 + 1.39525 S + 1.830815)

W2(S) = (S^2 + 28 S + 115)/(S^3 + 16 S^2 + 30 S + 110)

W3(S) = 1.0

Fig. 4.3-23 and Fig. 4.3-24 show the magnitude plots of these two functions.

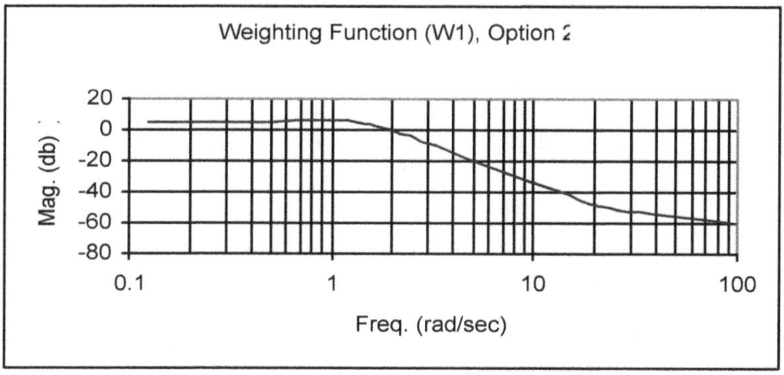

Fig. 4.3-23. Weighting Function, W1

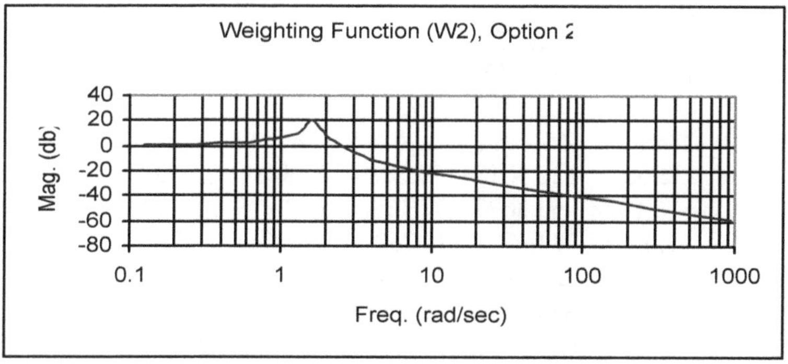

Fig. 4.3-24. Weighting Function, W2

The system matrices of the optimal dynamic feedback block, K(s), obtained by using "ncfsyn.m," are:

Afeedback =

Columns 1 through 6

19.5947	12.0686	-3.7622	7.7023	6.7710	-4.1811
-133.2636	-81.3601	29.7289	-54.3613	-36.2854	24.8438
-46.7851	-31.8817	8.7929	-14.3737	-12.6095	8.5470
-133.5946	-81.5792	21.4637	-54.6039	-35.3054	24.7822
-115.8128	-70.1261	21.5822	-48.0429	-32.1744	25.2708
-47.4318	-29.1598	8.9887	-19.6637	-17.2935	8.6218

-118.4524	-70.5996	21.8443	-47.8263	-32.7392	18.8194
106.8063	62.9077	-19.6788	42.1404	29.8548	-19.1798
-106.5078	-65.0057	20.4099	-43.6363	-29.7086	19.3542
133.1426	80.7087	-25.3372	54.1508	37.1132	-24.1498
-65.4590	-40.1790	12.6244	-26.9745	-18.2929	11.9443
37.3337	22.5886	-7.0906	15.1539	10.4004	-6.7649
-31.2964	-18.6202	5.8383	-12.4807	-8.6866	5.6250
8.8386	5.3556	-1.6813	3.5932	2.4630	-1.6027
-0.6477	-0.3779	0.1183	-0.2531	-0.1790	0.1153

Columns 7 through 12

14.9182	38.1821	19.7179	19.9269	15.7928	5.5047
-86.4198	-235.2462	-121.7501	-123.3440	-98.0154	-34.1578
-29.5882	-80.8882	-41.8229	-42.3475	-33.6417	-11.7240
-85.7023	-235.0003	-121.6330	-123.2438	-97.9396	-34.1313
-74.8080	-202.1754	-104.4021	-105.6954	-83.9568	-29.2589
-27.8920	-82.7209	-42.8374	-43.4291	-34.5102	-12.0267
-76.2136	-206.6423	-106.5130	-107.7738	-85.6030	-29.8324
69.2945	180.4935	92.2606	93.9391	74.6654	26.0177
-68.9450	-186.5591	-97.6990	-100.0744	-79.0846	-27.5750
86.1586	232.6831	122.3013	119.6089	92.8484	32.5284
-42.4715	-116.1848	-60.8672	-57.3352	-63.2600	-35.0713
24.1441	65.0268	34.0677	32.9259	42.8317	5.7416
-20.1625	-53.3004	-27.9061	-27.4400	-13.7407	6.4438
5.7179	15.4242	8.0803	7.7822	8.2735	0.4915
-0.4155	-1.0747	-0.5624	-0.5673	-0.1202	-0.0401

Columns 13 through 15

8.0918	1.3179	-0.3118
-50.2211	-8.1785	1.9354
-17.2370	-2.8071	0.6643
-50.1823	-8.1722	1.9339
-43.0170	-7.0055	1.6578
-17.6824	-2.8796	0.6814
-43.8598	-7.1428	1.6902
38.2489	6.2292	-1.4741
-40.5361	-6.6014	1.5621
47.7164	7.7740	-1.8396
-42.1937	-6.5502	1.5480
-3.7095	0.4995	-0.1146

```
-28.7172  -7.5264   1.7714
  5.8717 -23.2428  11.0553
 -1.5052   0.8051 -15.5606

Bfeedback =

   1.8934  -38.2539
   0.2523  233.1813
  -0.2165   80.3499
   0.3483  233.1477
  -0.5438  199.8992
   0.0762   82.4312
  -0.5860  203.4689
   0.0365 -180.9855
  -0.1269  185.9941
   0.1418 -231.1130
  -0.0711  114.7003
   0.0401  -64.7304
  -0.0332   53.6034
   0.0095  -15.3413
  -0.0007    1.0938

Cfeedback =

  Columns 1 through 6

 -252.5645 -152.2285   48.0056 -101.9783  -69.9110   45.8791
  -38.2234  -24.4774    7.7356  -16.4185  -10.9426    7.2453
    2.1971    1.3229   -0.4194    0.8852    0.6061   -0.4005

  Columns 7 through 12

 -162.7408 -440.2738 -228.1027 -230.8255 -183.4499  -63.9298
  -25.5209  -70.3915  -36.3600  -36.7206  -29.1900  -10.1716
    1.4144    3.8362    1.9861    2.0107    1.5978    0.5568

  Columns 13 through 15

  -93.9943  -15.3071    3.6223
  -14.9549   -2.4355    0.5763
    0.8187    0.1333   -0.0315
```

Dfeedback =

```
  -0.2711  437.3138

  -0.0431   69.5795

   0.0024   -3.8089
```

The gains are:

Gain1= 0.0114297

Gain2 = 1.0

The feedback dynamics that optimizes the control system has 15 states. Closed loop eigenvalues, the singular value vs. frequency plot, and step input response demonstrate its optimality.

The closed loop eigenvalues are:

```
   1.0e+002 *

  -0.2500
  -0.5824 + 1.3344i
  -0.5824 - 1.3344i
  -0.1052 + 0.2540i
  -0.1052 - 0.2540i
  -0.1052 + 0.2540i
  -0.1052 - 0.2540i
  -0.2245
  -0.2244
  -0.2035
  -0.1663
  -0.1662
  -0.1445 + 0.0000i
  -0.1445 - 0.0000i
  -0.0001 + 0.0597i
  -0.0001 - 0.0597i
  -0.0001 + 0.0597i
  -0.0001 - 0.0597i
  -0.0003 + 0.0514i
  -0.0003 - 0.0514i
```

-0.0003 + 0.0514i
-0.0003 - 0.0514i
-0.0176 + 0.0329i
-0.0176 - 0.0329i
-0.0176 + 0.0329i
-0.0176 - 0.0329i
-0.0254
-0.0252
-0.0102 + 0.0152i
-0.0102 - 0.0152i
-0.0101 + 0.0152i
-0.0101 - 0.0152i
-0.0100
-0.0100 |

All the eigenvalues are in the LHP (Left Half Plane) indicating stability.

The singular value vs. frequency plot is shown in Fig. 4.3-25.

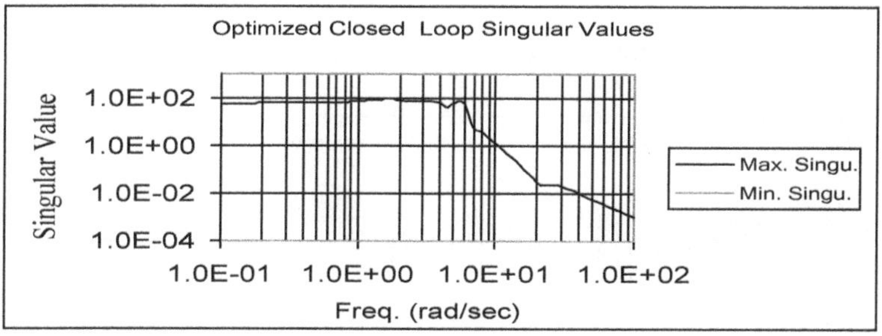

Fig. 4.3-25. Optimized Closed Loop System Singular Values

The singular value plot displays that the noise signals are attenuated considerably and the desired signals are amplified substantially, which improves the performance significantly. The maximum singular value is 87.6467 at 1.5999 rad/sec.

Unit step input response and state vector time histories of the optimal system are plotted below.

a) Step response

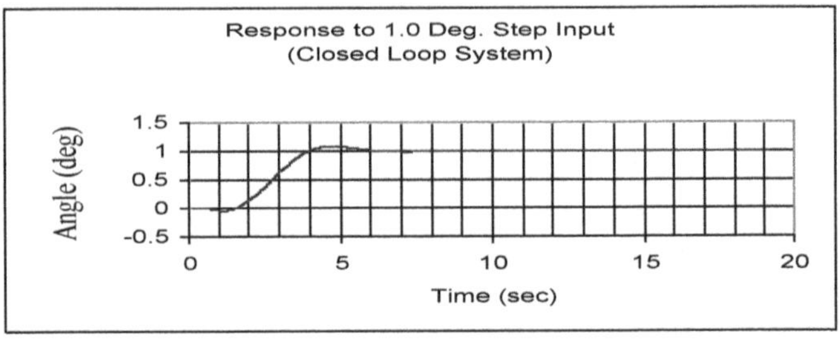

Fig. 4.3-26 Response to 1.0 Deg. Step Input

The response reaches the steady state at 4.00 sec.

b) Engine Deflection Angle

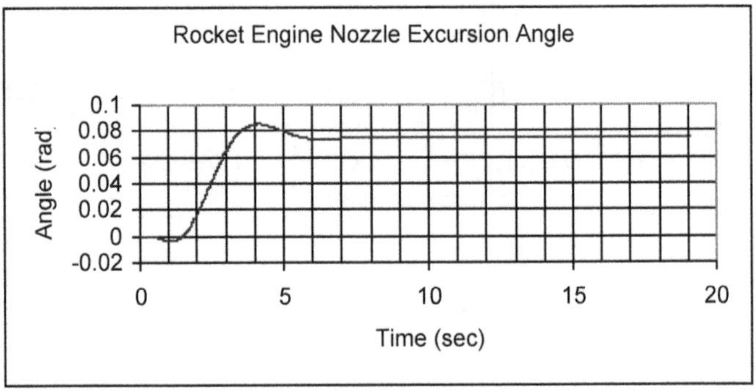

Fig. 4.3-27 Engine Nozzle Deflection History

The nozzle angle does not return to zero even after the attitude reaches steady state. The reason is that the LV's cg is no longer on the axial axis due to bending, liquid fuel sloshing, and, in addition, the non-zero angle of attack (to be addressed later).

c) Attitude Rate

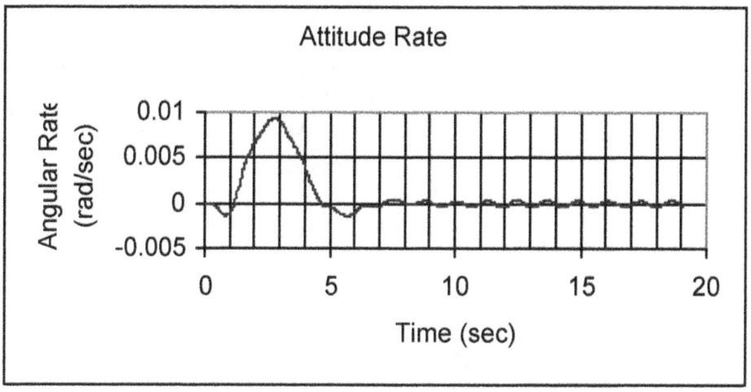

Fig. 4.3-28 Attitude Rate [x(2)]

d) Generalized Coordinate, Mode 1 (q1)

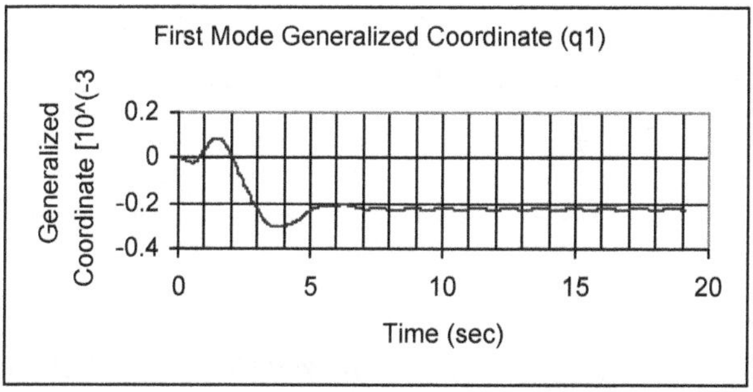

Fig. 4.3-29 Mode 1 Bending Generalized Coordinate (q1)

e) Generalized Coordinate Rate, Mode 1 (q1d)

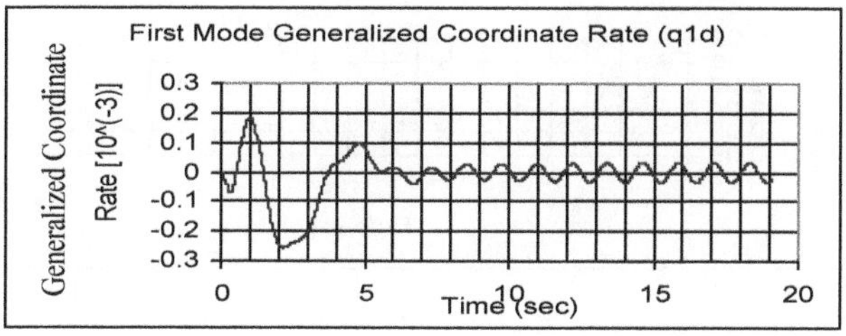

Fig. 4.3-30 Mode 1 Bending Generalized Coordinate Rate (q1d)

f) Generalized Coordinate , Mode 2 (q2)

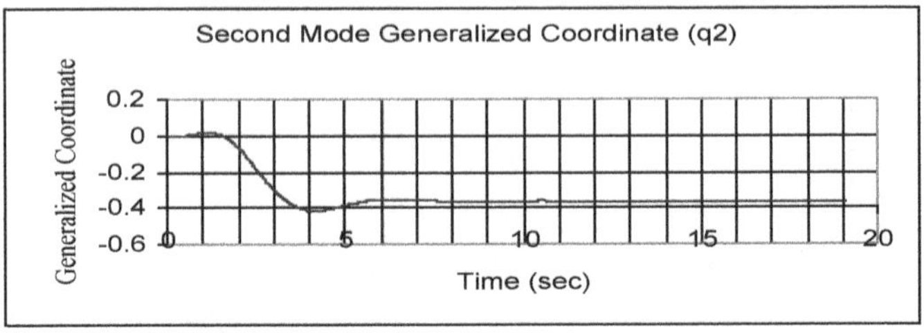

Fig. 4.3-31 Mode 2 Bending Generalized Coordinate (q2)

g) Generalized Coordinate Rate, Mode 2 (q2d)

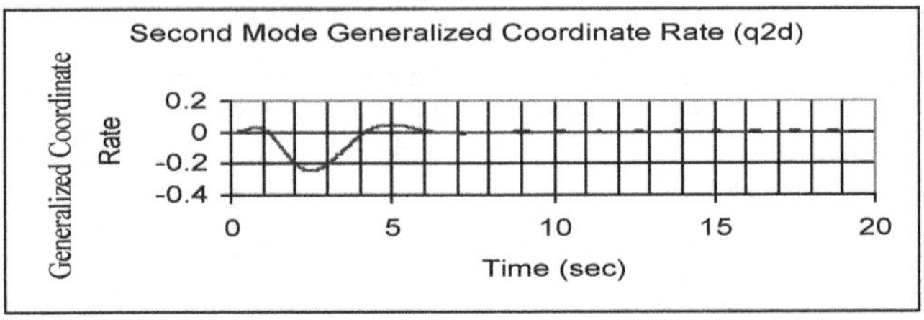

Fig. 4.3-32 Mode 2 Bending Generalized Coordinate Rate (q2d)

h) Fuel Sloshing 1

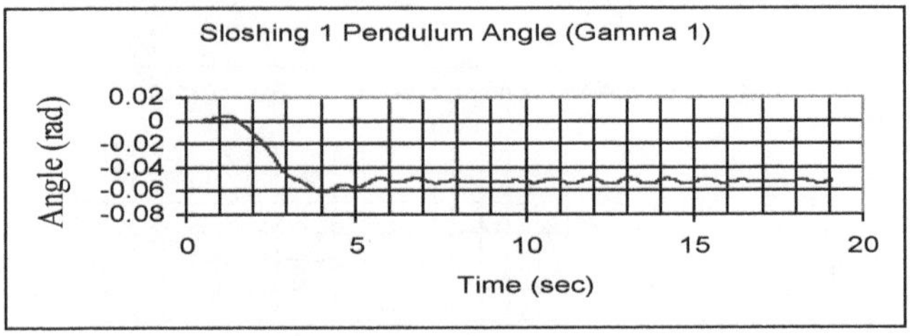

Fig. 4.3-33 Sloshing 1 Pendulum Angle (Γ1)

i) Fuel Sloshing Rate 1

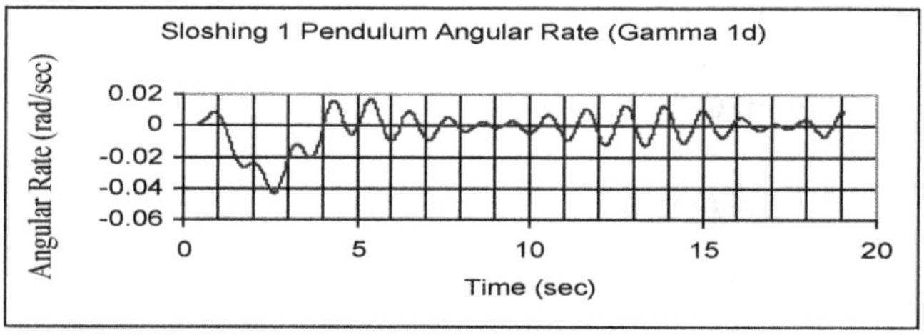

Fig. 4.3-34 Sloshing 1 Pendulum Angular Rate (Γ1d)

j) Fuel Sloshing 2

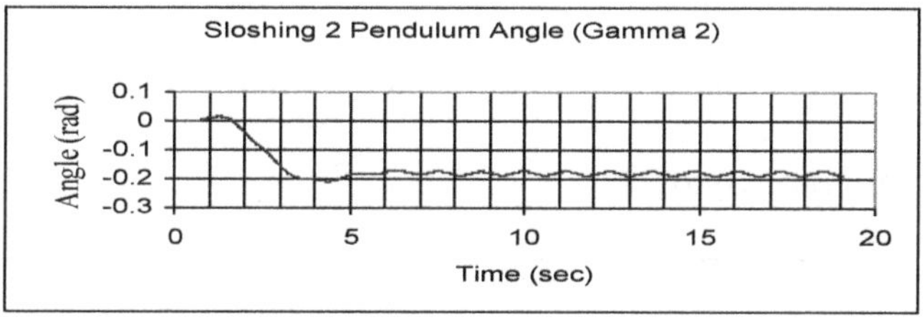

Fig. 4.3-35 Sloshing 2 Pendulum Angle (Γ2)

k) Fuel Sloshing Rate 2

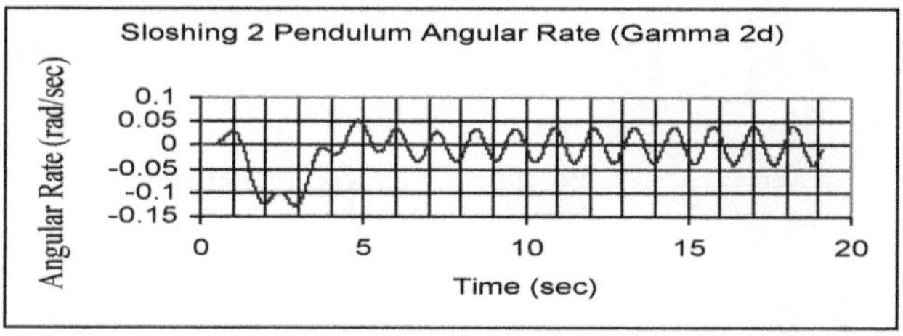

Fig. 4.3-36 Sloshing 2 Pendulum Angular Rate (Γ2d)

The amplitude seems to grow in the beginning. However, it reaches a finite steady state near 200 sec.

l) Position Gyro Read

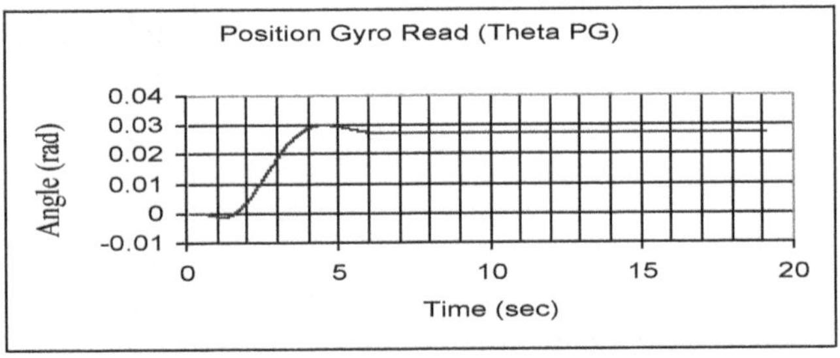

Fig. 4.3-37 Position Gyro Read (θPG)

m) Rate Gyro Read

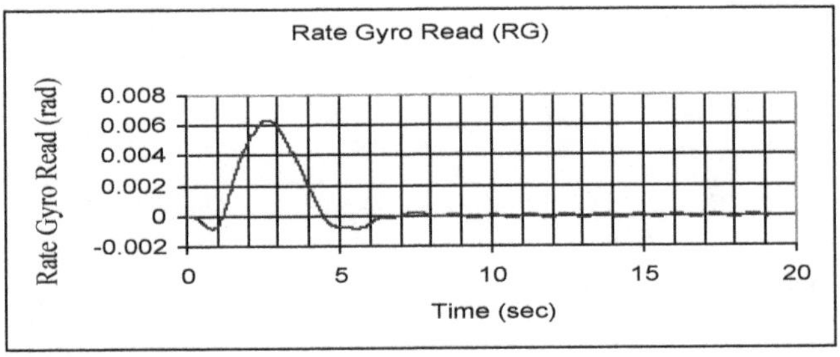

Fig. 4.3-38 Rate Gyro Read (θRG)

n) Rate Gyro Read Rate

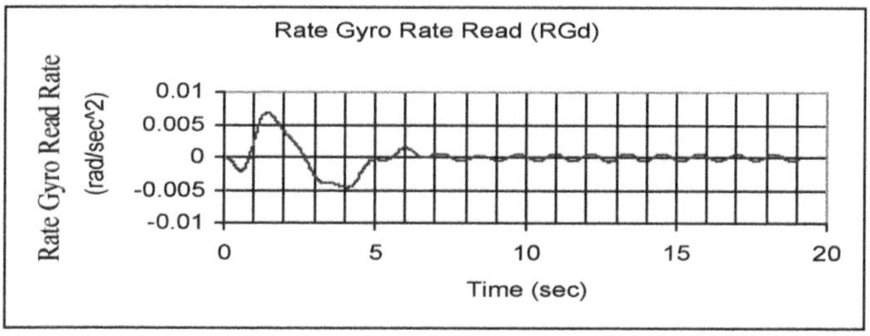

Fig. 4.3-39 Rate Gyro Rate Read (θRGd)

o) Plot of Attitude angle, Nozzle Angle, and the Angle of Attack

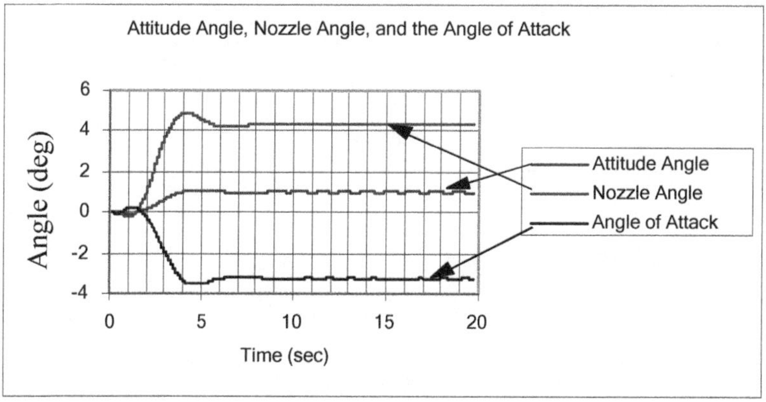

Fig. 4.3-40 Attitude angle, Nozzle Angle, and the Angle of Attack

In Fig. 4.3-40, it is shown that the attitude control system achieves the desired attitude of 1.0 deg. The nozzle angle and the angle of attack are optimally settled while the bending and sloshing are acting.

- Third Set

Actuator:

Act(S) = 16.67 /(S + 16.67)

The weighting functions are:

W1(S) = (0.059988 S^2 + 5.5188962 S +11.9976)/(S^2 + 1.3305 S + 1.8236)

W2(S) = (S^2 + 82 S + 100)/(S^3 + 16 S^2 + 25 S + 40)

W3(S) = 1.0

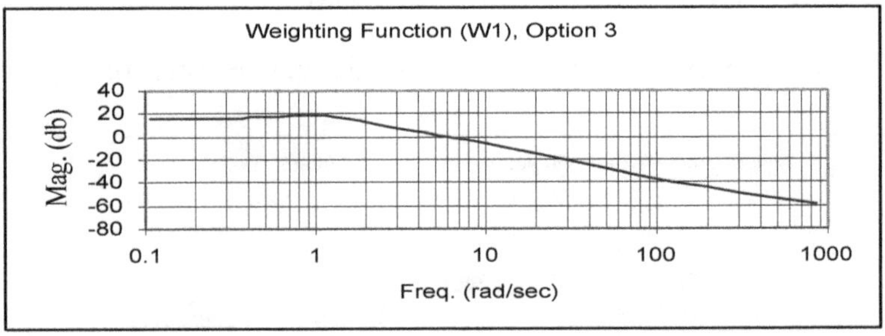

Fig. 4.3-41. Weighting Function, W1

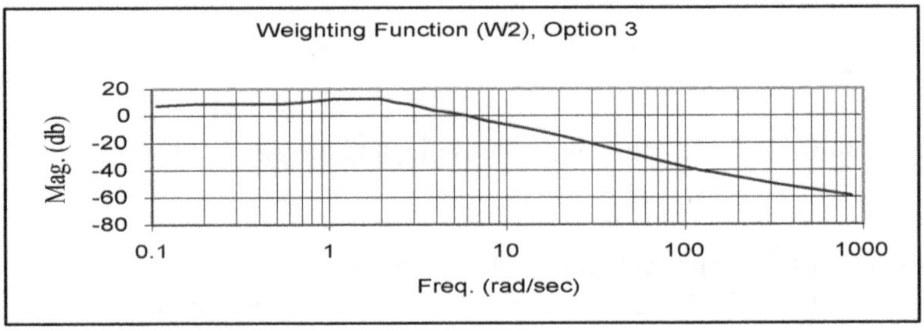

Fig. 4.3-42. Weighting Function, W2

Fig. 4-41 and Fig. 4-42 show the magnitude plots of these two functions.

The system matrices of the optimal dynamic feedback block, K(s), obtained by using "ncfsyn.m," are:

Afeedback =

-16.2845	19.9362	2.4273	16.4365	-24.7735	-9.0902
35.6806	-50.1261	-10.0950	-41.3132	66.1729	24.5990
-5.7176	11.7074	0.7877	2.0569	-9.3619	-3.5100
29.1493	-41.4818	-0.7918	-34.2900	52.9806	20.2653
-40.5223	55.6978	5.3062	46.8754	-74.9536	-27.8619
10.4734	-14.6517	-1.4924	-12.2059	20.4824	6.4518
-1.5902	1.7703	0.1798	1.4328	-2.3946	3.8813
39.5750	-51.1213	-5.2458	-42.1749	70.7792	22.9730
-23.1146	32.6212	3.3532	27.1065	-44.5777	-14.2259
-6.2002	8.9128	0.9162	7.4113	-12.1014	-3.7729
-24.3379	32.6746	3.3589	27.0128	-44.6393	-14.0340
-9.9420	13.7259	1.4115	11.3772	-18.7266	-5.9232
-4.0787	6.1377	0.6313	5.1236	-8.3152	-2.6169
-2.0808	2.4390	0.2507	1.9905	-3.3763	-1.0764
0.4624	-0.6312	-0.0649	-0.5226	0.8617	0.2723

Columns 7 through 12

-2.6116	-39.9836	22.2235	-6.5179	-22.1777	9.1083
7.0942	107.6845	-59.8243	17.5211	59.6114	-24.4937
-1.0244	-15.4110	8.5500	-2.5069	-8.5224	3.5029
5.8089	88.4817	-49.1756	14.3843	48.9550	-20.1184
-7.9554	-120.0230	66.3554	-19.3713	-65.8763	27.0977
-2.8242	27.8156	-15.5469	4.4433	15.1739	-6.2862
-0.2763	-5.0827	2.5009	-0.7207	-2.4258	1.0043
8.2207	95.7629	-49.3032	13.7431	46.9097	-20.1313
-4.6068	-65.5582	32.5704	-6.2375	-24.7623	11.8338
-1.1981	-16.0685	5.5454	-3.8782	-15.2858	7.2458
-4.6580	-56.6613	25.8979	-12.3602	-52.7937	33.4744
-1.9280	-25.3233	12.3707	-7.5299	-33.4120	-1.9601
-0.8055	-12.0770	5.1095	-3.6470	-12.7203	-0.8367
-0.3900	-3.7996	2.4682	-0.8857	-5.7802	-1.8114
0.0893	1.1452	-0.5605	0.3120	1.2939	0.4714

Columns 13 through 15

```
   5.4796   -3.2581    0.4188
 -14.7323    8.7598   -1.1259
   2.1062   -1.2525    0.1610
 -12.1012    7.1949   -0.9248
  16.2920   -9.6876    1.2451
  -3.7768    2.2440   -0.2884
   0.6026   -0.3586    0.0461
 -12.0340    7.1452   -0.9188
   7.0257   -4.1256    0.5304
   4.1644   -2.3768    0.3055
  17.7415   -9.9662    1.2867
  -4.3363    3.8521   -0.4768
  -4.2240    5.9681   -0.7789
   1.5173  -21.7616    5.4795
  0.4651      2.9619  -16.3669

Bfeedback =

  -1.0111  -22.0031
   0.7051   59.8657
   0.1979   -8.8238
   0.8633   49.1095
   0.5339  -65.5386
   0.0339   16.7592
   0.0103   -2.1446
   0.1010   59.1897
  -0.0569  -37.8175
  -0.0155  -10.3907
  -0.0577  -38.1712
  -0.0242  -15.9921
  -0.0108   -7.1279
  -0.0043   -2.8634
  0.11    0.7360

Cfeedback =

Columns 1 through 7

  24.9748  -33.3997   -3.6064  -27.4897   44.8549   15.7536    4.6598

   0.1865   -3.2348   -0.3489   -2.7919    4.1460    1.4664    0.3986
```

-0.0150 0.0203 0.0024 0.0168 -0.0277 -0.0101 -0.0029

Columns 8 through 14

69.7084 -38.8153 11.4365 38.8860 -15.9547 -9.5976 5.7078

6.7489 -3.6265 1.0826 3.6546 -1.5019 -0.9028 0.5373

-0.0445 0.0247 -0.0073 -0.0247 0.0101 0.0061 -0.0036

Column 15

-0.7336

-0.0691

0.0005

Dfeedback =

0.0600 39.6589
0.0057 3.7339
-0.0000 -0.0252

The gains are:

Gain1 = 0.044773
Gain2 = 1.0

The feedback dynamics that optimizes the control system has 15 states. The closed-loop eigenvalues, singular value vs. frequency plot, and step input response demonstrating its optimality are presented below.

The closed loop eigenvalues are

1.0e+002 *

-0.2500
-0.5824 + 1.3344i

-0.5824 - 1.3344i
-0.1040 + 0.2546i
-0.1040 - 0.2546i
-0.1046 + 0.2542i
-0.1046 - 0.2542i
-0.2284 + 0.0212i
-0.2284 - 0.0212i
-0.2293
-0.1353 + 0.0476i
-0.1353 - 0.0476i
-0.1446 + 0.0002i
-0.1446 - 0.0002i
-0.0868 + 0.0557i
-0.0868 - 0.0557i
-0.0002 + 0.0601i
-0.0002 - 0.0601i
-0.0002 + 0.0600i
-0.0002 - 0.0600i
-0.0009 + 0.0508i
-0.0009 - 0.0508i
-0.0009 + 0.0509i
-0.0009 - 0.0509i
-0.0347 + 0.0149i
-0.0347 - 0.0149i
-0.0204
-0.0052 + 0.0136i
-0.0052 - 0.0136i
-0.0078 + 0.0145i
-0.0078 - 0.0145i
-0.0134
-0.0106
-0.0100

All the eigenvalues are in LHP (Left Half Plane) indicating stability.

The singular value vs. frequency plot is shown in The singular value vs. frequency plot is shown in Fig. 4.3-43.

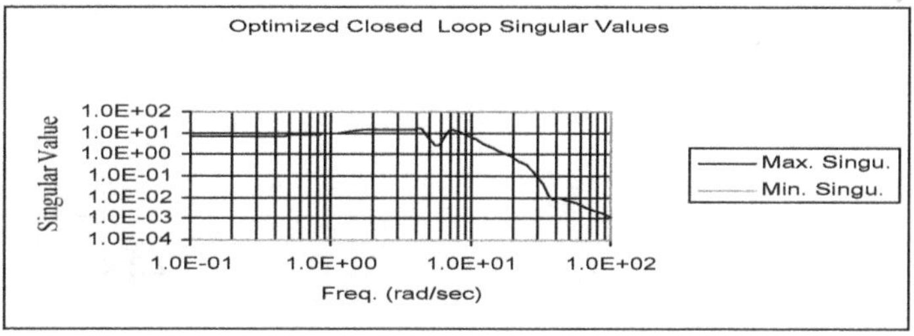

Fig. 4.3-43. Optimized, Closed-Loop System Singular Values

The singular value plot displays that the noise signals will be attenuated considerably while the desired signals will be amplified, which should improve performance significantly. The maximum singular value is 19.0845 at 1.9307 rad/sec.

Unit step input response and state vector time histories of the optimal attitude control system are plotted below.

a) Step response

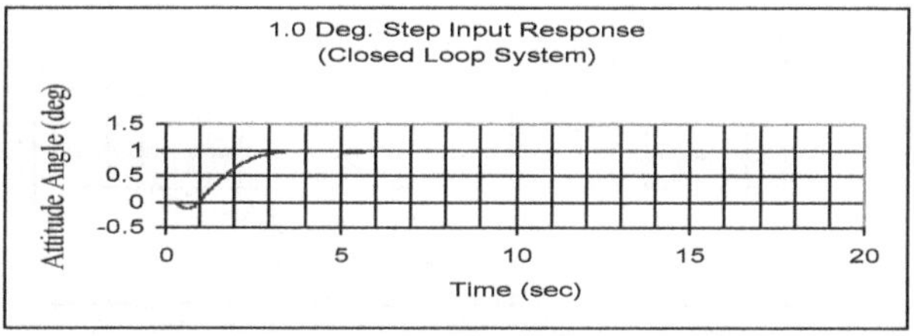

Fig. 4.3-44. Response to a 1.0 Deg Step Input

The response reaches steady state at 3.5 sec. Neither overshoot nor undershoot is displayed.

b) Engine Deflection Angle

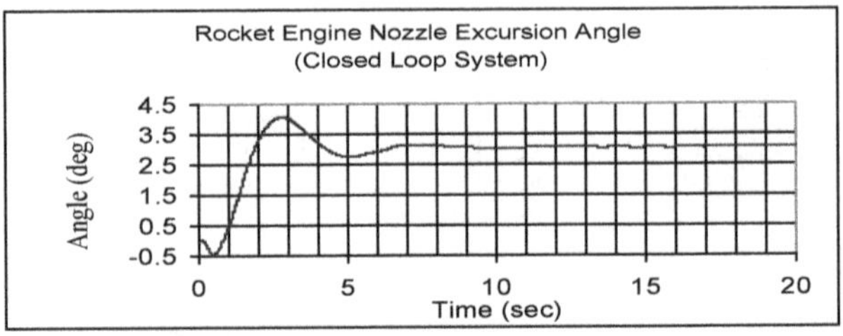

Fig. 4.3-45. Engine Nozzle Deflection Time History

The nozzle angle does not return to zero even after the attitude reaches steady state. The reason is that the LV's cg is no longer on the axial axis due to bending, liquid fuel sloshing, and, in addition, the non-zero angle of attack (to be shown later).

c) Attitude Rate

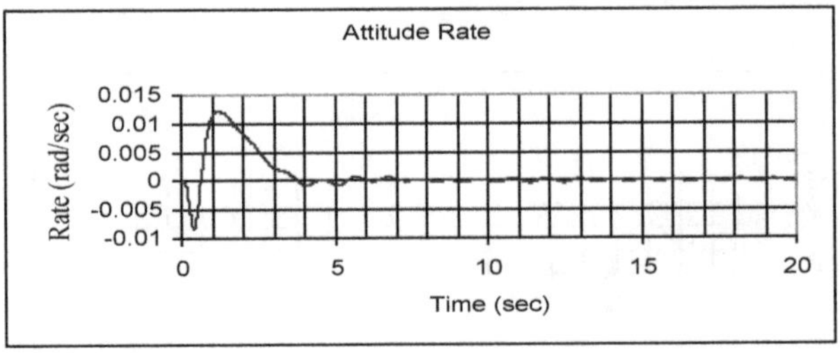

Fig. 4.3-46 Attitude Rate [x(2)]

d) Generalized Coordinate, Mode 1 (q1)

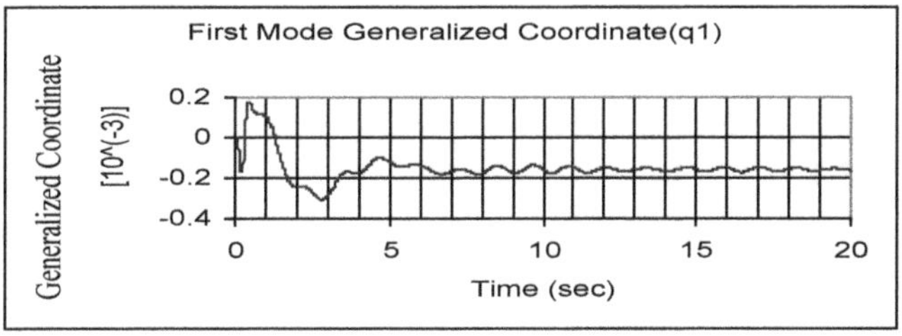

Fig. 4.3-47 Mode 1 Bending Generalized Coordinate (q1)

e) Generalized Coordinate Rate, Mode 1 (q1d)

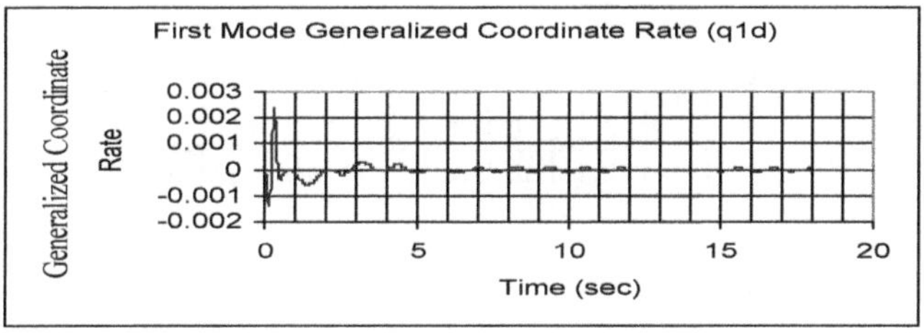

Fig. 4.3-48 Mode 1 Bending Generalized Coordinate Rate (q1d)

f) Generalized Coordinate , Mode 2 (q2)

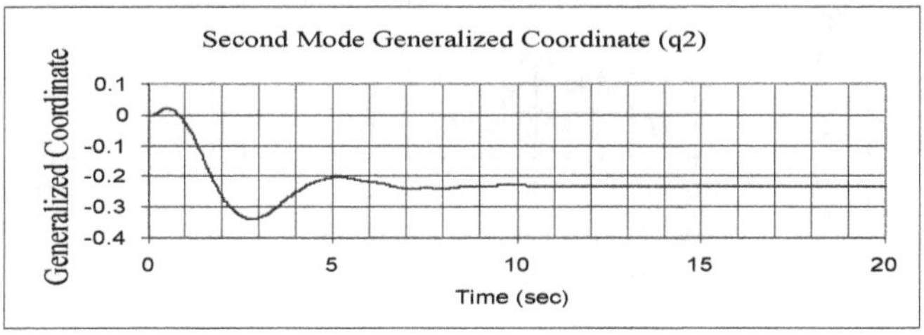

Fig. 4.3-49 Mode 2 Bending Generalized Coordinate Rate (q2)

g) Generalized Coordinate Rate, Mode 2 (q2d)

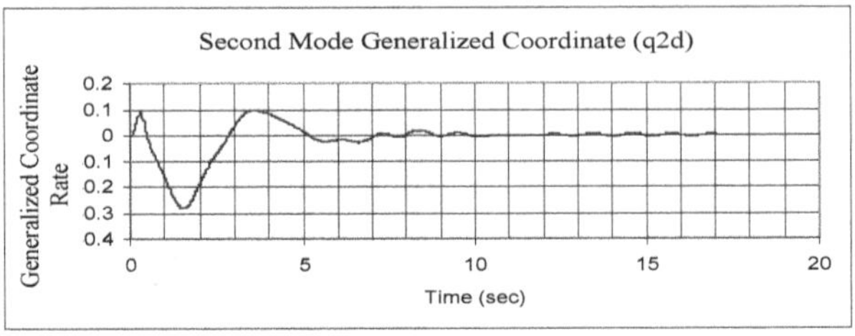

Fig. 4.3-50 Mode 2 Bending Generalized Coordinate Rate (q2d)

b) Fuel Sloshing 1

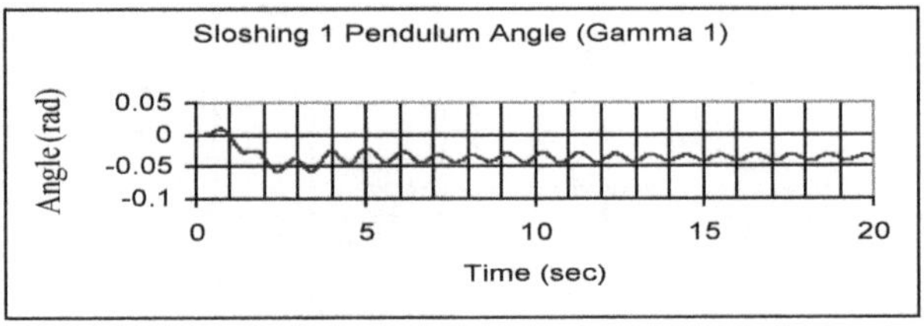

Fig. 4.3-51 Sloshing 1 Pendulum Angle (Γ1)

h) Fuel Sloshing Rate 1

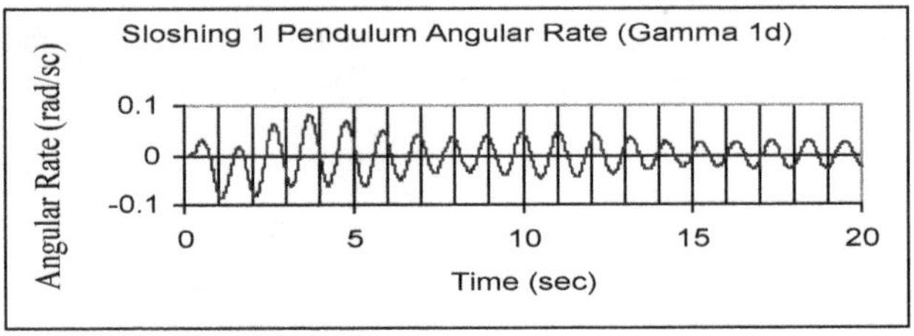

Fig. 4.3-52 Sloshing 1 Pendulum Angular Rate (Γ1d)

i) Fuel Sloshing 2

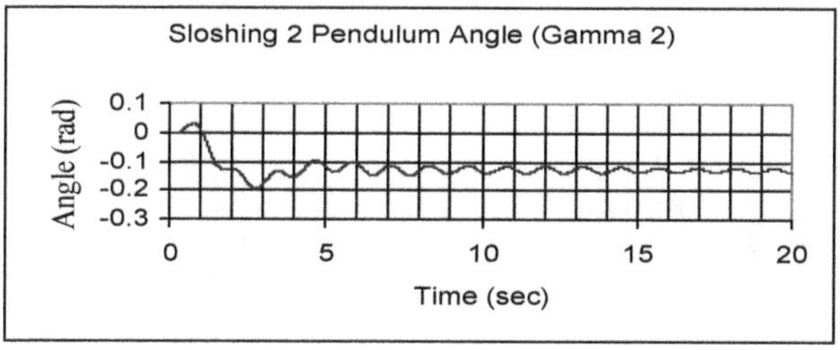

Fig. 4.3-53 Sloshing 2 Pendulum Angle (Γ2)

j) Fuel Sloshing Rate 2

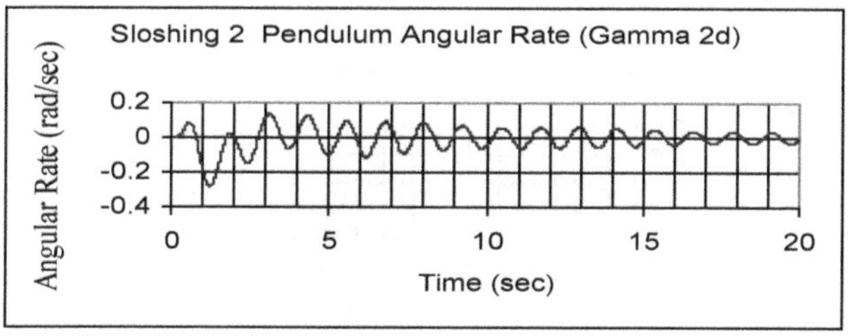

Fig. 4.3-54 Sloshing 2 Pendulum Angular Rate (Γ2d)

k) Position Gyro Read

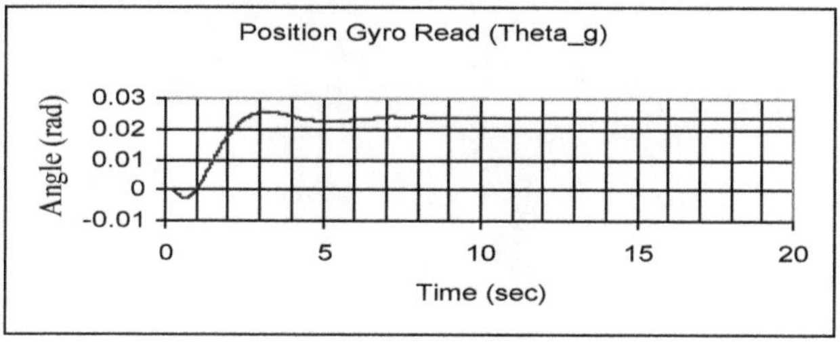

Fig. 4.3-55 Position Gyro Read (θPG)

m) Rate Gyro Read

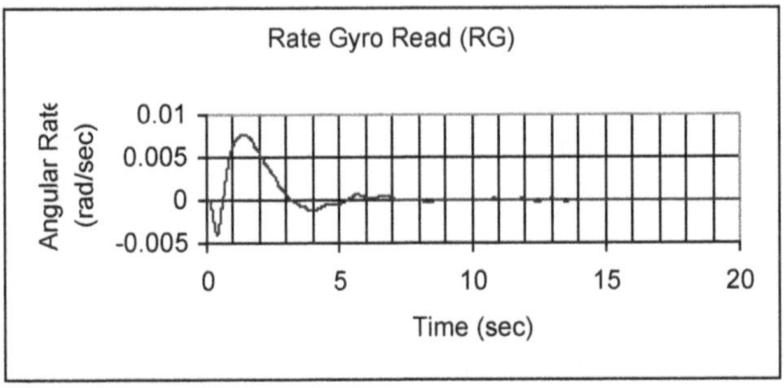

Fig. 4.3-56 Rate Gyro Read (θRG)

n) Rate Gyro Rate

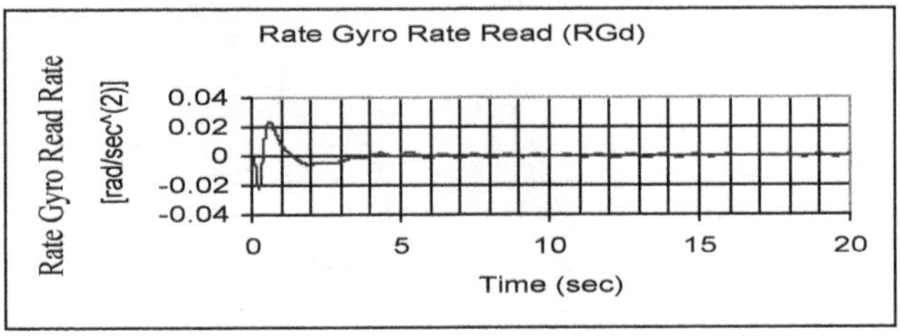

Fig. 4.3-57 Rate Gyro Rate Read (θRGd)

o) Plot of Attitude Angle, Nozzle Angle, and the Angle of Attack

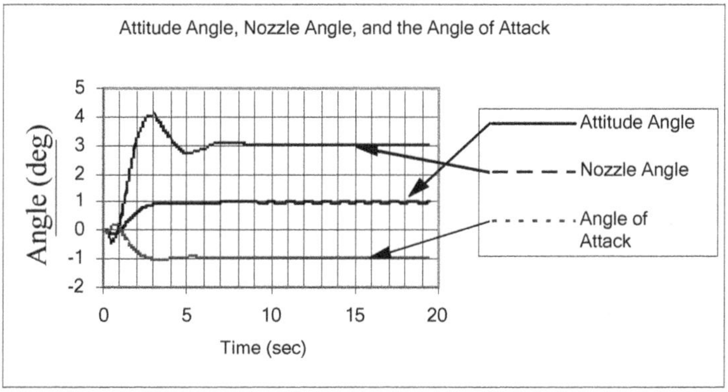

Fig. 4.3-58 Attitude Angle, Nozzle Angle, and the Angle of Attack

In Fig. 4.3-58, it is shown that the attitude control system achieves the desired attitude of 1.0 deg. The nozzle angle and the angle of attack are optimally settled while the bending and sloshing are acting.

	First Set	Second Set	Third Set
Attitude Angle(deg)	1 (Fig. 4.3-22)	1 (Fig. 4.3-40)	1 (Fig. 4.3-58)
Nozzle Angle (deg)	-1.4 (Fig. 4.3-22)	4.2 (Fig. 4.3-40)	3 (Fig. 4.3-58)
Angle of Attack (deg)	1.25 (Fig.4.3-22)	-3.2 (Fig. 4.3-40)	-1 (Fig. 4.3-58)
q1	5.00E-05 (Fig. 4.3-11)	-2.20E-04 (Fig. 4.3-29)	-1.70E-04 (Fig. 4.3-47)
q2	0.28 (Fig. 4.3-13)	-0.36 (Fig. 4.3-31)	-0.24 (Fig. 4.3-49)
Gamma 1(deg)	0.855 (Fig. 4.3-15)	-2.85 (Fig. 4.3-33)	-2.28 (Fig. 4.3-51)
Gamma 2(deg)	2.85 (Fig. 4.3-17)	-9.69 (Fig. 4.3-35)	-7.25 (Fig. 4.3-53)

Table 4.3-1. Comparison of the Three Sets of Weighting Functions

Table 4.3-1 summarizes the simulation results of the three cases studied in this chapter. The entries are steady state values. We notice that the attitude control responses are different depending on the weighting function. The desired attitudes are achieved, but the nozzle angles and the angles of attack, q1, q2, $\Gamma 1$, and $\Gamma 2$, are different. In general, this table shows the balance at steady state. When the nozzle angle is negative, the other parameters are positive, and when the nozzle angle is positive, the other parameters are negative. Numerical verification of the balance is not an easy task because it involves bending and sloshing. Fig. 4.3-59 shows a graphical explanation of bending, nozzle direction, and the angle of attack.

Note: The process of attaining an optimal set of weighting functions is not straight forward. It usually involves numerous trial and error processes. Lower order weighting functions are usually selected to start with because they have less number of coefficients that need to be determined. Primary objective of the weighting function is to meet the Loop Shaping requirements. Systematic way of searching for the optimal weighting function needs to be developed.

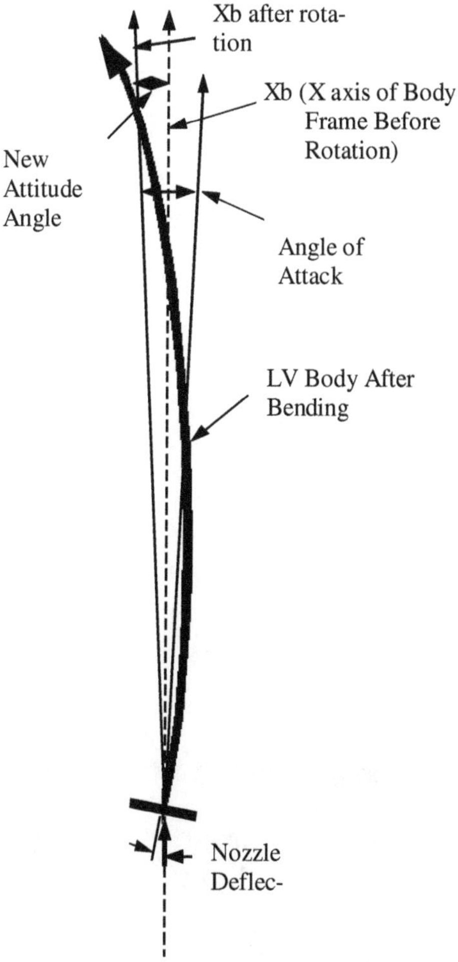

Fig. 4.3-59. The LV's Angles of Attack, Attitude, and Nozzle Deflection4)

- A case when Ka = 1.0 and Ki=1.5 in "En" block in Fig4.1-1a

In Fig. 4.1-1a, the "En" block generates the engine command signal, δc where δc is:

$$\delta c = Ka(Ki/S + 1)\,(\theta c - \theta f)$$

For this Fourth Set analysis, we choose;

$$Ka = 1.0$$
$$Ki = 1.5$$

Then,

$$\delta c = [(S + 1.5)/S]\,(\theta c - \theta f)$$
$$= (En)\,(\theta c - \theta f)$$

where

$$En = [(S + 1.5)/S]$$

Actuator:

$$Act(S) = 9.8 /(S + 14)$$

The weighting functions are:

$$W1(S) = (S + 5.57143)/(S + 1.2)$$

$$W2(S) = (S^2 + 82 S + 100)/(S^3 + 16 S^2 + 24.8 S + 81)$$

$$W3(S) = 1.0$$

Fig. 4.3-60 and Fig. 4.3-61 show the magnitude plots of these two functions.

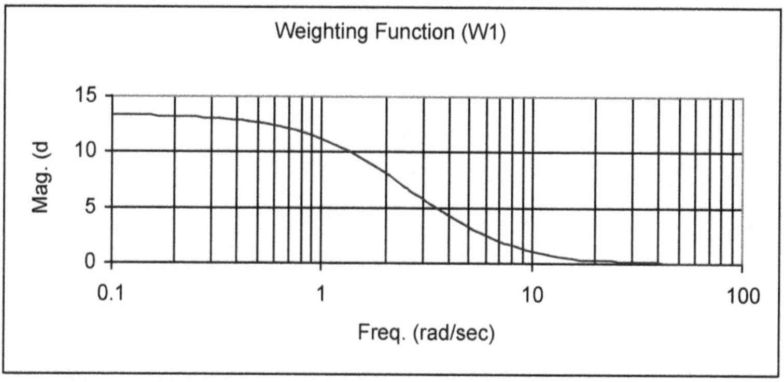

Fig. 4.3-60. Weighting Function, W1

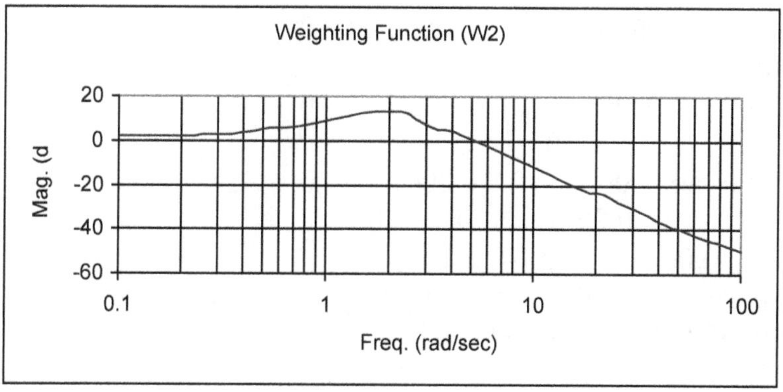

Fig. 4.3-61. Weighting Function, W2

System matrices

Ap =

1.0e+004 *

Columns 1 through 5

0	0.0001	0	0	0
0.0004	0.0002	-0.0000	0	0

0	0	0	0.0001	0
0.0227	0.0075	-0.0366	-0.0026	-0.0047
0	0	0	0	0
0.0099	0.0033	-0.3591	-0.0083	-0.0020
0	0	0	0	0
0	0	0	0	0
0	0	0	0	0
0	0	0	0	0
0.0025	0	0.0000	0	-0.0001
0	0	0	0	0
0	0.8480	0	0.0037	0

Columns 6 through 10

0	0	0	0	0
-0.0001	0	-0.0000	0.0000	0
0	0	0	0	0
0.0000	-0.0000	0	-0.0005	0
0.0001	0	0	0	0
0.0000	-0.0000	0	0.0002	0
0	0	0.0001	0	0
0	-0.0036	0	-0.0001	0
0	0	0	0	0.0001
0	-0.0001	0	-0.0027	0
0	0	0	0	0
0	0	0	0	0
-0.0234	0	0	0	0

Columns 11 through 13

0	0	0
0	0	0
0	0	0
0	0	0
0	0	0
0	0	0
0	0	0

```
        0         0         0
        0         0         0
        0         0         0
  -0.0025         0         0
        0         0    0.0001
        0   -2.1199   -0.0116

Bp =

        0         0         0
  -8.0500   -9.7660    0.0030
        0         0         0
-262.0970   38.5200    0.0100
        0         0         0
-114.7180   16.8600         0
        0         0         0
 -27.7200         0         0
        0         0         0
 -64.8150         0         0
        0         0         0
        0         0         0
        0         0         0

Cp =

  Columns 1 through 8

     1     0     0     0     0     0     0     0
     0     1     0     0     0     0     0     0

  Columns 9 through 13

     0     0     0     0     0
     0     0     0     0     0

Dp =
```

0	0	0
0	0	0

The system matrices of the optimal dynamic feedback block, K(s), obtained by using "ncfsyn.m," are:

Afeedback =

Columns 1 through 4

0.8985	3.7308	0.2576	-4.0358
-9.9292	-17.9041	-6.3054	19.6276
1.3503	6.9355	0.1265	1.1289
11.0695	20.4213	-1.7191	-22.4829
12.9871	21.6456	0.9682	-24.6312
8.1483	18.0321	0.9911	-20.2554
-2.3663	-2.7183	-0.1273	2.9364
1.3767	5.1965	0.3245	-5.8762
2.0475	2.3218	0.1145	-2.5118
-16.4336	-29.1315	-1.6795	32.4189
6.6941	12.7184	0.7455	-14.1558
-5.4741	-8.5745	-0.4827	9.4741
-1.3032	-2.8929	-0.1746	3.2377
-0.5393	-1.1529	-0.0691	1.2888
-0.1082	-0.1580	-0.0087	0.1741

Columns 5 through 8

-7.9630	-9.2832	2.5940	-5.2326
26.5493	33.6022	-9.1953	18.7114
-2.8103	-3.5164	0.9758	-2.0094
-28.6251	-37.4065	10.3117	-20.9363
-33.3952	-41.9732	11.1180	-23.0551
-26.1034	-35.2787	12.7897	-19.9542
4.3402	1.5503	-1.5650	6.1524
-7.2201	-10.8174	-0.6673	-6.3771
-3.7835	-3.1176	1.3618	0.5398
46.6695	57.2954	-15.5271	32.9776

-19.9008	-24.5680	6.1088	-13.3706
13.7182	15.7157	-4.5491	8.1425
4.4826	5.8466	-1.3298	3.3447
1.7919	2.3103	-0.5375	1.3114
0.2551	0.2833	-0.0871	0.1436

Columns 9 through 12

-2.4271	-13.9281	-6.3927	4.6118
8.6366	49.9089	22.6650	-16.4316
-0.9226	-5.3743	-2.4457	1.7684
-9.6965	-56.1519	-25.4266	18.4527
-10.3427	-60.8654	-27.2534	19.7524
-9.9033	-56.0689	-24.4450	17.8348
1.4918	9.9449	4.2088	-3.0316
-5.4436	-21.4608	-8.4977	6.2839
-1.3200	-11.7159	-4.0839	2.9024
17.4318	58.5403	12.0035	-13.9847
-5.6214	-11.2983	-21.6482	19.1244
4.4781	12.1810	15.3155	-20.6193
1.2269	4.6488	6.9068	-8.3613
0.4996	1.8385	2.6943	-3.2105
0.0876	0.2425	0.3104	-0.6834

Columns 13 through 15

-1.7434	-0.6599	0.0969
6.2072	2.3511	-0.3452
-0.6688	-0.2532	0.0372
-6.9709	-2.6405	0.3877
-7.4736	-2.8299	0.4156
-6.7484	-2.5552	0.3752
1.1521	0.4364	-0.0641
-2.3755	-0.8990	0.1320
-1.1124	-0.4214	0.0619
5.6960	2.1508	-0.3152
-7.8487	-2.9510	0.4315

```
  13.4358    4.9510   -0.7153
 -19.0120  -10.7462    2.1685
 -10.6602   -7.7364    2.3023
  -0.5888   -0.8307  -14.4999
```

Bfeedback =

```
   1.0543   -7.2699
   0.6223   23.2306
   0.1663   -2.6739
  -0.9587  -26.0535
   0.4882  -27.1912
   0.0416  -23.3589
  -0.0096    3.5368
   0.0089   -6.9115
  -0.0036   -2.9442
   0.0658   35.2443
  -0.0272  -15.7548
   0.0190   10.5139
   0.0065    3.5775
   0.0026    1.4246
   0.0003    0.1929
```

Cfeedback=

Columns 1 through 4

```
  -7.8735  -13.3333   -0.8863   14.6278
   2.1782    0.1236   -0.0213   -0.0347
   0.0020    0.0039    0.0003   -0.0044
```

Columns 5 through 8

```
  20.4402   24.5241   -6.5979   13.3937
  -0.3720    0.3612    0.1723    0.1962
  -0.0063   -0.0079    0.0021   -0.0044
```

Columns 9 through 12

6.1689	34.8310	16.1605	-11.6591
-0.1076	-0.1143	-0.0428	0.0463
-0.0020	-0.0116	-0.0053	0.0039

Columns 13 through 15

4.3988	1.6660	-0.2446
-0.0158	-0.0059	0.0009
-0.0015	-0.0006	0.0001

Dfeedback=

0.0309	17.0989
-0.0001	-0.0605
-0.0000	-0.0057

Gain1=0.134262
Gain2=1.0

The closed loop eigenvalues are:

1.0e+002 *

-0.2500
-0.5824 + 1.3344i
-0.5824 - 1.3344i
-0.1038 + 0.2650i
-0.1038 - 0.2650i
-0.1022 + 0.2613i
-0.1022 - 0.2613i
-0.2546
-0.1827 + 0.0833i
-0.1827 - 0.0833i
-0.2020 + 0.0320i
-0.2020 - 0.0320i
-0.1469

-0.1469
-0.0002 + 0.0600i
-0.0002 - 0.0600i
-0.0002 + 0.0600i
-0.0002 - 0.0600i
-0.0009 + 0.0509i
-0.0009 - 0.0509i
-0.0009 + 0.0509i
-0.0009 - 0.0509i
-0.0368 + 0.0212i
-0.0368 - 0.0212i
-0.0328 + 0.0108i
-0.0328 - 0.0108i
-0.0112 + 0.0168i
-0.0112 - 0.0168i
-0.0131 + 0.0166i
-0.0131 - 0.0166i
-0.0149
-0.0147
-0.0101
-0.0100

All the eigenvalues are in LHP (Left Half Plane) indicating stability.

The singular value vs. frequency plot is shown in Fig. 4.3-62.

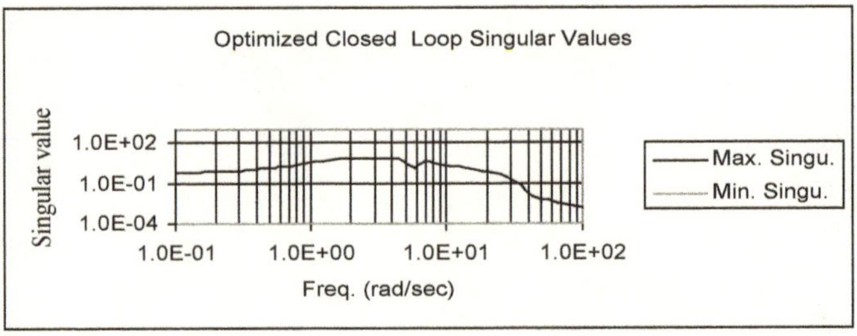

Fig. 4.3-62. Optimized, Closed-Loop System Singular Values

The singular value plot displays that the noise signals will be attenuated considerably while the desired signals will be amplified, which should improve performance significantly. The maximum singular is 8.3237 at 2.8118 rad/sec.

Unit step input response and state vector time histories of the optimal attitude control system are plotted below.

a) Step Response

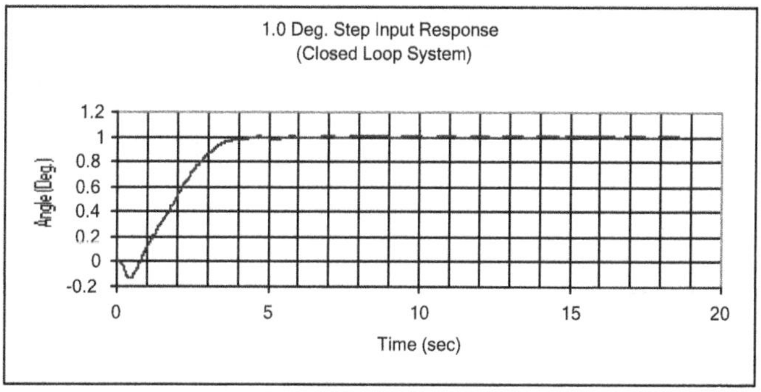

Fig. 4.3-63 Response to a 1.0 Deg Step Input

The response reaches steady state at 4.0 sec. Neither overshoot nor undershoot is displayed.

b) Engine Deflection Angle

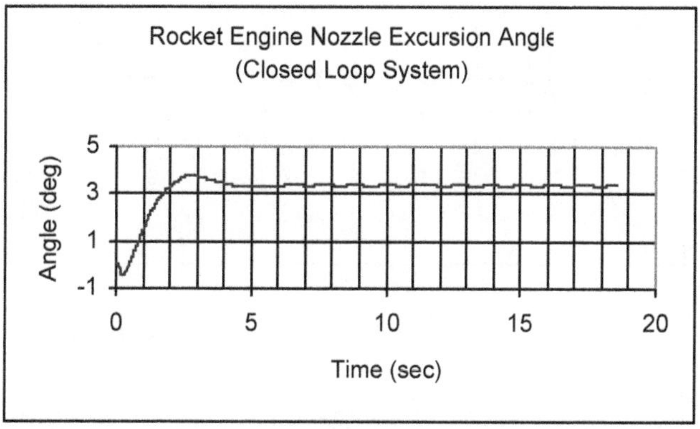

Fig. 4.3- 64 Engine Nozzle Deflection Time History

The nozzle angle does not return to zero even after the attitude reaches steady state. The reason is that the LV's cg is no longer on the axial axis due to bending, liquid fuel sloshing, and, in addition, the non-zero angle of attack.

c) Attitude Rate

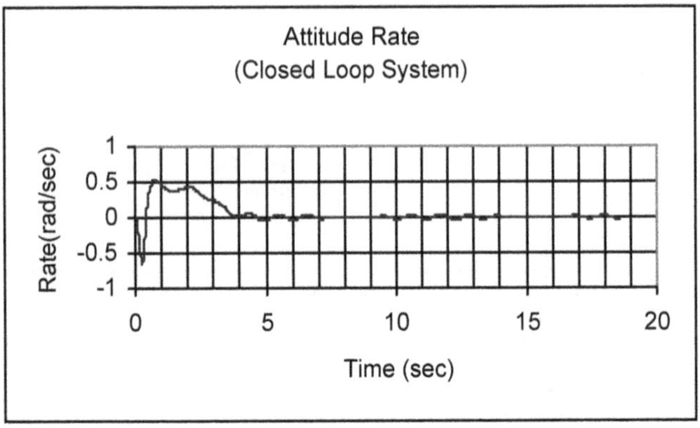

Fig. 4.3-65 Attitude Rate [x(2)]

d) Generalized Coordinate, Mode 1 (q1)

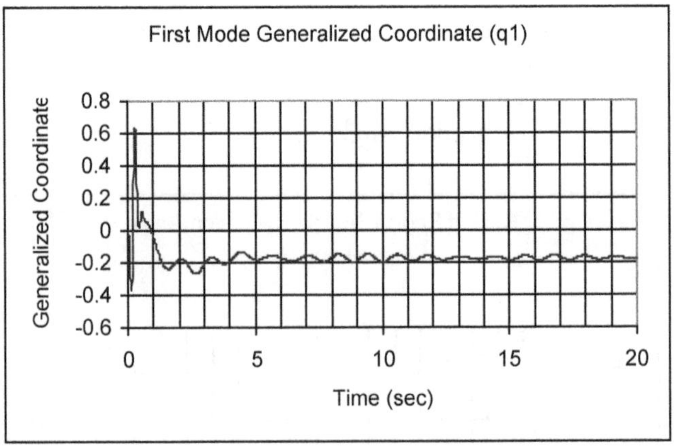

Fig. 4.3-66 Mode 1 Bending Generalized Coordinate (q1)

e) Generalized Coordinate Rate, Mode 1 (q1d)

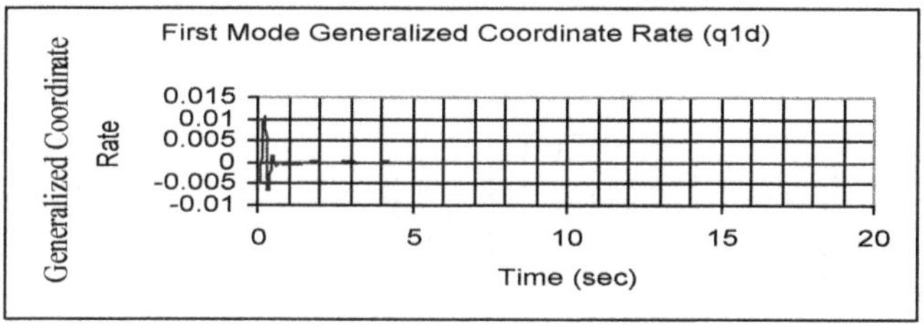

Fig. 4.3-67 Mode 1 Bending Generalized Coordinate Rate (q1d)

f) Generalized Coordinate , Mode 2 (q2)

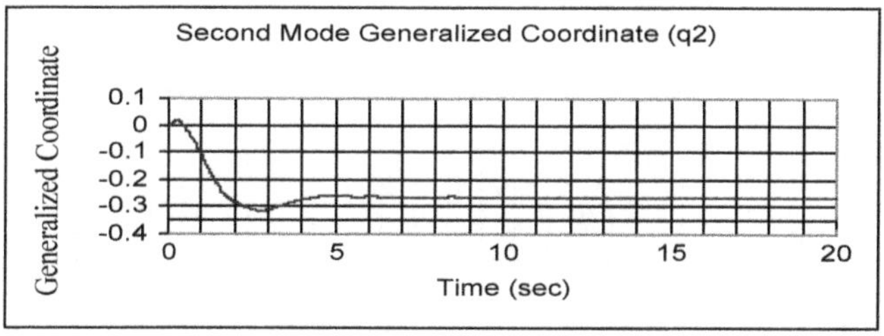

Fig. 4.3- 68 Mode 2 Bending Generalized Coordinate Rate (q2)

g) Generalized Coordinate Rate, Mode 2 (q2d)

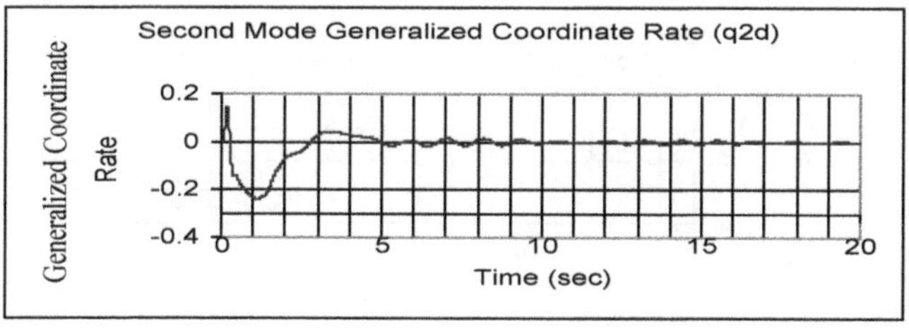

Fig. 4.3- 69 Mode 2 Bending Generalized Coordinate Rate (q2d)

b) Fuel Sloshing 1

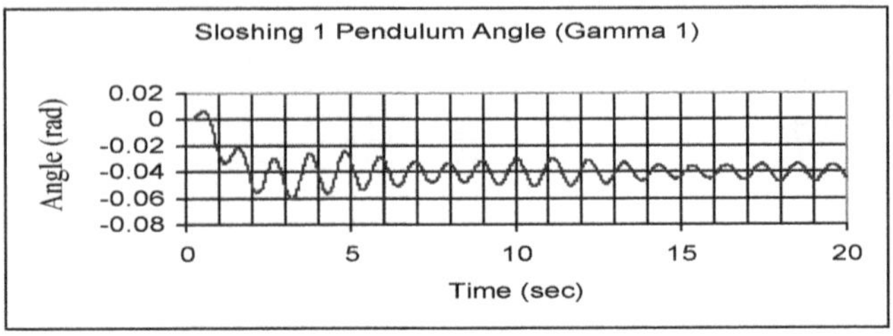

Fig. 4.3-70 Sloshing 1 Pendulum Angle (Γ1)

h) Fuel Sloshing Rate 1

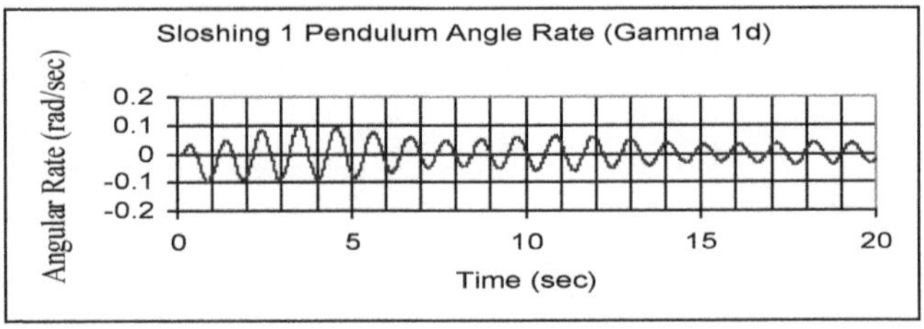

Fig. 4.3-71 Sloshing 1 Pendulum Angular Rate (Γ1d)

i) Fuel Sloshing 2

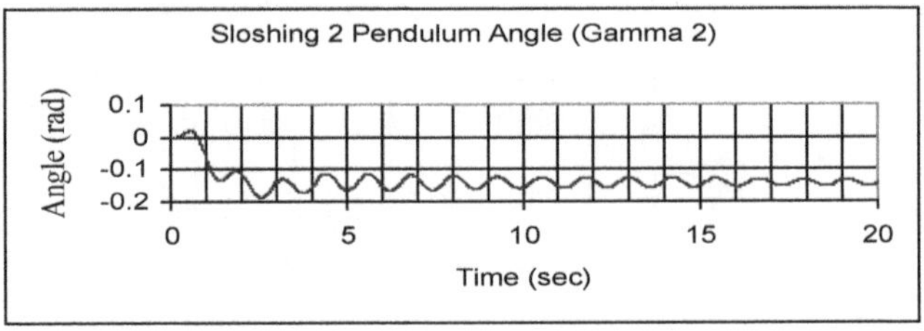

Fig. 4.3-72 Sloshing 2 Pendulum Angle (Γ2)

j) Fuel Sloshing Rate 2

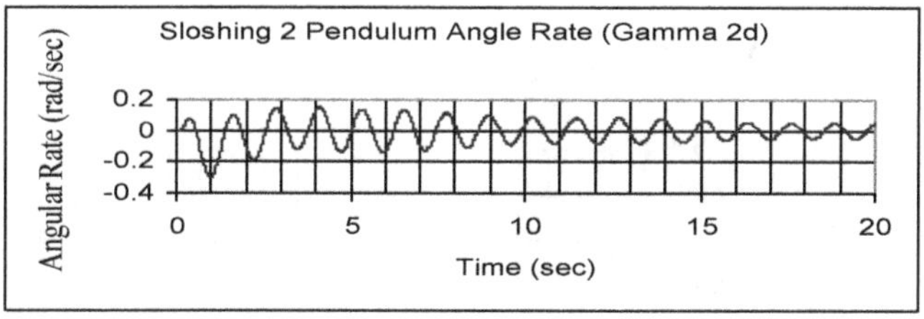

Fig. 4.3-73 Sloshing 2 Pendulum Angular Rate (Γ2d)

k) Position Gyro Read

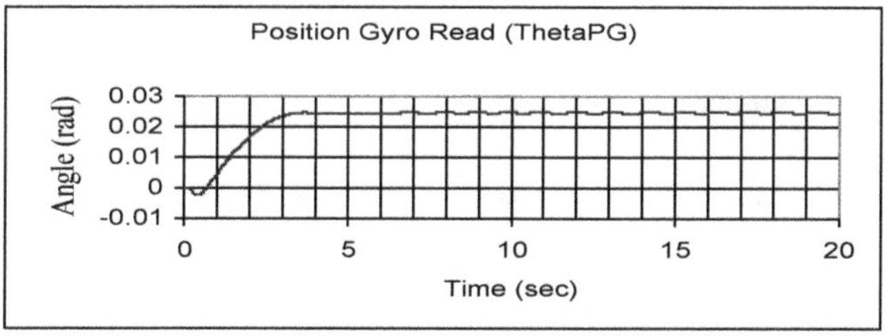

Fig. 4.3-74 Position Gyro Read (θPG)

e) Rate Gyro Read

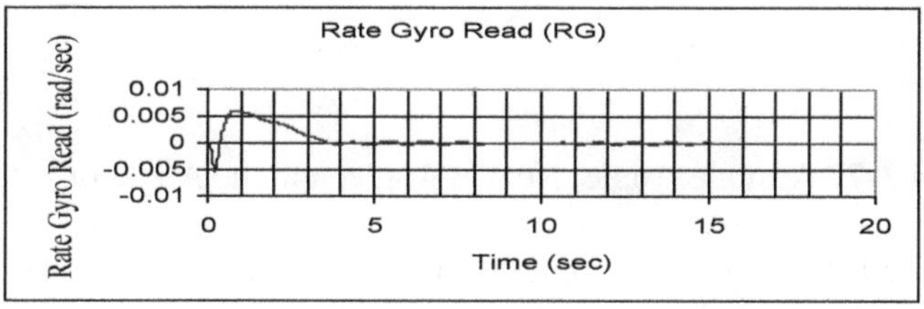

Fig. 4.3-75 Rate Gyro Read (θRG)

f) Rate Gyro Rate Read

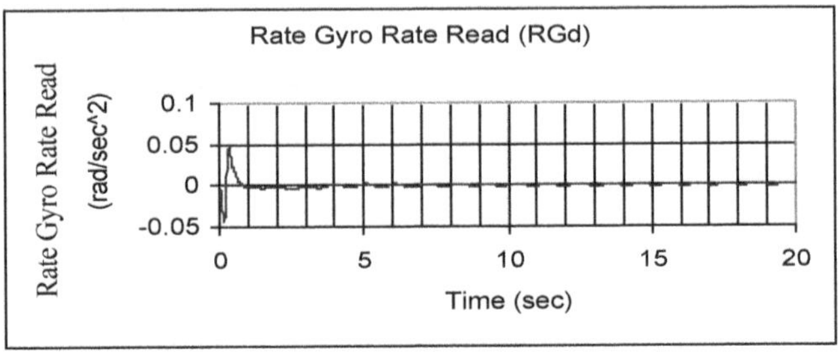

Fig. 4.3-76 Rate Gyro Rate Read (θRGd)

g) Plot of Attitude Angle, Nozzle Angle, and the Angle of Attack

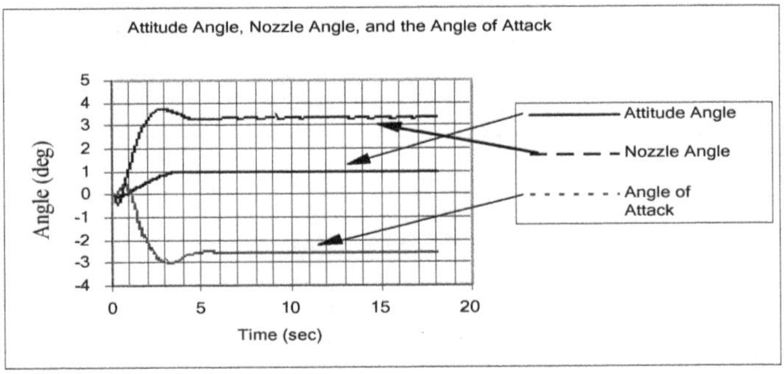

Fig. 4.3-77 Attitude Angle, Nozzle Angle, and the Angle of Attack

Fig. 4.3-59 shows how the angles of the LV's attitude, nozzle deflection, and attack are related.

4.4 Math Modeling Error Sensitivity Analysis

The modeling errors are generated in the following manner. First, an optimal feedback block, K(s), and associated weighting function, W(s), are designed for a given baseline system, F(s) (at top in Fig. 4.4-1), by using the "ncfsyn.m" from mu-Analysis and synthesis Tool Box, where

$$F(s) = \begin{vmatrix} A & B \\ C & D \end{vmatrix},$$

$$K(s) = \begin{vmatrix} Afeedback & Bfeedback \\ Cfeedback & Dfeedback \end{vmatrix}$$

$$W(s) = N(s)/D(s)$$

The optimally designed K(s) and W(s) are saved for future use.

The System B at the bottom in Fig. 4.4-1 represents the true system. The parameters of the true system are:

$$\begin{aligned} F(s)\ true &= F(s) + \Delta F \\ &= \begin{vmatrix} A + \Delta A & B + \Delta B \\ C + \Delta C & D + \Delta D \end{vmatrix} \end{aligned}$$

where $(A + \Delta A)$, $(B + \Delta B)$, $(C + \Delta C)$, and $(D + \Delta D)$ are real system parameters, and A, B, C, and D are math model parameters. The idea of setting $(A + \Delta A)$, $(B + \Delta B)$, $(C + \Delta C)$, and $(D + \Delta D)$ as real system parameters is that we can arbitrary create different real systems by simply changing ΔA, ΔB, ΔC, and ΔD.

In System B in Fig.4.4-1, we use the same optimally designed K(S) and W(S) from the System A, but we create a new real (true) system $(F(s) + \Delta F)$ by changing the values of ΔA, ΔB, ΔC, and ΔD. If $\Delta A = \Delta B = \Delta C = \Delta D = 0.0$, then the K(S) and W(S) remain as optimal functions. But if they are not equal to zero, then it creates a modeling error situation. That is, the K(S) and W(S) become no longer optimally designed functions because they are designed based on the parameters (A, B, C, and D) rather than the true parameter values $[(A + \Delta A)$, $(B + \Delta B)$, $(C + \Delta C)$, and $(D + \Delta D)]$. And the performance degradation will occur in the System B. The extent of the degradation increases as the parameter deviation, ΔA, ΔB, ΔC, and ΔD, increases.

In our analysis process, following simple procedure is taken. First, one parameter is selected at a time and the selected parameter is deviated gradually from the math model value. The rest 19 parameters out of total 20 parameters remain at the math model values. This process simulates a situation

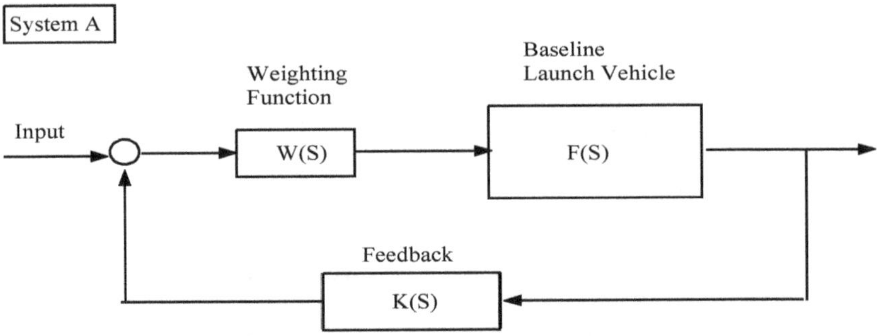

Optimal K(S) and W(S) are computed for this system

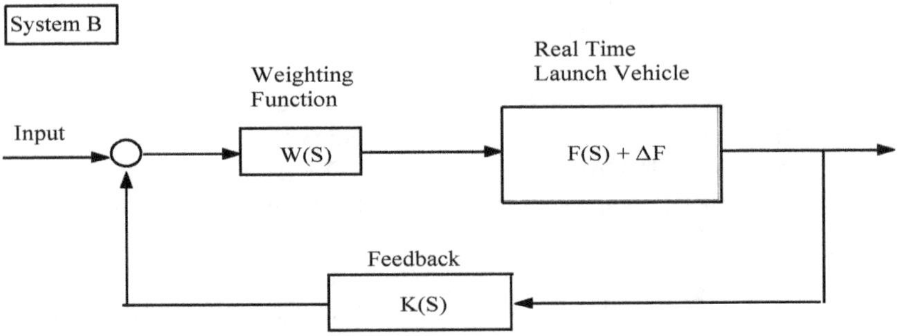

System used for modeling error Analysis

Fig. 4.4-1 Modeling Error Generation

that only one parameter modeling error occurs. The way of determining the sensitivity is as follows. The amount of deviation is increased positively (positive deviation) until the control system response displays instability, and then negatively decreased (negative deviation) until the system response displays instability again. The span between the parameter value at the positive end and that at the negative end represents the sensitivity. The larger the span is, the less sensitivity the system is. The same process is repeated for the rest of the 19 parameters.

The 20 parameters that are chosen for the sensitivity analysis are listed below:

1. Mass
2. Pitch MOI
3. Control Thrust
4. Aerodynamic Load
5. Rocket Engine Position
6. Center of Pressure Location
7. Fuel Sloshing 1 Pendulum Mass
8. Fuel Sloshing 2 Pendulum Mass
9. Fuel Sloshing 1 Pendulum Length
10. Fuel Sloshing 2 Pendulum Length
11. Fuel Sloshing 1 Pendulum Hinge Point
12. Fuel Sloshing 2 Pendulum Hinge Point
13. Bending Mode 1 Modal Mass
14. Bending Mode 2 Modal Mass
15. Bending Mode 1 Modal Frequency
16. Bending Mode 2 Modal Frequency
17. Bending Mode 1 Bending Slope at Tail
18. Bending Mode 1 Bending Slope at Middle
19. Bending Mode 2 Bending Slope at Tail
20. Bending Mode 2 Bending Slope at Middle

The format for the analysis is:

1. The parameter variation, ΔF, is generated as shown below.

   $\Delta F = a\ \%$ of F

   where

   1) Baseline Variation

   a = -20, -15, -10, -5 , 0, 5, 10, 15, 20

   2) Non-Baseline Variation

a : some values greater than -20 and less than 20

2. Response to a 1.0 deg step input is plotted for each variation.
3. Performance index is plotted for all variations. The performance index is defined as an integration of the absolute values of the difference between the ideal output and system response. (The smaller the index, the better the performance)
4. Singular value vs. frequency plot is generated for each variation.
5. Eigenvalues are tabulated for each variation.
6. The critical eigenvalues (related to stability) are plotted to determine at what value of the modeling error the system becomes unstable.
7. Comments and observations are made at the end of each case.

At the end of this chapter, a modeling error sensitivity table and a sensitivity bar chart are constructed to summarize the analysis. Since in real-time flight, the vehicle is likely to experience multiple parameter errors rather than just one parameter error, final multiple error simulation study is done for the case in which each of the 20 parameters is allowed to have 10 % error simultaneously. And then we observe whether the result is within the allowable modeling error, i.e., the amount of error that is allowed before the system becomes unstable.

To show the steps we take for our analysis, we select a case and demonstrate the procedure. The case we chose out of 20 parameters is LV mass modeling error. The analysis for the rest 19 parameters are presented in Appendix III.

Mass : mo (math model mass = 5,058 slugs)

Case 1 (Actual value is less then the value from the math model.)

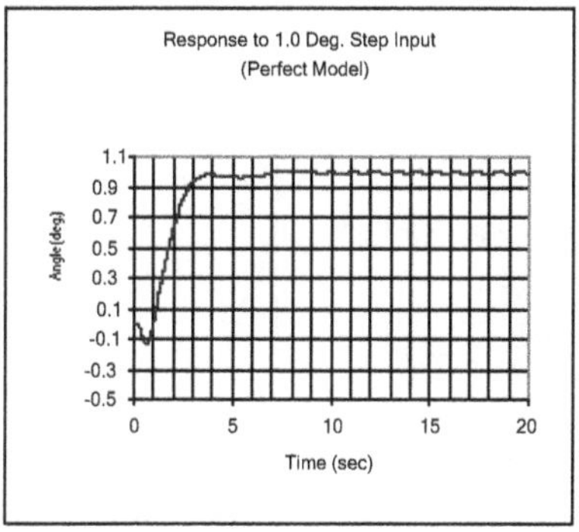

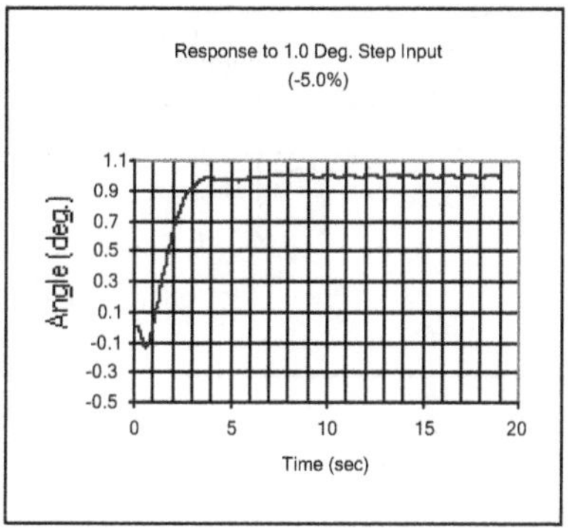

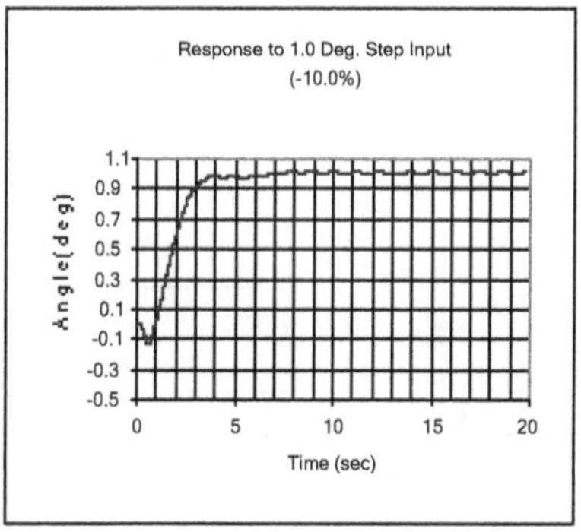

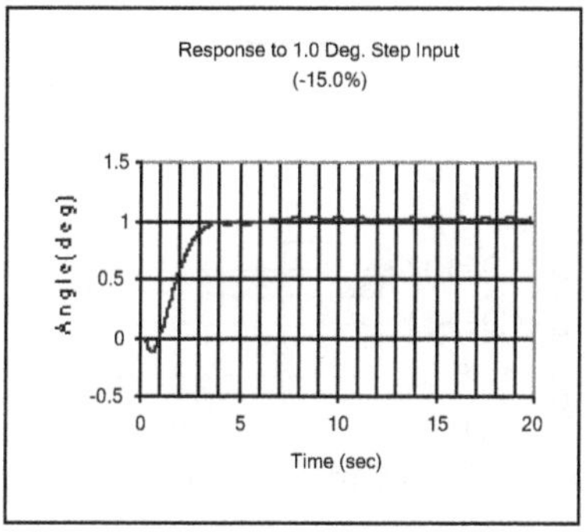

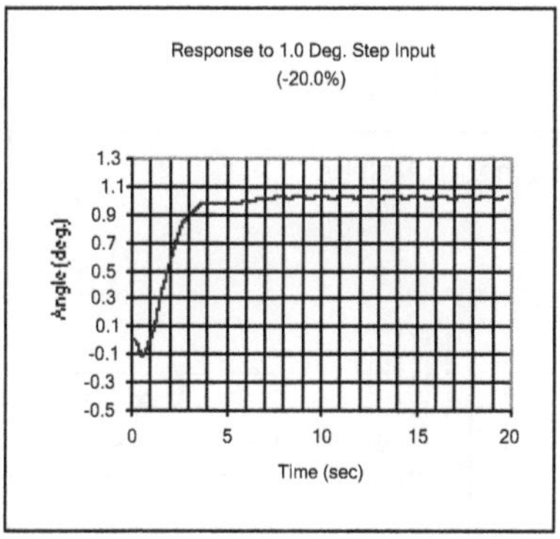

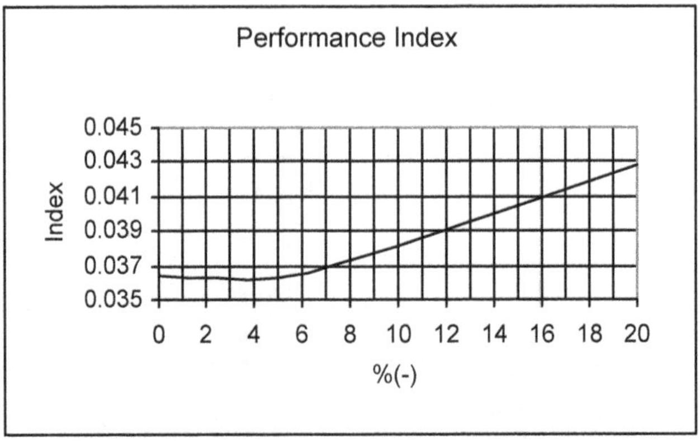

The following singular value vs. frequency plots appear to be identical but they are not (see page 455).

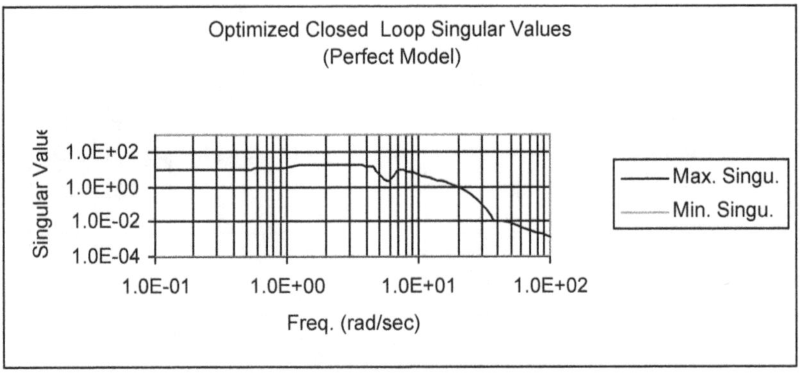

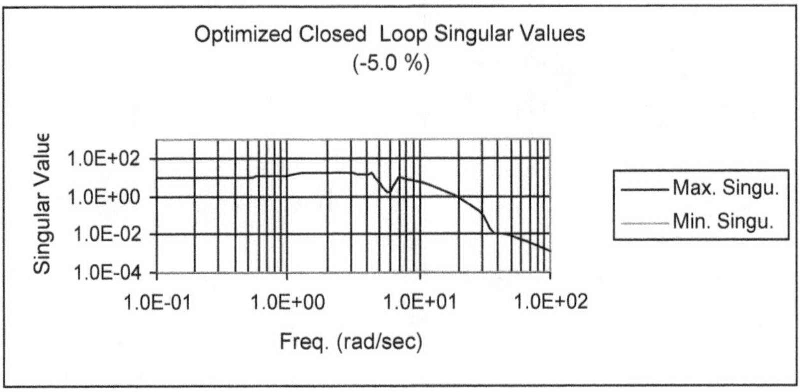

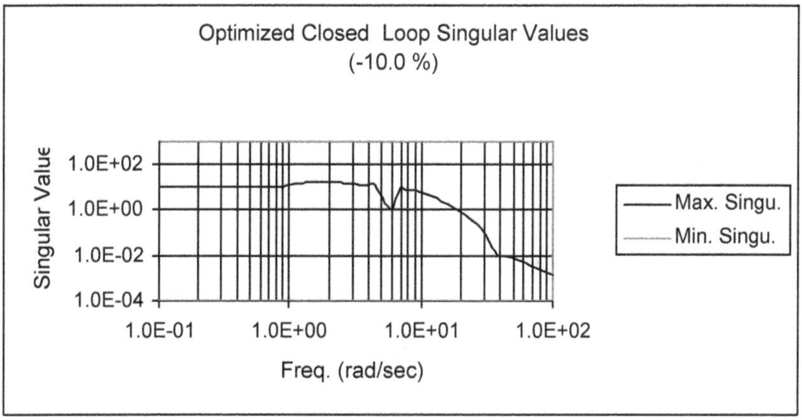
Optimized Closed Loop Singular Values
(-10.0 %)
Singular Value
1.0E+02
1.0E+00
1.0E-02
1.0E-04
1.0E-01
1.0E+00
1.0E+01
1.0E+02
Freq. (rad/sec)
Max. Singu.
Min. Singu.

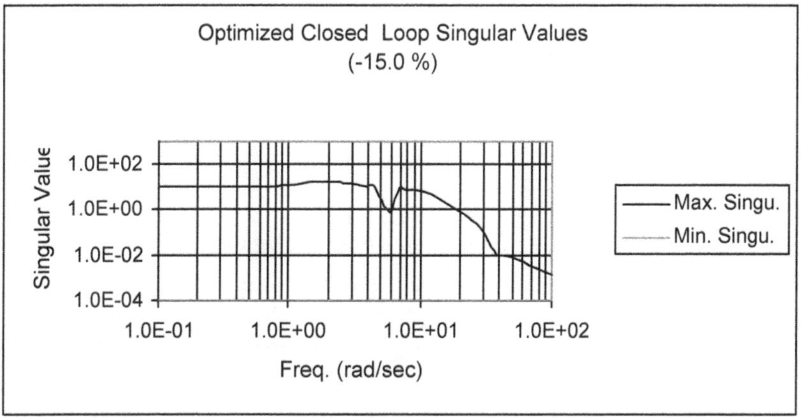
Optimized Closed Loop Singular Values
(-15.0 %)
Singular Value
1.0E+02
1.0E+00
1.0E-02
1.0E-04
1.0E-01
1.0E+00
1.0E+01
1.0E+02
Freq. (rad/sec)
Max. Singu.
Min. Singu.

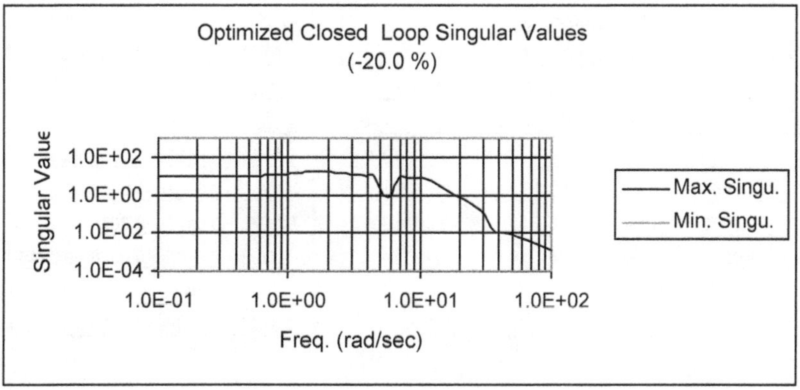
Optimized Closed Loop Singular Values
(-20.0 %)
Singular Value
1.0E+02
1.0E+00
1.0E-02
1.0E-04
1.0E-01
1.0E+00
1.0E+01
1.0E+02
Freq. (rad/sec)
Max. Singu.
Min. Singu.

Closed Loop System Eigenvalues					
	0%	-5%	-10.00%	-15.00%	-20.00%
	1.0e+002 *	1.0e+002 *	1.0e+002 *	1.0e+002 *	1.0e+002 *
1	-0.25	-0.25	-0.25	-0.25	-0.25
2	-0.5824 + 1.3344i	-0.5824 + 1.3344i	-0.5824 + 1.3344i	-0.5824 + 1.3344i	-0.5824 + 1.3344i
3	-0.5824 - 1.3344i	-0.5824 - 1.3344i	-0.5824 - 1.3344i	-0.5824 - 1.3344i	-0.5824 - 1.3344i
4	-0.1040 + 0.2546i	-0.1085 + 0.2642i	-0.1089 + 0.2680i	-0.1104 + 0.2703i	-0.1107 + 0.2724i
5	-0.1040 - 0.2546i	-0.1085 - 0.2642i	-0.1089 - 0.2680i	-0.1104 - 0.2703i	-0.1107 - 0.2724i
6	-0.1046 + 0.2542i	-0.0990 + 0.2406i	-0.0962 + 0.2327i	-0.0915 + 0.2254i	-0.0856 + 0.2192i
7	-0.1046 - 0.2542i	-0.0990 - 0.2406i	-0.0962 - 0.2327i	-0.0915 - 0.2254i	-0.0856 - 0.2192i
8	-0.2284 + 0.0212i	-0.2393 + 0.0445i	-0.2523	-0.2503 + 0.0544i	-0.2531 + 0.0570i
9	-0.2284 - 0.0212i	-0.2393 - 0.0445i	-0.2437 + 0.0513i	-0.2503 - 0.0544i	-0.2531 - 0.0570i
10	-0.2293	-0.245	-0.2437 - 0.0513i	-0.2522	-0.2549
11	-0.1353 + 0.0476i	-0.1165 + 0.1170i	-0.1200 + 0.1373i	-0.1228 + 0.1544i	-0.1289 + 0.1666i
12	-0.1353 - 0.0476i	-0.1165 - 0.1170i	-0.1200 - 0.1373i	-0.1228 - 0.1544i	-0.1289 - 0.1666i
13	-0.1446 + 0.0002i	-0.1609	-0.1636	-0.1646	-0.1651
14	-0.1446 - 0.0002i	-0.1451	-0.145	-0.145	-0.1449
15	-0.0868 + 0.0557i	-0.1446	-0.1446	-0.1446	-0.1446
16	-0.0868 - 0.0557i	-0.0281 + 0.0380i	-0.0018 + 0.0600i	-0.0024 + 0.0602i	-0.0028 + 0.0605i
17	-0.0002 + 0.0601i	-0.0281 - 0.0380i	-0.0018 - 0.0600i	-0.0024 - 0.0602i	-0.0028 - 0.0605i
18	-0.0002 - 0.0601i	-0.0012 + 0.0600i	0.0011 + 0.0597i	0.0014 + 0.0594i	0.0015 + 0.0591i
19	-0.0002 + 0.0600i	-0.0012 - 0.0600i	0.0011 - 0.0597i	0.0014 - 0.0594i	0.0015 - 0.0591i
20	-0.0002 - 0.0600i	0.0007 + 0.0599i	-0.0007 + 0.0524i	-0.0003 + 0.0530i	0.0004 + 0.0534i
21	-0.0009 + 0.0508i	0.0007 - 0.0599i	-0.0007 - 0.0524i	-0.0003 - 0.0530i	0.0004 - 0.0534i
22	-0.0009 - 0.0508i	-0.0009 + 0.0518i	-0.0011 + 0.0499i	-0.0012 + 0.0498i	-0.0014 + 0.0497i
23	-0.0009 + 0.0509i	-0.0009 - 0.0518i	-0.0011 - 0.0499i	-0.0012 - 0.0498i	-0.0014 - 0.0497i
24	-0.0009 - 0.0509i	-0.0010 + 0.0502i	-0.0194 + 0.0396i	-0.0140 + 0.0397i	-0.0100 + 0.0391i
25	-0.0347 + 0.0149i	-0.0010 - 0.0502i	-0.0194 - 0.0396i	-0.0140 - 0.0397i	-0.0100 - 0.0391i
26	-0.0347 - 0.0149i	-0.0315	-0.0303	-0.0298	-0.0297
27	-0.0204	-0.0052 + 0.0136i	-0.0053 + 0.0137i	-0.0053 + 0.0137i	-0.0053 + 0.0137i
28	-0.0052 + 0.0136i	-0.0052 - 0.0136i	-0.0053 - 0.0137i	-0.0053 - 0.0137i	-0.0053 - 0.0137i
29	-0.0052 - 0.0136i	-0.0078 + 0.0145i	-0.0078 + 0.0145i	-0.0078 + 0.0145i	-0.0078 + 0.0145i
30	-0.0078 + 0.0145i	-0.0078 - 0.0145i	-0.0078 - 0.0145i	-0.0078 - 0.0145i	-0.0078 - 0.0145i
31	-0.0078 - 0.0145i	-0.0172	-0.0163	-0.0161	-0.016
32	-0.0134	-0.0119 + 0.0026i	-0.0113 + 0.0040i	-0.0107 + 0.0048i	-0.0103 + 0.0053i

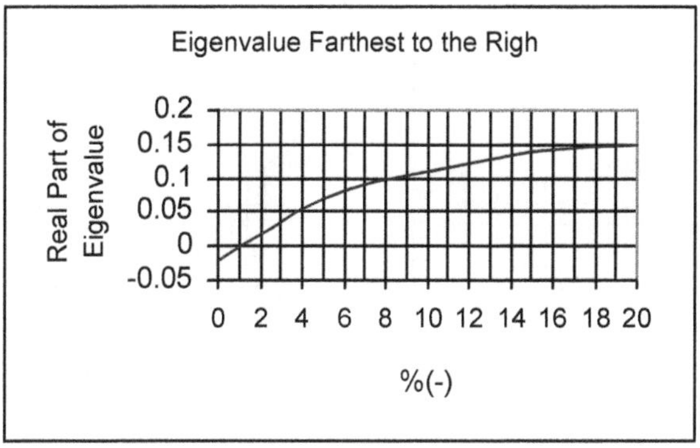

Observation:

Performance index is defined as an integration of absolute value of the difference between the command input and its response. Therefore, the smaller the value of the index is, the better the performance of the control system is. It is shown in Index figure that no significant performance degradation occurs until the difference between the actual mass value and the mass value used in math model is greater than 8.0 %.

The singular value plots show that there isn't any significant change in values until the frequency reaches about 10.0 rad/sec, which implies that noise is attenuated.

Case 2 (actual value is greater than the value in the math model)

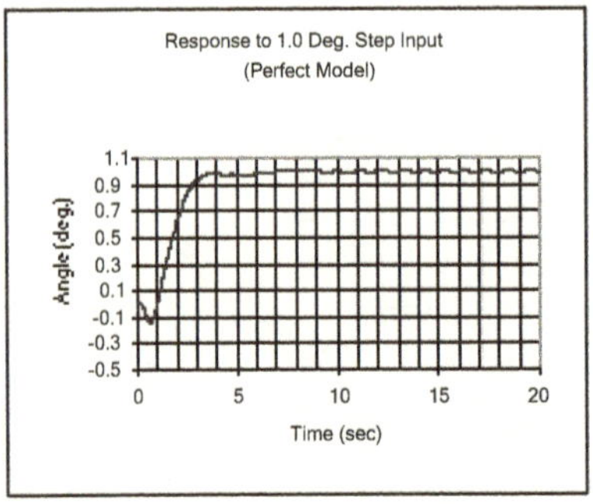

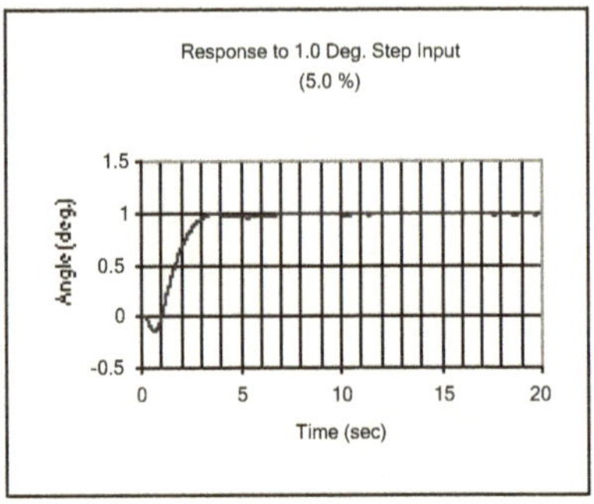

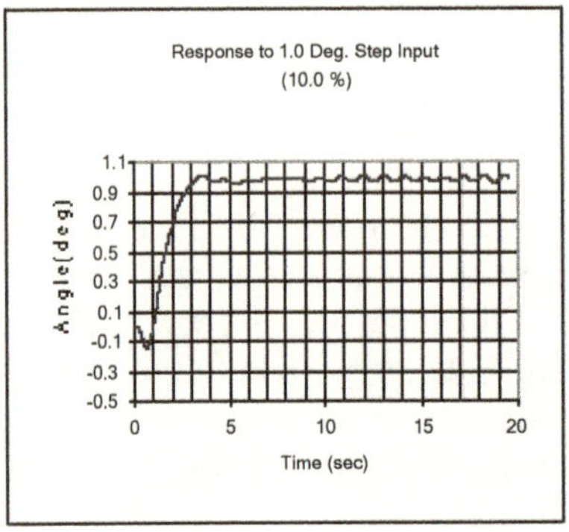

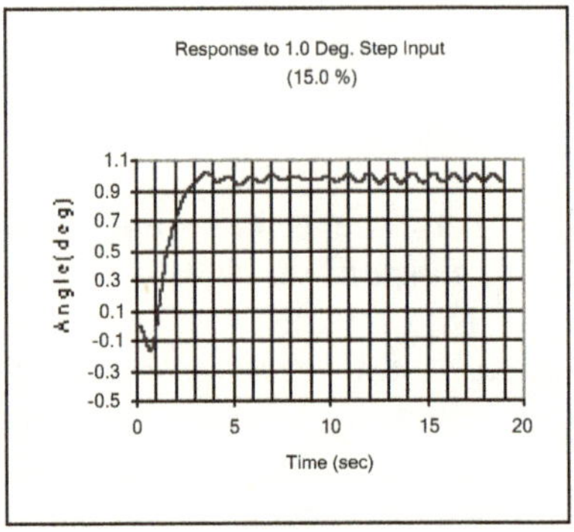

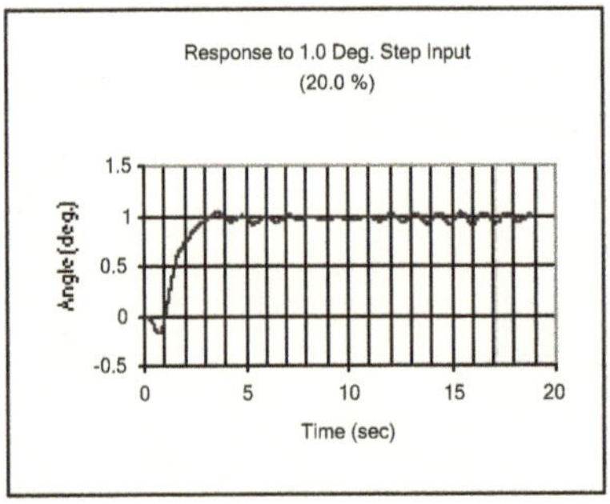

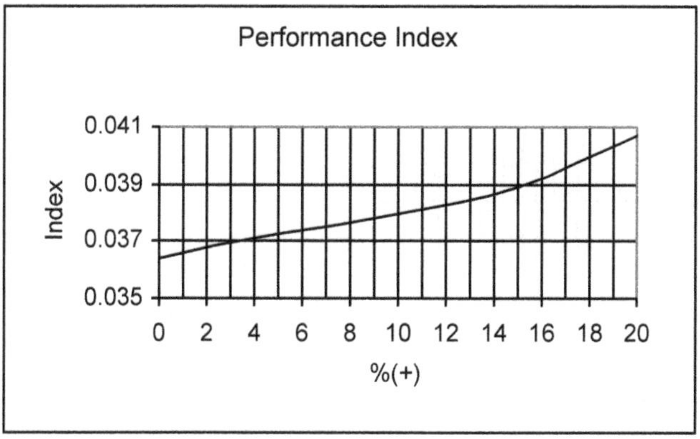
Performance Index
Index
0.041
0.039
0.037
0.035
0 2 4 6 8 10 12 14 16 18 20
%(+)

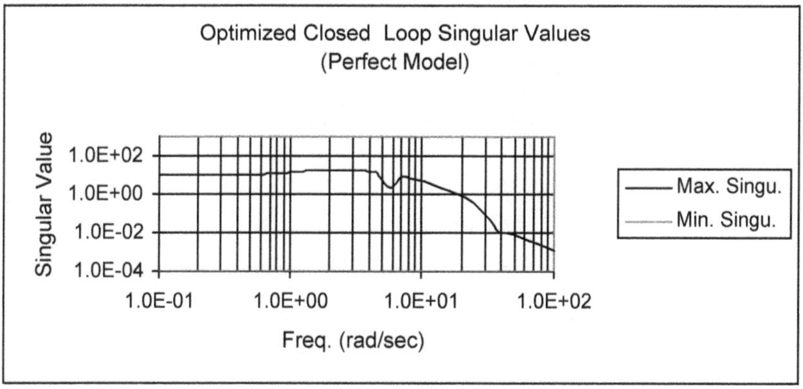
Optimized Closed Loop Singular Values
(Perfect Model)
Singular Value
1.0E+02
1.0E+00
1.0E-02
1.0E-04
1.0E-01 1.0E+00 1.0E+01 1.0E+02
Freq. (rad/sec)
Max. Singu.
Min. Singu.

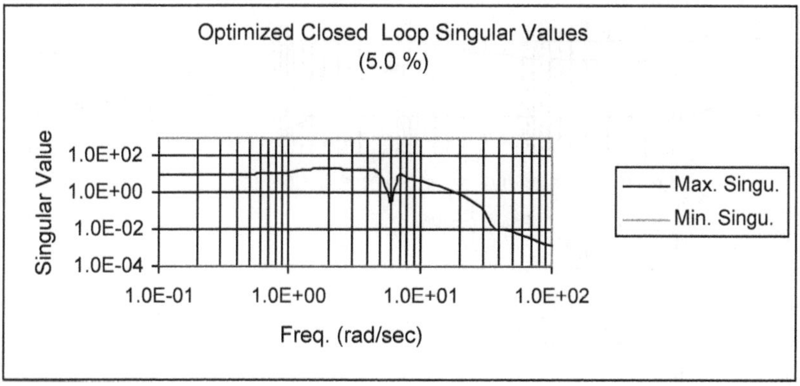
Optimized Closed Loop Singular Values
(5.0 %)
Singular Value
1.0E+02
1.0E+00
1.0E-02
1.0E-04
1.0E-01 1.0E+00 1.0E+01 1.0E+02
Freq. (rad/sec)
Max. Singu.
Min. Singu.

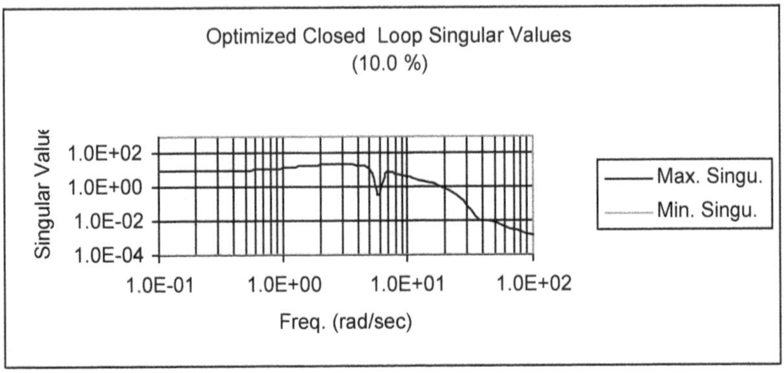
Optimized Closed Loop Singular Values
(10.0 %)
Singular Valu
1.0E+02
1.0E+00
1.0E-02
1.0E-04
1.0E-01
1.0E+00
1.0E+01
1.0E+02
Freq. (rad/sec)
Max. Singu.
Min. Singu.

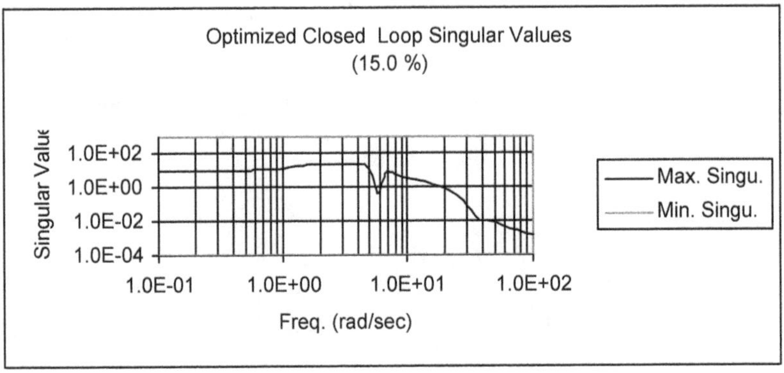
Optimized Closed Loop Singular Values
(15.0 %)
Singular Valu
1.0E+02
1.0E+00
1.0E-02
1.0E-04
1.0E-01
1.0E+00
1.0E+01
1.0E+02
Freq. (rad/sec)
Max. Singu.
Min. Singu.

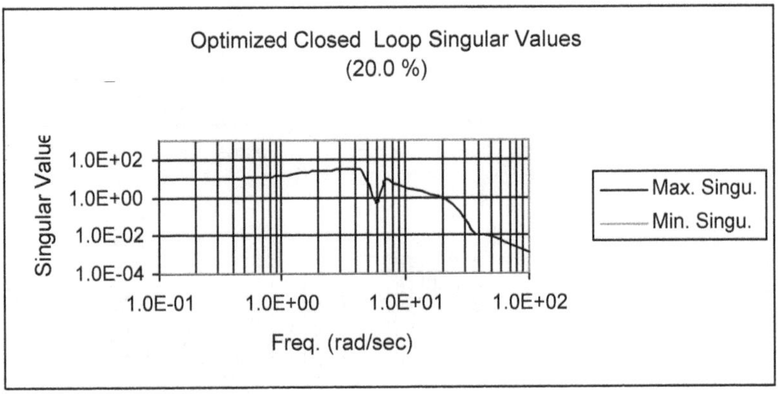
Optimized Closed Loop Singular Values
(20.0 %)
Singular Value
1.0E+02
1.0E+00
1.0E-02
1.0E-04
1.0E-01
1.0E+00
1.0E+01
1.0E+02
Freq. (rad/sec)
Max. Singu.
Min. Singu.

Closed Loop System Eigenvalues %(+)					
	0%	5%	10.00%	15.00%	20.00%
	1.0e+002 *	1.0e+002 *	1.0e+002 *	1.0e+002 *	1.0e+002 *
1	-0.25	-0.25	-0.25	-0.25	-0.25
2	-0.5824 + 1.3344i	-0.5824 + 1.3344i	-0.5824 + 1.3344i	-0.5824 + 1.3344i	-0.5824 + 1.3344i
3	-0.5824 - 1.3344i	-0.5824 - 1.3344i	-0.5824 - 1.3344i	-0.5824 - 1.3344i	-0.5824 - 1.3344i
4	-0.1040 + 0.2546i	-0.0934 + 0.2586i	-0.1211 + 0.2543i	-0.0870 + 0.2637i	-0.0848 + 0.2651i
5	-0.1040 - 0.2546i	-0.0934 - 0.2586i	-0.1211 - 0.2543i	-0.0870 - 0.2637i	-0.0848 - 0.2651i
6	-0.1046 + 0.2542i	-0.1164 + 0.2535i	-0.0895 + 0.2608i	-0.1240 + 0.2542i	-0.1268 + 0.2552i
7	-0.1046 - 0.2542i	-0.1164 - 0.2535i	-0.0895 - 0.2608i	-0.1240 - 0.2542i	-0.1268 - 0.2552i
8	-0.2284 + 0.0212i	-0.2496 + 0.0224i	-0.2551 + 0.0237i	-0.2580 + 0.0212i	-0.2608 + 0.0224i
9	-0.2284 - 0.0212i	-0.2496 - 0.0224i	-0.2551 - 0.0237i	-0.2580 - 0.0212i	-0.2608 - 0.0224i
10	-0.2293	-0.1853 + 0.0793i	-0.1957 + 0.0917i	-0.2037 + 0.1007i	-0.2080 + 0.1053i
11	-0.1353 + 0.0476i	-0.1853 - 0.0793i	-0.1957 - 0.0917i	-0.2037 - 0.1007i	-0.2080 - 0.1053i
12	-0.1353 - 0.0476i	-0.1736	-0.17	-0.1689	-0.1683
13	-0.1446 + 0.0002i	-0.0518 + 0.0863i	-0.0399 + 0.0954i	-0.0313 + 0.1004i	-0.0250 + 0.1044i
14	-0.1446 - 0.0002i	-0.0518 - 0.0863i	-0.0399 - 0.0954i	-0.0313 - 0.1004i	-0.0250 - 0.1044i
15	-0.0868 + 0.0557i	-0.1447 + 0.0001i	-0.1447 + 0.0000i	-0.1448	-0.1448
16	-0.0868 - 0.0557i	-0.1447 - 0.0001i	-0.1447 - 0.0000i	-0.1446	-0.1446
17	-0.0002 + 0.0601i	-0.0003 + 0.0609i	-0.0004 + 0.0612i	-0.0005 + 0.0613i	-0.0006 + 0.0614i
18	-0.0002 - 0.0601i	-0.0003 - 0.0609i	-0.0004 - 0.0612i	-0.0005 - 0.0613i	-0.0006 - 0.0614i
19	-0.0002 + 0.0600i	-0.0001 + 0.0592i	-0.0001 + 0.0590i	-0.0000 + 0.0589i	-0.0000 + 0.0588i
20	-0.0002 - 0.0600i	-0.0001 - 0.0592i	-0.0001 - 0.0590i	-0.0000 - 0.0589i	-0.0000 - 0.0588i
21	-0.0009 + 0.0508i	-0.0015 + 0.0507i	-0.0016 + 0.0506i	-0.0017 + 0.0505i	-0.0017 + 0.0505i
22	-0.0009 - 0.0508i	-0.0015 - 0.0507i	-0.0016 - 0.0506i	-0.0017 - 0.0505i	-0.0017 - 0.0505i
23	-0.0009 + 0.0509i	-0.0002 + 0.0507i	0.0000 + 0.0507i	0.0002 + 0.0506i	0.0003 + 0.0506i
24	-0.0009 - 0.0509i	-0.0002 - 0.0507i	0.0000 - 0.0507i	0.0002 - 0.0506i	0.0003 - 0.0506i
25	-0.0347 + 0.0149i	-0.0229 + 0.0140i	-0.0265	-0.0269	-0.0272
26	-0.0347 - 0.0149i	-0.0229 - 0.0140i	-0.0190 + 0.0144i	-0.0169 + 0.0148i	-0.0156 + 0.0148i
27	-0.0204	-0.0249	-0.0190 - 0.0144i	-0.0169 - 0.0148i	-0.0156 - 0.0148i
28	-0.0052 + 0.0136i	-0.0052 + 0.0136i	-0.0052 + 0.0135i	-0.0052 + 0.0135i	-0.0052 + 0.0135i
29	-0.0052 - 0.0136i	-0.0052 - 0.0136i	-0.0052 - 0.0135i	-0.0052 - 0.0135i	-0.0052 - 0.0135i
30	-0.0078 + 0.0145i	-0.0078 + 0.0145i	-0.0078 + 0.0145i	-0.0079 + 0.0145i	-0.0079 + 0.0145i
31	-0.0078 - 0.0145i	-0.0078 - 0.0145i	-0.0078 - 0.0145i	-0.0079 - 0.0145i	-0.0079 - 0.0145i
32	-0.0134	-0.0144	-0.0147	-0.0148	-0.0148
33	-0.0106	-0.0092	-0.0084	-0.0079	-0.0074
34	-0.01	-0.01	-0.01	-0.01	-0.01

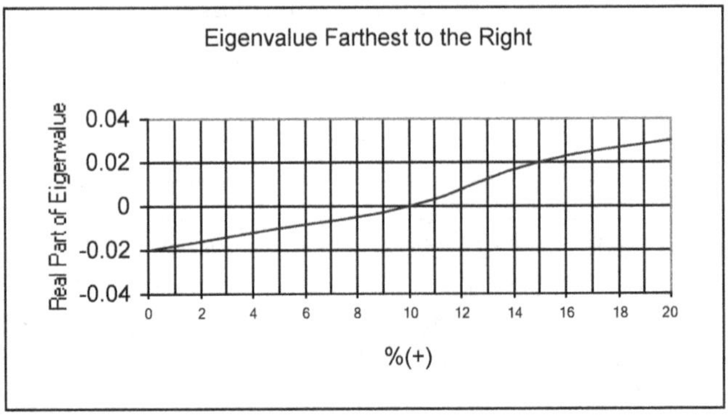

Observation:

The control system is fairly insensitive to mass modeling error.

****

NOTE: Similar analyses for the rest 19 parameters are presented in Appendix III.

****

Twenty system parameters are checked out to determine their sensitivities. The sensitivity measure is defined here as the maximum allowable percent error before the control system becomes unstable. Table 4.4.1 tabulates the sensitivity measures. Some of the LV's system parameters are relatively insensitive to modeling errors. For example, the LV's mass modeling error is allowed 20% negative error and 15 % positive error without causing instability. In the following specific example, the true mass is 5,058 slugs, and the math model mass is 4,046.4 slugs, which is 20 % less than the true value. Simulation analysis shows that the vehicle does not become unstable in spite of such a large modeling error

(Fig. 4.4-2a and Fig.4.4-2b)

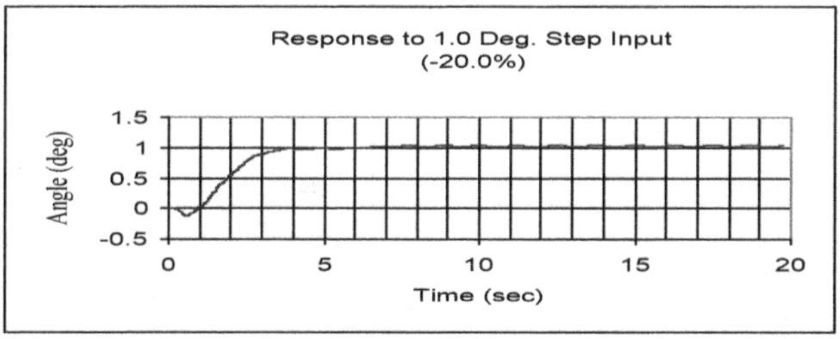

Fig. 4.4-2a Response to 1.0 Deg Step Input (-20.0 % Mass Modeling Error)

Simulation analysis of 15 % positive modeling error also shows robust response.

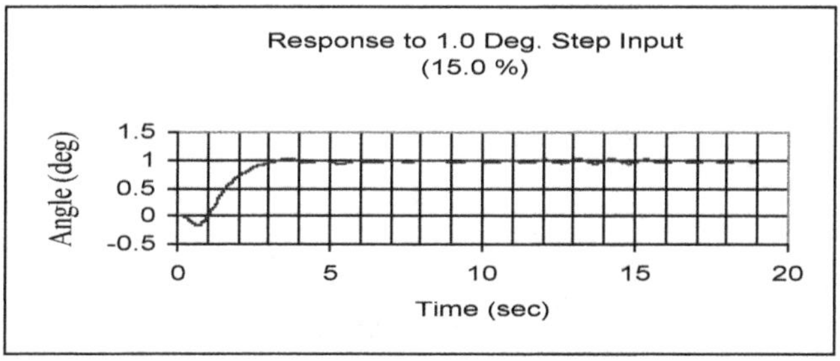

Fig. 4.4-2b Response to 1.0 Deg Step Input (-15.0 % Mass Modeling Error)

Table 4.4-1. Modeling Error Sensitivity Table

	LV System Parameter Name	Symbol	Math Model Value	Unit	Allowable Variation in % *1 Negative *2	Positive *3
1	LV Mass w/o Pendulum Masses	mo	5,058	slugs	20	20(*4)
2	Pitch Axis MOI	Iyy	1,694,000	slugs ft^2	15(*5)	10(*6)
3	Control Thrust	Tc	410,400	lbf	4 (*7)	6(*8)
4	Aerodynamic Load	Lalpha	198,000	lbf/rad	6(*8)	6(*7)
5	Rocket Engine Location	lc	-32.3	ft	1(*7)	2(*8)
6	Center of Pressure Location	lalpha	34.1	ft	1(*8)	1(*7)
7	Fuel Sloshing 1 Pendulum Mass	mp1	388	slugs	20(*9)	20
8	Fuel Sloshing 2 Pendulum Mass	mp2	280	slugs	20(*9)	20(*8)
9	Fuel Sloshing 1 Pendulum Length	Lp1	2	ft	10	1
10	Fuel Sloshing 2 Pendulum Length	Lp2	2.75	ft	0.5	1.5
11	Fuel Sloshing 1 Pendulum Hinge Loc.	lp1	5.56	ft	20	20
12	Fuel Sloshing 2 Pendulum Hinge Loc.	lp2	-20.62	ft	10(*9)	15(*8)
13	Bending Mode 1 Modal Mass	Mb1	1,590	slugs	20(*10)	(*11)
14	Bending mode 2 Modal Mass	Mb2	9,550	slugs	(*11)	(*11)
15	Bending Mode 1 Modal Frequency	FreqB1	18.9	rad/sec	10	(*11)
16	Bending Mode 2 Modal Frequency	FreqB2	60.7	rad/sec	(*11)	13
17	Bend. Mode 1 Bending Slope at Tail	SigG1t	-0.0648	rad/ft	(*11)	(*11)
18	Bend. Mode 1 Bending Slope at Middle	SigG1m	0.0044	rad/ft	(*11)	(*11)
19	Bend. Mode 2 Bending Slope at Tail	SigG2t	-0.171	rad/ft	(*11)	(*11)
20	Bend. Mode 2 Bending Slope at middle	SigG2m	-0.0276	rad/ft	(*11)	(*11)

*1 Maximum allowable variation from the true value before instability.
*2 When the true value is less than the math model value in absolute value sense.
*3 When the true value is greater than the math model value in absolute value sense.
*4 Mild instability when the math modeling error is greater than +20.0%.
*5 Stable until when the true value is less than 15% and then displays severe instability when the error increase to 18%.
*6 Stable until when the true value is greater than 10% and then displays severe instability when the error increase to 12.25%.
*7 Stable, but steady state occurs before the desired value is reached.
*8 Stable, but the desired value is exceeded in the steady state.
*9 Mild instability is shown even after the LV value is 50% greater than the math model

*10 Stable as long as the math model is within 20% of the true value

*11 No instability is shown up to 20 % modeling error

To visualize the modeling error sensitivity, a bar chart is generated below.

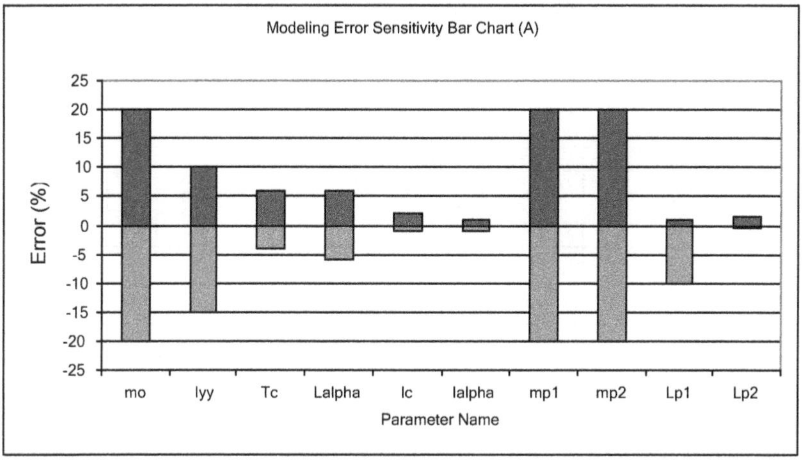

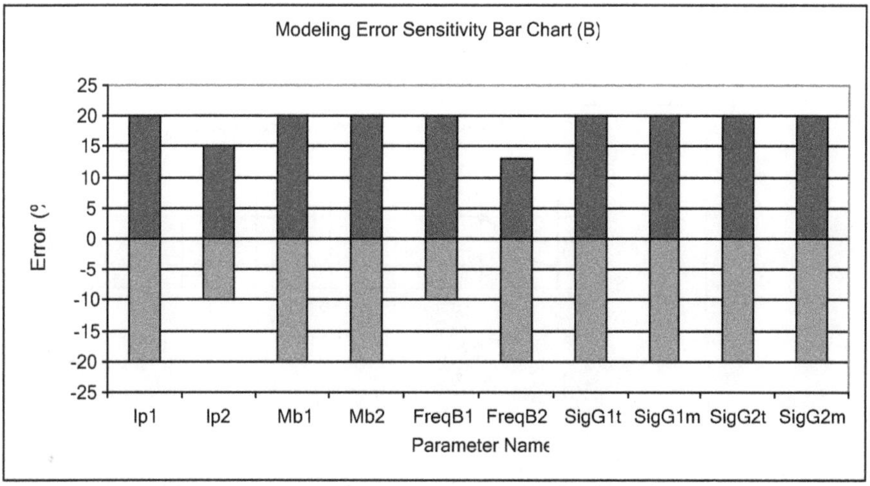

Fig. 4.4-3 Modeling Error Sensitivity Bar Chart

It is seen in Table 4.4-1 that Fuel Sloshing 2 Pendulum Length is the most sensitive parameter. When the true length is 0.4 % less than the length used in the math model, the control system becomes unstable. To observe how different sets of the weighting functions used in Chapter 4.3 influence the step response in this most sensitive case of Fuel Sloshing 2 Pendulum Length, we simulate the following cases of three different weighting functions.

Weighting Function Set 1

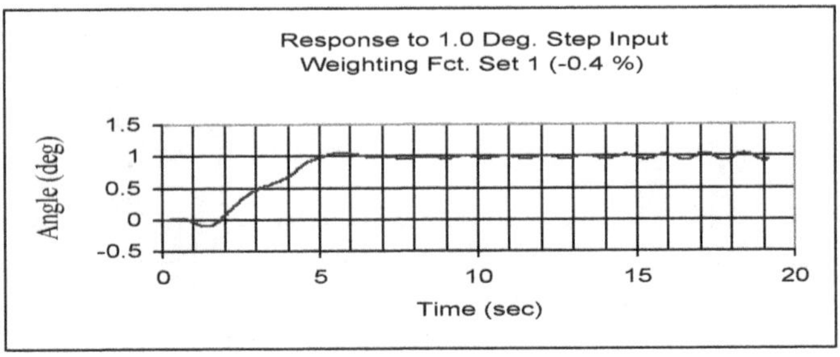

Fig. 4.4-4 Response to 1.0 Deg Step Input (-.40 % Fuel Sloshing 2 Pendulum Length Modeling Error and Weighting Function Set 1)

Weighting Function Set 2

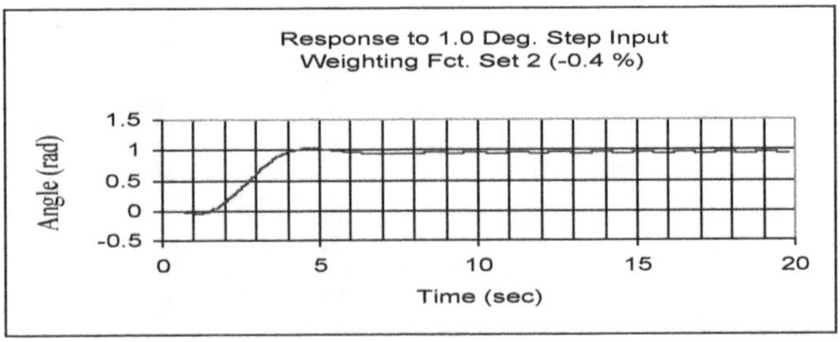

Fig. 4.4-5 Response to 1.0 Deg Step Input (-.40 % Fuel Sloshing 2 Pendulum Length Modeling Error and Weighting Function Set 2)

Weighting Function Set 3

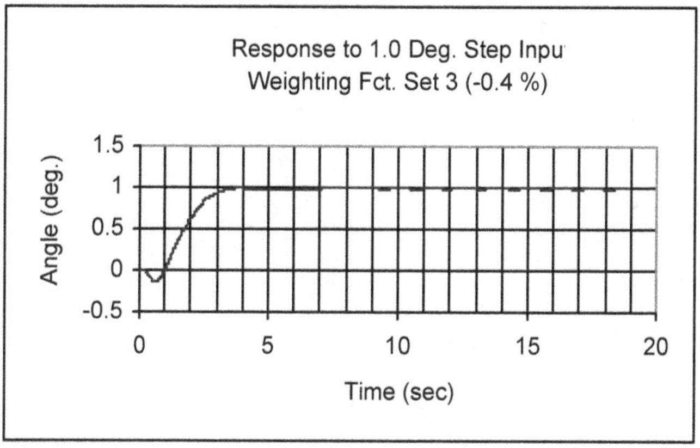

Fig. 4.4-6 Response to 1.0 Deg Step Input (-.40 % Fuel Sloshing 2 Pendulum Length Modeling Error and Weighting Function Set 3)

Weighting Function Set 3 shows that the desired steady state is reached faster than the other Sets. Set 3 also shows a stable response, while the Set 1 shows minor instability.

In the previous analysis, it is assumed that modeling error occurs for one parameter at a time. However, in reality, almost all parameters are subject to some degree of mismodeling. Here, we assumed that all 20 parameters have the modeling errors of 10% of the Allowable Errors tabulated in Table 4.4-1, and then a 1.0 deg step input is applied to the system. For example, mass is allowed to have 20% error. Therefore, the amount of mass error assigned in this study is 10 % of the 20 % of the mass, which is 2.0 % error in mass. Table 4.4-2 tabulates the amount of error assigned to each parameter in this analysis according to the 10 % error of the allowable error rule.

Table 4.4-2. Error Assigned to Each System Parameter

	LV System Parameter Name	Symbol	Math Model Value	Unit	Allowable Variation in % *1 Negative *2	Positive *3
1	LV Mass w/o Pendulum Masses	Mo	5,058	slugs	-2	0
2	Pitch Axis MOI	Iyy	1,694,000	slugs ft^2	0	1
3	Control Thrust	Tc	410,400	lbf	-0.4	0
4	Aerodynamic Load	Lalpha	198,000	lbf/rad	0	0.6
5	Rocket Engine Location	Lc	-32.3	ft	-0.1	0
6	Center of Pressure Location	Lalpha	34.1	ft	0	0.1
7	Fuel Sloshing 1 Pendulum Mass	mp1	388	slugs	-2	0

8	Fuel Sloshing 2 Pendulum Mass	mp2	280	slugs	0	2
9	Fuel Sloshing 1 Pendulum Length	Lp1	2	ft	0	0.1
10	Fuel Sloshing 2 Pendulum Length	Lp2	2.75	ft	-0.05	0
11	Fuel Sloshing 1 Pendulum Hinge Loc.	lp1	5.56	ft	0	2
12	Fuel Sloshing 2 Pendulum Hinge Loc.	lp2	-20.62	ft	-1	0
13	Bending Mode 1 Modal Mass	Mb1	1,590	slugs	0	2
14	Bending mode 2 Modal Mass	Mb2	9,550	slugs	-2	0
15	Bending Mode 1 Modal Frequency	FreqB1	18.9	rad/sec	-1	0
16	Bending Mode 2 Modal Frequency	FreqB2	60.7	rad/sec	0	1.3
17	Bend. Mode 1 Bending Slope at Tail	SigG1t	-0.0648	rad/ft	-2	0
18	Bend. Mode 1 Bending Slope at Middle	SigG1m	0.0044	rad/ft	0	2
19	Bend. Mode 2 Bending Slope at Tail	SigG2t	-0.171	rad/ft	-2	0
20	Bend. Mode 2 Bending Slope at middle	SigG2m	-0.0276	rad/ft	0	2

A 1.0 deg step input response is simulated for each of the three options of the weighting functions used in Chapter 4.3. The simulation results are shown below.

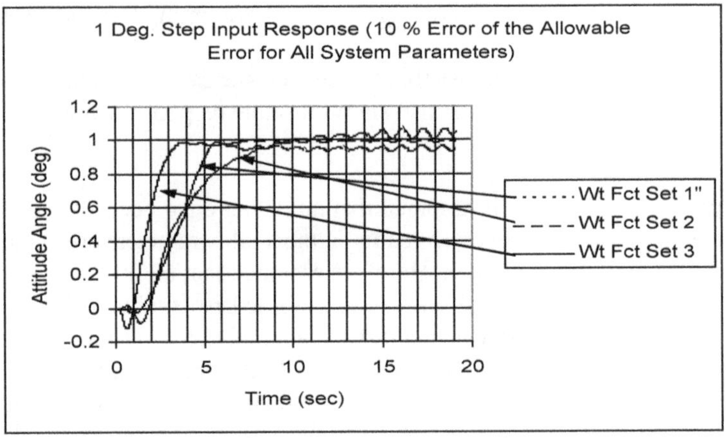

Fig. 4.4-7 Responses to a 1.0 Deg Step Input Applied to an LV System with10 % of the Maximum Allowable Error

Fig. 4.4-7 shows that the Weighting Function Set 3 offers the best response.

Singular value vs. frequency plots for three cases of the modeling errors, i.e., 0 % error, 15.0% error, and 30.0 % error are shown in Fig. 4.4-8 for Lp2 (Fuel Sloshing 2 Pendulum Length). It is seen here that because of the modeling error, the control system fails to adequately suppress some signals in the range of 1.5 to 8.0 rad/sec. Figs. 4.4-9, 4.4-10, and 4.4-11 show step responses when the modeling errors in Lp2 are 0.0 %, 15.0 %, and 30.0 %.

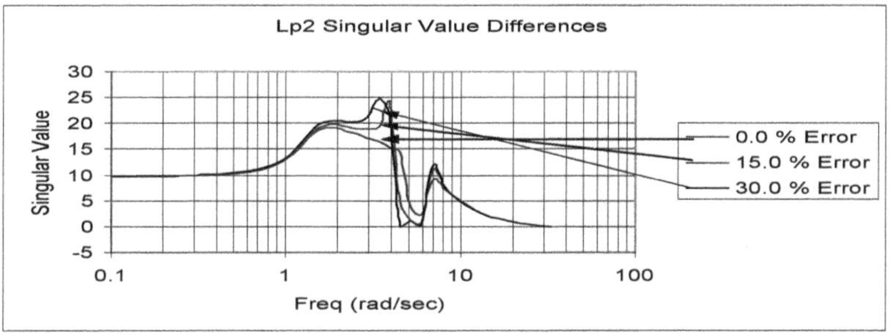

Fig. 4.4-8. Singular Value Comparison

The maximum singular value of 0.0% Error case is smaller than the other cases, 15.0 % Error and 30.0% Error, which suggests that 0.0 % case should generate better performance, and it is verified in Fig. 4.4-9, Fig. 4.4-10, and Fig. 4.4-11.

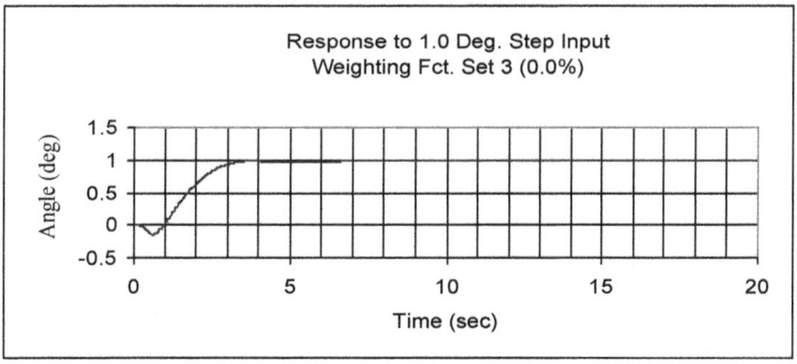

Fig. 4.4-9 Optimal Response

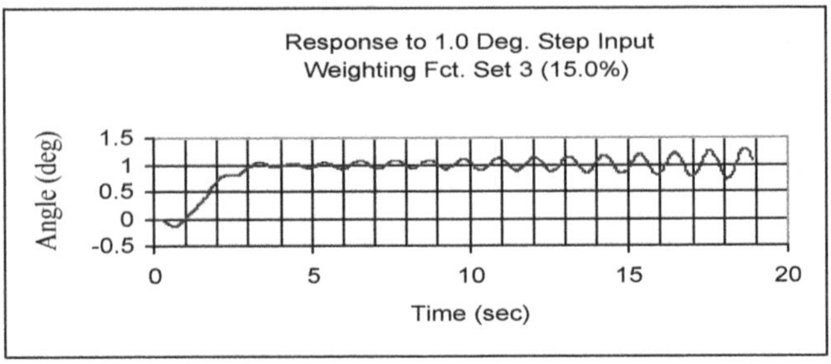

Fig. 4.4-10 Step Input Response (15.0 % Modeling Error)

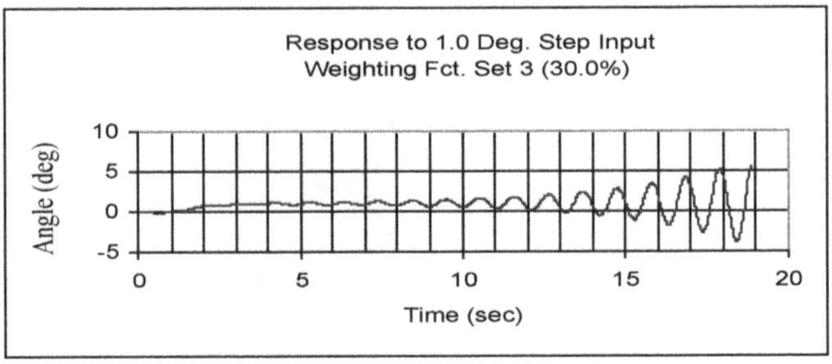

Fig. 4.4-11 Step Input Response (30.0 % Modeling Error)

4.5 Software Implementation Issue

It turns out that the dynamic feedback block, designed by using H-infinity, is usually a higher order system and takes a great deal of computation time, which is not recommended for flight control system. Since it is desirable to have a lower order feedback loop, it is investigated how much the feedback block system order can be reduced without losing stability. Starting from the full order of 15 states, the order of the feedback loop system is reduced one by one until the response becomes unstable. This exercise is repeated for all three Weighting Function sets to determine the effects of the weighting function on control system performance as the system order is reduced. For each set, Afeedback, Bfeedback, Cfeedback, Dfeedback, a 1.0 deg step input response, and the largest system eigenvalues are computed as the order is reduced.

a. Weighting Function Set 1

1) Feedback System Order: 14

Afeedback =

Columns 1 through 6

71.1392	23.0135	-2.9746	-0.4014	0.9478	-10.2110
-53.8266	-19.8974	4.5308	-0.3307	-1.8346	7.3692
-21.8304	-9.4368	0.9188	5.4973	-0.2381	2.9694
-31.9926	-11.6605	-4.1461	-0.2491	-1.0675	4.1008
2.9349	2.0466	-0.1934	0.6044	0.0393	-5.4396
25.7874	8.2522	-1.0909	0.1913	5.3832	-3.4750
68.0972	25.4255	-2.9827	0.9187	0.8946	-8.9057
1.2545	-0.3091	-0.0090	-0.1653	0.0357	-0.1815
-65.5059	-22.0915	2.7544	-0.2098	-0.9278	8.8229
49.6233	16.8656	-2.0944	0.1891	0.6998	-6.6815
-56.1547	-18.3374	2.3255	-0.0398	-0.8100	7.5758
45.0449	15.1911	-1.8941	0.1441	0.6381	-6.0673
-22.6192	-7.6073	0.9499	-0.0675	-0.3209	3.0471
-6.2451	-2.2405	0.2706	-0.0513	-0.0852	0.8385

Columns 7 through 12

-126.5575	-59.9853	34.9840	24.6600	-67.8688	-24.1449
94.9539	45.5110	-26.7671	-18.8248	51.8593	18.4462
37.5210	17.7916	-10.4154	-7.3351	20.1962	7.1843
55.6650	26.6143	-15.6327	-10.9987	30.2949	10.7760
-5.2169	-2.4885	1.4506	1.0225	-2.8139	-1.0011
-44.2156	-20.8068	12.1650	8.5709	-23.5951	-8.3935
-122.2004	-58.2959	34.0453	23.9782	-66.0039	-23.4821
0.1109	-1.7084	1.8352	1.2125	-3.4644	-1.1438
112.4143	53.0653	-32.0471	-23.5372	63.9676	22.5635
-85.5646	-40.6950	25.2055	15.4704	-38.8503	-13.3004
94.5956	48.1132	-29.1028	-19.3132	30.2389	-7.8940
-77.3124	-37.0669	22.5013	12.2359	-18.3426	-20.4422
38.7604	18.6899	-11.3536	-6.5075	14.8662	14.1408

```
  11.1197    4.7781   -3.0117   -1.7847   10.7361    1.5291

 Columns 13 through 14

 -12.8890   10.5247
   9.8493   -8.0416
   3.8355   -3.1318
   5.7536   -4.6977
  -0.5344    0.4364
  -4.4809    3.6589
 -12.5360   10.2361
  -0.6176    0.5043
  12.0852   -9.8644
  -7.4607    6.0677
   2.0530   -1.4835
 -14.6622   11.3749
 -10.2521   20.4527
  -0.5567  -34.9769

Bfeedback =

-2.3612  155.6045
  -0.7578 -118.3005
   0.1014  -46.1861
  -0.2661  -69.2361
  -0.0538    6.4410
  -0.2413   53.8345
  -0.3123  149.1294
  -0.0066    1.0065
   0.1034 -138.5882
  -0.0787  105.2740
   0.0879 -117.4180
  -0.0713   95.2925
   0.0358  -47.8029
  0.101 -13.5191

 Cfeedback =

 Columns 1 through 6

  309.8635  104.9641  -13.1099    1.0649    4.3791  -41.6462
   39.5921   14.6512   -1.7676    0.3561    0.5506   -5.4253
```

-4.1756	-1.4142	0.1774	-0.0137	-0.0591	0.5603

Columns 7 through 12

-533.1994	-252.9435	148.2120	104.3427	-287.3261	-102.2086
-71.8111	-33.8255	19.8312	13.9634	-38.4526	-13.6777
7.1865	3.4095	-1.9977	-1.4065	3.8730	1.3777

Columns 13 through 14

-54.5678	44.5553
-7.3024	5.9625
0.7355	-0.6006

Dfeedback=

-0.4914	656.6797
-0.0658	87.8786
0.66	-8.8517

2) Feedback System Order: 13

Afeedback =

Columns 1 through 5

71.1392	23.0135	-2.9746	-0.4014	0.9478
-53.8266	-19.8974	4.5308	-0.3307	-1.8346
-21.8304	-9.4368	0.9188	5.4973	-0.2381
-31.9926	-11.6605	-4.1461	-0.2491	-1.0675
2.9349	2.0466	-0.1934	0.6044	0.0393
25.7874	8.2522	-1.0909	0.1913	5.3832
68.0972	25.4255	-2.9827	0.9187	0.8946
1.2545	-0.3091	-0.0090	-0.1653	0.0357
-65.5059	-22.0915	2.7544	-0.2098	-0.9278
49.6233	16.8656	-2.0944	0.1891	0.6998
-56.1547	-18.3374	2.3255	-0.0398	-0.8100
45.0449	15.1911	-1.8941	0.1441	0.6381
-22.6192	-7.6073	0.9499	-0.0675	-0.3209

Columns 6 through 10

```
-10.2110 -126.5575  -59.9853   34.9840   24.6600
  7.3692   94.9539   45.5110  -26.7671  -18.8248
  2.9694   37.5210   17.7916  -10.4154   -7.3351
  4.1008   55.6650   26.6143  -15.6327  -10.9987
 -5.4396   -5.2169   -2.4885    1.4506    1.0225
 -3.4750  -44.2156  -20.8068   12.1650    8.5709
 -8.9057 -122.2004  -58.2959   34.0453   23.9782
 -0.1815    0.1109   -1.7084    1.8352    1.2125
  8.8229  112.4143   53.0653  -32.0471  -23.5372
 -6.6815  -85.5646  -40.6950   25.2055   15.4704
  7.5758   94.5956   48.1132  -29.1028  -19.3132
 -6.0673  -77.3124  -37.0669   22.5013   12.2359
  3.0471   38.7604   18.6899  -11.3536   -6.5075
```

Columns 11 through 13

```
-67.8688 -24.1449 -12.8890
 51.8593  18.4462   9.8493
 20.1962   7.1843   3.8355
 30.2949  10.7760   5.7536
 -2.8139  -1.0011  -0.5344
-23.5951  -8.3935  -4.4809
-66.0039 -23.4821 -12.5360
 -3.4644  -1.1438  -0.6176
 63.9676  22.5635  12.0852
-38.8503 -13.3004  -7.4607
 30.2389  -7.8940   2.0530
-18.3426 -20.4422 -14.6622
 14.8662           14.1408  -10.2521
```

Bfeedback =

```
 -2.3612  155.6045
 -0.7578 -118.3005
  0.1014  -46.1861
 -0.2661  -69.2361
 -0.0538    6.4410
 -0.2413   53.8345
 -0.3123  149.1294
 -0.0066    1.0065
  0.1034 -138.5882
```

```
   -0.0787  105.2740
    0.0879 -117.4180
   -0.0713   95.2925
    0.358 -47.8029

   Cfeedback =

  Columns 1 through 5

  309.8635  104.9641  -13.1099    1.0649    4.3791
   39.5921   14.6512   -1.7676    0.3561    0.5506
   -4.1756   -1.4142    0.1774   -0.0137   -0.0591

  Columns 6 through 10

  -41.6462 -533.1994 -252.9435  148.2120  104.3427
   -5.4253  -71.8111  -33.8255   19.8312   13.9634
    0.5603    7.1865    3.4095   -1.9977   -1.4065

  Columns 11 through 13

 -287.3261 -102.2086  -54.5678
  -38.4526  -13.6777   -7.3024
    3.8730    1.3777    0.7355

Dfeedback=

   -0.4914  656.6797
   -0.0658   87.8786
    0.0066   -8.8517
```

3) Feedback System Order: 12

```
Afeedback =

Columns 1 through 5

   71.1392   23.0135   -2.9746   -0.4014    0.9478
  -53.8266  -19.8974    4.5308   -0.3307   -1.8346
  -21.8304   -9.4368    0.9188    5.4973   -0.2381
```

-31.9926	-11.6605	-4.1461	-0.2491	-1.0675
2.9349	2.0466	-0.1934	0.6044	0.0393
25.7874	8.2522	-1.0909	0.1913	5.3832
68.0972	25.4255	-2.9827	0.9187	0.8946
1.2545	-0.3091	-0.0090	-0.1653	0.0357
-65.5059	-22.0915	2.7544	-0.2098	-0.9278
49.6233	16.8656	-2.0944	0.1891	0.6998
-56.1547	-18.3374	2.3255	-0.0398	-0.8100
45.0449	15.1911	-1.8941	0.1441	0.6381

Columns 6 through 10

-10.2110	-126.5575	-59.9853	34.9840	24.6600
7.3692	94.9539	45.5110	-26.7671	-18.8248
2.9694	37.5210	17.7916	-10.4154	-7.3351
4.1008	55.6650	26.6143	-15.6327	-10.9987
-5.4396	-5.2169	-2.4885	1.4506	1.0225
-3.4750	-44.2156	-20.8068	12.1650	8.5709
-8.9057	-122.2004	-58.2959	34.0453	23.9782
-0.1815	0.1109	-1.7084	1.8352	1.2125
8.8229	112.4143	53.0653	-32.0471	-23.5372
-6.6815	-85.5646	-40.6950	25.2055	15.4704
7.5758	94.5956	48.1132	-29.1028	-19.3132
-6.0673	-77.3124	-37.0669	22.5013	12.2359

Columns 11 through 12

-67.8688	-24.1449
51.8593	18.4462
20.1962	7.1843
30.2949	10.7760
-2.8139	-1.0011
-23.5951	-8.3935
-66.0039	-23.4821
-3.4644	-1.1438
63.9676	22.5635
-38.8503	-13.3004
30.2389	-7.8940
-18.3426	-20.4422

Bfeedback =

```
-2.3612  155.6045
-0.7578 -118.3005
 0.1014  -46.1861
-0.2661  -69.2361
-0.0538    6.4410
-0.2413   53.8345
-0.3123  149.1294
-0.0066    1.0065
 0.1034 -138.5882
-0.0787  105.2740
 0.0879 -117.4180
-0.0713   95.2925

Cfeedback =

  Columns 1 through 5

   309.8635  104.9641  -13.1099    1.0649    4.3791
    39.5921   14.6512   -1.7676    0.3561    0.5506
    -4.1756   -1.4142    0.1774   -0.0137   -0.0591

   Columns 6 through 10

   -41.6462 -533.1994 -252.9435  148.2120  104.3427
    -5.4253  -71.8111  -33.8255   19.8312   13.9634
     0.5603    7.1865    3.4095   -1.9977   -1.4065

   Columns 11 through 12

  -287.3261 -102.2086
   -38.4526  -13.6777
     3.8730    1.3777
```

Dfeedback=

```
 -0.4914  656.6797
 -0.0658   87.8786
  0.0066   -8.8517
```

4) Feedback System Order: 11

Afeedback =

Columns 1 through 5

71.1392	23.0135	-2.9746	-0.4014	0.9478
-53.8266	-19.8974	4.5308	-0.3307	-1.8346
-21.8304	-9.4368	0.9188	5.4973	-0.2381
-31.9926	-11.6605	-4.1461	-0.2491	-1.0675
2.9349	2.0466	-0.1934	0.6044	0.0393
25.7874	8.2522	-1.0909	0.1913	5.3832
68.0972	25.4255	-2.9827	0.9187	0.8946
1.2545	-0.3091	-0.0090	-0.1653	0.0357
-65.5059	-22.0915	2.7544	-0.2098	-0.9278
49.6233	16.8656	-2.0944	0.1891	0.6998
-56.1547	-18.3374	2.3255	-0.0398	-0.8100

Columns 6 through 10

-10.2110	-126.5575	-59.9853	34.9840	24.6600
7.3692	94.9539	45.5110	-26.7671	-18.8248
2.9694	37.5210	17.7916	-10.4154	-7.3351
4.1008	55.6650	26.6143	-15.6327	-10.9987
-5.4396	-5.2169	-2.4885	1.4506	1.0225
-3.4750	-44.2156	-20.8068	12.1650	8.5709
-8.9057	-122.2004	-58.2959	34.0453	23.9782
-0.1815	0.1109	-1.7084	1.8352	1.2125
8.8229	112.4143	53.0653	-32.0471	-23.5372
-6.6815	-85.5646	-40.6950	25.2055	15.4704
7.5758	94.5956	48.1132	-29.1028	-19.3132

Column 11

-67.8688
51.8593
20.1962
30.2949
-2.8139
-23.5951
-66.0039
-3.4644
63.9676

```
  -38.8503
   30.2389
Bfeedback =

   -2.3612  155.6045
   -0.7578 -118.3005
    0.1014  -46.1861
   -0.2661  -69.2361
   -0.0538    6.4410
   -0.2413   53.8345
   -0.3123  149.1294
   -0.0066    1.0065
    0.1034 -138.5882
   -0.0787  105.2740
    0.0879 -117.4180

Cfeedback =

  Columns 1 through 5

  309.8635  104.9641  -13.1099    1.0649    4.3791
   39.5921   14.6512   -1.7676    0.3561    0.5506
   -4.1756   -1.4142    0.1774   -0.0137   -0.0591

  Columns 6 through 10

  -41.6462 -533.1994 -252.9435  148.2120  104.3427
   -5.4253  -71.8111  -33.8255   19.8312   13.9634
    0.5603    7.1865    3.4095   -1.9977   -1.4065

  Column 11

 -287.3261
  -38.4526
    3.8730

Dfeedback=

   -0.4914  656.6797
   -0.0658   87.8786
    0.0066   -8.8517
```

Fig. 4.5-1 shows that the control system is stable and performs well as the order is reduced from 15 to 12. However, when the order is reduced to 11, a critical eigenvalue moves to the Right Half Plane, displaying instability (Fig. 4.5-2).

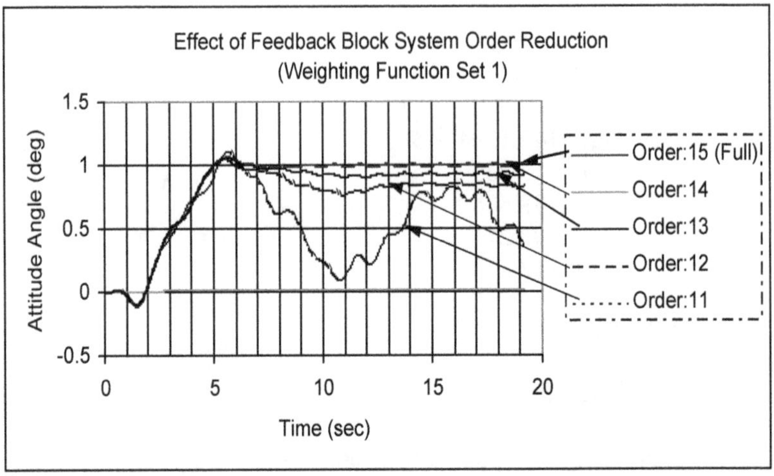

Fig. 4.5-1. Effect of Feedback Block System Order Reduction on Control System Performance, Weighting Function Set 1 (1.0 Deg Step Input)

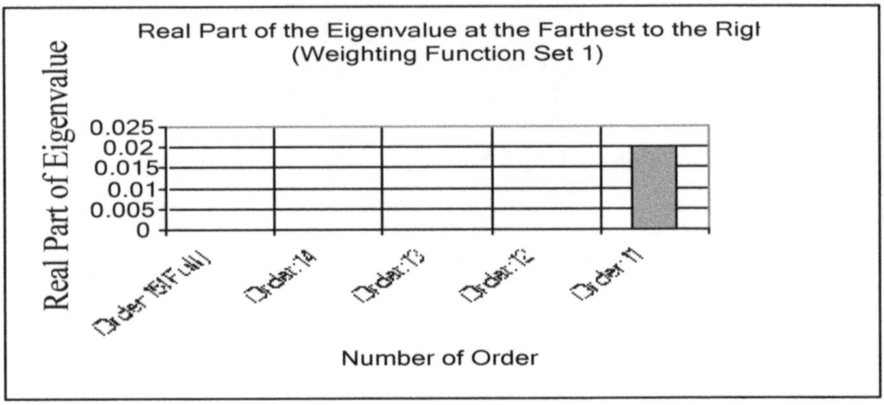

Fig. 4.5-2. Real Part of Eigenvalue at the Farthest to the Right, Weighting Function Set 1

b. Weighting Function Set 2

1) Feedback System Order: 14

Columns 1 through 5

19.5947	12.0686	-3.7622	7.7023	6.7710
-133.2636	-81.3601	29.7289	-54.3613	-36.2854
-46.7851	-31.8817	8.7929	-14.3737	-12.6095
-133.5946	-81.5792	21.4637	-54.6039	-35.3054
-115.8128	-70.1261	21.5822	-48.0429	-32.1744
-47.4318	-29.1598	8.9887	-19.6637	-17.2935
-118.4524	-70.5996	21.8443	-47.8263	-32.7392
106.8063	62.9077	-19.6788	42.1404	29.8548
-106.5078	-65.0057	20.4099	-43.6363	-29.7086
133.1426	80.7087	-25.3372	54.1508	37.1132
-65.4590	-40.1790	12.6244	-26.9745	-18.2929
37.3337	22.5886	-7.0906	15.1539	10.4004
-31.2964	-18.6202	5.8383	-12.4807	-8.6866
8.8386	5.3556	-1.6813	3.5932	2.4630

Columns 6 through 10

-4.1811	14.9182	38.1821	19.7179	19.9269
24.8438	-86.4198	-235.2462	-121.7501	-123.3440
8.5470	-29.5882	-80.8882	-41.8229	-42.3475
24.7822	-85.7023	-235.0003	-121.6330	-123.2438
25.2708	-74.8080	-202.1754	-104.4021	-105.6954
8.6218	-27.8920	-82.7209	-42.8374	-43.4291
18.8194	-76.2136	-206.6423	-106.5130	-107.7738
-19.1798	69.2945	180.4935	92.2606	93.9391
19.3542	-68.9450	-186.5591	-97.6990	-100.0744
-24.1498	86.1586	232.6831	122.3013	119.6089
11.9443	-42.4715	-116.1848	-60.8672	-57.3352
-6.7649	24.1441	65.0268	34.0677	32.9259
5.6250	-20.1625	-53.3004	-27.9061	-27.4400
-1.6027	5.7179	15.4242	8.0803	7.7822

Columns 11 through 14

15.7928	5.5047	8.0918	1.3179

```
 -98.0154 -34.1578 -50.2211  -8.1785
 -33.6417 -11.7240 -17.2370  -2.8071
 -97.9396 -34.1313 -50.1823  -8.1722
 -83.9568 -29.2589 -43.0170  -7.0055
 -34.5102 -12.0267 -17.6824  -2.8796
 -85.6030 -29.8324 -43.8598  -7.1428
  74.6654  26.0177  38.2489   6.2292
 -79.0846 -27.5750 -40.5361  -6.6014
  92.8484  32.5284  47.7164   7.7740
 -63.2600 -35.0713 -42.1937  -6.5502
  42.8317   5.7416  -3.7095   0.4995
 -13.7407   6.4438 -28.7172  -7.5264
   8.2735   0.4915   5.8717 -23.2428
```

Bfeedback =

```
   1.8934  -38.2539
   0.2523  233.1813
  -0.2165   80.3499
   0.3483  233.1477
  -0.5438  199.8992
   0.0762   82.4312
  -0.5860  203.4689
   0.0365 -180.9855
  -0.1269  185.9941
   0.1418 -231.1130
  -0.0711  114.7003
   0.0401  -64.7304
  -0.0332   53.6034
   0.0095  -15.3413
```

Cfeedback =

Columns 1 through 5

```
-252.5645 -152.2285   48.0056 -101.9783  -69.9110
 -38.2234  -24.4774    7.7356  -16.4185  -10.9426
   2.1971    1.3229   -0.4194    0.8852    0.6061
```

Columns 6 through 10

```
  45.8791 -162.7408 -440.2738 -228.1027 -230.8255
   7.2453  -25.5209  -70.3915  -36.3600  -36.7206
```

-0.4005 1.4144 3.8362 1.9861 2.0107

Columns 11 through 14

-183.4499 -63.9298 -93.9943 -15.3071
-29.1900 -10.1716 -14.9549 -2.4355
1.5978 0.5568 0.8187 0.1333

Dfeedback=

-0.2711 437.3138
-0.0431 69.5795
0.0024 -3.808917

2) Feedback System Order: 13

Afeedback =

Columns 1 through 5

19.5947 12.0686 -3.7622 7.7023 6.7710
-133.2636 -81.3601 29.7289 -54.3613 -36.2854
-46.7851 -31.8817 8.7929 -14.3737 -12.6095
-133.5946 -81.5792 21.4637 -54.6039 -35.3054
-115.8128 -70.1261 21.5822 -48.0429 -32.1744
-47.4318 -29.1598 8.9887 -19.6637 -17.2935
-118.4524 -70.5996 21.8443 -47.8263 -32.7392
106.8063 62.9077 -19.6788 42.1404 29.8548
-106.5078 -65.0057 20.4099 -43.6363 -29.7086
133.1426 80.7087 -25.3372 54.1508 37.1132
-65.4590 -40.1790 12.6244 -26.9745 -18.2929
37.3337 22.5886 -7.0906 15.1539 10.4004
-31.2964 -18.6202 5.8383 -12.4807 -8.6866

Columns 6 through 10

-4.1811 14.9182 38.1821 19.7179 19.9269
24.8438 -86.4198 -235.2462 -121.7501 -123.3440
8.5470 -29.5882 -80.8882 -41.8229 -42.3475
24.7822 -85.7023 -235.0003 -121.6330 -123.2438
25.2708 -74.8080 -202.1754 -104.4021 -105.6954

8.6218	-27.8920	-82.7209	-42.8374	-43.4291
18.8194	-76.2136	-206.6423	-106.5130	-107.7738
-19.1798	69.2945	180.4935	92.2606	93.9391
19.3542	-68.9450	-186.5591	-97.6990	-100.0744
-24.1498	86.1586	232.6831	122.3013	119.6089
11.9443	-42.4715	-116.1848	-60.8672	-57.3352
-6.7649	24.1441	65.0268	34.0677	32.9259
5.6250	-20.1625	-53.3004	-27.9061	-27.4400

Columns 11 through 13

15.7928	5.5047	8.0918
-98.0154	-34.1578	-50.2211
-33.6417	-11.7240	-17.2370
-97.9396	-34.1313	-50.1823
-83.9568	-29.2589	-43.0170
-34.5102	-12.0267	-17.6824
-85.6030	-29.8324	-43.8598
74.6654	26.0177	38.2489
-79.0846	-27.5750	-40.5361
92.8484	32.5284	47.7164
-63.2600	-35.0713	-42.1937
42.8317	5.7416	-3.7095
-13.7407	6.4438	-28.7172

Bfeedback =

1.8934	-38.2539
0.2523	233.1813
-0.2165	80.3499
0.3483	233.1477
-0.5438	199.8992
0.0762	82.4312
-0.5860	203.4689
0.0365	-180.9855
-0.1269	185.9941
0.1418	-231.1130
-0.0711	114.7003
0.0401	-64.7304
-0.0332	53.6034

Cfeedback =

Columns 1 through 5

-252.5645	-152.2285	48.0056	-101.9783	-69.9110
-38.2234	-24.4774	7.7356	-16.4185	-10.9426
2.1971	1.3229	-0.4194	0.8852	0.6061

Columns 6 through 10

45.8791	-162.7408	-440.2738	-228.1027	-230.8255
7.2453	-25.5209	-70.3915	-36.3600	-36.7206
-0.4005	1.4144	3.8362	1.9861	2.0107

Columns 11 through 13

-183.4499	-63.9298	-93.9943
-29.1900	-10.1716	-14.9549
1.5978	0.5568	0.8187

Dfeedback=

-0.2711	437.3138
-0.0431	69.5795
0.0024	-3.8089

3) Feedback System Order: 12

Afeedback =

Columns 1 through 5

19.5947	12.0686	-3.7622	7.7023	6.7710
-133.2636	-81.3601	29.7289	-54.3613	-36.2854
-46.7851	-31.8817	8.7929	-14.3737	-12.6095
-133.5946	-81.5792	21.4637	-54.6039	-35.3054
-115.8128	-70.1261	21.5822	-48.0429	-32.1744
-47.4318	-29.1598	8.9887	-19.6637	-17.2935
-118.4524	-70.5996	21.8443	-47.8263	-32.7392
106.8063	62.9077	-19.6788	42.1404	29.8548
-106.5078	-65.0057	20.4099	-43.6363	-29.7086
133.1426	80.7087	-25.3372	54.1508	37.1132
-65.4590	-40.1790	12.6244	-26.9745	-18.2929

37.3337	22.5886	-7.0906	15.1539	10.4004

Columns 6 through 10

-4.1811	14.9182	38.1821	19.7179	19.9269
24.8438	-86.4198	-235.2462	-121.7501	-123.3440
8.5470	-29.5882	-80.8882	-41.8229	-42.3475
24.7822	-85.7023	-235.0003	-121.6330	-123.2438
25.2708	-74.8080	-202.1754	-104.4021	-105.6954
8.6218	-27.8920	-82.7209	-42.8374	-43.4291
18.8194	-76.2136	-206.6423	-106.5130	-107.7738
-19.1798	69.2945	180.4935	92.2606	93.9391
19.3542	-68.9450	-186.5591	-97.6990	-100.0744
-24.1498	86.1586	232.6831	122.3013	119.6089
11.9443	-42.4715	-116.1848	-60.8672	-57.3352
-6.7649	24.1441	65.0268	34.0677	32.9259

Columns 11 through 12

15.7928	5.5047
-98.0154	-34.1578
-33.6417	-11.7240
-97.9396	-34.1313
-83.9568	-29.2589
-34.5102	-12.0267
-85.6030	-29.8324
74.6654	26.0177
-79.0846	-27.5750
92.8484	32.5284
-63.2600	-35.0713
42.8317	5.7416

Bfeedback =

1.8934	-38.2539
0.2523	233.1813
-0.2165	80.3499
0.3483	233.1477
-0.5438	199.8992
0.0762	82.4312
-0.5860	203.4689

```
 0.0365 -180.9855
-0.1269  185.9941
 0.1418 -231.1130
-0.0711  114.7003
 0.0401  -64.7304

Cfeedback =

Columns 1 through 5

-252.5645 -152.2285   48.0056 -101.9783  -69.9110
 -38.2234  -24.4774    7.7356  -16.4185  -10.9426
   2.1971    1.3229   -0.4194    0.8852    0.6061

Columns 6 through 10

  45.8791 -162.7408 -440.2738 -228.1027 -230.8255
   7.2453  -25.5209  -70.3915  -36.3600  -36.7206
  -0.4005    1.4144    3.8362    1.9861    2.0107

Columns 11 through 12

-183.4499  -63.9298
 -29.1900  -10.1716
   1.5978    0.5568
```

Dfeedback=

```
-0.2711  437.3138
-0.0431   69.5795
 0.0024   -3.8089
```

4) Feedback System Order: 11

Afeedback =

```
Columns 1 through 5

  19.5947   12.0686   -3.7622    7.7023    6.7710
-133.2636  -81.3601   29.7289  -54.3613  -36.2854
 -46.7851  -31.8817    8.7929  -14.3737  -12.6095
```

-133.5946	-81.5792	21.4637	-54.6039	-35.3054
-115.8128	-70.1261	21.5822	-48.0429	-32.1744
-47.4318	-29.1598	8.9887	-19.6637	-17.2935
-118.4524	-70.5996	21.8443	-47.8263	-32.7392
106.8063	62.9077	-19.6788	42.1404	29.8548
-106.5078	-65.0057	20.4099	-43.6363	-29.7086
133.1426	80.7087	-25.3372	54.1508	37.1132
-65.4590	-40.1790	12.6244	-26.9745	-18.2929

Columns 6 through 10

-4.1811	14.9182	38.1821	19.7179	19.9269
24.8438	-86.4198	-235.2462	-121.7501	-123.3440
8.5470	-29.5882	-80.8882	-41.8229	-42.3475
24.7822	-85.7023	-235.0003	-121.6330	-123.2438
25.2708	-74.8080	-202.1754	-104.4021	-105.6954
8.6218	-27.8920	-82.7209	-42.8374	-43.4291
18.8194	-76.2136	-206.6423	-106.5130	-107.7738
-19.1798	69.2945	180.4935	92.2606	93.9391
19.3542	-68.9450	-186.5591	-97.6990	-100.0744
-24.1498	86.1586	232.6831	122.3013	119.6089
11.9443	-42.4715	-116.1848	-60.8672	-57.3352

Column 11

15.7928
-98.0154
-33.6417
-97.9396
-83.9568
-34.5102
-85.6030
74.6654
-79.0846
92.8484
-63.2600

Bfeedback =

1.8934	-38.2539
0.2523	233.1813
-0.2165	80.3499

```
  0.3483  233.1477
 -0.5438  199.8992
  0.0762   82.4312
 -0.5860  203.4689
  0.0365 -180.9855
 -0.1269  185.9941
  0.1418 -231.1130
 -0.0711  114.7003

Cfeedback =

  Columns 1 through 5

-252.5645 -152.2285   48.0056 -101.9783  -69.9110
 -38.2234  -24.4774    7.7356  -16.4185  -10.9426
   2.1971    1.3229   -0.4194    0.8852    0.6061

  Columns 6 through 10

  45.8791 -162.7408 -440.2738 -228.1027 -230.8255
   7.2453  -25.5209  -70.3915  -36.3600  -36.7206
  -0.4005    1.4144    3.8362    1.9861    2.0107

  Column 11

-183.4499
 -29.1900
   1.5978
Dfeedback=

 -0.2711  437.3138
 -0.0431   69.5795
  0.0024   -3.8089
```

Fig. 4.5-3 shows that the control system is stable and performs well until the order is reduced down to 12. When the order is reduced to 12, a critical eigenvalue moves to the Right Half Plane, displaying instability (Fig. 4.5-4).

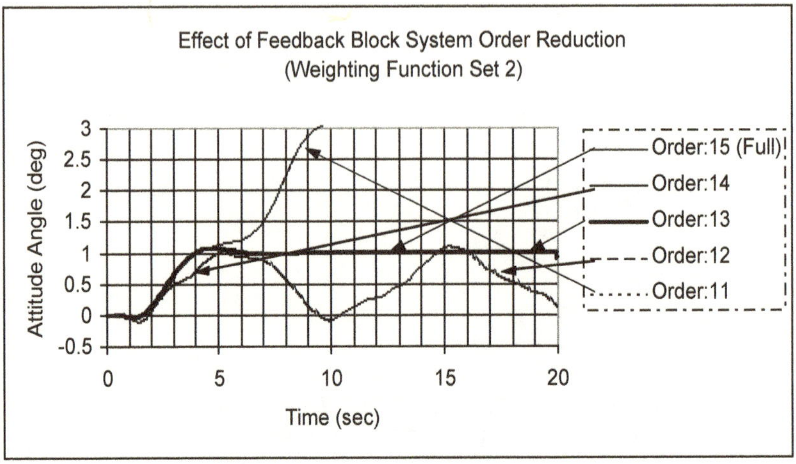

Fig. 4.5-3. Effect of Feedback Block System Order Reduction on Control System Performance, Weighting Function Set 2 (1.0 Deg Step Input)

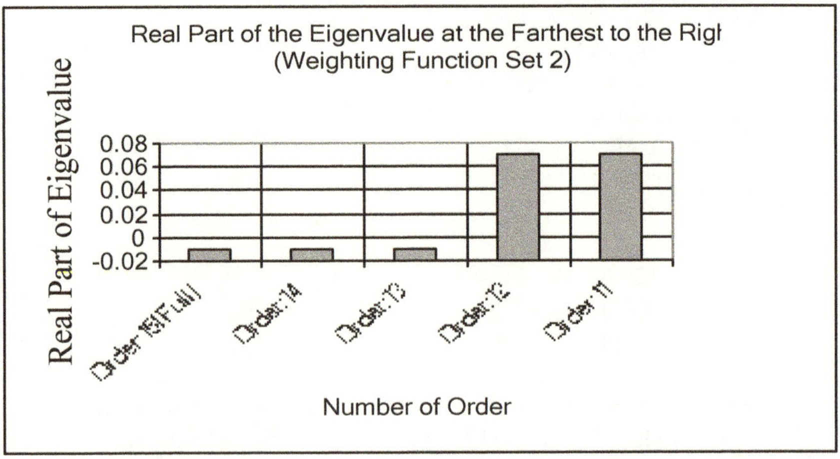

Fig. 4.5-4. Real Part of Eigenvalue at the Farthest Right, Weighting Function Set 2

c. Weighting Function Set 3

1) Feedback System Order: 14

Afeedback =

Columns 1 through 5

-16.2845	19.9362	2.4273	16.4365	-24.7735
35.6806	-50.1261	-10.0950	-41.3132	66.1729
-5.7176	11.7074	0.7877	2.0569	-9.3619
29.1493	-41.4818	-0.7918	-34.2900	52.9806
-40.5223	55.6978	5.3062	46.8754	-74.9536
10.4734	-14.6517	-1.4924	-12.2059	20.4824
-1.5902	1.7703	0.1798	1.4328	-2.3946
39.5750	-51.1213	-5.2458	-42.1749	70.7792
-23.1146	32.6212	3.3532	27.1065	-44.5777
-6.2002	8.9128	0.9162	7.4113	-12.1014
-24.3379	32.6746	3.3589	27.0128	-44.6393
-9.9420	13.7259	1.4115	11.3772	-18.7266
-4.0787	6.1377	0.6313	5.1236	-8.3152
-2.0808	2.4390	0.2507	1.9905	-3.3763

Columns 6 through 10

-9.0902	-2.6116	-39.9836	22.2235	-6.5179
24.5990	7.0942	107.6845	-59.8243	17.5211
-3.5100	-1.0244	-15.4110	8.5500	-2.5069
20.2653	5.8089	88.4817	-49.1756	14.3843
-27.8619	-7.9554	-120.0230	66.3554	-19.3713
6.4518	-2.8242	27.8156	-15.5469	4.4433
3.8813	-0.2763	-5.0827	2.5009	-0.7207
22.9730	8.2207	95.7629	-49.3032	13.7431
-14.2259	-4.6068	-65.5582	32.5704	-6.2375
-3.7729	-1.1981	-16.0685	5.5454	-3.8782
-14.0340	-4.6580	-56.6613	25.8979	-12.3602
-5.9232	-1.9280	-25.3233	12.3707	-7.5299
-2.6169	-0.8055	-12.0770	5.1095	-3.6470
-1.0764	-0.3900	-3.7996	2.4682	-0.8857

Columns 11 through 14

-22.1777	9.1083	5.4796	-3.2581
59.6114	-24.4937	-14.7323	8.7598
-8.5224	3.5029	2.1062	-1.2525
48.9550	-20.1184	-12.1012	7.1949
-65.8763	27.0977	16.2920	-9.6876
15.1739	-6.2862	-3.7768	2.2440
-2.4258	1.0043	0.6026	-0.3586
46.9097	-20.1313	-12.0340	7.1452
-24.7623	11.8338	7.0257	-4.1256
-15.2858	7.2458	4.1644	-2.3768
-52.7937	33.4744	17.7415	-9.9662
-33.4120	-1.9601	-4.3363	3.8521
-12.7203	-0.8367	-4.2240	5.9681
-5.7802	-1.8114	1.5173	-21.7616

Bfeedback =

-1.0111	-22.0031
0.7051	59.8657
0.1979	-8.8238
0.8633	49.1095
0.5339	-65.5386
0.0339	16.7592
0.0103	-2.1446
0.1010	59.1897
-0.0569	-37.8175
-0.0155	-10.3907
-0.0577	-38.1712
-0.0242	-15.9921
-0.0108	-7.1279
-0.0043	-2.8634

Cfeedback =

Columns 1 through 5

24.9748	-33.3997	-3.6064	-27.4897	44.8549
0.1865	-3.2348	-0.3489	-2.7919	4.1460
-0.0150	0.0203	0.0024	0.0168	-0.0277

Columns 6 through 10

15.7536	4.6598	69.7084	-38.8153	11.4365
1.4664	0.3986	6.7489	-3.6265	1.0826
-0.0101	-0.0029	-0.0445	0.0247	-0.0073

Columns 11 through 14

38.8860	-15.9547	-9.5976	5.7078
3.6546	-1.5019	-0.9028	0.5373
-0.0247	0.0101	0.0061	-0.0036

Dfeedback=

0.0600	39.6589
0.0057	3.7339
-0.0000	-0.0252

2) Feedback System Order: 13

Afeedback =

Columns 1 through 5

-16.2845	19.9362	2.4273	16.4365	-24.7735
35.6806	-50.1261	-10.0950	-41.3132	66.1729
-5.7176	11.7074	0.7877	2.0569	-9.3619
29.1493	-41.4818	-0.7918	-34.2900	52.9806
-40.5223	55.6978	5.3062	46.8754	-74.9536
10.4734	-14.6517	-1.4924	-12.2059	20.4824
-1.5902	1.7703	0.1798	1.4328	-2.3946
39.5750	-51.1213	-5.2458	-42.1749	70.7792
-23.1146	32.6212	3.3532	27.1065	-44.5777
-6.2002	8.9128	0.9162	7.4113	-12.1014
-24.3379	32.6746	3.3589	27.0128	-44.6393
-9.9420	13.7259	1.4115	11.3772	-18.7266
-4.0787	6.1377	0.6313	5.1236	-8.3152

Columns 6 through 10

-9.0902	-2.6116	-39.9836	22.2235	-6.5179
24.5990	7.0942	107.6845	-59.8243	17.5211
-3.5100	-1.0244	-15.4110	8.5500	-2.5069
20.2653	5.8089	88.4817	-49.1756	14.3843

```
-27.8619  -7.9554 -120.0230  66.3554 -19.3713
  6.4518  -2.8242  27.8156 -15.5469   4.4433
  3.8813  -0.2763  -5.0827   2.5009  -0.7207
 22.9730   8.2207  95.7629 -49.3032  13.7431
-14.2259  -4.6068 -65.5582  32.5704  -6.2375
 -3.7729  -1.1981 -16.0685   5.5454  -3.8782
-14.0340  -4.6580 -56.6613  25.8979 -12.3602
 -5.9232  -1.9280 -25.3233  12.3707  -7.5299
 -2.6169  -0.8055 -12.0770   5.1095  -3.6470

Columns 11 through 13

-22.1777   9.1083   5.4796
 59.6114 -24.4937 -14.7323
 -8.5224   3.5029   2.1062
 48.9550 -20.1184 -12.1012
-65.8763  27.0977  16.2920
 15.1739  -6.2862  -3.7768
 -2.4258   1.0043   0.6026
 46.9097 -20.1313 -12.0340
-24.7623  11.8338   7.0257
-15.2858   7.2458   4.1644
-52.7937  33.4744  17.7415
-33.4120  -1.9601  -4.3363
-12.7203  -0.8367  -4.2240

Bfeedback =

  -1.0111 -22.0031
   0.7051  59.8657
   0.1979  -8.8238
   0.8633  49.1095
   0.5339 -65.5386
   0.0339  16.7592
   0.0103  -2.1446
   0.1010  59.1897
  -0.0569 -37.8175
  -0.0155 -10.3907
  -0.0577 -38.1712
  -0.0242 -15.9921
  -0.0108  -7.1279
```

Cfeedback =

Columns 1 through 5

24.9748	-33.3997	-3.6064	-27.4897	44.8549
0.1865	-3.2348	-0.3489	-2.7919	4.1460
-0.0150	0.0203	0.0024	0.0168	-0.0277

Columns 6 through 10

15.7536	4.6598	69.7084	-38.8153	11.4365
1.4664	0.3986	6.7489	-3.6265	1.0826
-0.0101	-0.0029	-0.0445	0.0247	-0.0073

Columns 11 through 13

38.8860	-15.9547	-9.5976
3.6546	-1.5019	-0.9028
-0.0247	0.0101	0.0061

Dfeedback=

0.0600	39.6589
0.0057	3.7339
-0.0000	-0.0252

3) Feedback System Order: 12

Afeedback =

Columns 1 through 5

-16.2845	19.9362	2.4273	16.4365	-24.7735
35.6806	-50.1261	-10.0950	-41.3132	66.1729
-5.7176	11.7074	0.7877	2.0569	-9.3619
29.1493	-41.4818	-0.7918	-34.2900	52.9806
-40.5223	55.6978	5.3062	46.8754	-74.9536
10.4734	-14.6517	-1.4924	-12.2059	20.4824
-1.5902	1.7703	0.1798	1.4328	-2.3946
39.5750	-51.1213	-5.2458	-42.1749	70.7792
-23.1146	32.6212	3.3532	27.1065	-44.5777

-6.2002	8.9128	0.9162	7.4113	-12.1014
-24.3379	32.6746	3.3589	27.0128	-44.6393
-9.9420	13.7259	1.4115	11.3772	-18.7266

Columns 6 through 10

-9.0902	-2.6116	-39.9836	22.2235	-6.5179
24.5990	7.0942	107.6845	-59.8243	17.5211
-3.5100	-1.0244	-15.4110	8.5500	-2.5069
20.2653	5.8089	88.4817	-49.1756	14.3843
-27.8619	-7.9554	-120.0230	66.3554	-19.3713
6.4518	-2.8242	27.8156	-15.5469	4.4433
3.8813	-0.2763	-5.0827	2.5009	-0.7207
22.9730	8.2207	95.7629	-49.3032	13.7431
-14.2259	-4.6068	-65.5582	32.5704	-6.2375
-3.7729	-1.1981	-16.0685	5.5454	-3.8782
-14.0340	-4.6580	-56.6613	25.8979	-12.3602
-5.9232	-1.9280	-25.3233	12.3707	-7.5299

Columns 11 through 12

-22.1777	9.1083
59.6114	-24.4937
-8.5224	3.5029
48.9550	-20.1184
-65.8763	27.0977
15.1739	-6.2862
-2.4258	1.0043
46.9097	-20.1313
-24.7623	11.8338
-15.2858	7.2458
-52.7937	33.4744
-33.4120	-1.9601

Bfeedback =

-1.0111	-22.0031
0.7051	59.8657
0.1979	-8.8238
0.8633	49.1095

0.5339	-65.5386
0.0339	16.7592
0.0103	-2.1446
0.1010	59.1897
-0.0569	-37.8175
-0.0155	-10.3907
-0.0577	-38.1712
-0.0242	-15.9921

Cfeedback =

Columns 1 through 5

24.9748	-33.3997	-3.6064	-27.4897	44.8549
0.1865	-3.2348	-0.3489	-2.7919	4.1460
-0.0150	0.0203	0.0024	0.0168	-0.0277

Columns 6 through 10

15.7536	4.6598	69.7084	-38.8153	11.4365
1.4664	0.3986	6.7489	-3.6265	1.0826
-0.0101	-0.0029	-0.0445	0.0247	-0.0073

Columns 11 through 12

38.8860	-15.9547
3.6546	-1.5019
-0.0247	0.0101

Dfeedback=

0.0600	39.6589
0.0057	3.7339
-0.0000	-0.0252

4) Feedback System Order: 11

Afeedback =

Columns 1 through 5

-16.2845	19.9362	2.4273	16.4365	-24.7735

```
  35.6806  -50.1261  -10.0950  -41.3132   66.1729
  -5.7176   11.7074    0.7877    2.0569   -9.3619
  29.1493  -41.4818   -0.7918  -34.2900   52.9806
 -40.5223   55.6978    5.3062   46.8754  -74.9536
  10.4734  -14.6517   -1.4924  -12.2059   20.4824
  -1.5902    1.7703    0.1798    1.4328   -2.3946
  39.5750  -51.1213   -5.2458  -42.1749   70.7792
 -23.1146   32.6212    3.3532   27.1065  -44.5777
  -6.2002    8.9128    0.9162    7.4113  -12.1014
 -24.3379   32.6746    3.3589   27.0128  -44.6393

 Columns 6 through 10

  -9.0902   -2.6116  -39.9836   22.2235   -6.5179
  24.5990    7.0942  107.6845  -59.8243   17.5211
  -3.5100   -1.0244  -15.4110    8.5500   -2.5069
  20.2653    5.8089   88.4817  -49.1756   14.3843
 -27.8619   -7.9554 -120.0230   66.3554  -19.3713
   6.4518   -2.8242   27.8156  -15.5469    4.4433
   3.8813   -0.2763   -5.0827    2.5009   -0.7207
  22.9730    8.2207   95.7629  -49.3032   13.7431
 -14.2259   -4.6068  -65.5582   32.5704   -6.2375
  -3.7729   -1.1981  -16.0685    5.5454   -3.8782
 -14.0340   -4.6580  -56.6613   25.8979  -12.3602

 Column 11

 -22.1777
  59.6114
  -8.5224
  48.9550
 -65.8763
  15.1739
  -2.4258
  46.9097
 -24.7623
 -15.2858
 -52.7937

Bfeedback =

  -1.0111  -22.0031
```

```
    0.7051   59.8657
    0.1979   -8.8238
    0.8633   49.1095
    0.5339  -65.5386
    0.0339   16.7592
    0.0103   -2.1446
    0.1010   59.1897
   -0.0569  -37.8175
   -0.0155  -10.3907
   -0.0577  -38.1712


Cfeedback =

  Columns 1 through 5

   24.9748  -33.3997   -3.6064  -27.4897   44.8549
    0.1865   -3.2348   -0.3489   -2.7919    4.1460
   -0.0150    0.0203    0.0024    0.0168   -0.0277

  Columns 6 through 10

   15.7536    4.6598   69.7084  -38.8153   11.4365
    1.4664    0.3986    6.7489   -3.6265    1.0826
   -0.0101   -0.0029   -0.0445    0.0247   -0.0073

  Column 11

   38.8860
    3.6546
   -0.0247

Dfeedback=

    0.0600   39.6589
    0.0057    3.7339
   -0.0000   -0.0252
```

5) Feedback System Order: 10

Afeedback =

Columns 1 through 4

-16.2845	19.9362	2.4273	16.4365
35.6806	-50.1261	-10.0950	-41.3132
-5.7176	11.7074	0.7877	2.0569
29.1493	-41.4818	-0.7918	-34.2900
-40.5223	55.6978	5.3062	46.8754
10.4734	-14.6517	-1.4924	-12.2059
-1.5902	1.7703	0.1798	1.4328
39.5750	-51.1213	-5.2458	-42.1749
-23.1146	32.6212	3.3532	27.1065
-6.2002	8.9128	0.9162	7.4113

Columns 5 through 8

-24.7735	-9.0902	-2.6116	-39.9836
66.1729	24.5990	7.0942	107.6845
-9.3619	-3.5100	-1.0244	-15.4110
52.9806	20.2653	5.8089	88.4817
-74.9536	-27.8619	-7.9554	-120.0230
20.4824	6.4518	-2.8242	27.8156
-2.3946	3.8813	-0.2763	-5.0827
70.7792	22.9730	8.2207	95.7629
-44.5777	-14.2259	-4.6068	-65.5582
-12.1014	-3.7729	-1.1981	-16.0685

Columns 9 through 10

22.2235	-6.5179
-59.8243	17.5211
8.5500	-2.5069
-49.1756	14.3843
66.3554	-19.3713
-15.5469	4.4433
2.5009	-0.7207
-49.3032	13.7431
32.5704	-6.2375
5.5454	-3.8782

Bfeedback =

-1.0111	-22.0031
0.7051	59.8657
0.1979	-8.8238
0.8633	49.1095
0.5339	-65.5386
0.0339	16.7592
0.0103	-2.1446
0.1010	59.1897
-0.0569	-37.8175
-0.0155	-10.3907

Cfeedback =

columns 1 through 4

24.9748	-33.3997	-3.6064	-27.4897
0.1865	-3.2348	-0.3489	-2.7919
-0.0150	0.0203	0.0024	0.0168

Columns 5 through 8

44.8549	15.7536	4.6598	69.7084
4.1460	1.4664	0.3986	6.7489
-0.0277	-0.0101	-0.0029	-0.0445

Columns 9 through 10

-38.8153	11.4365
-3.6265	1.0826
0.0247	-0.0073

Dfeedback=

0.0600	39.6589
0.0057	3.7339
-0.0000	-0.0252

Fig. 4.5-5 shows that the control system is stable all the way down to the 11th order, but performance degrades considerably when the order is reduced below 11. Notice that all the critical eigenvalues remain in the Left Half Plane until the order is reduced to 11. (Fig. 4.5-6).

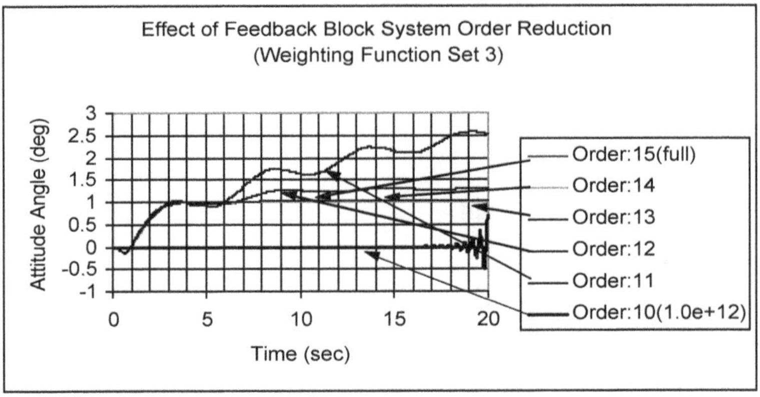

Fig. 4.5-5. Effect of Feedback Block System Order Reduction on Control System Performance, Weighting Function Set 3

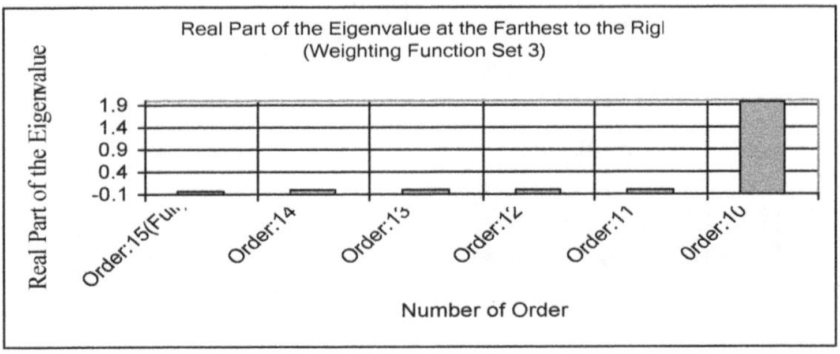

Fig. 4.5-6. Real Part of Eigenvalue at the Farthest Right, Weighting Function Set 3

Here again, this analysis shows that the weighting function must be designed properly to achieve a robust and high performance control system design. The findings are tabulated below.

Weighting Function Set 1:	system becomes unstable when the order is reduced to 11
Weighting Function Set 2:	system becomes unstable when the order is reduced to 12
Weighting Function Set 3:	system becomes unstable when the order is reduced to 11

Chapter 5. Wind Influence on LV Attitude and AOA in Flight

Since the wind affects significantly rocket's aerodynamic response and behavior when the rocket is flying through the atmosphere, it is necessary to analyze the wind effects before launch to ensure that the flight system performs optimally and avoid potential problems such as exceeding structural and thermal safety limits, and/or losing attitude stability primarily due to the increase of the Angle of Attack. The attitude control system performance during the atmospheric flight is one of the most serious concerns in our current flight system. The reason is that the current launch vehicle control systems were designed using classical control theories, e.g., Bode, Nyquist, and Root Locus, and, therefore, it is not possible to deal with multi-input multi-output systems. More specifically, classical control theory is not the right theory when we need to deal with attitude input command and wind "input" at the same time.

MIMO (Multi-Input Multi-Output) control theory can deal with more than one input. It is more preferable theory which is applicable to launch vehicle attitude control systems design. By taking MIMO system design approach, a feedback block can be constructed that responds to wind input, AOA, as well as attitude command input, which, in turn, enables the control system to contain the wind effect and reduce the AOA to some degree. Once a stable control system is designed for multiple inputs, e.g., attitude command and wind, no matter how strong the wind is, the control system can manage the wind disturbance because it is treated as a part of inputs in the design. It should be noted that this new control system may require a new set of control surfaces or on-and-off reaction jets discussed in Chapter 4 to respond to feedback signals. It also has another capability of reducing the AOA by providing an appropriate AOA command input. This capability is demonstrated in the later part of this chapter. The H-infinity MIMO approach used here truly demonstrates a design technique that can revolutionize LV attitude control systems design.

While a strong wind cannot drive the control system unstable, it may cause excessive drift from the nominal trajectory. Guidance must include a drift compensation scheme and eliminate the excessive drift by providing appropriate trajectory. Appropriate trajectory design requires the wind analysis, and therefore, the wind analysis is still required for successful flights.

5.1 Simulations of Wind Influence on LV Attititude

To demonstrate robust performance of the control system designed by use of H-infinity MIMO control theory, several wind cases are fabricated and a 1.0 deg step input is applied to observe how the system responds to these winds. The system under test is the one from Section 4.3 (Weighting Function Option 3).

From Fig. 5.1-1, four cases of wind are selected:

1) 0.0 ft/sec lateral wind
2) 50.0 ft/sec lateral wind
3) 100.0 ft/sec lateral wind
4) 200.0 ft/sec lateral wind

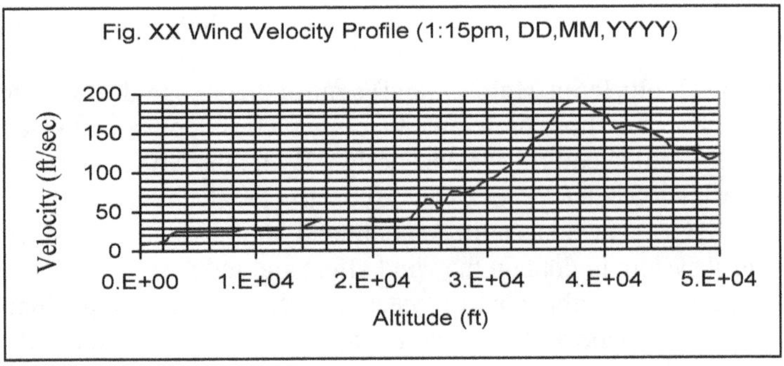

Fig. 5.1-1. Wind Velocity Profile

1) 0.0 ft/sec. Lateral Wind

To start with, it is assumed that there is no wind. In this case, there will be zero wind variation (Fig. 5.1-2), and the step response is not affected by the wind (Fig. 5.1-3). The angle of attack reaches steady state (Fig. 5.1-4). Fig. 5.1-5 shows the lateral thrust that is needed to counteract the moment due to the non-zero angle of attack. The angle of attack is generated because the cg is not on the X-body axis due to structural bending and sloshing. Fig. 5.1-6 shows a graphical explanation of what is happening.

a) Lateral Wind Velocity

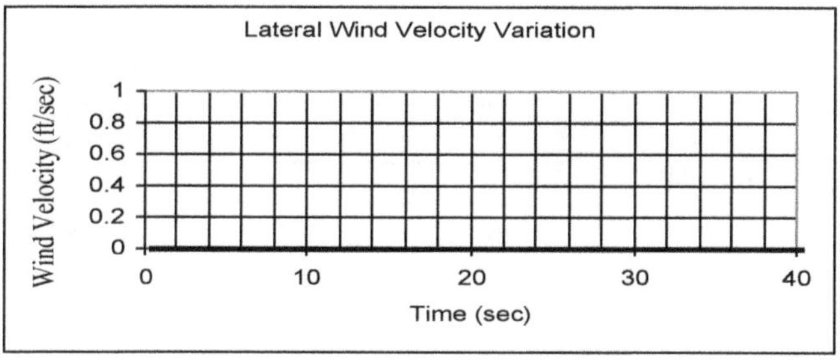

Fig. 5.1-2. Lateral Wind Velocity Variation (0.0 Vari.)

a) 1.0-Deg Step Input Attitude Response

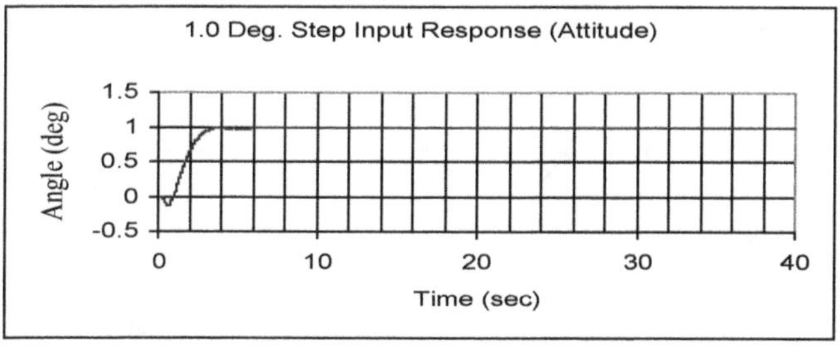

Fig. 5.1-3. 1.0-Deg Step Input Attitude Response

c) The Angle of Attack

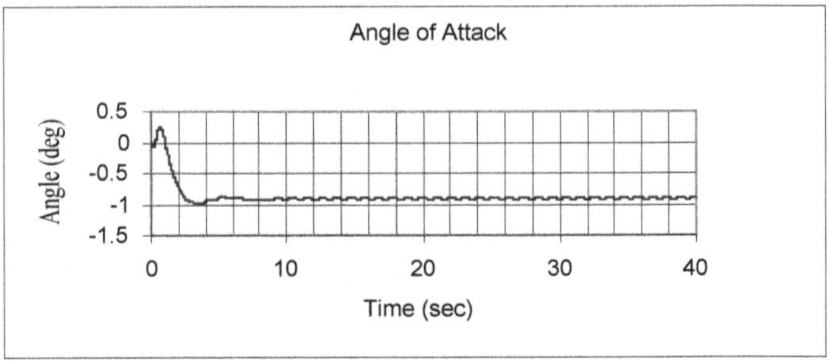

Fig. 5.1-4. Angle of Attack

d) Lateral Thrust

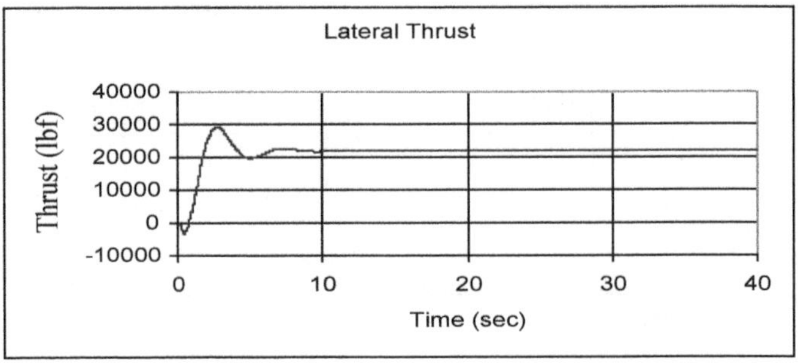

Fig. 5.1-5. Lateral Thrust

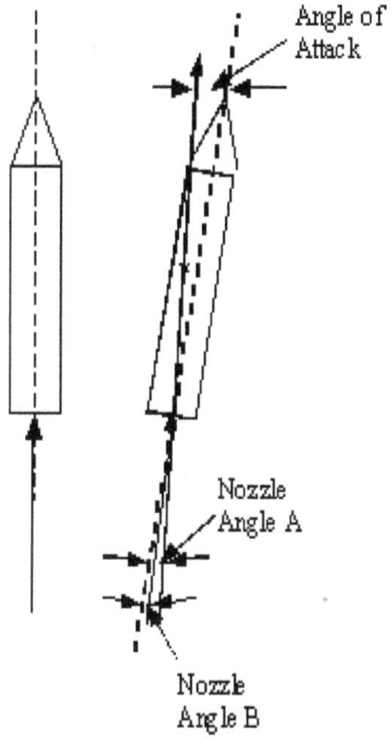

Fig. 5.1-6. Launch Vehicle Rotation

In Fig. 5.1-6, the Nozzle Angle A is the nozzle angle at t ≈ 2.8 sec. and the Nozzle Angle B is the nozzle angle after t ≈ 4.2 sec. The nozzle angle swings from A to B to create the lateral thrust needed for balancing the vehicle.

2) Up To 50 ft/sec Lateral Wind Velocity Variation

The wind velocity profile is shown in Fig. 5.1-7. The wind velocity starts with 0.0 ft/sec., and then increases to a maximum velocity of 50 ft/sec, and gradually decreases to 0.0 ft/sec at t = 20.0 sec (Fig. 5.1-7). The direction of the velocity is perpendicular to the LV X-body axis. The step input response reflects the wind effect applied (Fig. 5.1-8). Initially, the attitude begins to approach the desired attitude of 1.0 degree. However, as the velocity increases, attitude stops increasing and, instead, decreases until it becomes negative. When the wind velocity reaches its maximum, the attitude goes down to -1.32 deg at t = 11 sec. But the control system does not lose its stability and controllability. As the wind velocity decreases, the attitude recovers to the desired attitude, i.e., 1.0 degree. Fig. 5.1-9 and Fig. 5.1-10 show the time histories of the angle of attack and lateral thrust. The complexities in the figures come from bending and sloshing. Fig 5.1-11 shows how the angle of attack increases as the wind velocity increases.

a) Lateral Wind Velocity Variation profile

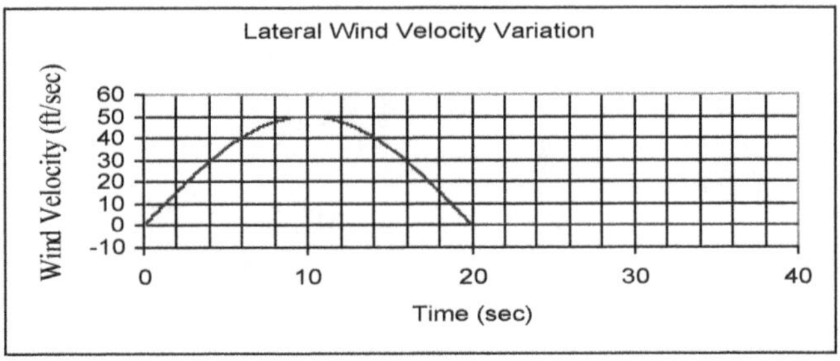

Fig. 5.1-7. Lateral Wind Velocity Variation

b) A 1.0-Deg Step Input Attitude Response

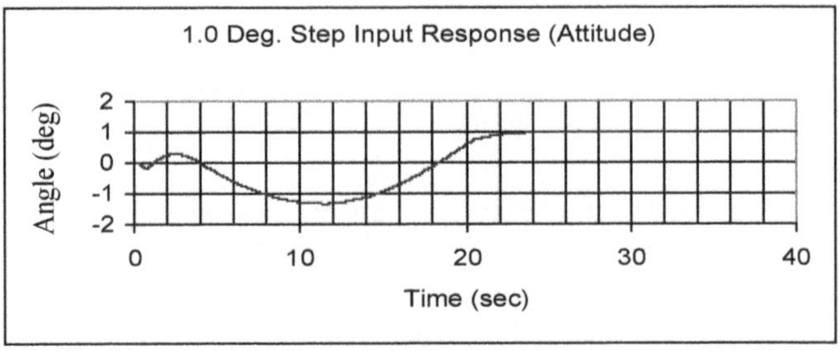

Fig. 5.1-8. A 1.0-Deg Step Input Attitude Response

c) The Angle of Attack

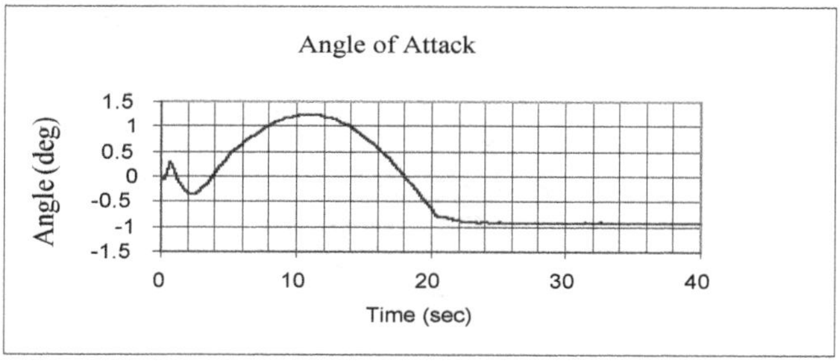

Fig. 5.1-9. The Angle of Attack

d) Lateral Thrust

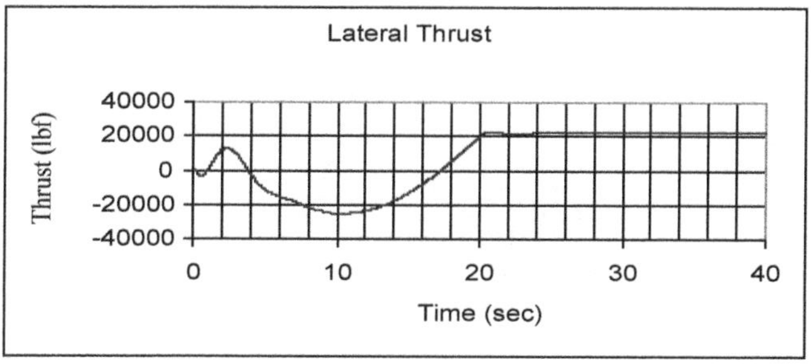

Fig. 5.1-10. Lateral Thrust

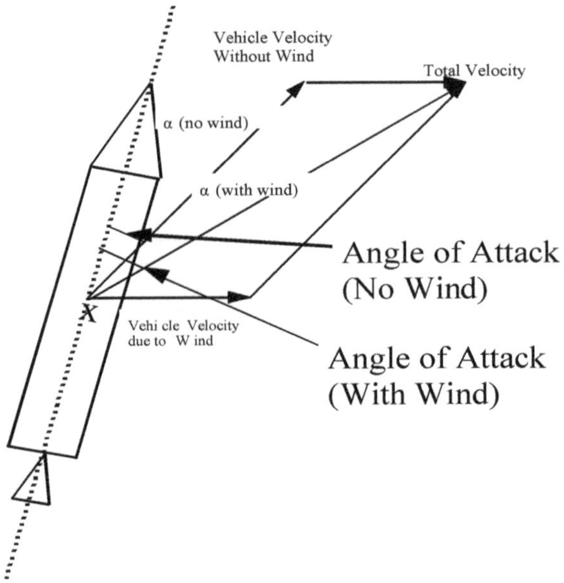

Fig. 5.1-11. Angle of Attack and Wind

3) Lateral Wind Velocity Variation up to 100 ft/sec

A wind velocity profile is shown in Fig. 5.1-12. The wind starts with 0.0 ft/sec. and increases to a maximum velocity of 100 ft/sec. at t = 10.0 sec and gradually decreases to 0.0 ft/sec. at t = 20.0 sec (Fig. 5.1-12). The direction of the velocity is perpendicular to the LV X-body axis. The step input response reflects the wind effect applied (Fig. 5.1-13). The attitude cannot turn toward the desired attitude of 1.0 degree, because the wind velocity increases immediately. As the velocity increases, the attitude actually decreases until it becomes negative. When the wind velocity reaches its maximum, the attitude decreases to –3.64 deg at t = 11.5 sec. But the control system does not lose its ability.. As the wind velocity decreases, the attitude recovers to the desired attitude, 1.0 deg.

a) Lateral Wind Velocity Variation profile

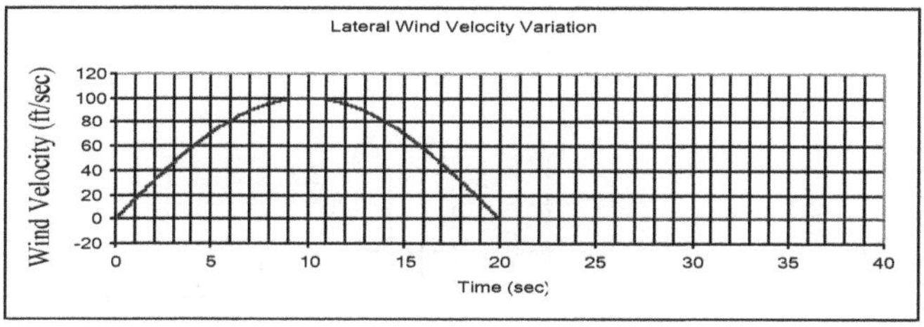

Fig. 5.1-12. Lateral Wind Velocity Variation

b) 1.0 Deg Step Input Attitude Response

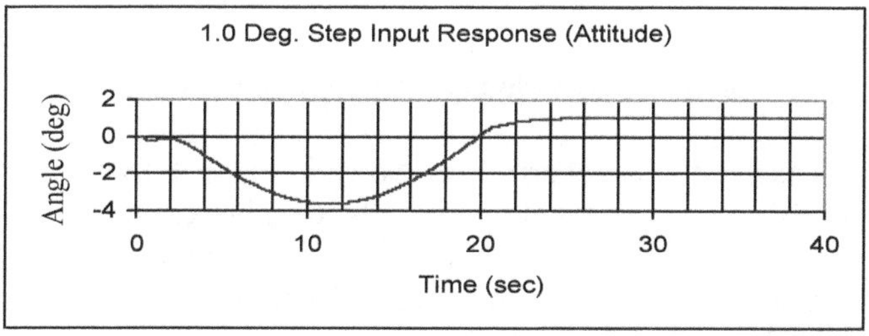

Fig. 5.1-13. 1.0 Deg Step Input Attitude Response

c) The Angle of Attack

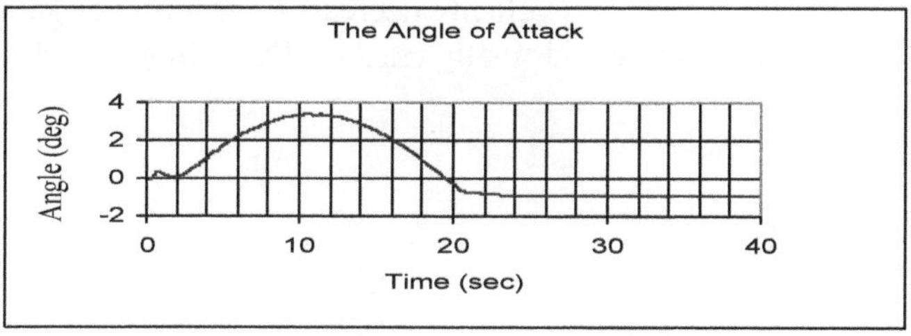

Fig. 5.1-14 The Angle of Attack

d) Lateral Thrust

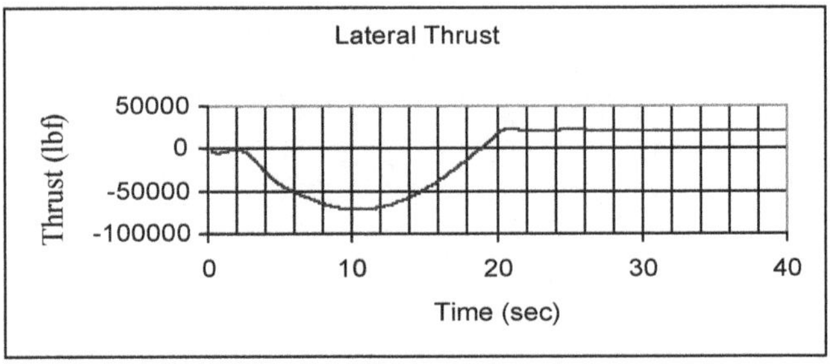

Fig. 5.1-15. Lateral Thrust

5) Lateral Wind Velocity Variation up to 200 ft/sec

The wind velocity profile is shown in Fig. 5.1-16. The direction of the wind is normal to the LV's X-body axis. Its velocity starts with 0.0 ft/sec. and then grows to its maximum velocity, 200 ft/sec, at t = 10.0 sec and gradually decreases to 0.0 ft/sec at t = 20.0 sec. (Fig. 5.1-16). The step input response reflects the wind effect applied (Fig. 5.1-17). In the beginning, the attitude cannot turn toward the desired attitude of 1.0 deg because the wind velocity increases immediately. As the wind velocity increases, the attitude actually decreases until it becomes negative. When the wind velocity reaches its maximum, the attitude goes down to -8.27 deg at t = 11.2 sec. But the control system does not lose its ability. As the wind velocity decreases, the attitude again becomes positive and finally reaches the desired attitude of 1.0 deg.

a) Lateral Wind Velocity Variation profile

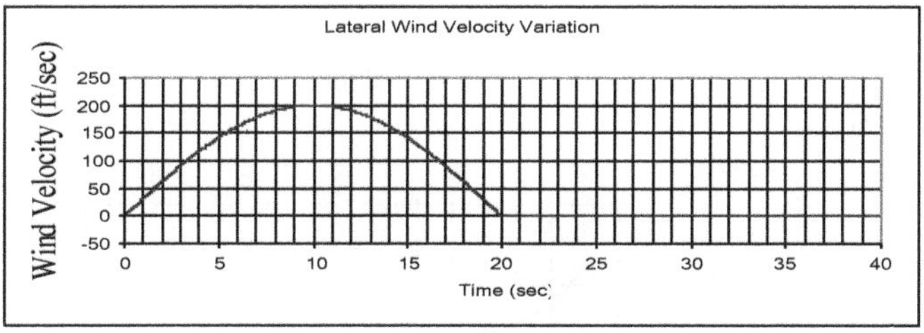

Fig. 5.1-16. Lateral Wind Velocity Variation

b) 1.0 Deg. Step Input Response in Attitude

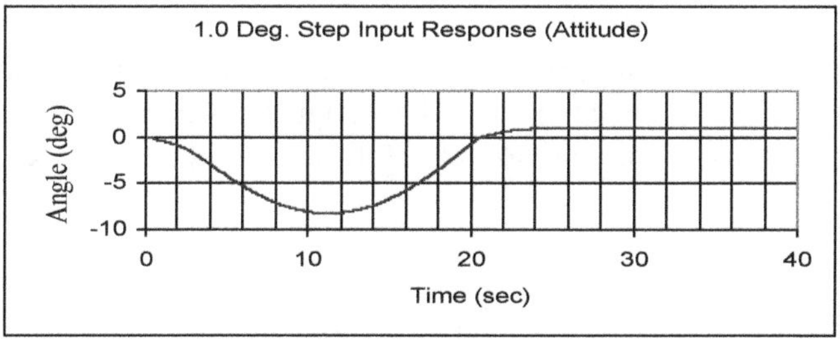

Fig. 5.1-17. 1.0 Deg Step Input Attitude Response

c) Angle of Attack

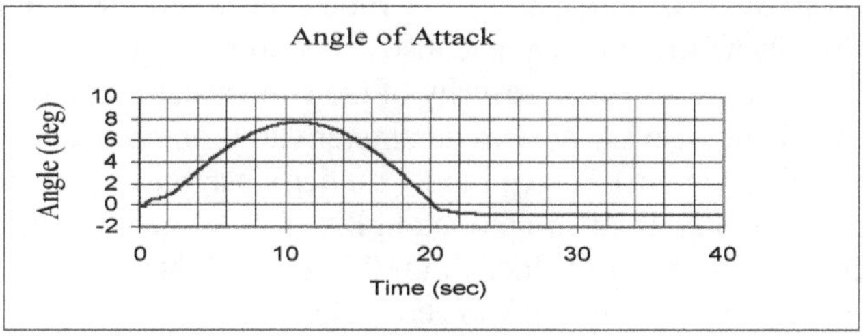

Fig. 5.1-18. Angle of Attack

d) Lateral Thrust

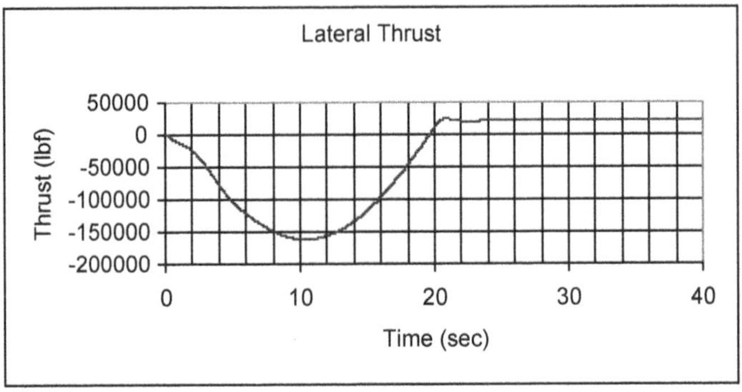

Fig. 5.1-19. Lateral Thrust

It is shown that the control system designed by using the H-infinity MIMO control theory is robust, and the desired attitude can be obtained regardless how high the lateral wind velocity is as long as an appropriate lateral thrust is provided.

5.2 Random Wind Velocity Variation Cases

In the random velocity variation case, two wind profiles, A and B, in Fig. 5.2-1 are chosen for the wind analysis. In case A, the average velocity is 8.4 ft/sec and its standard deviation is 85.5 ft/sec. In case B, the average velocity is 0.5 ft/sec and its standard deviation is 4.2 ft/sec. It is seen here that case A has more severe wind shear than case B. The objective of this simulation study is to demonstrate how robust the control system is under a severe wind-shear environment and also show how the quality of the system response varies as the severity of the wind shear increases. Here we assume that there is no LV system parameter variation for 50 sec (time-invariant flight segment). A step input (1.0 deg) is applied, and then the responses to this step input are plotted in Fig. 5.2-2. The wind velocity starts increasing from zero velocity at t=0.0 sec, and then after 29.0 sec velocity variation, it comes back to zero velocity and

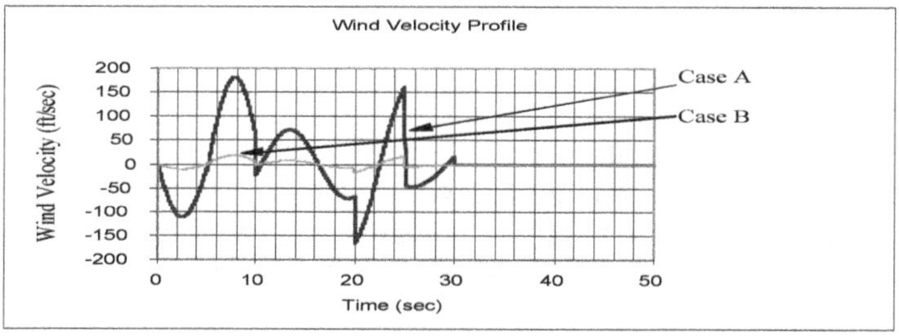

Fig. 5.2-1 Lateral Wind Velocity Variation

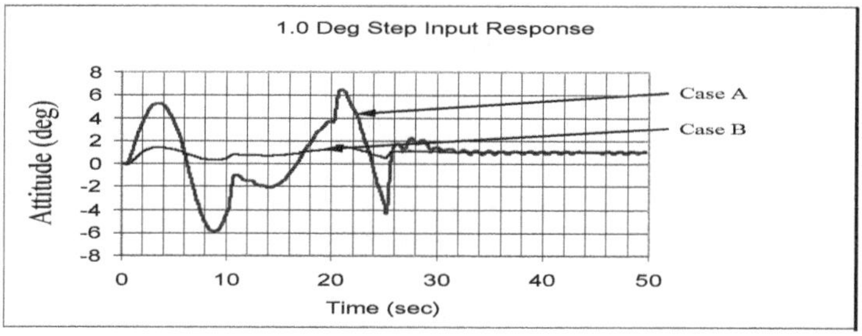

Fig. 5.2-2 1.0 Deg. Step Input Responses under Wind Shear Disturbances

remains at that velocity. In the case B, the attitude reaches the desired 1.0 deg at t=26 sec as the wind velocity goes down below 5 ft/sec. But in case A, it reaches the desired 1.0 deg attitude at t=31.0, and remains oscillatory until t=42 sec but does not diverge.

This analysis shows that the vehicle does not lose its controllability in the midst of strong lateral wind accompanied with severe shear winds. In fact, a further simulation study confirms that the control systems never become unstable regardless of how strong the wind is. While strong lateral wind cannot cause instability, there could be excessive position drift from the nominal trajectory. The guidance and navigation systems must incorporate a proper trajectory compensation scheme in order to accomplish satellite launch mission successfully.

Modeling errors are one of the concerns in attitude control systems design. In order to assess how much the modeling error affects the system response, a case of -15% modeling error in pitch moment of inertia is investigated to observe the effect. Fig. 5.2-3 shows plots of the differences between the nominal response and the responses when there are modeling error and wind disturbance. It is seen here that there is a significant difference in step input responses between the perfect math modeling case (Case B) and -15% modeling error case (Case A) when the wind shear forces are large, but when the wind velocity comes down to zero, the modeling error effect becomes insignificant. Thus, it demonstrates that the control system is so robust that there is no significant degradation in the system response in spite of such a large modeling error as large as -15.0 % as long as there is no severe wind disturbance. However, even in the severe wind environment, the system responses robustly and achieves the control objective.

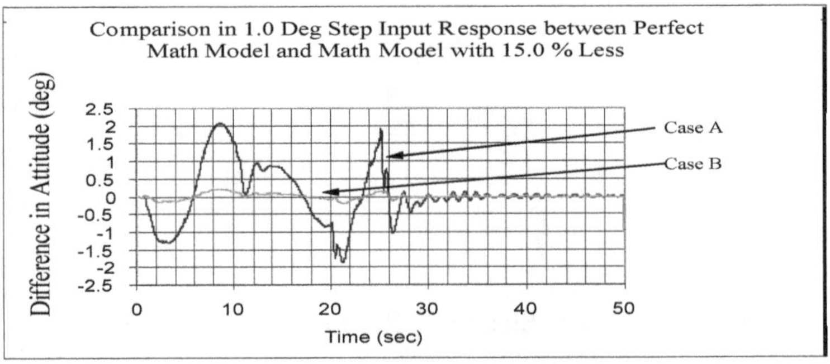

Fig. 5.2-3 Modeling Error Effect on System Response Under Large Wind Shear Disturbances

5.3 AOA Reduction Demonstration

To demonstrate the AOA reduction capability, several cases are simulated. For this analysis, we used the system designed in the section of "A case when Ka = 1.0 and Ki=1.5 in "En" block in Fig4.1-1" in Chapter 4.

First we show that the attitude control system designed is stable. To demonstrate the stability, 1.0 degree step input is applied to the system and the response is simulated. Its output attitude is plotted in Fig. 5.3-1. The steady state is reached within 4 sec without overshoot, and it demonstrates

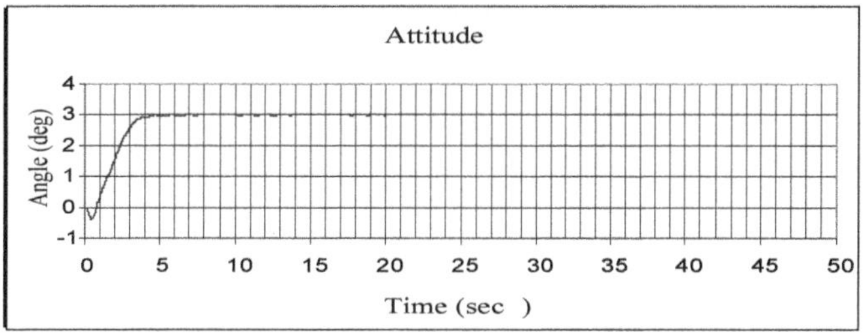

Fig. 5.3-1 1 Degree Step Response

the stability. Since our concern here is the AOA (Angle of Attack), it is plotted in Fig. 5.3-2. It shows that the AOA increases from 0.0 degree at t=0.0 sec. to -2.15 degrees at t=4.2 sec and remains at the same angle for

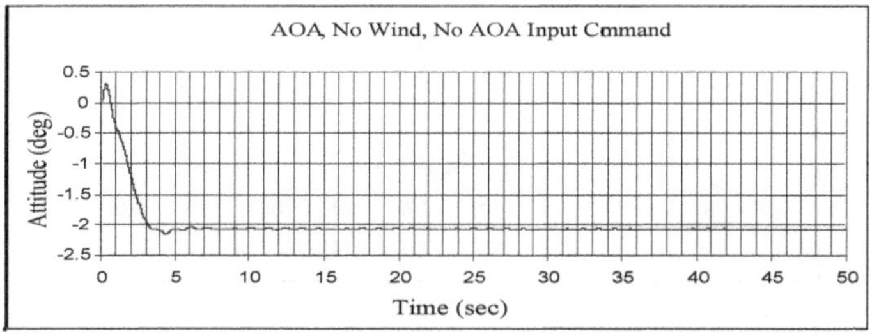

Fig. 5.3-2 AOA When There Is No Wind

the rest of the simulation.

After checking the system stability, we repeat the same simulation but with wind applied. The wind velocity profile is shown in Fig. 5.3-3. It starts at t=15.0 sec. and lasts until t=35.0 sec. The velocity profile is a sinusoidal and the peak velocity is 200 ft/sec. It is seen here that the wind is applied when the system is in the steady state. The attitude plot in Fig. 5.3-4 shows how the wind disturbance affects the attitude. The wind causes the attitude to decrease from 1.0 degree to 0.1 degree at t=18.5 sec and

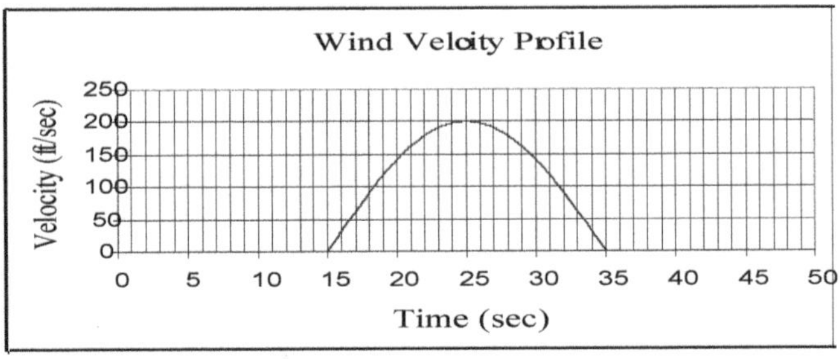

Fig. 5.3-3 Wind Velocity Profile

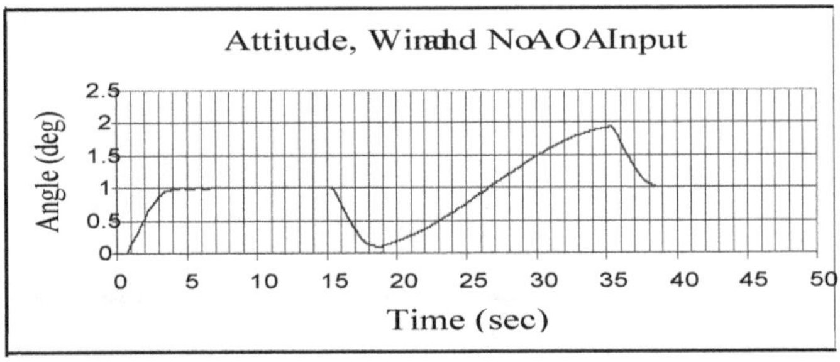

Fig. 5.3-4 Attitude When Wind Is Applied

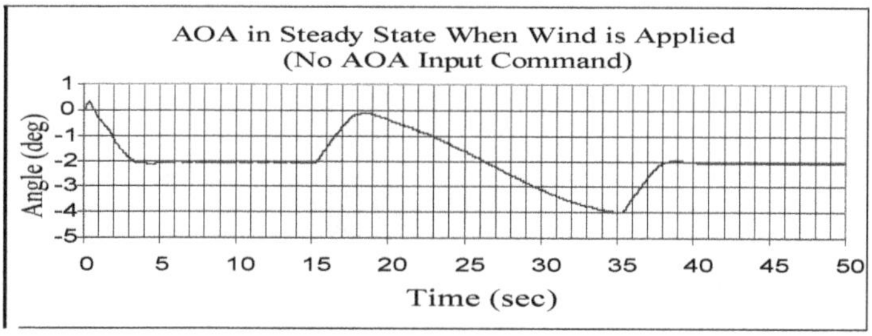

Fig. 5.3-5 AOA When Wind Is Applied in The Steady State

then back up to 1.9 deg at t=35.2 sec. About 3.0 second later, the attitude recovers its desired attitude, 1.0 degree. The wind also causes the AOA to increase from 2.15 degree to 4.0 degrees (Fig. 5.3-5), which is almost 100% increase. The current LV attitude control system is not able to control the AOA. However, using the H-infinity control concept, it is possible to control the AOA to some extent. If it gets out of the boundary, then we can bring it back to inside the boundary by commanding the AOA. This is possible because in this MIMO control system, the AOA is a part of input set. As we design an appropriate AOA input command, we can reduce the AOA. In the following, the AOA reduction is demonstrated. Fig. 5.3-6 shows an example of AOA input command which is designed for this particular simulation. Note that the design process used is a

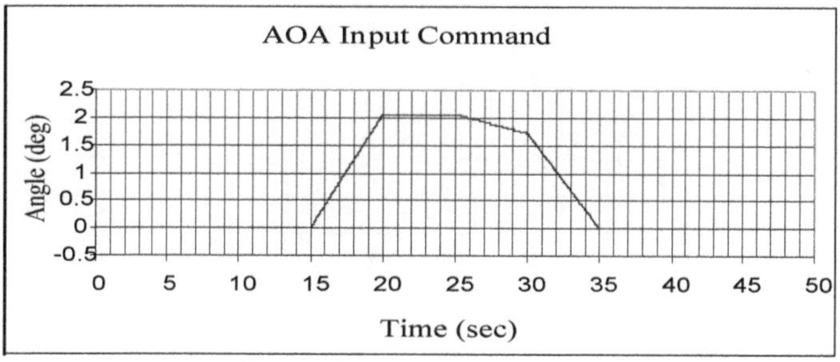

Fig. 5.3-6 Designed AOA Input Command

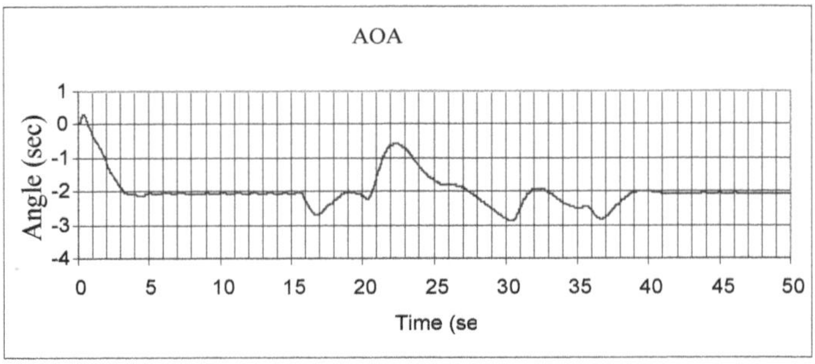

Fig. 5.3-7 AOA when Wind and Designed AOA Input Command Are Applied.

trial-and-error approach. The AOA is plotted when the designed AOA input command is applied in Fig. 5.3-7. It is seen here that the maximum AOA is reduced from 4.0 degrees (Fig. 5.3-5) to 2.9 degree (Fig. 5.3-7), which is about 28% decrease. Fig. 5.3-8 shows the attitude when the wind is applied from t=15.0 sec to t=35.0 sec. During this period the attitude deviates from the desired angle, but it recovers its desired attitude immediately as the wind disappears.

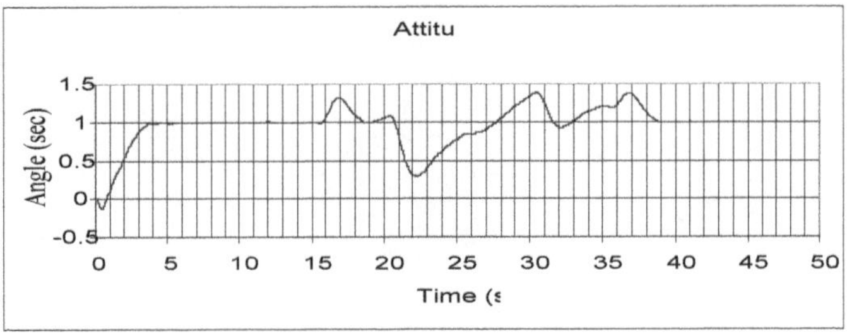

Fig. 5.3-8 Attitude When The Wind and Designed AOA Input Command Are Applied.

In order to investigate how the system responses during the transient period, we shift the wind starting time and AOA commanding time from t=15.0 sec to t=0.0 sec. Fig. 5.3-9 shows the new wind velocity profile and Fig. 5.3-10 shows the new designed AOA input command. Fig. 5.3-11

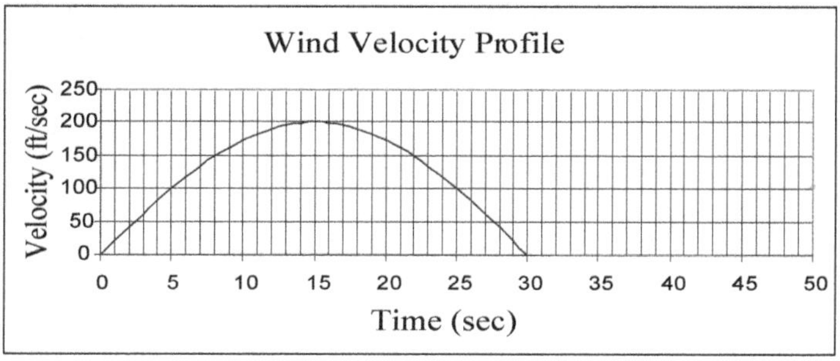

Fig. 5.3-9 Wind Velocity Profile

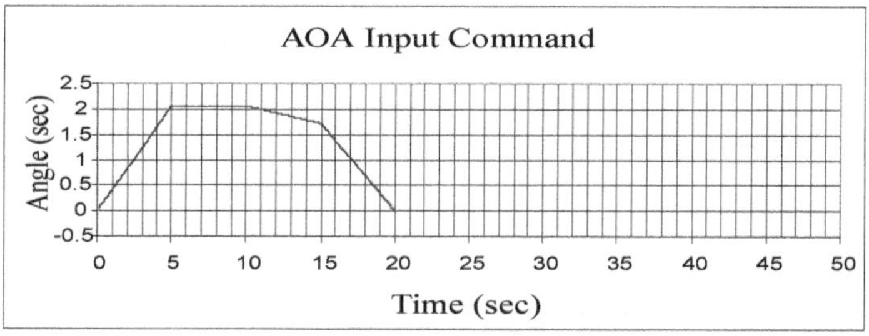

Fig. 5.3-10 Designed AOA Input Command

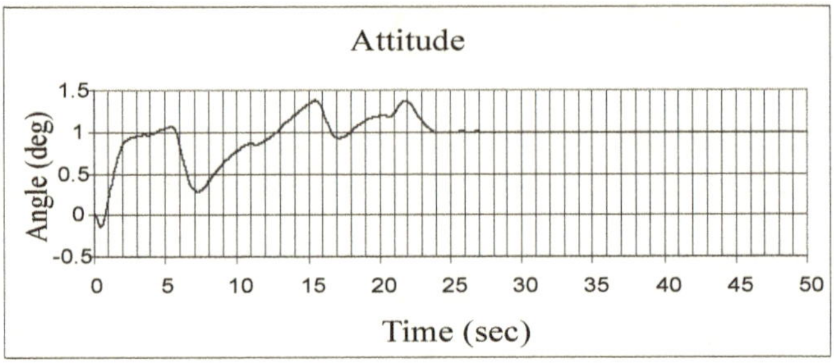

F ig. 5.3-11 Attitude When Wind and Designed AOA Input Command Are Applied.

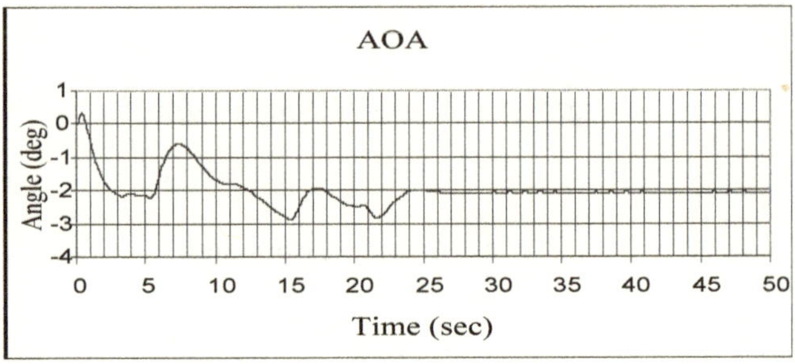

Fig. 5.3-12 AOA When Wind and Designed AOA Input Command Are Applied

shows the attitude. It is seen in Fig. 5.3-12 that the designed AOA input command is as effective as in the case when the wind occurs in the middle of the steady state (Fig. 5.3-7). It appears that the wind velocity profile and the desired AOA input command profile are closely related.

5.4 Wind Analysis

The fundamental problem with the wind is the highly unpredictability. Because of this nature of unpredictability, we need to depend on statistical analysis. In order to perform the statistical wind analysis, we must collect data at the launch site by flying balloons in a systematic manner throughout the year, possibly for many years. Some balloons are equipped with a simple navigation system that transmits their positions to the ground station. By tracking the balloon, we compute the balloon velocity, thus wind velocity. Another approach to measure the balloon velocity is to add an attachment to the balloon that reflects radar signals from the ground. Then, the ground station can track azimuth and slant angles of the balloon and compute their position and velocity.

Currently available wind data are from the data obtained by collecting 150 profiles per month for one year. A sample profile generated at a certain date and time is shown in Fig. 5.4-1. It is seen here that the wind velocity increases as the altitude increases up to about 39,000 ft, where the wind velocity is about 190 ft/sec (129 mph).

NASA systematically collected a reasonable amount of such data. These data are used to predict, with a reasonable level of confidence, what the wind velocity is likely to be at any given altitude and time. If we collect these data for many years, then we will have an even more reliable data base that will provide valuable information needed for successful launch.

Before each flight, the balloon measurements will be made according to a prescheduled timetable until the measurements are no longer needed. Fig. 5.4-1 shows when the last wind measurement should be made. It is seen in Fig. 5.4-1 that it takes about 440 seconds for the balloon to reach 50,000 ft. Therefore, if we need wind data up to 50,000 ft for the LV, then we need to have at least 440 seconds for collecting the data. Assuming the data processing and implementation time required is 120 sec, we need to launch the last balloon at least 560 sec before the launch. Ideally, we desire to make some statistical predictions of the wind effect on the LV from t = 0.0 to t = 70 sec from wind data measured from t = -1,800 sec to t = -560.0 sec.

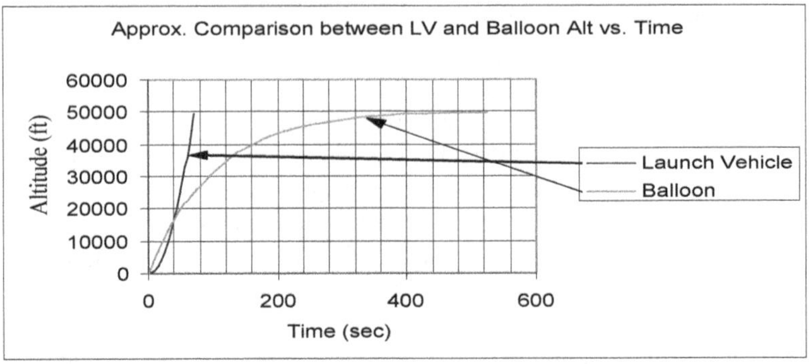

Fig. 5.4-1 Approximate Altitude Comparison between Balloon and Launch Vehicle

Since wind data usually have high frequency content, we need to perform curve fitting of the data for wind analysis. In the following, we discuss briefly a curve-fitting technique.

Least Square Curve Fitting

Mathematical derivation of the curve-fitting technique is simple, and it is shown below. A measurement equation can be written in a matrix format:

$$Z = HX + V \tag{5.4-1}$$

Where

Z	:	measurement data
H	:	measurement matrix
X	:	independent variable
V	:	measurement noise

The measurement error is,

$$E = Z - HX \tag{5.4-2}$$

where H is a measurement matrix math model.

The square of the error with weighting matrix, W, is,

$$J = E'WE$$
$$= (Z - HX)'W(Z - HX) \qquad (5.4\text{-}3)$$

( )' means transpose of ( ).

Taking a derivative of J with respect to X to minimize J (error), and equating the result equal to zero, we have,

$$\delta J/\delta X = (-H)'R(Z - HX) + (Z - HX)'R(-H)$$
$$= -2H'RZ + 2H'RHX$$
$$= 0 \qquad (5.4\text{-}4)$$

Solving for X,

$$X = [\text{Inverse of } (H'RH)]H'RZ \qquad (5.4\text{-}5)$$

Example

A data table is shown below and Fig. 5.4-2 shows a curve connecting these data points.

Time (sec)	Velocity (ft/sec)
0	0.0
1	3.0
2	7.9
3	6.2
4	9.4
5	6.9
6	8.5
7	3.7
8	3.0

Table 5.4-1. Wind Velocity Data

We are curve fitting these points to a second order polynomial.

```
Y = a*t**2 + b*t + c
  = [t**2    t        1.0] | a|
                           | b|
                           | c|

  = H X

     H = [t**2     t        1.0]
     X'= [ a   b  c ]
```

Now by use of Eqn. (5.4-5), we obtain a, b, and c. Note that Z in Eqn. (5.4-5) represents the velocity measurements in Table 5.4-1. For this example, we assume R = [I], a unity matrix and " t " is from the first column of Table 5.4-1. Thus, we have,

```
H = |0    0    1| : t=0
    |1    1    1| : t=1
    |4    2    1| : t=2
    |9    3    1| : t=3
    |16   4    1| : t=4
    |25   5    1| : t=5
    |36   6    1| : t=6
    |49   7    1| : t=7
    |64   8    1| : t=8
```

And,

```
                                                     (inverse)
|a| = | 1  1  1  1  1  1  1  1  1     0   0   1 | (-1)| 1  1  1  1  1   1   1   1  1 | |0.0|
|b|   | 0  1  2  3  4  5  6  7  8     1   1   1 |     | 0  1  2  3  4   5   6   7  8 | |3.0|
|c|   | 0  1  4  9 16 25 36 49 64     4   2   1 |     | 0  1  4  9  16  25  36  49 64| |7.9|
                                      9   3   1 |                                      |6.2|
                                     16   4   1 |                                      |9.4|
                                     25   5   1 |                                      |6.9|
                                     36   6   1 |                                      |8.5|
                                     49   7   1 |                                      |3.7|
                                     64   8   1 |                                      |3.0|
```

```
|a| = | 204    36     9   | (-1)| 1  1  1  1  1   1   1   1  1 | |0.0|
|b|   | 1296   204    36  |     | 0  1  2  3  4   5   6   7  8 | |3.0|
|c|   | 8772   1296   204 |     | 0  1  4  9  16  25  36  49 64| |7.9|
```

$$\left|\begin{matrix}6.2\\9.4\\6.9\\8.5\\3.7\\3.0\end{matrix}\right|$$

$$\left|\begin{matrix}a\\b\\c\end{matrix}\right| = \left|\begin{matrix}0.030 & -0.026 & 0.003\\-0.309 & 0.224 & -0.026\\0.661 & -0.309 & 0.030\end{matrix}\right| \left|\begin{matrix}1 & 1 & 1 & 1 & 1 & 1 & 1 & 1 & 1\\0 & 1 & 2 & 3 & 4 & 5 & 6 & 7 & 8\\0 & 1 & 4 & 9 & 16 & 25 & 36 & 49 & 64\end{matrix}\right| \left|\begin{matrix}0.0\\3.0\\7.9\\6.2\\9.4\\6.9\\8.5\\3.7\\3.0\end{matrix}\right|$$

$$\left|\begin{matrix}a\\b\\c\end{matrix}\right| = \left|\begin{matrix}-0.445\\3.825\\0.182\end{matrix}\right|$$

Now, the 2nd order curve fitting equation is,

$$Vel. = -0.445*t^2 + 3.825*t + 0.182$$

Fig. 5.4-2 demonstrates the curve fitting.

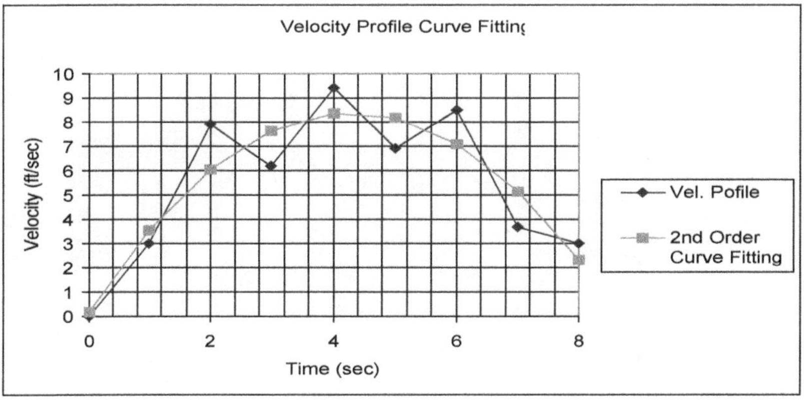

Fig. 5.4-2. Curve Fitting

The curve fitting technique is one of the tools used to analyze wind data. The statistical wind analysis enables us to predict wind characteristics. The current flight control system needs the wind prediction to ensure that the engine deflection angle required to fly through the wind (predicted) is less than the maximum vehicle engine deflection angle with some safety margin. Otherwise, the launch would have to be cancelled because the strong wind could cause not only excessive vehicle position drift but also its incapability of propulsion system to direct the nozzle angle to the direction that compensates for the wind effect. Unfortunately, the classical control theory used to design most of the current launch vehicle attitude control systems has the inherent deficiency that it does not provide a means of computing multi-variable feedback signals, implying that it does not compute proper feedbacks to deal with the attitude command, AOA, and the strong wind at the same time. On the other hand, as has been claimed, the H-infinity with Loop Shaping, and Coprime factorization can be used to design optimal feedback block that computes optimal feedback signals that reflect not only the attitude but also AOA as well as the wind effects.

Appendix I : Launch Vehicle System Parameter Table

Appendix I: <u>Launch Vehicle System Parameter (Symbol) Table</u>

(Blank Page)

Appendix I : Launch Vehicle System Parameter Table

Parameters (Symbol)					unit	
t	flight time	0.50	72.00	152.00	sec	
Rigid Body	-					
mOR	sum of rigid body and engine		5.089E+03		slugs	
la (=2.0*lalp)	dist. from LV nose to the origin	6.820E+01	6.820E+01	6.820E+01	ft	
TT (or FT)	total thrust	3.661E+05	4.189E+05	4.364E+05	lbf	
Ts (or Fs) = TT - Tc	total thrust - control thrust	5.810E+04	7.690E+04	8.040E+04	lbf	
muC	control moment coefficient	3.210E+00	4.560E+00	1.110E+01	sec^(-2)	
Lalp	aerodynamic load	0.000E+00	1.980E+05	9.000E+03	lb/rad	
m0	total mass - sloshing masses	8.225E+03	5.058E+03	1.670E+03	slugs	
mp1	mass of fuel tank 1 sloshing	1.130E+02	3.880E+02	2.390E+02	slugs	
mp2	mass of fuel tank 2 sloshing	2.000E+02	2.800E+02	1.280E+02	slugs	
mT	total mass	8.538E+03	5.726E+03	2.037E+03	slugs	mT=m0+sum of mpi
Uo	steady state axial velocity	2.050E+01	3.837E+03	7.023E+03	ft/sec	
Uodot	axial acceleration	4.050E+01	6.630E+01	2.152E+02	ft/sec^2	
lalp	dist. from CP to the origin	-	3.410E+01	1.860E+01	ft	
lc (el C)	dist. from the origin to engine hinge	3.390E+01	3.230E+01	4.680E+01	ft	
mualp	aerodynamic moment coefficient	0.000E+00	2.800E+00	1.900E-01	sec^(-2)	(Lalp*lalp/Iyy)
Theta0 (deg.)	pitch angle	9.000E+01	6.000E+01	1.000E+01	degree	assumed
Theta0 (rad)	pitch angle	1.571E+00	1.047E+00	1.745E-01	rad/sec	
lavg	dist from the origin to middle of LV		1.795E+01		ft	
lT(el T)	tail location from the origin	3.390E+01	3.230E+01	4.680E+01	ft	
Mass per unit length(η)			- 5.698E+01			from note computing C1
Iyy	pitch moment of inertia		2.420E+06		slug ft^2	
Engine Inertia	-					
mR	rocket engine mass	3.080E+01	3.080E+01	3.080E+01	slugs	
lR (el r)	dist from rocket cg to hinge point	2.520E+00	2.520E+00	2.520E+00	ft	
Tc (or FC)	control thrust	3.080E+05	3.420E+05	3.560E+05	lbf	
muCmRLR/Tc		8.089E-04	1.03E-03	2.42E-03		
IR	rocket MOI		377		slug ft^2	

Appendix I : Launch Vehicle System Parameter Table

	about hinge point					
Bending Mode 1	-	-	-	-	-	
M1	1st bending mode modal mass	2.540E+03	1.590E+03	1.200E+03	slugs	
Th1	damping ratio	7.000E-01	7.000E-01	7.000E-01		
W1	1st bending mode modal frequency	1.650E+01	1.890E+01	4.110E+01	rad/sec	
SigG1(lc,Tail Loc)	negative slope of 1st bending mode	-5.500E-01	-6.480E-02	-1.375E-01	rad/ft	
SigG1(lavg)	negative slope of 1st bending mode	3.84E-02	4.40E-03	6.23E-02	rad/ft	
mR*IR*/M1			4.882E-02			
Tc/M1			2.151E+02			
2*Th1*W1			2.646E+01			
W1^2			3.572E+02			
Bending Mode 2	-	-	-	-	-	
M2	2nd bending mode modal mass	9.550E+03	3.420E+05	3.560E+05	slugs	
Th2	damping ratio	7.000E-01	7.000E-01	7.000E-01		
W2	2nd bending mode modal frequency	4.160E+01	6.070E+01	7.290E+01	rad/sec	
SigG2(lc,Tail Loc)	negative slope of 1st bending mode	-1.218E-01	-1.710E-01	-1.218E-01	rad/ft	
SigG2(lavg)	negative slope of 1st bending mode	0.00192	-0.0276	0.0309	rad/ft	
mR*IR*/M2			8.127E-03			
Tc/M2			3.581E+01			
2*Th2*W2			8.498E+01			
W2^2			3.684E+03			
Sloshing 1	-					
Lp1	1st pendulum length		2.730E+00		ft	
lp1	1st pendulum hinge location		5.560E+00		ft	
a1	function of many variables		-2.569E+01			
Wp1^2	1st square of pendulum Frequency		2.621E+01		(rad/sec)^2	
mp1	1st pendulum mass		3.880E+02		slugs	
muP1	1st pendulum velocity		5.910E-02		sec^(-2)	
Sloshing 2	-					
Lp2	2nd pendulum length		2.750E+00		ft	
lp2	2nd pendulum hinge location		-2.062E+01		ft	
a2	function of many variables					

Appendix I : Launch Vehicle System Parameter Table

Wp2^2	2nd square of pendulum Frequency		2.679E+01		(rad/sec)^2	
mp2	2nd pendulum mass		2.800E+02		slugs	
muP2	2nd pendulum velocity		-1.582E-01		sec^(-2)	
Rate Gyro	-					
Kr	rate gyro gain	4.000E-01			sec^(-1)	
WR	natural frequency	1.456E+02			rad/sec	
Theta R	damping ratio	4.000E-01				
Sigma 1(IG)	see Eqn. (3.7-8)	4.400E-03			rad/ft	
Sigma 2(IG)	see Eqn. (3.7-8)	-2.760E-02			rad/ft	
Position Gyro	-					
Tau	time constant	4.000E-02			sec	
Sigma 1(IG)	see Eqn. (3.7-7)	4.400E-03			rad/ft	
Sigma 2(IG)	see Eqn. (3.7-7)	-2.760E-02			rad/ft	

(Ref. 18)

(Blank Page)

Appendix II : LV and Feedback Dynamic System Matrices Computation

(Blank Page)

Appendix II: LV and Feedback Dynamic System Matrices Computation

1) Computation of Launch Vehicle Attitude Dynamic System Matrices

The CD attached at the back of this book computes Ap and Bp matrices in the following equations. (See Chapter 3 for details.)

$$\underline{\mathbf{xd}} = Ap^*\underline{\mathbf{x}} + Bp^*\underline{\mathbf{u}}$$

$$\underline{\mathbf{y}} = Cp^*\underline{\mathbf{x}} + Dp^*\underline{\mathbf{u}}$$

where

xd = **dx/dt**
x = [**θ θd q1 q1d q2 q2d Γ1 Γ1d Γ2 Γ2d θPG θRG (θRG)d**]

θ	: attitude angle in pitch (rad)
θd	: attitude angular rate in pitch (rad/sec)
q1	: generalized coordinate of the 1st bending mode in pitch (ft)
q1d	: rate of generalized coordinate of the 1st bending mode in pitch (ft/sec)
q2	: generalized coordinate of the 2nd bending mode in pitch (ft)
q2d	: rate of generalized coordinate of the 2nd bending mode in pitch (ft/sec)
Γ1	: pendulum angle of fuel tank 1 sloshing (rad)
Γ1d	: pendulum angular rate of fuel tank 1 sloshing (rad/sec)
Γ2	: pendulum angle of fuel tank 2 sloshing (rad)
Γ2d	: pendulum angular rate of fuel tank 2 sloshing (rad/sec)
θPG	: position gyro output (rad)
θRG	: rate gyro output (rad/sec)
θRG)d	: time derivative rate gyro output (rad/sec^(2))

u = [**δ** ; **α** ; **Ww**], (transpose of [**δ α Ww**])

δ	: engine deflection angle (rad)

α : angle of attack (rad)
Ww : wind velocity (ft/sec)

$\underline{y} = [\theta \;\; \theta d\,]$

where

$$\theta d = d\theta/dt$$

The CD provides a table just like the one shown below. It is the input table and has 49 entries to be filled with data from LV dynamics. From the data entered in the table, an Excel program coded in the CD computes Ap and Bp matrices.

INPUT

	Parameters			unit
1	Time	flight time		sec
2	Gravity			ft/sec^2
	Rigid Body			
3	la	dist. from LV nose to the origin		ft
4	TT (or FT)	total thrust		lbf
	Ts (or Fs) = TT - Tc	total thrust - control thrust		lbf
5	muC	control moment coefficient		sec^(-2)
6	Lalp	aerodynamic load		lb/rad
	m0	total mass - sloshing masses		slugs
7	mp1	mass of fuel tank 1 sloshing		slugs
8	mp2	mass of fuel tank 2 sloshing		slugs
9	mT	total mass		slugs
10	Uo	steady state axial velocity		ft/sec
11	Uodot	axial acceleration		ft/sec^2
12	lalp	dist. from CP to the origin		ft
13	lc (el C)	dist. from the origin to engine hinge		ft
14	mualp	aerodynamic moment coefficient		sec^(-2)
15	Theta0 (deg.)	pitch angle		deg
	Theta0 (rad)	pitch angle		rad/sec
16	lavg	dist from the origin to middle of LV		ft
17	lT(el T)	tail location from the origin		ft

	Mass per unit length(η)			
18	Iyy	pitch moment of inertia		slug ft^2
	Engine Inertia			
19	mR	rocket engine mass		slugs
20	lR (el r)	dist from rocket cg to hinge point		ft
21	Tc (or FC)	control thrust		lbf
	muCmRLR/Tc			
22	IR	rocket MOI about hinge point		slug ft^2
	Bending Mode 1			
23	M1	1st bending mode modal mass		slugs
24	Th1	damping ratio		
25	W1	1st bending mode modal frequency		rad/sec
26	SigG1(lc,Tail Loc)	negative slope of 1st bending mode		rad/ft
27	SigG1(lavg)	negative slope of 1st bending mode		rad/ft
	mR*lR/M1			
	Tc/M1			
	2*Th1*W1			
	W1^2			
	Bending Mode 2			
28	M2	2nd bending mode modal mass		slugs
29	Th2	damping ratio		
30	W2	2nd bending mode modal frequency		rad/sec
31	SigG2(lc,Tail Loc)	negative slope of 1st bending mode		rad/ft
32	SigG2(lavg)	negative slope of 1st bending mode		rad/ft
	mR*lR/M2			
	Tc/M2			
	2*Th2*W2			
	W2^2			
	Sloshing 1			
33	Lp1	1st pendulum length		ft
34	lp1	1st pendulum hinge location		ft
	a1	function of many variables		
35	Wp1^2	1st square of pend. Frequency		(rad/sec)^2
	mp1	1st pendulum mass		slugs
36	muP1	1st pendulum velocity		sec^(-2)
	Sloshing 2			
37	Lp2	2nd pendulum length		ft

38	lp2	2nd pendulum hinge location		ft
	a2	function of many variables		
39	Wp2^2	2nd square of pend. Frequency		(rad/sec)^2
	mp2	2nd pendulum mass		slugs
40	muP2	2nd pendulum velocity		sec^(-2)
	Rate Gyro			
41	Kr	rate gyro gain		sec^(-1)
42	WR(rad/sec)	natural frequency		rad/sec
43	TheR	damping ratio		
44	Sigma 1(IG)	see Eqn. (3.7-8)		rad/ft
45	Sigma 2(IG)	see Eqn. (3.7-8)		rad/ft
	Position Gyro			
46	Tau	time constant		sec
47	Sigma 1(IG)	see Eqn. (3.7-7)		rad/ft
48	Sigma 2(IG)	see Eqn. (3.7-7)		rad/ft

OUTPUT

Ap Matrix

	1	2	3	4	5	6	7
1							
2							
3							
4							
5							
6							
7							
8							
9							
10							
11							
12							
13							

	8	9	10	11	12	13
1						
2						

3						
4						
5						
6						
7						
8						
9						
10						
11						
12						
13						

Bp Matrix

	1	2	3
1			
2			
3			
4			
5			
6			
7			
8			
9			
10			
11			
12			
13			

Cp Matrix

	1	2	3	4	5	6	7
1							
2							

	8	9	10	11	12	13
1						
2						

Dp Matrix

	1	2	3
1			
2			

Example

INPUT

	Parameters			unit
1	t	flight time	72	sec
2	Gravity		32.174	ft/sec^2
	Rigid Body			
3	la	dist. from LV nose to the origin	68.2	ft
4	TT (or FT)	total thrust	418900	lbf
	Ts (or Fs) = TT - Tc	total thrust - control thrust	8500	lbf
5	muC	control moment coefficient	4.56	sec^(-2)
6	Lalp	aerodynamic load	198000	lb/rad
	m0	total mass - sloshing masses	5058	slugs
7	mp1	mass of fuel tank 1 sloshing	388	slugs
8	mp2	mass of fuel tank 2 sloshing	280	slugs
9	mT	total mass	5726	slugs
10	Uo	steady state axial velocity	3836.5	ft/sec
11	Uodot	axial acceleration	66.3	ft/sec^2
12	lalp	dist. from CP to the origin	34.1	ft

13	lc (el C)	dist. from the origin to engine hinge	-32.3	ft
14	mualp	aerodynamic moment coefficient	2.8	sec^(-2)
15	Theta0 (deg.)	pitch angle	60	deg
	Theta0 (rad)	pitch angle	1.05	rad/sec
16	lavg	dist from the origin to middle of LV	17.95	ft
17	lT(el T)	tail location from the origin	-32.3	ft
	Mass per unit length(η)		56.97512438	
18	Iyy	pitch moment of inertia	1,694,000	slug ft^2
	Engine Inertia			
19	mR	rocket engine mass	30.8	slugs
20	lR (el r)	dist from rocket cg to hinge point	2.52	ft
21	Tc (or FC)	control thrust	410400	lbf
	muCmRLR/Tc		0.0008624	
22	IR	rocket MOI about hinge point	377	slug ft^2
	Bending Mode 1			
23	M1	1st bending mode modal mass	1590	slugs
24	Th1	damping ratio	0.7	
25	W1	1st bending mode modal frequency	18.9	rad/sec
26	SigG1(lc,Tail Loc)	negative slope of 1st bending mode	-0.0648	rad/ft
27	SigG1(lavg)	negative slope of 1st bending mode	0.0044	rad/ft
	mR*lR/M1		0.048815094	
	Tc/M1		258.1132075	
	2*Th1*W1		26.46	
	W1^2		357.21	
	Bending Mode 2			
28	M2	2nd bending mode modal mass	9550	slugs

29	Th2	damping ratio	0.7	
30	W2	2nd bending mode modal frequency	60.7	rad/sec
31	SigG2(lc,Tail Loc)	negative slope of 1st bending mode	-0.171	rad/ft
32	SigG2(lavg)	negative slope of 1st bending mode	-0.0276	rad/ft
	mR*IR/M2		0.00812733	
	Tc/M2		42.97382199	
	2*Th2*W2		84.98	
	W2^2		3684.49	
	Sloshing 1			
33	Lp1	1st pendulum length	2	ft
34	lp1	1st pendulum hinge location	5.56	ft
	a1	function of many variables	-32.452595	
35	Wp1^2	1st square of pend. Frequency	26.2099402	(rad/sec)^2
	mp1	1st pendulum mass	388	slugs
36	muP1	1st pendulum velocity	0.0591	sec^(-2)
	Sloshing 2			
37	Lp2	2nd pendulum length	2.75	ft
38	lp2	2nd pendulum hinge location	-20.62	ft
	a2	function of many variables	-68.256724	
39	Wp2^2	2nd square of pend. Frequency	26.7881307	(rad/sec)^2
	mp2	2nd pendulum mass	280	slugs
40	muP2	2nd pendulum velocity	-0.1582	sec^(-2)
	Rate Gyro			
41	Kr	rate gyro gain	0.4	sec^(-1)
42	WR(rad/sec)	natural frequency	145.6	rad/sec
43	TheR	damping ratio	0.4	

44	Sigma 1(IG)	see Eqn. (3.7-8)	0.0044	rad/ft
45	Sigma 2(IG)	see Eqn. (3.7-8)	-0.0276	rad/ft
	Position Gyro			
46	Tau	time constant	0.04	sec
47	Sigma 1(IG)	see Eqn. (3.7-7)	0.0044	rad/ft
48	Sigma 2(IG)	see Eqn. (3.7-7)	-0.0276	rad/ft

OUTPUT

Ap Matrix

	1	2	3	4	5	6	7
1	0.00E+00	1.00E+00	0.00E+00	0.00E+00	0.00E+00	0.00E+00	0.00E+00
2	3.87E+00	2.02E+00	-2.44E-01	-1.58E-04	3.90E-06	-7.15E-01	-4.08E-04
3	0.00E+00	0.00E+00	0.00E+00	1.00E+00	0.00E+00	0.00E+00	0.00E+00
4	2.27E+02	7.52E+01	-3.66E+02	-2.58E+01	-4.68E+01	2.85E-02	-3.90E-01
5	0.00E+00	0.00E+00	0.00E+00	0.00E+00	0.00E+00	1.00E+00	0.00E+00
6	9.92E+01	3.29E+01	-3.59E+03	-8.26E+01	-2.05E+01	1.53E-02	-1.71E-01
7	0.00E+00	0.00E+00	0.00E+00	0.00E+00	0.00E+00	0.00E+00	0.00E+00
8	0.00E+00	0.00E+00	0.00E+00	0.00E+00	0.00E+00	0.00E+00	-3.55E+01
9	0.00E+00	0.00E+00	0.00E+00	0.00E+00	0.00E+00	0.00E+00	0.00E+00
10	0.00E+00	0.00E+00	0.00E+00	0.00E+00	0.00E+00	0.00E+00	-1.13E+00
11	2.50E+01	0.00E+00	1.10E-01	0.00E+00	-6.90E-01	0.00E+00	0.00E+00
12	0.00E+00	0.00E+00	0.00E+00	0.00E+00	0.00E+00	0.00E+00	0.00E+00
13	0.00E+00	8.48E+03	0.00E+00	3.73E+01	0.00E+00	-2.34E+02	0.00E+00

	8	9	10	11	12	13
1	0.00E+00	0.00E+00	0.00E+00	0.00E+00	0.00E+00	0.00E+00
2	-8.04E-02	2.24E-01	0.00E+00	0.00E+00	0.00E+00	0.00E+00
3	0.00E+00	0.00E+00	0.00E+00	0.00E+00	0.00E+00	0.00E+00
4	0.00E+00	-4.79E+00	0.00E+00	0.00E+00	0.00E+00	0.00E+00
5	0.00E+00	0.00E+00	0.00E+00	0.00E+00	0.00E+00	0.00E+00
6	0.00E+00	2.16E+00	0.00E+00	0.00E+00	0.00E+00	0.00E+00
7	1.00E+00	0.00E+00	0.00E+00	0.00E+00	0.00E+00	0.00E+00
8	0.00E+00	-1.34E+00	0.00E+00	0.00E+00	0.00E+00	0.00E+00
9	0.00E+00	0.00E+00	1.00E+00	0.00E+00	0.00E+00	0.00E+00

10	0.00E+00	-2.66E+01	0.00E+00	0.00E+00	0.00E+00	0.00E+00
11	0.00E+00	0.00E+00	0.00E+00	-2.50E+01	0.00E+00	0.00E+00
12	0.00E+00	0.00E+00	0.00E+00	0.00E+00	0.00E+00	1.00E+00
13	0.00E+00	0.00E+00	0.00E+00	0.00E+00	-2.12E+04	-1.16E+02

Bp Matrix

	1	2	3
1	0.00E+00	0.00E+00	0.00E+00
2	-7.65E+00	-9.29E+00	2.41E-03
3	0.00E+00	0.00E+00	0.00E+00
4	-2.62E+02	3.85E+01	1.02E-02
5	0.00E+00	0.00E+00	0.00E+00
6	-1.15E+02	1.69E+01	8.69E-05
7	0.00E+00	0.00E+00	0.00E+00
8	-2.77E+01	0.00E+00	0.00E+00
9	0.00E+00	0.00E+00	0.00E+00
10	-6.48E+01	0.00E+00	0.00E+00
11	0.00E+00	0.00E+00	0.00E+00
12	0.00E+00	0.00E+00	0.00E+00
13	0.00E+00	0.00E+00	0.00E+00

Cp Matrix

	1	2	3	4	5	6	7
1	1.00E+00	0.00E+00	0.00E+00	0.00E+00	0.00E+00	0.00E+00	0.00E+00
2	0.00E+00	1.00E+00	0.00E+00	0.00E+00	0.00E+00	0.00E+00	0.00E+00

	8	9	10	11	12	13
1	0.00E+00	0.00E+00	0.00E+00	0.00E+00	0.00E+00	0.00E+00
2	0.00E+00	0.00E+00	0.00E+00	0.00E+00	0.00E+00	0.00E+00

Dp MARIX

	1	2	3
1	0.00E+00	0.00E+00	0.00E+00

2	0.00E+00	0.00E+00	0.00E+00

2) Computation of Control System Feedback Dynamic System Matrices

a. Copy Ap, Bp, Cp, and Dp computed in section 1) and paste them in Matlab Command Window as shown below.

```
Ap =[                                   ];
Bp =[                                   ];
Cp =[                                   ];
Dp =[                                   ];
```

b. Execute c_ncf_lv.m (in Matlab Command Window as show below)

c_ncf_lv.m code is listed below.

```
% c_ncf_lv.m code:

Function[af, bf, cf, df]=c_ncf_lv(Ap, Bp, Cp, Dp)
%
sys =pck(Ap, Bp, Cp, Dp);
% Weighting Function 1
w1a = nd2sys ([              ],[              ]); % include actuator
dynamics
w1b = nd2sys ([              ],[              ]);
w1c = nd2sys ([              ],[              ]);
% Weighting Function 2
w2a = nd2sys ([ 1.0 ],[ 1.0 ]);
w2b = nd2sys ([ 1.0 ],[ 1.0 ]);
%
w1 = daug(w1a, w1b, w1c);
w2 = daug(w2a, w2b);  %Note that w2=I (identity matrix) for launch
vehicle
%
sysgw=mmult(sys, w1);
```

```
sysgw=mmult(w2,sysgw);
[Kfeedback, emax]=ncfsyn(sysgw,1);
[af, bf, cf, df]=unpack(Kfeedback);
%
% af, bf, cf, and df are control system feedback dynamic system matrices.
%
```

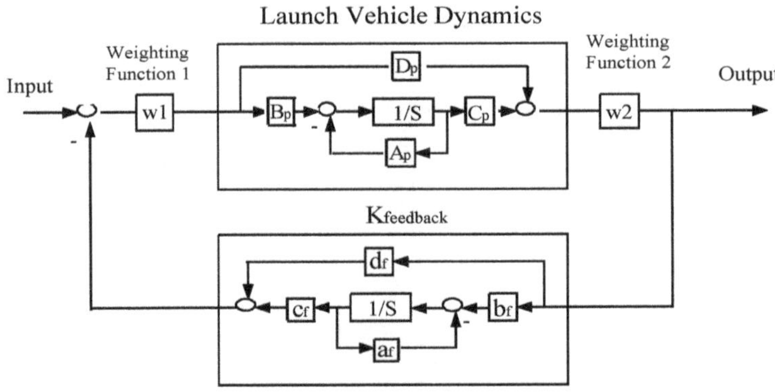

Fig. Appendix II-1 Launch Vehicle Dynamics Block Diagram

Fig. Appendix II-1 shows launch vehicle attitude control system block diagram.
W1 set is defined by designer, and w2 is an identity matrix. af, bf,cf, and df are computed by using ncfsyn.m.

Appendix III : Math Modeling Error Sensitivity Response and Singular Value Plots, and Eigenvalue Tables

Note: In the following pages, numerous plots of singular value vs. frequency are presented. Author observed that some of these plots look very similar, but not identical. The reason for the apparent similarity is due to the log scale in singular value axis. The step responses corresponding to each singular value vs. frequency plot and corresponding eigenvalues, which are also presented, are different. It is recommended that the singular value plots near the frequency of interest should be either expanded or tabulated to observe the differences.

A case is shown below that two singular value vs. frequency plots look identical but they are actually different. The difference is shown in Fig. III-3. The first two plots are from page 434.

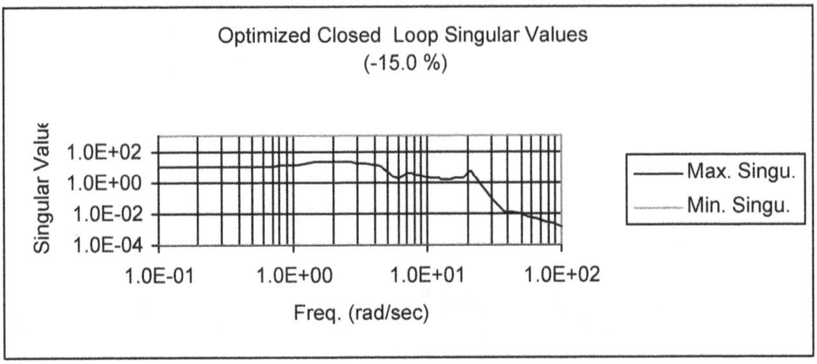

Fig. III-1 Optimized Closed Loop Singular Value
When The Pitch MOI Modeling Error is -15%

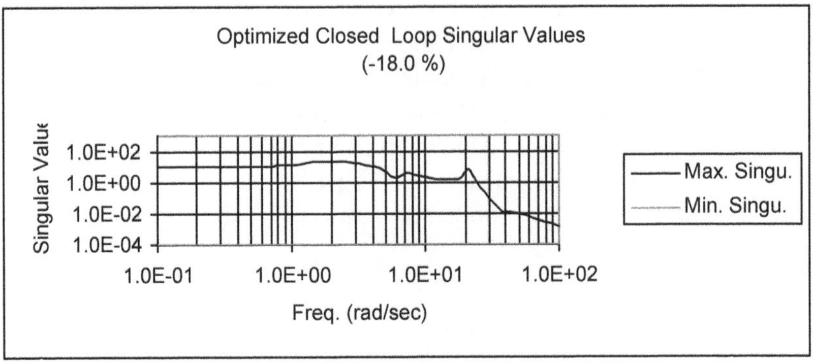

Fig. III-2 Optimized Closed Loop Singular Value
When The Pitch MOI Modeling Error is -18%

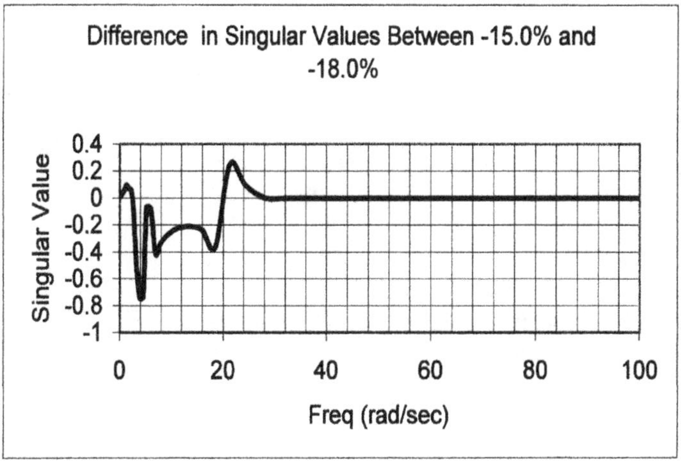

Fig. III-3 Difference in Singular Value between -15.0% Modeling Error and -18.0 % Modeling Error

It is seen here that the singular value plots in Fig. III-1 and Fig. III-2 look identical. But they are not identical as verified in Fig. III-3 that the difference between these plots are not zero..

Appendix III: Collection of Math Modeling Error Sensitivity Response and Singular Value Plots, and Eigenvalue Tables

Appendix III: Collection of Math Modeling Error Sensitivity Response and Singular Value Plots, and Eigenvalue Tables

1) **Pitch MOI : Iyy (math model MOI in y axis = 1,694,000 slugs ft^2)**

Case 1 (actual value is less than the value in the math model)

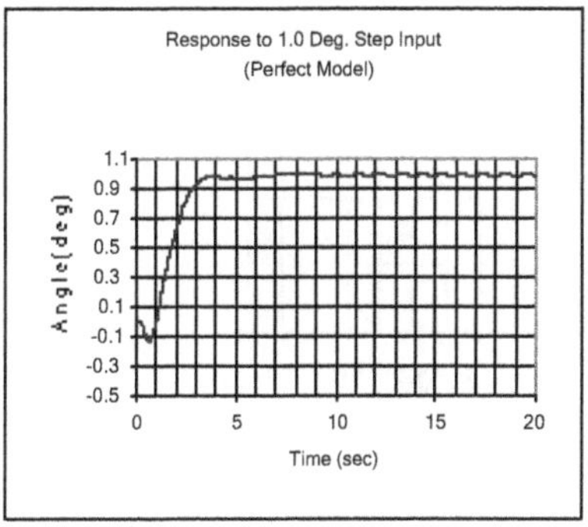

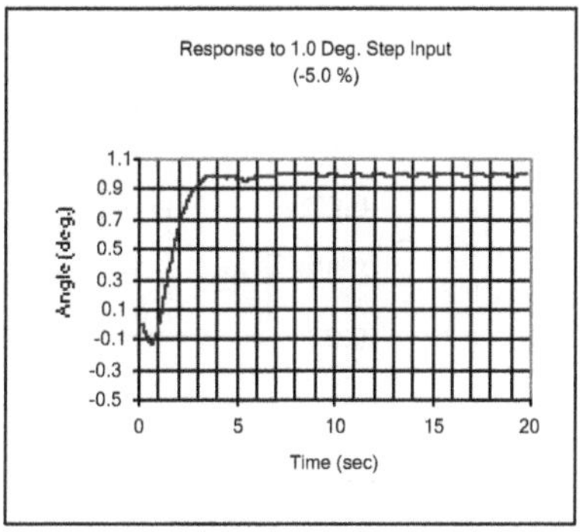

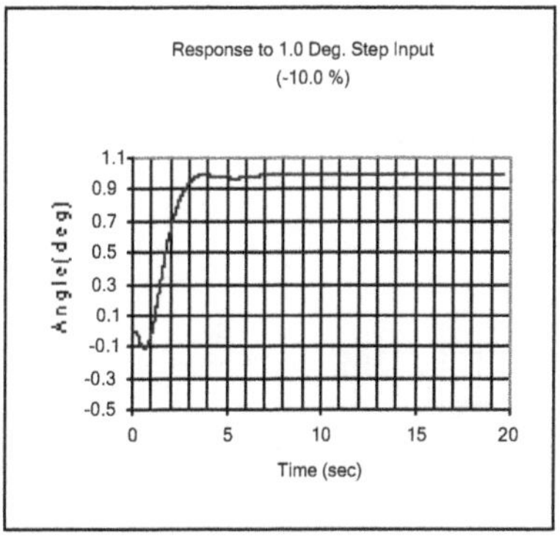

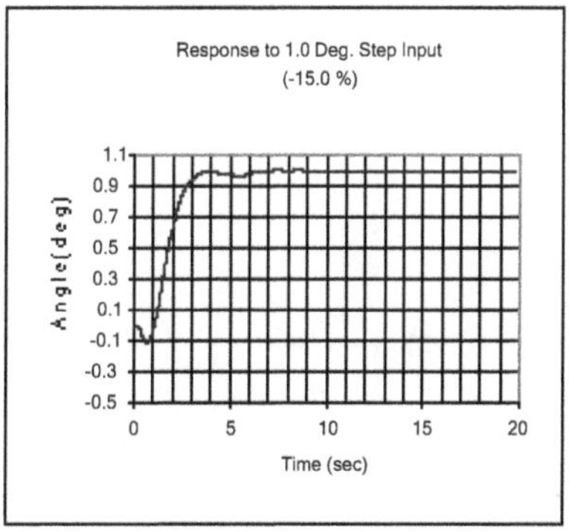

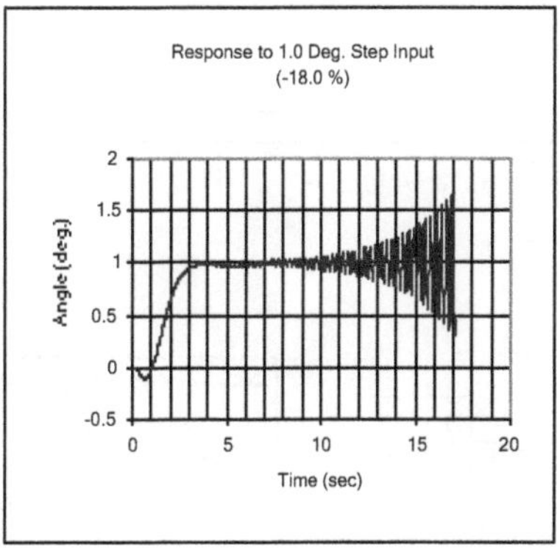

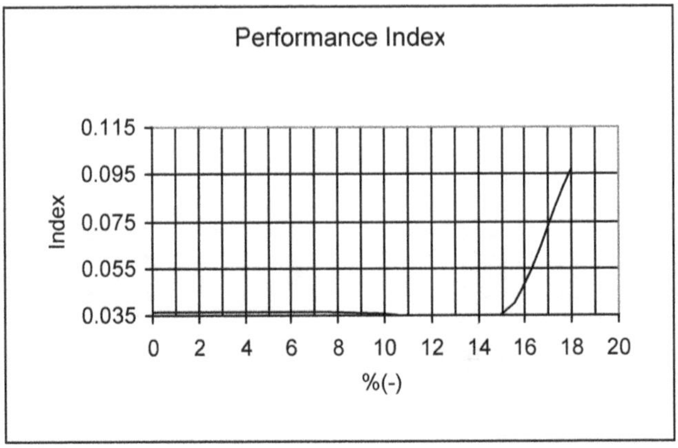
Performance Index
Index
0.115
0.095
0.075
0.055
0.035
0 2 4 6 8 10 12 14 16 18 20
%(-)

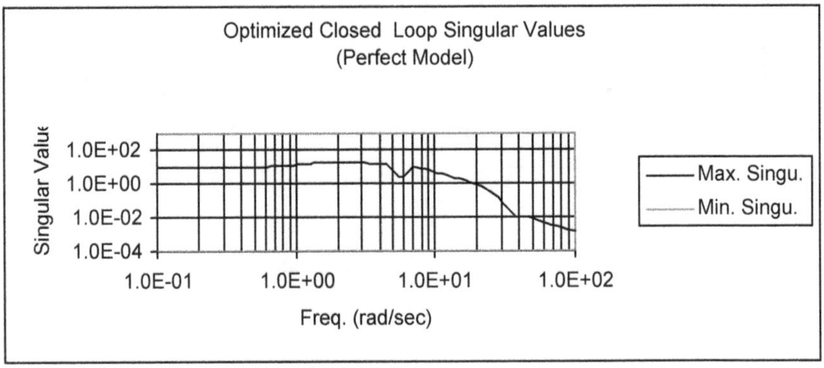
Optimized Closed Loop Singular Values
(Perfect Model)
Singular Valu
1.0E+02
1.0E+00
1.0E-02
1.0E-04
1.0E-01 1.0E+00 1.0E+01 1.0E+02
Freq. (rad/sec)
Max. Singu.
Min. Singu.

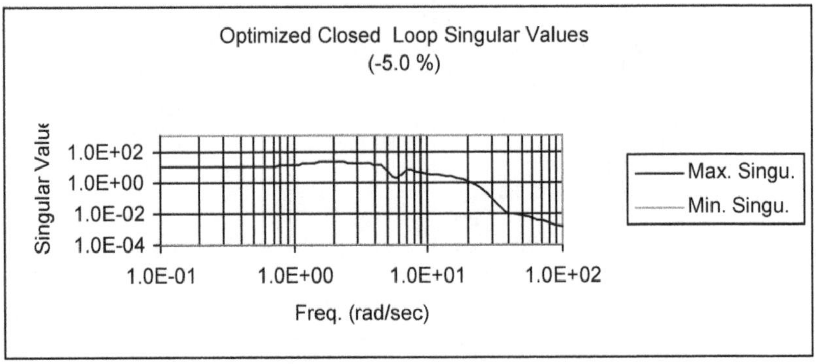
Optimized Closed Loop Singular Values
(-5.0 %)
Singular Valu
1.0E+02
1.0E+00
1.0E-02
1.0E-04
1.0E-01 1.0E+00 1.0E+01 1.0E+02
Freq. (rad/sec)
Max. Singu.
Min. Singu.

Appendix III: Collection of Math Modeling Error Sensitivity Response and Singular Value Plots, and Eigenvalue Tables

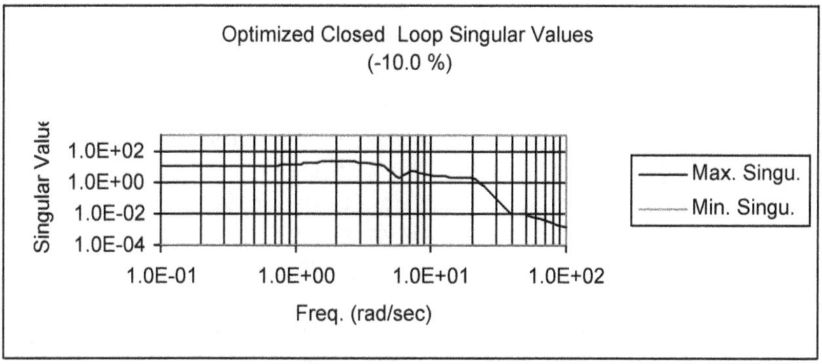

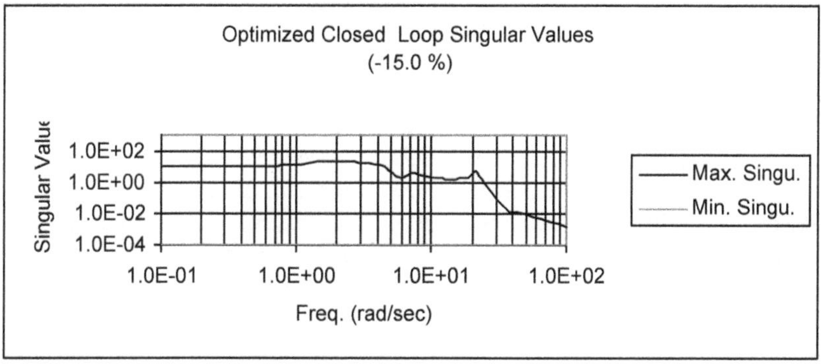

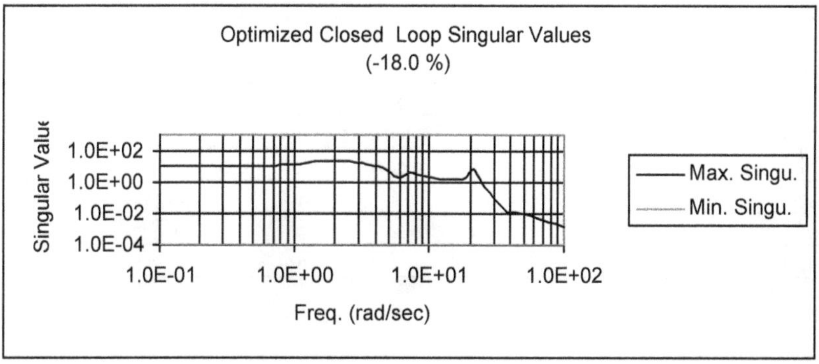

Closed Loop System Eigenvalues					
	0%	-5%	-10.00%	-18.00%	
	1.0e+002 *	1.0e+002 *	1.0e+002 *	1.0e+002 *	
1	-0.25	-0.25	-0.25	-0.25	
2	-0.5824 + 1.3344i	-0.5824 + 1.3344i	-0.5824 + 1.3344i	-0.5824 + 1.3344i	

3	-0.5824 - 1.3344i	-0.5824 - 1.3344i	-0.5824 - 1.3344i	-0.5824 - 1.3344i	
4	-0.1040 + 0.2546i	-0.0830 + 0.2893i	-0.0749 + 0.3043i	-0.0718 + 0.3103i	
5	-0.1040 - 0.2546i	-0.0830 - 0.2893i	-0.0749 - 0.3043i	-0.0718 - 0.3103i	
6	-0.1046 + 0.2542i	-0.2450 + 0.1472i	-0.2749 + 0.1723i	-0.2866 + 0.1805i	
7	-0.1046 - 0.2542i	-0.2450 - 0.1472i	-0.2749 - 0.1723i	-0.2866 - 0.1805i	
8	-0.2284 + 0.0212i	-0.2912	-0.2998	-0.3024	
9	-0.2284 - 0.0212i	-0.1118 + 0.2266i	-0.1101 + 0.2215i	-0.1095 + 0.2202i	
10	-0.2293	-0.1118 - 0.2266i	-0.1101 - 0.2215i	-0.1095 - 0.2202i	
11	-0.1353 + 0.0476i	-0.1907	-0.192	-0.1922	
12	-0.1353 - 0.0476i	-0.17	-0.1683	-0.168	
13	-0.1446 + 0.0002i	-0.1446 + 0.0001i	-0.1446 + 0.0000i	-0.1446 + 0.0000i	
14	-0.1446 - 0.0002i	-0.1446 - 0.0001i	-0.1446 - 0.0000i	-0.1446 - 0.0000i	
15	-0.0868 + 0.0557i	-0.0212 + 0.0761i	-0.0045 + 0.0761i	0.0011 + 0.0755i	
16	-0.0868 - 0.0557i	-0.0212 - 0.0761i	-0.0045 - 0.0761i	0.0011 - 0.0755i	
17	-0.0002 + 0.0601i	-0.0006 + 0.0599i	-0.0007 + 0.0596i	-0.0000 + 0.0600i	
18	-0.0002 - 0.0601i	-0.0006 - 0.0599i	-0.0007 - 0.0596i	-0.0000 - 0.0600i	
19	-0.0002 + 0.0600i	-0.0000 + 0.0600i	-0.0001 + 0.0600i	-0.0006 + 0.0594i	
20	-0.0002 - 0.0600i	-0.0000 - 0.0600i	-0.0001 - 0.0600i	-0.0006 - 0.0594i	
21	-0.0009 + 0.0508i	-0.0011 + 0.0508i	-0.0011 + 0.0507i	-0.0011 + 0.0507i	
22	-0.0009 - 0.0508i	-0.0011 - 0.0508i	-0.0011 - 0.0507i	-0.0011 - 0.0507i	
23	-0.0009 + 0.0509i	-0.0007 + 0.0506i	-0.0005 + 0.0503i	-0.0003 + 0.0503i	
24	-0.0009 - 0.0509i	-0.0007 - 0.0506i	-0.0005 - 0.0503i	-0.0003 - 0.0503i	
25	-0.0347 + 0.0149i	-0.0239 + 0.0112i	-0.0264	-0.0269	
26	-0.0347 - 0.0149i	-0.0239 - 0.0112i	-0.0203 + 0.0112i	-0.0193 + 0.0113i	
27	-0.0204	-0.0241	-0.0203 - 0.0112i	-0.0193 - 0.0113i	
28	-0.0052 + 0.0136i	-0.0053 + 0.0136i	-0.0053 + 0.0136i	-0.0053 + 0.0136i	
29	-0.0052 - 0.0136i	-0.0053 - 0.0136i	-0.0053 - 0.0136i	-0.0053 - 0.0136i	
30	-0.0078 + 0.0145i	-0.0078 + 0.0145i	-0.0078 + 0.0145i	-0.0078 + 0.0145i	
31	-0.0078 - 0.0145i	-0.0078 - 0.0145i	-0.0078 - 0.0145i	-0.0078 - 0.0145i	
32	-0.0134	-0.0141	-0.0144	-0.0146	
33	-0.0106	-0.01	-0.0096	-0.0094	
34	-0.01	-0.01	-0.01	-0.01	

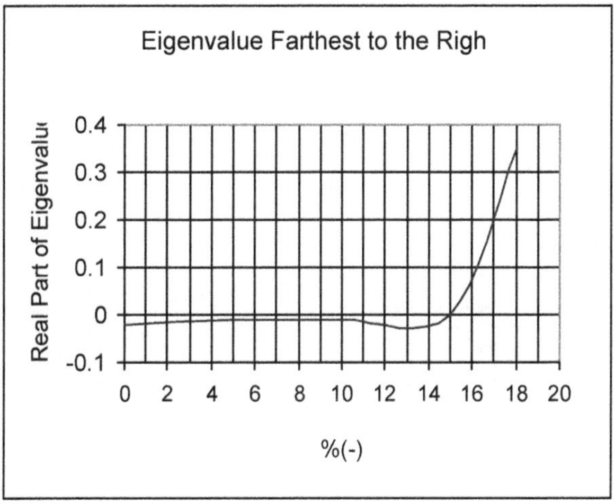
Eigenvalue Farthest to the Righ
Real Part of Eigenvalu
0.4
0.3
0.2
0.1
0
-0.1
0 2 4 6 8 10 12 14 16 18 20
%(-)

Case 2 (actual value is greater than the math model value)

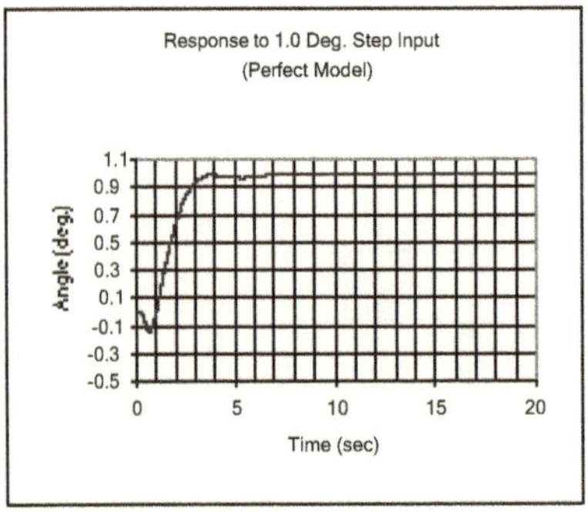

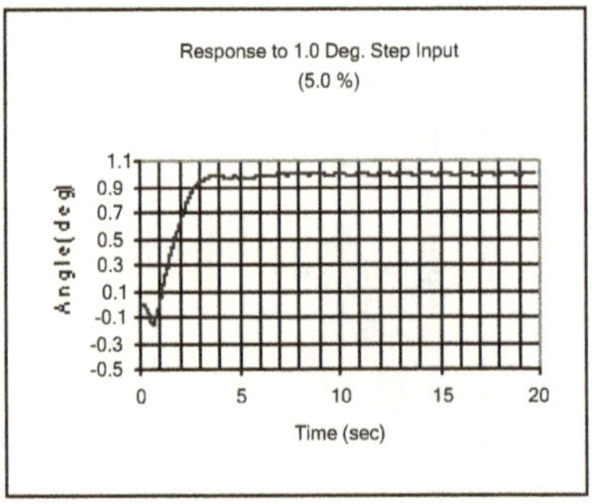

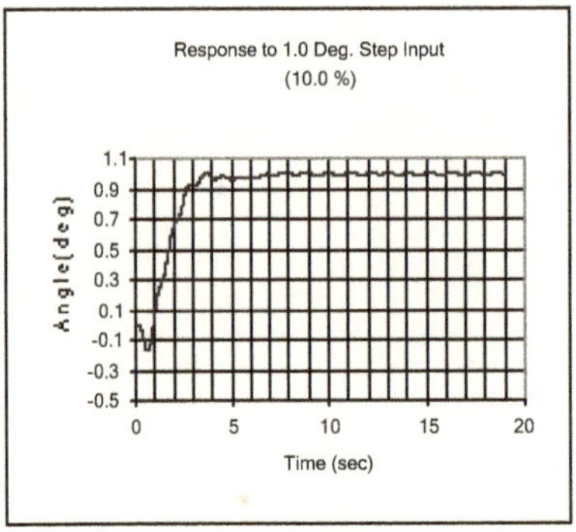

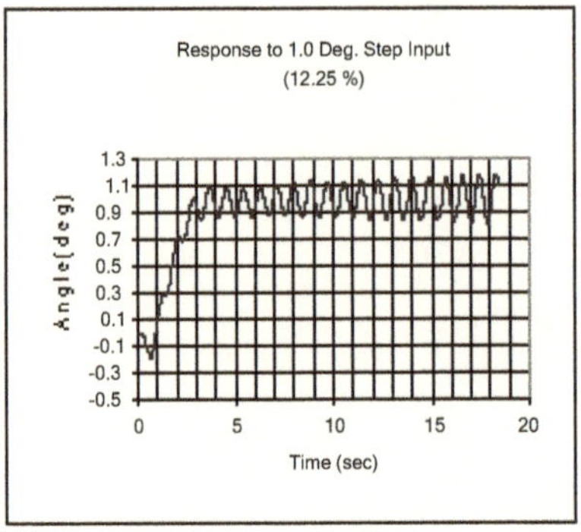

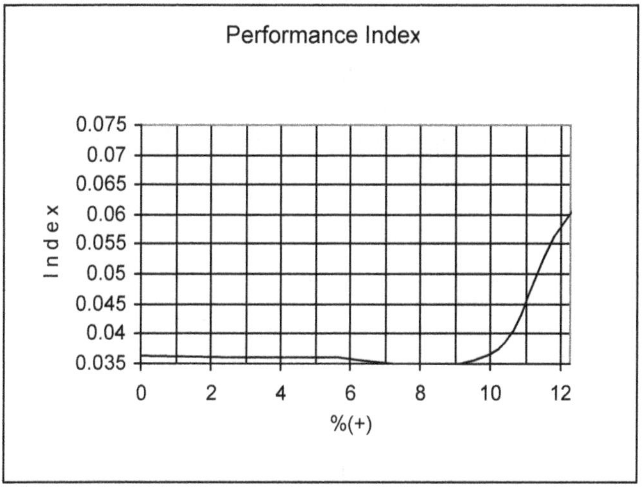
Performance Index
Index
0.075
0.07
0.065
0.06
0.055
0.05
0.045
0.04
0.035
0
2
4
6
8
10
12
%(+)

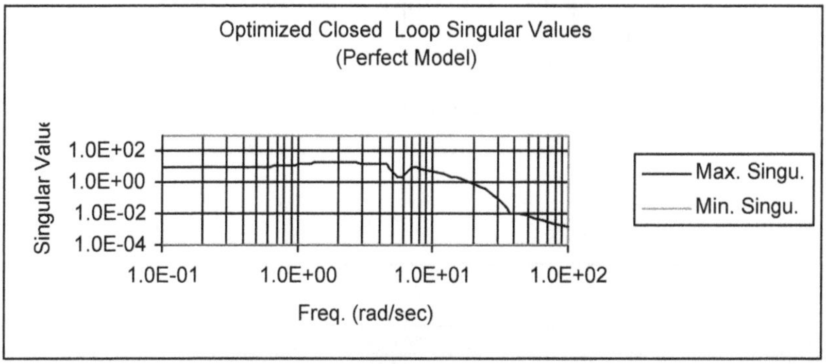
Optimized Closed Loop Singular Values
(Perfect Model)
Singular Valu
1.0E+02
1.0E+00
1.0E-02
1.0E-04
1.0E-01
1.0E+00
1.0E+01
1.0E+02
Freq. (rad/sec)
Max. Singu.
Min. Singu.

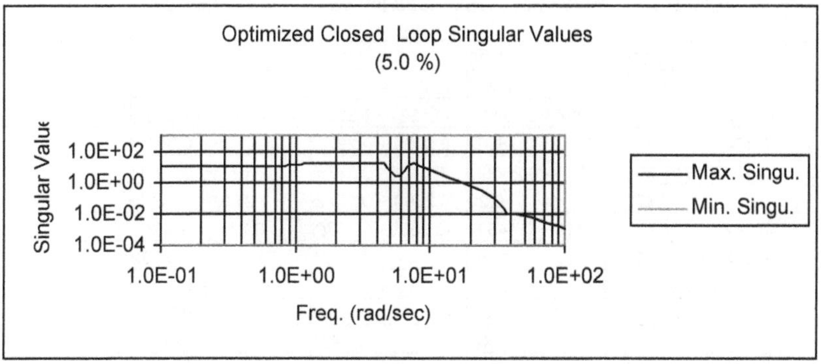
Optimized Closed Loop Singular Values
(5.0 %)
Singular Valu
1.0E+02
1.0E+00
1.0E-02
1.0E-04
1.0E-01
1.0E+00
1.0E+01
1.0E+02
Freq. (rad/sec)
Max. Singu.
Min. Singu.

Appendix III: Collection of Math Modeling Error Sensitivity Response and Singular Value Plots, and Eigenvalue Tables

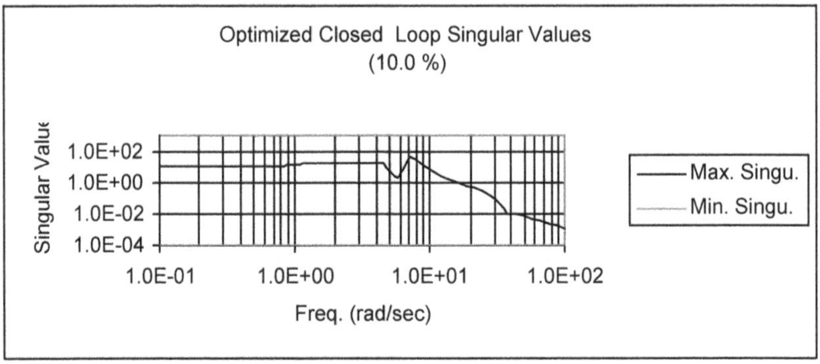

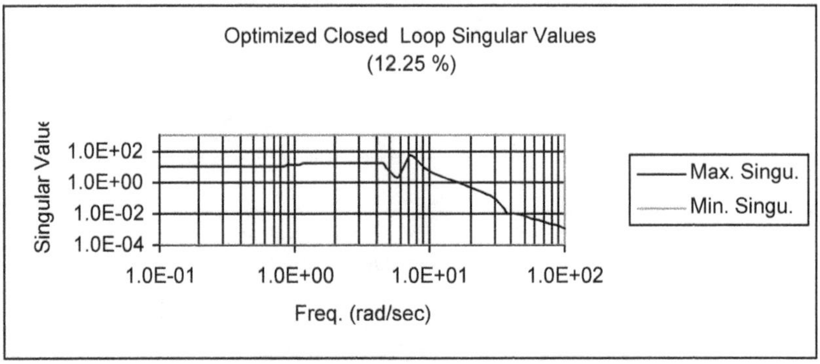

Closed Loop System Eigenvalues

	0%	5.00%	10.00%	12.25%	
	1.0e+002 *	1.0e+002 *	1.0e+002 *	1.0e+002 *	
1	-0.25	-0.25	-0.25	-0.25	
2	-0.5824 + 1.3344i	-0.5824 + 1.3344i	-0.5824 + 1.3344i	-0.5824 + 1.3344i	
3	-0.5824 - 1.3344i	-0.5824 - 1.3344i	-0.5824 - 1.3344i	-0.5824 - 1.3344i	
4	-0.1040 + 0.2546i	-0.0833 + 0.2866i	-0.0760 + 0.2976i	-0.0735 + 0.3014i	
5	-0.1040 - 0.2546i	-0.0833 - 0.2866i	-0.0760 - 0.2976i	-0.0735 - 0.3014i	
6	-0.1046 + 0.2542i	-0.29	-0.2660 + 0.1672i	-0.2740 + 0.1735i	
7	-0.1046 - 0.2542i	-0.2416 + 0.1454i	-0.2660 - 0.1672i	-0.2740 - 0.1735i	
8	-0.2284 + 0.0212i	-0.2416 - 0.1454i	-0.2971	-0.2992	
9	-0.2284 - 0.0212i	-0.1118 + 0.2263i	-0.1102 + 0.2213i	-0.1097 + 0.2201i	
10	-0.2293	-0.1118 - 0.2263i	-0.1102 - 0.2213i	-0.1097 - 0.2201i	
11	-0.1353 + 0.0476i	-0.1909	-0.1921	-0.1923	
12	-0.1353 - 0.0476i	-0.1699	-0.1682	-0.1679	
13	-0.1446 + 0.0002i	-0.1447 + 0.0000i	-0.1446 + 0.0000i	-0.1446 + 0.0000i	

14	-0.1446 - 0.0002i	-0.1447 - 0.0000i	-0.1446 - 0.0000i	-0.1446 - 0.0000i	
15	-0.0868 + 0.0557i	-0.0210 + 0.0746i	-0.0051 + 0.0743i	-0.0002 + 0.0734i	
16	-0.0868 - 0.0557i	-0.0210 - 0.0746i	-0.0051 - 0.0743i	-0.0002 - 0.0734i	
17	-0.0002 + 0.0601i	-0.0005 + 0.0602i	-0.0008 + 0.0597i	-0.0008 + 0.0595i	
18	-0.0002 - 0.0601i	-0.0005 - 0.0602i	-0.0008 - 0.0597i	-0.0008 - 0.0595i	
19	-0.0002 + 0.0600i	-0.0002 + 0.0600i	-0.0001 + 0.0600i	-0.0001 + 0.0600i	
20	-0.0002 - 0.0600i	-0.0002 - 0.0600i	-0.0001 - 0.0600i	-0.0001 - 0.0600i	
21	-0.0009 + 0.0508i	-0.0011 + 0.0506i	-0.0010 + 0.0507i	-0.0010 + 0.0507i	
22	-0.0009 - 0.0508i	-0.0011 - 0.0506i	-0.0010 - 0.0507i	-0.0010 - 0.0507i	
23	-0.0009 + 0.0509i	-0.0007 + 0.0507i	-0.0005 + 0.0503i	-0.0004 + 0.0502i	
24	-0.0009 - 0.0509i	-0.0007 - 0.0507i	-0.0005 - 0.0503i	-0.0004 - 0.0502i	
25	-0.0347 + 0.0149i	-0.0237 + 0.0115i	-0.0264	-0.0268	
26	-0.0347 - 0.0149i	-0.0237 - 0.0115i	-0.0201 + 0.0112i	-0.0192 + 0.0114i	
27	-0.0204	-0.0241	-0.0201 - 0.0112i	-0.0192 - 0.0114i	
28	-0.0052 + 0.0136i	-0.0053 + 0.0136i	-0.0053 + 0.0136i	-0.0053 + 0.0136i	
29	-0.0052 - 0.0136i	-0.0053 - 0.0136i	-0.0053 - 0.0136i	-0.0053 - 0.0136i	
30	-0.0078 + 0.0145i	-0.0078 + 0.0145i	-0.0078 + 0.0145i	-0.0078 + 0.0145i	
31	-0.0078 - 0.0145i	-0.0078 - 0.0145i	-0.0078 - 0.0145i	-0.0078 - 0.0145i	
32	-0.0134	-0.0141	-0.0144	-0.0145	
33	-0.0106	-0.01	-0.0096	-0.0094	
34	-0.01	-0.01	-0.01	-0.01	

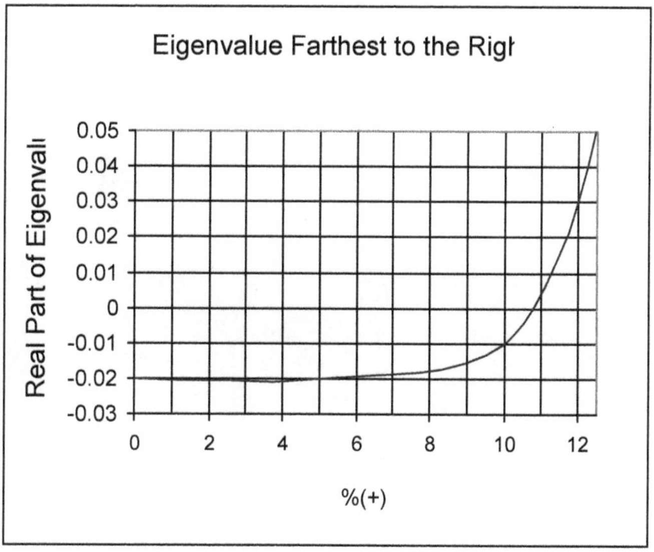

Observation:

The control system is fairly stable when the modeling Iyy is in the range of 15% greater than the true value and less than 10 % of the true value. But significant instability is displayed when a slight increase or decrease in error occurs.

2) Control Thrust : Tc (math model Control Thrust = 410,400 lbf)

Case 1 (actual value is less than the value in the math model)

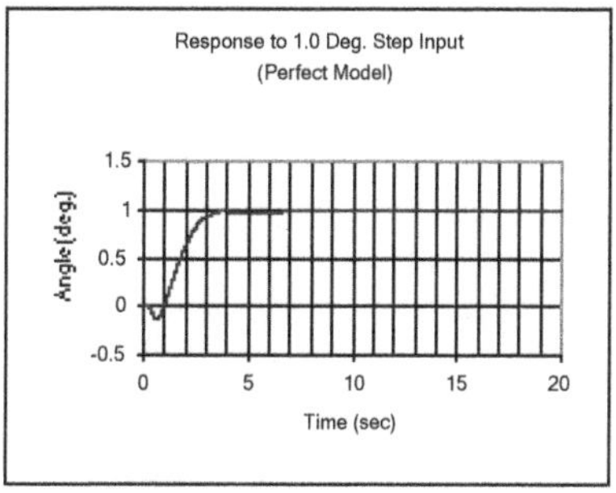

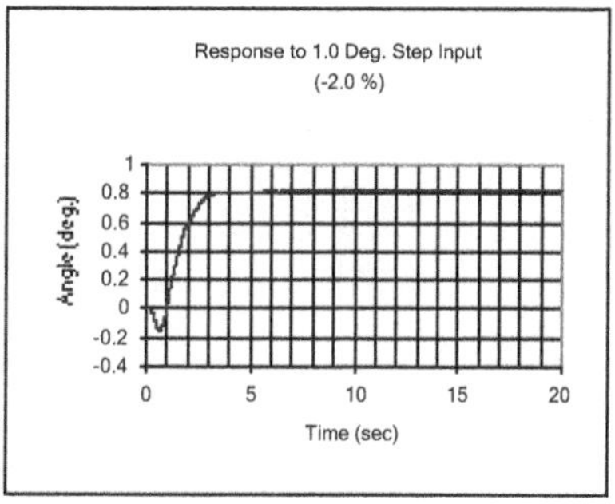

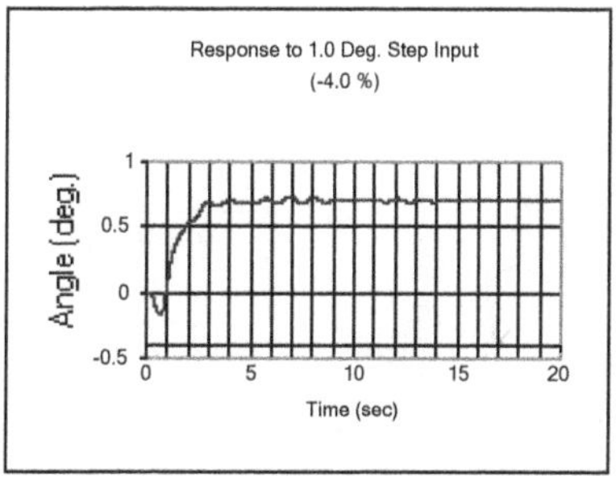

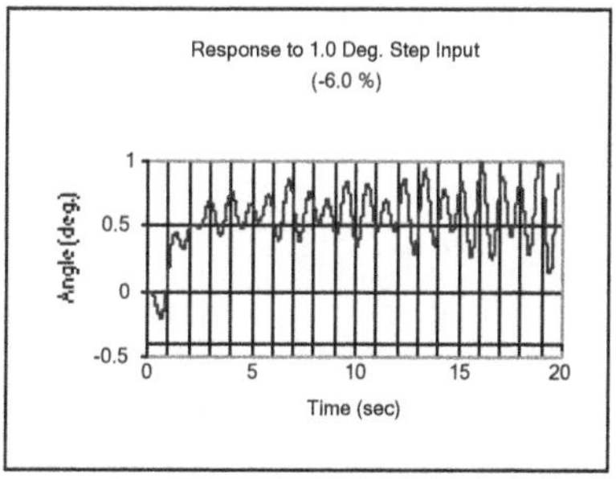

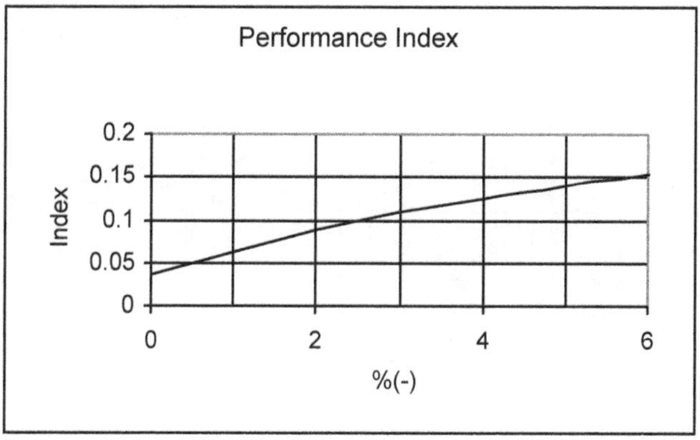

Appendix III: Collection of Math Modeling Error Sensitivity Response and Singular Value Plots, and Eigenvalue Tables

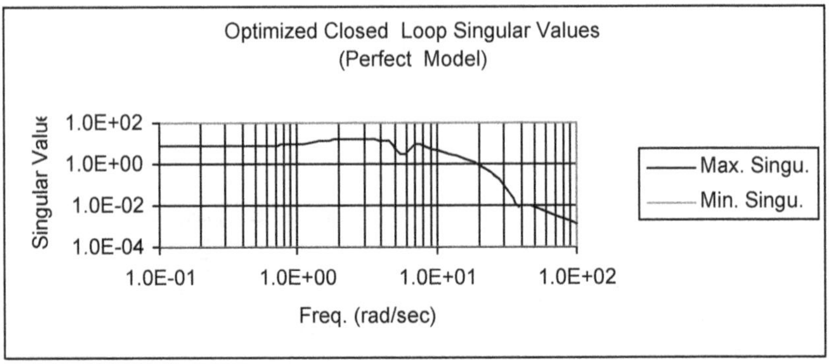

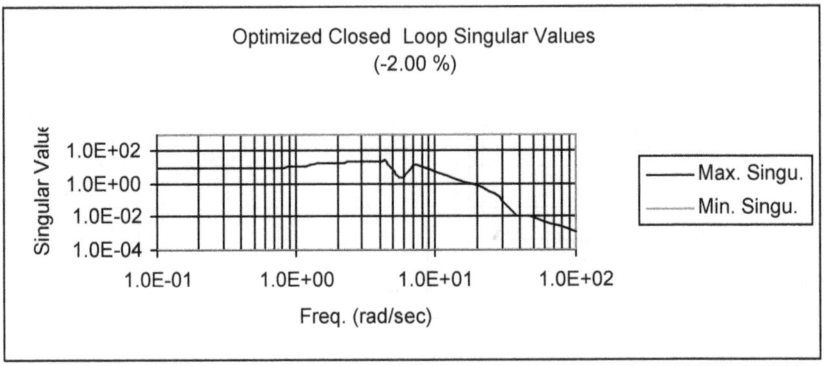

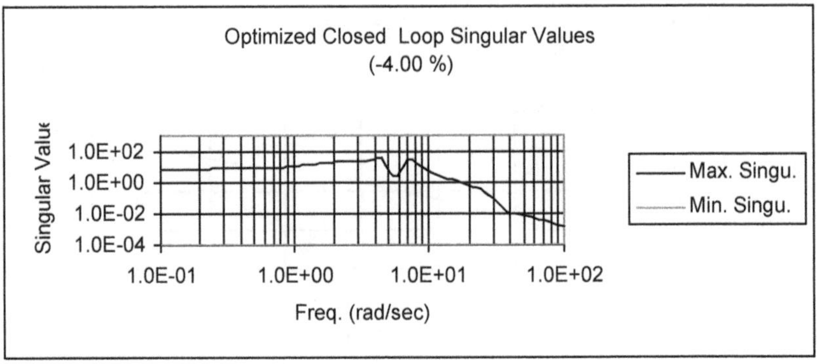

Appendix III: Collection of Math Modeling Error Sensitivity Response and Singular Value Plots, and Eigenvalue Tables

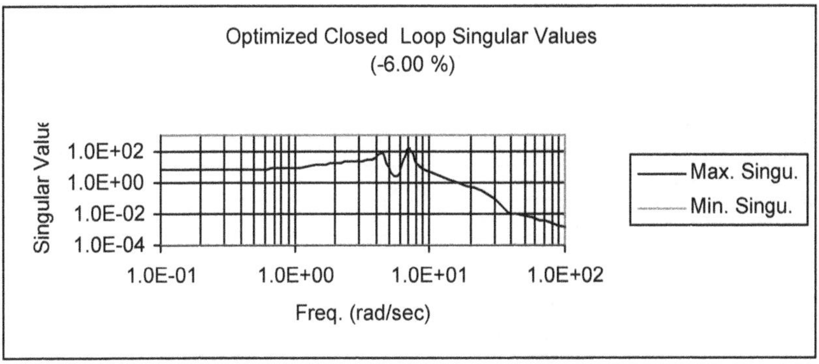

Closed Loop System Eigenvalues					
	0%	-2.00%	-4.00%	-6.00%	
	1.0e+002 *	1.0e+002 *	1.0e+002 *	1.0e+002 *	
1	-0.25	-0.25	-0.25	-0.25	
2	-0.5824 + 1.3344i	-0.5824 + 1.3344i	-0.5824 + 1.3344i	-0.5824 + 1.3344i	
3	-0.5824 - 1.3344i	-0.5824 - 1.3344i	-0.5824 - 1.3344i	-0.5824 - 1.3344i	
4	-0.1040 + 0.2546i	-0.0916 + 0.2648i	-0.0845 + 0.2732i	-0.2588 + 0.1596i	
5	-0.1040 - 0.2546i	-0.0916 - 0.2648i	-0.0845 - 0.2732i	-0.2588 - 0.1596i	
6	-0.1046 + 0.2542i	-0.1052 + 0.2540i	-0.1052 + 0.2540i	-0.2924	
7	-0.1046 - 0.2542i	-0.1052 - 0.2540i	-0.1052 - 0.2540i	-0.0795 + 0.2802i	
8	-0.2284 + 0.0212i	-0.2777	-0.2446 + 0.1481i	-0.0795 - 0.2802i	
9	-0.2284 - 0.0212i	-0.2236 + 0.1286i	-0.2446 - 0.1481i	-0.1052 + 0.2540i	
10	-0.2293	-0.2236 - 0.1286i	-0.2871	-0.1052 - 0.2540i	
11	-0.1353 + 0.0476i	-0.2245	-0.2246	-0.2246	
12	-0.1353 - 0.0476i	-0.1687	-0.1677	-0.1673	
13	-0.1446 + 0.0002i	-0.1442	-0.1442	-0.1442	
14	-0.1446 - 0.0002i	-0.1446	-0.1446	-0.1446	
15	-0.0868 + 0.0557i	-0.0266 + 0.0688i	-0.0092 + 0.0681i	0.0007 + 0.0674i	
16	-0.0868 - 0.0557i	-0.0266 - 0.0688i	-0.0092 - 0.0681i	0.0007 - 0.0674i	
17	-0.0002 + 0.0601i	-0.0006 + 0.0600i	-0.0011 + 0.0597i	0.0000 + 0.0600i	
18	-0.0002 - 0.0601i	-0.0006 - 0.0600i	-0.0011 - 0.0597i	0.0000 - 0.0600i	
19	-0.0002 + 0.0600i	-0.0001 + 0.0600i	-0.0000 + 0.0600i	-0.0009 + 0.0586i	
20	-0.0002 - 0.0600i	-0.0001 - 0.0600i	-0.0000 - 0.0600i	-0.0009 - 0.0586i	

21	-0.0009 + 0.0508i	-0.0015 + 0.0500i	-0.0015 + 0.0484i	0.0004 + 0.0472i
22	-0.0009 - 0.0508i	-0.0015 - 0.0500i	-0.0015 - 0.0484i	0.0004 - 0.0472i
23	-0.0009 + 0.0509i	-0.0007 + 0.0510i	-0.0006 + 0.0510i	-0.0006 + 0.0511i
24	-0.0009 - 0.0509i	-0.0007 - 0.0510i	-0.0006 - 0.0510i	-0.0006 - 0.0511i
25	-0.0347 + 0.0149i	-0.0247 + 0.0216i	-0.0230 + 0.0252i	-0.0231 + 0.0274i
26	-0.0347 - 0.0149i	-0.0247 - 0.0216i	-0.0230 - 0.0252i	-0.0231 - 0.0274i
27	-0.0204	-0.0216	-0.0219	-0.022
28	-0.0052 + 0.0136i	-0.0059 + 0.0137i	-0.0066 + 0.0135i	-0.0068 + 0.0131i
29	-0.0052 - 0.0136i	-0.0059 - 0.0137i	-0.0066 - 0.0135i	-0.0068 - 0.0131i
30	-0.0078 + 0.0145i	-0.0077 + 0.0144i	-0.0075 + 0.0144i	-0.0075 + 0.0146i
31	-0.0078 - 0.0145i	-0.0077 - 0.0144i	-0.0075 - 0.0144i	-0.0075 - 0.0146i
32	-0.0134	-0.0124	-0.0118 + 0.0006i	-0.0117 + 0.0009i
33	-0.0106	-0.0112	-0.0118 - 0.0006i	-0.0117 - 0.0009i
34	-0.01	-0.01	-0.01	-0.01

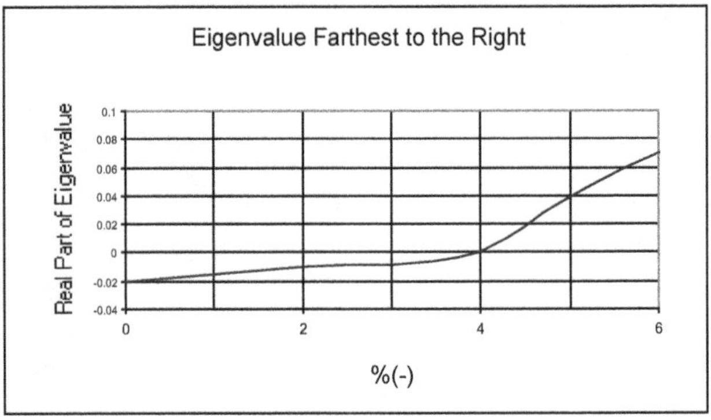

Observation:

The control system is sensitive to Tc modeling error. When the actual value of Tc is 2.0% less than the value in the math model, the response reaches only about 80.0% of the desired output. When the error is as large as 6.0%, the system becomes unstable and the instability rapidly becomes worse as the error increases beyond 6.0%.

Case 2 (actual value is greater than the value in the math model)

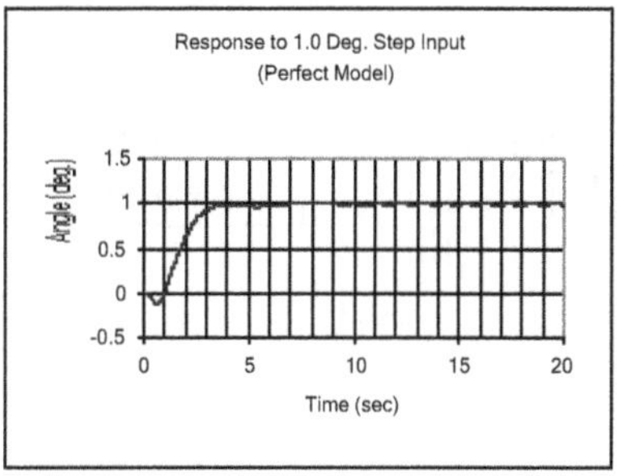

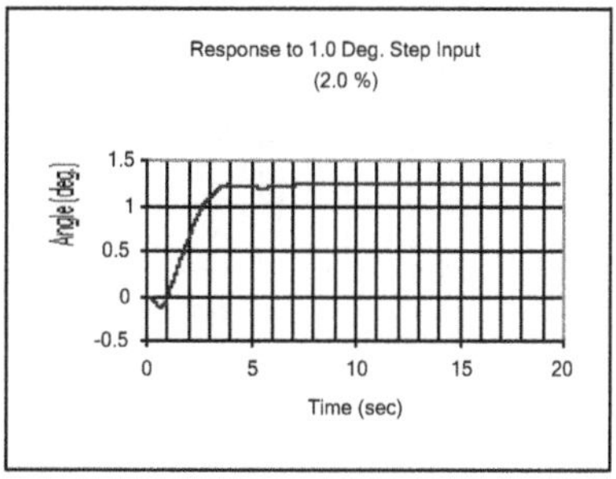

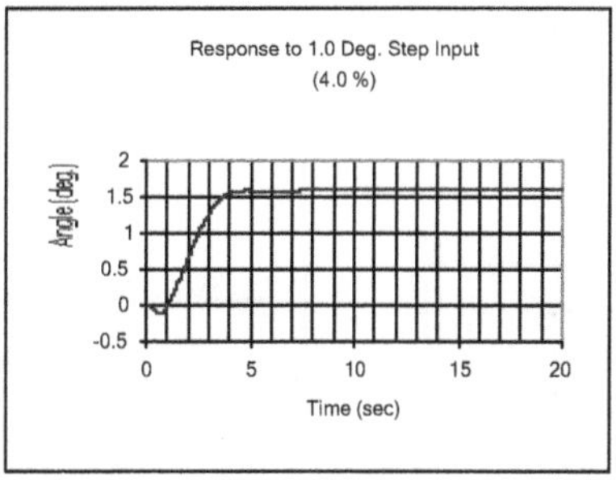

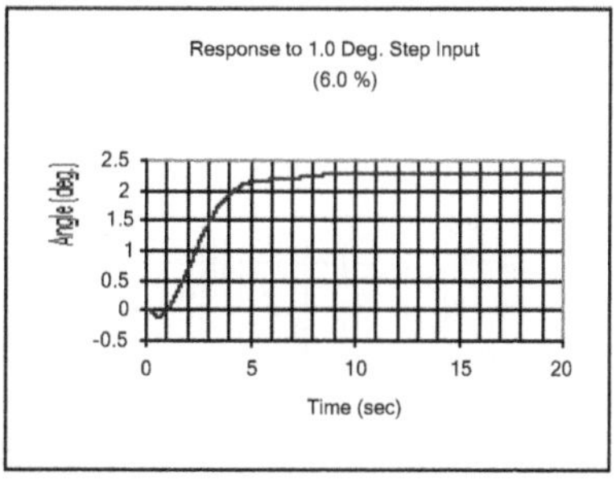

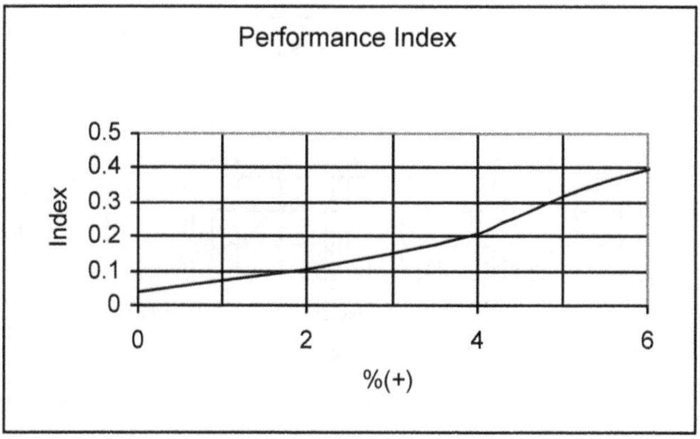

Appendix III: Collection of Math Modeling Error Sensitivity Response and Singular Value Plots, and Eigenvalue Tables

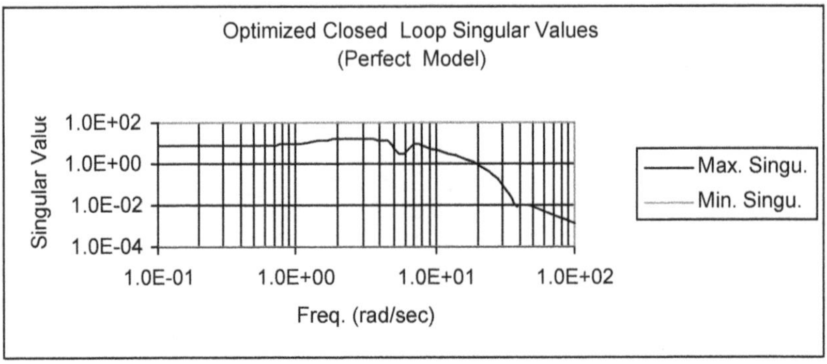

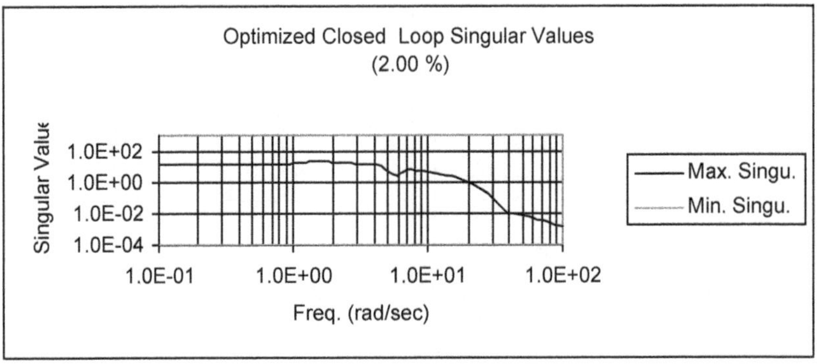

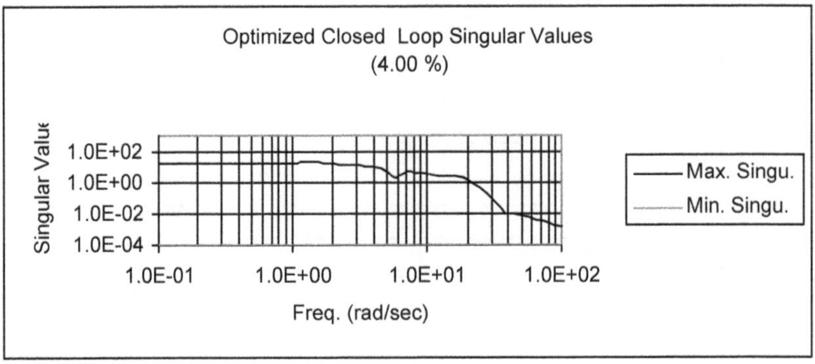

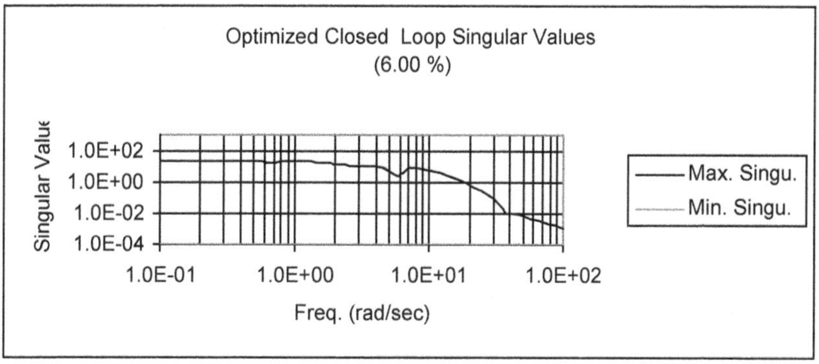

Closed Loop System Eigenvalues

	0%	2.00%	4.00%	6.00%	
	1.0e+002 *	1.0e+002 *	1.0e+002 *	1.0e+002 *	
1	-0.25	-0.25	-0.25	-0.25	
2	-0.5824 + 1.3344i	-0.5824 + 1.3344i	-0.5824 + 1.3344i	-0.5824 + 1.3344i	
3	-0.5824 - 1.3344i	-0.5824 - 1.3344i	-0.5824 - 1.3344i	-0.5824 - 1.3344i	
4	-0.1040 + 0.2546i	-0.1247 + 0.2503i	-0.1424 + 0.2586i	-0.0808 + 0.2950i	
5	-0.1040 - 0.2546i	-0.1247 - 0.2503i	-0.1424 - 0.2586i	-0.0808 - 0.2950i	
6	-0.1046 + 0.2542i	-0.1053 + 0.2540i	-0.1053 + 0.2540i	-0.2924	
7	-0.1046 - 0.2542i	-0.1053 - 0.2540i	-0.1053 - 0.2540i	-0.2376 + 0.1377i	
8	-0.2284 + 0.0212i	-0.2737 + 0.0500i	-0.2889 + 0.0525i	-0.2376 - 0.1377i	
9	-0.2284 - 0.0212i	-0.2737 - 0.0500i	-0.2889 - 0.0525i	-0.1351 + 0.1901i	
10	-0.2293	-0.2248	-0.0413 + 0.1752i	-0.1351 - 0.1901i	
11	-0.1353 + 0.0476i	-0.0682 + 0.1562i	-0.0413 - 0.1752i	-0.1649	
12	-0.1353 - 0.0476i	-0.0682 - 0.1562i	-0.2247	-0.0455 + 0.0954i	
13	-0.1446 + 0.0002i	-0.1647	-0.1657	-0.0455 - 0.0954i	
14	-0.1446 - 0.0002i	-0.1441	-0.1442	-0.1419	
15	-0.0868 + 0.0557i	-0.1446	-0.1446	-0.1446	
16	-0.0868 - 0.0557i	-0.0001 + 0.0602i	-0.0001 + 0.0602i	-0.0810 + 0.0228i	
17	-0.0002 + 0.0601i	-0.0001 - 0.0602i	-0.0001 - 0.0602i	-0.0810 - 0.0228i	
18	-0.0002 - 0.0601i	-0.0002 + 0.0599i	-0.0001 + 0.0598i	-0.0002 + 0.0604i	
19	-0.0002 + 0.0600i	-0.0002 - 0.0599i	-0.0001 - 0.0598i	-0.0002 - 0.0604i	
20	-0.0002 - 0.0600i	-0.0010 + 0.0513i	-0.0009 + 0.0514i	-0.0002 + 0.0598i	
21	-0.0009 + 0.0508i	-0.0010 - 0.0513i	-0.0009 - 0.0514i	-0.0002 - 0.0598i	

22	-0.0009 - 0.0508i	-0.0006 + 0.0507i	-0.0005 + 0.0507i	-0.0011 + 0.0514i	
23	-0.0009 + 0.0509i	-0.0006 - 0.0507i	-0.0005 - 0.0507i	-0.0011 - 0.0514i	
24	-0.0009 - 0.0509i	-0.0337 + 0.0313i	-0.0307 + 0.0318i	-0.0002 + 0.0507i	
25	-0.0347 + 0.0149i	-0.0337 - 0.0313i	-0.0307 - 0.0318i	-0.0002 - 0.0507i	
26	-0.0347 - 0.0149i	-0.0238	-0.0229	-0.0206	
27	-0.0204	-0.0047 + 0.0131i	-0.0044 + 0.0125i	-0.0044 + 0.0119i	
28	-0.0052 + 0.0136i	-0.0047 - 0.0131i	-0.0044 - 0.0125i	-0.0044 - 0.0119i	
29	-0.0052 - 0.0136i	-0.0078 + 0.0146i	-0.0078 + 0.0147i	-0.0078 + 0.0147i	
30	-0.0078 + 0.0145i	-0.0078 - 0.0146i	-0.0078 - 0.0147i	-0.0078 - 0.0147i	
31	-0.0078 - 0.0145i	-0.0147 + 0.0026i	-0.0129 + 0.0032i	-0.0127 + 0.0042i	
32	-0.0134	-0.0147 - 0.0026i	-0.0129 - 0.0032i	-0.0127 - 0.0042i	
33	-0.0106	-0.0099	-0.0087	-0.0061	
34	-0.01	-0.01	-0.01	-0.01	

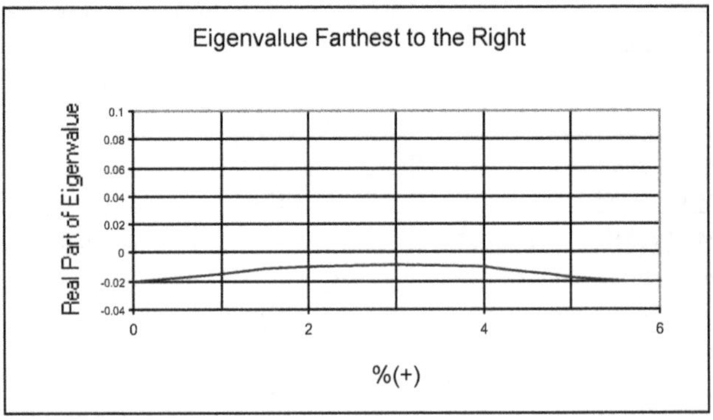

Observation:

When actual control thrust is larger than the value in the math model, the attitude control system is less sensitive. However, the response is larger than the commanded input value. Simulation study shows that as the difference exceeds 13 %, it becomes unstable.

3) **Aerodynamic Load: L-alpha (math model Aerodynamic Load = 198,000 lbf/rad)**

Case 1 (actual value is less than the value in the math model)

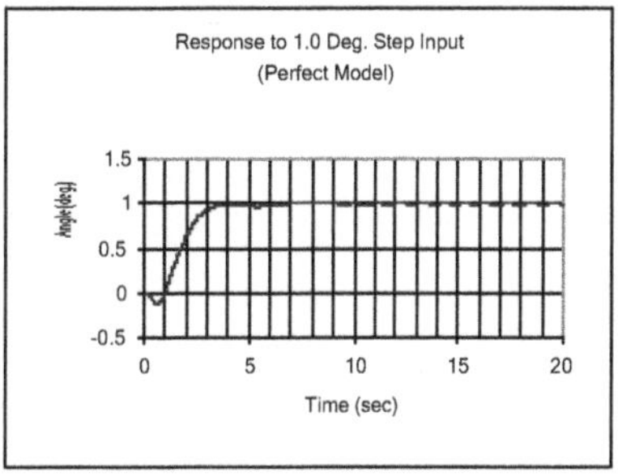

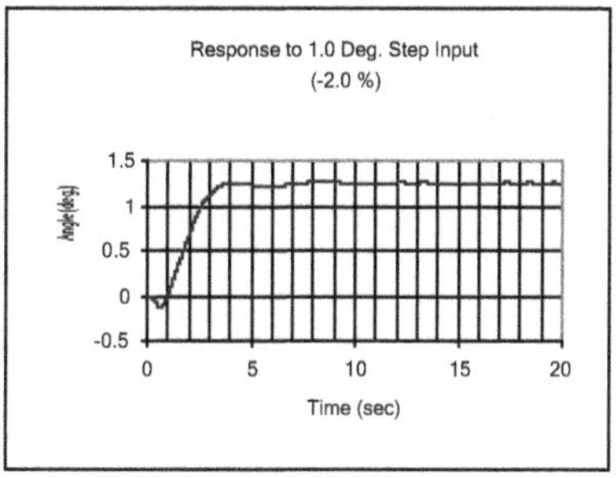

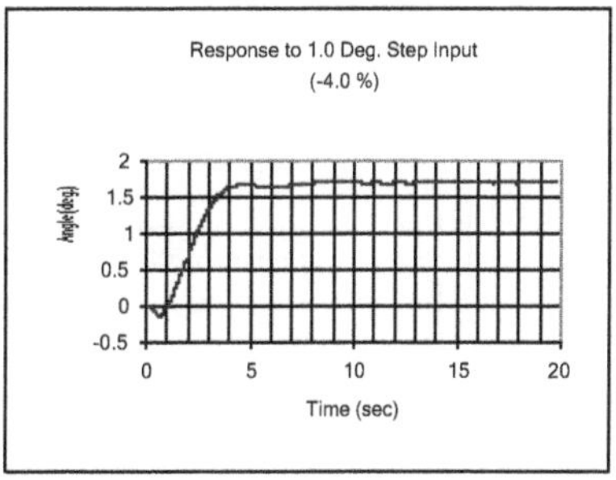

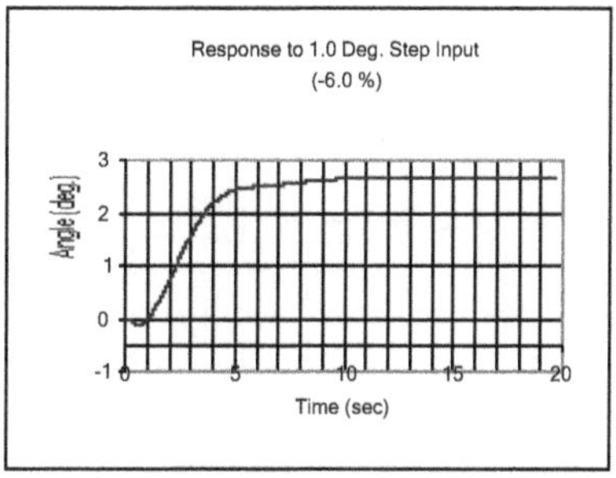

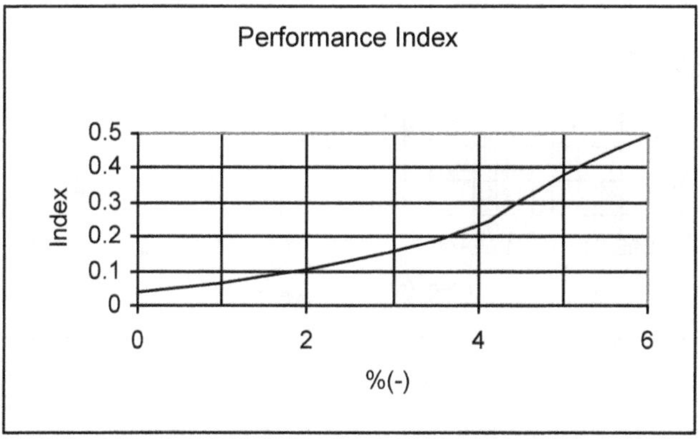

Appendix III: Collection of Math Modeling Error Sensitivity Response and Singular Value Plots, and Eigenvalue Tables

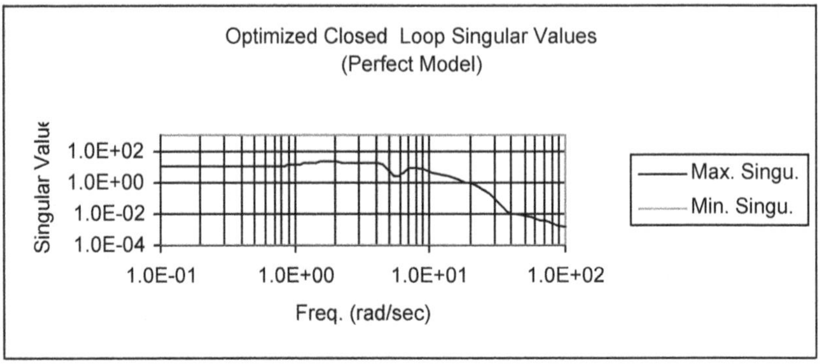

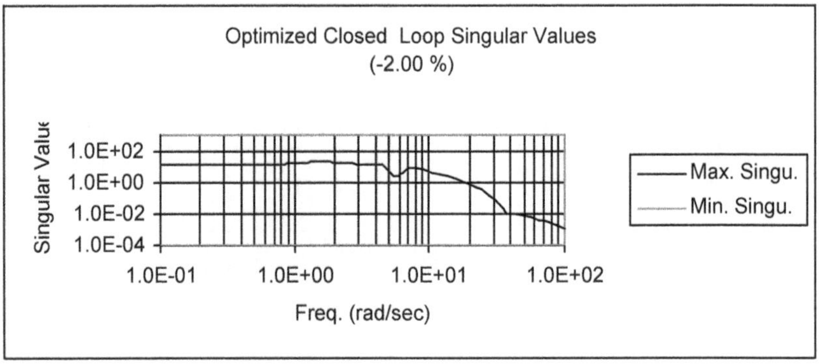

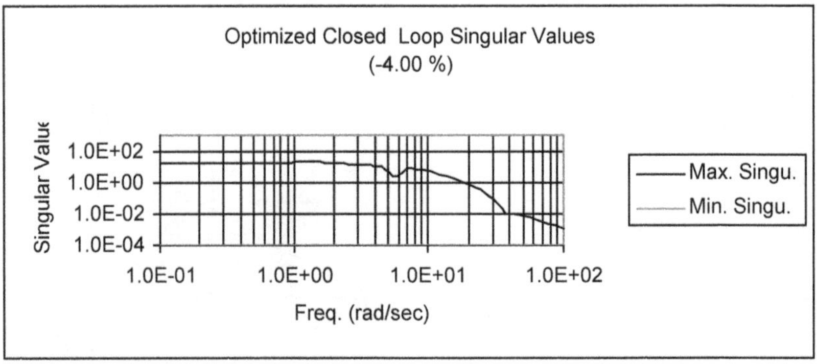

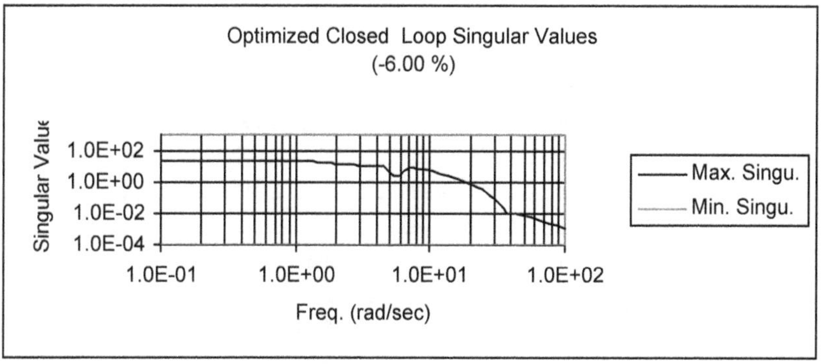

Closed Loop System Eigenvalues

	0%	-2.00%	-4.00%	-6.00%	
	1.0e+002 *	1.0e+002 *	1.0e+002 *	1.0e+002 *	
1	-0.25	-0.25	-0.25	-0.25	
2	-0.5824 + 1.3344i	-0.5824 + 1.3344i	-0.5824 + 1.3344i	-0.5824 + 1.3344i	
3	-0.5824 - 1.3344i	-0.5824 - 1.3344i	-0.5824 - 1.3344i	-0.5824 - 1.3344i	
4	-0.1040 + 0.2546i	-0.1012 + 0.2609i	-0.0994 + 0.2637i	-0.0986 + 0.2656i	
5	-0.1040 - 0.2546i	-0.1012 - 0.2609i	-0.0994 - 0.2637i	-0.0986 - 0.2656i	
6	-0.1046 + 0.2542i	-0.1075 + 0.2478i	-0.1098 + 0.2449i	-0.1107 + 0.2429i	
7	-0.1046 - 0.2542i	-0.1075 - 0.2478i	-0.1098 - 0.2449i	-0.1107 - 0.2429i	
8	-0.2284 + 0.0212i	-0.2526	-0.2605	-0.2647	
9	-0.2284 - 0.0212i	-0.2135 + 0.0340i	-0.2075 + 0.0467i	-0.2022 + 0.0566i	
10	-0.2293	-0.2135 - 0.0340i	-0.2075 - 0.0467i	-0.2022 - 0.0566i	
11	-0.1353 + 0.0476i	-0.1499 + 0.0528i	-0.1574 + 0.0501i	-0.1659 + 0.0468i	
12	-0.1353 - 0.0476i	-0.1499 - 0.0528i	-0.1574 - 0.0501i	-0.1659 - 0.0468i	
13	-0.1446 + 0.0002i	-0.1492	-0.152	-0.1538	
14	-0.1446 - 0.0002i	-0.1446	-0.1446	-0.1446	
15	-0.0868 + 0.0557i	-0.0630 + 0.0630i	-0.0543 + 0.0717i	-0.0488 + 0.0775i	
16	-0.0868 - 0.0557i	-0.0630 - 0.0630i	-0.0543 - 0.0717i	-0.0488 - 0.0775i	
17	-0.0002 + 0.0601i	-0.0729	-0.083	-0.0861	
18	-0.0002 - 0.0601i	-0.0002 + 0.0602i	-0.0002 + 0.0602i	-0.0002 + 0.0602i	
19	-0.0002 + 0.0600i	-0.0002 - 0.0602i	-0.0002 - 0.0602i	-0.0002 - 0.0602i	
20	-0.0002 - 0.0600i	-0.0002 + 0.0599i	-0.0002 + 0.0599i	-0.0002 + 0.0598i	
21	-0.0009 + 0.0508i	-0.0002 - 0.0599i	-0.0002 - 0.0599i	-0.0002 - 0.0598i	
22	-0.0009 - 0.0508i	-0.0010 + 0.0507i	-0.0010 + 0.0506i	-0.0009 + 0.0506i	

23	-0.0009 + 0.0509i	-0.0010 - 0.0507i	-0.0010 - 0.0506i	-0.0009 - 0.0506i	
24	-0.0009 - 0.0509i	-0.0007 + 0.0510i	-0.0007 + 0.0510i	-0.0006 + 0.0510i	
25	-0.0347 + 0.0149i	-0.0007 - 0.0510i	-0.0007 - 0.0510i	-0.0006 - 0.0510i	
26	-0.0347 - 0.0149i	-0.0047 + 0.0132i	-0.0213	-0.0212	
27	-0.0204	-0.0047 - 0.0132i	-0.0043 + 0.0125i	-0.0044 + 0.0118i	
28	-0.0052 + 0.0136i	-0.0078 + 0.0146i	-0.0043 - 0.0125i	-0.0044 - 0.0118i	
29	-0.0052 - 0.0136i	-0.0078 - 0.0146i	-0.0078 + 0.0147i	-0.0078 + 0.0147i	
30	-0.0078 + 0.0145i	-0.0219	-0.0078 - 0.0147i	-0.0078 - 0.0147i	
31	-0.0078 - 0.0145i	-0.0155 + 0.0030i	-0.0132 + 0.0040i	-0.006	
32	-0.0134	-0.0155 - 0.0030i	-0.0132 - 0.0040i	-0.0121 + 0.0036i	
33	-0.0106	-0.0097	-0.0083	-0.0121 - 0.0036i	
34	-0.01	-0.01	-0.01	-0.01	

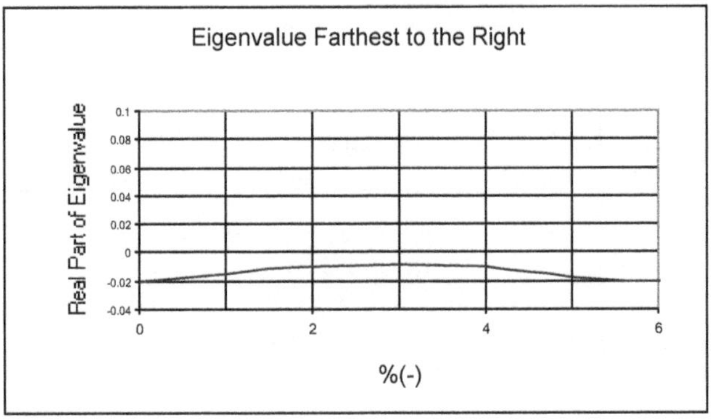

Observation:

L-alpha error does not affect system stability, but it does cause inaccurate responses. The output is larger than the input value, and it increases as the error increases. The real parts of the eigenvalues farthest to the Right Half Plane remain in the Left Half Plane.

Case 2 (actual value is greater than the value in the math model)

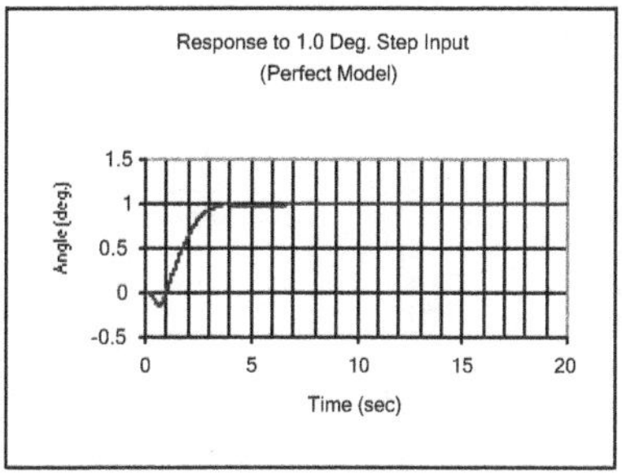

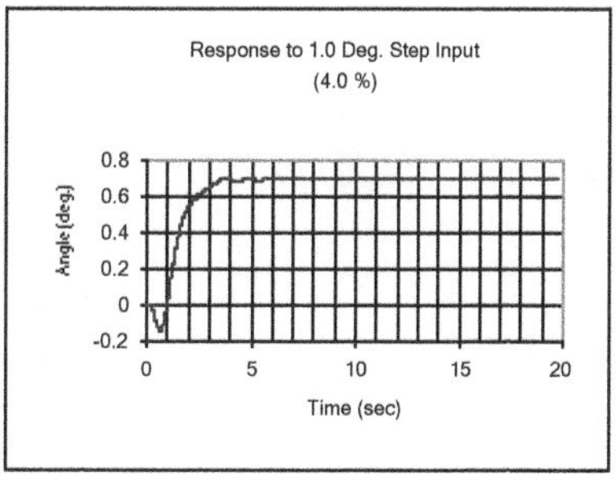

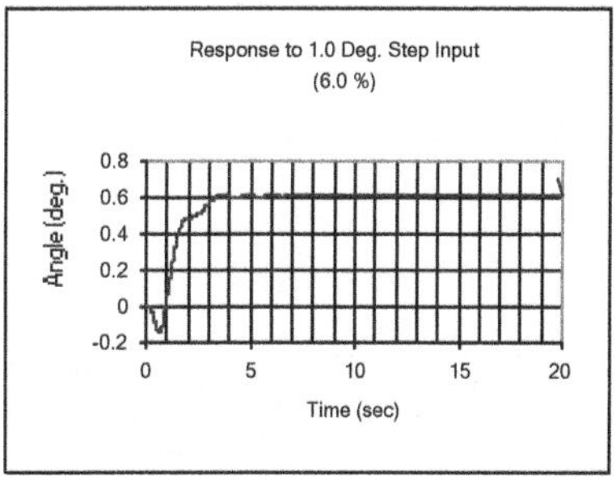

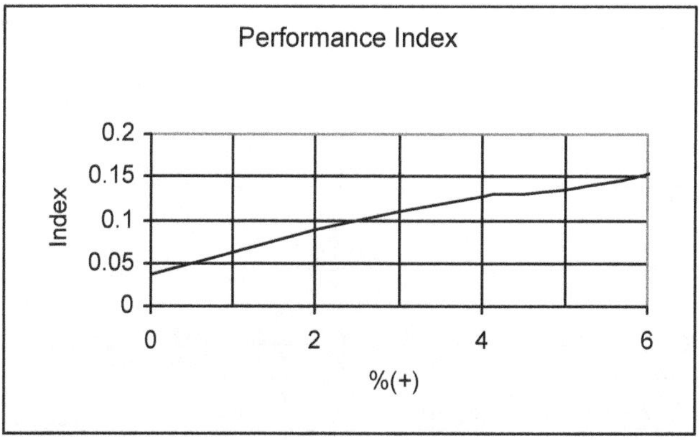

Appendix III: Collection of Math Modeling Error Sensitivity Response and Singular Value Plots, and Eigenvalue Tables

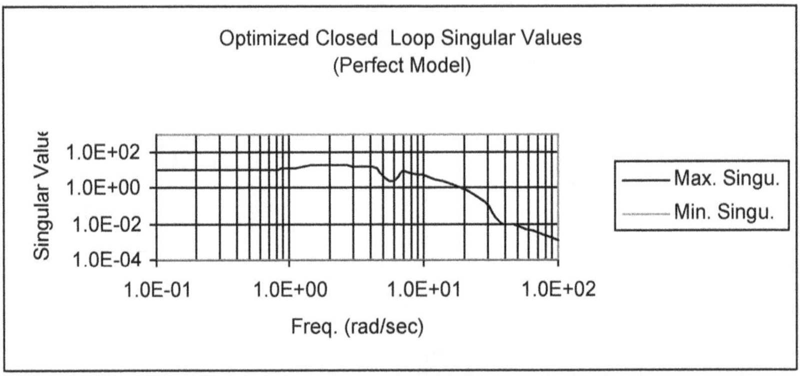

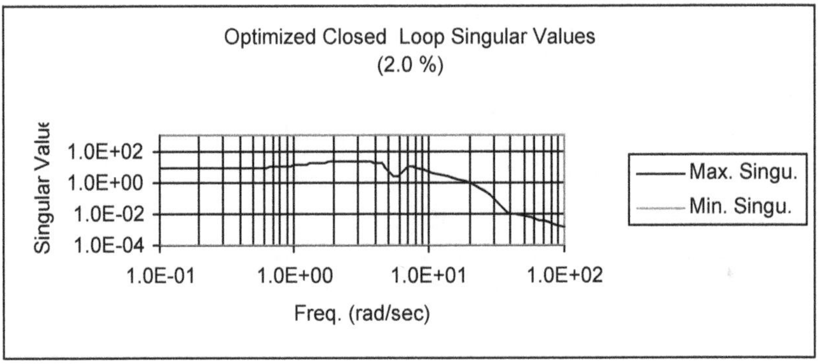

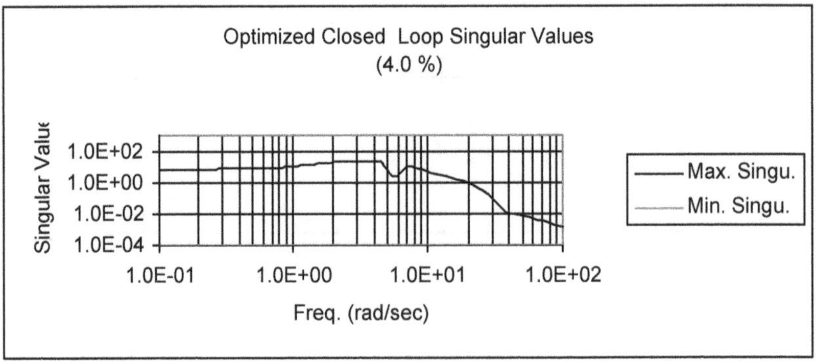

Appendix III: Collection of Math Modeling Error Sensitivity Response and Singular Value Plots, and Eigenvalue Tables

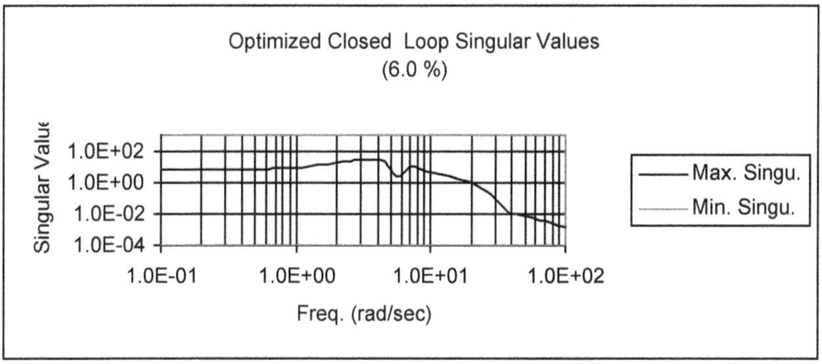

Closed Loop System Eigenvalues

	0%	2.00%	4.00%	6.00%	
	1.0e+002 *	1.0e+002 *	1.0e+002 *	1.0e+002 *	
1	-0.25	-0.25	-0.25	-0.25	
2	-0.5824 + 1.3344i	-0.5824 + 1.3344i	-0.5824 + 1.3344i	-0.5824 + 1.3344i	
3	-0.5824 - 1.3344i	-0.5824 - 1.3344i	-0.5824 - 1.3344i	-0.5824 - 1.3344i	
4	-0.1040 + 0.2546i	-0.1108 + 0.2587i	-0.1135 + 0.2605i	-0.1155 + 0.2614i	
5	-0.1040 - 0.2546i	-0.1108 - 0.2587i	-0.1135 - 0.2605i	-0.1155 - 0.2614i	
6	-0.1046 + 0.2542i	-0.0974 + 0.2502i	-0.0943 + 0.2485i	-0.0921 + 0.2479i	
7	-0.1046 - 0.2542i	-0.0974 - 0.2502i	-0.0943 - 0.2485i	-0.0921 - 0.2479i	
8	-0.2284 + 0.0212i	-0.2453 + 0.0290i	-0.2516 + 0.0334i	-0.2555 + 0.0363i	
9	-0.2284 - 0.0212i	-0.2453 - 0.0290i	-0.2516 - 0.0334i	-0.2555 - 0.0363i	
10	-0.2293	-0.1995	-0.1909	-0.1875	
11	-0.1353 + 0.0476i	-0.1075 + 0.0741i	-0.1108 + 0.0879i	-0.1112 + 0.0976i	
12	-0.1353 - 0.0476i	-0.1075 - 0.0741i	-0.1108 - 0.0879i	-0.1112 - 0.0976i	
13	-0.1446 + 0.0002i	-0.1295 + 0.0277i	-0.1415 + 0.0225i	-0.1475 + 0.0219i	
14	-0.1446 - 0.0002i	-0.1295 - 0.0277i	-0.1415 - 0.0225i	-0.1475 - 0.0219i	
15	-0.0868 + 0.0557i	-0.1328	-0.1446	-0.1446	
16	-0.0868 - 0.0557i	-0.1446	-0.1119	-0.1038	
17	-0.0002 + 0.0601i	-0.0003 + 0.0600i	-0.0004 + 0.0601i	-0.0005 + 0.0601i	
18	-0.0002 - 0.0601i	-0.0003 - 0.0600i	-0.0004 - 0.0601i	-0.0005 - 0.0601i	
19	-0.0002 + 0.0600i	-0.0001 + 0.0600i	-0.0001 + 0.0600i	-0.0000 + 0.0600i	
20	-0.0002 - 0.0600i	-0.0001 - 0.0600i	-0.0001 - 0.0600i	-0.0000 - 0.0600i	
21	-0.0009 + 0.0508i	-0.0011 + 0.0510i	-0.0013 + 0.0511i	-0.0015 + 0.0512i	
22	-0.0009 - 0.0508i	-0.0011 - 0.0510i	-0.0013 - 0.0511i	-0.0015 - 0.0512i	
23	-0.0009 + 0.0509i	-0.0008 + 0.0507i	-0.0008 + 0.0507i	-0.0008 + 0.0506i	

24	-0.0009 - 0.0509i	-0.0008 - 0.0507i	-0.0008 - 0.0507i	-0.0008 - 0.0506i	
25	-0.0347 + 0.0149i	-0.0234 + 0.0279i	-0.0166 + 0.0325i	-0.0119 + 0.0352i	
26	-0.0347 - 0.0149i	-0.0234 - 0.0279i	-0.0166 - 0.0325i	-0.0119 - 0.0352i	
27	-0.0204	-0.0208	-0.0209	-0.0209	
28	-0.0052 + 0.0136i	-0.0059 + 0.0137i	-0.0065 + 0.0135i	-0.0067 + 0.0132i	
29	-0.0052 - 0.0136i	-0.0059 - 0.0137i	-0.0065 - 0.0135i	-0.0067 - 0.0132i	
30	-0.0078 + 0.0145i	-0.0077 + 0.0144i	-0.0076 + 0.0144i	-0.0075 + 0.0146i	
31	-0.0078 - 0.0145i	-0.0077 - 0.0144i	-0.0076 - 0.0144i	-0.0075 - 0.0146i	
32	-0.0134	-0.0118 + 0.0006i	-0.0117 + 0.0012i	-0.0117 + 0.0014i	
33	-0.0106	-0.0118 - 0.0006i	-0.0117 - 0.0012i	-0.0117 - 0.0014i	
34	-0.01	-0.01	-0.01	-0.01	

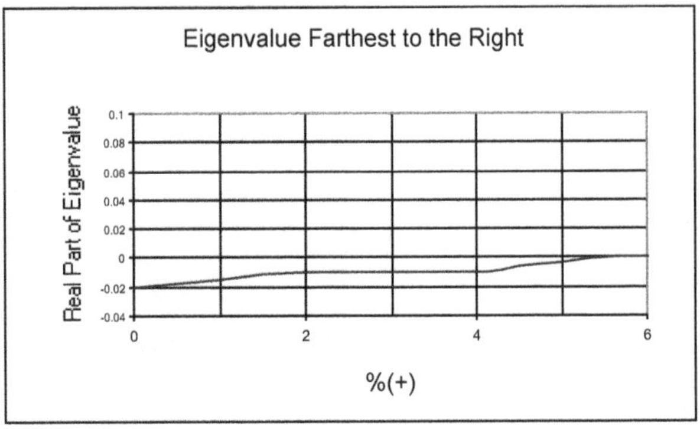

Observation:

Here again, the output is less than the input, and the system remains stable. The output decreases to a greater extent as the positive error gets larger. The real parts of the eigenvalues remain in the Left Half Plane, indicating system stability.

4) **Rocket Engine Position: lc(The distance of the engine location from LV's body frame origin in the math model = -32.3 ft.)**

Case 1 (actual value is less than the value in the math model)

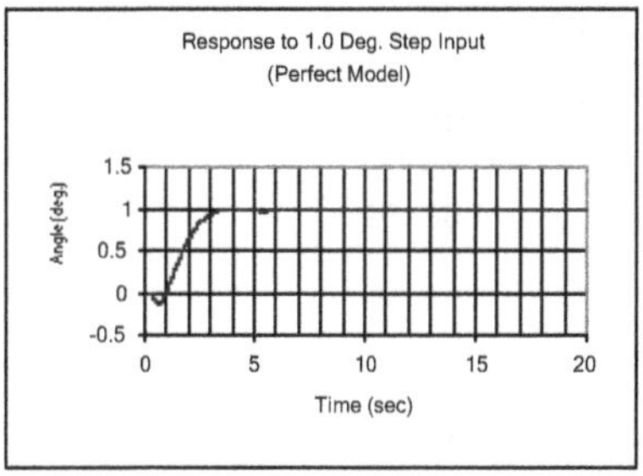

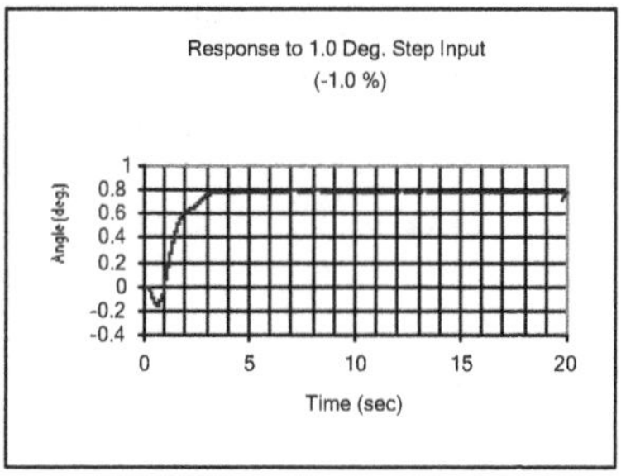

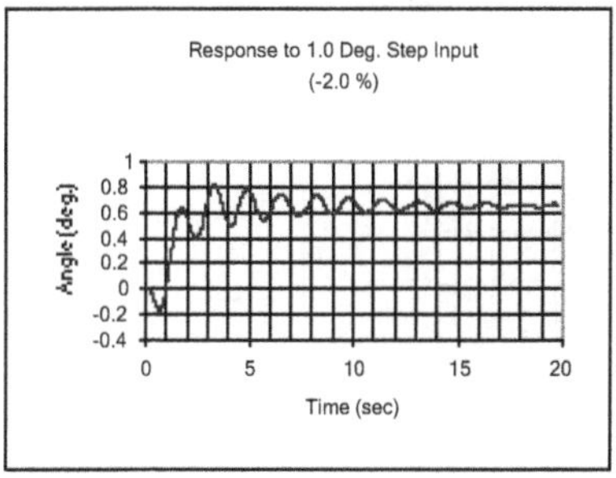

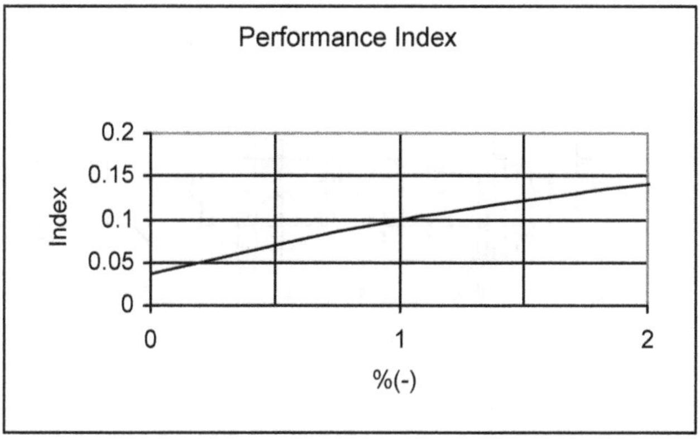

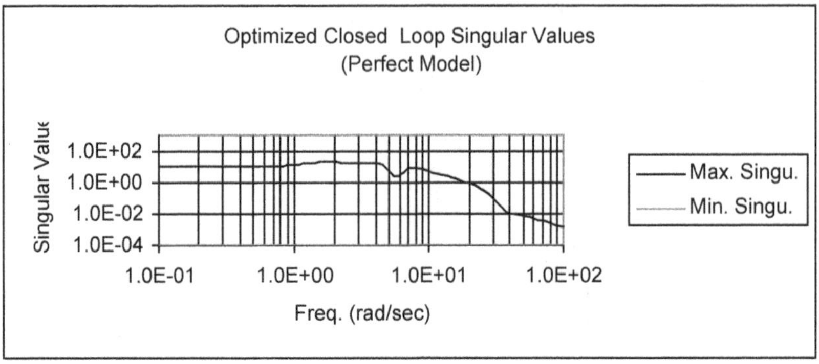
Optimized Closed Loop Singular Values
(Perfect Model)
Singular Valu
1.0E+02
1.0E+00
1.0E-02
1.0E-04
1.0E-01
1.0E+00
1.0E+01
1.0E+02
Freq. (rad/sec)
Max. Singu.
Min. Singu.

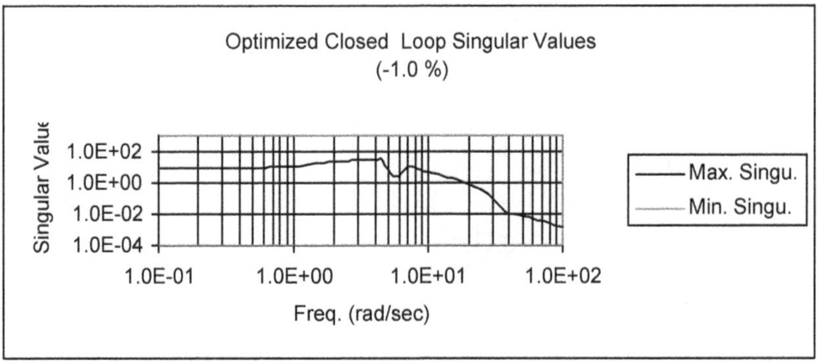
Optimized Closed Loop Singular Values
(-1.0 %)
Singular Valu
1.0E+02
1.0E+00
1.0E-02
1.0E-04
1.0E-01
1.0E+00
1.0E+01
1.0E+02
Freq. (rad/sec)
Max. Singu.
Min. Singu.

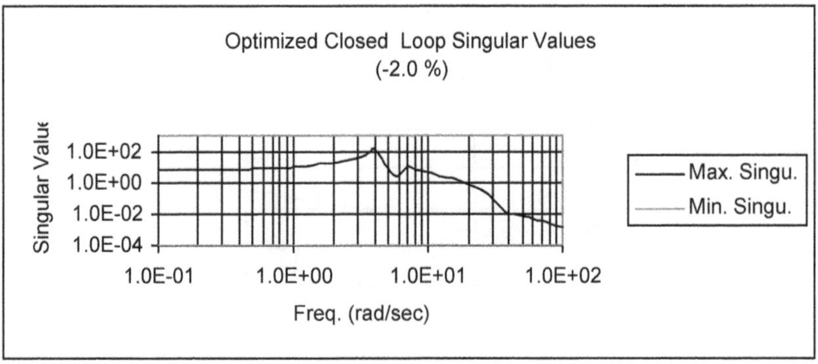
Optimized Closed Loop Singular Values
(-2.0 %)
Singular Valu
1.0E+02
1.0E+00
1.0E-02
1.0E-04
1.0E-01
1.0E+00
1.0E+01
1.0E+02
Freq. (rad/sec)
Max. Singu.
Min. Singu.

Closed Loop System Eigenvalues					
	0%	-1.00%	-2.00%		
	1.0e+002 *	1.0e+002 *	1.0e+002 *		
1	-0.25	-0.25	-0.25		
2	-0.5824 + 1.3344i	-0.5824 + 1.3344i	-0.5824 + 1.3344i		
3	-0.5824 - 1.3344i	-0.5824 - 1.3344i	-0.5824 - 1.3344i		
4	-0.1040 + 0.2546i	-0.0922 + 0.2796i	-0.0870 + 0.2888i		
5	-0.1040 - 0.2546i	-0.0922 - 0.2796i	-0.0870 - 0.2888i		
6	-0.1046 + 0.2542i	-0.2832	-0.2932		
7	-0.1046 - 0.2542i	-0.1098 + 0.2218i	-0.2157 + 0.1220i		
8	-0.2284 + 0.0212i	-0.1098 - 0.2218i	-0.2157 - 0.1220i		
9	-0.2284 - 0.0212i	-0.2102 + 0.0909i	-0.1043 + 0.2072i		
10	-0.2293	-0.2102 - 0.0909i	-0.1043 - 0.2072i		
11	-0.1353 + 0.0476i	-0.1315 + 0.0905i	-0.1441 + 0.0933i		
12	-0.1353 - 0.0476i	-0.1315 - 0.0905i	-0.1441 - 0.0933i		
13	-0.1446 + 0.0002i	-0.1597	-0.1637		
14	-0.1446 - 0.0002i	-0.1518	-0.1502		
15	-0.0868 + 0.0557i	-0.1446	-0.1446		
16	-0.0868 - 0.0557i	-0.0484	-0.0464		
17	-0.0002 + 0.0601i	-0.0003 + 0.0603i	-0.0000 + 0.0607i		
18	-0.0002 - 0.0601i	-0.0003 - 0.0603i	-0.0000 - 0.0607i		
19	-0.0002 + 0.0600i	-0.0001 + 0.0600i	-0.0001 + 0.0599i		
20	-0.0002 - 0.0600i	-0.0001 - 0.0600i	-0.0001 - 0.0599i		
21	-0.0009 + 0.0508i	-0.0020 + 0.0504i	-0.0029 + 0.0517i		
22	-0.0009 - 0.0508i	-0.0020 - 0.0504i	-0.0029 - 0.0517i		
23	-0.0009 + 0.0509i	-0.0007 + 0.0511i	-0.0007 + 0.0513i		
24	-0.0009 - 0.0509i	-0.0007 - 0.0511i	-0.0007 - 0.0513i		
25	-0.0347 + 0.0149i	-0.0148 + 0.0379i	-0.0017 + 0.0385i		
26	-0.0347 - 0.0149i	-0.0148 - 0.0379i	-0.0017 - 0.0385i		
27	-0.0204	-0.0194	-0.0191		
28	-0.0052 + 0.0136i	-0.0062 + 0.0137i	-0.0075 + 0.0146i		
29	-0.0052 - 0.0136i	-0.0062 - 0.0137i	-0.0075 - 0.0146i		
30	-0.0078 + 0.0145i	-0.0077 + 0.0144i	-0.0068 + 0.0132i		
31	-0.0078 - 0.0145i	-0.0077 - 0.0144i	-0.0068 - 0.0132i		
32	-0.0134	-0.0117 + 0.0022i	-0.0116 + 0.0029i		
33	-0.0106	-0.0117 - 0.0022i	-0.0116 - 0.0029i		
34	-0.01	-0.01	-0.01		

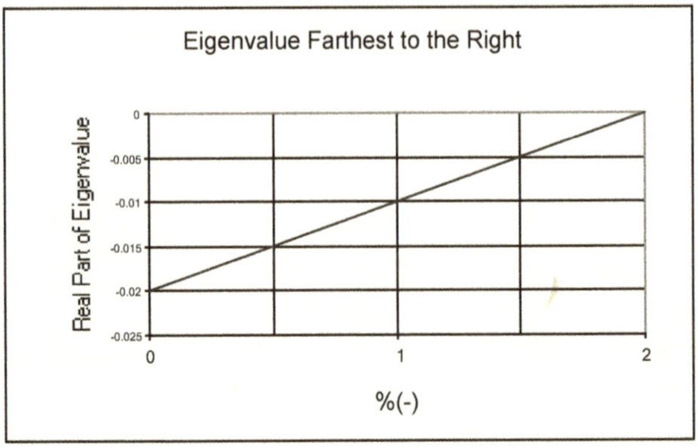

Observation:

This study shows that the location of the engine with respect to the Body Frame origin needs to be known very accurately. The control system approaches instability even when the distance is only 2.0 % less than the true distance.

Case 2 (actual value is greater than the value in the math model)

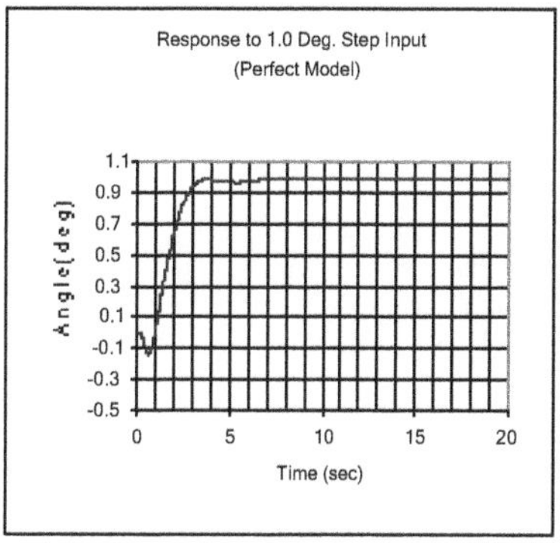

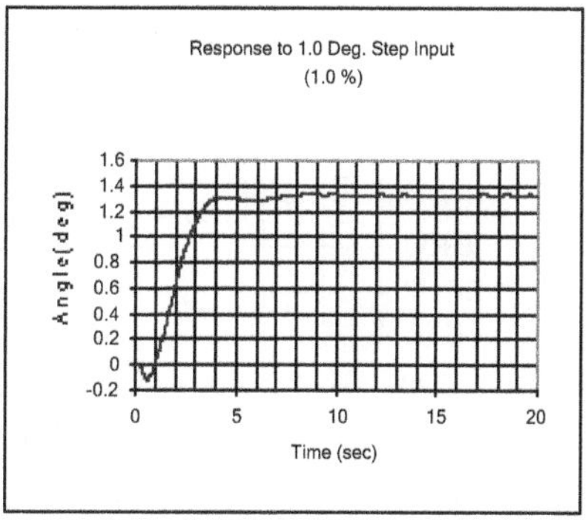

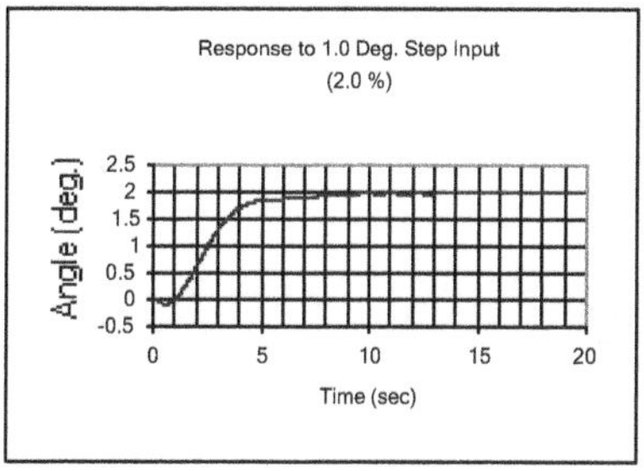

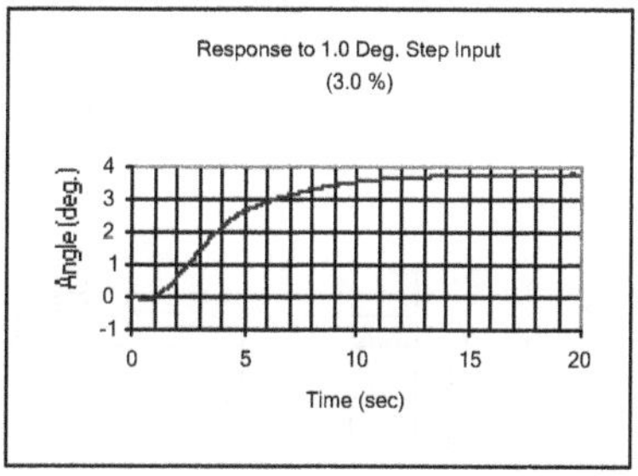

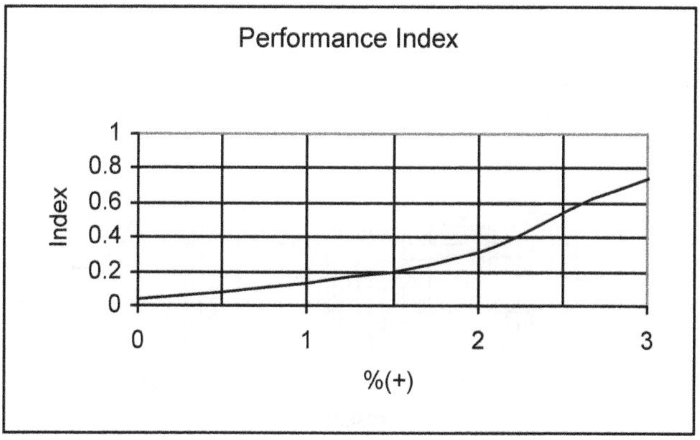

Appendix III: Collection of Math Modeling Error Sensitivity Response and Singular Value Plots, and Eigenvalue Tables

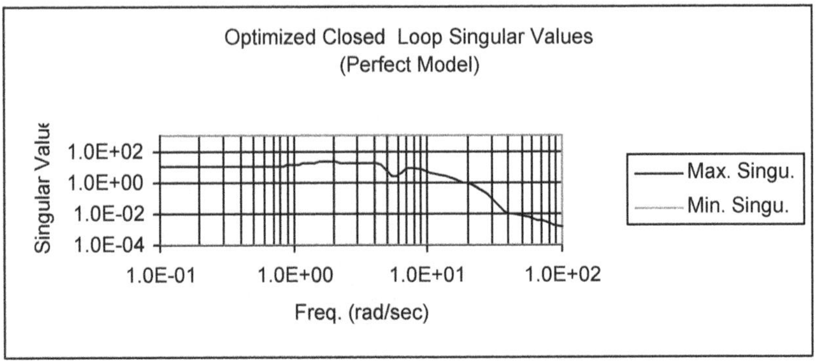

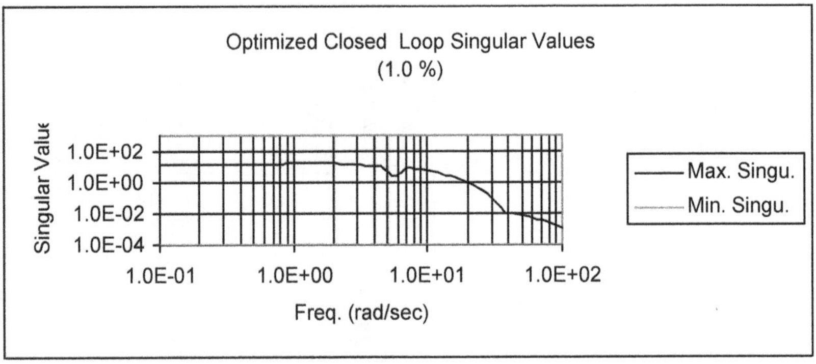

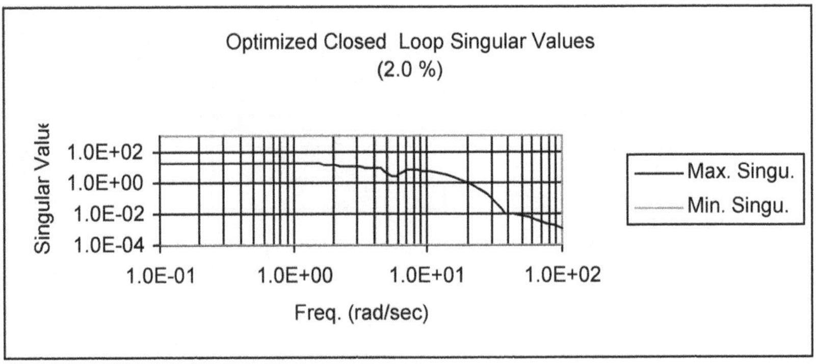

Appendix III: Collection of Math Modeling Error Sensitivity Response and Singular Value Plots, and Eigenvalue Tables

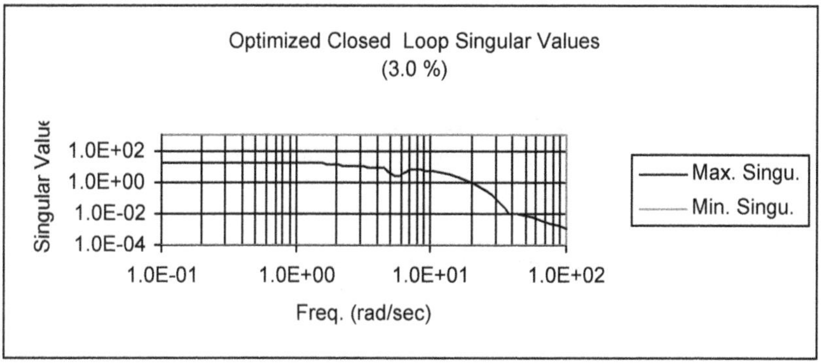

Closed Loop System Eigenvalues					
	0%	1.00%	2.00%	3.00%	
	1.0e+002 *	1.0e+002 *	1.0e+002 *	1.0e+002 *	
1	-0.25	-0.25	-0.25	-0.25	
2	-0.5824 + 1.3344i	-0.5824 + 1.3344i	-0.5824 + 1.3344i	-0.5824 + 1.3344i	
3	-0.5824 - 1.3344i	-0.5824 - 1.3344i	-0.5824 - 1.3344i	-0.5824 - 1.3344i	
4	-0.1040 + 0.2546i	-0.1329 + 0.2701i	-0.1432 + 0.2787i	-0.1432 + 0.2787i	
5	-0.1040 - 0.2546i	-0.1329 - 0.2701i	-0.1432 - 0.2787i	-0.1432 - 0.2787i	
6	-0.1046 + 0.2542i	-0.0790 + 0.2454i	-0.0705 + 0.2425i	-0.0705 + 0.2425i	
7	-0.1046 - 0.2542i	-0.0790 - 0.2454i	-0.0705 - 0.2425i	-0.0705 - 0.2425i	
8	-0.2284 + 0.0212i	-0.2738 + 0.0472i	-0.2858 + 0.0520i	-0.2858 + 0.0520i	
9	-0.2284 - 0.0212i	-0.2738 - 0.0472i	-0.2858 - 0.0520i	-0.2858 - 0.0520i	
10	-0.2293	-0.1509 + 0.0719i	-0.1525 + 0.0769i	-0.1525 + 0.0769i	
11	-0.1353 + 0.0476i	-0.1509 - 0.0719i	-0.1525 - 0.0769i	-0.1525 - 0.0769i	
12	-0.1353 - 0.0476i	-0.1722	-0.0450 + 0.1102i	-0.0450 + 0.1102i	
13	-0.1446 + 0.0002i	-0.0562 + 0.0955i	-0.0450 - 0.1102i	-0.0450 - 0.1102i	
14	-0.1446 - 0.0002i	-0.0562 - 0.0955i	-0.1697	-0.1697	
15	-0.0868 + 0.0557i	-0.1484	-0.1489	-0.1489	
16	-0.0868 - 0.0557i	-0.1446	-0.1446	-0.1446	
17	-0.0002 + 0.0601i	-0.0001 + 0.0601i	-0.0001 + 0.0601i	-0.0001 + 0.0601i	
18	-0.0002 - 0.0601i	-0.0001 - 0.0601i	-0.0001 - 0.0601i	-0.0001 - 0.0601i	
19	-0.0002 + 0.0600i	-0.0003 + 0.0598i	-0.0002 + 0.0598i	-0.0002 + 0.0598i	
20	-0.0002 - 0.0600i	-0.0003 - 0.0598i	-0.0002 - 0.0598i	-0.0002 - 0.0598i	
21	-0.0009 + 0.0508i	-0.0011 + 0.0511i	-0.0011 + 0.0512i	-0.0011 + 0.0512i	
22	-0.0009 - 0.0508i	-0.0011 - 0.0511i	-0.0011 - 0.0512i	-0.0011 - 0.0512i	
23	-0.0009 + 0.0509i	-0.0004 + 0.0508i	-0.0002 + 0.0509i	-0.0002 + 0.0509i	

24	-0.0009 - 0.0509i	-0.0004 - 0.0508i	-0.0002 - 0.0509i	-0.0002 - 0.0509i	
25	-0.0347 + 0.0149i	-0.0434	-0.0444	-0.0444	
26	-0.0347 - 0.0149i	-0.0178	-0.0045 + 0.0120i	-0.0045 + 0.0120i	
27	-0.0204	-0.0162 + 0.0058i	-0.0045 - 0.0120i	-0.0045 - 0.0120i	
28	-0.0052 + 0.0136i	-0.0162 - 0.0058i	-0.0078 + 0.0147i	-0.0078 + 0.0147i	
29	-0.0052 - 0.0136i	-0.0046 + 0.0129i	-0.0078 - 0.0147i	-0.0078 - 0.0147i	
30	-0.0078 + 0.0145i	-0.0046 - 0.0129i	-0.0185	-0.0185	
31	-0.0078 - 0.0145i	-0.0078 + 0.0146i	-0.0136 + 0.0059i	-0.0136 + 0.0059i	
32	-0.0134	-0.0078 - 0.0146i	-0.0136 - 0.0059i	-0.0136 - 0.0059i	
33	-0.0106	-0.0083	-0.0059	-0.0059	
34	-0.01	-0.01	-0.01	-0.01	

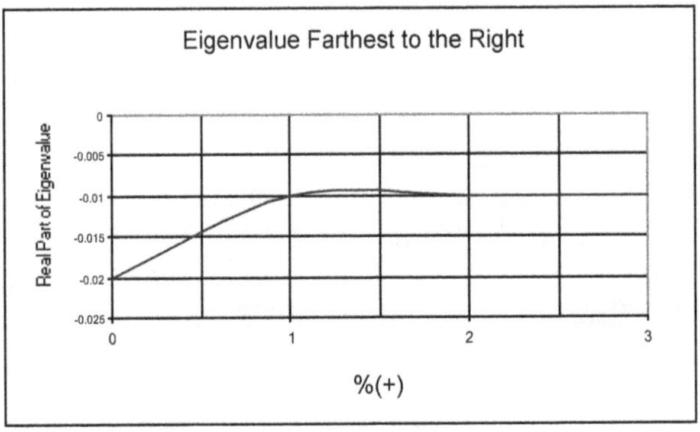

Observation :

As the positive error increases, the positive discrepancy in step response increases, but it remains stable until the modeling error increases too much. This instability becomes evident when the modeling error reaches 20%.

(Blank Page)

5) **Center of pressure: lalpha (The distance of the center of pressure from the Body Frame origin in the math model = 34.1 ft.)**

Case 1 (actual value is less than the value in the math model)

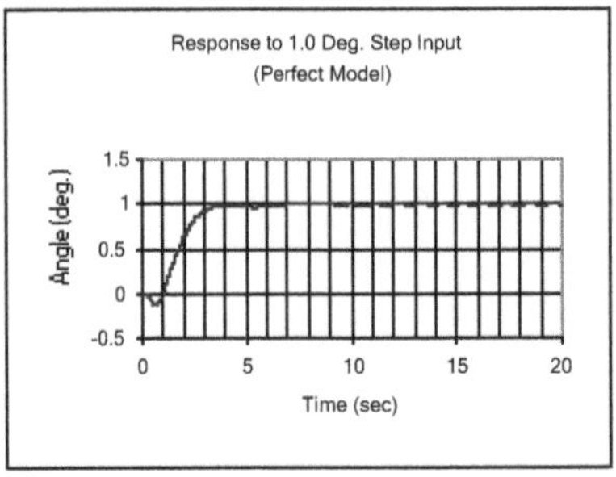

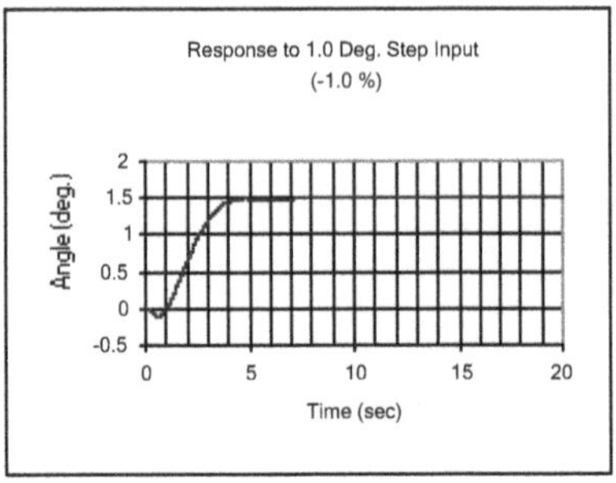

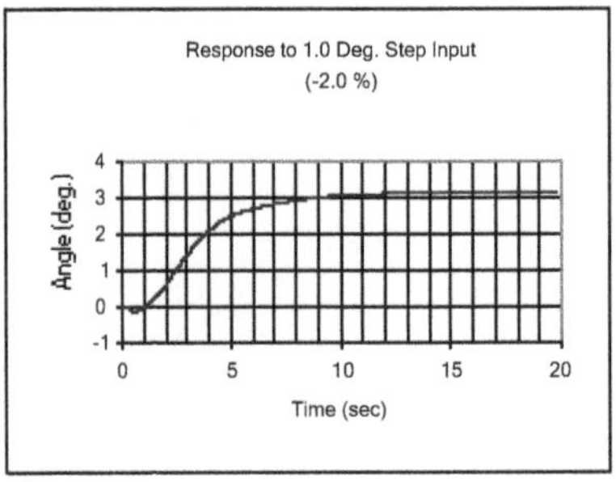

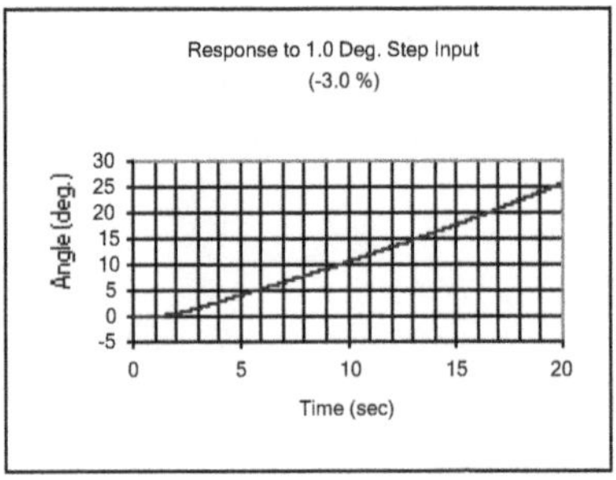

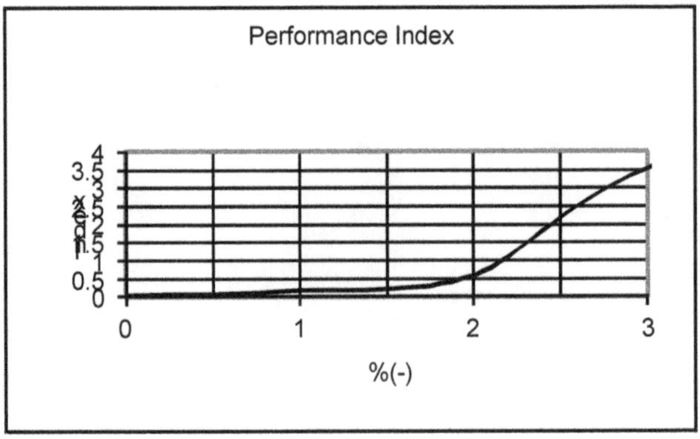

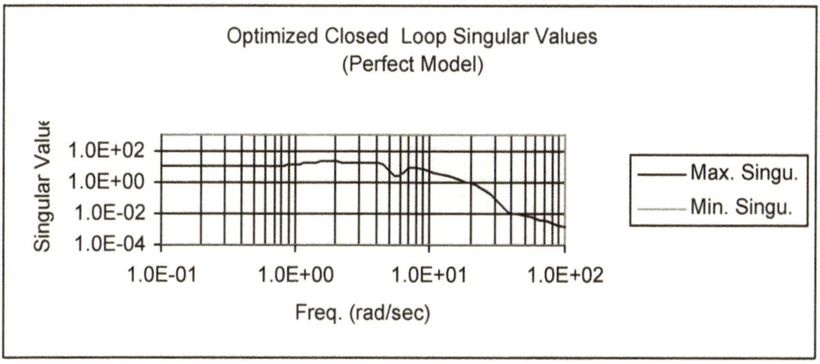
Optimized Closed Loop Singular Values
(Perfect Model)
Singular Value
1.0E+02
1.0E+00
1.0E-02
1.0E-04
1.0E-01
1.0E+00
1.0E+01
1.0E+02
Freq. (rad/sec)
Max. Singu.
Min. Singu.

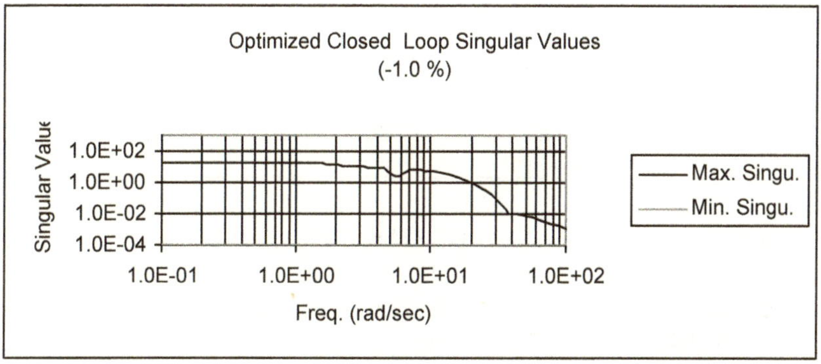
Optimized Closed Loop Singular Values
(-1.0 %)
Singular Value
1.0E+02
1.0E+00
1.0E-02
1.0E-04
1.0E-01
1.0E+00
1.0E+01
1.0E+02
Freq. (rad/sec)
Max. Singu.
Min. Singu.

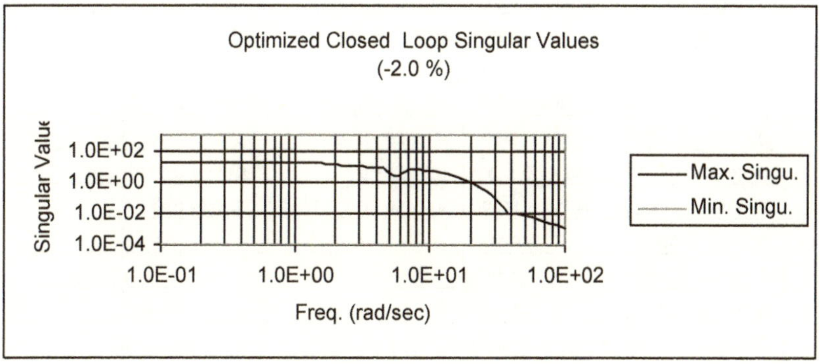
Optimized Closed Loop Singular Values
(-2.0 %)
Singular Value
1.0E+02
1.0E+00
1.0E-02
1.0E-04
1.0E-01
1.0E+00
1.0E+01
1.0E+02
Freq. (rad/sec)
Max. Singu.
Min. Singu.

Appendix III: Collection of Math Modeling Error Sensitivity Response and Singular Value Plots, and Eigenvalue Tables

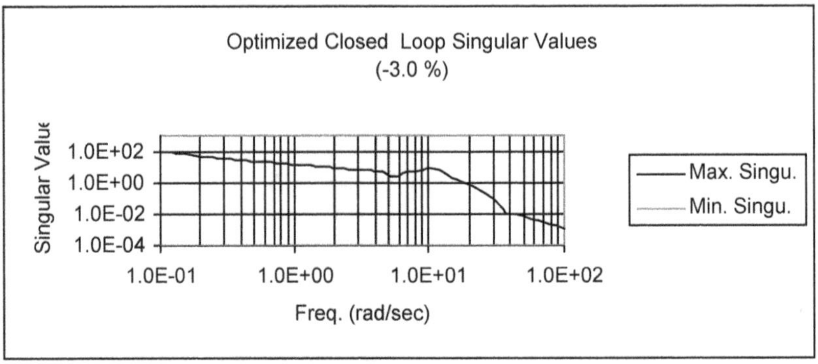

Closed Loop System Eigenvalues

	0%	-1.00%	-2.00%	-3.00%	
	1.0e+002 *	1.0e+002 *	1.0e+002 *	1.0e+002 *	
1	-0.25	-0.25	-0.25	-0.25	
2	-0.5824 + 1.3344i	-0.5824 + 1.3344i	-0.5824 + 1.3344i	-0.5824 + 1.3344i	
3	-0.5824 - 1.3344i	-0.5824 - 1.3344i	-0.5824 - 1.3344i	-0.5824 - 1.3344i	
4	-0.1040 + 0.2546i	-0.1432 + 0.2787i	-0.1432 + 0.2787i	-0.1306 + 0.2561i	
5	-0.1040 - 0.2546i	-0.1432 - 0.2787i	-0.1432 - 0.2787i	-0.1306 - 0.2561i	
6	-0.1046 + 0.2542i	-0.0705 + 0.2425i	-0.0705 + 0.2425i	-0.0853 + 0.2644i	
7	-0.1046 - 0.2542i	-0.0705 - 0.2425i	-0.0705 - 0.2425i	-0.0853 - 0.2644i	
8	-0.2284 + 0.0212i	-0.2858 + 0.0520i	-0.2858 + 0.0520i	-0.2591 + 0.0276i	
9	-0.2284 - 0.0212i	-0.2858 - 0.0520i	-0.2858 - 0.0520i	-0.2591 - 0.0276i	
10	-0.2293	-0.1525 + 0.0769i	-0.1525 + 0.0769i	-0.2096 + 0.1017i	
11	-0.1353 + 0.0476i	-0.1525 - 0.0769i	-0.1525 - 0.0769i	-0.2096 - 0.1017i	
12	-0.1353 - 0.0476i	-0.0450 + 0.1102i	-0.0450 + 0.1102i	-0.0167 + 0.1119i	
13	-0.1446 + 0.0002i	-0.0450 - 0.1102i	-0.0450 - 0.1102i	-0.0167 - 0.1119i	
14	-0.1446 - 0.0002i	-0.1697	-0.1697	-0.1685	
15	-0.0868 + 0.0557i	-0.1489	-0.1489	-0.1464	
16	-0.0868 - 0.0557i	-0.1446	-0.1446	-0.1446	
17	-0.0002 + 0.0601i	-0.0001 + 0.0601i	-0.0001 + 0.0601i	-0.0471	
18	-0.0002 - 0.0601i	-0.0001 - 0.0601i	-0.0001 - 0.0601i	-0.0001 + 0.0602i	
19	-0.0002 + 0.0600i	-0.0002 + 0.0598i	-0.0002 + 0.0598i	-0.0001 - 0.0602i	
20	-0.0002 - 0.0600i	-0.0002 - 0.0598i	-0.0002 - 0.0598i	-0.0002 + 0.0597i	
21	-0.0009 + 0.0508i	-0.0011 + 0.0512i	-0.0011 + 0.0512i	-0.0002 - 0.0597i	
22	-0.0009 - 0.0508i	-0.0011 - 0.0512i	-0.0011 - 0.0512i	-0.0008 + 0.0506i	

23	-0.0009 + 0.0509i	-0.0002 + 0.0509i	-0.0002 + 0.0509i	-0.0008 - 0.0506i	
24	-0.0009 - 0.0509i	-0.0002 - 0.0509i	-0.0002 - 0.0509i	-0.0003 + 0.0511i	
25	-0.0347 + 0.0149i	-0.0444	-0.0444	-0.0003 - 0.0511i	
26	-0.0347 - 0.0149i	-0.0045 + 0.0120i	-0.0045 + 0.0120i	-0.0192	
27	-0.0204	-0.0045 - 0.0120i	-0.0045 - 0.0120i	0.0002	
28	-0.0052 + 0.0136i	-0.0078 + 0.0147i	-0.0078 + 0.0147i	-0.0078 + 0.0147i	
29	-0.0052 - 0.0136i	-0.0078 - 0.0147i	-0.0078 - 0.0147i	-0.0078 - 0.0147i	
30	-0.0078 + 0.0145i	-0.0185	-0.0185	-0.0054 + 0.0111i	
31	-0.0078 - 0.0145i	-0.0136 + 0.0059i	-0.0136 + 0.0059i	-0.0054 - 0.0111i	
32	-0.0134	-0.0136 - 0.0059i	-0.0136 - 0.0059i	-0.0119 + 0.0046i	
33	-0.0106	-0.0059	-0.0059	-0.0119 - 0.0046i	
34	-0.01	-0.01	-0.01	-0.01	

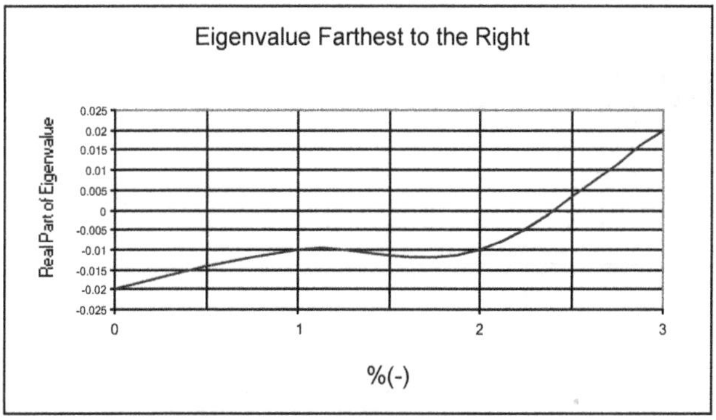

Observation:

This system parameter is sensitive to modeling error. System instability results when the value used in the model is only 2.4% less than the true value.

Case 2 (actual value is greater than the value in the math model)

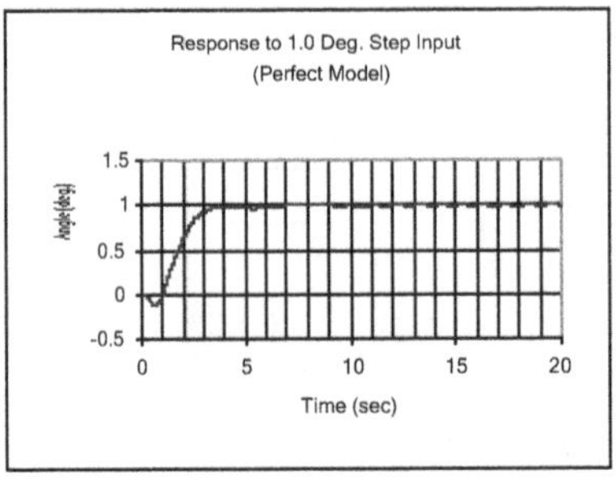

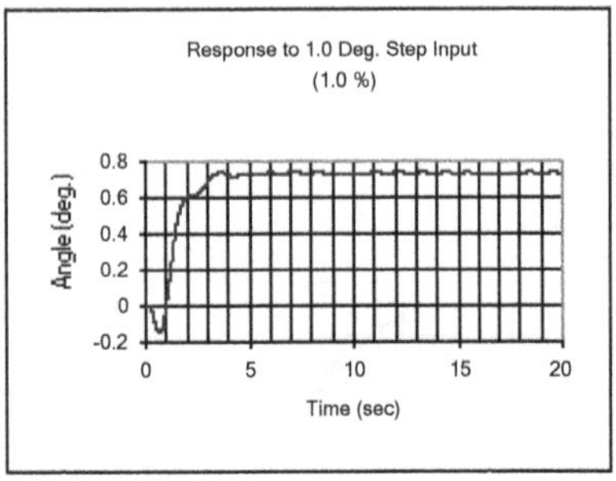

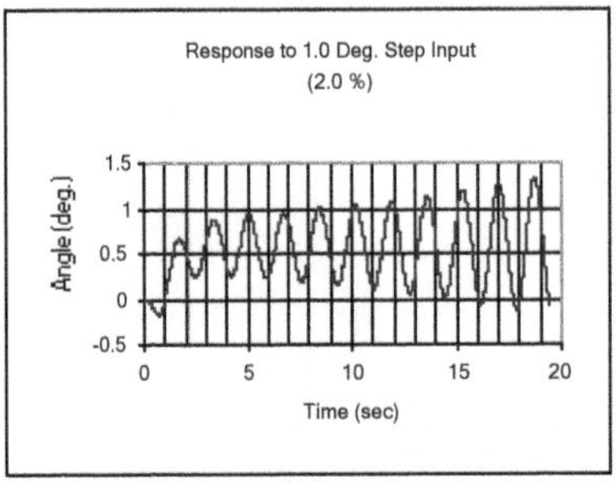

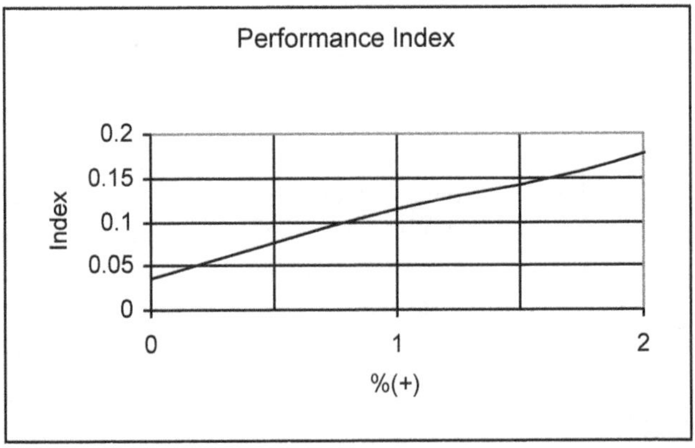

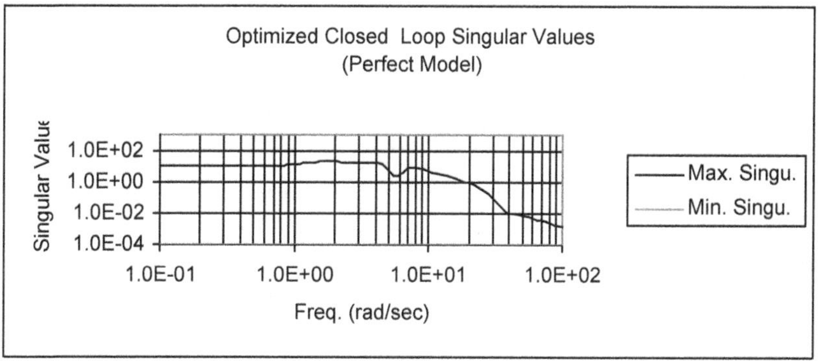
Optimized Closed Loop Singular Values
(Perfect Model)
Singular Value
1.0E+02
1.0E+00
1.0E-02
1.0E-04
1.0E-01
1.0E+00
1.0E+01
1.0E+02
Freq. (rad/sec)
Max. Singu.
Min. Singu.

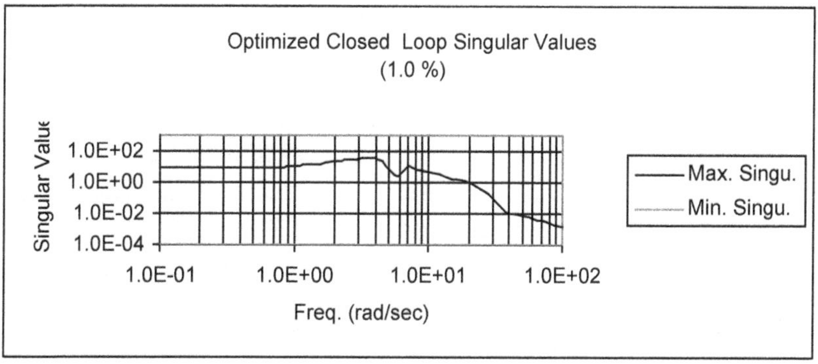
Optimized Closed Loop Singular Values
(1.0 %)
Singular Value
1.0E+02
1.0E+00
1.0E-02
1.0E-04
1.0E-01
1.0E+00
1.0E+01
1.0E+02
Freq. (rad/sec)
Max. Singu.
Min. Singu.

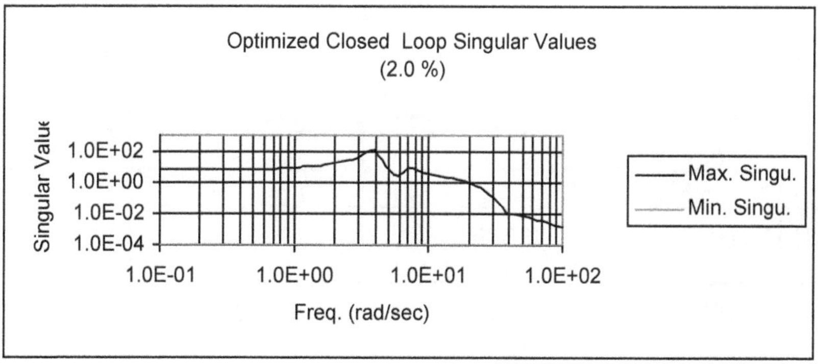
Optimized Closed Loop Singular Values
(2.0 %)
Singular Value
1.0E+02
1.0E+00
1.0E-02
1.0E-04
1.0E-01
1.0E+00
1.0E+01
1.0E+02
Freq. (rad/sec)
Max. Singu.
Min. Singu.

Appendix III: Collection of Math Modeling Error Sensitivity Response and Singular Value Plots, and Eigenvalue Tables

Closed Loop System Eigenvalues					
	0%	1.00%	2.00%		
	1.0e+002 *	1.0e+002 *	1.0e+002 *		
1	-0.25	-0.25	-0.25		
2	-0.5824 + 1.3344i	-0.5824 + 1.3344i	-0.5824 + 1.3344i		
3	-0.5824 - 1.3344i	-0.5824 - 1.3344i	-0.5824 - 1.3344i		
4	-0.1040 + 0.2546i	-0.1071 + 0.2679i	-0.1074 + 0.2726i		
5	-0.1040 - 0.2546i	-0.1071 - 0.2679i	-0.1074 - 0.2726i		
6	-0.1046 + 0.2542i	-0.0962 + 0.2354i	-0.0873 + 0.2255i		
7	-0.1046 - 0.2542i	-0.0962 - 0.2354i	-0.0873 - 0.2255i		
8	-0.2284 + 0.0212i	-0.2556	-0.2621		
9	-0.2284 - 0.0212i	-0.2360 + 0.0471i	-0.2427 + 0.0531i		
10	-0.2293	-0.2360 - 0.0471i	-0.2427 - 0.0531i		
11	-0.1353 + 0.0476i	-0.1239 + 0.1294i	-0.1335 + 0.1553i		
12	-0.1353 - 0.0476i	-0.1239 - 0.1294i	-0.1335 - 0.1553i		
13	-0.1446 + 0.0002i	-0.1619	-0.1644		
14	-0.1446 - 0.0002i	-0.1468	-0.1465		
15	-0.0868 + 0.0557i	-0.1446	-0.1446		
16	-0.0868 - 0.0557i	-0.049	-0.0477		
17	-0.0002 + 0.0601i	-0.0003 + 0.0603i	-0.0002 + 0.0604i		
18	-0.0002 - 0.0601i	-0.0003 - 0.0603i	-0.0002 - 0.0604i		
19	-0.0002 + 0.0600i	-0.0001 + 0.0600i	-0.0000 + 0.0600i		
20	-0.0002 - 0.0600i	-0.0001 - 0.0600i	-0.0000 - 0.0600i		
21	-0.0009 + 0.0508i	-0.0008 + 0.0507i	-0.0011 + 0.0523i		
22	-0.0009 - 0.0508i	-0.0008 - 0.0507i	-0.0011 - 0.0523i		
23	-0.0009 + 0.0509i	-0.0015 + 0.0515i	-0.0008 + 0.0507i		
24	-0.0009 - 0.0509i	-0.0015 - 0.0515i	-0.0008 - 0.0507i		
25	-0.0347 + 0.0149i	-0.0104 + 0.0360i	0.0006 + 0.0368i		
26	-0.0347 - 0.0149i	-0.0104 - 0.0360i	0.0006 - 0.0368i		
27	-0.0204	-0.0195	-0.0193		
28	-0.0052 + 0.0136i	-0.0065 + 0.0136i	-0.0075 + 0.0146i		
29	-0.0052 - 0.0136i	-0.0065 - 0.0136i	-0.0075 - 0.0146i		
30	-0.0078 + 0.0145i	-0.0076 + 0.0144i	-0.0069 + 0.0129i		
31	-0.0078 - 0.0145i	-0.0076 - 0.0144i	-0.0069 - 0.0129i		

32	-0.0134	-0.0116 + 0.0023i	-0.0116 + 0.0028i		
33	-0.0106	-0.0116 - 0.0023i	-0.0116 - 0.0028i		
34	-0.01	-0.01	-0.01		

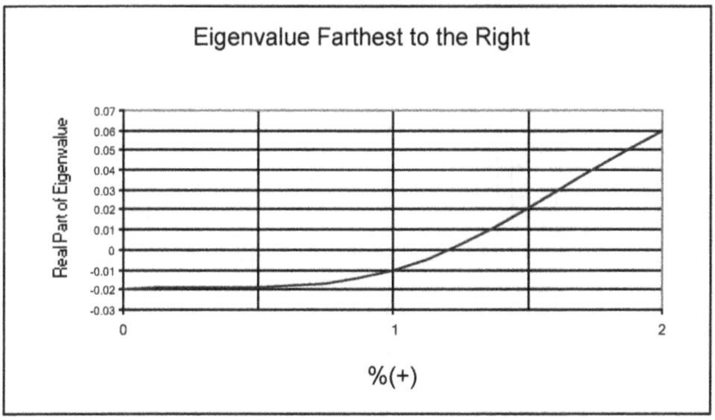

Observation:

The system is more sensitive to positive modeling errors than negative errors. Modeling errors larger than about 1.2 % cause system instability.

6) Fuel Sloshing 1 Pendulum Mass: mp1 [math model Pendulum Mass (Sloshing) 1= 388 slugs]

Case 1 (actual value is less than the value in the math model)

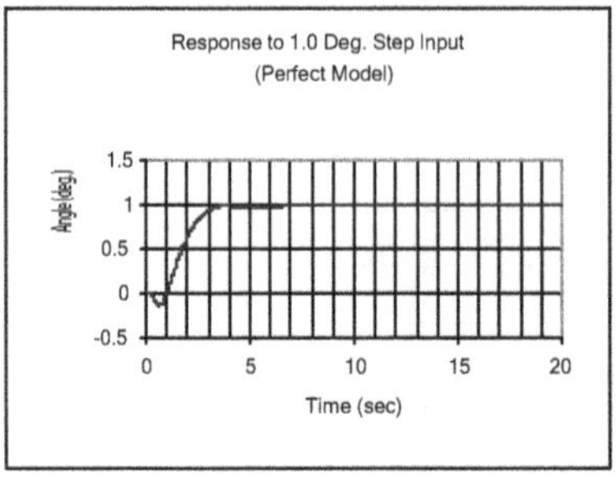

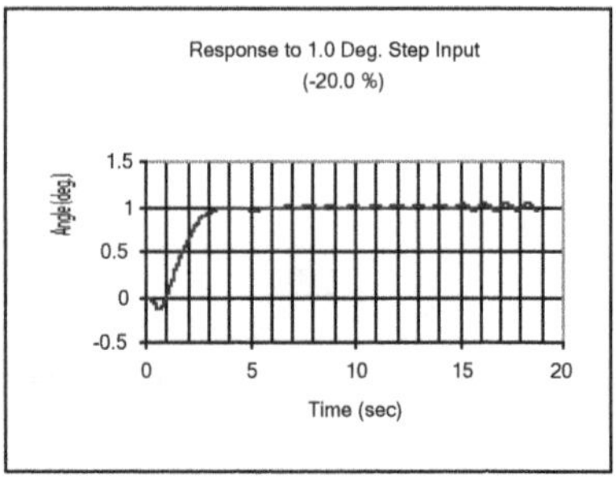

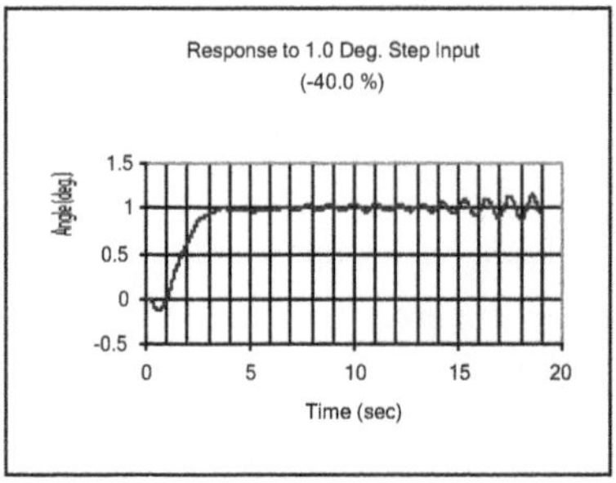

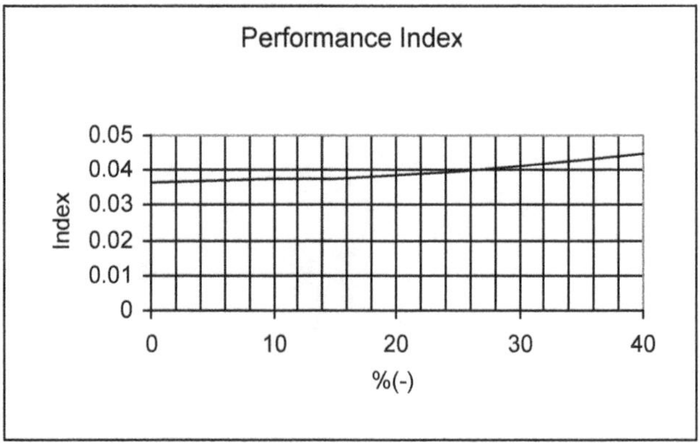

Appendix III: Collection of Math Modeling Error Sensitivity Response and Singular Value Plots, and Eigenvalue Tables

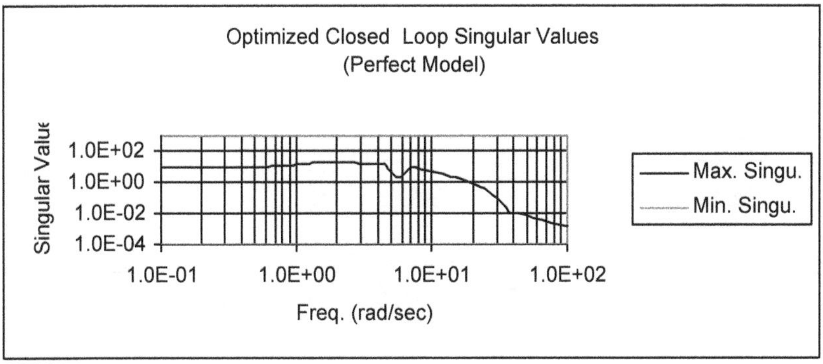

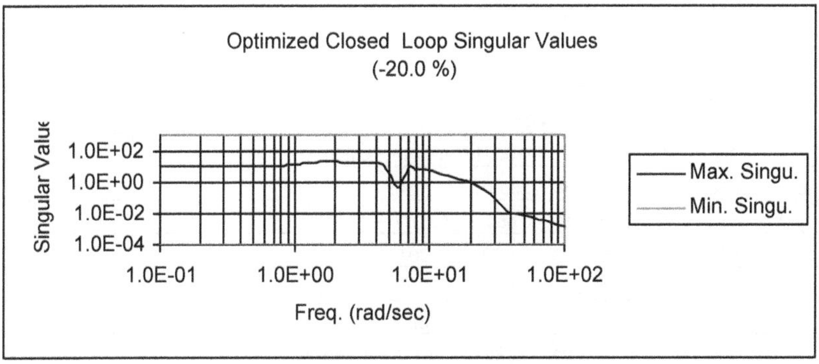

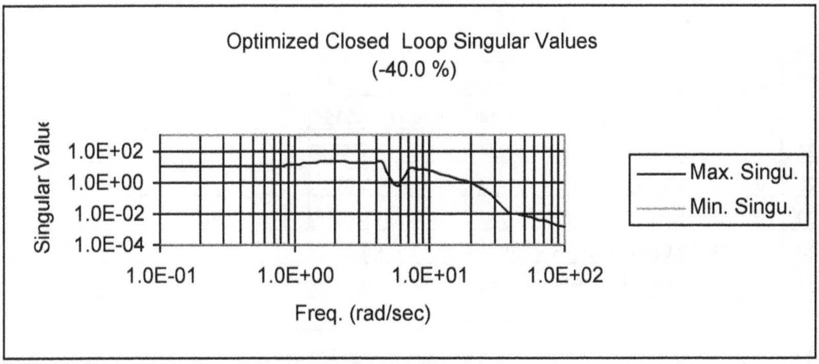

Appendix III: Collection of Math Modeling Error Sensitivity Response and Singular Value Plots, and Eigenvalue Tables

Closed Loop System Eigenvalues					
	0%	-20.00%	-40.00%		
	1.0e+002 *	1.0e+002 *	1.0e+002 *		
1	-0.25	-0.25	-0.25		
2	-0.5824 + 1.3344i	-0.5824 + 1.3344i	-0.5824 + 1.3344i		
3	-0.5824 - 1.3344i	-0.5824 - 1.3344i	-0.5824 - 1.3344i		
4	-0.1040 + 0.2546i	-0.1112 + 0.2538i	-0.1140 + 0.2539i		
5	-0.1040 - 0.2546i	-0.1112 - 0.2538i	-0.1140 - 0.2539i		
6	-0.1046 + 0.2542i	-0.0978 + 0.2561i	-0.0952 + 0.2570i		
7	-0.1046 - 0.2542i	-0.0978 - 0.2561i	-0.0952 - 0.2570i		
8	-0.2284 + 0.0212i	-0.2411 + 0.0223i	-0.2450 + 0.0248i		
9	-0.2284 - 0.0212i	-0.2411 - 0.0223i	-0.2450 - 0.0248i		
10	-0.2293	-0.1697 + 0.0589i	-0.1825 + 0.0691i		
11	-0.1353 + 0.0476i	-0.1697 - 0.0589i	-0.1825 - 0.0691i		
12	-0.1353 - 0.0476i	-0.1887	-0.1776		
13	-0.1446 + 0.0002i	-0.0652 + 0.0773i	-0.0567 + 0.0868i		
14	-0.1446 - 0.0002i	-0.0652 - 0.0773i	-0.0567 - 0.0868i		
15	-0.0868 + 0.0557i	-0.1446 + 0.0001i	-0.1446 + 0.0001i		
16	-0.0868 - 0.0557i	-0.1446 - 0.0001i	-0.1446 - 0.0001i		
17	-0.0002 + 0.0601i	0.0013 + 0.0594i	0.0017 + 0.0587i		
18	-0.0002 - 0.0601i	0.0013 - 0.0594i	0.0017 - 0.0587i		
19	-0.0002 + 0.0600i	-0.0017 + 0.0590i	-0.0017 + 0.0582i		
20	-0.0002 - 0.0600i	-0.0017 - 0.0590i	-0.0017 - 0.0582i		
21	-0.0009 + 0.0508i	-0.0022 + 0.0508i	-0.0029 + 0.0505i		
22	-0.0009 - 0.0508i	-0.0022 - 0.0508i	-0.0029 - 0.0505i		
23	-0.0009 + 0.0509i	-0.0000 + 0.0507i	0.0004 + 0.0506i		
24	-0.0009 - 0.0509i	-0.0000 - 0.0507i	0.0004 - 0.0506i		
25	-0.0347 + 0.0149i	-0.0281 + 0.0148i	-0.0247 + 0.0151i		
26	-0.0347 - 0.0149i	-0.0281 - 0.0148i	-0.0247 - 0.0151i		
27	-0.0204	-0.0219	-0.0232		
28	-0.0052 + 0.0136i	-0.0053 + 0.0136i	-0.0053 + 0.0137i		
29	-0.0052 - 0.0136i	-0.0053 - 0.0136i	-0.0053 - 0.0137i		
30	-0.0078 + 0.0145i	-0.0078 + 0.0145i	-0.0078 + 0.0145i		
31	-0.0078 - 0.0145i	-0.0078 - 0.0145i	-0.0078 - 0.0145i		
32	-0.0134	-0.0139	-0.0141		
33	-0.0106	-0.01	-0.0096		
34	-0.01	-0.01	-0.01		

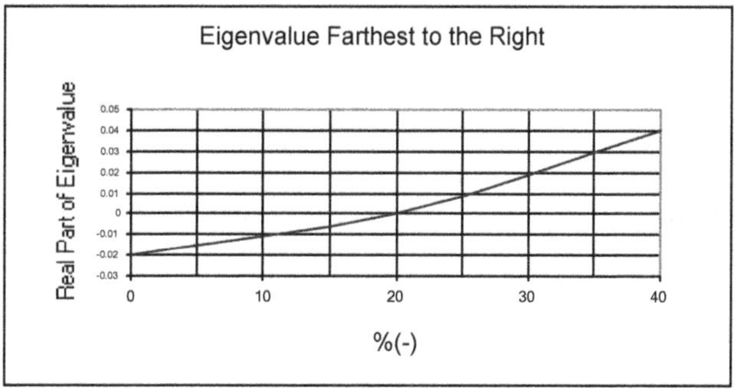

Observation:

The control system is insensitive to pendulum mass 1 modeling error. When the error is - 20.0%, the step input response shows some instability, but it is not significant.

Case 2 (actual value is greater than the value in the math model)

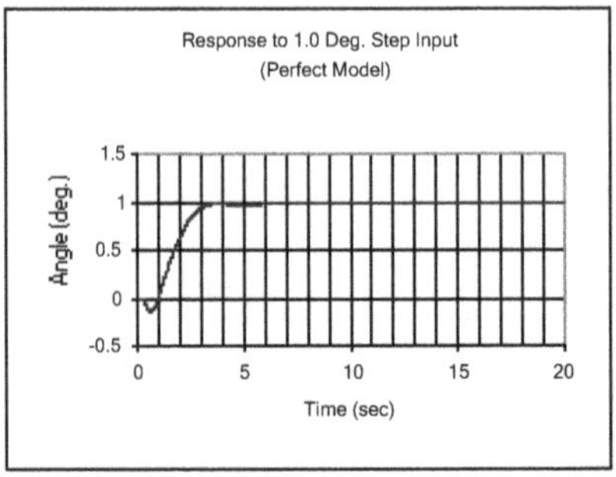

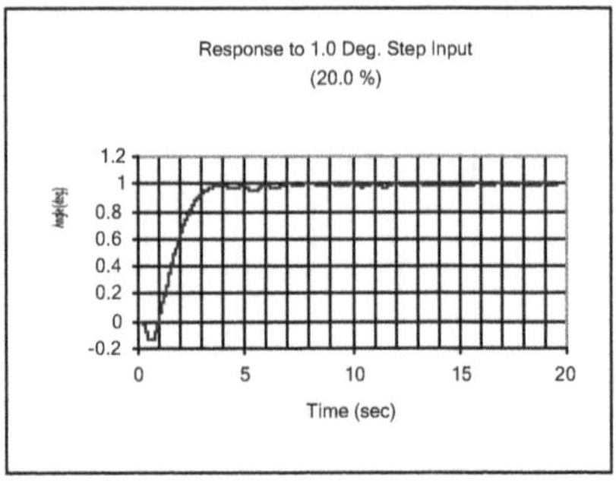

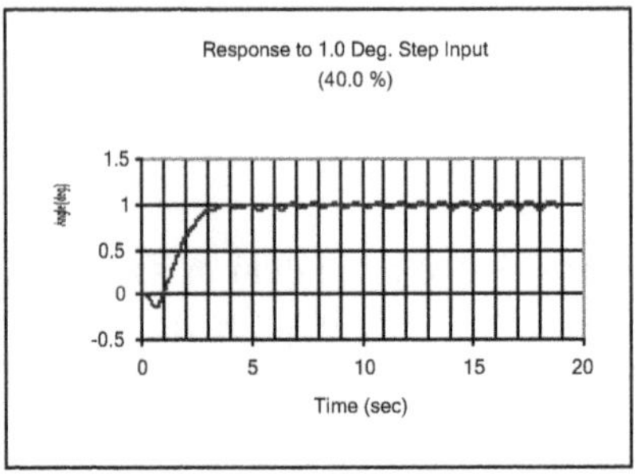

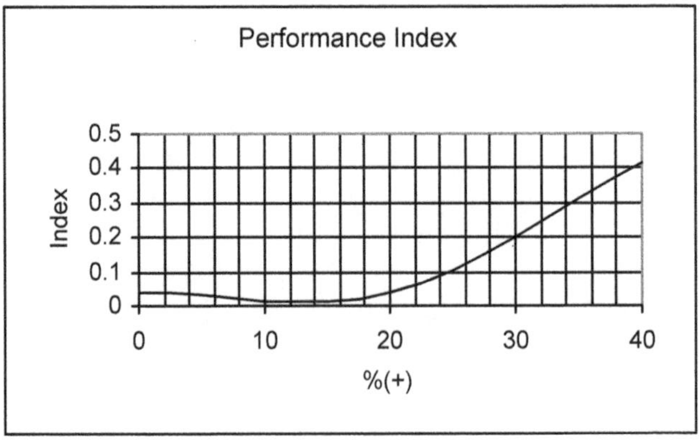

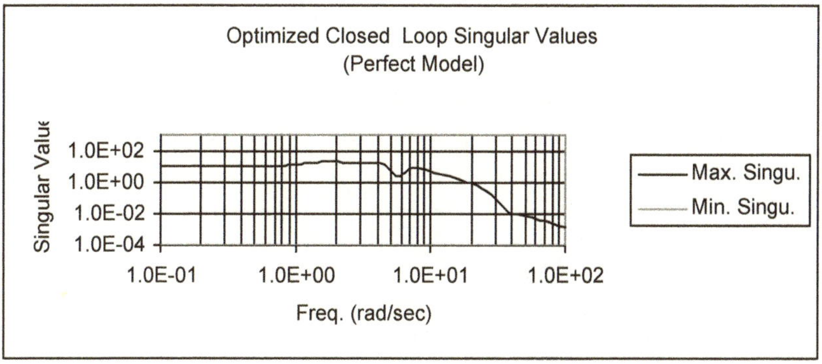
Optimized Closed Loop Singular Values
(Perfect Model)
Singular Valu
1.0E+02
1.0E+00
1.0E-02
1.0E-04
1.0E-01
1.0E+00
1.0E+01
1.0E+02
Freq. (rad/sec)
Max. Singu.
Min. Singu.

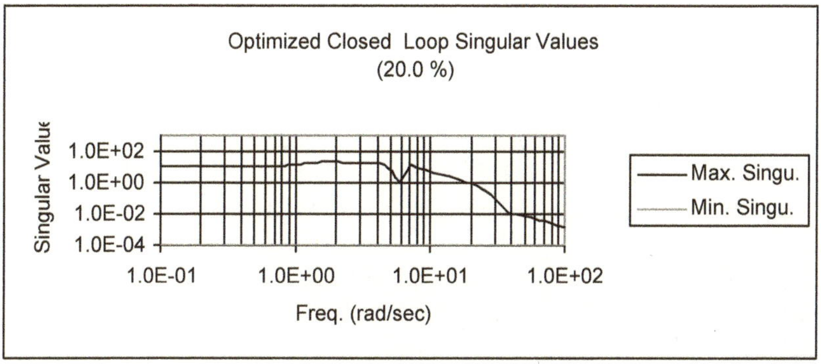
Optimized Closed Loop Singular Values
(20.0 %)
Singular Valu
1.0E+02
1.0E+00
1.0E-02
1.0E-04
1.0E-01
1.0E+00
1.0E+01
1.0E+02
Freq. (rad/sec)
Max. Singu.
Min. Singu.

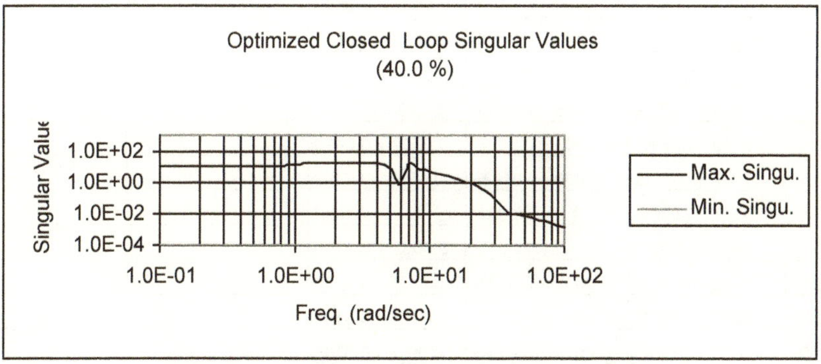
Optimized Closed Loop Singular Values
(40.0 %)
Singular Valu
1.0E+02
1.0E+00
1.0E-02
1.0E-04
1.0E-01
1.0E+00
1.0E+01
1.0E+02
Freq. (rad/sec)
Max. Singu.
Min. Singu.

Appendix III: Collection of Math Modeling Error Sensitivity Response and Singular Value Plots, and Eigenvalue Tables

	Closed Loop System Eigenvalues				
	0%	20.00%	40.00%		
	1.0e+002 *	1.0e+002 *	1.0e+002 *		
1	-0.25	-0.25	-0.25		
2	-0.5824 + 1.3344i	-0.5824 + 1.3344i	-0.5824 + 1.3344i		
3	-0.5824 - 1.3344i	-0.5824 - 1.3344i	-0.5824 - 1.3344i		
4	-0.1040 + 0.2546i	-0.1053 + 0.2607i	-0.1057 + 0.2631i		
5	-0.1040 - 0.2546i	-0.1053 - 0.2607i	-0.1057 - 0.2631i		
6	-0.1046 + 0.2542i	-0.1029 + 0.2471i	-0.1021 + 0.2436i		
7	-0.1046 - 0.2542i	-0.1029 - 0.2471i	-0.1021 - 0.2436i		
8	-0.2284 + 0.0212i	-0.2472	-0.253		
9	-0.2284 - 0.0212i	-0.2267 + 0.0341i	-0.2289 + 0.0393i		
10	-0.2293	-0.2267 - 0.0341i	-0.2289 - 0.0393i		
11	-0.1353 + 0.0476i	-0.1188 + 0.0905i	-0.1211 + 0.1058i		
12	-0.1353 - 0.0476i	-0.1188 - 0.0905i	-0.1211 - 0.1058i		
13	-0.1446 + 0.0002i	-0.151	-0.158		
14	-0.1446 - 0.0002i	-0.1443	-0.1444		
15	-0.0868 + 0.0557i	-0.1449	-0.1447		
16	-0.0868 - 0.0557i	-0.0393 + 0.0309i	0.0007 + 0.0647i		
17	-0.0002 + 0.0601i	-0.0393 - 0.0309i	0.0007 - 0.0647i		
18	-0.0002 - 0.0601i	0.0000 + 0.0628i	-0.0321 + 0.0351i		
19	-0.0002 + 0.0600i	0.0000 - 0.0628i	-0.0321 - 0.0351i		
20	-0.0002 - 0.0600i	-0.0002 + 0.0589i	-0.0002 + 0.0587i		
21	-0.0009 + 0.0508i	-0.0002 - 0.0589i	-0.0002 - 0.0587i		
22	-0.0009 - 0.0508i	-0.0006 + 0.0518i	-0.0004 + 0.0521i		
23	-0.0009 + 0.0509i	-0.0006 - 0.0518i	-0.0004 - 0.0521i		
24	-0.0009 - 0.0509i	-0.0009 + 0.0499i	-0.0009 + 0.0496i		
25	-0.0347 + 0.0149i	-0.0009 - 0.0499i	-0.0009 - 0.0496i		
26	-0.0347 - 0.0149i	-0.0354	-0.0316		
27	-0.0204	-0.0189	-0.0052 + 0.0136i		
28	-0.0052 + 0.0136i	-0.0052 + 0.0136i	-0.0052 - 0.0136i		
29	-0.0052 - 0.0136i	-0.0052 - 0.0136i	-0.0078 + 0.0145i		
30	-0.0078 + 0.0145i	-0.0078 + 0.0145i	-0.0078 - 0.0145i		
31	-0.0078 - 0.0145i	-0.0078 - 0.0145i	-0.0178		
32	-0.0134	-0.0120 + 0.0003i	-0.0120 + 0.0017i		
33	-0.0106	-0.0120 - 0.0003i	-0.0120 - 0.0017i		
34	-0.01	-0.01	-0.01		

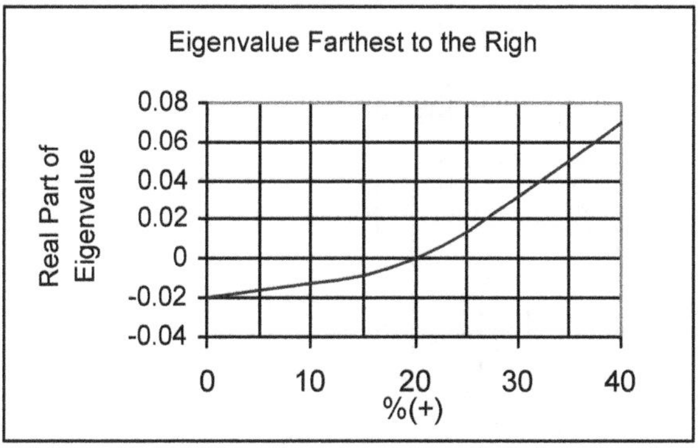

Observation:

When the error is +20.0%, the system displays some instability.

7) **Fuel Sloshing 2 Pendulum Mass: mp2 (math model Pendulum Mass (Sloshing 2 = 280 slugs)**

Case 1 (actual value is less than the value in the math model)

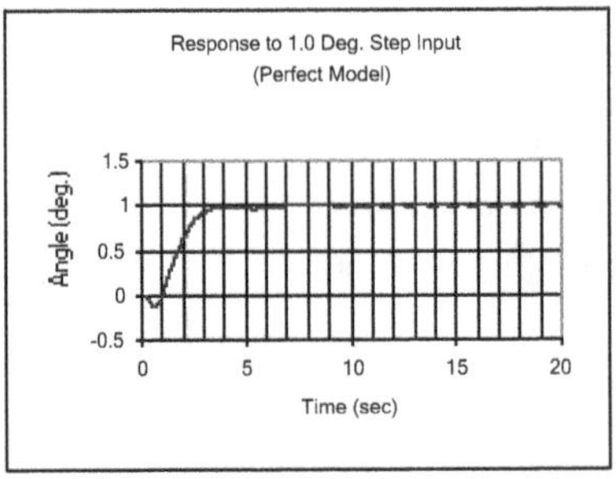

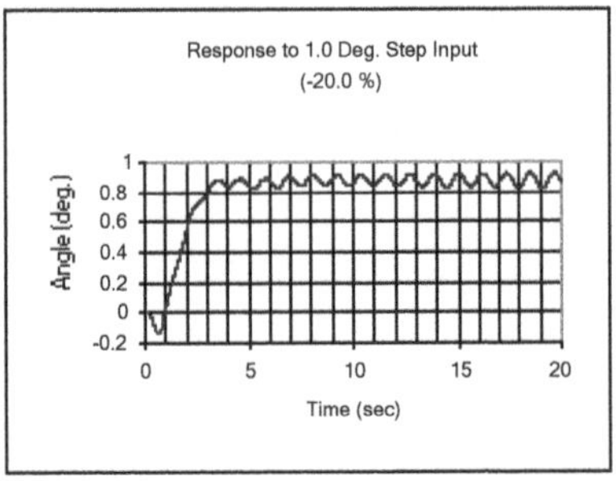

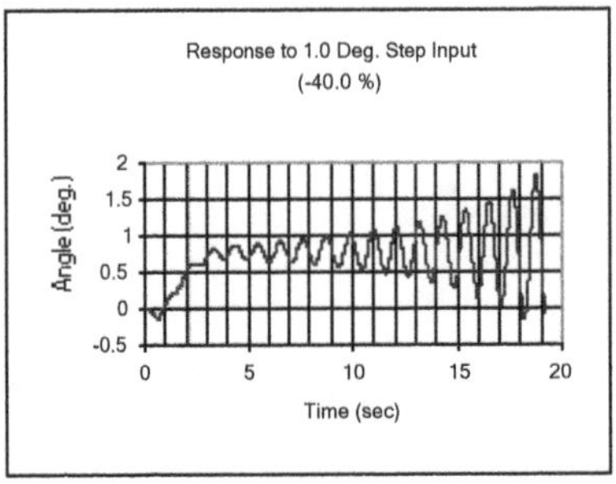

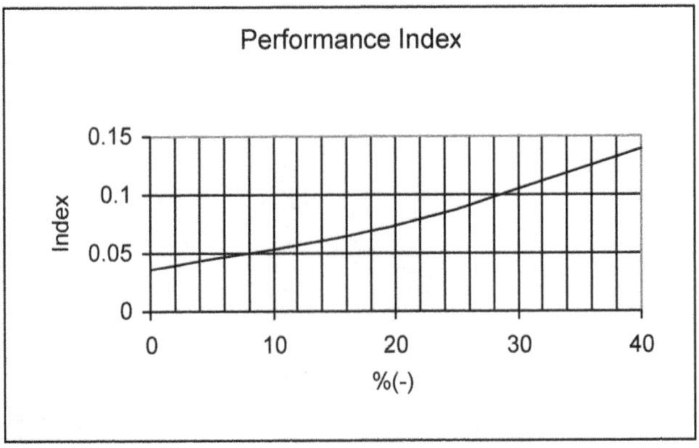

Appendix III: Collection of Math Modeling Error Sensitivity Response and Singular Value Plots, and Eigenvalue Tables

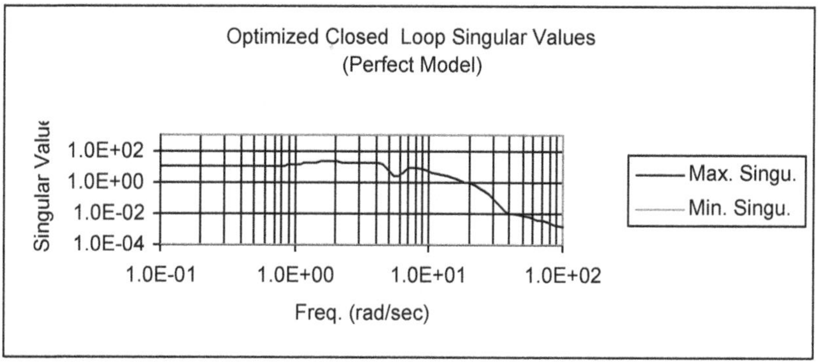

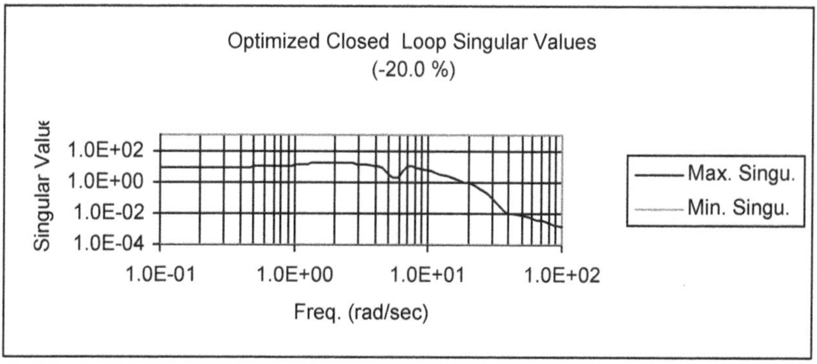

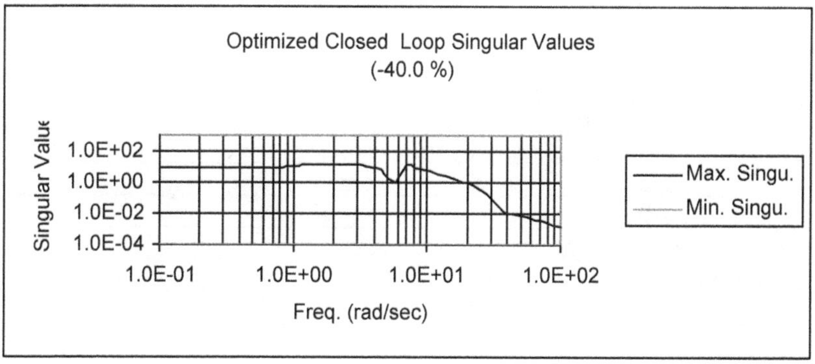

Appendix III: Collection of Math Modeling Error Sensitivity Response and Singular Value Plots, and Eigenvalue Tables

Closed Loop System Eigenvalues					
	0%	-20.00%	-40.00%		
	1.0e+002 *	1.0e+002 *	1.0e+002 *		
1	-0.25	-0.25	-0.25		
2	-0.5824 + 1.3344i	-0.5824 + 1.3344i	-0.5824 + 1.3344i		
3	-0.5824 - 1.3344i	-0.5824 - 1.3344i	-0.5824 - 1.3344i		
4	-0.1040 + 0.2546i	-0.1101 + 0.2538i	-0.1124 + 0.2538i		
5	-0.1040 - 0.2546i	-0.1101 - 0.2538i	-0.1124 - 0.2538i		
6	-0.1046 + 0.2542i	-0.0986 + 0.2558i	-0.0963 + 0.2566i		
7	-0.1046 - 0.2542i	-0.0986 - 0.2558i	-0.0963 - 0.2566i		
8	-0.2284 + 0.0212i	-0.2403 + 0.0236i	-0.2448 + 0.0264i		
9	-0.2284 - 0.0212i	-0.2403 - 0.0236i	-0.2448 - 0.0264i		
10	-0.2293	-0.202	-0.1912		
11	-0.1353 + 0.0476i	-0.1460 + 0.0559i	-0.1529 + 0.0624i		
12	-0.1353 - 0.0476i	-0.1460 - 0.0559i	-0.1529 - 0.0624i		
13	-0.1446 + 0.0002i	-0.1447 + 0.0001i	-0.1449		
14	-0.1446 - 0.0002i	-0.1447 - 0.0001i	-0.1447		
15	-0.0868 + 0.0557i	-0.0811 + 0.0559i	-0.0778 + 0.0559i		
16	-0.0868 - 0.0557i	-0.0811 - 0.0559i	-0.0778 - 0.0559i		
17	-0.0002 + 0.0601i	0.0002 + 0.0602i	-0.0006 + 0.0602i		
18	-0.0002 - 0.0601i	0.0002 - 0.0602i	-0.0006 - 0.0602i		
19	-0.0002 + 0.0600i	-0.0005 + 0.0601i	0.0009 + 0.0600i		
20	-0.0002 - 0.0600i	-0.0005 - 0.0601i	0.0009 - 0.0600i		
21	-0.0009 + 0.0508i	0.0003 + 0.0543i	0.0017 + 0.0570i		
22	-0.0009 - 0.0508i	0.0003 - 0.0543i	0.0017 - 0.0570i		
23	-0.0009 + 0.0509i	-0.0001 + 0.0504i	-0.0000 + 0.0505i		
24	-0.0009 - 0.0509i	-0.0001 - 0.0504i	-0.0000 - 0.0505i		
25	-0.0347 + 0.0149i	-0.0328 + 0.0146i	-0.0315 + 0.0143i		
26	-0.0347 - 0.0149i	-0.0328 - 0.0146i	-0.0315 - 0.0143i		
27	-0.0204	-0.0215	-0.0226		
28	-0.0052 + 0.0136i	-0.0055 + 0.0134i	-0.0056 + 0.0132i		
29	-0.0052 - 0.0136i	-0.0055 - 0.0134i	-0.0056 - 0.0132i		
30	-0.0078 + 0.0145i	-0.0078 + 0.0145i	-0.0078 + 0.0145i		
31	-0.0078 - 0.0145i	-0.0078 - 0.0145i	-0.0078 - 0.0145i		
32	-0.0134	-0.0136	-0.0137		
33	-0.0106	-0.0103	-0.0101		
34	-0.01	-0.01	-0.01		

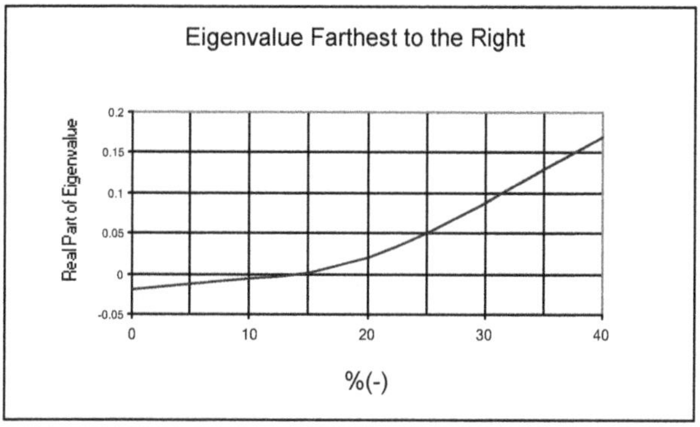

Observation:

Instability occurs even when the actual mass is only 15.0% less than the mass in the math model. The pendulum 1 case did not show significant instability even when the actual mass was 20.0% less or greater than the mass in the math model. The pendulum 1 is located in the positive side of the Body Frame origin, and the pendulum 2 is located in the other side. The sensitivity to modeling error depends on the location of the pendulum hinge point.

Case 2 (actual value is greater than the value in the math model)

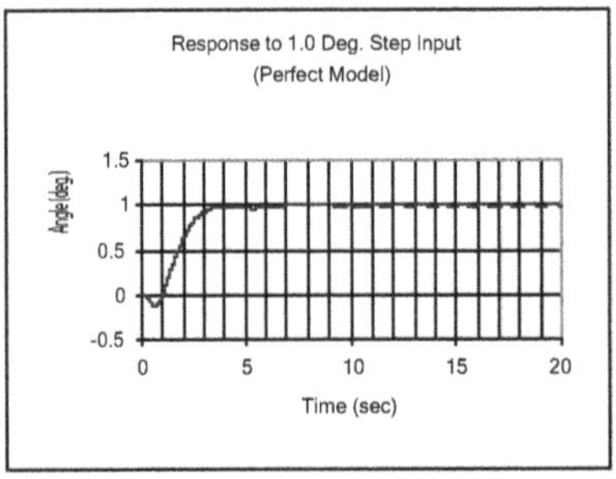

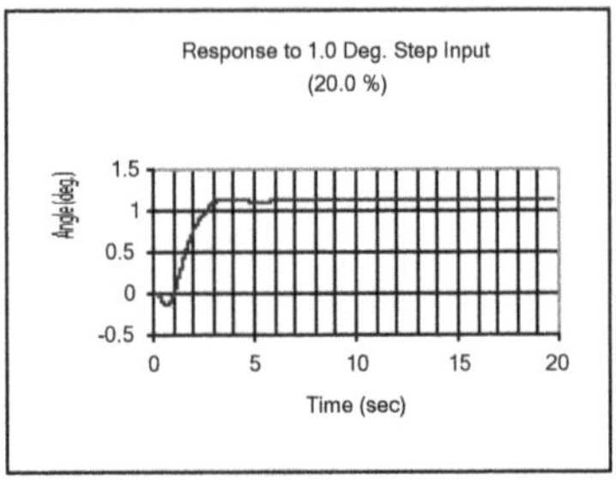

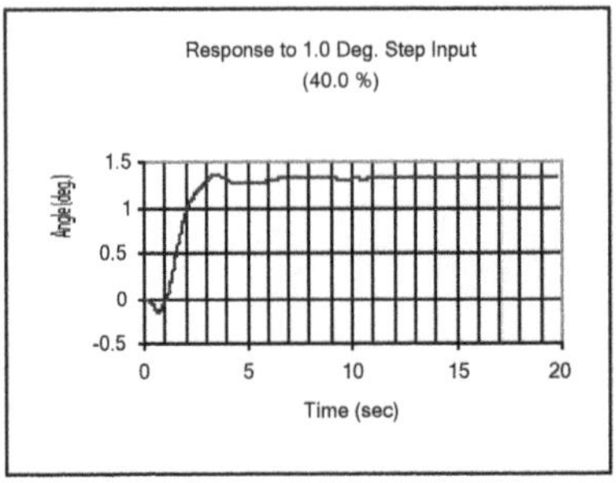

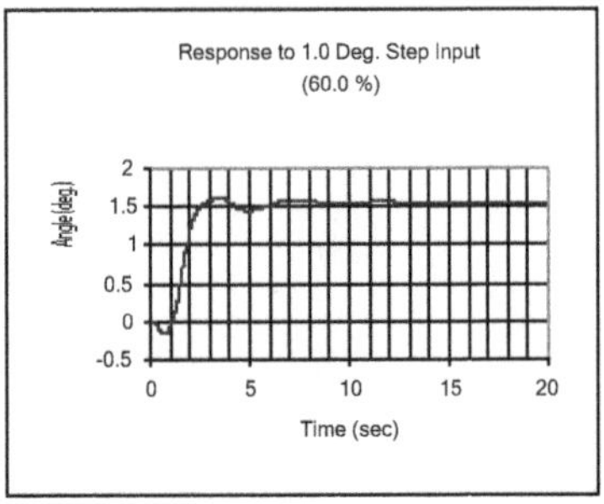

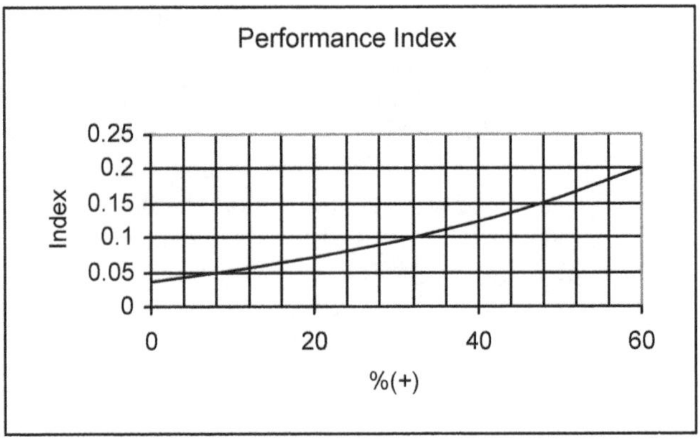

Appendix III: Collection of Math Modeling Error Sensitivity Response and Singular Value Plots, and Eigenvalue Tables

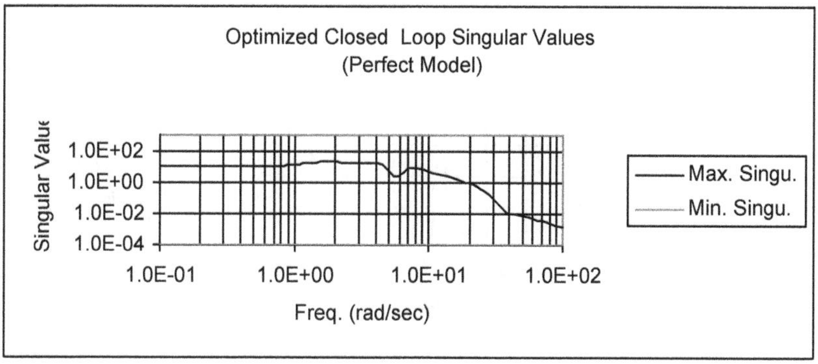

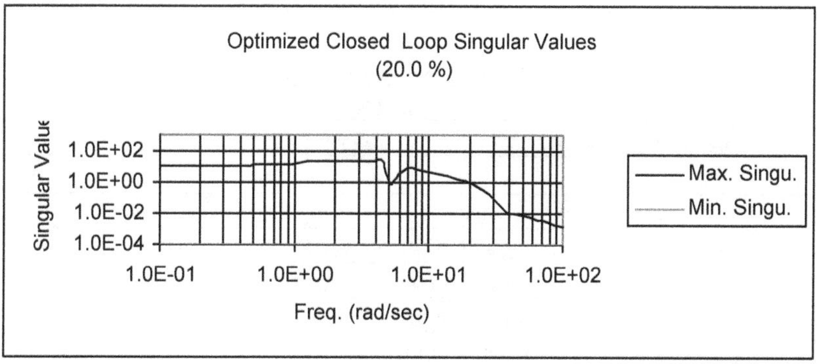

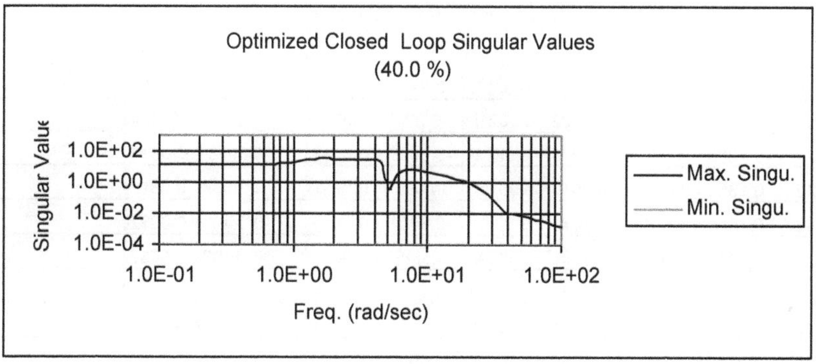

Appendix III: Collection of Math Modeling Error Sensitivity Response and Singular Value Plots, and Eigenvalue Tables

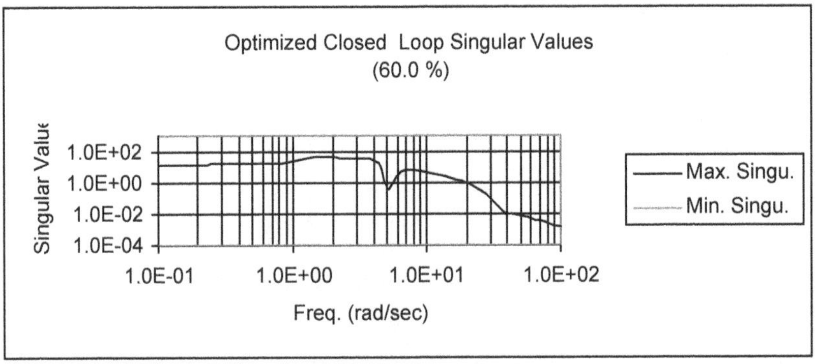

Closed Loop System Eigenvalues

	0%	20.00%	40.00%	60.00%	
	1.0e+002 *	1.0e+002 *	1.0e+002 *	1.0e+002 *	
1	-0.25	-0.25	-0.25	-0.25	
2	-0.5824 + 1.3344i	-0.5824 + 1.3344i	-0.5824 + 1.3344i	-0.5824 + 1.3344i	
3	-0.5824 - 1.3344i	-0.5824 - 1.3344i	-0.5824 - 1.3344i	-0.5824 - 1.3344i	
4	-0.1040 + 0.2546i	-0.1053 + 0.2598i	-0.1057 + 0.2619i	-0.1060 + 0.2634i	
5	-0.1040 - 0.2546i	-0.1053 - 0.2598i	-0.1057 - 0.2619i	-0.1060 - 0.2634i	
6	-0.1046 + 0.2542i	-0.1033 + 0.2482i	-0.1029 + 0.2453i	-0.1025 + 0.2429i	
7	-0.1046 - 0.2542i	-0.1033 - 0.2482i	-0.1029 - 0.2453i	-0.1025 - 0.2429i	
8	-0.2284 + 0.0212i	-0.2463	-0.2522	-0.256	
9	-0.2284 - 0.0212i	-0.2222 + 0.0321i	-0.2217 + 0.0388i	-0.2221 + 0.0434i	
10	-0.2293	-0.2222 - 0.0321i	-0.2217 - 0.0388i	-0.2221 - 0.0434i	
11	-0.1353 + 0.0476i	-0.0979 + 0.0613i	-0.0998 + 0.0733i	-0.0988 + 0.0813i	
12	-0.1353 - 0.0476i	-0.0979 - 0.0613i	-0.0998 - 0.0733i	-0.0988 - 0.0813i	
13	-0.1446 + 0.0002i	-0.1176 + 0.0269i	-0.1421 + 0.0028i	-0.1467 + 0.0032i	
14	-0.1446 - 0.0002i	-0.1176 - 0.0269i	-0.1421 - 0.0028i	-0.1467 - 0.0032i	
15	-0.0868 + 0.0557i	-0.1444 + 0.0002i	-0.1445	-0.1446	
16	-0.0868 - 0.0557i	-0.1444 - 0.0002i	-0.0716	-0.0519 + 0.0280i	
17	-0.0002 + 0.0601i	-0.0003 + 0.0602i	-0.0003 + 0.0602i	-0.0519 - 0.0280i	
18	-0.0002 - 0.0601i	-0.0003 - 0.0602i	-0.0003 - 0.0602i	-0.0003 + 0.0602i	
19	-0.0002 + 0.0600i	-0.0002 + 0.0597i	-0.0002 + 0.0596i	-0.0003 - 0.0602i	
20	-0.0002 - 0.0600i	-0.0002 - 0.0597i	-0.0002 - 0.0596i	-0.0002 + 0.0596i	
21	-0.0009 + 0.0508i	-0.0001 + 0.0507i	-0.0000 + 0.0507i	-0.0002 - 0.0596i	

22	-0.0009 - 0.0508i	-0.0001 - 0.0507i	-0.0000 - 0.0507i	-0.0000 + 0.0507i	
23	-0.0009 + 0.0509i	-0.0039 + 0.0474i	-0.0061 + 0.0433i	-0.0000 - 0.0507i	
24	-0.0009 - 0.0509i	-0.0039 - 0.0474i	-0.0061 - 0.0433i	-0.0080 + 0.0385i	
25	-0.0347 + 0.0149i	-0.0377 + 0.0152i	-0.0442 + 0.0157i	-0.0080 - 0.0385i	
26	-0.0347 - 0.0149i	-0.0377 - 0.0152i	-0.0442 - 0.0157i	-0.0429	
27	-0.0204	-0.0049 + 0.0138i	-0.0045 + 0.0141i	-0.0039 + 0.0145i	
28	-0.0052 + 0.0136i	-0.0049 - 0.0138i	-0.0045 - 0.0141i	-0.0039 - 0.0145i	
29	-0.0052 - 0.0136i	-0.0079 + 0.0145i	-0.0079 + 0.0146i	-0.0080 + 0.0146i	
30	-0.0078 + 0.0145i	-0.0079 - 0.0145i	-0.0079 - 0.0146i	-0.0080 - 0.0146i	
31	-0.0078 - 0.0145i	-0.0191	-0.0177	-0.0164	
32	-0.0134	-0.0131	-0.0122 + 0.0009i	-0.0122 + 0.0020i	
33	-0.0106	-0.0111	-0.0122 - 0.0009i	-0.0122 - 0.0020i	
34	-0.01	-0.01	-0.01	-0.01	

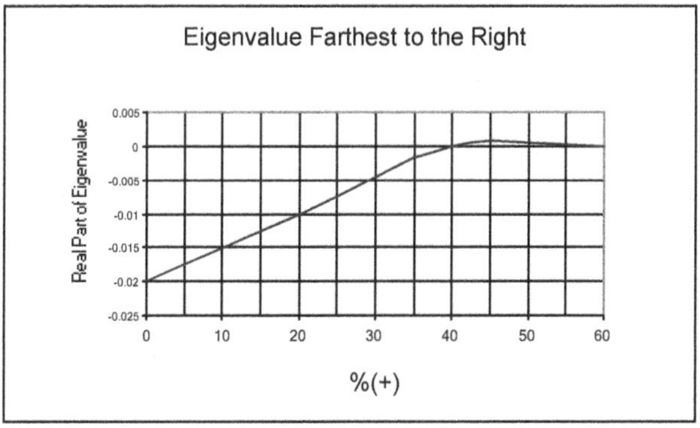

Observation:

The real parts of the system eigenvalues remain in the Left-Half Plane even after the error increases to 60.0%, implying that the control system is insensitive to mp2 modeling error.

8) Fuel Sloshing 1 Pendulum Length: Lp1 (The length of pendulum 1 in the math model = 2.0 ft.)

Case 1 (actual value is less than the value in the math model)

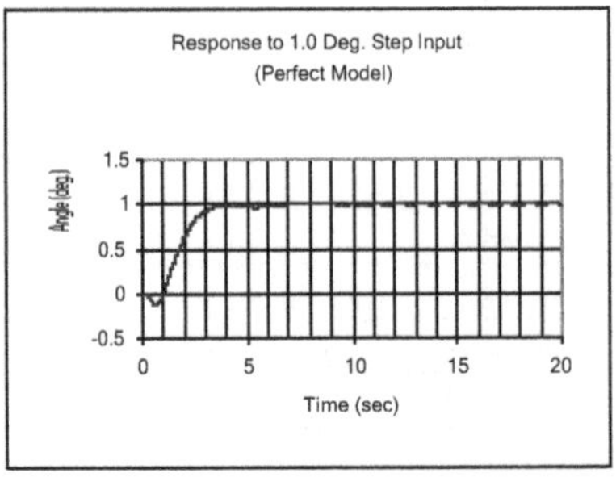

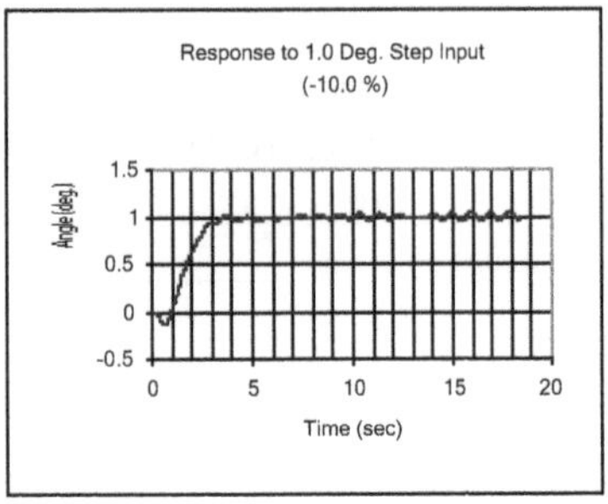

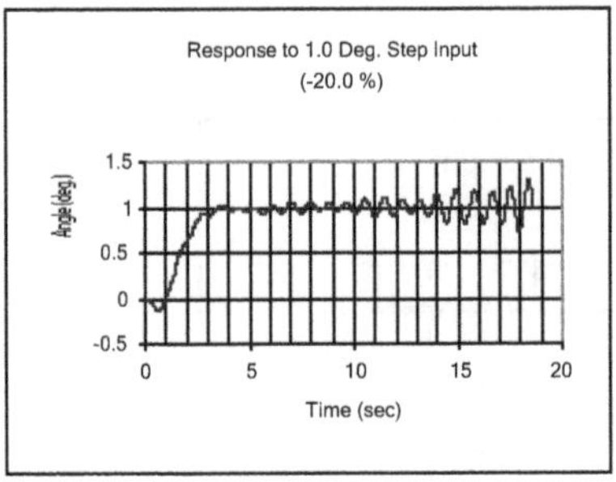

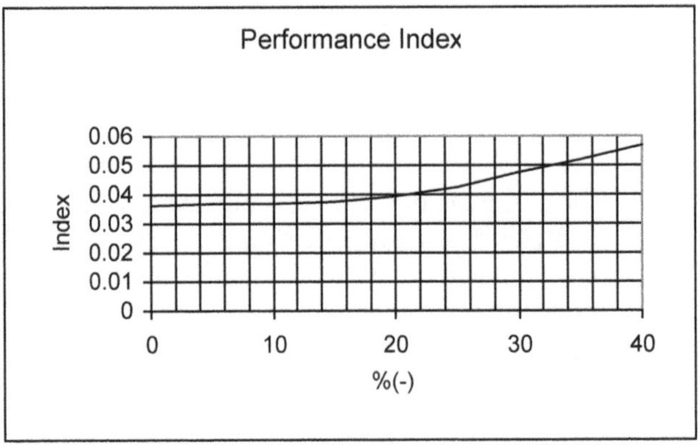

Appendix III: Collection of Math Modeling Error Sensitivity Response and Singular Value Plots, and Eigenvalue Tables

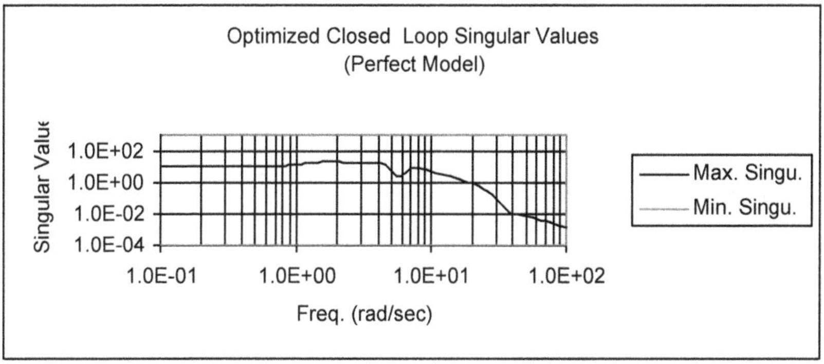

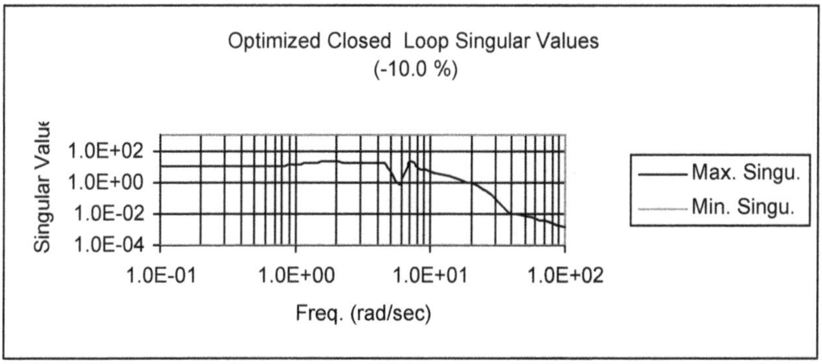

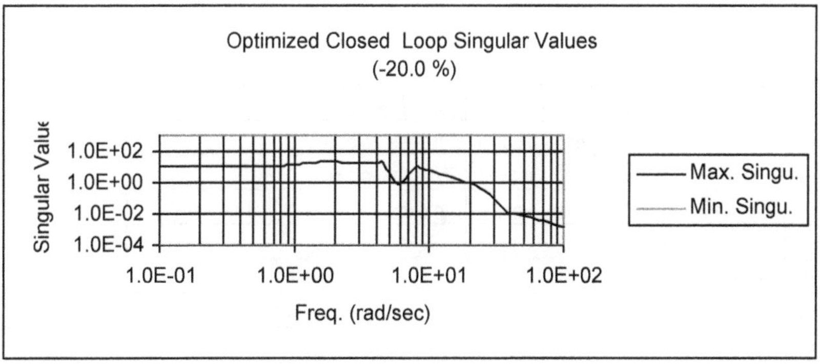

Appendix III: Collection of Math Modeling Error Sensitivity Response and Singular Value Plots, and Eigenvalue Tables

Closed Loop System Eigenvalues					
	0%	-10.00%	-20.00%		
	1.0e+002 *	1.0e+002 *	1.0e+002 *		
1	-0.25	-0.25	-0.25		
2	-0.5824 + 1.3344i	-0.5824 + 1.3344i	-0.5824 + 1.3344i		
3	-0.5824 - 1.3344i	-0.5824 - 1.3344i	-0.5824 - 1.3344i		
4	-0.1040 + 0.2546i	-0.1042 + 0.2547i	-0.1043 + 0.2549i		
5	-0.1040 - 0.2546i	-0.1042 - 0.2547i	-0.1043 - 0.2549i		
6	-0.1046 + 0.2542i	-0.1045 + 0.2542i	-0.1045 + 0.2541i		
7	-0.1046 - 0.2542i	-0.1045 - 0.2542i	-0.1045 - 0.2541i		
8	-0.2284 + 0.0212i	-0.23	-0.2313		
9	-0.2284 - 0.0212i	-0.2283 + 0.0195i	-0.2280 + 0.0174i		
10	-0.2293	-0.2283 - 0.0195i	-0.2280 - 0.0174i		
11	-0.1353 + 0.0476i	-0.1301 + 0.0495i	-0.1224 + 0.0561i		
12	-0.1353 - 0.0476i	-0.1301 - 0.0495i	-0.1224 - 0.0561i		
13	-0.1446 + 0.0002i	-0.1447 + 0.0002i	-0.1447 + 0.0002i		
14	-0.1446 - 0.0002i	-0.1447 - 0.0002i	-0.1447 - 0.0002i		
15	-0.0868 + 0.0557i	-0.0922 + 0.0509i	-0.1002 + 0.0404i		
16	-0.0868 - 0.0557i	-0.0922 - 0.0509i	-0.1002 - 0.0404i		
17	-0.0002 + 0.0601i	0.0006 + 0.0670i	0.0016 + 0.0720i		
18	-0.0002 - 0.0601i	0.0006 - 0.0670i	0.0016 - 0.0720i		
19	-0.0002 + 0.0600i	0.0002 + 0.0570i	0.0005 + 0.0564i		
20	-0.0002 - 0.0600i	0.0002 - 0.0570i	0.0005 - 0.0564i		
21	-0.0009 + 0.0508i	-0.0022 + 0.0507i	-0.0027 + 0.0505i		
22	-0.0009 - 0.0508i	-0.0022 - 0.0507i	-0.0027 - 0.0505i		
23	-0.0009 + 0.0509i	-0.0001 + 0.0507i	0.0001 + 0.0507i		
24	-0.0009 - 0.0509i	-0.0001 - 0.0507i	0.0001 - 0.0507i		
25	-0.0347 + 0.0149i	-0.0349 + 0.0151i	-0.0352 + 0.0153i		
26	-0.0347 - 0.0149i	-0.0349 - 0.0151i	-0.0352 - 0.0153i		
27	-0.0204	-0.0204	-0.0204		
28	-0.0052 + 0.0136i	-0.0052 + 0.0136i	-0.0052 + 0.0136i		
29	-0.0052 - 0.0136i	-0.0052 - 0.0136i	-0.0052 - 0.0136i		
30	-0.0078 + 0.0145i	-0.0078 + 0.0145i	-0.0078 + 0.0145i		
31	-0.0078 - 0.0145i	-0.0078 - 0.0145i	-0.0078 - 0.0145i		
32	-0.0134	-0.0134	-0.0134		
33	-0.0106	-0.0106	-0.0106		
34	-0.01	-0.01	-0.01		

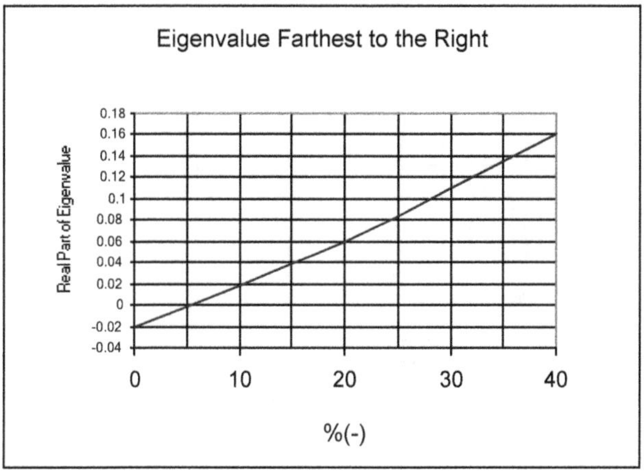

Observation:

When the actual pendulum length of sloshing 1 is about 5.0 % less than the length in the math model, the control system becomes unstable.

Case 2 (actual value is greater than the value in the math model)

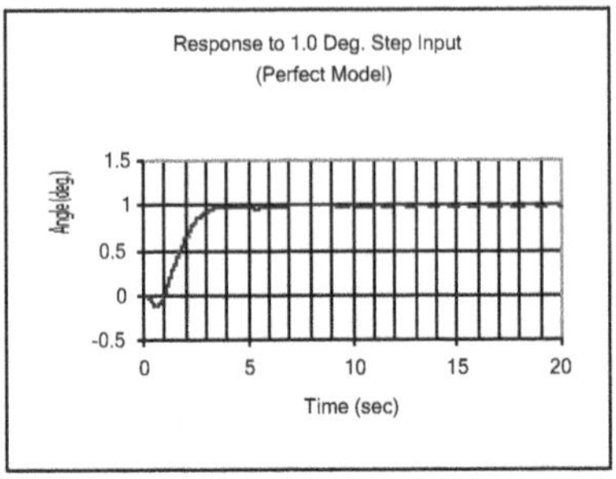

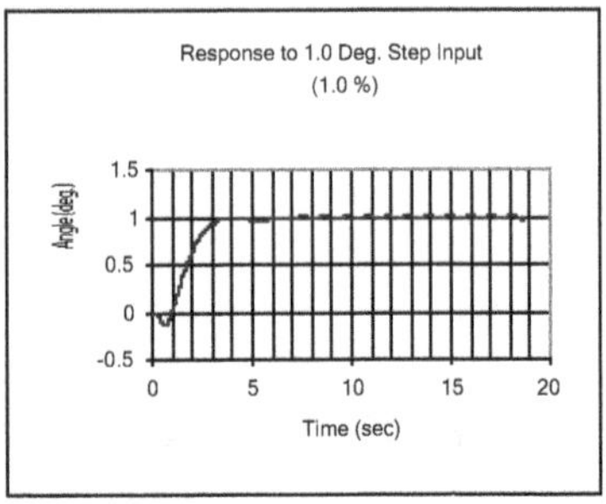

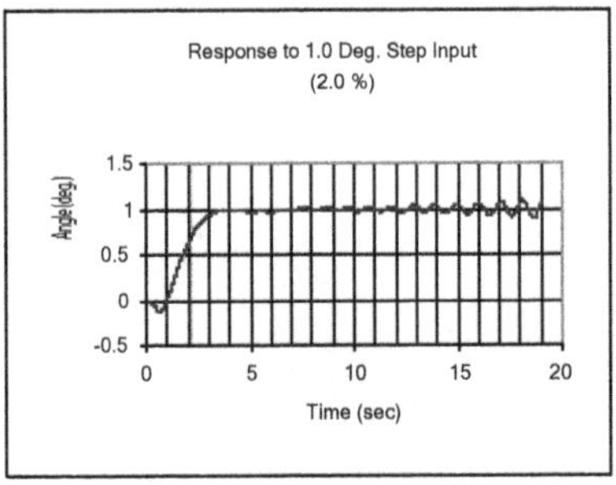

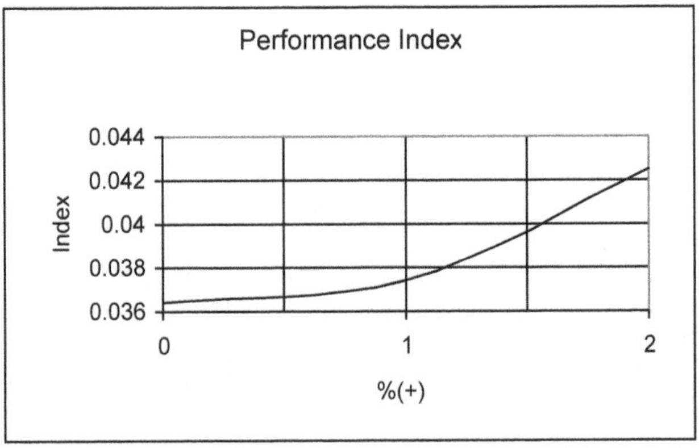

Appendix III: Collection of Math Modeling Error Sensitivity Response and Singular Value Plots, and Eigenvalue Tables

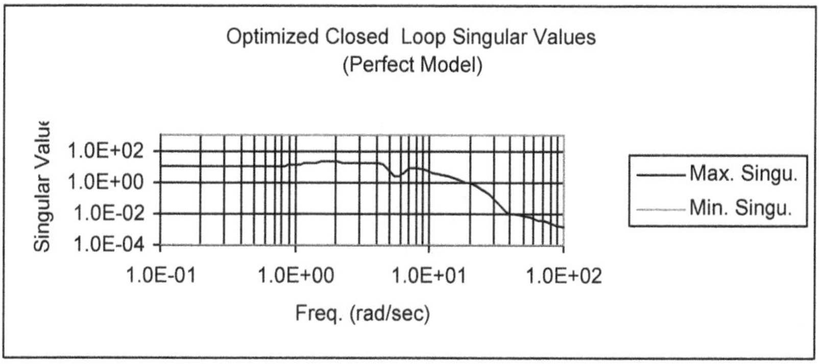

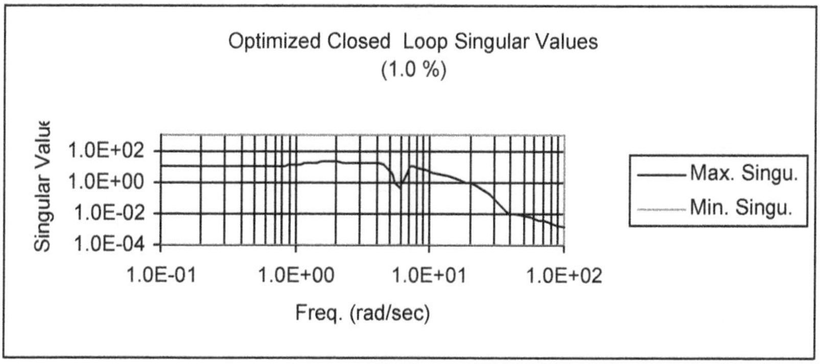

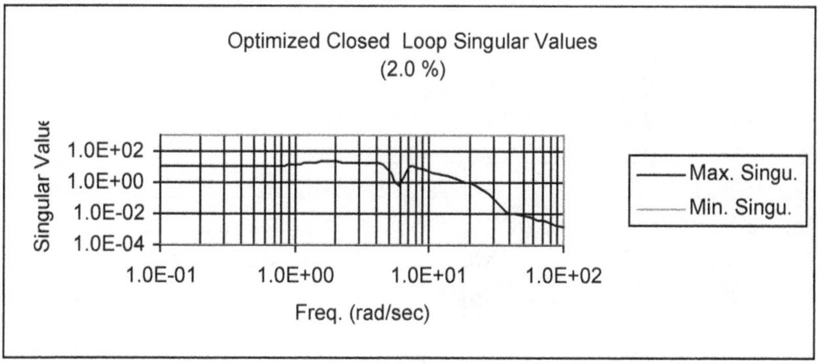

Appendix III: Collection of Math Modeling Error Sensitivity Response and Singular Value Plots, and Eigenvalue Tables

Closed Loop System Eigenvalues					
	0%	1.00%	2.00%		
	1.0e+002 *	1.0e+002 *	1.0e+002 *		
1	-0.25	-0.25	-0.25		
2	-0.5824 + 1.3344i	-0.5824 + 1.3344i	-0.5824 + 1.3344i		
3	-0.5824 - 1.3344i	-0.5824 - 1.3344i	-0.5824 - 1.3344i		
4	-0.1040 + 0.2546i	-0.1040 + 0.2546i	-0.1040 + 0.2546i		
5	-0.1040 - 0.2546i	-0.1040 - 0.2546i	-0.1040 - 0.2546i		
6	-0.1046 + 0.2542i	-0.1046 + 0.2543i	-0.1046 + 0.2543i		
7	-0.1046 - 0.2542i	-0.1046 - 0.2543i	-0.1046 - 0.2543i		
8	-0.2284 + 0.0212i	-0.2284 + 0.0214i	-0.2284 + 0.0215i		
9	-0.2284 - 0.0212i	-0.2284 - 0.0214i	-0.2284 - 0.0215i		
10	-0.2293	-0.2293	-0.2292		
11	-0.1353 + 0.0476i	-0.1357 + 0.0474i	-0.1361 + 0.0473i		
12	-0.1353 - 0.0476i	-0.1357 - 0.0474i	-0.1361 - 0.0473i		
13	-0.1446 + 0.0002i	-0.1446 + 0.0002i	-0.1446 + 0.0002i		
14	-0.1446 - 0.0002i	-0.1446 - 0.0002i	-0.1446 - 0.0002i		
15	-0.0868 + 0.0557i	-0.0864 + 0.0561i	-0.0860 + 0.0564i		
16	-0.0868 - 0.0557i	-0.0864 - 0.0561i	-0.0860 - 0.0564i		
17	-0.0002 + 0.0601i	0.0012 + 0.0599i	0.0017 + 0.0597i		
18	-0.0002 - 0.0601i	0.0012 - 0.0599i	0.0017 - 0.0597i		
19	-0.0002 + 0.0600i	-0.0017 + 0.0599i	-0.0024 + 0.0597i		
20	-0.0002 - 0.0600i	-0.0017 - 0.0599i	-0.0024 - 0.0597i		
21	-0.0009 + 0.0508i	-0.0009 + 0.0505i	-0.0009 + 0.0504i		
22	-0.0009 - 0.0508i	-0.0009 - 0.0505i	-0.0009 - 0.0504i		
23	-0.0009 + 0.0509i	-0.0008 + 0.0512i	-0.0008 + 0.0513i		
24	-0.0009 - 0.0509i	-0.0008 - 0.0512i	-0.0008 - 0.0513i		
25	-0.0347 + 0.0149i	-0.0347 + 0.0149i	-0.0347 + 0.0148i		
26	-0.0347 - 0.0149i	-0.0347 - 0.0149i	-0.0347 - 0.0148i		
27	-0.0204	-0.0204	-0.0204		
28	-0.0052 + 0.0136i	-0.0052 + 0.0136i	-0.0052 + 0.0136i		
29	-0.0052 - 0.0136i	-0.0052 - 0.0136i	-0.0052 - 0.0136i		
30	-0.0078 + 0.0145i	-0.0078 + 0.0145i	-0.0078 + 0.0145i		
31	-0.0078 - 0.0145i	-0.0078 - 0.0145i	-0.0078 - 0.0145i		
32	-0.0134	-0.0134	-0.0134		
33	-0.0106	-0.0106	-0.0106		
34	-0.01	-0.01	-0.01		

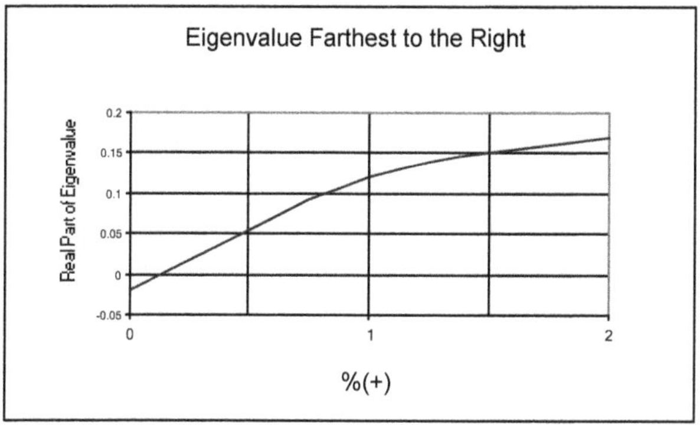

Observation:

The control system is sensitive to positive modeling error (pendulum length of the sloshing 1). A slight error causes instability.

9) **Fuel Sloshing 2 Pendulum Length: Lp2 (The value of pendulum length 2 in the math model = 2.75 ft.)**

Case 1 (actual value is less than the value in the math model)

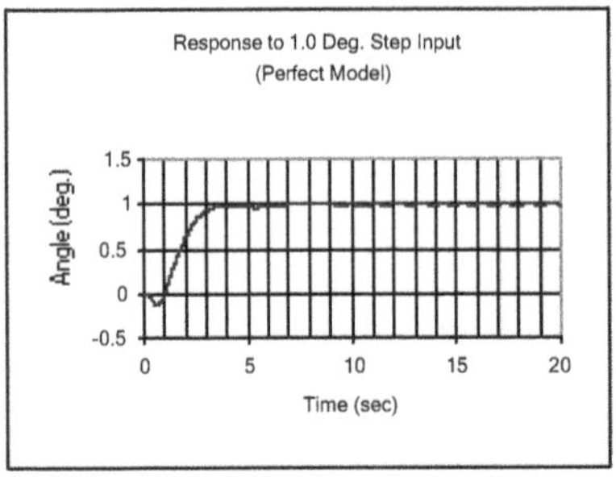

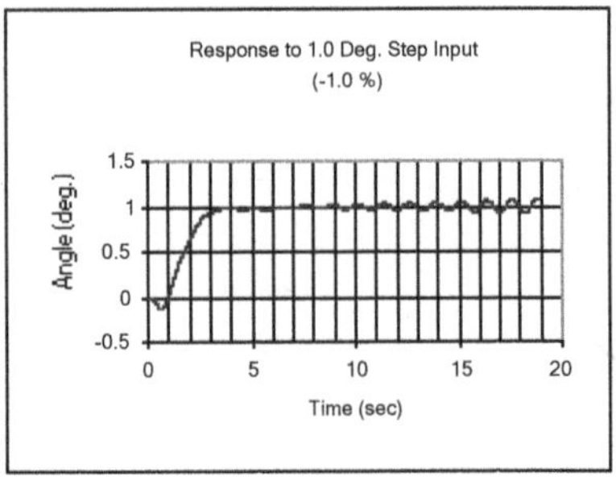

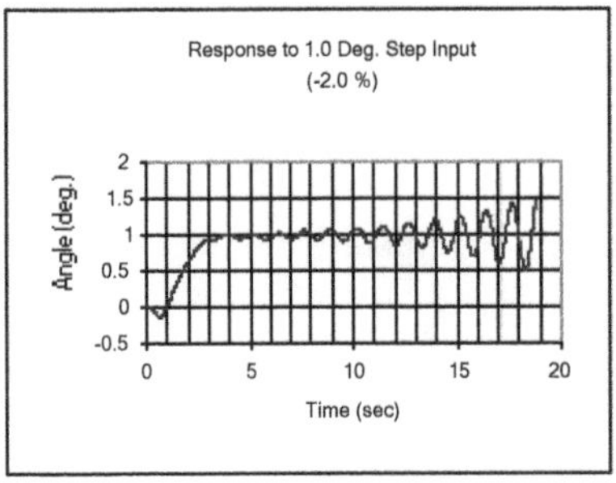

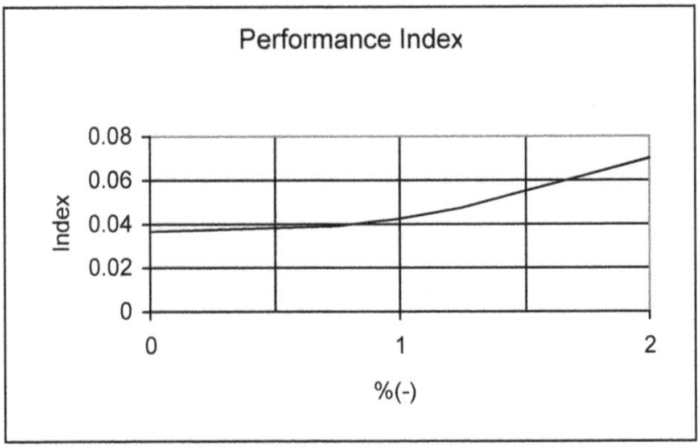

Appendix III: Collection of Math Modeling Error Sensitivity Response and Singular Value Plots, and Eigenvalue Tables

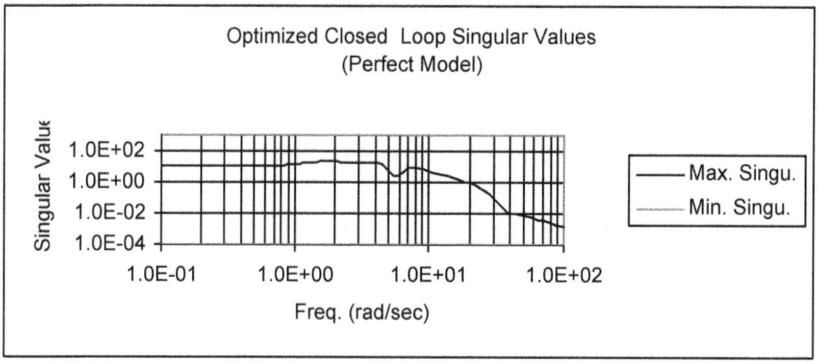

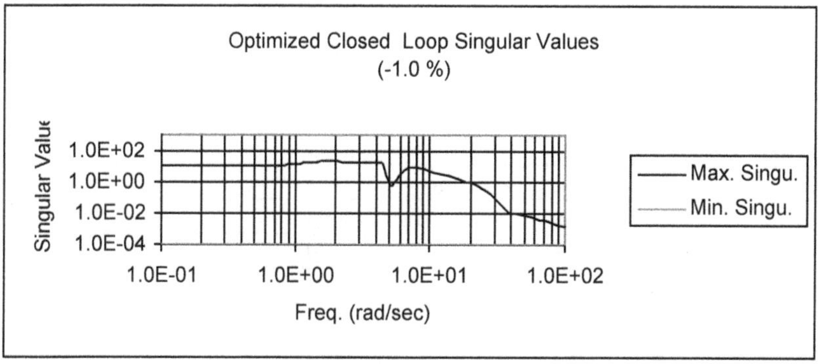

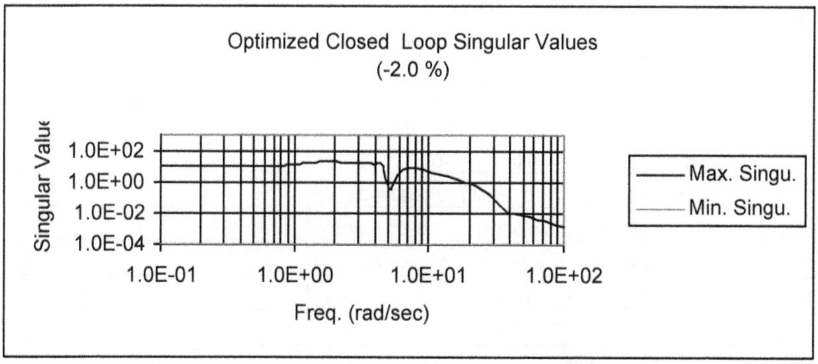

Appendix III: Collection of Math Modeling Error Sensitivity Response and Singular Value Plots, and Eigenvalue Tables

Closed Loop System Eigenvalues					
	0%	-1.00%	-2.00%		
	1.0e+002 *	1.0e+002 *	1.0e+002 *		
1	-0.25	-0.25	-0.25		
2	-0.5824 + 1.3344i	-0.5824 + 1.3344i	-0.5824 + 1.3344i		
3	-0.5824 - 1.3344i	-0.5824 - 1.3344i	-0.5824 - 1.3344i		
4	-0.1040 + 0.2546i	-0.1041 + 0.2546i	-0.1041 + 0.2546i		
5	-0.1040 - 0.2546i	-0.1041 - 0.2546i	-0.1041 - 0.2546i		
6	-0.1046 + 0.2542i	-0.1046 + 0.2542i	-0.1045 + 0.2542i		
7	-0.1046 - 0.2542i	-0.1046 - 0.2542i	-0.1045 - 0.2542i		
8	-0.2284 + 0.0212i	-0.2282 + 0.0209i	-0.2297		
9	-0.2284 - 0.0212i	-0.2282 - 0.0209i	-0.2279 + 0.0206i		
10	-0.2293	-0.2295	-0.2279 - 0.0206i		
11	-0.1353 + 0.0476i	-0.1366 + 0.0477i	-0.1379 + 0.0478i		
12	-0.1353 - 0.0476i	-0.1366 - 0.0477i	-0.1379 - 0.0478i		
13	-0.1446 + 0.0002i	-0.1446 + 0.0002i	-0.1446 + 0.0002i		
14	-0.1446 - 0.0002i	-0.1446 - 0.0002i	-0.1446 - 0.0002i		
15	-0.0868 + 0.0557i	-0.0855 + 0.0565i	-0.0843 + 0.0572i		
16	-0.0868 - 0.0557i	-0.0855 - 0.0565i	-0.0843 - 0.0572i		
17	-0.0002 + 0.0601i	-0.0002 + 0.0602i	-0.0003 + 0.0603i		
18	-0.0002 - 0.0601i	-0.0002 - 0.0602i	-0.0003 - 0.0603i		
19	-0.0002 + 0.0600i	-0.0002 + 0.0598i	-0.0002 + 0.0598i		
20	-0.0002 - 0.0600i	-0.0002 - 0.0598i	-0.0002 - 0.0598i		
21	-0.0009 + 0.0508i	-0.0031 + 0.0515i	-0.0041 + 0.0518i		
22	-0.0009 - 0.0508i	-0.0031 - 0.0515i	-0.0041 - 0.0518i		
23	-0.0009 + 0.0509i	0.0012 + 0.0504i	0.0020 + 0.0502i		
24	-0.0009 - 0.0509i	0.0012 - 0.0504i	0.0020 - 0.0502i		
25	-0.0347 + 0.0149i	-0.0347 + 0.0148i	-0.0346 + 0.0148i		
26	-0.0347 - 0.0149i	-0.0347 - 0.0148i	-0.0346 - 0.0148i		
27	-0.0204	-0.0204	-0.0204		
28	-0.0052 + 0.0136i	-0.0052 + 0.0136i	-0.0052 + 0.0136i		
29	-0.0052 - 0.0136i	-0.0052 - 0.0136i	-0.0052 - 0.0136i		
30	-0.0078 + 0.0145i	-0.0078 + 0.0145i	-0.0078 + 0.0145i		
31	-0.0078 - 0.0145i	-0.0078 - 0.0145i	-0.0078 - 0.0145i		
32	-0.0134	-0.0134	-0.0134		
33	-0.0106	-0.0106	-0.0106		
34	-0.01	-0.01	-0.01		

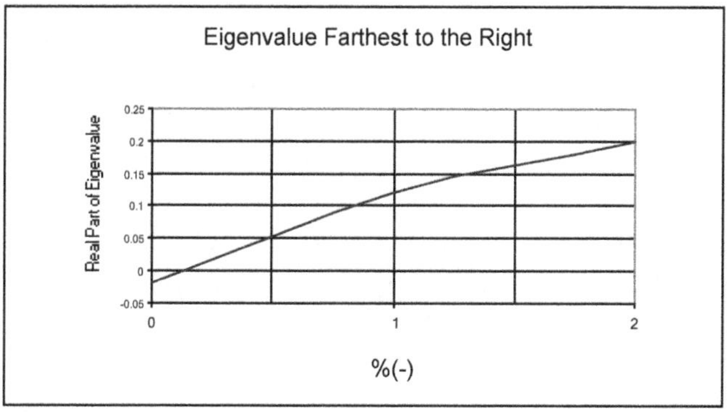

Observation:

The control system is sensitive to negative modeling error (pendulum length of sloshing 2). A slight error causes instability.

Case 2 (actual value is greater than the value in the math model)

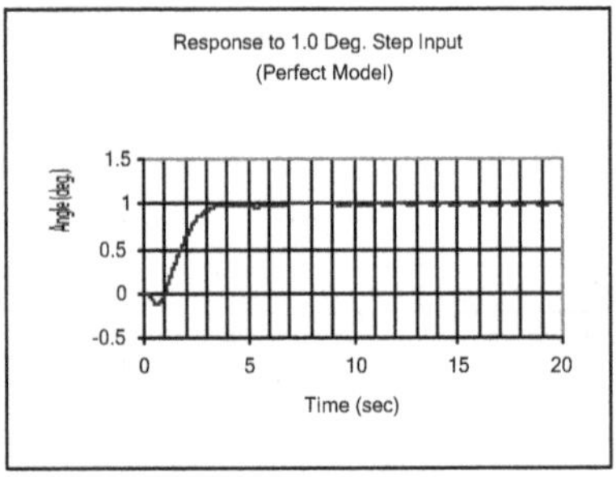

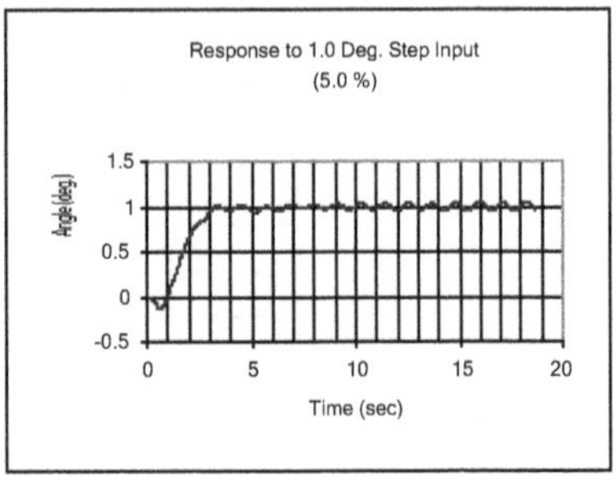

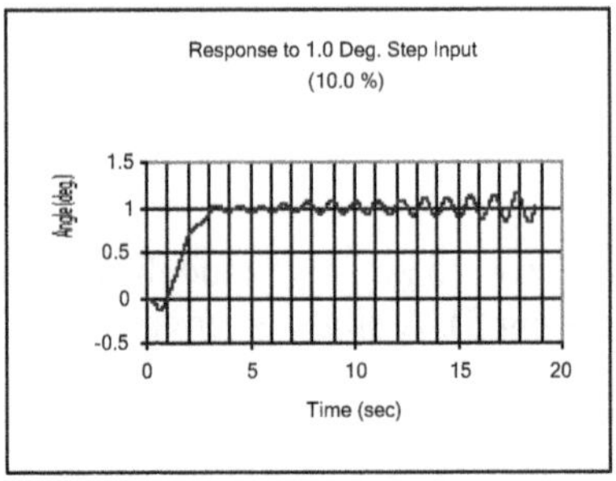

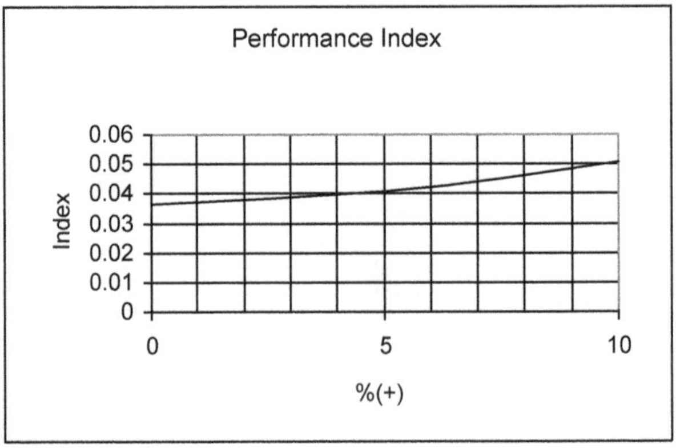

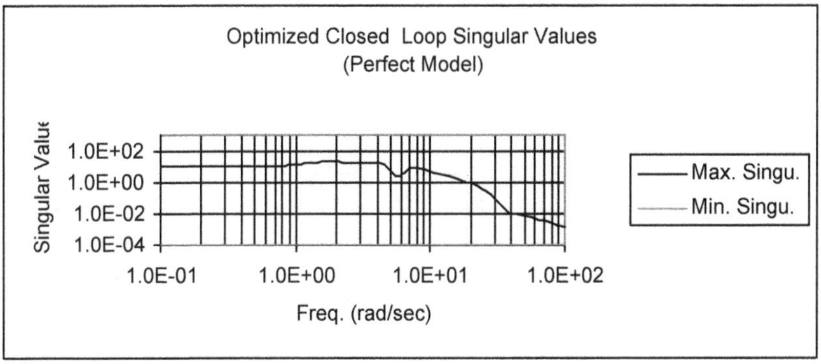

Optimized Closed Loop Singular Values
(Perfect Model)

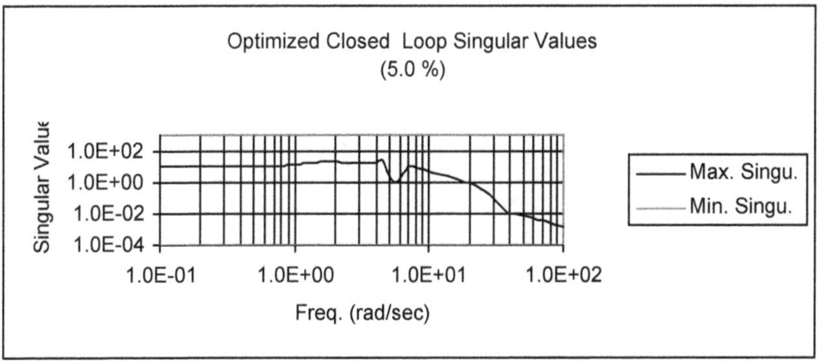

Optimized Closed Loop Singular Values
(5.0 %)

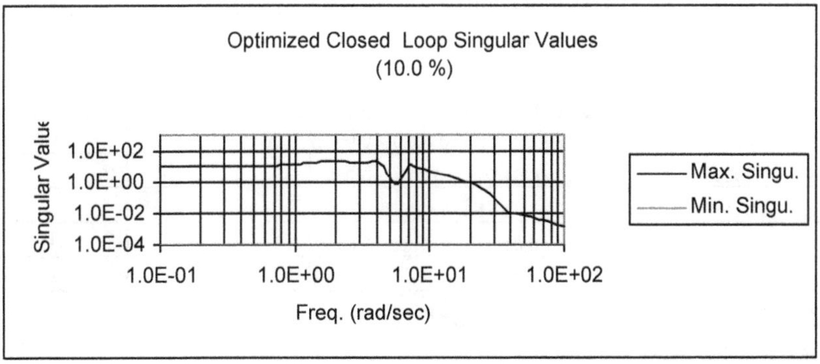

Optimized Closed Loop Singular Values
(10.0 %)

Appendix III: Collection of Math Modeling Error Sensitivity Response and Singular Value Plots, and Eigenvalue Tables

Closed Loop System Eigenvalues					
	0%	5.00%	10.00%		
	1.0e+002 *	1.0e+002 *	1.0e+002 *		
1	-0.25	-0.25	-0.25		
2	-0.5824 + 1.3344i	-0.5824 + 1.3344i	-0.5824 + 1.3344i		
3	-0.5824 - 1.3344i	-0.5824 - 1.3344i	-0.5824 - 1.3344i		
4	-0.1040 + 0.2546i	-0.1039 + 0.2546i	-0.1038 + 0.2546i		
5	-0.1040 - 0.2546i	-0.1039 - 0.2546i	-0.1038 - 0.2546i		
6	-0.1046 + 0.2542i	-0.1046 + 0.2543i	-0.1047 + 0.2543i		
7	-0.1046 - 0.2542i	-0.1046 - 0.2543i	-0.1047 - 0.2543i		
8	-0.2284 + 0.0212i	-0.2295 + 0.0225i	-0.2304 + 0.0236i		
9	-0.2284 - 0.0212i	-0.2295 - 0.0225i	-0.2304 - 0.0236i		
10	-0.2293	-0.2287	-0.2282		
11	-0.1353 + 0.0476i	-0.1270 + 0.0474i	-0.1150 + 0.0563i		
12	-0.1353 - 0.0476i	-0.1270 - 0.0474i	-0.1150 - 0.0563i		
13	-0.1446 + 0.0002i	-0.1447 + 0.0002i	-0.1447 + 0.0002i		
14	-0.1446 - 0.0002i	-0.1447 - 0.0002i	-0.1447 - 0.0002i		
15	-0.0868 + 0.0557i	-0.0947 + 0.0510i	-0.1065 + 0.0362i		
16	-0.0868 - 0.0557i	-0.0947 - 0.0510i	-0.1065 - 0.0362i		
17	-0.0002 + 0.0601i	-0.0006 + 0.0601i	-0.0007 + 0.0602i		
18	-0.0002 - 0.0601i	-0.0006 - 0.0601i	-0.0007 - 0.0602i		
19	-0.0002 + 0.0600i	0.0003 + 0.0601i	0.0006 + 0.0601i		
20	-0.0002 - 0.0600i	0.0003 - 0.0601i	0.0006 - 0.0601i		
21	-0.0009 + 0.0508i	0.0005 + 0.0549i	0.0011 + 0.0562i		
22	-0.0009 - 0.0508i	0.0005 - 0.0549i	0.0011 - 0.0562i		
23	-0.0009 + 0.0509i	-0.0018 + 0.0457i	-0.0020 + 0.0435i		
24	-0.0009 - 0.0509i	-0.0018 - 0.0457i	-0.0020 - 0.0435i		
25	-0.0347 + 0.0149i	-0.0350 + 0.0151i	-0.0352 + 0.0154i		
26	-0.0347 - 0.0149i	-0.0350 - 0.0151i	-0.0352 - 0.0154i		
27	-0.0204	-0.0204	-0.0204		
28	-0.0052 + 0.0136i	-0.0052 + 0.0136i	-0.0052 + 0.0136i		
29	-0.0052 - 0.0136i	-0.0052 - 0.0136i	-0.0052 - 0.0136i		
30	-0.0078 + 0.0145i	-0.0078 + 0.0145i	-0.0078 + 0.0145i		
31	-0.0078 - 0.0145i	-0.0078 - 0.0145i	-0.0078 - 0.0145i		
32	-0.0134	-0.0134	-0.0134		
33	-0.0106	-0.0106	-0.0106		
34	-0.01	-0.01	-0.01		

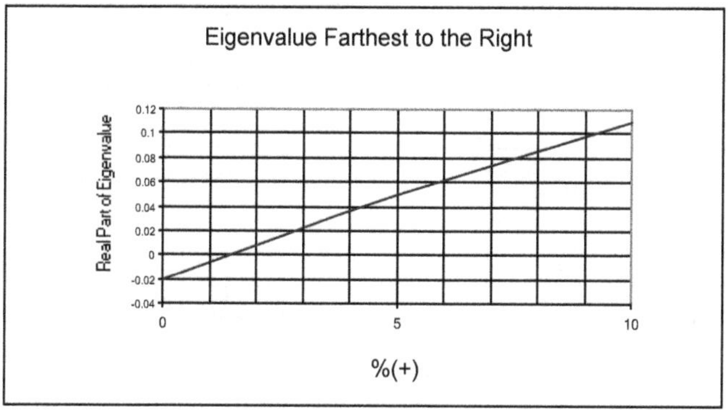

Observation:

When the actual pendulum length of sloshing 2 is about 1.5% greater than the length in the math model, the real part of a complex root crosses the borderline and moves to the Right-Half Plane, indicating that the system becomes unstable.

10) **Fuel Sloshing 1 Pendulum Hinge Point: lp1 (The value of Fuel Sloshing 2 Pendulum Hinge Point in the math model = 5.56 ft.)**

Case 1 (actual value is less than the value in the math model)

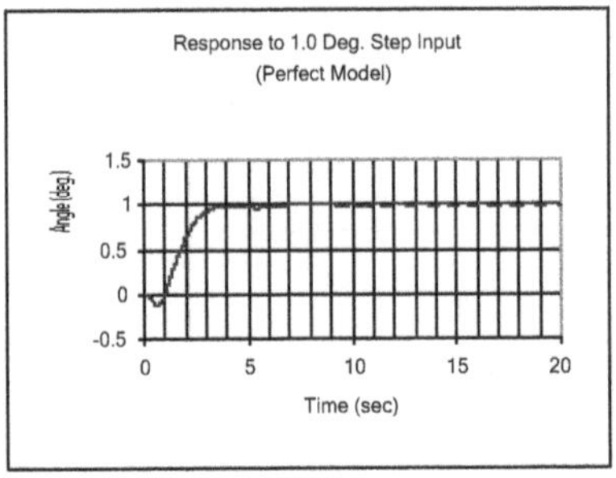

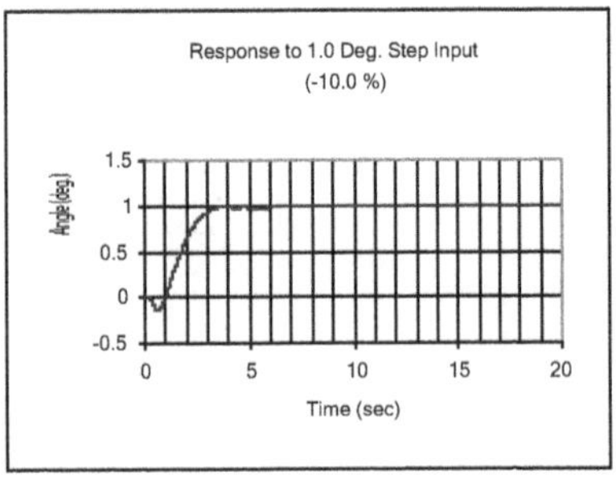

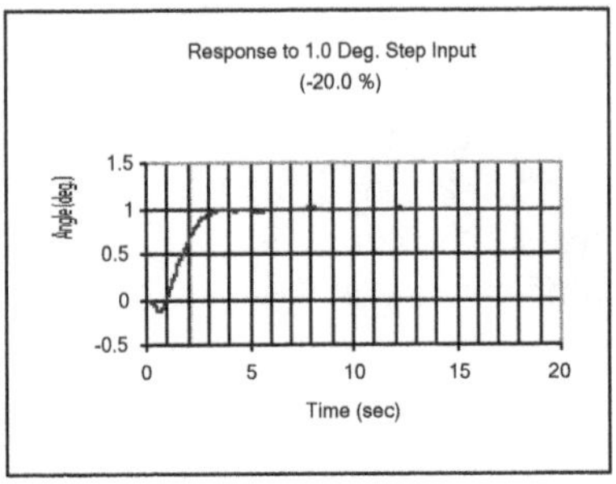

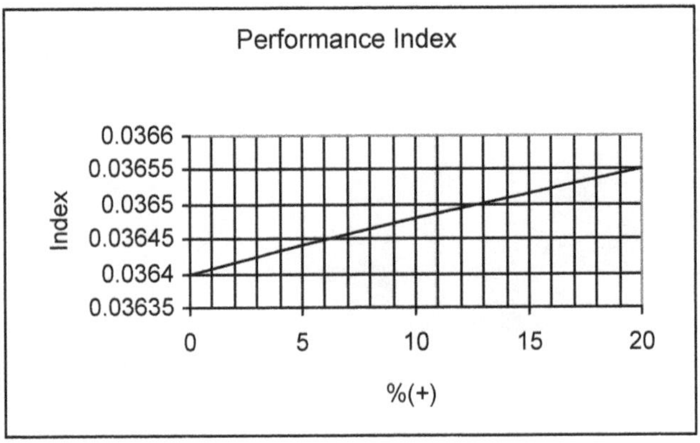

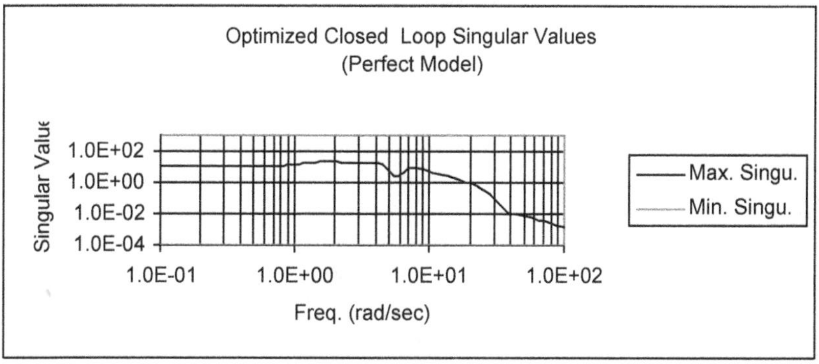
Optimized Closed Loop Singular Values
(Perfect Model)
Singular Valu
1.0E+02
1.0E+00
1.0E-02
1.0E-04
1.0E-01
1.0E+00
1.0E+01
1.0E+02
Freq. (rad/sec)
Max. Singu.
Min. Singu.

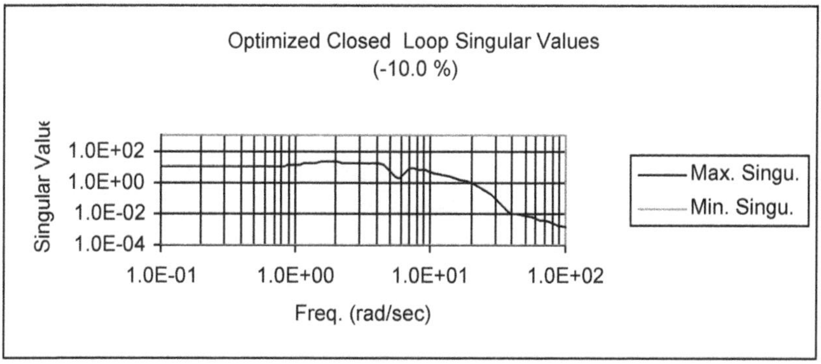
Optimized Closed Loop Singular Values
(-10.0 %)
Singular Valu
1.0E+02
1.0E+00
1.0E-02
1.0E-04
1.0E-01
1.0E+00
1.0E+01
1.0E+02
Freq. (rad/sec)
Max. Singu.
Min. Singu.

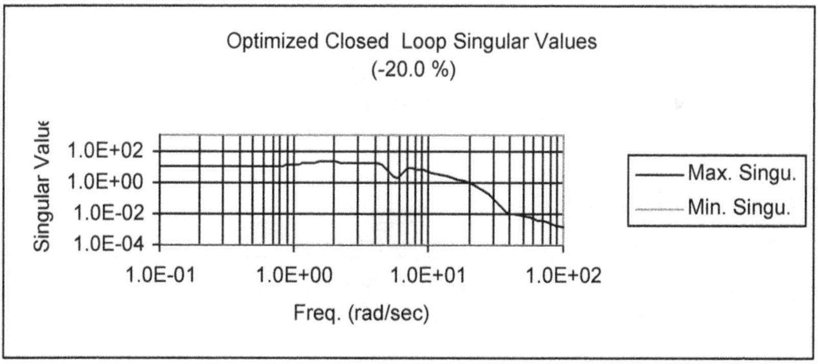
Optimized Closed Loop Singular Values
(-20.0 %)
Singular Valu
1.0E+02
1.0E+00
1.0E-02
1.0E-04
1.0E-01
1.0E+00
1.0E+01
1.0E+02
Freq. (rad/sec)
Max. Singu.
Min. Singu.

Appendix III: Collection of Math Modeling Error Sensitivity Response and Singular Value Plots, and Eigenvalue Tables

Closed Loop System Eigenvalues (% off from the math model)					
	0%	-10%	-20.00%		
	1.0e+002 *	1.0e+002 *	1.0e+002 *		
1	-0.25	-0.25	-0.25		
2	-0.5824 + 1.3344i	-0.5824 + 1.3344i	-0.5824 + 1.3344i		
3	-0.5824 - 1.3344i	-0.5824 - 1.3344i	-0.5824 - 1.3344i		
4	-0.0973 + 0.2537i	-0.0972 + 0.2537i	-0.0971 + 0.2537i		
5	-0.0973 - 0.2537i	-0.0972 - 0.2537i	-0.0971 - 0.2537i		
6	-0.0978 + 0.2534i	-0.0979 + 0.2534i	-0.0979 + 0.2535i		
7	-0.0978 - 0.2534i	-0.0979 - 0.2534i	-0.0979 - 0.2535i		
8	-0.2272 + 0.0221i	-0.2275 + 0.0234i	-0.2277 + 0.0248i		
9	-0.2272 - 0.0221i	-0.2275 - 0.0234i	-0.2277 - 0.0248i		
10	-0.2245	-0.2239	-0.2233		
11	-0.1110 + 0.0554i	-0.1040 + 0.0552i	-0.1305 + 0.0362i		
12	-0.1110 - 0.0554i	-0.1040 - 0.0552i	-0.1305 - 0.0362i		
13	-0.1172 + 0.0325i	-0.1244 + 0.0356i	-0.0983 + 0.0575i		
14	-0.1172 - 0.0325i	-0.1244 - 0.0356i	-0.0983 - 0.0575i		
15	-0.1452	-0.1451	-0.1449		
16	-0.1448	-0.1448	-0.145		
17	-0.0002 + 0.0600i	-0.0008 + 0.0596i	-0.0009 + 0.0591i		
18	-0.0002 - 0.0600i	-0.0008 - 0.0596i	-0.0009 - 0.0591i		
19	-0.0002 + 0.0600i	0.0002 + 0.0599i	0.0002 + 0.0597i		
20	-0.0002 - 0.0600i	0.0002 - 0.0599i	0.0002 - 0.0597i		
21	-0.0009 + 0.0508i	-0.0011 + 0.0512i	-0.0013 + 0.0514i		
22	-0.0009 - 0.0508i	-0.0011 - 0.0512i	-0.0013 - 0.0514i		
23	-0.0009 + 0.0509i	-0.0007 + 0.0507i	-0.0006 + 0.0507i		
24	-0.0009 - 0.0509i	-0.0007 - 0.0507i	-0.0006 - 0.0507i		
25	-0.0357	-0.0336 + 0.0033i	-0.0331 + 0.0051i		
26	-0.0323	-0.0336 - 0.0033i	-0.0331 - 0.0051i		
27	-0.0075 + 0.0138i	-0.0074 + 0.0139i	-0.0074 + 0.0140i		
28	-0.0075 - 0.0138i	-0.0074 - 0.0139i	-0.0074 - 0.0140i		
29	-0.0075 + 0.0138i	-0.0076 + 0.0137i	-0.0076 + 0.0137i		
30	-0.0075 - 0.0138i	-0.0076 - 0.0137i	-0.0076 - 0.0137i		
31	-0.0182	-0.0186	-0.019		
32	-0.0173	-0.017	-0.0168		
33	-0.0098	-0.0098	-0.0098		
34	-0.0098	-0.0098	-0.0098		

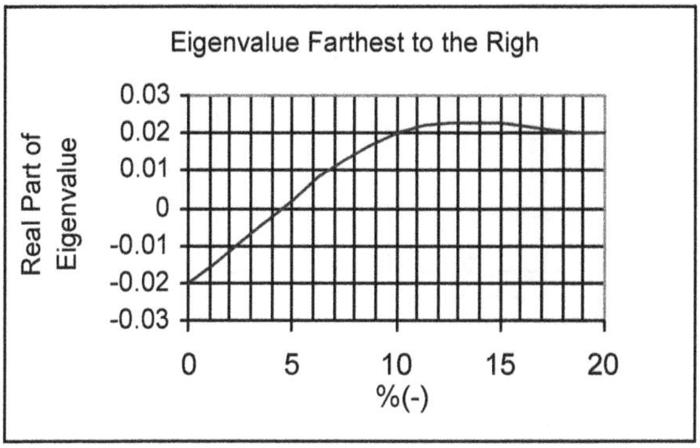

Observation:

The system appears to be stable even when the actual value is 20.0 % less than the value in the math model. The eigenvalue computation reveals that there is one complex root with a real value of 0.02, implying instability. Further investigation is needed to understand why the step response does not show instability. One conjecture would be that the coefficients associated with this root might be close to zero.

Case 2 (actual value is greater than the value in the math model)

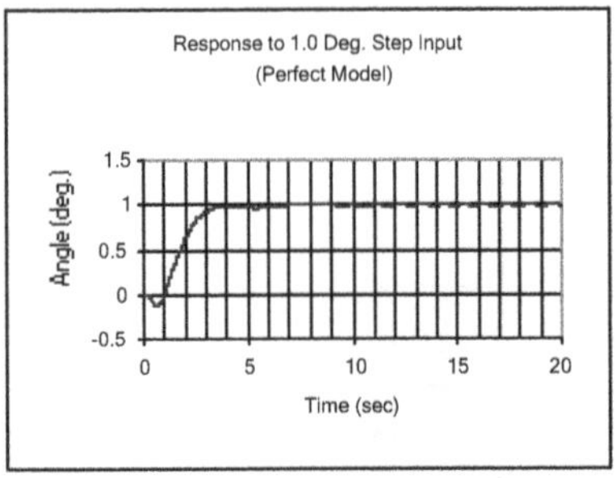

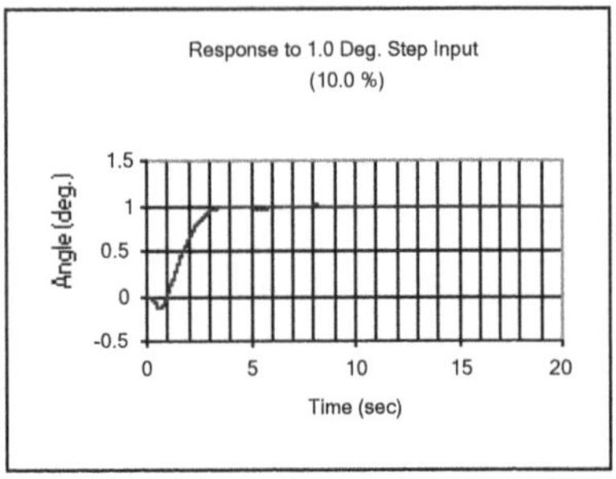

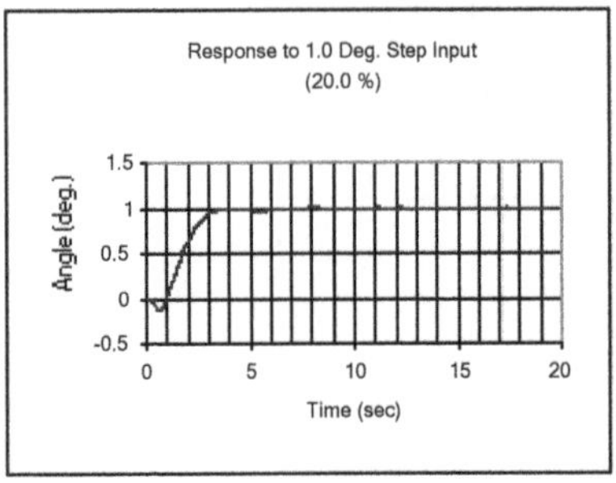

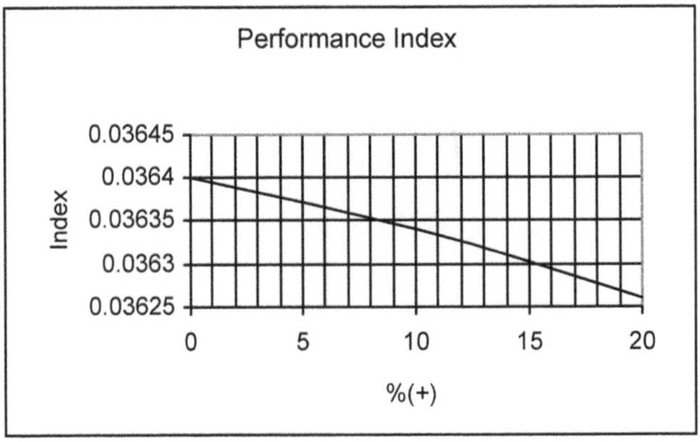

Appendix III: Collection of Math Modeling Error Sensitivity Response and Singular Value Plots, and Eigenvalue Tables

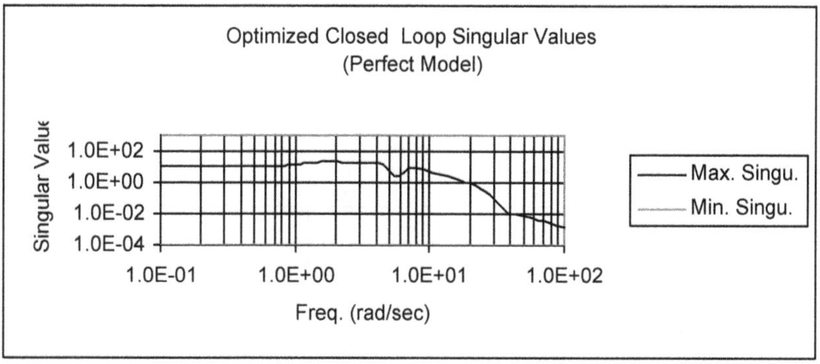

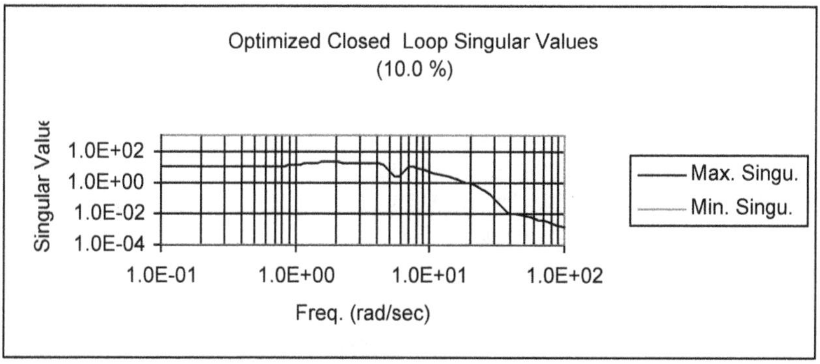

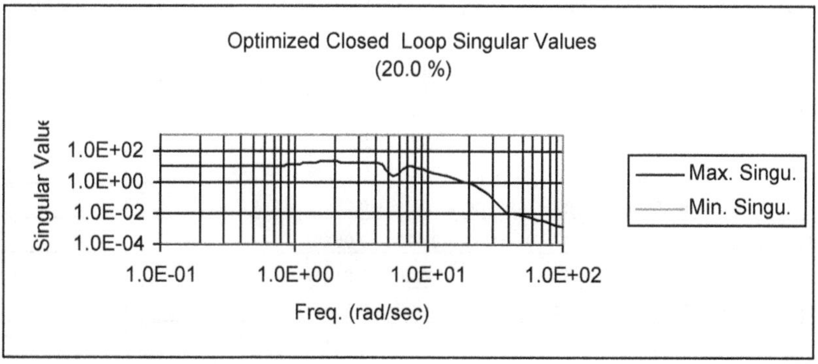

Appendix III: Collection of Math Modeling Error Sensitivity Response and Singular Value Plots, and Eigenvalue Tables

Closed Loop System Eigenvalues					
	0%	10.00%	20.00%		
	1.0e+002 *	1.0e+002 *	1.0e+002 *		
1	-0.25	-0.25	-0.25		
2	-0.5824 + 1.3344i	-0.5824 + 1.3344i	-0.5824 + 1.3344i		
3	-0.5824 - 1.3344i	-0.5824 - 1.3344i	-0.5824 - 1.3344i		
4	-0.1040 + 0.2546i	-0.1041 + 0.2547i	-0.1042 + 0.2547i		
5	-0.1040 - 0.2546i	-0.1041 - 0.2547i	-0.1042 - 0.2547i		
6	-0.1046 + 0.2542i	-0.1045 + 0.2542i	-0.1045 + 0.2541i		
7	-0.1046 - 0.2542i	-0.1045 - 0.2542i	-0.1045 - 0.2541i		
8	-0.2284 + 0.0212i	-0.2299	-0.2304		
9	-0.2284 - 0.0212i	-0.2282 + 0.0199i	-0.2280 + 0.0190i		
10	-0.2293	-0.2282 - 0.0199i	-0.2280 - 0.0190i		
11	-0.1353 + 0.0476i	-0.1326 + 0.0493i	-0.1313 + 0.0505i		
12	-0.1353 - 0.0476i	-0.1326 - 0.0493i	-0.1313 - 0.0505i		
13	-0.1446 + 0.0002i	-0.1447 + 0.0002i	-0.1447 + 0.0002i		
14	-0.1446 - 0.0002i	-0.1447 - 0.0002i	-0.1447 - 0.0002i		
15	-0.0868 + 0.0557i	-0.0888 + 0.0518i	-0.0898 + 0.0493i		
16	-0.0868 - 0.0557i	-0.0888 - 0.0518i	-0.0898 - 0.0493i		
17	-0.0002 + 0.0601i	-0.0002 + 0.0608i	-0.0002 + 0.0612i		
18	-0.0002 - 0.0601i	-0.0002 - 0.0608i	-0.0002 - 0.0612i		
19	-0.0002 + 0.0600i	-0.0001 + 0.0597i	-0.0001 + 0.0597i		
20	-0.0002 - 0.0600i	-0.0001 - 0.0597i	-0.0001 - 0.0597i		
21	-0.0009 + 0.0508i	-0.0011 + 0.0505i	-0.0011 + 0.0504i		
22	-0.0009 - 0.0508i	-0.0011 - 0.0505i	-0.0011 - 0.0504i		
23	-0.0009 + 0.0509i	-0.0007 + 0.0510i	-0.0006 + 0.0510i		
24	-0.0009 - 0.0509i	-0.0007 - 0.0510i	-0.0006 - 0.0510i		
25	-0.0347 + 0.0149i	-0.0354 + 0.0146i	-0.0357 + 0.0144i		
26	-0.0347 - 0.0149i	-0.0354 - 0.0146i	-0.0357 - 0.0144i		
27	-0.0204	-0.0203	-0.0203		
28	-0.0052 + 0.0136i	-0.0052 + 0.0136i	-0.0052 + 0.0136i		
29	-0.0052 - 0.0136i	-0.0052 - 0.0136i	-0.0052 - 0.0136i		
30	-0.0078 + 0.0145i	-0.0078 + 0.0145i	-0.0078 + 0.0145i		
31	-0.0078 - 0.0145i	-0.0078 - 0.0145i	-0.0078 - 0.0145i		
32	-0.0134	-0.0135	-0.0135		
33	-0.0106	-0.0106	-0.0106		
34	-0.01	-0.01	-0.01		

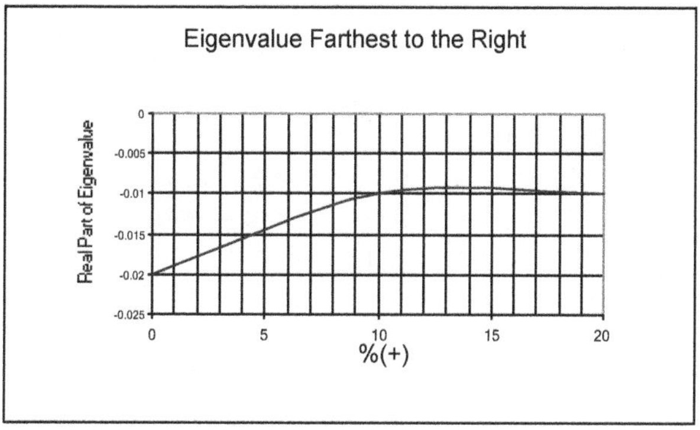

Observation:

The system is stable even after the actual value is 20 % greater than the value in the math model.

11) **Fuel Sloshing 2 Pendulum Hinge Point: lp2 (The value of the Fuel Sloshing 2 Pendulum Hinge Point = -20.26 ft.)**

Case 1 (actual value is less than the value in the math model)

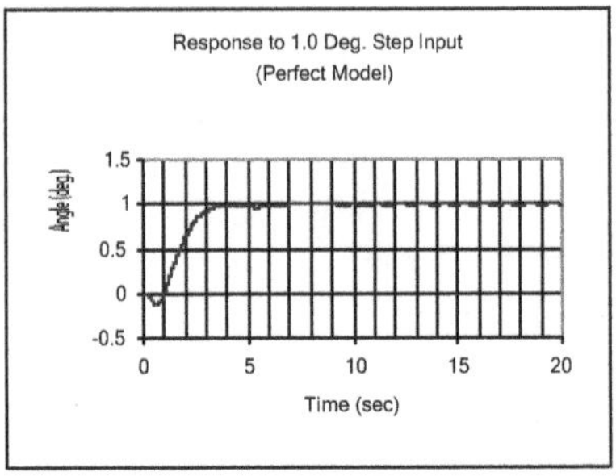

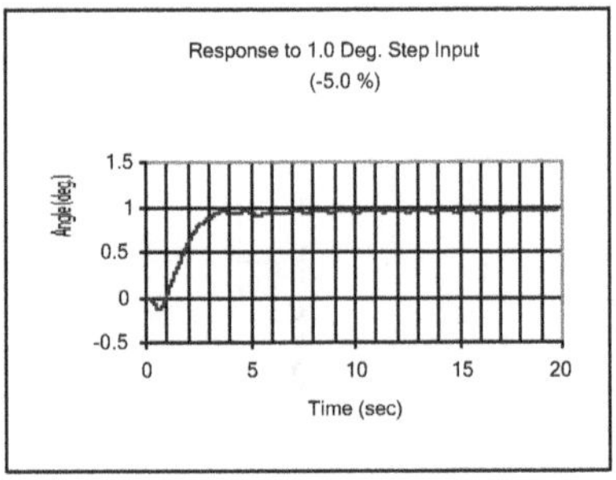

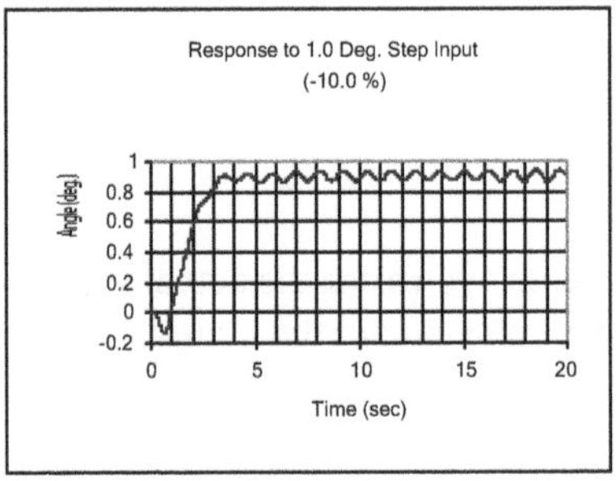

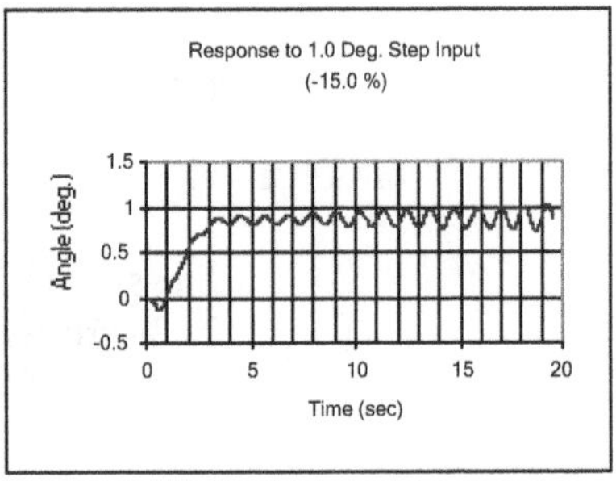

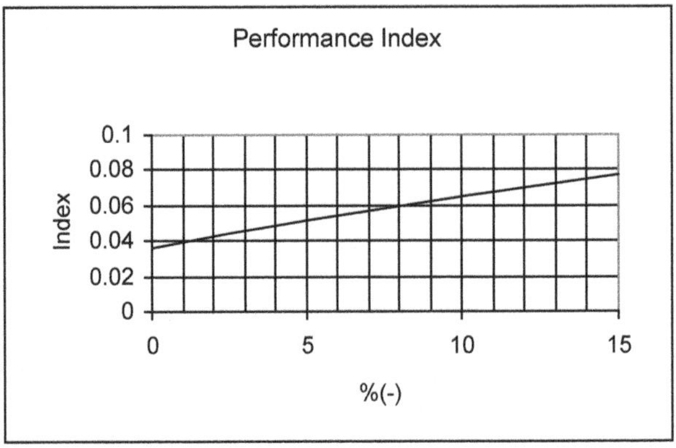

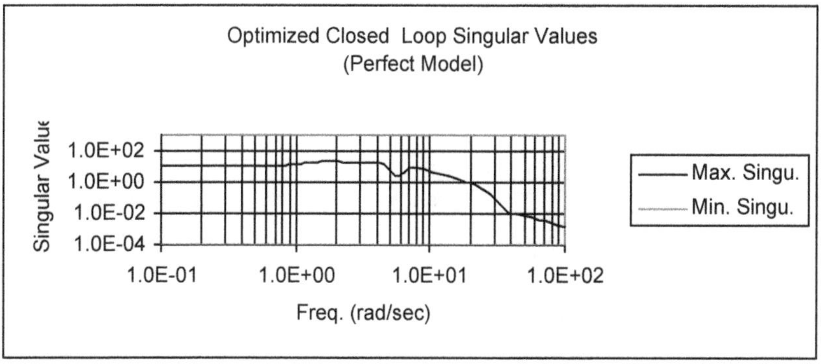
Optimized Closed Loop Singular Values
(Perfect Model)
Singular Valu
1.0E+02
1.0E+00
1.0E-02
1.0E-04
1.0E-01
1.0E+00
1.0E+01
1.0E+02
Freq. (rad/sec)
Max. Singu.
Min. Singu.

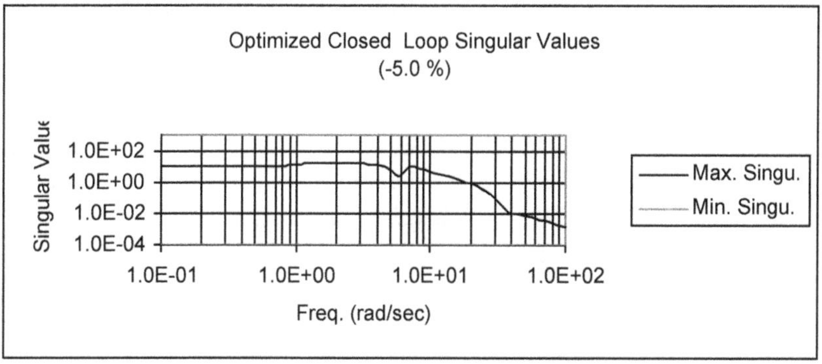
Optimized Closed Loop Singular Values
(-5.0 %)
Singular Valu
1.0E+02
1.0E+00
1.0E-02
1.0E-04
1.0E-01
1.0E+00
1.0E+01
1.0E+02
Freq. (rad/sec)
Max. Singu.
Min. Singu.

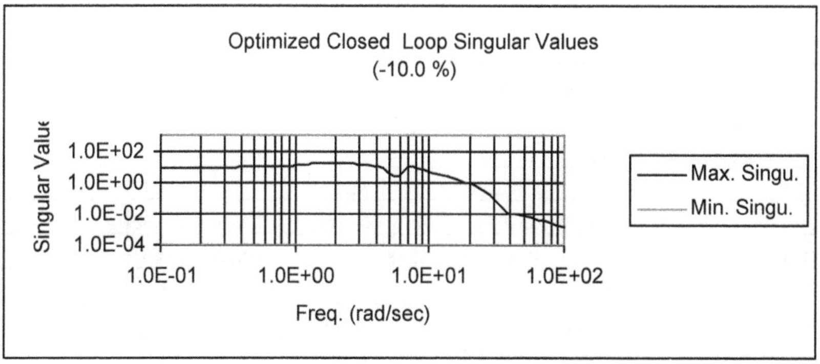
Optimized Closed Loop Singular Values
(-10.0 %)
Singular Valu
1.0E+02
1.0E+00
1.0E-02
1.0E-04
1.0E-01
1.0E+00
1.0E+01
1.0E+02
Freq. (rad/sec)
Max. Singu.
Min. Singu.

Appendix III: Collection of Math Modeling Error Sensitivity Response and Singular Value Plots, and Eigenvalue Tables

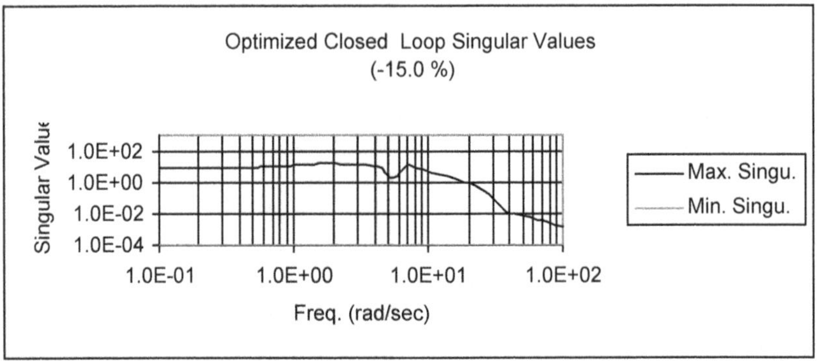

Closed Loop System Eigenvalues					
	0%	-5.00%	-10.00%	-15.00%	
	1.0e+002 *	1.0e+002 *	1.0e+002 *	1.0e+002 *	
1	-0.25	-0.25	-0.25	-0.25	
2	-0.5824 + 1.3344i	-0.5824 + 1.3344i	-0.5824 + 1.3344i	-0.5824 + 1.3344i	
3	-0.5824 - 1.3344i	-0.5824 - 1.3344i	-0.5824 - 1.3344i	-0.5824 - 1.3344i	
4	-0.1040 + 0.2546i	-0.1037 + 0.2546i	-0.1035 + 0.2546i	-0.1032 + 0.2546i	
5	-0.1040 - 0.2546i	-0.1037 - 0.2546i	-0.1035 - 0.2546i	-0.1032 - 0.2546i	
6	-0.1046 + 0.2542i	-0.1047 + 0.2543i	-0.1048 + 0.2543i	-0.1048 + 0.2542i	
7	-0.1046 - 0.2542i	-0.1047 - 0.2543i	-0.1048 - 0.2543i	-0.1048 - 0.2542i	
8	-0.2284 + 0.0212i	-0.2313 + 0.0244i	-0.2334 + 0.0268i	-0.2350 + 0.0286i	
9	-0.2284 - 0.0212i	-0.2313 - 0.0244i	-0.2334 - 0.0268i	-0.2350 - 0.0286i	
10	-0.2293	-0.2279	-0.2271	-0.2267	
11	-0.1353 + 0.0476i	-0.1195 + 0.0674i	-0.1218 + 0.0788i	-0.1234 + 0.0853i	
12	-0.1353 - 0.0476i	-0.1195 - 0.0674i	-0.1218 - 0.0788i	-0.1234 - 0.0853i	
13	-0.1446 + 0.0002i	-0.1448 + 0.0002i	-0.1376	-0.1477	
14	-0.1446 - 0.0002i	-0.1448 - 0.0002i	-0.1456	-0.142	
15	-0.0868 + 0.0557i	-0.1181	-0.1448	-0.1447	
16	-0.0868 - 0.0557i	-0.0734	-0.0001 + 0.0608i	-0.0419 + 0.0277i	
17	-0.0002 + 0.0601i	-0.0002 + 0.0605i	-0.0001 - 0.0608i	-0.0419 - 0.0277i	
18	-0.0002 - 0.0601i	-0.0002 - 0.0605i	-0.0002 + 0.0598i	0.0002 + 0.0612i	
19	-0.0002 + 0.0600i	-0.0002 + 0.0599i	-0.0002 - 0.0598i	0.0002 - 0.0612i	
20	-0.0002 - 0.0600i	-0.0002 - 0.0599i	0.0001 + 0.0543i	-0.0002 + 0.0598i	
21	-0.0009 + 0.0508i	-0.0007 + 0.0529i	0.0001 - 0.0543i	-0.0002 - 0.0598i	

22	-0.0009 - 0.0508i	-0.0007 - 0.0529i	-0.0001 + 0.0507i	0.0008 + 0.0554i	
23	-0.0009 + 0.0509i	-0.0001 + 0.0506i	-0.0001 - 0.0507i	0.0008 - 0.0554i	
24	-0.0009 - 0.0509i	-0.0001 - 0.0506i	-0.0450 + 0.0238i	-0.0001 + 0.0508i	
25	-0.0347 + 0.0149i	-0.0403 + 0.0140i	-0.0450 - 0.0238i	-0.0001 - 0.0508i	
26	-0.0347 - 0.0149i	-0.0403 - 0.0140i	-0.0382	-0.0342	
27	-0.0204	-0.0203	-0.0203	-0.0202	
28	-0.0052 + 0.0136i	-0.0053 + 0.0135i	-0.0054 + 0.0134i	-0.0054 + 0.0133i	
29	-0.0052 - 0.0136i	-0.0053 - 0.0135i	-0.0054 - 0.0134i	-0.0054 - 0.0133i	
30	-0.0078 + 0.0145i	-0.0078 + 0.0145i	-0.0078 + 0.0145i	-0.0078 + 0.0145i	
31	-0.0078 - 0.0145i	-0.0078 - 0.0145i	-0.0078 - 0.0145i	-0.0078 - 0.0145i	
32	-0.0134	-0.0134	-0.0134	-0.0133	
33	-0.0106	-0.0106	-0.0107	-0.0107	
34	-0.01	-0.01	-0.01	-0.01	

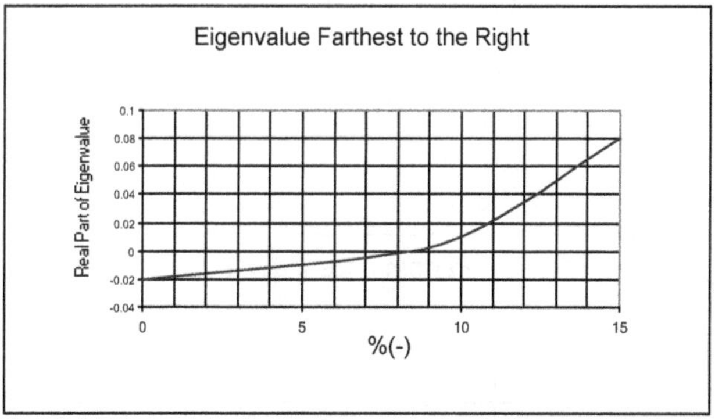

Observation:

The control system is stable until the actual value is 9.0 % less than the value in the math model.

Case 2 (actual value is greater than the value in the math model)

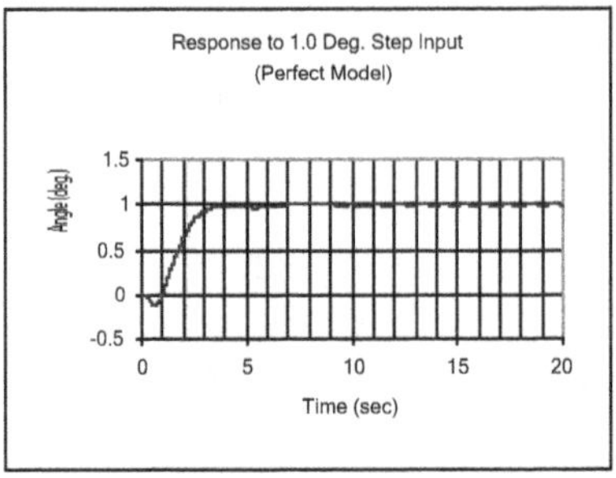

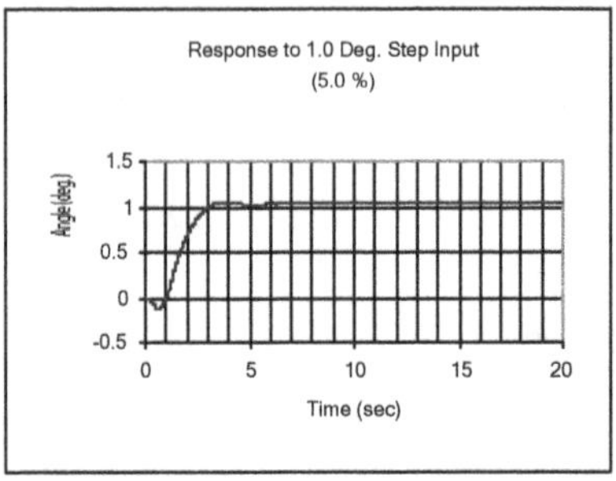

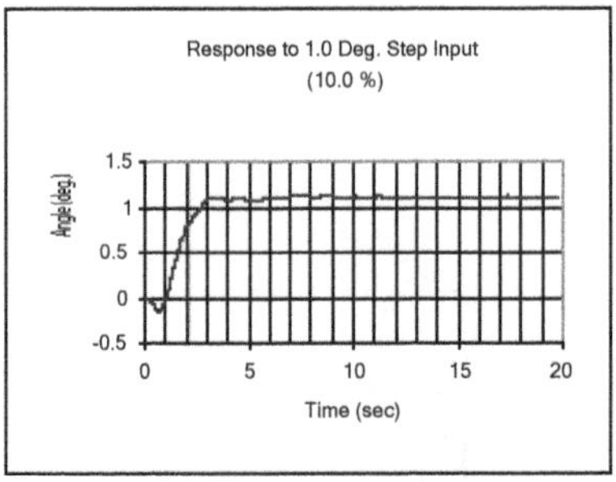

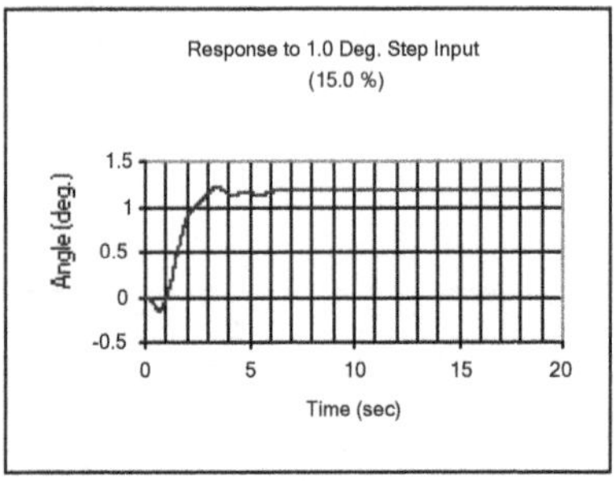

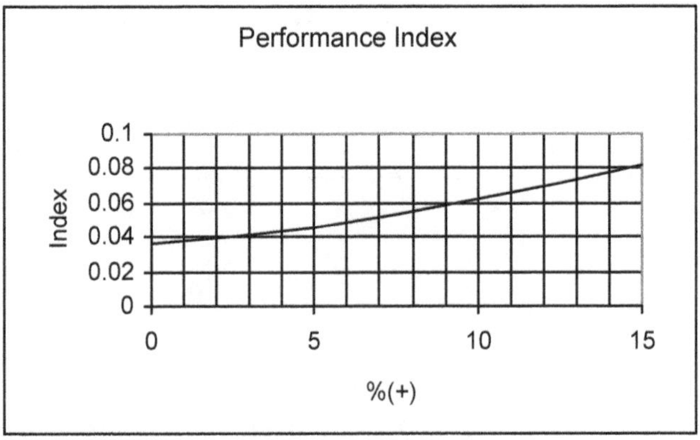

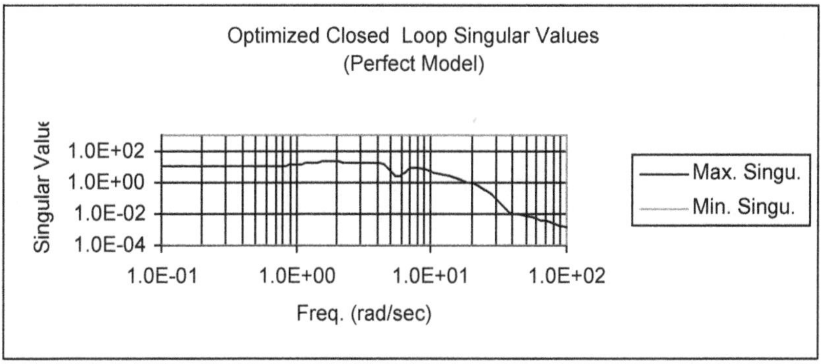
Optimized Closed Loop Singular Values
(Perfect Model)
Singular Value
1.0E+02
1.0E+00
1.0E-02
1.0E-04
1.0E-01
1.0E+00
1.0E+01
1.0E+02
Freq. (rad/sec)
Max. Singu.
Min. Singu.

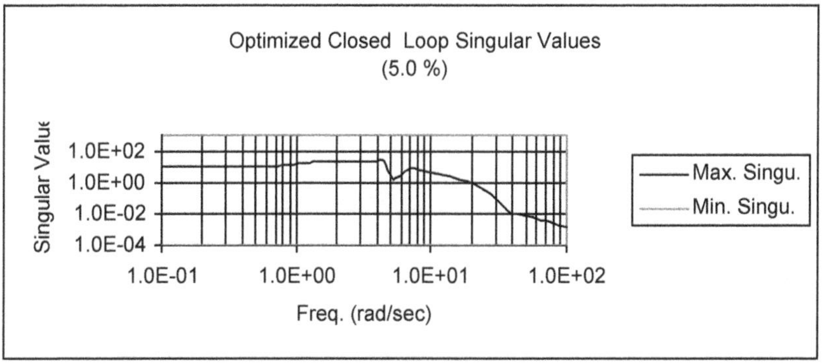
Optimized Closed Loop Singular Values
(5.0 %)
Singular Value
1.0E+02
1.0E+00
1.0E-02
1.0E-04
1.0E-01
1.0E+00
1.0E+01
1.0E+02
Freq. (rad/sec)
Max. Singu.
Min. Singu.

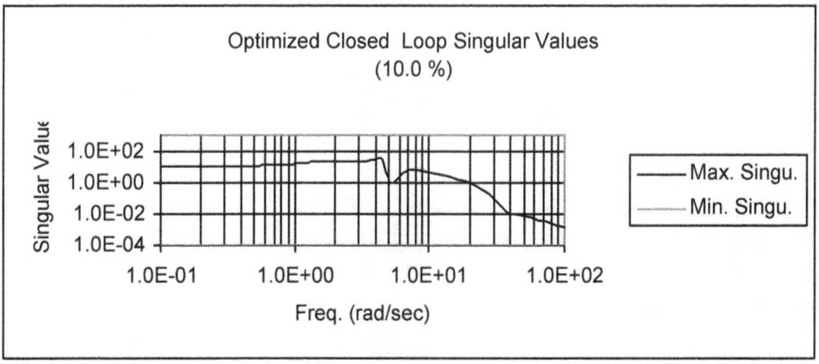
Optimized Closed Loop Singular Values
(10.0 %)
Singular Value
1.0E+02
1.0E+00
1.0E-02
1.0E-04
1.0E-01
1.0E+00
1.0E+01
1.0E+02
Freq. (rad/sec)
Max. Singu.
Min. Singu.

Appendix III: Collection of Math Modeling Error Sensitivity Response and Singular Value Plots, and Eigenvalue Tables

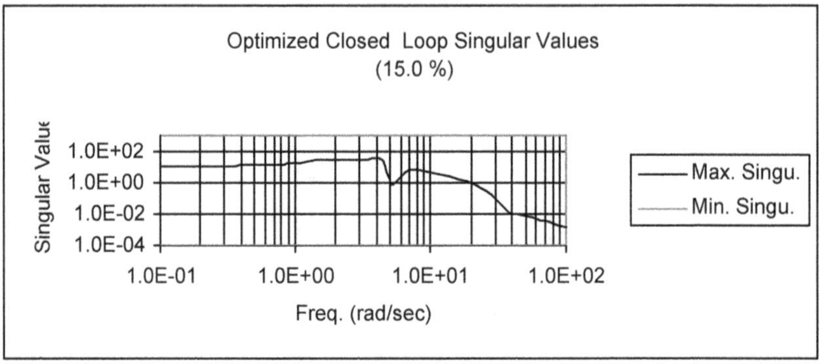

Closed Loop System Eigenvalues

	0%	5.00%	10.00%	15.00%	
	1.0e+002 *	1.0e+002 *	1.0e+002 *	1.0e+002 *	
1	-0.25	-0.25	-0.25	-0.25	
2	-0.5824 + 1.3344i	-0.5824 + 1.3344i	-0.5824 + 1.3344i	-0.5824 + 1.3344i	
3	-0.5824 - 1.3344i	-0.5824 - 1.3344i	-0.5824 - 1.3344i	-0.5824 - 1.3344i	
4	-0.1040 + 0.2546i	-0.1044 + 0.2548i	-0.1046 + 0.2550i	-0.1047 + 0.2552i	
5	-0.1040 - 0.2546i	-0.1044 - 0.2548i	-0.1046 - 0.2550i	-0.1047 - 0.2552i	
6	-0.1046 + 0.2542i	-0.1044 + 0.2540i	-0.1045 + 0.2539i	-0.1045 + 0.2537i	
7	-0.1046 - 0.2542i	-0.1044 - 0.2540i	-0.1045 - 0.2539i	-0.1045 - 0.2537i	
8	-0.2284 + 0.0212i	-0.2327	-0.2382	-0.243	
9	-0.2284 - 0.0212i	-0.2241 + 0.0169i	-0.2181 + 0.0117i	-0.2166	
10	-0.2293	-0.2241 - 0.0169i	-0.2181 - 0.0117i	-0.2075	
11	-0.1353 + 0.0476i	-0.1509 + 0.0483i	-0.1618 + 0.0501i	-0.1713 + 0.0529i	
12	-0.1353 - 0.0476i	-0.1509 - 0.0483i	-0.1618 - 0.0501i	-0.1713 - 0.0529i	
13	-0.1446 + 0.0002i	-0.1446 + 0.0002i	-0.0695 + 0.0776i	-0.0650 + 0.0846i	
14	-0.1446 - 0.0002i	-0.1446 - 0.0002i	-0.0695 - 0.0776i	-0.0650 - 0.0846i	
15	-0.0868 + 0.0557i	-0.0758 + 0.0690i	-0.1445 + 0.0001i	-0.1445 + 0.0001i	
16	-0.0868 - 0.0557i	-0.0758 - 0.0690i	-0.1445 - 0.0001i	-0.1445 - 0.0001i	
17	-0.0002 + 0.0601i	-0.0004 + 0.0599i	-0.0004 + 0.0598i	-0.0004 + 0.0597i	
18	-0.0002 - 0.0601i	-0.0004 - 0.0599i	-0.0004 - 0.0598i	-0.0004 - 0.0597i	
19	-0.0002 + 0.0600i	-0.0000 + 0.0600i	0.0000 + 0.0600i	0.0001 + 0.0599i	
20	-0.0002 - 0.0600i	-0.0000 - 0.0600i	0.0000 - 0.0600i	0.0001 - 0.0599i	
21	-0.0009 + 0.0508i	-0.0024 + 0.0484i	-0.0002 + 0.0510i	-0.0002 + 0.0510i	
22	-0.0009 - 0.0508i	-0.0024 - 0.0484i	-0.0002 - 0.0510i	-0.0002 - 0.0510i	

23	-0.0009 + 0.0509i	-0.0003 + 0.0510i	-0.0031 + 0.0458i	-0.0036 + 0.0431i	
24	-0.0009 - 0.0509i	-0.0003 - 0.0510i	-0.0031 - 0.0458i	-0.0036 - 0.0431i	
25	-0.0347 + 0.0149i	-0.0318 + 0.0156i	-0.0296 + 0.0163i	-0.0277 + 0.0170i	
26	-0.0347 - 0.0149i	-0.0318 - 0.0156i	-0.0296 - 0.0163i	-0.0277 - 0.0170i	
27	-0.0204	-0.0205	-0.0205	-0.0207	
28	-0.0052 + 0.0136i	-0.0052 + 0.0137i	-0.0050 + 0.0139i	-0.0049 + 0.0141i	
29	-0.0052 - 0.0136i	-0.0052 - 0.0137i	-0.0050 - 0.0139i	-0.0049 - 0.0141i	
30	-0.0078 + 0.0145i	-0.0079 + 0.0145i	-0.0079 + 0.0145i	-0.0079 + 0.0145i	
31	-0.0078 - 0.0145i	-0.0079 - 0.0145i	-0.0079 - 0.0145i	-0.0079 - 0.0145i	
32	-0.0134	-0.0135	-0.0135	-0.0135	
33	-0.0106	-0.0106	-0.0105	-0.0105	
34	-0.01	-0.01	-0.01	-0.01	

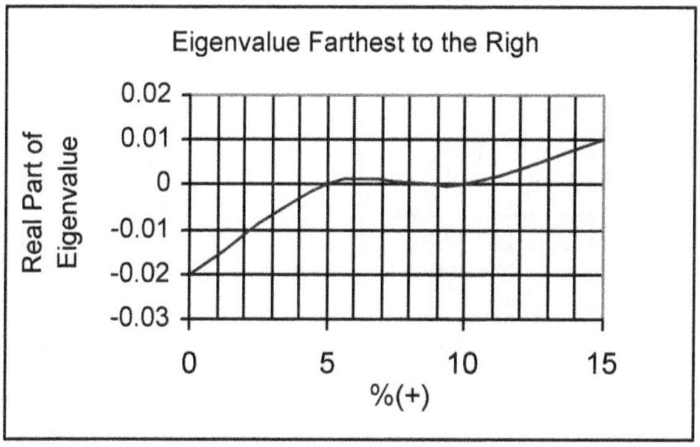

Observation:

The response is not oscillatory like the previous case and appears stable even when the actual value is 10.0 % greater than the value in the math model. The real part of one eigenvalue is on the borderline. Any further increase in error may cause instability, but the instability would be insignificant.

12) **Bending Mode 1 Modal Mass: Mb1 (The value of Bending Mode 1 Modal Mass in the math model = 1,590 slugs.)**

Case 1 (actual value is less than the value in the math model)

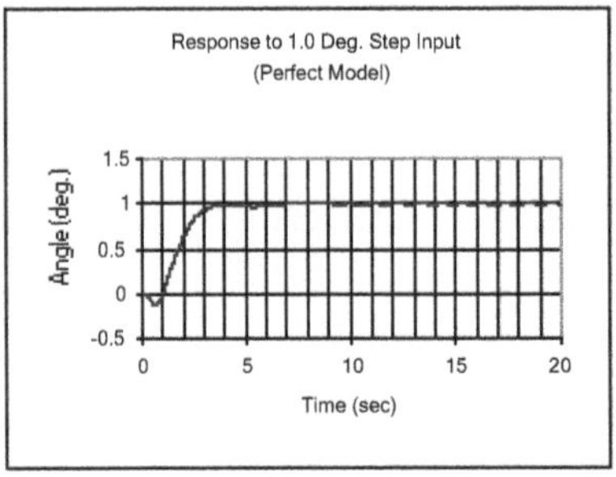

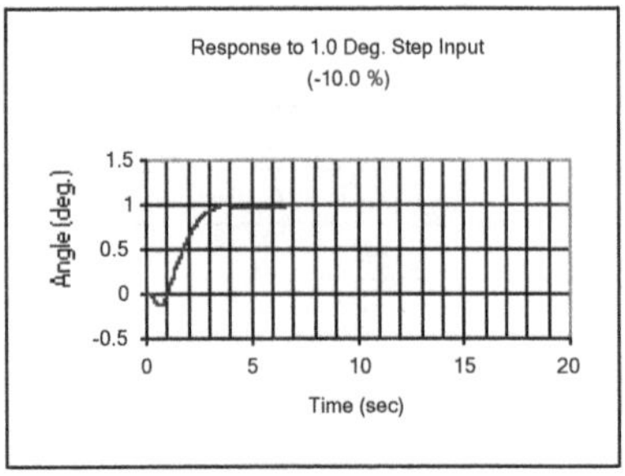

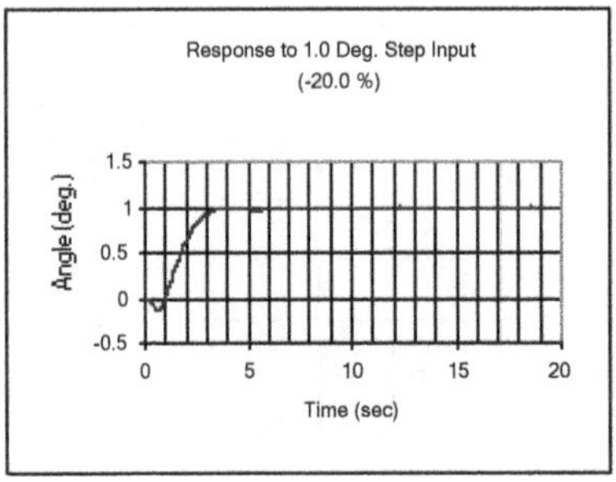

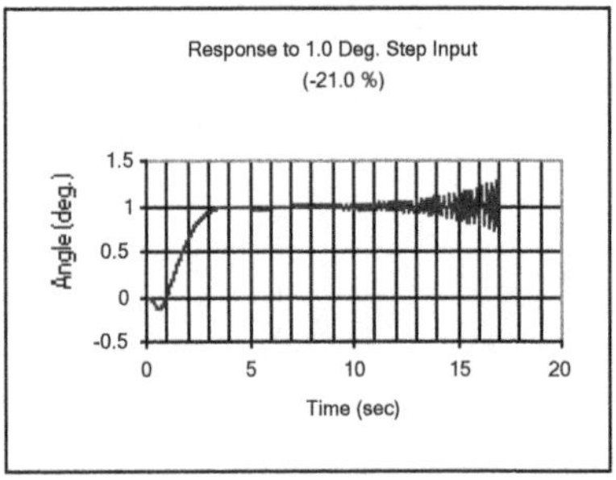

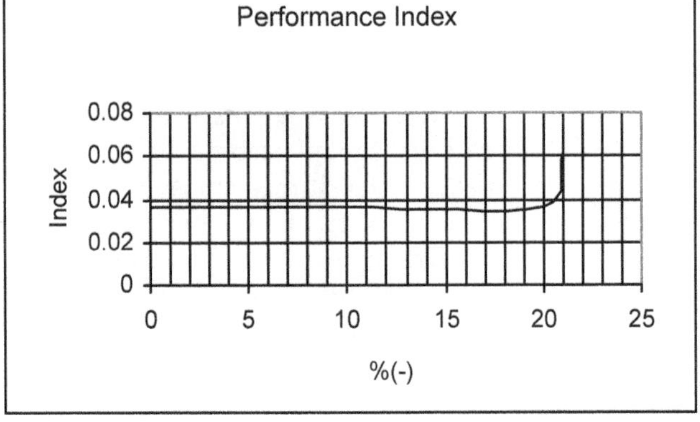

Appendix III: Collection of Math Modeling Error Sensitivity Response and Singular Value Plots, and Eigenvalue Tables

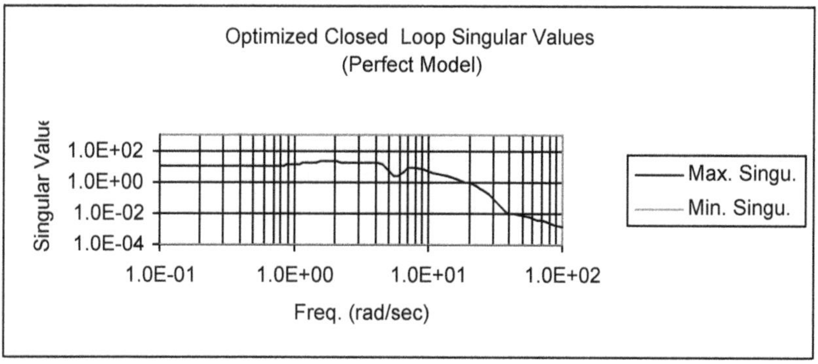

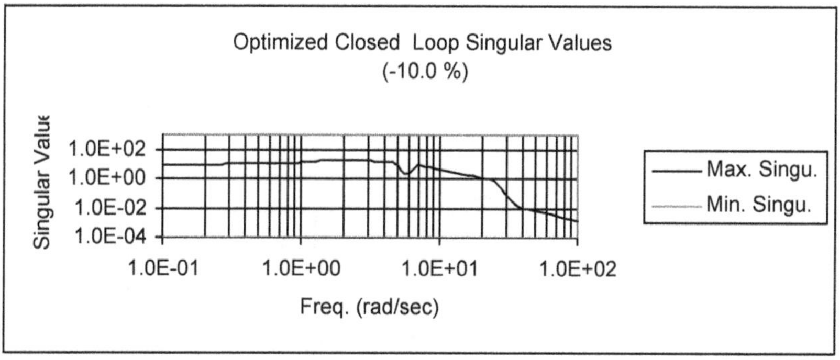

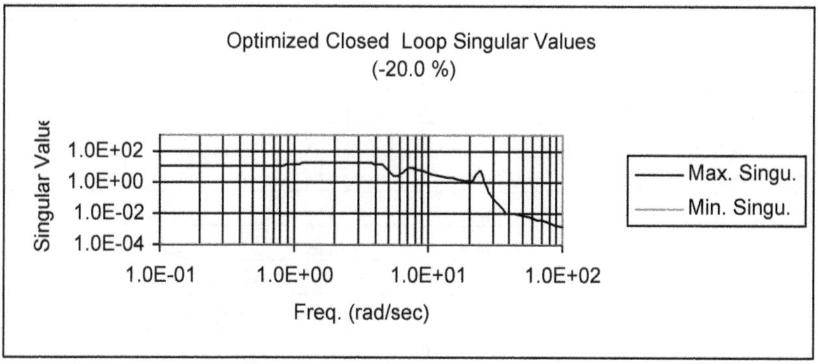

Appendix III: Collection of Math Modeling Error Sensitivity Response and Singular Value Plots, and Eigenvalue Tables

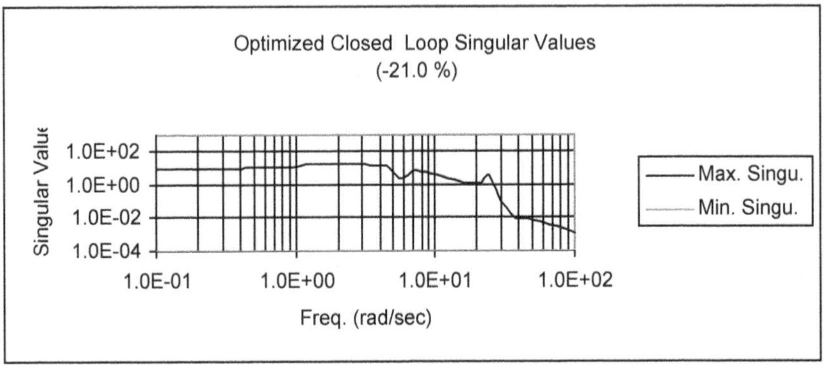

Closed Loop System Eigenvalues

	0%	-10.00%	-20.00%	-21.00%	
	1.0e+002 *	1.0e+002 *	1.0e+002 *	1.0e+002 *	
1	-0.25	-0.25	-0.25	-0.25	
2	-0.5824 + 1.3344i	-0.5824 + 1.3344i	-0.5824 + 1.3344i	-0.5824 + 1.3344i	
3	-0.5824 - 1.3344i	-0.5824 - 1.3344i	-0.5824 - 1.3344i	-0.5824 - 1.3344i	
4	-0.1040 + 0.2546i	-0.1633 + 0.3002i	-0.1823 + 0.3243i	-0.1838 + 0.3265i	
5	-0.1040 - 0.2546i	-0.1633 - 0.3002i	-0.1823 - 0.3243i	-0.1838 - 0.3265i	
6	-0.1046 + 0.2542i	-0.3106 + 0.0710i	-0.3323 + 0.0779i	-0.3340 + 0.0783i	
7	-0.1046 - 0.2542i	-0.3106 - 0.0710i	-0.3323 - 0.0779i	-0.3340 - 0.0783i	
8	-0.2284 + 0.0212i	-0.0304 + 0.2392i	0.0007 + 0.2410i	0.0035 + 0.2413i	
9	-0.2284 - 0.0212i	-0.0304 - 0.2392i	0.0007 - 0.2410i	0.0035 - 0.2413i	
10	-0.2293	-0.1337 + 0.1167i	-0.1327 + 0.1242i	-0.1328 + 0.1247i	
11	-0.1353 + 0.0476i	-0.1337 - 0.1167i	-0.1327 - 0.1242i	-0.1328 - 0.1247i	
12	-0.1353 - 0.0476i	-0.1679	-0.1673	-0.1673	
13	-0.1446 + 0.0002i	-0.1471	-0.1472	-0.1472	
14	-0.1446 - 0.0002i	-0.1446	-0.1446	-0.1446	
15	-0.0868 + 0.0557i	-0.0538 + 0.0509i	-0.0471 + 0.0512i	-0.0467 + 0.0513i	
16	-0.0868 - 0.0557i	-0.0538 - 0.0509i	-0.0471 - 0.0512i	-0.0467 - 0.0513i	
17	-0.0002 + 0.0601i	-0.0001 + 0.0601i	-0.0000 + 0.0601i	-0.0000 + 0.0601i	
18	-0.0002 - 0.0601i	-0.0001 - 0.0601i	-0.0000 - 0.0601i	-0.0000 - 0.0601i	
19	-0.0002 + 0.0600i	-0.0003 + 0.0599i	-0.0003 + 0.0599i	-0.0003 + 0.0599i	
20	-0.0002 - 0.0600i	-0.0003 - 0.0599i	-0.0003 - 0.0599i	-0.0003 - 0.0599i	
21	-0.0009 + 0.0508i	-0.0011 + 0.0511i	-0.0012 + 0.0512i	-0.0012 + 0.0513i	
22	-0.0009 - 0.0508i	-0.0011 - 0.0511i	-0.0012 - 0.0512i	-0.0012 - 0.0513i	

23	-0.0009 + 0.0509i	-0.0006 + 0.0507i	-0.0005 + 0.0507i	-0.0005 + 0.0507i	
24	-0.0009 - 0.0509i	-0.0006 - 0.0507i	-0.0005 - 0.0507i	-0.0005 - 0.0507i	
25	-0.0347 + 0.0149i	-0.0312 + 0.0141i	-0.0293 + 0.0137i	-0.0291 + 0.0137i	
26	-0.0347 - 0.0149i	-0.0312 - 0.0141i	-0.0293 - 0.0137i	-0.0291 - 0.0137i	
27	-0.0204	-0.0205	-0.0205	-0.0205	
28	-0.0052 + 0.0136i	-0.0052 + 0.0136i	-0.0052 + 0.0136i	-0.0052 + 0.0136i	
29	-0.0052 - 0.0136i	-0.0052 - 0.0136i	-0.0052 - 0.0136i	-0.0052 - 0.0136i	
30	-0.0078 + 0.0145i	-0.0078 + 0.0145i	-0.0078 + 0.0145i	-0.0078 + 0.0145i	
31	-0.0078 - 0.0145i	-0.0078 - 0.0145i	-0.0078 - 0.0145i	-0.0078 - 0.0145i	
32	-0.0134	-0.0134	-0.0135	-0.0135	
33	-0.0106	-0.0106	-0.0106	-0.0106	
34	-0.01	-0.01	-0.01	-0.01	

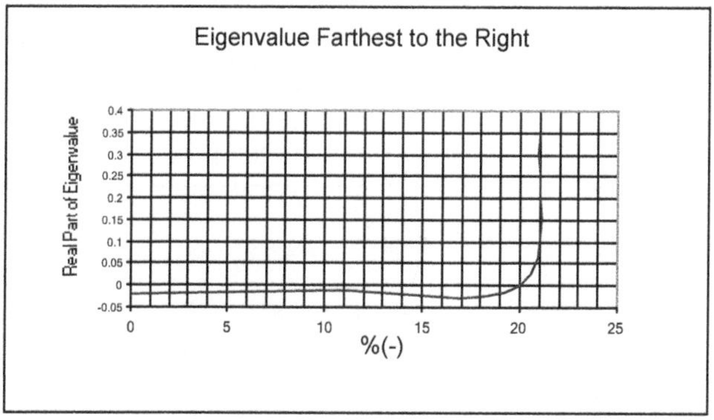

Observation:

The step response displays stability until the error reaches –20.0%, but it shows severe instability when the error increases slightly above 20%. Further analysis is needed to determine the causes of this sudden change.

Case 2 (actual value is greater than the value in the math model)

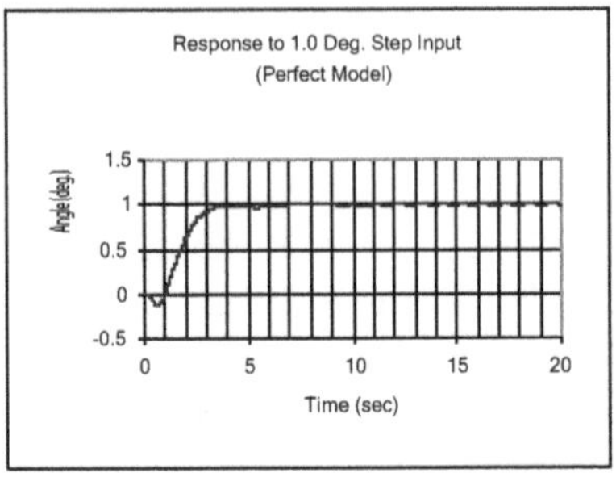

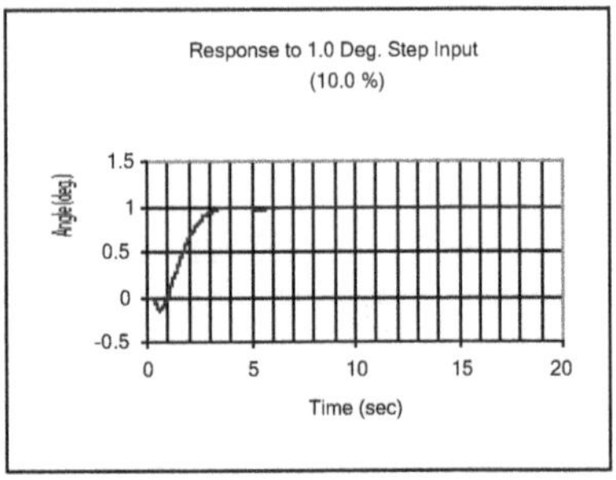

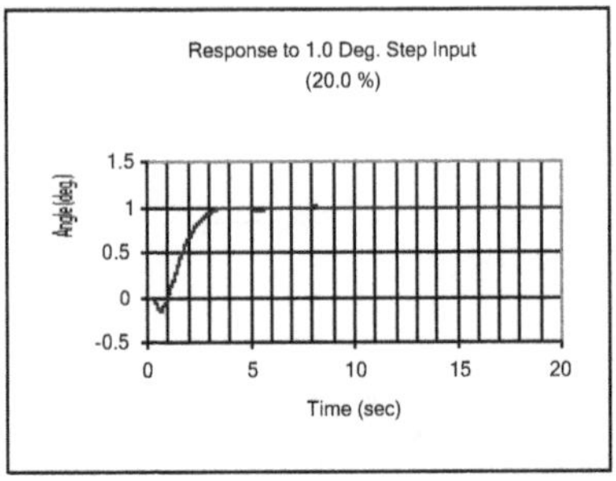

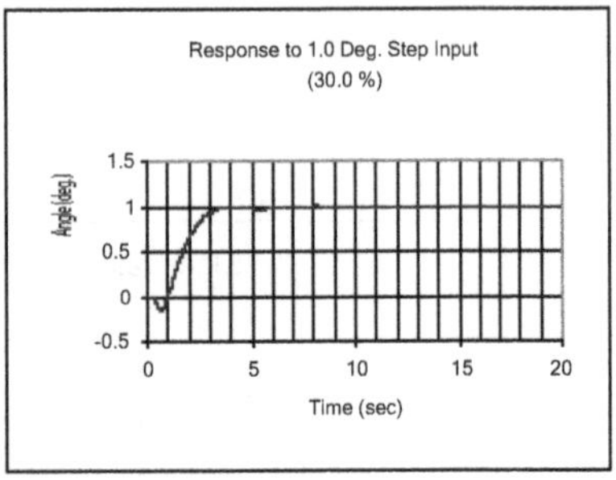

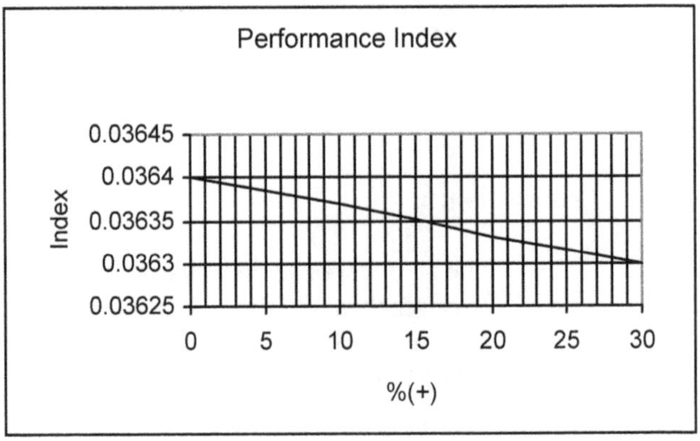

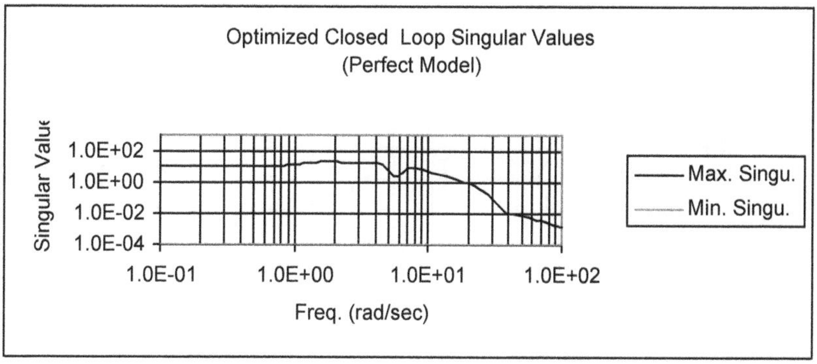
Optimized Closed Loop Singular Values
(Perfect Model)
Singular Value
1.0E+02
1.0E+00
1.0E-02
1.0E-04
1.0E-01
1.0E+00
1.0E+01
1.0E+02
Freq. (rad/sec)
Max. Singu.
Min. Singu.

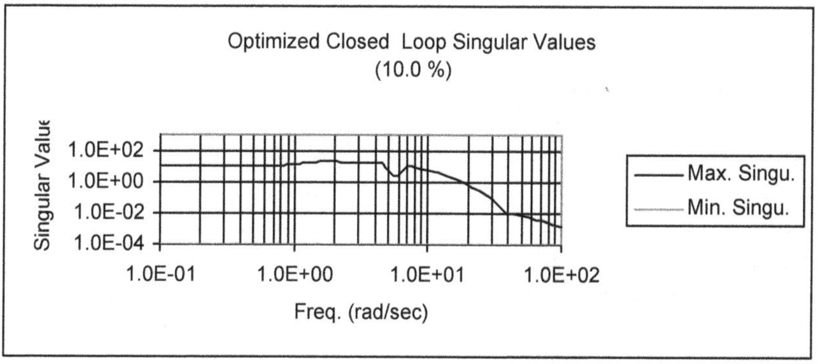
Optimized Closed Loop Singular Values
(10.0 %)
Singular Value
1.0E+02
1.0E+00
1.0E-02
1.0E-04
1.0E-01
1.0E+00
1.0E+01
1.0E+02
Freq. (rad/sec)
Max. Singu.
Min. Singu.

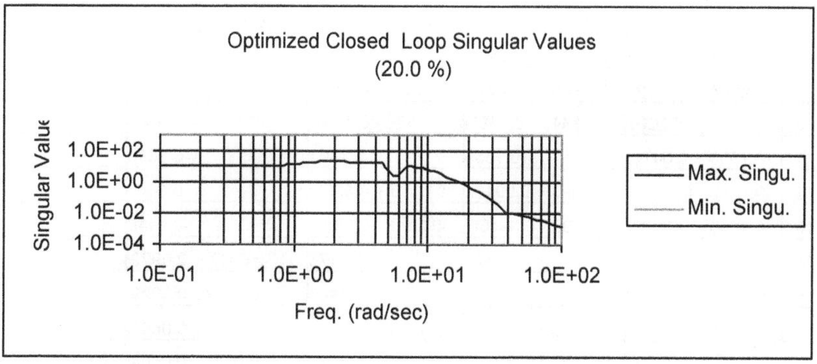
Optimized Closed Loop Singular Values
(20.0 %)
Singular Value
1.0E+02
1.0E+00
1.0E-02
1.0E-04
1.0E-01
1.0E+00
1.0E+01
1.0E+02
Freq. (rad/sec)
Max. Singu.
Min. Singu.

Appendix III: Collection of Math Modeling Error Sensitivity Response and Singular Value Plots, and Eigenvalue Tables

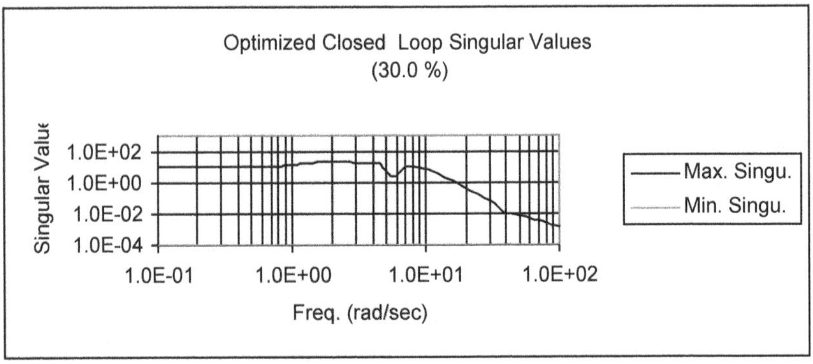

Closed Loop System Eigenvalues				
	0%	10.00%	20.00%	30.00%
	1.0e+002 *	1.0e+002 *	1.0e+002 *	1.0e+002 *
1	-0.25	-0.25	-0.25	-0.25
2	-0.5824 + 1.3344i	-0.5824 + 1.3344i	-0.5824 + 1.3344i	-0.5824 + 1.3344i
3	-0.5824 - 1.3344i	-0.5824 - 1.3344i	-0.5824 - 1.3344i	-0.5824 - 1.3344i
4	-0.1040 + 0.2546i	-0.0790 + 0.3098i	-0.0716 + 0.3260i	-0.0670 + 0.3362i
5	-0.1040 - 0.2546i	-0.0790 - 0.3098i	-0.0716 - 0.3260i	-0.0670 - 0.3362i
6	-0.1046 + 0.2542i	-0.3194	-0.3325	-0.3399
7	-0.1046 - 0.2542i	-0.2338 + 0.1550i	-0.2591 + 0.1794i	-0.2740 + 0.1898i
8	-0.2284 + 0.0212i	-0.2338 - 0.1550i	-0.2591 - 0.1794i	-0.2740 - 0.1898i
9	-0.2284 - 0.0212i	-0.1340 + 0.1685i	-0.1329 + 0.1501i	-0.1326 + 0.1446i
10	-0.2293	-0.1340 - 0.1685i	-0.1329 - 0.1501i	-0.1326 - 0.1446i
11	-0.1353 + 0.0476i	-0.1655	-0.1661	-0.1663
12	-0.1353 - 0.0476i	-0.0593 + 0.0946i	-0.0418 + 0.0957i	-0.1473
13	-0.1446 + 0.0002i	-0.0593 - 0.0946i	-0.0418 - 0.0957i	-0.1446
14	-0.1446 - 0.0002i	-0.1474	-0.1473	-0.0317 + 0.0929i
15	-0.0868 + 0.0557i	-0.1446	-0.1446	-0.0317 - 0.0929i
16	-0.0868 - 0.0557i	-0.0003 + 0.0602i	-0.0003 + 0.0603i	-0.0004 + 0.0604i
17	-0.0002 + 0.0601i	-0.0003 - 0.0602i	-0.0003 - 0.0603i	-0.0004 - 0.0604i
18	-0.0002 - 0.0601i	-0.0002 + 0.0599i	-0.0001 + 0.0599i	-0.0001 + 0.0599i
19	-0.0002 + 0.0600i	-0.0002 - 0.0599i	-0.0001 - 0.0599i	-0.0001 - 0.0599i
20	-0.0002 - 0.0600i	-0.0012 + 0.0504i	-0.0013 + 0.0502i	-0.0015 + 0.0500i
21	-0.0009 + 0.0508i	-0.0012 - 0.0504i	-0.0013 - 0.0502i	-0.0015 - 0.0500i

22	-0.0009 - 0.0508i	-0.0007 + 0.0511i	-0.0007 + 0.0511i	-0.0006 + 0.0512i	
23	-0.0009 + 0.0509i	-0.0007 - 0.0511i	-0.0007 - 0.0511i	-0.0006 - 0.0512i	
24	-0.0009 - 0.0509i	-0.0383 + 0.0213i	-0.0356 + 0.0257i	-0.0334 + 0.0278i	
25	-0.0347 + 0.0149i	-0.0383 - 0.0213i	-0.0356 - 0.0257i	-0.0334 - 0.0278i	
26	-0.0347 - 0.0149i	-0.0409	-0.0344	-0.0318	
27	-0.0204	-0.0203	-0.0202	-0.0201	
28	-0.0052 + 0.0136i	-0.0053 + 0.0136i	-0.0053 + 0.0136i	-0.0053 + 0.0136i	
29	-0.0052 - 0.0136i	-0.0053 - 0.0136i	-0.0053 - 0.0136i	-0.0053 - 0.0136i	
30	-0.0078 + 0.0145i	-0.0078 + 0.0145i	-0.0078 + 0.0145i	-0.0078 + 0.0145i	
31	-0.0078 - 0.0145i	-0.0078 - 0.0145i	-0.0078 - 0.0145i	-0.0078 - 0.0145i	
32	-0.0134	-0.0134	-0.0134	-0.0134	
33	-0.0106	-0.0106	-0.0106	-0.0106	
34	-0.01	-0.01	-0.01	-0.01	

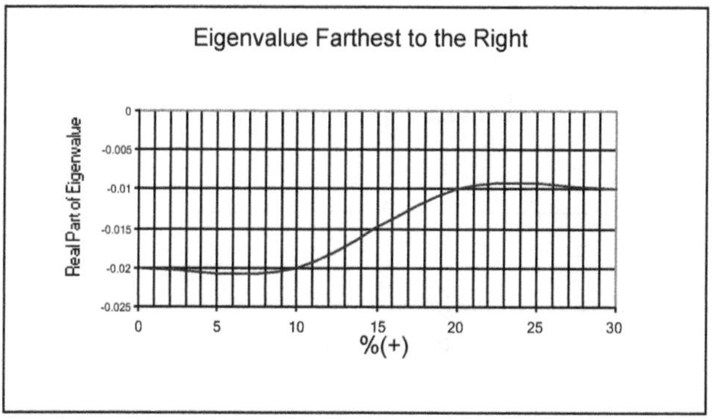

Observation:

When the actual Bending Mode 1 Modal mass is greater than the modal mass used in the math model, the real parts of the complex roots remain in the Left Half Plane, indicating system stability.

13) Bending Mode 2 Modal Mass: Mb2 (The value of Bending Mode 2 Modal Mass in the math model = 9,550 slugs.)

Case 1 (actual value is less than the value in the math model)

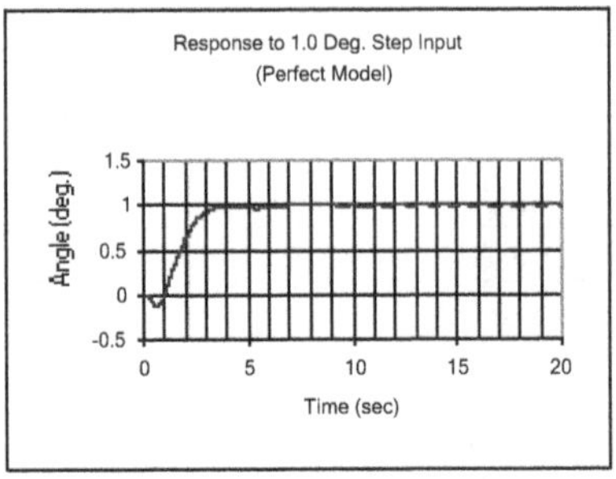

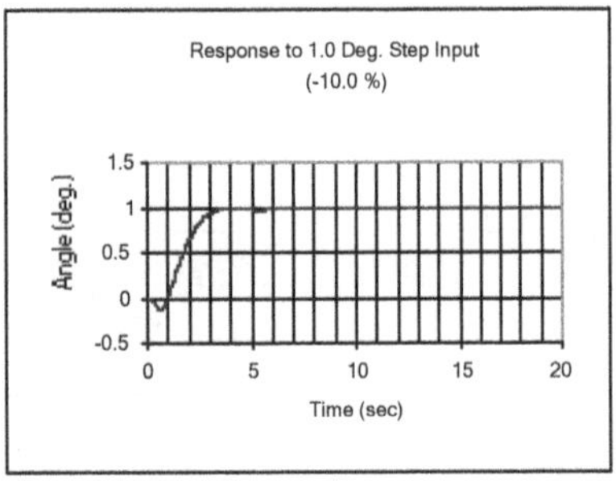

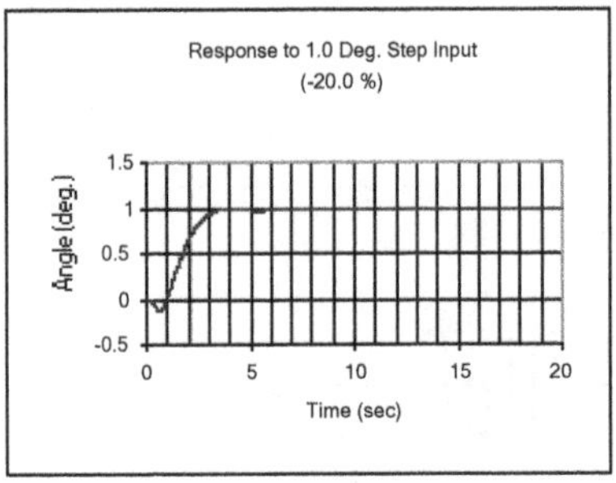

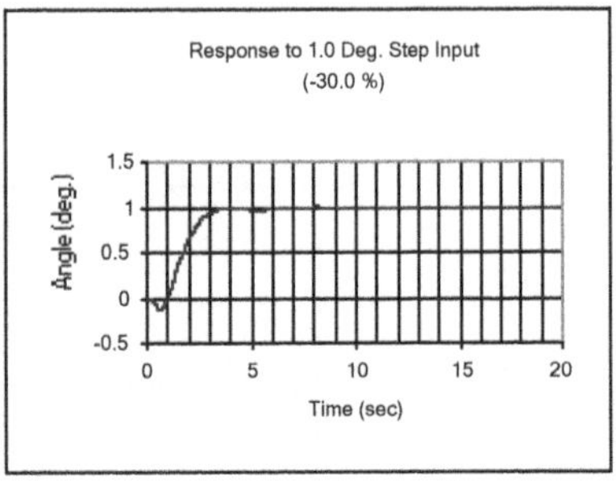

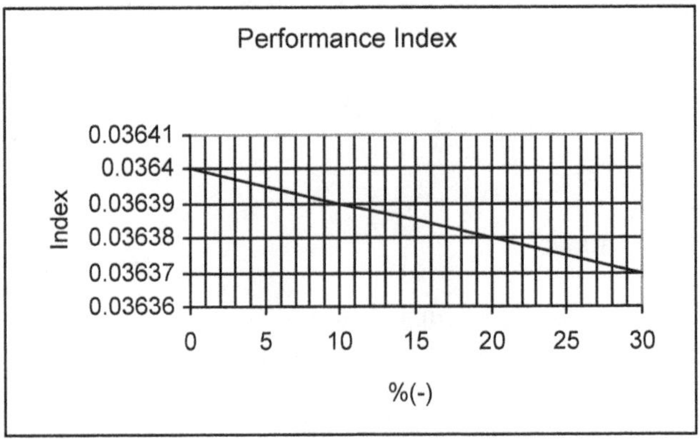

Appendix III: Collection of Math Modeling Error Sensitivity Response and Singular Value Plots, and Eigenvalue Tables

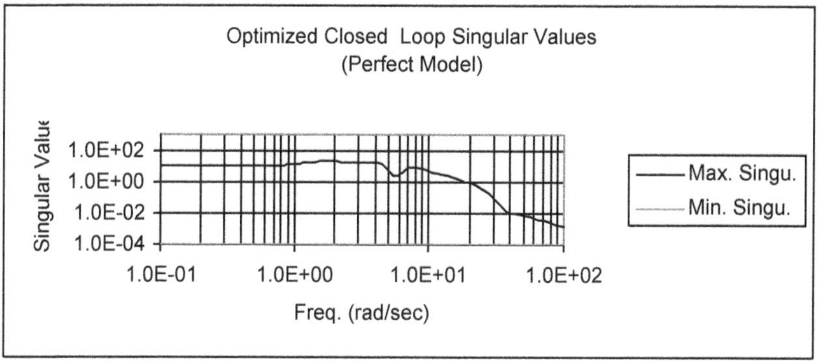

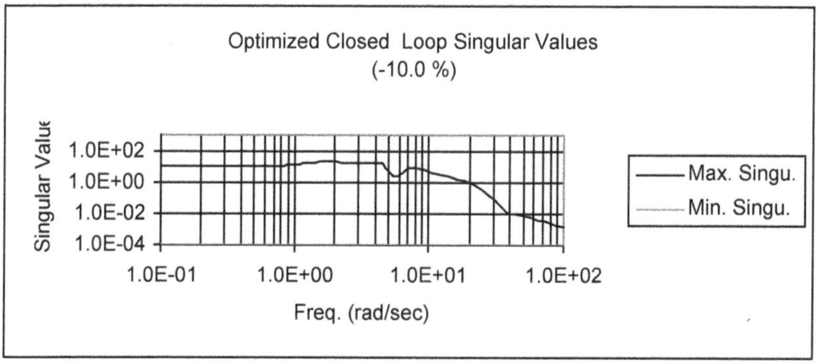

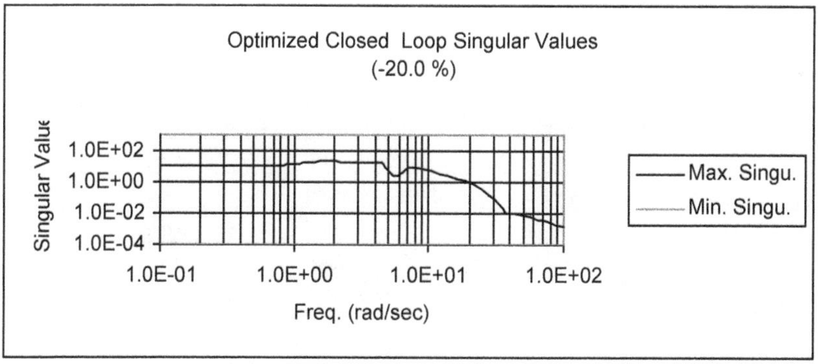

Appendix III: Collection of Math Modeling Error Sensitivity Response and Singular Value Plots, and Eigenvalue Tables

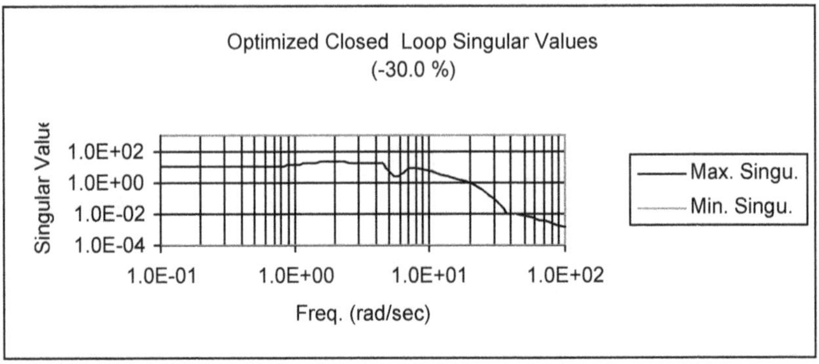

Closed Loop System Eigenvalues				
	0%	-10.00%	-20.00%	-30.00%
	1.0e+002 *	1.0e+002 *	1.0e+002 *	1.0e+002 *
1	-0.25	-0.25	-0.25	-0.25
2	-0.5824 + 1.3344i	-0.5824 + 1.3344i	-0.5824 + 1.3344i	-0.5824 + 1.3344i
3	-0.5824 - 1.3344i	-0.5824 - 1.3344i	-0.5824 - 1.3344i	-0.5824 - 1.3344i
4	-0.1040 + 0.2546i	-0.1126 + 0.2705i	-0.1172 + 0.2787i	-0.1217 + 0.2866i
5	-0.1040 - 0.2546i	-0.1126 - 0.2705i	-0.1172 - 0.2787i	-0.1217 - 0.2866i
6	-0.1046 + 0.2542i	-0.2927	-0.3124	-0.3281
7	-0.1046 - 0.2542i	-0.0975 + 0.2387i	-0.0949 + 0.2308i	-0.0929 + 0.2234i
8	-0.2284 + 0.0212i	-0.0975 - 0.2387i	-0.0949 - 0.2308i	-0.0929 - 0.2234i
9	-0.2284 - 0.0212i	-0.1986 + 0.0697i	-0.1957 + 0.0887i	-0.1935 + 0.1015i
10	-0.2293	-0.1986 - 0.0697i	-0.1957 - 0.0887i	-0.1935 - 0.1015i
11	-0.1353 + 0.0476i	-0.0885 + 0.0663i	-0.0854 + 0.0743i	-0.1625
12	-0.1353 - 0.0476i	-0.0885 - 0.0663i	-0.0854 - 0.0743i	-0.0816 + 0.0807i
13	-0.1446 + 0.0002i	-0.1488 + 0.0084i	-0.1583	-0.0816 - 0.0807i
14	-0.1446 - 0.0002i	-0.1488 - 0.0084i	-0.1496	-0.1482
15	-0.0868 + 0.0557i	-0.1446	-0.1446	-0.1446
16	-0.0868 - 0.0557i	-0.1073	-0.0852	-0.074
17	-0.0002 + 0.0601i	-0.0002 + 0.0601i	-0.0001 + 0.0601i	-0.0001 + 0.0601i
18	-0.0002 - 0.0601i	-0.0002 - 0.0601i	-0.0001 - 0.0601i	-0.0001 - 0.0601i
19	-0.0002 + 0.0600i	-0.0003 + 0.0600i	-0.0003 + 0.0600i	-0.0003 + 0.0599i
20	-0.0002 - 0.0600i	-0.0003 - 0.0600i	-0.0003 - 0.0600i	-0.0003 - 0.0599i
21	-0.0009 + 0.0508i	-0.0011 + 0.0504i	-0.0011 + 0.0500i	-0.0011 + 0.0497i

22	-0.0009 - 0.0508i	-0.0011 - 0.0504i	-0.0011 - 0.0500i	-0.0011 - 0.0497i	
23	-0.0009 + 0.0509i	-0.0007 + 0.0511i	-0.0007 + 0.0512i	-0.0006 + 0.0513i	
24	-0.0009 - 0.0509i	-0.0007 - 0.0511i	-0.0007 - 0.0512i	-0.0006 - 0.0513i	
25	-0.0347 + 0.0149i	-0.0348 + 0.0152i	-0.0348 + 0.0156i	-0.0349 + 0.0162i	
26	-0.0347 - 0.0149i	-0.0348 - 0.0152i	-0.0348 - 0.0156i	-0.0349 - 0.0162i	
27	-0.0204	-0.0204	-0.0204	-0.0204	
28	-0.0052 + 0.0136i	-0.0052 + 0.0136i	-0.0053 + 0.0136i	-0.0053 + 0.0136i	
29	-0.0052 - 0.0136i	-0.0052 - 0.0136i	-0.0053 - 0.0136i	-0.0053 - 0.0136i	
30	-0.0078 + 0.0145i	-0.0078 + 0.0145i	-0.0078 + 0.0145i	-0.0078 + 0.0145i	
31	-0.0078 - 0.0145i	-0.0078 - 0.0145i	-0.0078 - 0.0145i	-0.0078 - 0.0145i	
32	-0.0134	-0.0134	-0.0134	-0.0134	
33	-0.0106	-0.0106	-0.0106	-0.0106	
34	-0.01	-0.01	-0.01	-0.01	

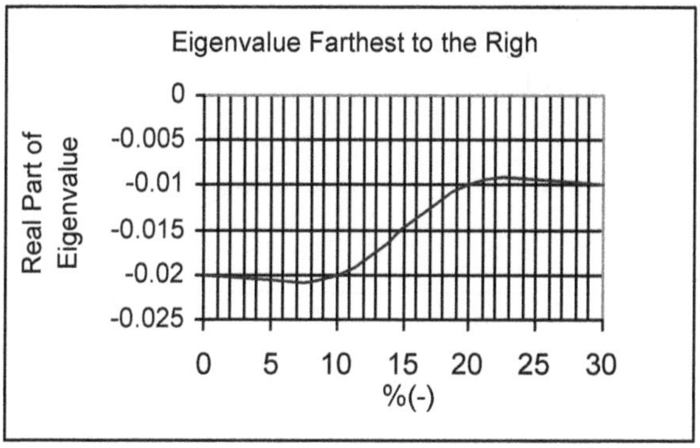

Observation:

When the actual Bending Mode 2 Modal Mass is less than the modal mass used in the math model, the real parts of the complex roots remain in the Left Half Plane, indicating system stability.

Case 2 (actual value is greater than the values in the math model)

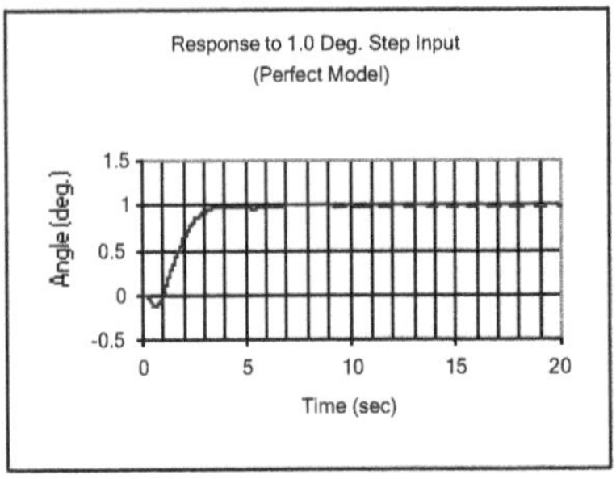

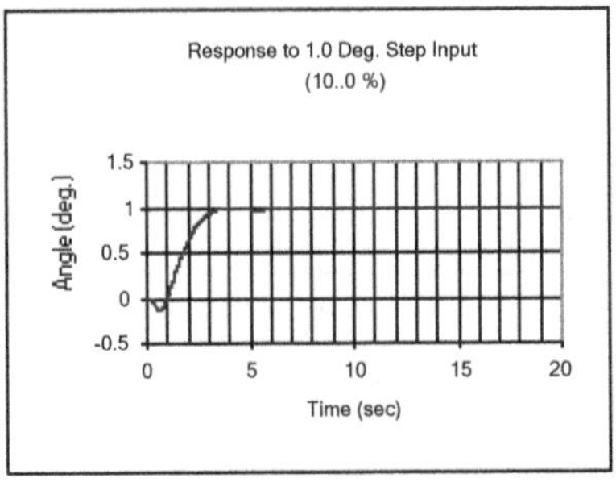

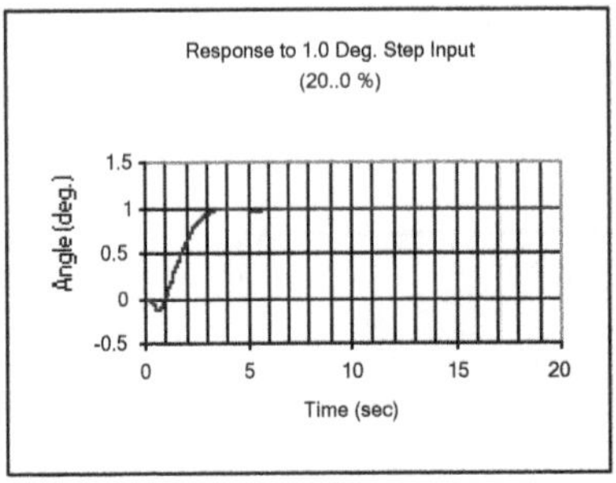

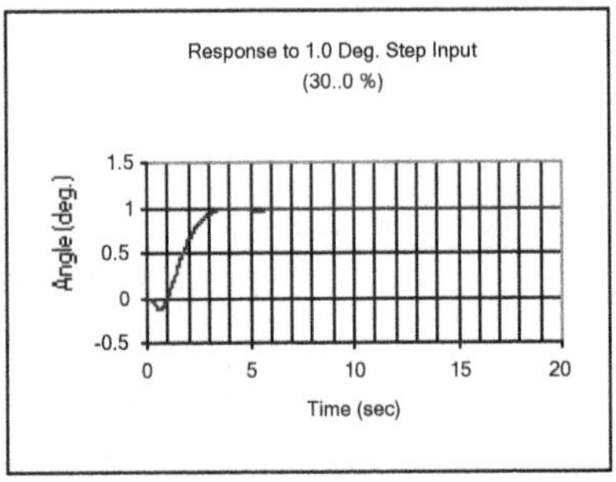

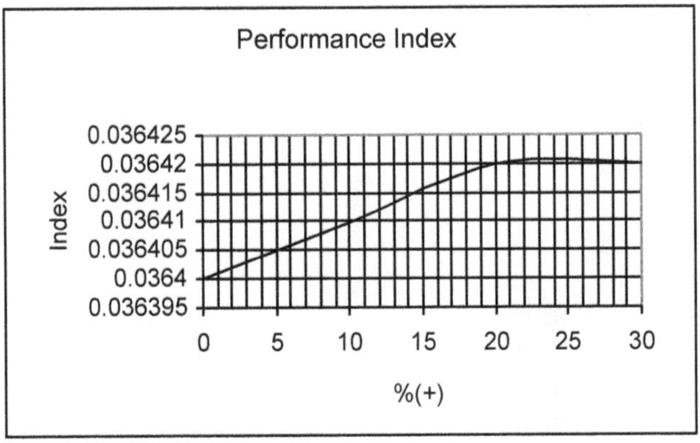

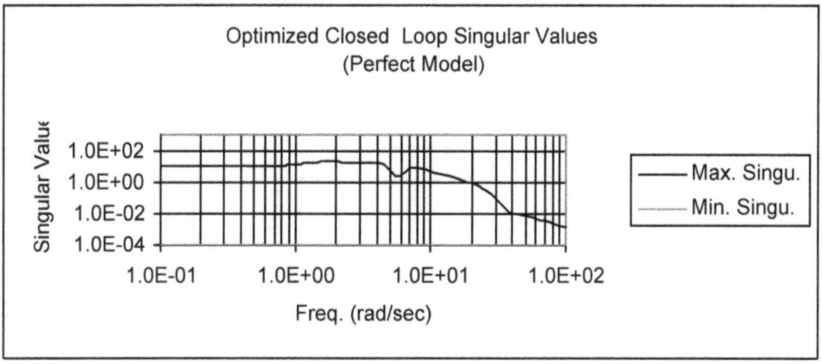
Optimized Closed Loop Singular Values
(Perfect Model)
Singular Value
1.0E+02
1.0E+00
1.0E-02
1.0E-04
1.0E-01
1.0E+00
1.0E+01
1.0E+02
Freq. (rad/sec)
Max. Singu.
Min. Singu.

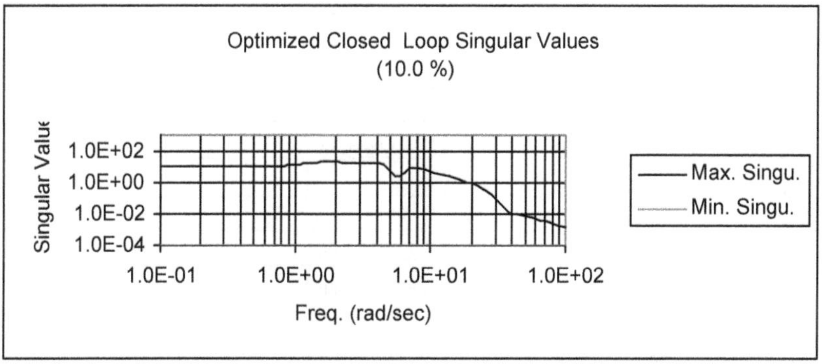
Optimized Closed Loop Singular Values
(10.0 %)
Singular Value
1.0E+02
1.0E+00
1.0E-02
1.0E-04
1.0E-01
1.0E+00
1.0E+01
1.0E+02
Freq. (rad/sec)
Max. Singu.
Min. Singu.

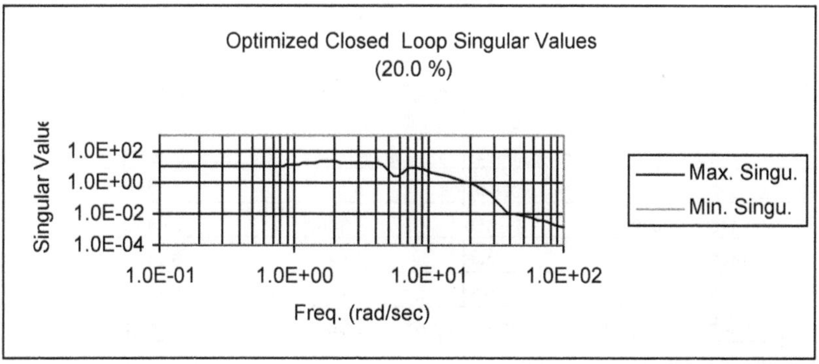
Optimized Closed Loop Singular Values
(20.0 %)
Singular Value
1.0E+02
1.0E+00
1.0E-02
1.0E-04
1.0E-01
1.0E+00
1.0E+01
1.0E+02
Freq. (rad/sec)
Max. Singu.
Min. Singu.

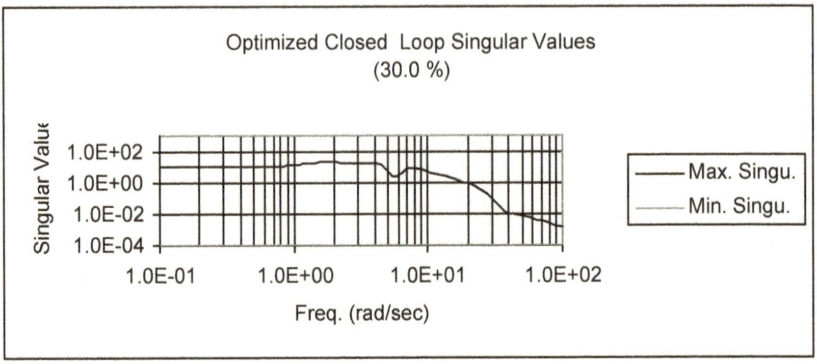

Closed Loop System Eigenvalues

	0%	10.00%	20.00%	30.00%	
	1.0e+002 *	1.0e+002 *	1.0e+002 *	1.0e+002 *	
1	-0.25	-0.25	-0.25	-0.25	
2	-0.5824 + 1.3344i	-0.5824 + 1.3344i	-0.5824 + 1.3344i	-0.5824 + 1.3344i	
3	-0.5824 - 1.3344i	-0.5824 - 1.3344i	-0.5824 - 1.3344i	-0.5824 - 1.3344i	
4	-0.1040 + 0.2546i	-0.0894 + 0.2612i	-0.2693 + 0.0666i	-0.2743 + 0.0735i	
5	-0.1040 - 0.2546i	-0.0894 - 0.2612i	-0.2693 - 0.0666i	-0.2743 - 0.0735i	
6	-0.1046 + 0.2542i	-0.1179 + 0.2474i	-0.1223 + 0.2447i	-0.0801 + 0.2653i	
7	-0.1046 - 0.2542i	-0.1179 - 0.2474i	-0.1223 - 0.2447i	-0.0801 - 0.2653i	
8	-0.2284 + 0.0212i	-0.2613 + 0.0557i	-0.0839 + 0.2637i	-0.1252 + 0.2429i	
9	-0.2284 - 0.0212i	-0.2613 - 0.0557i	-0.0839 - 0.2637i	-0.1252 - 0.2429i	
10	-0.2293	-0.1794	-0.1339 + 0.0771i	-0.1335 + 0.0822i	
11	-0.1353 + 0.0476i	-0.1344 + 0.0684i	-0.1339 - 0.0771i	-0.1335 - 0.0822i	
12	-0.1353 - 0.0476i	-0.1344 - 0.0684i	-0.1747	-0.1728	
13	-0.1446 + 0.0002i	-0.1459	-0.1463	-0.1465	
14	-0.1446 - 0.0002i	-0.1446	-0.1446	-0.1446	
15	-0.0868 + 0.0557i	-0.0805 + 0.0509i	-0.0762 + 0.0495i	-0.0734 + 0.0489i	
16	-0.0868 - 0.0557i	-0.0805 - 0.0509i	-0.0762 - 0.0495i	-0.0734 - 0.0489i	
17	-0.0002 + 0.0601i	-0.0002 + 0.0601i	-0.0003 + 0.0601i	-0.0003 + 0.0601i	
18	-0.0002 - 0.0601i	-0.0002 - 0.0601i	-0.0003 - 0.0601i	-0.0003 - 0.0601i	
19	-0.0002 + 0.0600i	-0.0002 + 0.0600i	-0.0002 + 0.0600i	-0.0002 + 0.0600i	
20	-0.0002 - 0.0600i	-0.0002 - 0.0600i	-0.0002 - 0.0600i	-0.0002 - 0.0600i	
21	-0.0009 + 0.0508i	-0.0012 + 0.0510i	-0.0014 + 0.0511i	-0.0014 + 0.0512i	

22	-0.0009 - 0.0508i	-0.0012 - 0.0510i	-0.0014 - 0.0511i	-0.0014 - 0.0512i	
23	-0.0009 + 0.0509i	-0.0006 + 0.0507i	-0.0005 + 0.0507i	-0.0004 + 0.0507i	
24	-0.0009 - 0.0509i	-0.0006 - 0.0507i	-0.0005 - 0.0507i	-0.0004 - 0.0507i	
25	-0.0347 + 0.0149i	-0.0347 + 0.0146i	-0.0346 + 0.0144i	-0.0346 + 0.0142i	
26	-0.0347 - 0.0149i	-0.0347 - 0.0146i	-0.0346 - 0.0144i	-0.0346 - 0.0142i	
27	-0.0204	-0.0204	-0.0203	-0.0203	
28	-0.0052 + 0.0136i	-0.0052 + 0.0136i	-0.0052 + 0.0136i	-0.0052 + 0.0136i	
29	-0.0052 - 0.0136i	-0.0052 - 0.0136i	-0.0052 - 0.0136i	-0.0052 - 0.0136i	
30	-0.0078 + 0.0145i	-0.0078 + 0.0145i	-0.0078 + 0.0145i	-0.0078 + 0.0145i	
31	-0.0078 - 0.0145i	-0.0078 - 0.0145i	-0.0078 - 0.0145i	-0.0078 - 0.0145i	
32	-0.0134	-0.0134	-0.0135	-0.0135	
33	-0.0106	-0.0106	-0.0106	-0.0106	
34	-0.01	-0.01	-0.01	-0.01	

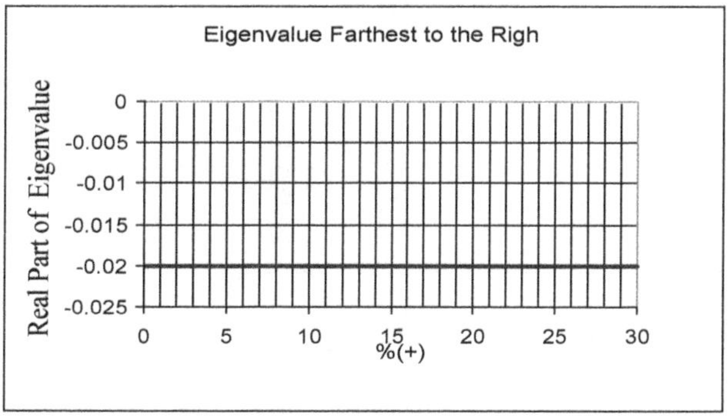

Observation:

When the actual Bending Mode 2 Modal mass is greater than the modal mass used in the math model, the real parts of the complex roots still remain in the Left Half Plane, indicating system stability.

14) Bending Mode 1 Modal Frequency: FreqB1 (The value of Bending Mode 1 Frequency in the math model = 18.9 rad/sec.)

Case 1 (actual value is less than the value in the math model)

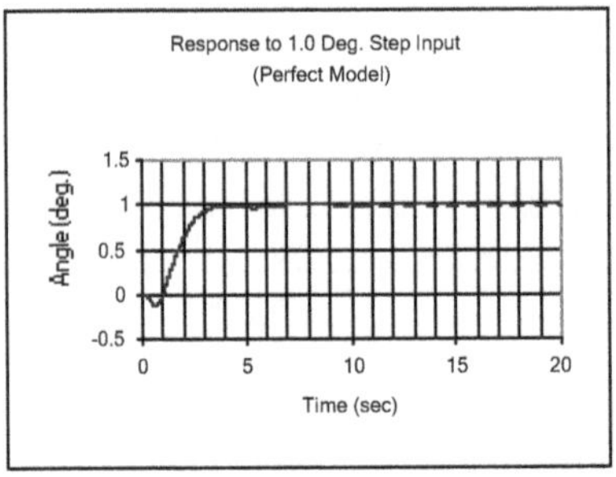

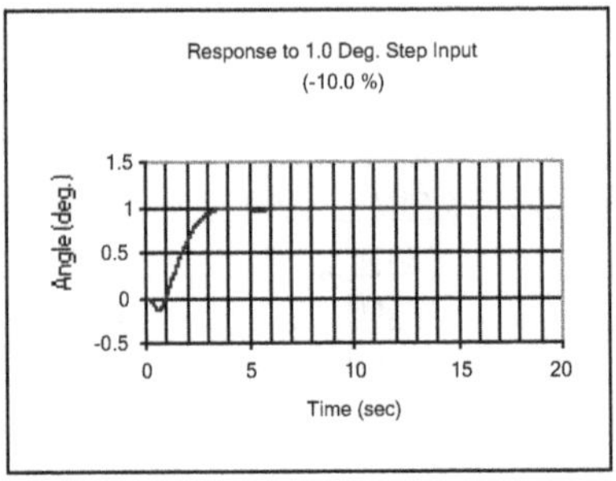

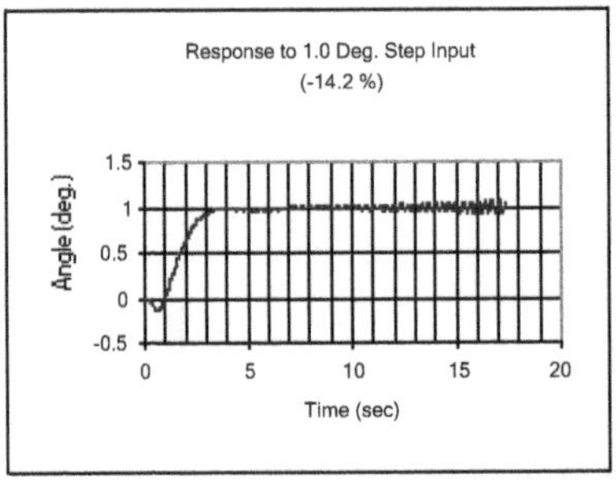

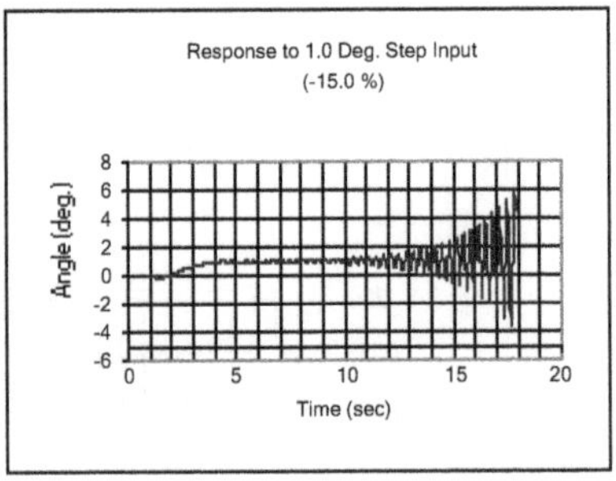

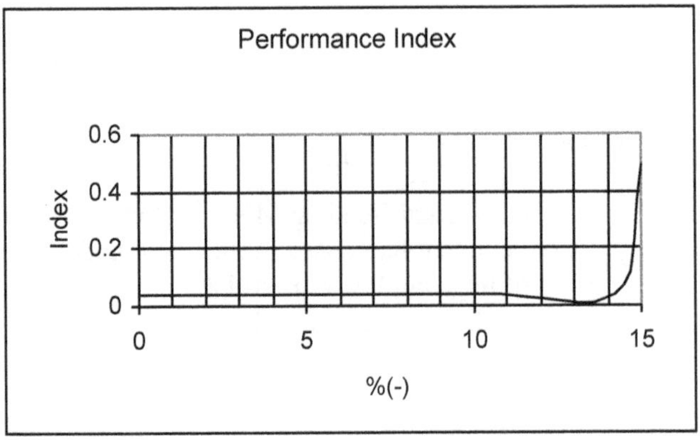

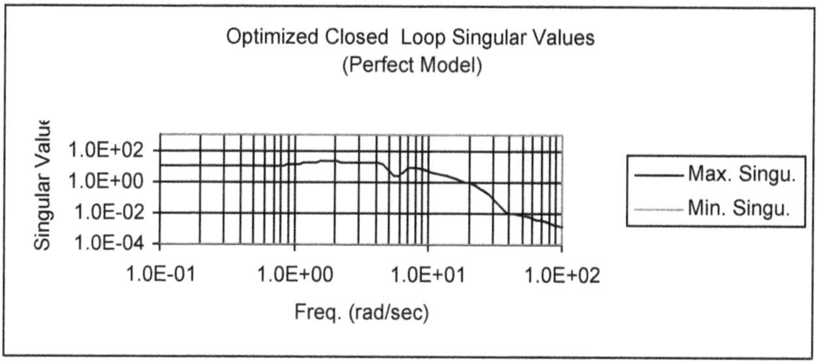
Optimized Closed Loop Singular Values
(Perfect Model)
Singular Value
1.0E+02
1.0E+00
1.0E-02
1.0E-04
1.0E-01
1.0E+00
1.0E+01
1.0E+02
Freq. (rad/sec)
Max. Singu.
Min. Singu.

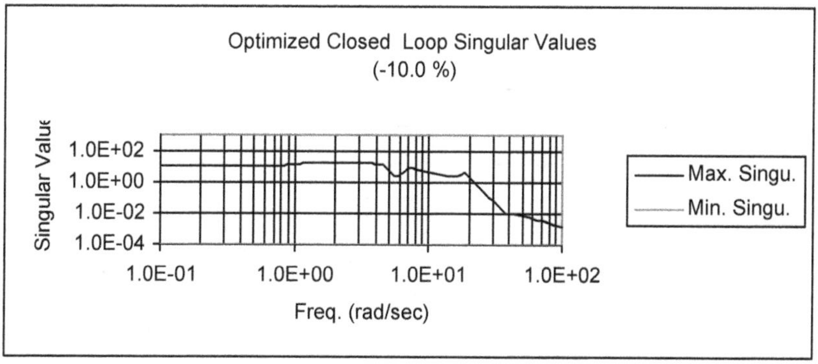
Optimized Closed Loop Singular Values
(-10.0 %)
Singular Value
1.0E+02
1.0E+00
1.0E-02
1.0E-04
1.0E-01
1.0E+00
1.0E+01
1.0E+02
Freq. (rad/sec)
Max. Singu.
Min. Singu.

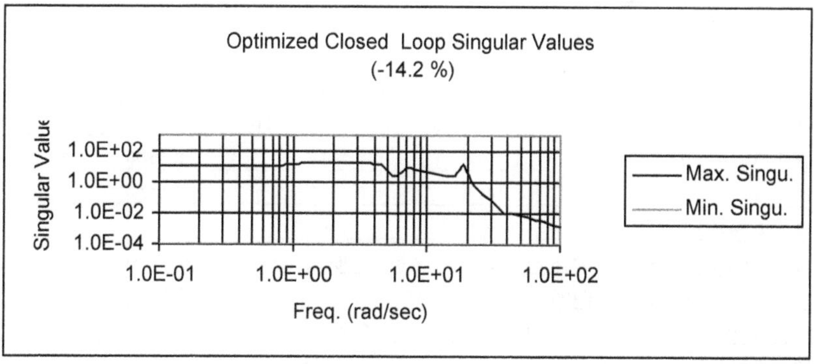
Optimized Closed Loop Singular Values
(-14.2 %)
Singular Value
1.0E+02
1.0E+00
1.0E-02
1.0E-04
1.0E-01
1.0E+00
1.0E+01
1.0E+02
Freq. (rad/sec)
Max. Singu.
Min. Singu.

Appendix III: Collection of Math Modeling Error Sensitivity Response and Singular Value Plots, and Eigenvalue Tables

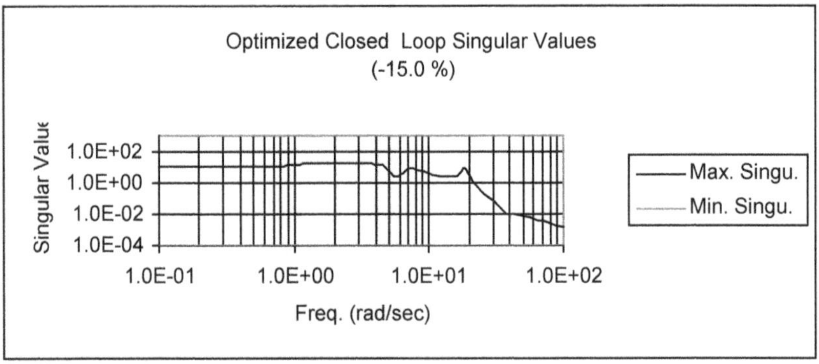

Closed Loop System Eigenvalues					
	0%	-10.00%	-14.20%	-15.00%	
	1.0e+002 *	1.0e+002 *	1.0e+002 *	1.0e+002 *	
1	-0.25	-0.25	-0.25	-0.25	
2	-0.5824 + 1.3344i	-0.5824 + 1.3344i	-0.5824 + 1.3344i	-0.5824 + 1.3344i	
3	-0.5824 - 1.3344i	-0.5824 - 1.3344i	-0.5824 - 1.3344i	-0.5824 - 1.3344i	
4	-0.1040 + 0.2546i	-0.1099 + 0.3202i	-0.1095 + 0.3289i	-0.1095 + 0.3303i	
5	-0.1040 - 0.2546i	-0.1099 - 0.3202i	-0.1095 - 0.3289i	-0.1095 - 0.3303i	
6	-0.1046 + 0.2542i	-0.2371 + 0.1694i	-0.2493 + 0.1745i	-0.2512 + 0.1752i	
7	-0.1046 - 0.2542i	-0.2371 - 0.1694i	-0.2493 - 0.1745i	-0.2512 - 0.1752i	
8	-0.2284 + 0.0212i	-0.2911	-0.3028	-0.3045	
9	-0.2284 - 0.0212i	-0.2593	-0.2517	-0.2505	
10	-0.2293	-0.0135 + 0.1892i	0.0018 + 0.1876i	0.0043 + 0.1874i	
11	-0.1353 + 0.0476i	-0.0135 - 0.1892i	0.0018 - 0.1876i	0.0043 - 0.1874i	
12	-0.1353 - 0.0476i	-0.1677	-0.1674	-0.1674	
13	-0.1446 + 0.0002i	-0.1471	-0.1472	-0.1472	
14	-0.1446 - 0.0002i	-0.1446	-0.1446	-0.1446	
15	-0.0868 + 0.0557i	-0.0458 + 0.0501i	-0.0430 + 0.0503i	-0.0426 + 0.0503i	
16	-0.0868 - 0.0557i	-0.0458 - 0.0501i	-0.0430 - 0.0503i	-0.0426 - 0.0503i	
17	-0.0002 + 0.0601i	-0.0001 + 0.0601i	-0.0000 + 0.0600i	-0.0000 + 0.0600i	
18	-0.0002 - 0.0601i	-0.0001 - 0.0601i	-0.0000 - 0.0600i	-0.0000 - 0.0600i	
19	-0.0002 + 0.0600i	-0.0003 + 0.0599i	-0.0004 + 0.0599i	-0.0004 + 0.0599i	
20	-0.0002 - 0.0600i	-0.0003 - 0.0599i	-0.0004 - 0.0599i	-0.0004 - 0.0599i	
21	-0.0009 + 0.0508i	-0.0012 + 0.0512i	-0.0012 + 0.0513i	-0.0012 + 0.0513i	
22	-0.0009 - 0.0508i	-0.0012 - 0.0512i	-0.0012 - 0.0513i	-0.0012 - 0.0513i	

23	-0.0009 + 0.0509i	-0.0005 + 0.0506i	-0.0004 + 0.0506i	-0.0004 + 0.0506i	
24	-0.0009 - 0.0509i	-0.0005 - 0.0506i	-0.0004 - 0.0506i	-0.0004 - 0.0506i	
25	-0.0347 + 0.0149i	-0.0292 + 0.0136i	-0.0281 + 0.0134i	-0.0279 + 0.0134i	
26	-0.0347 - 0.0149i	-0.0292 - 0.0136i	-0.0281 - 0.0134i	-0.0279 - 0.0134i	
27	-0.0204	-0.0205	-0.0206	-0.0206	
28	-0.0052 + 0.0136i	-0.0052 + 0.0136i	-0.0052 + 0.0136i	-0.0052 + 0.0136i	
29	-0.0052 - 0.0136i	-0.0052 - 0.0136i	-0.0052 - 0.0136i	-0.0052 - 0.0136i	
30	-0.0078 + 0.0145i	-0.0078 + 0.0145i	-0.0078 + 0.0145i	-0.0078 + 0.0145i	
31	-0.0078 - 0.0145i	-0.0078 - 0.0145i	-0.0078 - 0.0145i	-0.0078 - 0.0145i	
32	-0.0134	-0.0135	-0.0135	-0.0135	
33	-0.0106	-0.0106	-0.0106	-0.0106	
34	-0.01	-0.01	-0.01	-0.01	

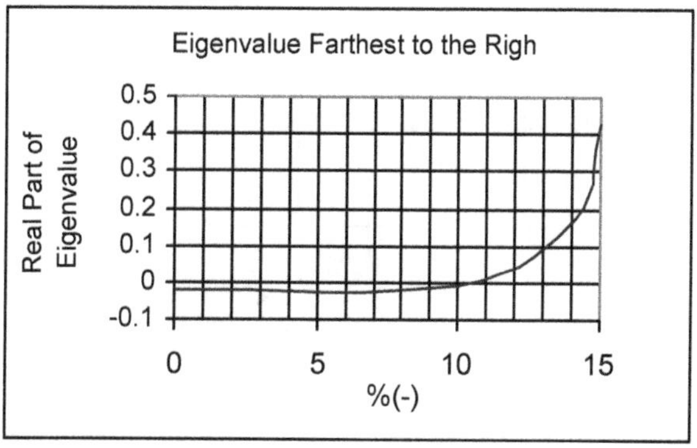

Observation:

When the actual Bending Mode 1 frequency is about 10.0% less than the frequency used in the math model, the real part of a complex root crosses the borderline and moves to the Right Half Plane, indicating that the system becomes unstable. Once the move into the Right Half Plane occurs, instability increases at a rapid rate.

Case 2 (actual value is greater than the value in the math model)

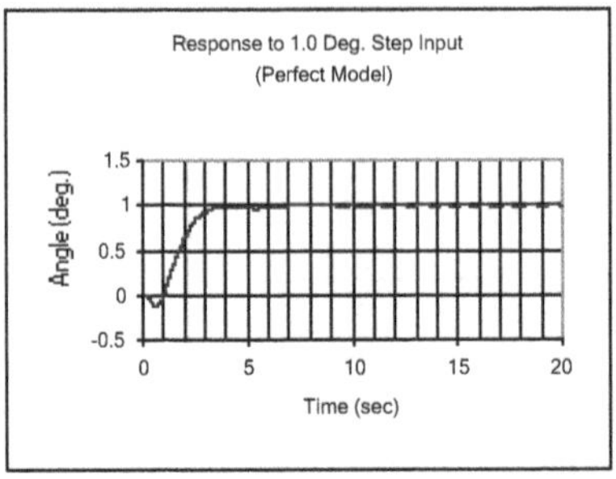

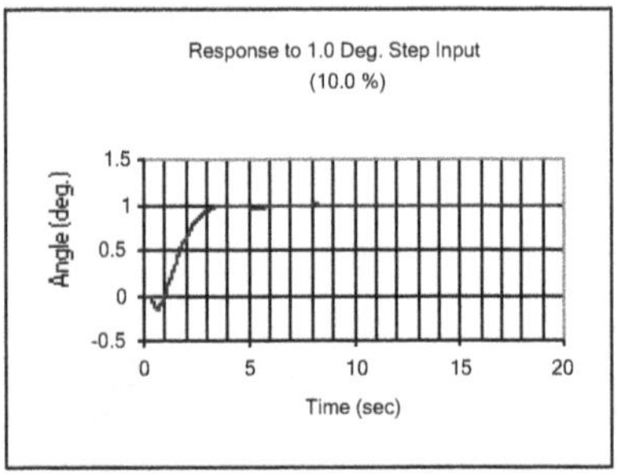

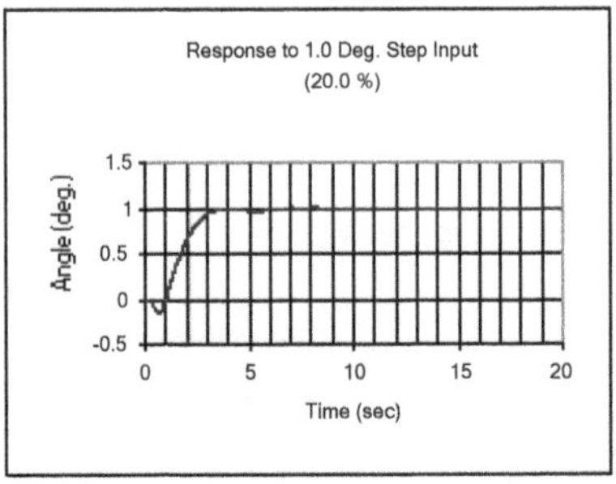

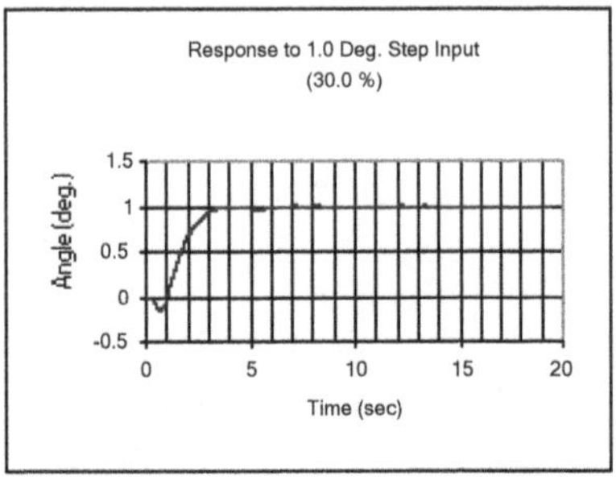

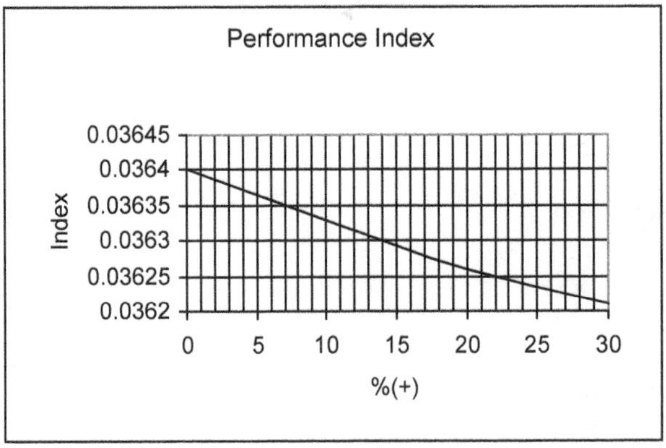

Appendix III: Collection of Math Modeling Error Sensitivity Response and Singular Value Plots, and Eigenvalue Tables

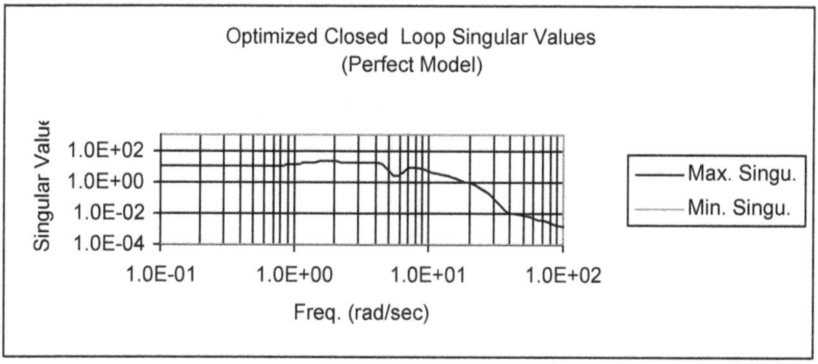

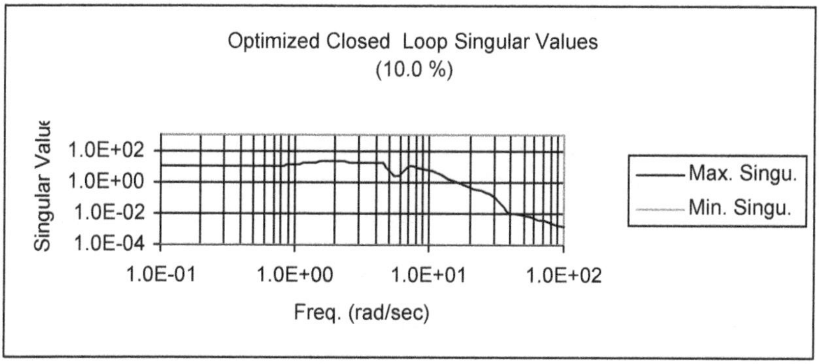

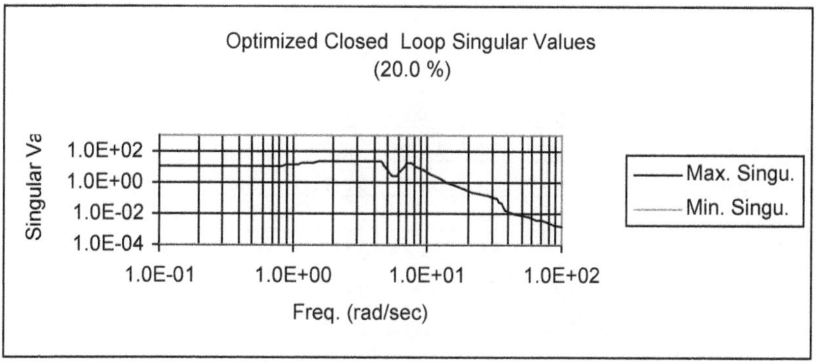

Appendix III: Collection of Math Modeling Error Sensitivity Response and Singular Value Plots, and Eigenvalue Tables

Closed Loop System Eigenvalues					
	0%	10.00%	20.00%	30.00%	
	1.0e+002 *	1.0e+002 *	1.0e+002 *	1.0e+002 *	
1	-0.25	-0.25	-0.25	-0.25	
2	-0.5824 + 1.3344i	-0.5824 + 1.3344i	-0.5824 + 1.3344i	-0.5824 + 1.3344i	
3	-0.5824 - 1.3344i	-0.5824 - 1.3344i	-0.5824 - 1.3344i	-0.5824 - 1.3344i	
4	-0.1040 + 0.2546i	-0.1877 + 0.2813i	-0.2087 + 0.2986i	-0.2232 + 0.3113i	
5	-0.1040 - 0.2546i	-0.1877 - 0.2813i	-0.2087 - 0.2986i	-0.2232 - 0.3113i	
6	-0.1046 + 0.2542i	-0.0431 + 0.2917i	-0.0297 + 0.3072i	-0.0220 + 0.3185i	
7	-0.1046 - 0.2542i	-0.0431 - 0.2917i	-0.0297 - 0.3072i	-0.0220 - 0.3185i	
8	-0.2284 + 0.0212i	-0.2816 + 0.0854i	-0.2992 + 0.0947i	-0.3104 + 0.1015i	
9	-0.2284 - 0.0212i	-0.2816 - 0.0854i	-0.2992 - 0.0947i	-0.3104 - 0.1015i	
10	-0.2293	-0.2737	-0.2853	-0.2953	
11	-0.1353 + 0.0476i	-0.166	-0.1664	-0.1665	
12	-0.1353 - 0.0476i	-0.1473	-0.1473	-0.1473	
13	-0.1446 + 0.0002i	-0.1446	-0.1446	-0.1446	
14	-0.1446 - 0.0002i	-0.0369 + 0.0861i	-0.0246 + 0.0799i	-0.0183 + 0.0743i	
15	-0.0868 + 0.0557i	-0.0369 - 0.0861i	-0.0246 - 0.0799i	-0.0183 - 0.0743i	
16	-0.0868 - 0.0557i	-0.0003 + 0.0603i	-0.0003 + 0.0605i	-0.0004 + 0.0608i	
17	-0.0002 + 0.0601i	-0.0003 - 0.0603i	-0.0003 - 0.0605i	-0.0004 - 0.0608i	
18	-0.0002 - 0.0601i	-0.0002 + 0.0599i	-0.0001 + 0.0599i	-0.0001 + 0.0598i	
19	-0.0002 + 0.0600i	-0.0002 - 0.0599i	-0.0001 - 0.0599i	-0.0001 - 0.0598i	
20	-0.0002 - 0.0600i	-0.0014 + 0.0503i	-0.0019 + 0.0499i	-0.0024 + 0.0495i	
21	-0.0009 + 0.0508i	-0.0014 - 0.0503i	-0.0019 - 0.0499i	-0.0024 - 0.0495i	
22	-0.0009 - 0.0508i	-0.0006 + 0.0511i	-0.0005 + 0.0512i	-0.0005 + 0.0512i	
23	-0.0009 + 0.0509i	-0.0006 - 0.0511i	-0.0005 - 0.0512i	-0.0005 - 0.0512i	
24	-0.0009 - 0.0509i	-0.0342 + 0.0263i	-0.0297 + 0.0295i	-0.0266 + 0.0313i	
25	-0.0347 + 0.0149i	-0.0342 - 0.0263i	-0.0297 - 0.0295i	-0.0266 - 0.0313i	
26	-0.0347 - 0.0149i	-0.0332	-0.0293	-0.0274	
27	-0.0204	-0.0202	-0.0053 + 0.0137i	-0.0053 + 0.0137i	
28	-0.0052 + 0.0136i	-0.0053 + 0.0136i	-0.0053 - 0.0137i	-0.0053 - 0.0137i	
29	-0.0052 - 0.0136i	-0.0053 - 0.0136i	-0.0078 + 0.0145i	-0.0078 + 0.0145i	
30	-0.0078 + 0.0145i	-0.0078 + 0.0145i	-0.0078 - 0.0145i	-0.0078 - 0.0145i	
31	-0.0078 - 0.0145i	-0.0078 - 0.0145i	-0.02	-0.0197	
32	-0.0134	-0.0134	-0.0134	-0.0134	
33	-0.0106	-0.0106	-0.0107	-0.0107	
34	-0.01	-0.01	-0.01	-0.01	

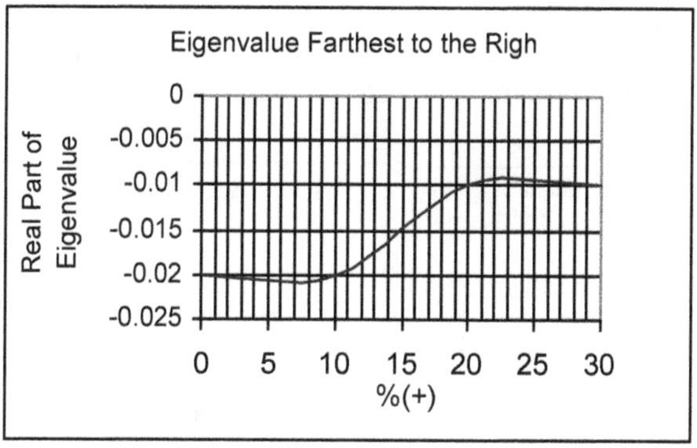

Observation:

When the actual Bending Mode 1 frequency is greater than the modal frequency used in the math model, the real parts of complex roots remain in the Left Half Plane, indicating system stability.

15) **Bending Mode 2 Modal Frequency: FreqB2 (The value of Bending Mode 2 Frequency in the math model = 60.7 rad/sec.)**

Case 1 (actual value is less than the value in the math model)

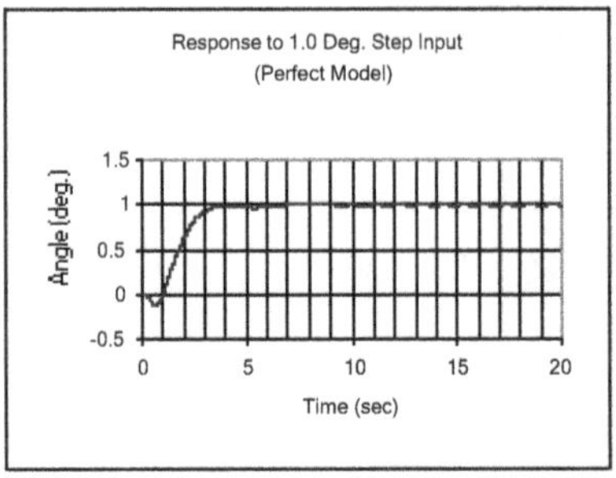

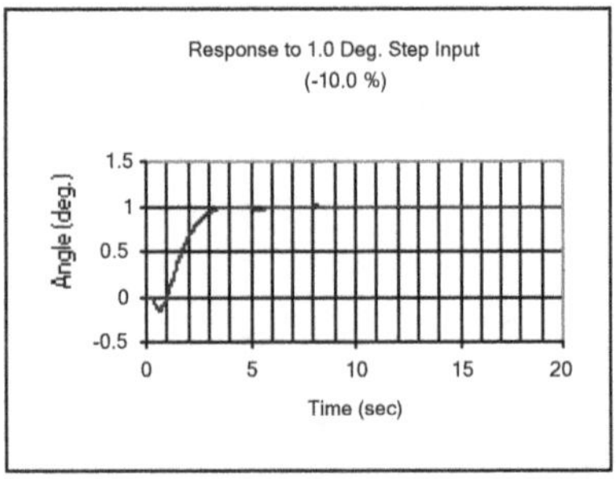

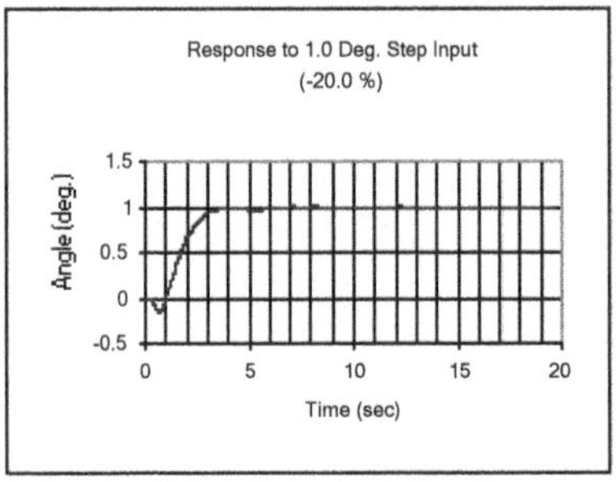

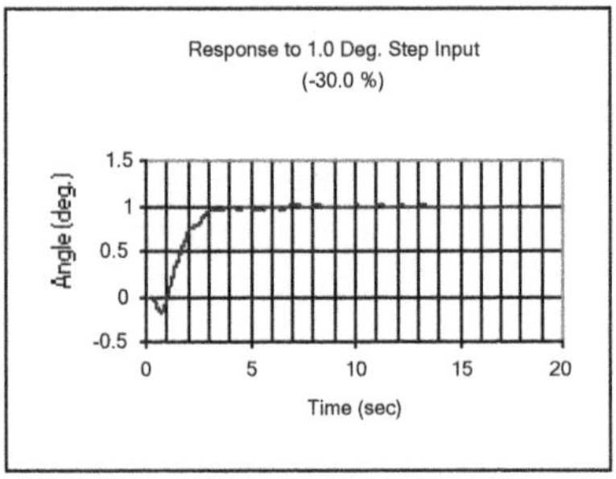

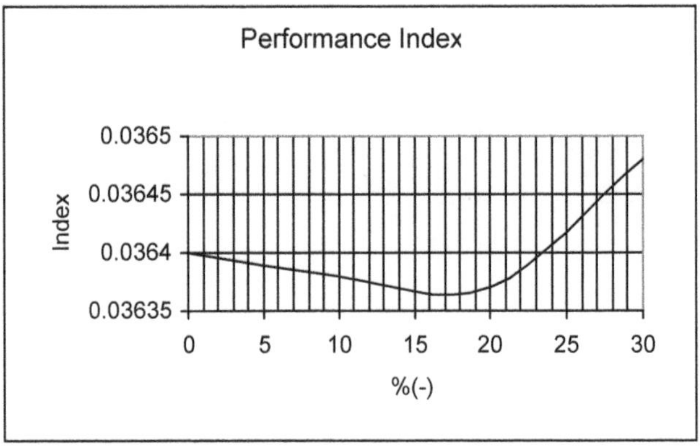

Appendix III: Collection of Math Modeling Error Sensitivity Response and Singular Value Plots, and Eigenvalue Tables

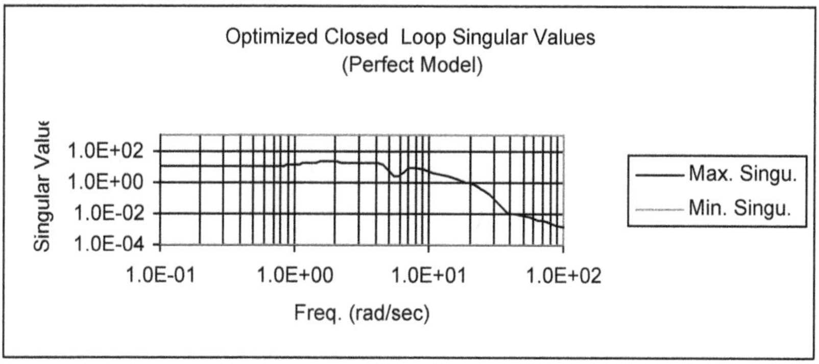

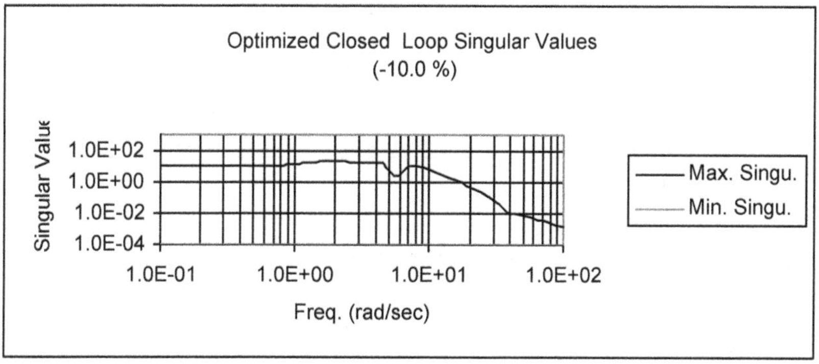

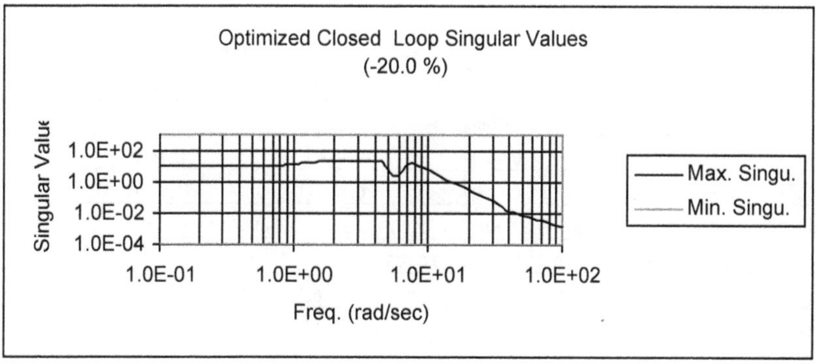

Appendix III: Collection of Math Modeling Error Sensitivity Response and Singular Value Plots, and Eigenvalue Tables

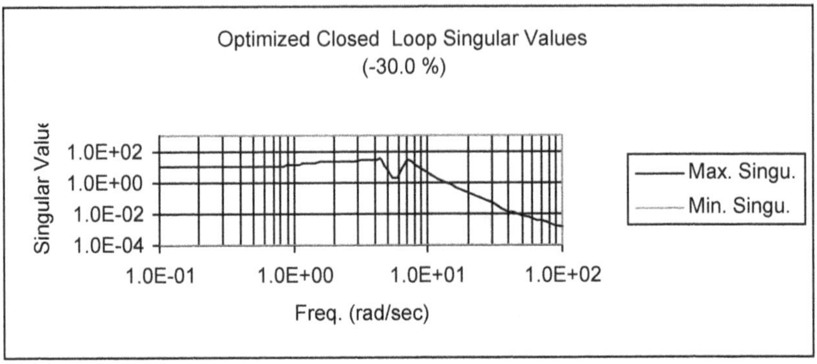

Closed Loop System Eigenvalues

	0%	-10.00%	-20.00%	-30.00%	
	1.0e+002 *	1.0e+002 *	1.0e+002 *	1.0e+002 *	
1	-0.25	-0.25	-0.25	-0.25	
2	-0.5824 + 1.3344i	-0.5824 + 1.3344i	-0.5824 + 1.3344i	-0.5824 + 1.3344i	
3	-0.5824 - 1.3344i	-0.5824 - 1.3344i	-0.5824 - 1.3344i	-0.5824 - 1.3344i	
4	-0.1040 + 0.2546i	-0.3644	-0.3835	-0.3933	
5	-0.1040 - 0.2546i	-0.2584 + 0.2120i	-0.2864 + 0.2321i	-0.3024 + 0.2412i	
6	-0.1046 + 0.2542i	-0.2584 - 0.2120i	-0.2864 - 0.2321i	-0.3024 - 0.2412i	
7	-0.1046 - 0.2542i	-0.0627 + 0.3217i	-0.0518 + 0.3420i	-0.0453 + 0.3555i	
8	-0.2284 + 0.0212i	-0.0627 - 0.3217i	-0.0518 - 0.3420i	-0.0453 - 0.3555i	
9	-0.2284 - 0.0212i	-0.1333 + 0.1518i	-0.1310 + 0.1454i	-0.1304 + 0.1434i	
10	-0.2293	-0.1333 - 0.1518i	-0.1310 - 0.1454i	-0.1304 - 0.1434i	
11	-0.1353 + 0.0476i	-0.1664	-0.1666	-0.1666	
12	-0.1353 - 0.0476i	-0.1473	-0.1473	-0.1473	
13	-0.1446 + 0.0002i	-0.1446	-0.1446	-0.1446	
14	-0.1446 - 0.0002i	-0.0370 + 0.0903i	-0.0198 + 0.0811i	-0.0110 + 0.0706i	
15	-0.0868 + 0.0557i	-0.0370 - 0.0903i	-0.0198 - 0.0811i	-0.0110 - 0.0706i	
16	-0.0868 - 0.0557i	-0.0003 + 0.0603i	-0.0005 + 0.0606i	-0.0006 + 0.0614i	
17	-0.0002 + 0.0601i	-0.0003 - 0.0603i	-0.0005 - 0.0606i	-0.0006 - 0.0614i	
18	-0.0002 - 0.0601i	-0.0001 + 0.0599i	-0.0001 + 0.0599i	-0.0001 + 0.0598i	
19	-0.0002 + 0.0600i	-0.0001 - 0.0599i	-0.0001 - 0.0599i	-0.0001 - 0.0598i	
20	-0.0002 - 0.0600i	-0.0014 + 0.0502i	-0.0020 + 0.0495i	-0.0032 + 0.0483i	
21	-0.0009 + 0.0508i	-0.0014 - 0.0502i	-0.0020 - 0.0495i	-0.0032 - 0.0483i	

22	-0.0009 - 0.0508i	-0.0006 + 0.0511i	-0.0005 + 0.0512i	-0.0004 + 0.0513i	
23	-0.0009 + 0.0509i	-0.0006 - 0.0511i	-0.0005 - 0.0512i	-0.0004 - 0.0513i	
24	-0.0009 - 0.0509i	-0.0339 + 0.0269i	-0.0284 + 0.0307i	-0.0236 + 0.0338i	
25	-0.0347 + 0.0149i	-0.0339 - 0.0269i	-0.0284 - 0.0307i	-0.0236 - 0.0338i	
26	-0.0347 - 0.0149i	-0.0325	-0.0282	-0.0259	
27	-0.0204	-0.0202	-0.0053 + 0.0137i	-0.0054 + 0.0137i	
28	-0.0052 + 0.0136i	-0.0053 + 0.0136i	-0.0053 - 0.0137i	-0.0054 - 0.0137i	
29	-0.0052 - 0.0136i	-0.0053 - 0.0136i	-0.0078 + 0.0145i	-0.0078 + 0.0145i	
30	-0.0078 + 0.0145i	-0.0078 + 0.0145i	-0.0078 - 0.0145i	-0.0078 - 0.0145i	
31	-0.0078 - 0.0145i	-0.0078 - 0.0145i	-0.0198	-0.0193	
32	-0.0134	-0.0134	-0.0134	-0.0133	
33	-0.0106	-0.0106	-0.0107	-0.0107	
34	-0.01	-0.01	-0.01	-0.01	

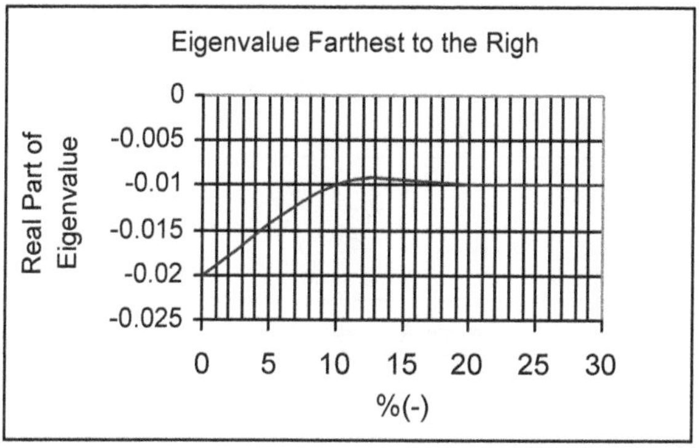

Observation:

When the actual Bending Mode 2 frequency is less than the value of modal frequency in the math model, the real parts of complex roots remain in the Left Half Plane, indicating system stability.

Case 2 (actual value is greater than the value in the math model)

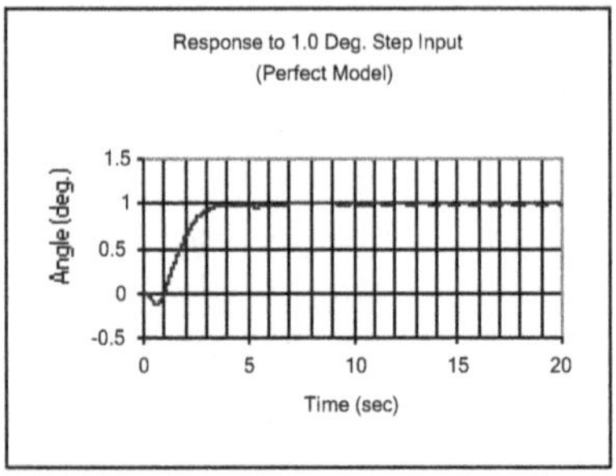

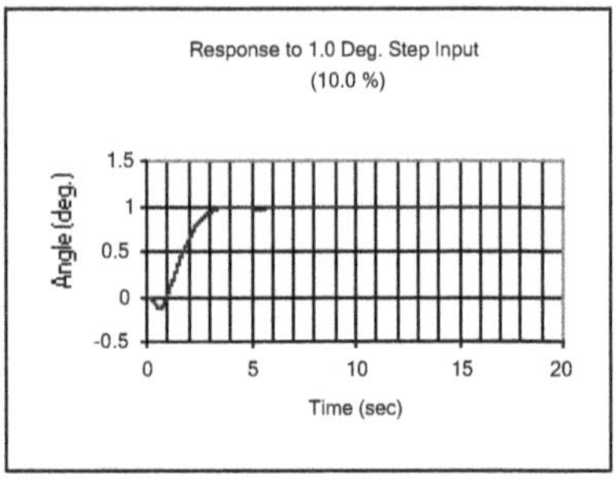

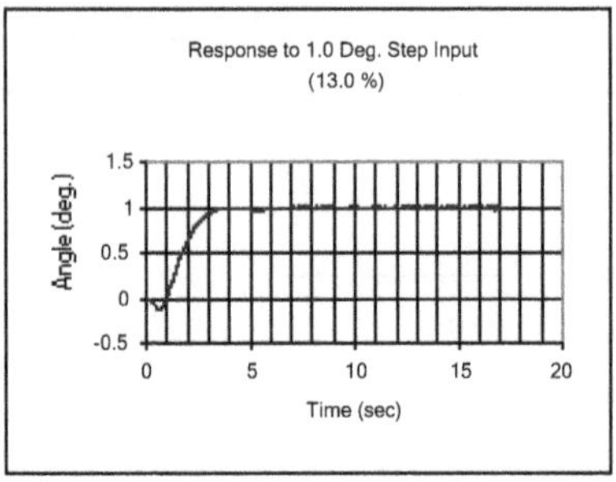

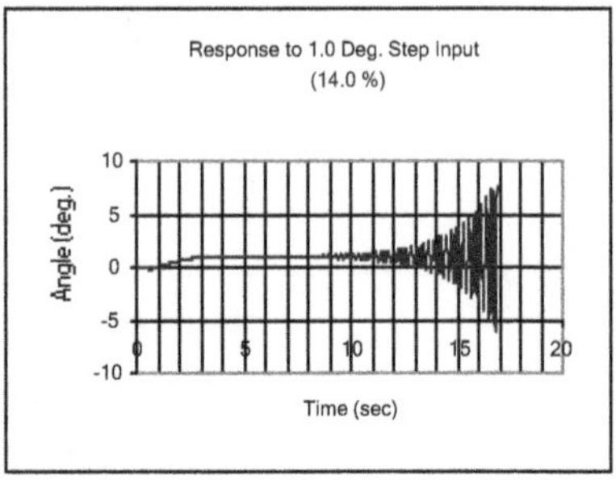

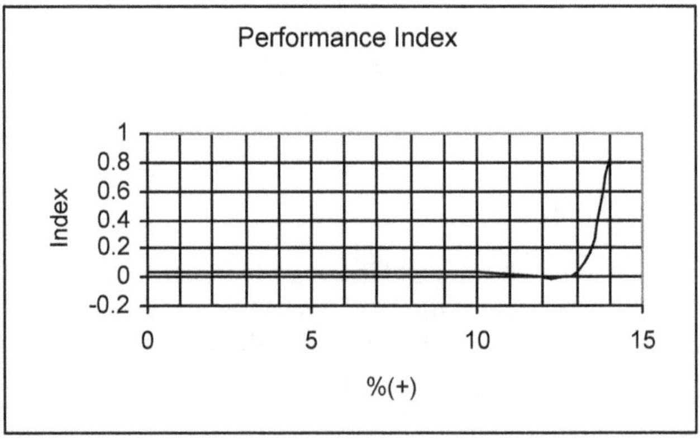

Appendix III: Collection of Math Modeling Error Sensitivity Response and Singular Value Plots, and Eigenvalue Tables

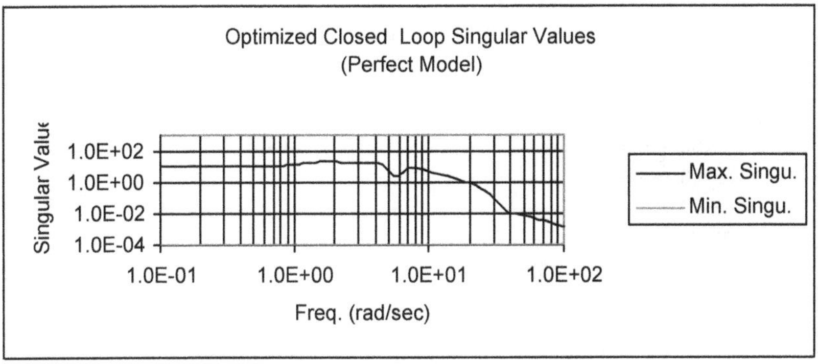

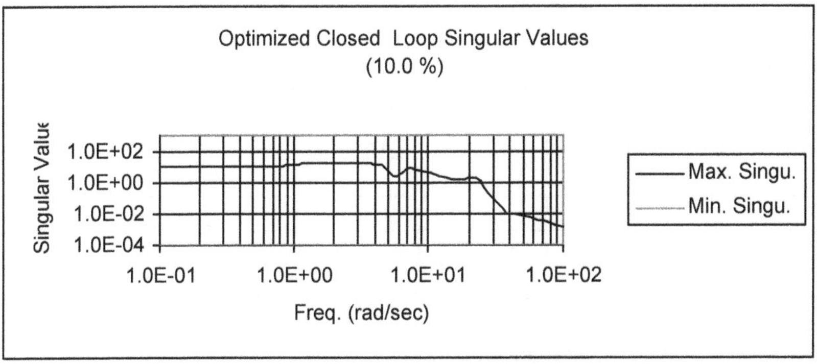

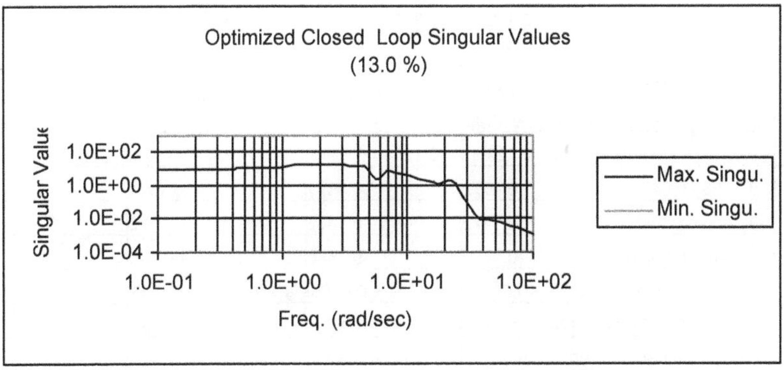

Appendix III: Collection of Math Modeling Error Sensitivity Response and Singular Value Plots, and Eigenvalue Tables

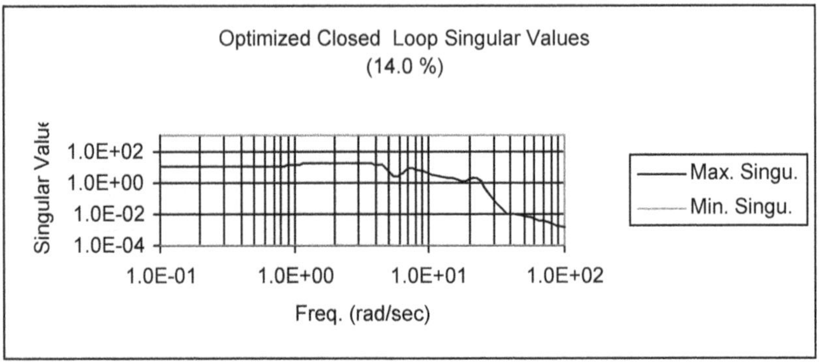

Closed Loop System Eigenvalues

	0%	10.00%	13.00%	14.00%	
	1.0e+002 *	1.0e+002 *	1.0e+002 *	1.0e+002 *	
1	-0.25	-0.25	-0.25	-0.25	
2	-0.5824 + 1.3344i	-0.5824 + 1.3344i	-0.5824 + 1.3344i	-0.5824 + 1.3344i	
3	-0.5824 - 1.3344i	-0.5824 - 1.3344i	-0.5824 - 1.3344i	-0.5824 - 1.3344i	
4	-0.1040 + 0.2546i	-0.1623 + 0.3252i	-0.1668 + 0.3349i	-0.1681 + 0.3379i	
5	-0.1040 - 0.2546i	-0.1623 - 0.3252i	-0.1668 - 0.3349i	-0.1681 - 0.3379i	
6	-0.1046 + 0.2542i	-0.3465 + 0.1031i	-0.3563 + 0.1077i	-0.3592 + 0.1090i	
7	-0.1046 - 0.2542i	-0.3465 - 0.1031i	-0.3563 - 0.1077i	-0.3592 - 0.1090i	
8	-0.2284 + 0.0212i	-0.0100 + 0.2297i	0.0017 + 0.2300i	0.0052 + 0.2302i	
9	-0.2284 - 0.0212i	-0.0100 - 0.2297i	0.0017 - 0.2300i	0.0052 - 0.2302i	
10	-0.2293	-0.1282 + 0.1295i	-0.1284 + 0.1315i	-0.1284 + 0.1320i	
11	-0.1353 + 0.0476i	-0.1282 - 0.1295i	-0.1284 - 0.1315i	-0.1284 - 0.1320i	
12	-0.1353 - 0.0476i	-0.1672	-0.1671	-0.1671	
13	-0.1446 + 0.0002i	-0.1472	-0.1472	-0.1472	
14	-0.1446 - 0.0002i	-0.1446	-0.1446	-0.1446	
15	-0.0868 + 0.0557i	-0.0474 + 0.0504i	-0.0454 + 0.0506i	-0.0448 + 0.0506i	
16	-0.0868 - 0.0557i	-0.0474 - 0.0504i	-0.0454 - 0.0506i	-0.0448 - 0.0506i	
17	-0.0002 + 0.0601i	-0.0001 + 0.0601i	-0.0000 + 0.0601i	-0.0000 + 0.0601i	
18	-0.0002 - 0.0601i	-0.0001 - 0.0601i	-0.0000 - 0.0601i	-0.0000 - 0.0601i	
19	-0.0002 + 0.0600i	-0.0003 + 0.0599i	-0.0003 + 0.0599i	-0.0003 + 0.0599i	
20	-0.0002 - 0.0600i	-0.0003 - 0.0599i	-0.0003 - 0.0599i	-0.0003 - 0.0599i	
21	-0.0009 + 0.0508i	-0.0012 + 0.0512i	-0.0012 + 0.0513i	-0.0012 + 0.0513i	

22	-0.0009 - 0.0508i	-0.0012 - 0.0512i	-0.0012 - 0.0513i	-0.0012 - 0.0513i	
23	-0.0009 + 0.0509i	-0.0005 + 0.0506i	-0.0004 + 0.0506i	-0.0004 + 0.0506i	
24	-0.0009 - 0.0509i	-0.0005 - 0.0506i	-0.0004 - 0.0506i	-0.0004 - 0.0506i	
25	-0.0347 + 0.0149i	-0.0294 + 0.0137i	-0.0287 + 0.0136i	-0.0285 + 0.0136i	
26	-0.0347 - 0.0149i	-0.0294 - 0.0137i	-0.0287 - 0.0136i	-0.0285 - 0.0136i	
27	-0.0204	-0.0205	-0.0206	-0.0206	
28	-0.0052 + 0.0136i	-0.0052 + 0.0136i	-0.0052 + 0.0136i	-0.0052 + 0.0136i	
29	-0.0052 - 0.0136i	-0.0052 - 0.0136i	-0.0052 - 0.0136i	-0.0052 - 0.0136i	
30	-0.0078 + 0.0145i	-0.0078 + 0.0145i	-0.0078 + 0.0145i	-0.0078 + 0.0145i	
31	-0.0078 - 0.0145i	-0.0078 - 0.0145i	-0.0078 - 0.0145i	-0.0078 - 0.0145i	
32	-0.0134	-0.0135	-0.0135	-0.0135	
33	-0.0106	-0.0106	-0.0106	-0.0106	
34	-0.01	-0.01	-0.01	-0.01	

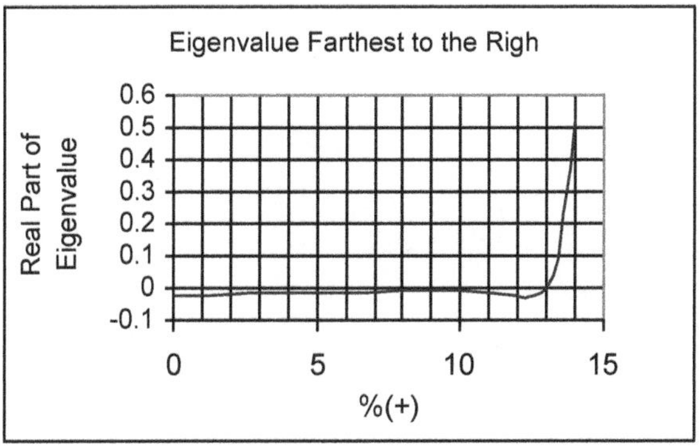

Observation:

When the actual Bending Mode 2 modal frequency is about 12.0% less than the modal frequency in the math model, the real part of a complex root crosses the borderline and moves to the right half plane, indicating that the system has become unstable. A notable observation that needs further analysis is that the system response displays severe instability even if the error is slightly less than 12.0%. A similar phenomenon was observed earlier when we analyzed modal frequency in the Bending Mode 1 modeling error case. In this case, however, severe instability was displayed when the actual value of modal frequency in Bending Mode 1 is slightly larger than 10% modeling error.

16) **Bending Mode 1 Bending Slope at Tail: SigG1t (The value used for Bending Mode 1 bending slope at the tail = -0.0648 rad/ft in the math model.)**

Case 1 (actual value is less than the value in the math model)

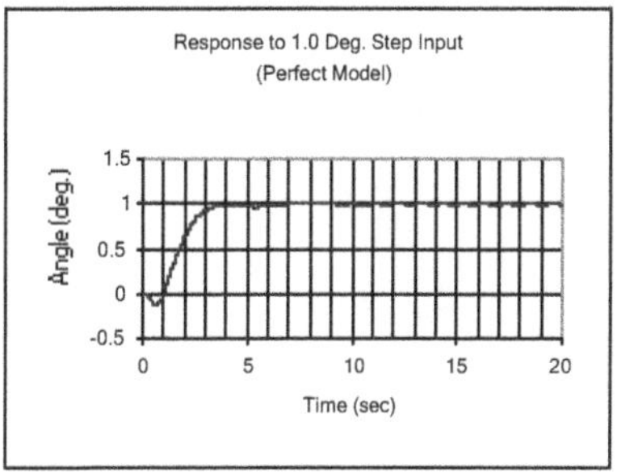

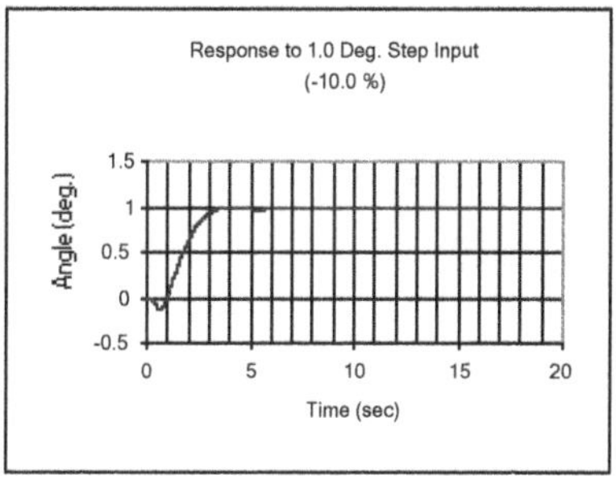

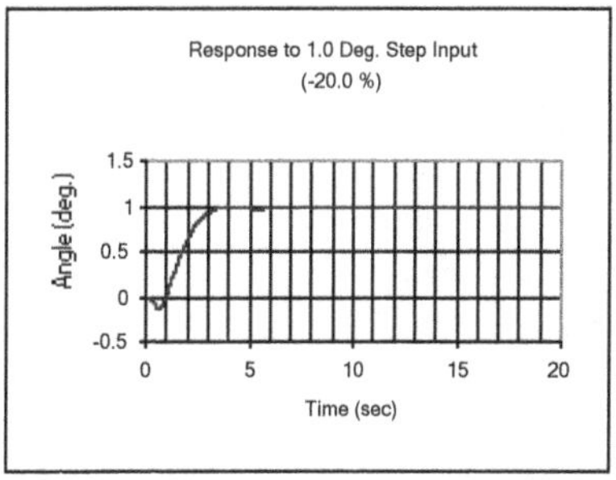

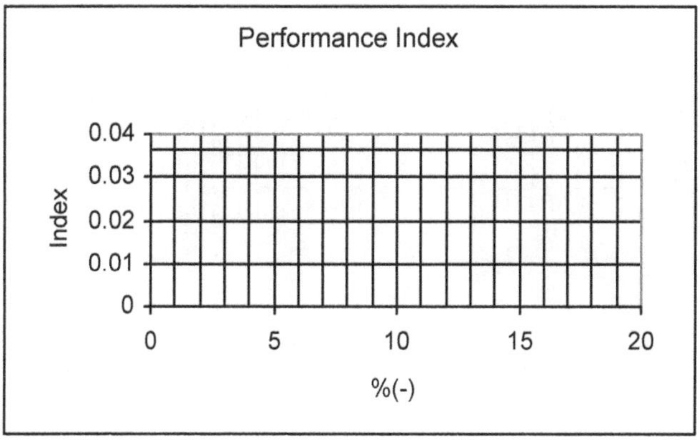

Appendix III: Collection of Math Modeling Error Sensitivity Response and Singular Value Plots, and Eigenvalue Tables

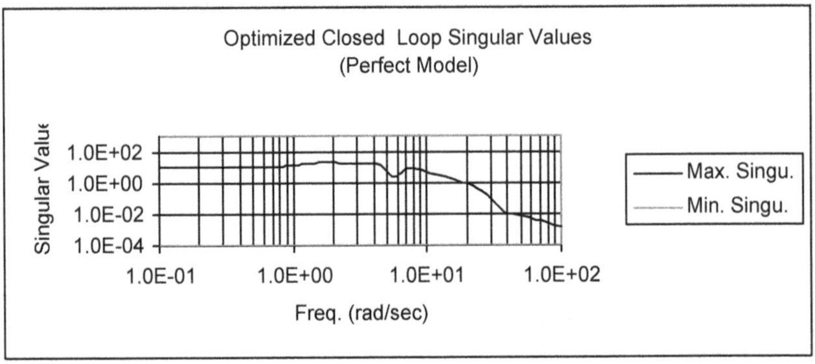

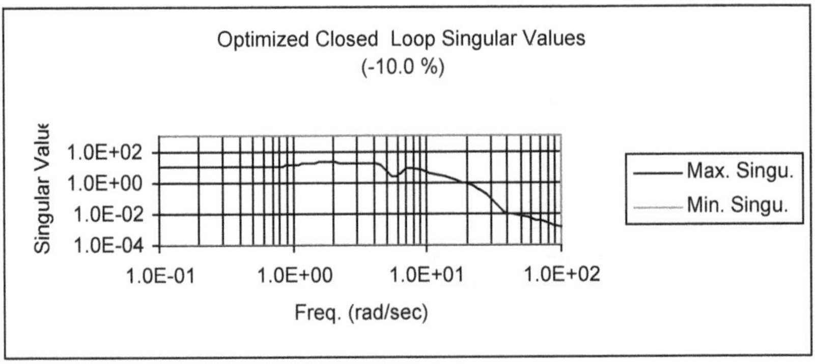

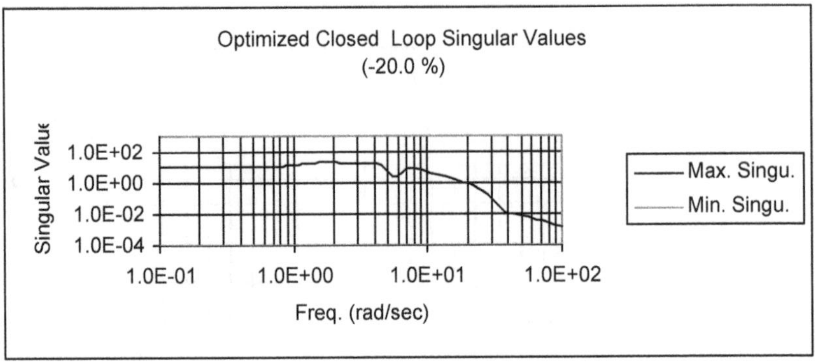

Closed Loop System Eigenvalues					
	0%	-10.00%	-20.00%		

	1.0e+002 *	1.0e+002 *	1.0e+002 *		
1	-0.25	-0.25	-0.25		
2	-0.5824 + 1.3344i	-0.5824 + 1.3344i	-0.5824 + 1.3344i		
3	-0.5824 - 1.3344i	-0.5824 - 1.3344i	-0.5824 - 1.3344i		
4	-0.1040 + 0.2546i	-0.1040 + 0.2546i	-0.1040 + 0.2546i		
5	-0.1040 - 0.2546i	-0.1040 - 0.2546i	-0.1040 - 0.2546i		
6	-0.1046 + 0.2542i	-0.1046 + 0.2542i	-0.1046 + 0.2542i		
7	-0.1046 - 0.2542i	-0.1046 - 0.2542i	-0.1046 - 0.2542i		
8	-0.2284 + 0.0212i	-0.2284 + 0.0212i	-0.2284 + 0.0212i		
9	-0.2284 - 0.0212i	-0.2284 - 0.0212i	-0.2284 - 0.0212i		
10	-0.2293	-0.2293	-0.2293		
11	-0.1353 + 0.0476i	-0.1353 + 0.0476i	-0.1353 + 0.0476i		
12	-0.1353 - 0.0476i	-0.1353 - 0.0476i	-0.1353 - 0.0476i		
13	-0.1446 + 0.0002i	-0.1446 + 0.0002i	-0.1446 + 0.0002i		
14	-0.1446 - 0.0002i	-0.1446 - 0.0002i	-0.1446 - 0.0002i		
15	-0.0868 + 0.0557i	-0.0868 + 0.0557i	-0.0868 + 0.0557i		
16	-0.0868 - 0.0557i	-0.0868 - 0.0557i	-0.0868 - 0.0557i		
17	-0.0002 + 0.0601i	-0.0002 + 0.0601i	-0.0002 + 0.0601i		
18	-0.0002 - 0.0601i	-0.0002 - 0.0601i	-0.0002 - 0.0601i		
19	-0.0002 + 0.0600i	-0.0002 + 0.0600i	-0.0002 + 0.0600i		
20	-0.0002 - 0.0600i	-0.0002 - 0.0600i	-0.0002 - 0.0600i		
21	-0.0009 + 0.0508i	-0.0009 + 0.0508i	-0.0009 + 0.0508i		
22	-0.0009 - 0.0508i	-0.0009 - 0.0508i	-0.0009 - 0.0508i		
23	-0.0009 + 0.0509i	-0.0009 + 0.0509i	-0.0009 + 0.0509i		
24	-0.0009 - 0.0509i	-0.0009 - 0.0509i	-0.0009 - 0.0509i		
25	-0.0347 + 0.0149i	-0.0347 + 0.0149i	-0.0347 + 0.0149i		
26	-0.0347 - 0.0149i	-0.0347 - 0.0149i	-0.0347 - 0.0149i		
27	-0.0204	-0.0204	-0.0204		
28	-0.0052 + 0.0136i	-0.0052 + 0.0136i	-0.0052 + 0.0136i		
29	-0.0052 - 0.0136i	-0.0052 - 0.0136i	-0.0052 - 0.0136i		
30	-0.0078 + 0.0145i	-0.0078 + 0.0145i	-0.0078 + 0.0145i		
31	-0.0078 - 0.0145i	-0.0078 - 0.0145i	-0.0078 - 0.0145i		
32	-0.0134	-0.0134	-0.0134		
33	-0.0106	-0.0106	-0.0106		
34	-0.01	-0.01	-0.01		

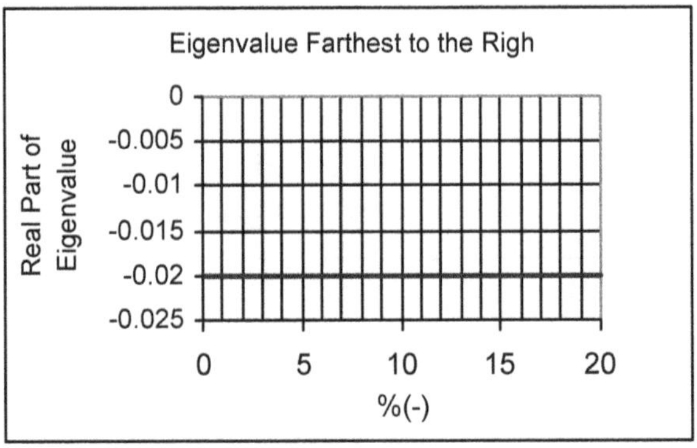

Observation:

Modeling error associated with bending slope does not cause control system instability.

Case 2 (actual value is greater than the value in the math model)

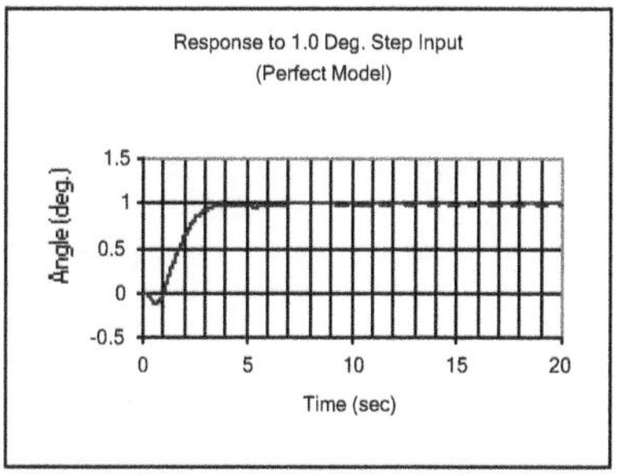

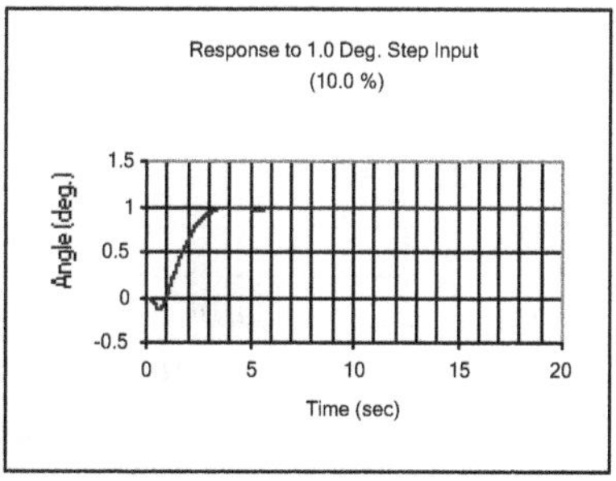

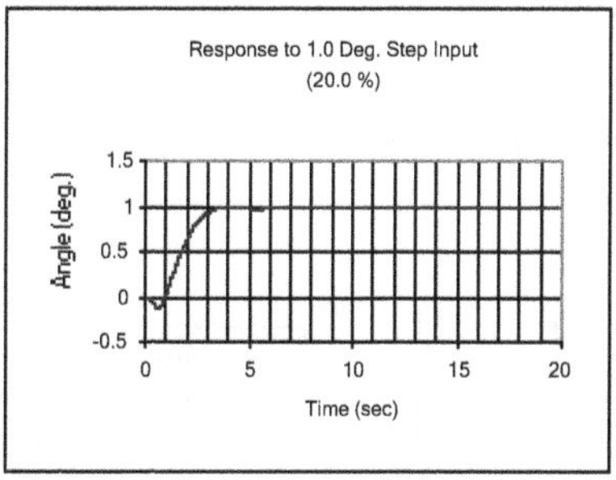

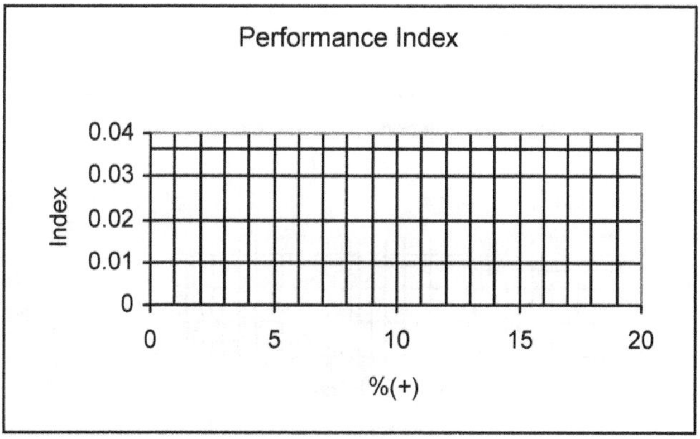

Appendix III: Collection of Math Modeling Error Sensitivity Response and Singular Value Plots, and Eigenvalue Tables

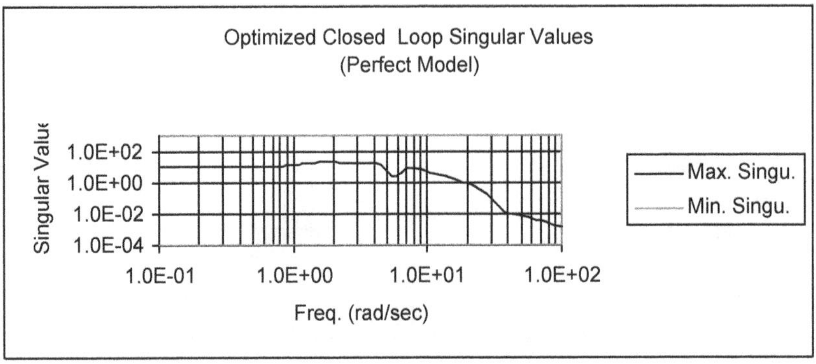

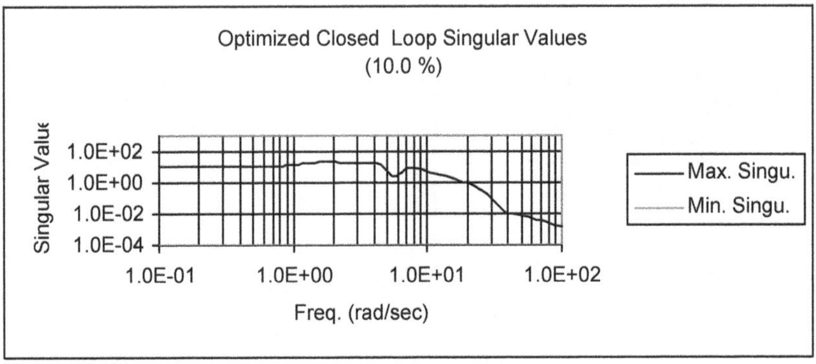

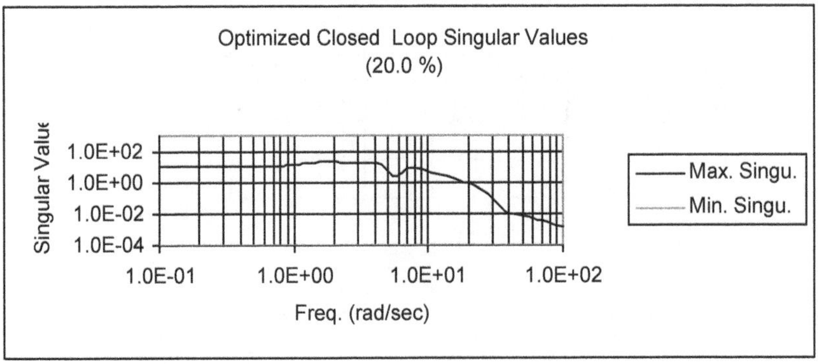

Appendix III: Collection of Math Modeling Error Sensitivity Response and Singular Value Plots, and Eigenvalue Tables

Closed Loop System Eigenvalues					
	0%	10.00%	20.00%		
	1.0e+002 *	1.0e+002 *	1.0e+002 *		
1	-0.25	-0.25	-0.25		
2	-0.5824 + 1.3344i	-0.5824 + 1.3344i	-0.5824 + 1.3344i		
3	-0.5824 - 1.3344i	-0.5824 - 1.3344i	-0.5824 - 1.3344i		
4	-0.1040 + 0.2546i	-0.1040 + 0.2546i	-0.1040 + 0.2546i		
5	-0.1040 - 0.2546i	-0.1040 - 0.2546i	-0.1040 - 0.2546i		
6	-0.1046 + 0.2542i	-0.1046 + 0.2542i	-0.1046 + 0.2542i		
7	-0.1046 - 0.2542i	-0.1046 - 0.2542i	-0.1046 - 0.2542i		
8	-0.2284 + 0.0212i	-0.2284 + 0.0212i	-0.2284 + 0.0212i		
9	-0.2284 - 0.0212i	-0.2284 - 0.0212i	-0.2284 - 0.0212i		
10	-0.2293	-0.2293	-0.2293		
11	-0.1353 + 0.0476i	-0.1353 + 0.0476i	-0.1353 + 0.0476i		
12	-0.1353 - 0.0476i	-0.1353 - 0.0476i	-0.1353 - 0.0476i		
13	-0.1446 + 0.0002i	-0.1446 + 0.0002i	-0.1446 + 0.0002i		
14	-0.1446 - 0.0002i	-0.1446 - 0.0002i	-0.1446 - 0.0002i		
15	-0.0868 + 0.0557i	-0.0868 + 0.0557i	-0.0868 + 0.0557i		
16	-0.0868 - 0.0557i	-0.0868 - 0.0557i	-0.0868 - 0.0557i		
17	-0.0002 + 0.0601i	-0.0002 + 0.0601i	-0.0002 + 0.0601i		
18	-0.0002 - 0.0601i	-0.0002 - 0.0601i	-0.0002 - 0.0601i		
19	-0.0002 + 0.0600i	-0.0002 + 0.0600i	-0.0002 + 0.0600i		
20	-0.0002 - 0.0600i	-0.0002 - 0.0600i	-0.0002 - 0.0600i		
21	-0.0009 + 0.0508i	-0.0009 + 0.0508i	-0.0009 + 0.0508i		
22	-0.0009 - 0.0508i	-0.0009 - 0.0508i	-0.0009 - 0.0508i		
23	-0.0009 + 0.0509i	-0.0009 + 0.0509i	-0.0009 + 0.0509i		
24	-0.0009 - 0.0509i	-0.0009 - 0.0509i	-0.0009 - 0.0509i		
25	-0.0347 + 0.0149i	-0.0347 + 0.0149i	-0.0347 + 0.0149i		
26	-0.0347 - 0.0149i	-0.0347 - 0.0149i	-0.0347 - 0.0149i		
27	-0.0204	-0.0204	-0.0204		
28	-0.0052 + 0.0136i	-0.0052 + 0.0136i	-0.0052 + 0.0136i		
29	-0.0052 - 0.0136i	-0.0052 - 0.0136i	-0.0052 - 0.0136i		
30	-0.0078 + 0.0145i	-0.0078 + 0.0145i	-0.0078 + 0.0145i		
31	-0.0078 - 0.0145i	-0.0078 - 0.0145i	-0.0078 - 0.0145i		
32	-0.0134	-0.0134	-0.0134		
33	-0.0106	-0.0106	-0.0106		
34	-0.01	-0.01	-0.01		

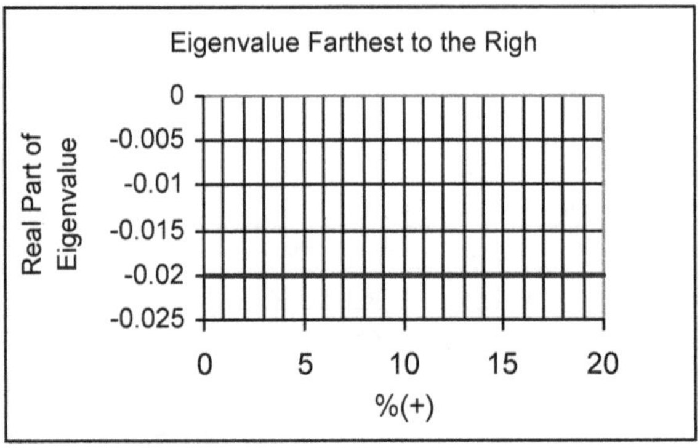

Observation:

Modeling error associated with bending slope does not cause control system instability.

17) **Bending Mode 1 Bending Slope at Middle: SigG1m (The value of Bending Mode 1 bending slope at middle = 0.0044 rad/ft in the math model.)**

Case 1 (actual value is less than the value in the math model)

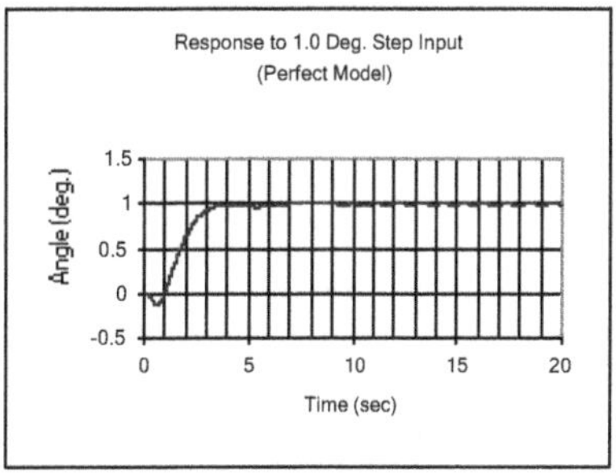

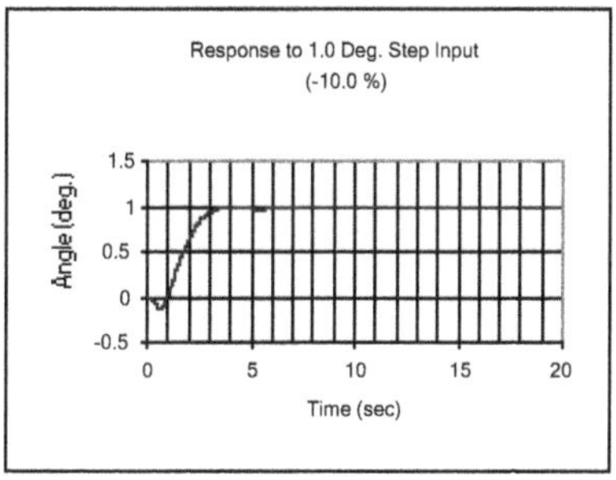

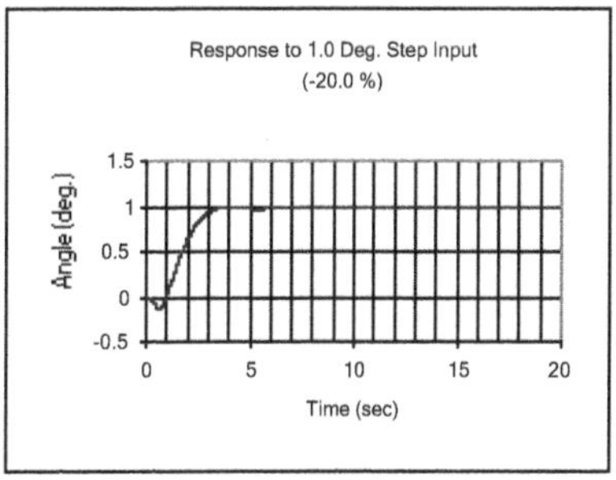

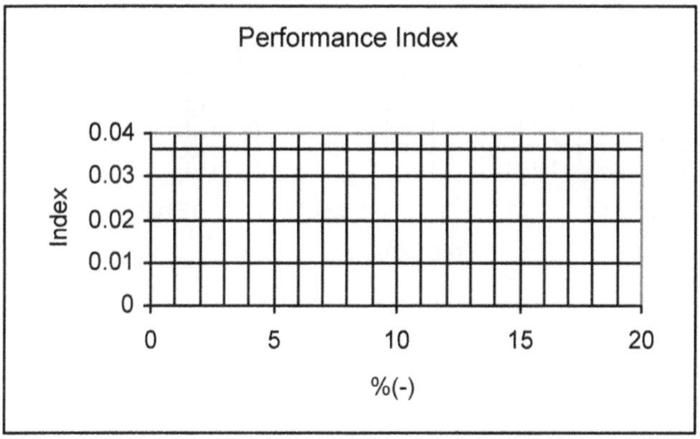

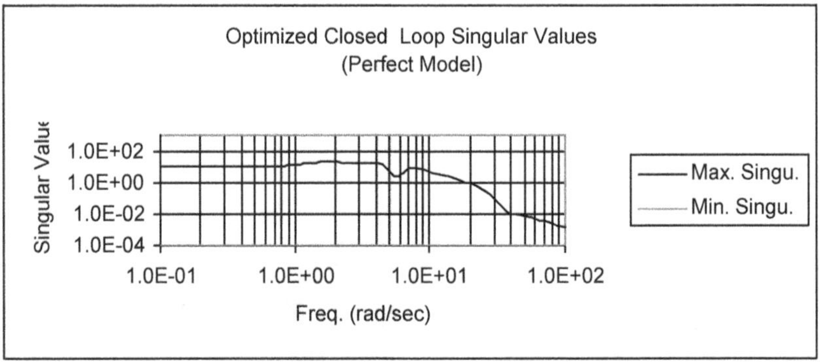
Optimized Closed Loop Singular Values
(Perfect Model)
Singular Valu
1.0E+02
1.0E+00
1.0E-02
1.0E-04
1.0E-01
1.0E+00
1.0E+01
1.0E+02
Freq. (rad/sec)
Max. Singu.
Min. Singu.

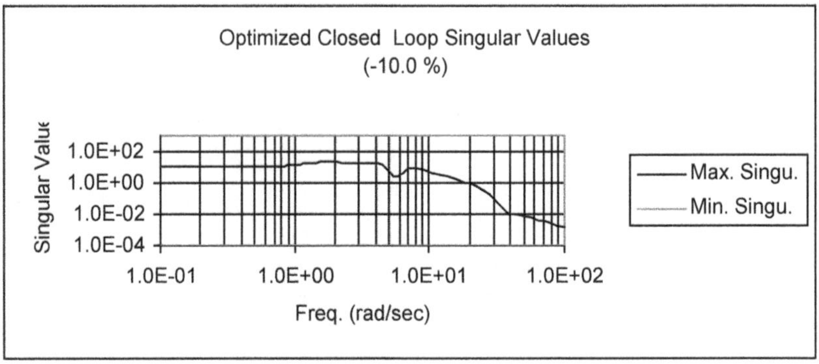
Optimized Closed Loop Singular Values
(-10.0 %)
Singular Valu
1.0E+02
1.0E+00
1.0E-02
1.0E-04
1.0E-01
1.0E+00
1.0E+01
1.0E+02
Freq. (rad/sec)
Max. Singu.
Min. Singu.

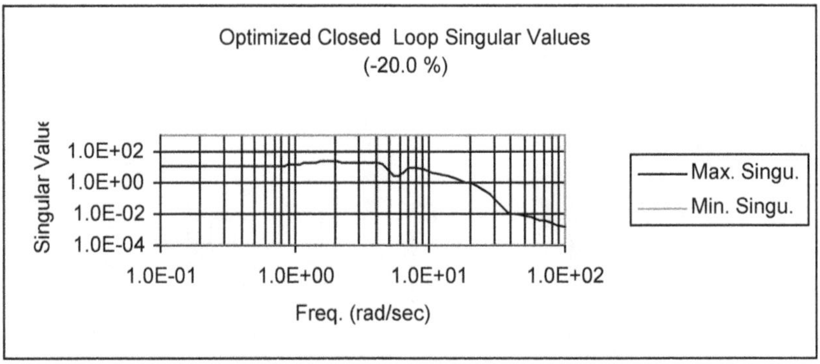
Optimized Closed Loop Singular Values
(-20.0 %)
Singular Valu
1.0E+02
1.0E+00
1.0E-02
1.0E-04
1.0E-01
1.0E+00
1.0E+01
1.0E+02
Freq. (rad/sec)
Max. Singu.
Min. Singu.

Closed Loop System Eigenvalues					
	0%	-10.00%	-20.00%		
	1.0e+002 *	1.0e+002 *	1.0e+002 *		
1	-0.25	-0.25	-0.25		
2	-0.5824 + 1.3344i	-0.5824 + 1.3344i	-0.5824 + 1.3344i		
3	-0.5824 - 1.3344i	-0.5824 - 1.3344i	-0.5824 - 1.3344i		
4	-0.1040 + 0.2546i	-0.1057 + 0.2548i	-0.1062 + 0.2551i		
5	-0.1040 - 0.2546i	-0.1057 - 0.2548i	-0.1062 - 0.2551i		
6	-0.1046 + 0.2542i	-0.1029 + 0.2540i	-0.1024 + 0.2538i		
7	-0.1046 - 0.2542i	-0.1029 - 0.2540i	-0.1024 - 0.2538i		
8	-0.2284 + 0.0212i	-0.2325	-0.2344		
9	-0.2284 - 0.0212i	-0.2269 + 0.0221i	-0.2260 + 0.0231i		
10	-0.2293	-0.2269 - 0.0221i	-0.2260 - 0.0231i		
11	-0.1353 + 0.0476i	-0.1352 + 0.0470i	-0.1350 + 0.0463i		
12	-0.1353 - 0.0476i	-0.1352 - 0.0470i	-0.1350 - 0.0463i		
13	-0.1446 + 0.0002i	-0.1446 + 0.0002i	-0.1446 + 0.0002i		
14	-0.1446 - 0.0002i	-0.1446 - 0.0002i	-0.1446 - 0.0002i		
15	-0.0868 + 0.0557i	-0.0868 + 0.0558i	-0.0870 + 0.0560i		
16	-0.0868 - 0.0557i	-0.0868 - 0.0558i	-0.0870 - 0.0560i		
17	-0.0002 + 0.0601i	-0.0002 + 0.0601i	-0.0002 + 0.0601i		
18	-0.0002 - 0.0601i	-0.0002 - 0.0601i	-0.0002 - 0.0601i		
19	-0.0002 + 0.0600i	-0.0002 + 0.0600i	-0.0002 + 0.0600i		
20	-0.0002 - 0.0600i	-0.0002 - 0.0600i	-0.0002 - 0.0600i		
21	-0.0009 + 0.0508i	-0.0009 + 0.0508i	-0.0009 + 0.0508i		
22	-0.0009 - 0.0508i	-0.0009 - 0.0508i	-0.0009 - 0.0508i		
23	-0.0009 + 0.0509i	-0.0009 + 0.0509i	-0.0009 + 0.0509i		
24	-0.0009 - 0.0509i	-0.0009 - 0.0509i	-0.0009 - 0.0509i		
25	-0.0347 + 0.0149i	-0.0347 + 0.0149i	-0.0347 + 0.0149i		
26	-0.0347 - 0.0149i	-0.0347 - 0.0149i	-0.0347 - 0.0149i		
27	-0.0204	-0.0204	-0.0204		
28	-0.0052 + 0.0136i	-0.0052 + 0.0136i	-0.0052 + 0.0136i		
29	-0.0052 - 0.0136i	-0.0052 - 0.0136i	-0.0052 - 0.0136i		
30	-0.0078 + 0.0145i	-0.0078 + 0.0145i	-0.0078 + 0.0145i		
31	-0.0078 - 0.0145i	-0.0078 - 0.0145i	-0.0078 - 0.0145i		
32	-0.0134	-0.0134	-0.0134		
33	-0.0106	-0.0106	-0.0106		
34	-0.01	-0.01	-0.01		

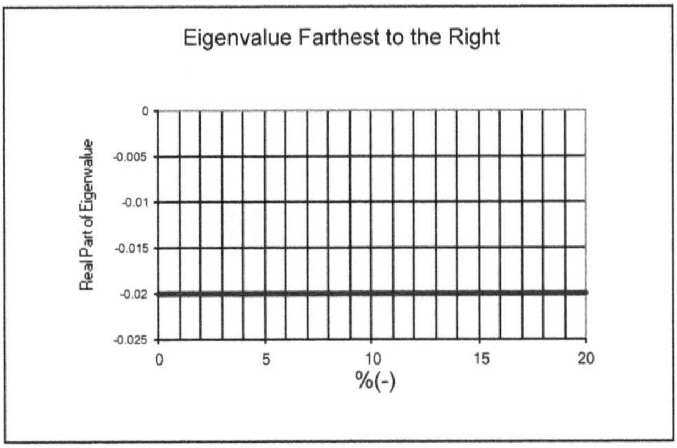

Observation:

Modeling error associated with bending slope does not cause control system instability.

Case 2 (actual value is greater than the value in the math model)

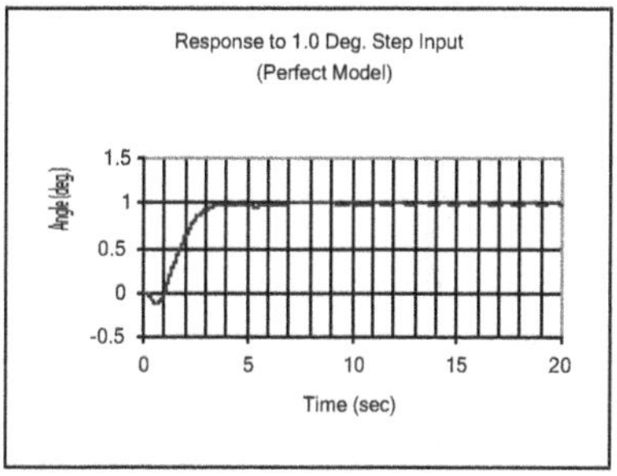

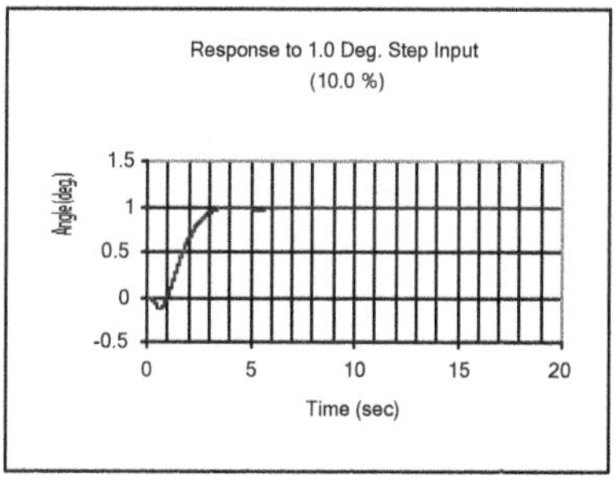

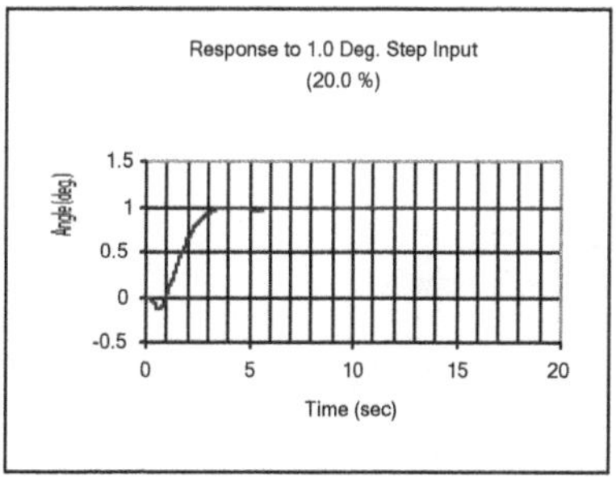

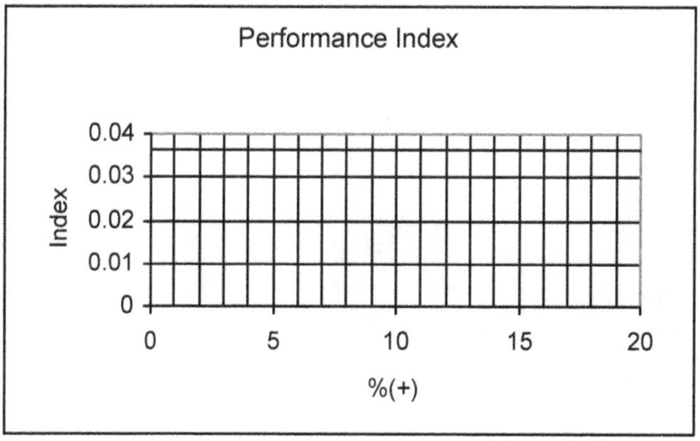

Appendix III: Collection of Math Modeling Error Sensitivity Response and Singular Value Plots, and Eigenvalue Tables

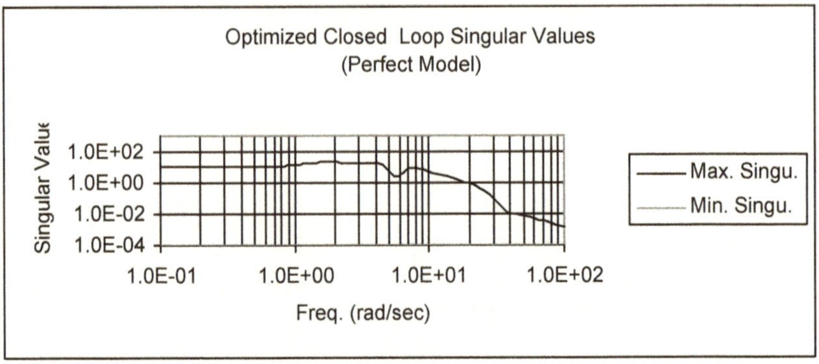

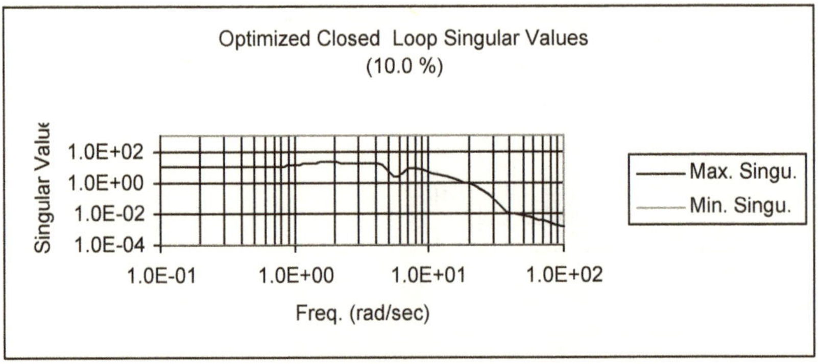

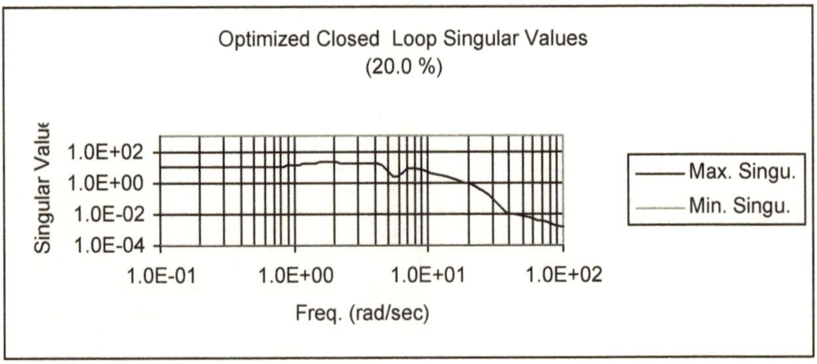

Closed Loop System Eigenvalues					
	0%	10.00%	20.00%		
	1.0e+002 *	1.0e+002 *	-0.1		
1	-0.25	-0.25	-0.25		
2	-0.5824 + 1.3344i	-0.5824 + 1.3344i	-0.5824 + 1.3344i		
3	-0.5824 - 1.3344i	-0.5824 - 1.3344i	-0.5824 - 1.3344i		
4	-0.1040 + 0.2546i	-0.1038 + 0.2557i	-0.1036 + 0.2563i		
5	-0.1040 - 0.2546i	-0.1038 - 0.2557i	-0.1036 - 0.2563i		
6	-0.1046 + 0.2542i	-0.1049 + 0.2531i	-0.1051 + 0.2525i		
7	-0.1046 - 0.2542i	-0.1049 - 0.2531i	-0.1051 - 0.2525i		
8	-0.2284 + 0.0212i	-0.2298 + 0.0205i	-0.2313 + 0.0202i		
9	-0.2284 - 0.0212i	-0.2298 - 0.0205i	-0.2313 - 0.0202i		
10	-0.2293	-0.2264	-0.2232		
11	-0.1353 + 0.0476i	-0.1354 + 0.0483i	-0.1356 + 0.0489i		
12	-0.1353 - 0.0476i	-0.1354 - 0.0483i	-0.1356 - 0.0489i		
13	-0.1446 + 0.0002i	-0.1447 + 0.0002i	-0.1447 + 0.0002i		
14	-0.1446 - 0.0002i	-0.1447 - 0.0002i	-0.1447 - 0.0002i		
15	-0.0868 + 0.0557i	-0.0867 + 0.0555i	-0.0865 + 0.0553i		
16	-0.0868 - 0.0557i	-0.0867 - 0.0555i	-0.0865 - 0.0553i		
17	-0.0002 + 0.0601i	-0.0002 + 0.0601i	-0.0002 + 0.0601i		
18	-0.0002 - 0.0601i	-0.0002 - 0.0601i	-0.0002 - 0.0601i		
19	-0.0002 + 0.0600i	-0.0002 + 0.0600i	-0.0002 + 0.0600i		
20	-0.0002 - 0.0600i	-0.0002 - 0.0600i	-0.0002 - 0.0600i		
21	-0.0009 + 0.0508i	-0.0009 + 0.0508i	-0.0009 + 0.0508i		
22	-0.0009 - 0.0508i	-0.0009 - 0.0508i	-0.0009 - 0.0508i		
23	-0.0009 + 0.0509i	-0.0009 + 0.0509i	-0.0009 + 0.0509i		
24	-0.0009 - 0.0509i	-0.0009 - 0.0509i	-0.0009 - 0.0509i		
25	-0.0347 + 0.0149i	-0.0347 + 0.0149i	-0.0347 + 0.0149i		
26	-0.0347 - 0.0149i	-0.0347 - 0.0149i	-0.0347 - 0.0149i		
27	-0.0204	-0.0204	-0.0204		
28	-0.0052 + 0.0136i	-0.0052 + 0.0136i	-0.0052 + 0.0136i		
29	-0.0052 - 0.0136i	-0.0052 - 0.0136i	-0.0052 - 0.0136i		
30	-0.0078 + 0.0145i	-0.0078 + 0.0145i	-0.0078 + 0.0145i		
31	-0.0078 - 0.0145i	-0.0078 - 0.0145i	-0.0078 - 0.0145i		
32	-0.0134	-0.0134	-0.0134		
33	-0.0106	-0.0106	-0.0106		
34	-0.01	-0.01	-0.01		

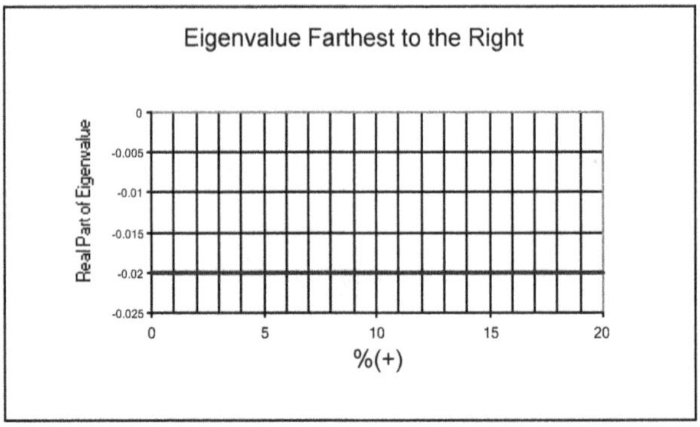

Observation:

Modeling error associated with bending slope does not cause control system instability.

18) **Bending Mode 2 Bending Slope at Tail: SigG2t (The value of Bending Mode 2 bending slope at tail = -0.171 rad/ft in the math model.)**

Case 1 (actual value is less than the value in the math model)

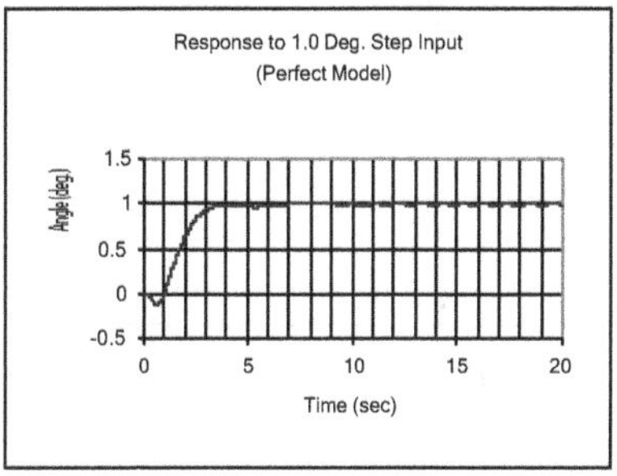

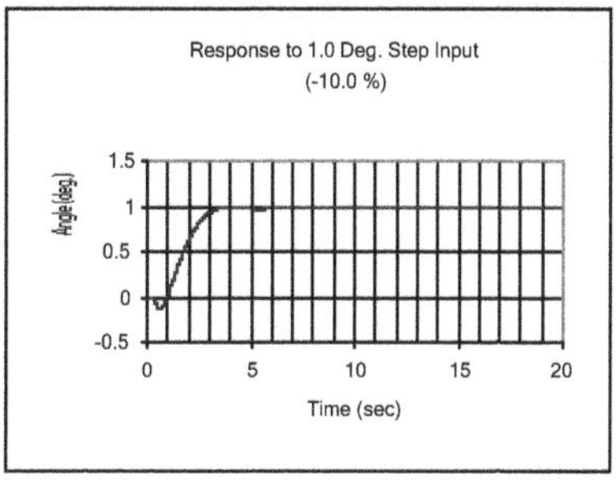

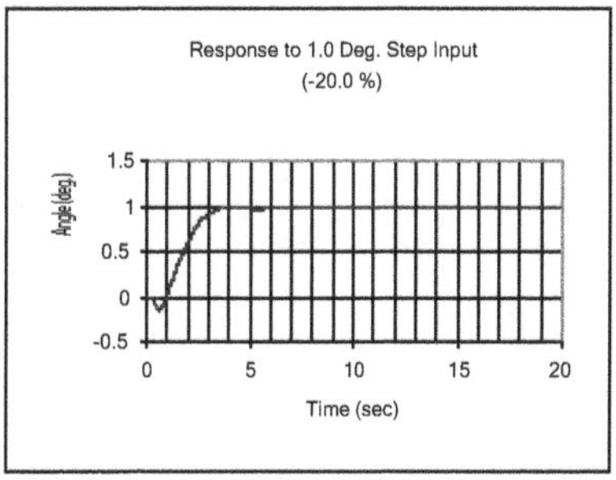

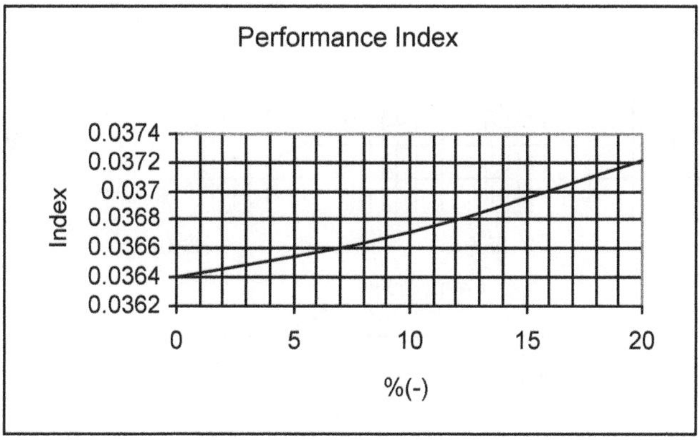

Appendix III: Collection of Math Modeling Error Sensitivity Response and Singular Value Plots, and Eigenvalue Tables

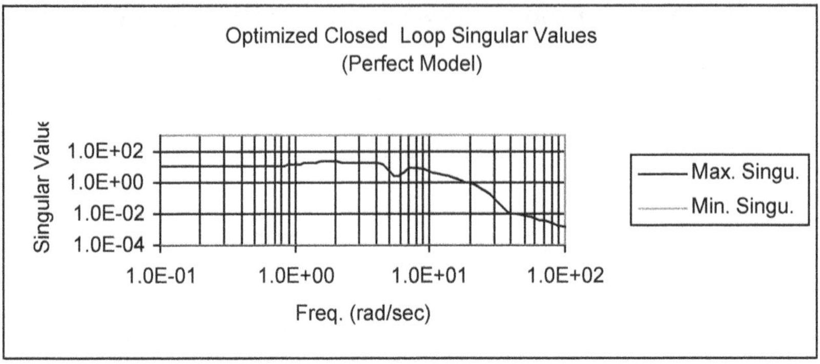

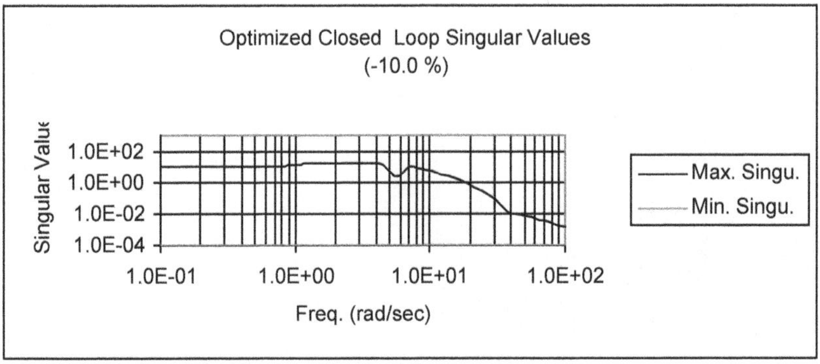

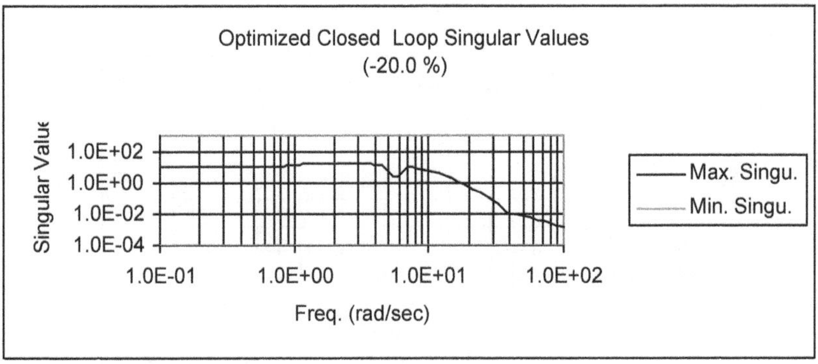

Appendix III: Collection of Math Modeling Error Sensitivity Response and Singular Value Plots, and Eigenvalue Tables

Closed Loop System Eigenvalues					
	0%	-10.00%	-20.00%		
	1.0e+002 *	1.0e+002 *	-0.1		
1	-0.25	-0.25	-0.25		
2	-0.5824 + 1.3344i	-0.5824 + 1.3344i	-0.5824 + 1.3344i		
3	-0.5824 - 1.3344i	-0.5824 - 1.3344i	-0.5824 - 1.3344i		
4	-0.1040 + 0.2546i	-0.0648 + 0.3285i	-0.1036 + 0.2563i		
5	-0.1040 - 0.2546i	-0.0648 - 0.3285i	-0.1036 - 0.2563i		
6	-0.1046 + 0.2542i	-0.2762 + 0.1787i	-0.1051 + 0.2525i		
7	-0.1046 - 0.2542i	-0.2762 - 0.1787i	-0.1051 - 0.2525i		
8	-0.2284 + 0.0212i	-0.3204	-0.2313 + 0.0202i		
9	-0.2284 - 0.0212i	-0.1350 + 0.1515i	-0.2313 - 0.0202i		
10	-0.2293	-0.1350 - 0.1515i	-0.2232		
11	-0.1353 + 0.0476i	-0.0405 + 0.1005i	-0.1356 + 0.0489i		
12	-0.1353 - 0.0476i	-0.0405 - 0.1005i	-0.1356 - 0.0489i		
13	-0.1446 + 0.0002i	-0.1662	-0.1447 + 0.0002i		
14	-0.1446 - 0.0002i	-0.1473	-0.1447 - 0.0002i		
15	-0.0868 + 0.0557i	-0.1446	-0.0865 + 0.0553i		
16	-0.0868 - 0.0557i	-0.0355 + 0.0339i	-0.0865 - 0.0553i		
17	-0.0002 + 0.0601i	-0.0355 - 0.0339i	-0.0002 + 0.0601i		
18	-0.0002 - 0.0601i	-0.0004 + 0.0602i	-0.0002 - 0.0601i		
19	-0.0002 + 0.0600i	-0.0004 - 0.0602i	-0.0002 + 0.0600i		
20	-0.0002 - 0.0600i	-0.0001 + 0.0599i	-0.0002 - 0.0600i		
21	-0.0009 + 0.0508i	-0.0001 - 0.0599i	-0.0009 + 0.0508i		
22	-0.0009 - 0.0508i	-0.0014 + 0.0511i	-0.0009 - 0.0508i		
23	-0.0009 + 0.0509i	-0.0014 - 0.0511i	-0.0009 + 0.0509i		
24	-0.0009 - 0.0509i	-0.0005 + 0.0508i	-0.0009 - 0.0509i		
25	-0.0347 + 0.0149i	-0.0005 - 0.0508i	-0.0347 + 0.0149i		
26	-0.0347 - 0.0149i	-0.0257	-0.0347 - 0.0149i		
27	-0.0204	-0.0052 + 0.0134i	-0.0204		
28	-0.0052 + 0.0136i	-0.0052 - 0.0134i	-0.0052 + 0.0136i		
29	-0.0052 - 0.0136i	-0.0078 + 0.0146i	-0.0052 - 0.0136i		
30	-0.0078 + 0.0145i	-0.0078 - 0.0146i	-0.0078 + 0.0145i		
31	-0.0078 - 0.0145i	-0.0177	-0.0078 - 0.0145i		
32	-0.0134	-0.0136	-0.0134		
33	-0.0106	-0.011	-0.0106		
34	-0.01	-0.01	-0.01		

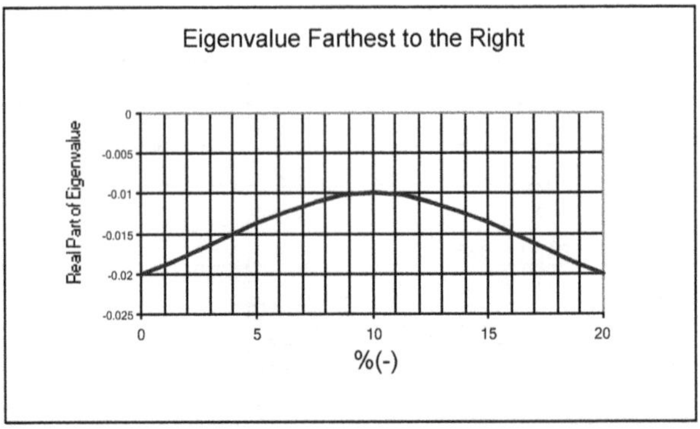

Observation:

Modeling error associated with bending slope does not cause control system instability.

Case 2 (actual value is greater than the value in the math model)

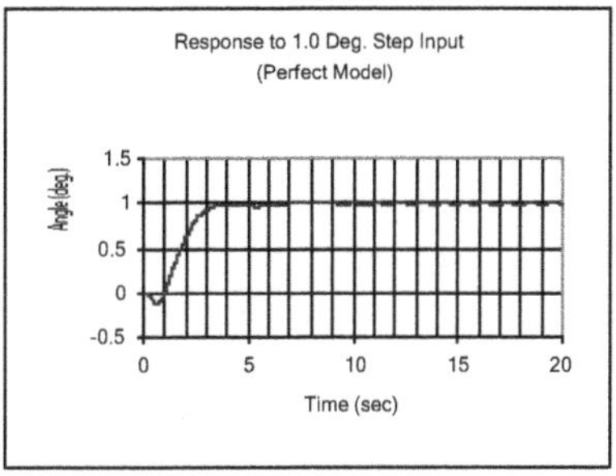

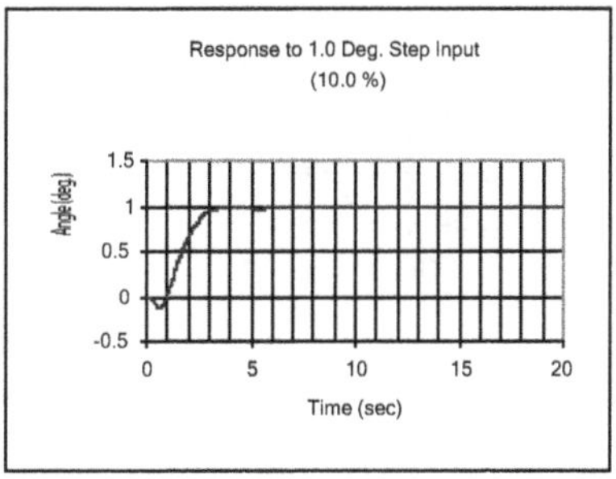

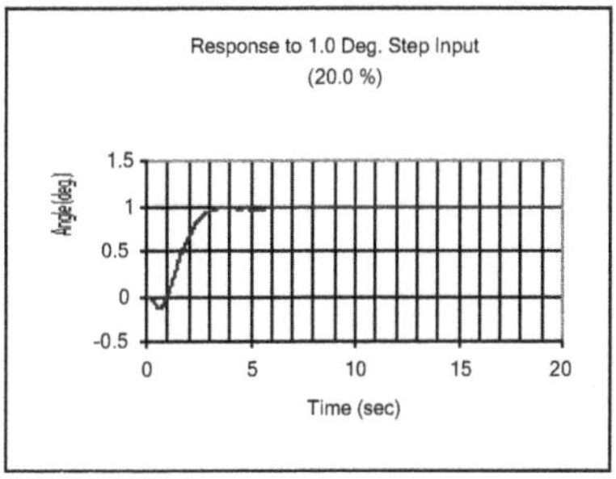

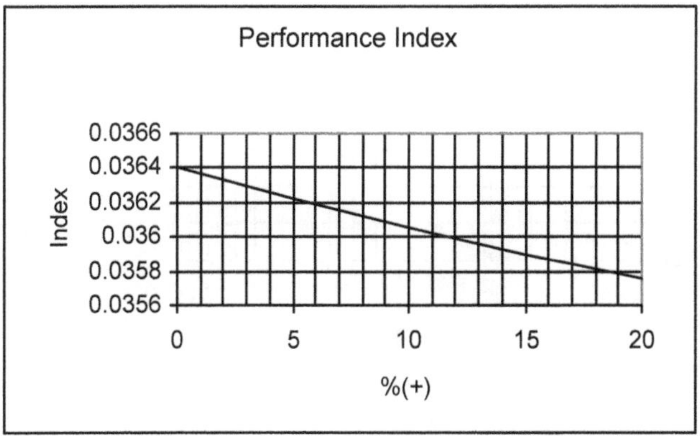

Appendix III: Collection of Math Modeling Error Sensitivity Response and Singular Value Plots, and Eigenvalue Tables

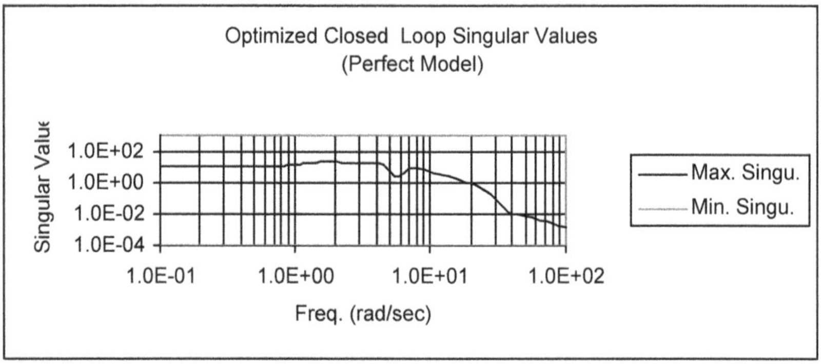

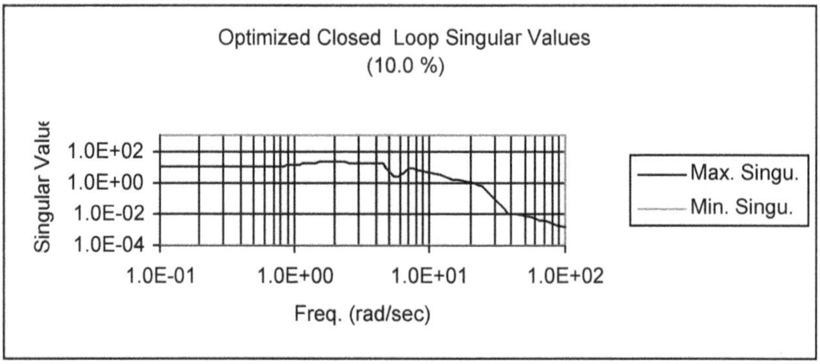

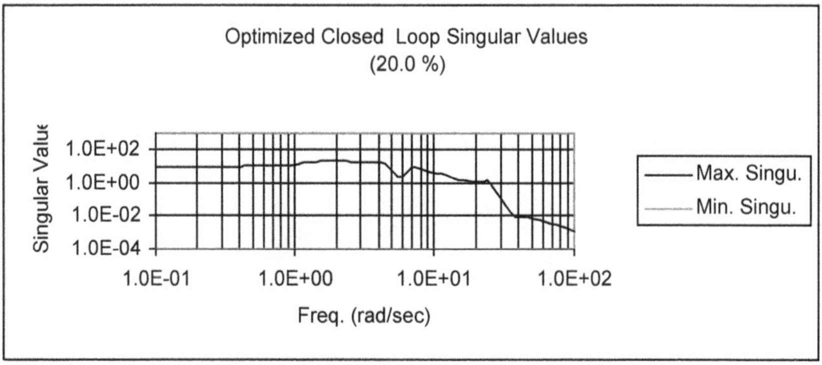

Appendix III: Collection of Math Modeling Error Sensitivity Response and Singular Value Plots, and Eigenvalue Tables

Closed Loop System Eigenvalues					
	0%	10.00%	20.00%		
	1.0e+002 *	1.0e+002 *	1.0e+002 *		
1	-0.25	-0.25	-0.25		
2	-0.5824 + 1.3344i	-0.5824 + 1.3344i	-0.5824 + 1.3344i		
3	-0.5824 - 1.3344i	-0.5824 - 1.3344i	-0.5824 - 1.3344i		
4	-0.1040 + 0.2546i	-0.1602 + 0.2999i	-0.1760 + 0.3212i		
5	-0.1040 - 0.2546i	-0.1602 - 0.2999i	-0.1760 - 0.3212i		
6	-0.1046 + 0.2542i	-0.3045 + 0.0567i	-0.3215 + 0.0574i		
7	-0.1046 - 0.2542i	-0.3045 - 0.0567i	-0.3215 - 0.0574i		
8	-0.2284 + 0.0212i	-0.0383 + 0.2364i	-0.0150 + 0.2352i		
9	-0.2284 - 0.0212i	-0.0383 - 0.2364i	-0.0150 - 0.2352i		
10	-0.2293	-0.1381 + 0.1173i	-0.1374 + 0.1250i		
11	-0.1353 + 0.0476i	-0.1381 - 0.1173i	-0.1374 - 0.1250i		
12	-0.1353 - 0.0476i	-0.168	-0.1674		
13	-0.1446 + 0.0002i	-0.1471	-0.1471		
14	-0.1446 - 0.0002i	-0.1446	-0.1446		
15	-0.0868 + 0.0557i	-0.0543 + 0.0561i	-0.0485 + 0.0570i		
16	-0.0868 - 0.0557i	-0.0543 - 0.0561i	-0.0485 - 0.0570i		
17	-0.0002 + 0.0601i	-0.0001 + 0.0601i	-0.0001 + 0.0601i		
18	-0.0002 - 0.0601i	-0.0001 - 0.0601i	-0.0001 - 0.0601i		
19	-0.0002 + 0.0600i	-0.0003 + 0.0600i	-0.0003 + 0.0599i		
20	-0.0002 - 0.0600i	-0.0003 - 0.0600i	-0.0003 - 0.0599i		
21	-0.0009 + 0.0508i	-0.0010 + 0.0504i	-0.0011 + 0.0502i		
22	-0.0009 - 0.0508i	-0.0010 - 0.0504i	-0.0011 - 0.0502i		
23	-0.0009 + 0.0509i	-0.0008 + 0.0511i	-0.0007 + 0.0512i		
24	-0.0009 - 0.0509i	-0.0008 - 0.0511i	-0.0007 - 0.0512i		
25	-0.0347 + 0.0149i	-0.0277 + 0.0169i	-0.0248 + 0.0178i		
26	-0.0347 - 0.0149i	-0.0277 - 0.0169i	-0.0248 - 0.0178i		
27	-0.0204	-0.0209	-0.0213		
28	-0.0052 + 0.0136i	-0.0053 + 0.0137i	-0.0053 + 0.0138i		
29	-0.0052 - 0.0136i	-0.0053 - 0.0137i	-0.0053 - 0.0138i		
30	-0.0078 + 0.0145i	-0.0078 + 0.0145i	-0.0079 + 0.0145i		
31	-0.0078 - 0.0145i	-0.0078 - 0.0145i	-0.0079 - 0.0145i		
32	-0.0134	-0.0134	-0.0134		
33	-0.0106	-0.0105	-0.0104		
34	-0.01	-0.01	-0.01		

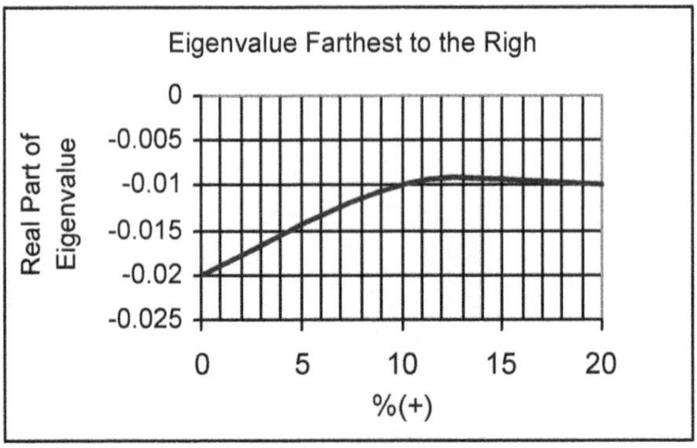

Observation:

Modeling error associated with bending slope does not cause control system instability.

17) **Bending Mode 1 Bending Slope at Middle: SigG1m (The value of Bending Mode 1 bending slope at middle = 0.0044 rad/ft in the math model.)**

Case 1 (actual value is less than the value in the math model)

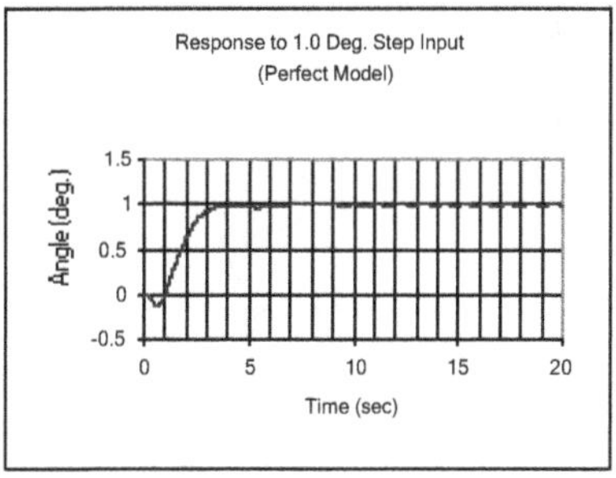

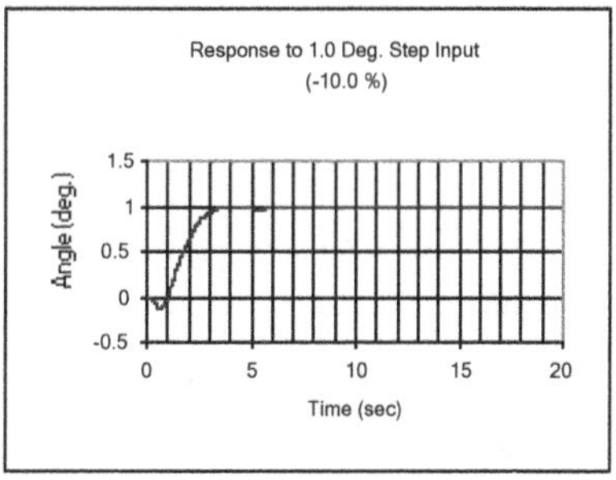

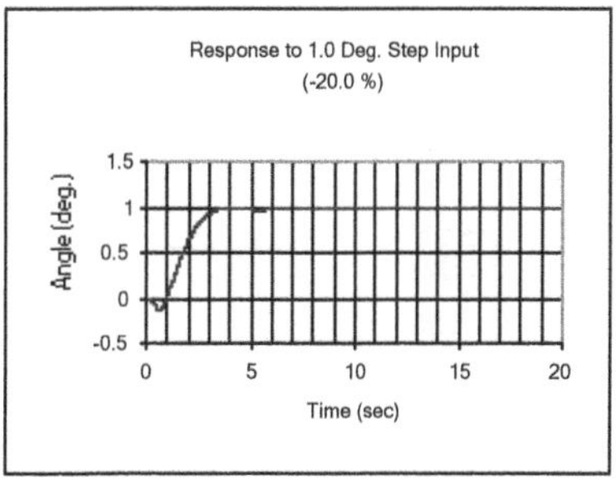

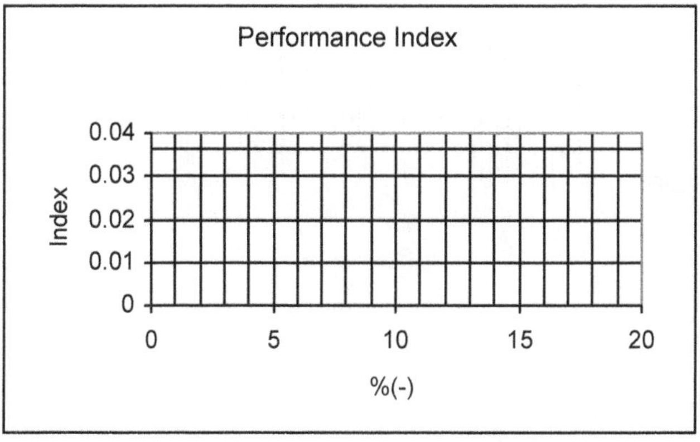

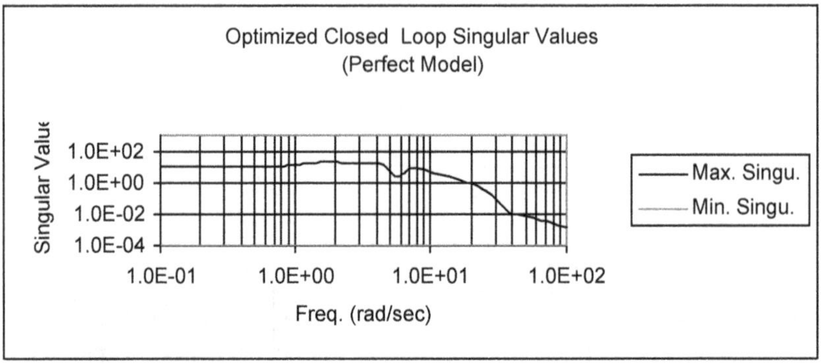
Optimized Closed Loop Singular Values
(Perfect Model)
Singular Valu
1.0E+02
1.0E+00
1.0E-02
1.0E-04
1.0E-01
1.0E+00
1.0E+01
1.0E+02
Freq. (rad/sec)
Max. Singu.
Min. Singu.

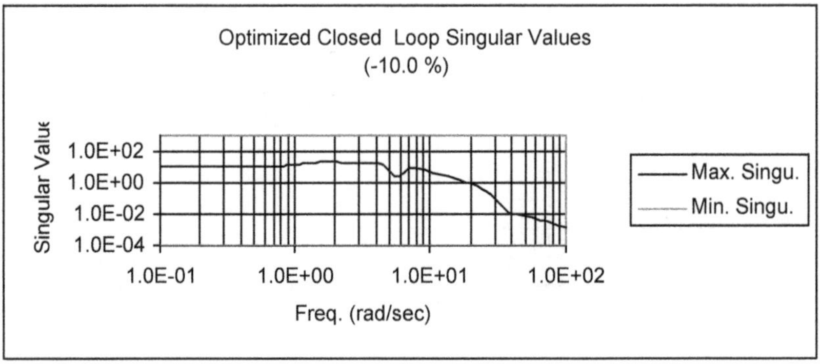
Optimized Closed Loop Singular Values
(-10.0 %)
Singular Valu
1.0E+02
1.0E+00
1.0E-02
1.0E-04
1.0E-01
1.0E+00
1.0E+01
1.0E+02
Freq. (rad/sec)
Max. Singu.
Min. Singu.

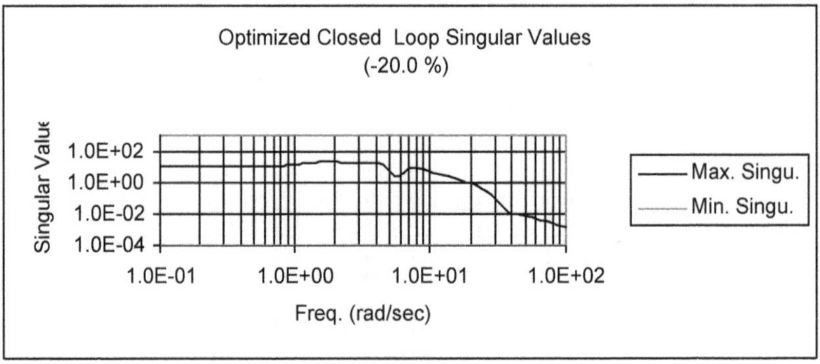
Optimized Closed Loop Singular Values
(-20.0 %)
Singular Valu
1.0E+02
1.0E+00
1.0E-02
1.0E-04
1.0E-01
1.0E+00
1.0E+01
1.0E+02
Freq. (rad/sec)
Max. Singu.
Min. Singu.

Closed Loop System Eigenvalues					
	0%	-10.00%	-20.00%		
	1.0e+002 *	1.0e+002 *	1.0e+002 *		
1	-0.25	-0.25	-0.25		
2	-0.5824 + 1.3344i	-0.5824 + 1.3344i	-0.5824 + 1.3344i		
3	-0.5824 - 1.3344i	-0.5824 - 1.3344i	-0.5824 - 1.3344i		
4	-0.1040 + 0.2546i	-0.1057 + 0.2548i	-0.1062 + 0.2551i		
5	-0.1040 - 0.2546i	-0.1057 - 0.2548i	-0.1062 - 0.2551i		
6	-0.1046 + 0.2542i	-0.1029 + 0.2540i	-0.1024 + 0.2538i		
7	-0.1046 - 0.2542i	-0.1029 - 0.2540i	-0.1024 - 0.2538i		
8	-0.2284 + 0.0212i	-0.2325	-0.2344		
9	-0.2284 - 0.0212i	-0.2269 + 0.0221i	-0.2260 + 0.0231i		
10	-0.2293	-0.2269 - 0.0221i	-0.2260 - 0.0231i		
11	-0.1353 + 0.0476i	-0.1352 + 0.0470i	-0.1350 + 0.0463i		
12	-0.1353 - 0.0476i	-0.1352 - 0.0470i	-0.1350 - 0.0463i		
13	-0.1446 + 0.0002i	-0.1446 + 0.0002i	-0.1446 + 0.0002i		
14	-0.1446 - 0.0002i	-0.1446 - 0.0002i	-0.1446 - 0.0002i		
15	-0.0868 + 0.0557i	-0.0868 + 0.0558i	-0.0870 + 0.0560i		
16	-0.0868 - 0.0557i	-0.0868 - 0.0558i	-0.0870 - 0.0560i		
17	-0.0002 + 0.0601i	-0.0002 + 0.0601i	-0.0002 + 0.0601i		
18	-0.0002 - 0.0601i	-0.0002 - 0.0601i	-0.0002 - 0.0601i		
19	-0.0002 + 0.0600i	-0.0002 + 0.0600i	-0.0002 + 0.0600i		
20	-0.0002 - 0.0600i	-0.0002 - 0.0600i	-0.0002 - 0.0600i		
21	-0.0009 + 0.0508i	-0.0009 + 0.0508i	-0.0009 + 0.0508i		
22	-0.0009 - 0.0508i	-0.0009 - 0.0508i	-0.0009 - 0.0508i		
23	-0.0009 + 0.0509i	-0.0009 + 0.0509i	-0.0009 + 0.0509i		
24	-0.0009 - 0.0509i	-0.0009 - 0.0509i	-0.0009 - 0.0509i		
25	-0.0347 + 0.0149i	-0.0347 + 0.0149i	-0.0347 + 0.0149i		
26	-0.0347 - 0.0149i	-0.0347 - 0.0149i	-0.0347 - 0.0149i		
27	-0.0204	-0.0204	-0.0204		
28	-0.0052 + 0.0136i	-0.0052 + 0.0136i	-0.0052 + 0.0136i		
29	-0.0052 - 0.0136i	-0.0052 - 0.0136i	-0.0052 - 0.0136i		
30	-0.0078 + 0.0145i	-0.0078 + 0.0145i	-0.0078 + 0.0145i		
31	-0.0078 - 0.0145i	-0.0078 - 0.0145i	-0.0078 - 0.0145i		
32	-0.0134	-0.0134	-0.0134		
33	-0.0106	-0.0106	-0.0106		
34	-0.01	-0.01	-0.01		

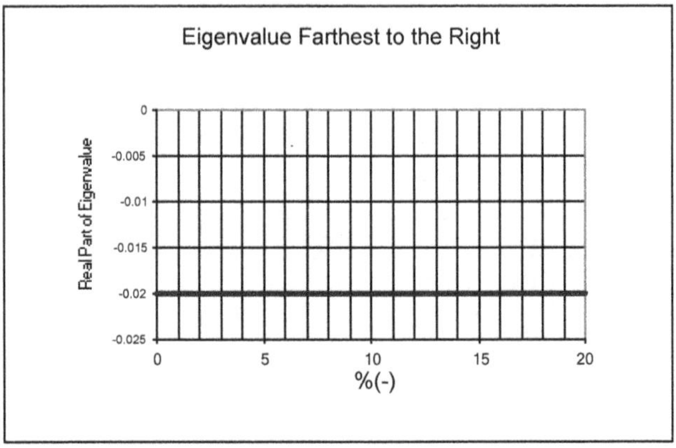

Observation:

Modeling error associated with bending slope does not cause control system instability.

Case 2 (actual value is greater than the value in the math model)

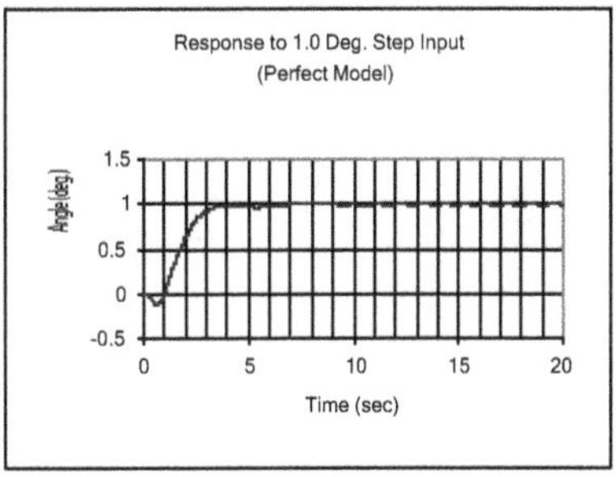

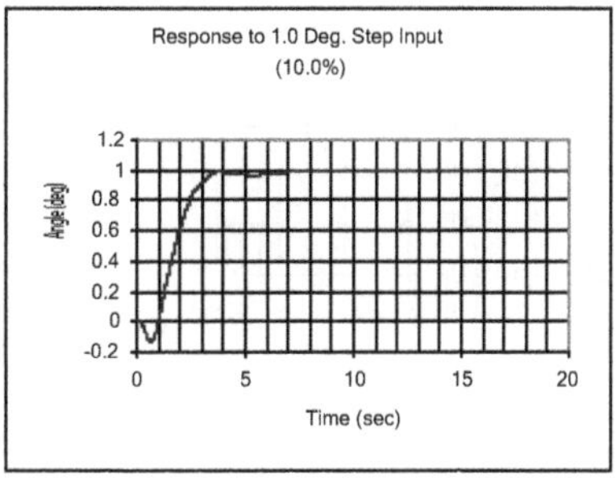

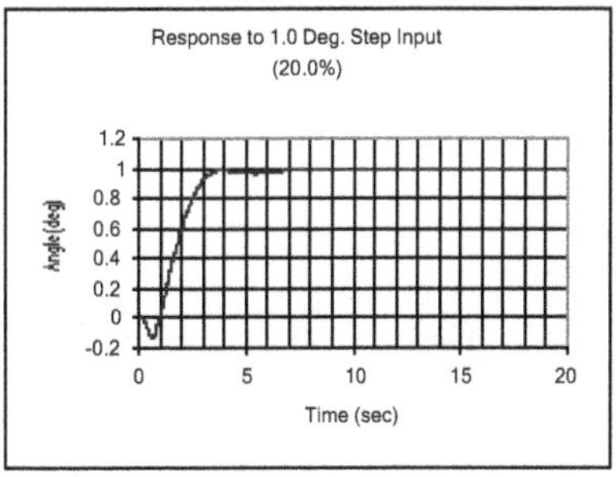

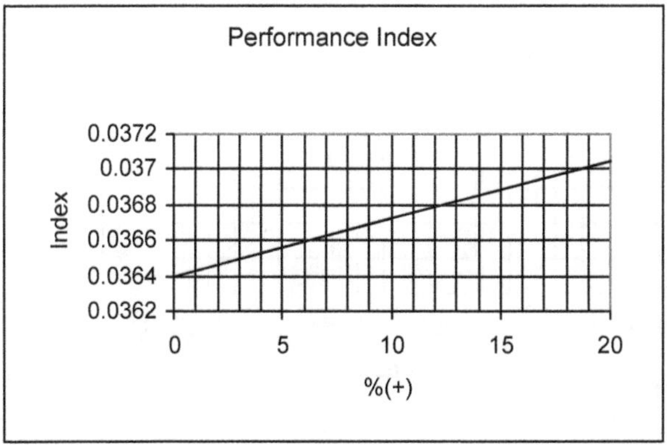

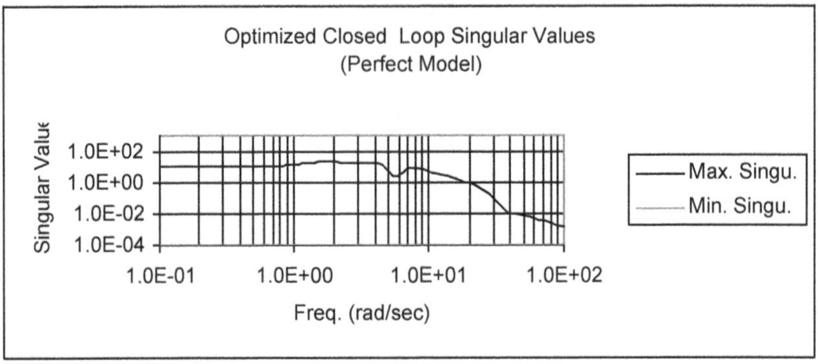
Optimized Closed Loop Singular Values
(Perfect Model)
Singular Valu
1.0E+02
1.0E+00
1.0E-02
1.0E-04
1.0E-01
1.0E+00
1.0E+01
1.0E+02
Freq. (rad/sec)
Max. Singu.
Min. Singu.

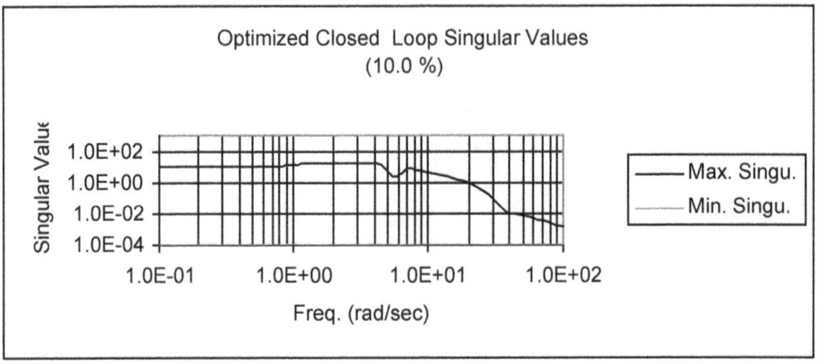
Optimized Closed Loop Singular Values
(10.0 %)
Singular Valu
1.0E+02
1.0E+00
1.0E-02
1.0E-04
1.0E-01
1.0E+00
1.0E+01
1.0E+02
Freq. (rad/sec)
Max. Singu.
Min. Singu.

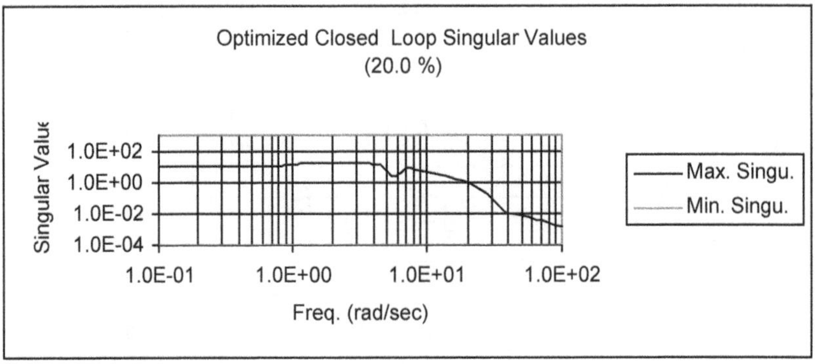
Optimized Closed Loop Singular Values
(20.0 %)
Singular Valu
1.0E+02
1.0E+00
1.0E-02
1.0E-04
1.0E-01
1.0E+00
1.0E+01
1.0E+02
Freq. (rad/sec)
Max. Singu.
Min. Singu.

Appendix III: Collection of Math Modeling Error Sensitivity Response and Singular Value Plots, and Eigenvalue Tables

Closed Loop System Eigenvalues					
	0%	10.00%	20.00%		
	1.0e+002 *	1.0e+002 *	1.0e+002 *		
1	-0.25	-0.25	-0.25		
2	-0.5824 + 1.3344i	-0.5824 + 1.3344i	-0.5824 + 1.3344i		
3	-0.5824 - 1.3344i	-0.5824 - 1.3344i	-0.5824 - 1.3344i		
4	-0.1040 + 0.2546i	-0.1199 + 0.2623i	-0.1266 + 0.2660i		
5	-0.1040 - 0.2546i	-0.1199 - 0.2623i	-0.1266 - 0.2660i		
6	-0.1046 + 0.2542i	-0.0898 + 0.2468i	-0.0839 + 0.2433i		
7	-0.1046 - 0.2542i	-0.0898 - 0.2468i	-0.0839 - 0.2433i		
8	-0.2284 + 0.0212i	-0.2584 + 0.0423i	-0.2682 + 0.0490i		
9	-0.2284 - 0.0212i	-0.2584 - 0.0423i	-0.2682 - 0.0490i		
10	-0.2293	-0.1922	-0.1026 + 0.1085i		
11	-0.1353 + 0.0476i	-0.1067 + 0.0922i	-0.1026 - 0.1085i		
12	-0.1353 - 0.0476i	-0.1067 - 0.0922i	-0.187		
13	-0.1446 + 0.0002i	-0.1446 + 0.0058i	-0.1487 + 0.0046i		
14	-0.1446 - 0.0002i	-0.1446 - 0.0058i	-0.1487 - 0.0046i		
15	-0.0868 + 0.0557i	-0.1446	-0.1446		
16	-0.0868 - 0.0557i	-0.0491 + 0.0260i	-0.0440 + 0.0335i		
17	-0.0002 + 0.0601i	-0.0491 - 0.0260i	-0.0440 - 0.0335i		
18	-0.0002 - 0.0601i	-0.0001 + 0.0601i	-0.0001 + 0.0601i		
19	-0.0002 + 0.0600i	-0.0001 - 0.0601i	-0.0001 - 0.0601i		
20	-0.0002 - 0.0600i	-0.0003 + 0.0600i	-0.0003 + 0.0599i		
21	-0.0009 + 0.0508i	-0.0003 - 0.0600i	-0.0003 - 0.0599i		
22	-0.0009 - 0.0508i	-0.0009 + 0.0505i	-0.0008 + 0.0504i		
23	-0.0009 + 0.0509i	-0.0009 - 0.0505i	-0.0008 - 0.0504i		
24	-0.0009 - 0.0509i	-0.0009 + 0.0511i	-0.0008 + 0.0511i		
25	-0.0347 + 0.0149i	-0.0009 - 0.0511i	-0.0008 - 0.0511i		
26	-0.0347 - 0.0149i	-0.0332	-0.028		
27	-0.0204	-0.0052 + 0.0135i	-0.0052 + 0.0134i		
28	-0.0052 + 0.0136i	-0.0052 - 0.0135i	-0.0052 - 0.0134i		
29	-0.0052 - 0.0136i	-0.0078 + 0.0145i	-0.0078 + 0.0146i		
30	-0.0078 + 0.0145i	-0.0078 - 0.0145i	-0.0078 - 0.0146i		
31	-0.0078 - 0.0145i	-0.0197	-0.0188		
32	-0.0134	-0.0135	-0.0135		
33	-0.0106	-0.0107	-0.0109		
34	-0.01	-0.01	-0.01		

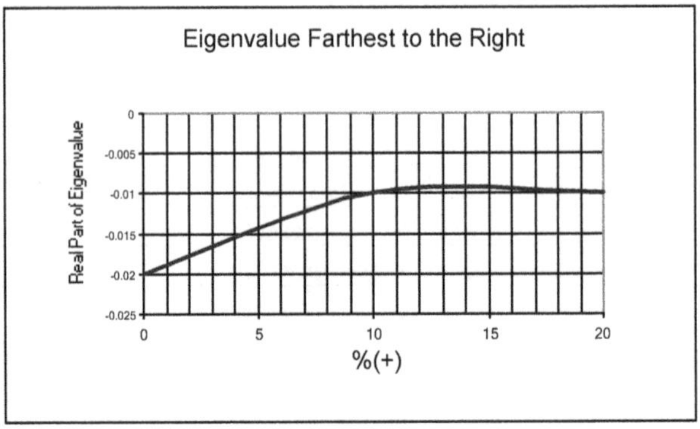

Observation:

Modeling error associated with bending slope does not cause control system instability.

Launch Vehicle System Matrices (A, B, C, D) Computation

(CD in Excel is available upon request, chkimca@gmail.com)

(Blank Page)

Launch Vehicle System Matrices (A, B, C, D) Computation

1) Columns: A, B, C, D, E
Rows: 1 through 81

Launch Vehicle System Matrices (A, B, C, D) Computation

Enter Data in the Input Column
(Fill the blanks in dark square only)
(A, B, C, D Matrices Computed
Are on Your Right Side)

	Code B38	80378		
	Parameters		Input	Unit
1	Time	flight time		sec
2	Gravity			ft/sec^2
	Rigid Body			
3	la	dist. frm LV nose to the origin		ft
4	TT (or FT)	total thrust		lbf
	Ts (or Fs) = TT - Tc	total thrust - control thrust		lbf
5	muC	control moment coefficent		sec^(-2)
6	Lalp	aerodynamic load		lb/rad
	m0	total mass - sloshing masses		slugs
7	mp1	mass of fuel tank 1 sloshing		slugs
8	mp2	mass of fuel tank 2 sloshing		slugs
9	mT	total mass		slugs
10	Uo	steady state axial velocity		ft/sec
11	Uodot	axial acceleration		ft/sec^2
12	lalp	dist. from CP to the origin		ft
13	lc (el C)	dist. from the origin to engine hinge		ft
14	mualp	aerodynamic moment coefficient		sec^(-2)
15	Theta0 (deg.)	pitch angle		degree
	Theta0 (rad)	pitch angle		rad/sec
16	lavg	dist frm the origin to middle of LV		ft
17	lT(el T)	tail location from the origin		ft
	Mass per unit length(η)			
18	Iyy	pitch moment of inertia		slug ft^2
	Engine Inertia			
19	mR	rocket engin mass		slugs
20	lR (el r)	dist frm rocket cg to hinge point		ft
21	Tc (or FC)	control thrust		lbf
	muCmRLR/Tc			
22	IR	rocket MOI about hinge point		slug ft^2
	Bending Mode 1			
23	M1	1st bending mode modal mass		slugs
24	Th1	damping ratio		
25	W1	1st bending mode modal frequency		rad/sec
26	SigG1(lc,Tail Loc)	negative slope of 1st bending mode		rad/ft
27	SigG1(lavg)	negative slope of 1st bending mode		rad/ft
	mR*lR*/M1			
	Tc/M1			
	2*Th1*W1			
	W1^2			

	Bending Mode 2			
28	M2	2nd bending mode modal mass		slugs
29	Th2	damping ratio		
39	W2	2nd bending mode modal frequency		rad/sec
31	SigG2(lc,Tail Loc)	negative slope of 1st bending mode		rad/ft
32	SigG2(lavg)	negative slope of 1st bending mode		rad/ft
	mR*IR*/M2			
	Tc/M2			
	2*Th2*W2			
	W2^2			
	Sloshing 1			
33	Lp1	1st pendulum length		ft
34	lp1	1st pendulum hinge location		ft
	a1	function of many variables		
35	Wp1^2	1st square of pend. Frequency		(rad/sec)^2
	mp1	1st mendulum mass		slugs
36	muP1	1st pendulum velocity		sec^(-2)
	Sloshing 2			
37	Lp2	2nd pendulum length		ft
38	lp2	2nd pendulum hinge location		ft
	a2	function of many variables		
39	Wp2^2	2nd square of pend. Frequency		(rad/sec)^2
	mp2	2nd mendulum mass		slugs
40	muP2	2nd pendulum velocity		sec^(-2)
	Rate Gyro			
41	Kr	rate gyro gain		sec^(-1)
42	WR(rad/sec)	natural frequency		rad/sec
43	TheR	damping ratio		
44	Sigma 1(IG)	see Eqn. (3.7-8)		rad/ft
45	Sigma 2(IG)	see Eqn. (3.7-8)		rad/ft
	Position Gyro			
46	Tau	time constant		sec
47	Sigma 1(IG)	see Eqn. (3.7-7)		rad/ft
48	Sigma 2(IG)	see Eqn. (3.7-7)		rad/ft

(Blank Page)

Launch Vehicle System Matrices (A, B, C, D) Computation

2) Columns: F, G, H
   Rows: 53 through 223

Launch Vehicle System Matrices (A, B, C, D) Computation

KA	=G86/G83	
KI		
g	=D9	ft/sec^2
Actuator		
KCb		
set it zero		
WC		

Launch Vehicle System Matrices (A, B, C, D) Computation

Kcb*Wc^2	=G11*G29^2	
2.0 eta We	=M51+M45/M32	
Wc^2	=G29^2	
Rigid Body		
Time	0.5 sec	72.0 sec
mOR		
		=D12
TT	366100	=D13
Ts = TT - Tc	=G60-G87	=D13-D35
muC	3.21	=D15
Lalp(pp220,347)	0	=D16
mo	8225	=D17
mp1	113	=D18
mp2	200	=D19
mT	=G64+G65+G66	=D20
Uo	20.5	=D21
Uodot	40.5	=D22
lalp	-	=D23
lc	-33.9	=D24
mualp	0	=D25
Theta0	90	=D26
	=G74/57.3	=H74/57.3
lavg		=D28
IT(el T)	=G71	=D29
Mass per unit length(η)		=D30
Iyy		=D31
Engine Inertia		
Time	0.5 sec	72.0 sec
muC	=G62	=H62
mR	=D33	=D33
IR (el r)	2.52	=D34
mualp	=G73	=H73
Tc	308000	=D35
muCmRLR/Tc	=G83*G84*G85/G87	=D36
IR		=D37

Launch Vehicle System Matrices (A, B, C, D) Computation

Bending Mode 1		
mR	=D33	slugs
lR	2.52	ft
M1	2540	slugs at 0.5 sec
	=D39	slugs at 72 sec
	1200	slugs at 152 sec
Tc		lb at 0.5 sec
		lb at 72 sec
		lb at 152 sec
Th1	0.7	
W1		
t=0.5 sec	16.5	rad/sec
t=72.0 sec.	=D41	rad/sec
t=152 sec	41.1	rad/sec
SigG1(lc and lT(Tail Location))		
t=0.5 sec	-0.55	rad/ft
t=72.0 sec.	=D42	rad/ft
t=152 sec	-0.1375	rad/ft
mR*lR*/M1	=G97*G98/G100	
Tc/M1	=G104/G100	
2*Th1*W1	=2*G106*G109	
W1^2	=G109^2	
Bending Mode 2		
mR	=G97	slugs
lR	=G98	ft
M2	=D49	slugs
Tc	308000	lb at 0.5 sec
	342000	lb at 72 sec
	356000	lb at 152 sec
Th2	=G106	
W2		
t=0.5 sec	41.6	rad/sec
t=72.0 sec.	=D51	rad/sec
t=152 sec		rad/sec
SigG2 (lc and lT(Tail Location))		
t=0.5 sec	-0.1218	rad/ft
t=72.0 sec.	=D52	rad/ft
t=152 sec	-0.368	rad/ft
mR*lR*/M2	=G123*G124/G125	
Tc/M2	=G128/G125	
2*Th2*W2	=2*G130*G133	
W2^2	=G133^2	

Launch Vehicle System Matrices (A, B, C, D) Computation

Bending Mode 3		
mR	30.8	slugs
lR	2.52	ft
M3	1840	slugs
Tc	308000	lb at 0.5 sec
	=H87	lb at 72 sec
	356000	lb at 152 sec
Th3	0.7	
W3		
t=0.5 sec	67.4	rad/sec
t=72.0 sec.	87.1	rad/sec
t=152 sec	238	rad/sec
SigG3		
t=0.5 sec	-0.272	rad/ft
t=72.0 sec.	-0.565	rad/ft
t=152 sec	0.34	rad/ft
mR*lR*/M3	=G147*G148/G149	
Tc/M3	=G151/G149	
2*Th3*W3	=2*G153*G156	
W3^2	=G156^2	
Sloshing 1		
muC(pp287)	=G83	pp347
a1	=-(1/H172)*((H173/H64)-(H175-H172)*H176)	
	Lp1(72 sec)	=D59
	Tc(75 sec)	=G104
	Tt(75 sec)	418900
	lp1(72 sec)	=D60
	muC	=H83
Wp1^2	=(1/H172)*((1+(H180/H64))*H181+(H175-H172)*H182)	
5.12	mp1(75 sec)	=H65
	Uodot (75 sec)	66.3
pp298	muP1(75 sec)	0.0591
Sloshing 2		
muC	=G83	
a2	=-(1/H187)*((H188/H64)-(H190-H187)*H191)	
	Lp2(72 sec)	=D66
	Tc(75 sec)	=G104
	Tt(75 sec)	=H174
	lp2(72 sec)	=D67
	muC	=H83
Wp2^2	=(1/H187)*((1+(H195/H64))*H69+(H190-H187)*H197)	
5.17	mp2(75 sec)	=H66

Launch Vehicle System Matrices (A, B, C, D) Computation

	muP2(75 sec)	-0.1582
Rate Gyro		
Kr	=D73	sec^(-1)
WR(rad/sec)	=D74	Wr>2.0*1.4*WB
		WB(rad/sec)
		WR=c*WB, c>1.0
Damping Ratio	=D75	
Sigma 1(IG)	=D76	rad/ft
Sigma 2(IG)	=D77	rad/ft
Position Gyro		
Tau	=D79	sec
Sigma 1(IG)	=D80	rad/ft
Sigma 2(IG)	=D81	rad/ft
Wind Coefficients		
G1	1	
G2	1	
G3	1	
G4	1	
G5	1	
G6	1	
G7	1	
G8	1	

Launch Vehicle System Matrices (A, B, C, D) Computation

3) Columns: I, J, K
   Rows: 7 through 211

Launch Vehicle System Matrices (A, B, C, D) Computation

CHECK		Servo Gain (KA=mualp/muc, stability border line)
		Integrator Gain
	rad/sec/rad	Effective (at load) open loop velocity gain
		KCb=alp*KC
		alp=1+(4/3pi)*(PLh/(Ps-Pr))cos (phi)
		KC=(Kap*KF*Kv*sqrt(PS-PR))/A
		WC=sqrt(1/(1/Wa^2 + 1/Wp^2 + 1/Wm^2))
	Wa^2	=4*M34*M30^2*M31^2/(M32*M33)
	Wp^2	=M31^2*M35/M32
	Wm^2	=M31^2*M35/M32
		SQRT(WC^2 + Cfb*K0b/IR)

Launch Vehicle System Matrices (A, B, C, D) Computation

		K0b + Cfb/Ir
152 sec		
	rigid body + engine	
436400		
=I60-I87		
11.1	sec^(-2)	
9000	lb/rad	
1670	slugs	
239	slugs	
128	slugs	
=I64+I65+I66	slugs	mT=m0+sum of mpi
7023.3	ft/sec	
215.2	ft/sec^2	
18.6	ft	
-46.8	ft	
0.19	sec^(-2)	(Lalp*lalp/Iyy)
10	degree	assumed
=I74/57.3	rad	
	ft	dist from orign to middle of lv
=I71		Tail Location
		from note computing C1
	slug ft^2	
152 sec		
=I62	sec^(-2)	
=D33	slugs	
2.52	ft	pp349
=I73	sec^(-2)	
356000	lb	
=I83*I84*I85/I87		
	slug ft^2	pp349

SigG1(Imiddle)		
0.0384	rad/ft	
=D43	rad/ft	
0.0623	rad/ft	
SigG2(Imiddle)		
0.00192	rad/ft	
=D53	rad/ft	
0.0309	rad/ft	

ft	originally 2.73.	
lb		
lb		
ft		
sec(-2)		
slugs		
ft/sec^2		
sec^(-2)		
ft		
lb		
lb		
ft		
sec(-2)		
slugs		

sec^(-2)		
=2.8*I202		
40		
pp348, SN400		
pp348, SN400		
pp320		
pp348, SN400		
pp348, SN400		

Launch Vehicle System Matrices (A, B, C, D) Computation

4) Columns: L, M
   Rows: 7 through 115

Launch Vehicle System Matrices (A, B, C, D) Computation

mualp/muC
Approxi
2.7
70
0.000007125
432000
7200
0.0247
115
0.26
0.0247
1.77
377
0.0076
38900000
1120000
=O49+(4/3.14157)*(O46/(O48*O49))

Launch Vehicle System Matrices (A, B, C, D) Computation

=(O52+G11*(M30^2/M35))/(M33/(4*M34)+M30^2/M35)
CL
=H71
=-$G$113*$M$59^2/(2*$H$71)
=-$G$113*((2*H70+H71)/2)^2/(2*$H$71)
=-$G$113*$H$175^2/(2*$H$71)
=-$G$113*$H$190^2/(2*$H$71)
=-$G$138*$M$59^2/(2*$H$71)
=-$G$138*((2*H70+H71)/2)^2/(2*$H$71)
=-$G$138*$H$175^2/(2*$H$71)
=-$G$138*$H$190^2/(2*$H$71)
=(G113/H71)*(H76)
=(G138/H71)*(H76)
=(-H84*H69*M61+G9*COS(H75)*(G113*H79/(2*H71))*((0.25*(2*H70+H71)^(4) -0)-((1/3)*(2*H70+H71)^(3) -0)))/G100
=-(((H70-H76)*H63*G113*H71/(4*H68)))/G100
=(H87*M61)/G100
=-(H63*M62)/G100
= (H69*H180*M63)/G100
=(H69*H195*M64)/G100
=-((H87+H61)*M61*G113+H63*M62*I113)/G100
=-((H87+H61)*M61*G138+H63*M62*I138)/G100
=(H63*M62^(2))/(H68*G100)
=H63*M62*M66/(H68*G100)
=(-H84*H69*M66+G9*COS(H75)*(G138*H79/(2*H71))*((0.25*(2*H70+H71)^(4) -0)-((1/3)*(2*H70+H71)^(3) -0)))/G125
=-(((H70-H76)*H63*G138*H71/(4*H68)))/G125
=(H87*M66)/G125
=-(H63*M67)/G125
= (H69*H180*M68)/G125
=-(H69*H195*M69)/G125
=-((H87+H61)*M66*G113+H63*M67*I113)/G125
=-((H87+H61)*M66*G138+H63*M67*I138)/G125

Launch Vehicle System Matrices (A, B, C, D) Computation

=(H63*M66^(2))/(H68*G125)
=H63*M66*M61/(H68*G125)
=-(1/H172)*((1+(H180/H67))*H69+(H175-H172)*H182)
=-(1/H172)*((H87/H67)-(H175-H172)*H62)
=-(1/H187)*((H69*H180/H67)+H182*(H190-H187))
=-(1/H187)*((H87/H67)-(H190-H187)*H62)

Launch Vehicle System Matrices (A, B, C, D) Computation

5) Columns: N, O, P, Q
Rows: 7 through 54

Launch Vehicle System Matrices (A, B, C, D) Computation

KA=2.0, pp271at Max q		
ma/rad		
volts/ft		
(ft^3/sec)/ma(lb/ft^2)		
lb/ft^2		
lb/ft^2		
ft^2		
see sheet2		
assumed lag (15.0 Deg)		
ft^2		
ft		
slug ft^2		
ft^3		
lb/ft^2		
lb/ft		
Cv+(4/pi)*(CB/omega*delb)		
CB	565	lb/ft
omega	10	

Launch Vehicle System Matrices (A, B, C, D) Computation

delb	0.4		
Cv	3600	lb ft/(rad/sec)	
(CLb+KCb*(A^2/Km))/(VT/4B +A^2/Km)			
CLb	=1.11*N54/M27	1.11 CL/sqrt(PLb)	
0.00000478	(ft^3/sec)/(lb/ft^(0.5))		

(Blank Page)

6) Columns: R, S, T
   Rows: 7 through 137

Launch Vehicle System Matrices (A, B, C, D) Computation

=H80+H93+H84*(H71+ H85)^(2)		
=(H84*H71*H69+H70*H63)/R7		
= -(1/H68)*((H70-H76)^(2))*H63/R7		
= (-H71*(H87+H61)*G113-(H87+H61)*M61 - H84*H85*H69*G113 +I113*(H70-H76)*H63)/R7		
=(-(1/H68)*((H70-H76)*H63*G113*H76^(2))/(2*H71))/R7		
=((-H71*(H87+H61)*G138-(H87+H61)*M66)-H84*H85*H69*G138+I138*(H70-H76)*H63)/R7		
=(-(1/H68)*(H70-H76)*H63*G138*H76^(2)/(2*H71))/R7		
=(-H180*H175*H69)/R7		
=(-H195*H190*H69)/R7		1
=(H71*H87 + H84*H85*H69)/R7		2
=(H93+H84*H85*H71)/R7		3
=(H63*(H70-H76)*H76^2/(2*H71))/R7		4
=-H84*(H85+H71)/R7		5
=(-(1/H68)*((H70-H76)*H63*H76^(2))/(2*H71))/R7		6
=H64+H84		7
=(H84*H69-H67*G9*COS(H75))/R26		8
=(H64*H68+(1/H68)*(H70-H76)*H63)/R26		9
=-H84*(H71+H85)/R26		10
=(-(H87+H61)*G113-I113*H63)/R26		11
=(-(1/H68)*((G113*H76^(2))/(2*H71))*H63)/R26		12
=(-H84*(M61+H85*G113))/R26		13
=(-(H87+H61)*G138-I138*H63)/R26		
=((1/H68)*H63*G138*H76^(2)/(2*H71))/R26		
=(-H84*(M66+H85*G138))/R26		
=H180*H69/R26		
=H195*H69/R26		
=H87/R26		
=H84*H85/R26		
=-H63/R26		1
=(-(1/H67)*((H69-H75)*H62*H75^(2))/(2*H70))/R26		2
=1- R24*R29		3
=(R8+ R24*R27)/R42		4
=(R9+R24*R28)/R42		5
=(R10+R24*R30)/R42		6
=(R11+R24*R31)/R42		7
=R24*R32/R42		8

Launch Vehicle System Matrices (A, B, C, D) Computation

=(R12+R24*R33)/R42		9
=(R15+R24*R34)/R42		10
=R24*R35/R42		11
=(R19+R24*R36)/R42		12
=(R20+R24*R37)/R42		13
=(R21+R24*R38)/R42		
=(R22+R24*R39)/R42		
=(R23+R24*R40)/R42		
=(R25 + R24*R41)/R42		
=$M$76		
=$M$77		
=-(H84*M61*((H71+H85)))/G100	1	1
=$M$82	2	2
=$M$85		
=-((H84*M61*(M61+H85*G113)))/G100		
=$M$83		
=$M$86		
=-(H84*M61*(M66+H85*G138))/G100		
=$M$80	1	1
=$M$81	2	2
=$M$78		
=H84*H85*M61/G100		
=$M$79*80378/C5		
=-H84*M61*80378/(G100*C5)		
=R70/H68		
=1-(R62+R71*R32)		
=(R57+R71*R27)/R73		
=(R58+R71*R28)/R73		
=(R59+R71*R29)/R73		
=(-(G109^(2))+R60+R71*R30)/R73		
=(-2*G106*G109+R61+R71*R31)/R73		
=(R63+R71*R33)/R73		
=(R64+R71*R34)*(80378)/(R73*C5)		
=(R65+R71*R35)/R73		
=(R66+R71*R36)/R73		
=(R67+R71*R37)/R73		
=(R68+R71*R38)/R73		
=(R69+R71*R39)/R73		
=(R70+R71*R40)/R73		
=(R72+R71*R41)/R73		
=$M$88		
=$M$89		
=-(H84*M66*(H71+H85))/G125		
=$M$94		
=$M$97		
=-(H84*M66*(M61+H85*G113))/G125		
=$M$95		
=$M$98		

Launch Vehicle System Matrices (A, B, C, D) Computation

=-(H84*M66*(M66+H85*G138))/G125		
=$M$92		
=$M$93		
=$M$90		
=H84*H85*M66/G125		
=$M$91		
=-H84*M66/G125		
=R101/H63		
=1-(R96+R102*R35)		
=(R88+R102*R27)/R104		
=(R89+R102*R28)/R104		
=(R90+R102*R29)/R104		
=(-(G133^(2))+R91+R102*R30)/R104		
=(-2*G130*G133+R92+R102*R31)/R104		
=(R93+R102*R32)/R104		
=(R94+R102*R33)/R104		
=(R95+R102*R34)/R104		
=(R97+R102*R36)/R104		
=(R98+R102*R37)/R104		
=(R99+R102*R38)/R104		
=(R100+R102*R39)/R104		
=(R101+R102*R40)/R104		
=(R103+R102*R41)/R104		
=$M$112		
=-(1/H172)*(H69*H195/H67 +H197*(H175-H172))		
=$M$113		
=$M$114		
=(-1/H187)*((1+H195/H67)*H69 + H197*(H190-H187))		
=$M$115		
=1/G209		
=G210/G209		
=G211/G209		
=-1/G209		
= (G201^(2))*G200		
=G201^2*G200*G205		
=(G201^(2))*G200*G206		
=-(G201^(2))		
=-2*G204*G201		

7) Columns: U, V, W, X, Y, Z
   Rows: 18 through 67

Launch Vehicle System Matrices (A, B, C, D) Computation

A Matrix

1	2	3	4	5	6	7
0	1	0	0	0	0	0
=R43	=R44	=R45	=R46	=R47	=R48	=R49
0	0	0	1	0	0	0
=R74	=R75	=R77	=R78	=R79	=R80	=R82
0	0	0	0	0	1	0
=R105	=R106	=R108	=R109	=R111	=R112	=R113
0	0	0	0	0	0	0
0	0	0	0	0	0	=R120
0	0	0	0	0	0	0
0	0	0	0	0	0	=R124
=R128	0	=R129	0	=R130	0	0
0	0	0	0	0	0	0
0	=R133	0	=R134	0	=R135	0

B Matrix

1	2	3
0	0	0
=R53	=R55	=R56
0	0	0
=R84	=R86	=R87
0	0	0
=R115	=R117	=R118
0	0	0
=R122	0	=R142

Launch Vehicle System Matrices (A, B, C, D) Computation

0	0	0
=R126	0	=R143
0	0	=R144
0	0	=R145
0	0	=R146

C Matrix

1	2	3	4	5	6	7
1	0	0	0	0	0	0
0	1	0	0	0	0	0

D Matrix

1	2	3
0	0	0
0	0	0

(Blank Page)

Launch Vehicle System Matrices (A, B, C, D) Computation

8) Columns: AB, AC, AD, AE, AF, AG
   Rows: 19 through 60

8	9	10	11	12	13
0	0	0	0	0	0
=R51	=R52	0	0	0	0
0	0	0	0	0	0
0	=R83	0	0	0	0
0	0	0	0	0	0
0	=R114	0	0	0	0
1	0	0	0	0	0
0	=R121	0	0	0	0
0	0	1	0	0	0
0	=R125	0	0	0	0
0	0	0	=R131	0	0
0	0	0	0	0	1
0	0	0	0	=R136	=R137

References

(Blank Page)

Reference

1 Bate, Roger R., Donald D. Mueller, and Jerry E. White, "*Fundamentals of Astrodynamics*," Dover Publication, Inc., New York, 1971.

2 Blakelock, John H., "*Automatic Control of Aircraft and Missile*," John Wiley & Sons, Inc., New York, 1991.

3 Britting, Kenneth R., "*Inertial Navigation Systems Analysis*, " Wiley-Interscience, New York, 1971.

4 Bryson, Arthur E. Jr., Yu-Chi Ho, " *Applied Optimal Control*," Ginn and Company, Waltham, MA 1969.

5 Chen, C. T., "*Introduction to Linear System Theory*," Holt, Rinehart and Winston, Inc., New York, 1970.

6 "Control Design via mu Synthesis, Xmath Control Design Module, Integrated Systems," Santa Clara, CA Feb. 1995.

7 "Control System Toolbox for use with MATLAB," The MathWorks, Inc., Natick, MA.

8 Derusso, Paul M., Rob J. Roy, and Charles M. Close, "*State Variables for Engineers,* " John Wiley & Sons, Inc., New York, 1965.

9 Doyle, John C., Gunter Stein, Yeh Banda, and His-Han, "Lecture Notes for the Workshop on H-infinity and mu Methods for Robust Control," American Control Conference, San Diego, CA 1990.

10 Doyle, John C., Keith Glover, Pramod Khargonekar, and Bruce A Francis, "State-Space Solutions to Standard H-2 and H-infinity Control Problems," IEEE Trans. Automat. Contr., Vol. AC-34, No. 8, Aug. 1989.

11 Dwight, Herbert Bristol, "*Tables of Integrals and Other Mathematical Data*," The MacMillan Company, New York, 1961.

12 Gelb, Arthur, "*Applied Optimal Estimation*," The M.I.T. Press, Cambridge, MA, 1974.

13 Glover, Keith, "All Optimal Hankel-Norm Approximation of Linear Multivariables Systems and Their H-infinity - Error Bounds," Int. J. Control, Vol. 39, No. 6, 1984.

14 Glover, Keith, and John C. Doyle, "State-Space Formulae for All Stabilizing Controllers That Satisfy an H-infinity Norm Bounded Relations to Risk Sensitivity," Systems and Control Letters, Vol. 11, 1988.

15 Glover, Keith, J. C. Doyle, "A State Space Approach to H-infinity Optimal Control," Springer-Verlag, Lecture Notes in Control and Information Sciences, Vol. 135, 1989.

16 Goldstein, Herbert, "*Classical Mechanics*," Addison-Wesley Publishing Company, Inc., Reading, MA 1950.

17 Greensite, Arthur L., "*Control Theory: Vol. I Elements of Modern Control Theory*," Spartan Books, New York, 1970.

18 Greensite, Arthur L., "*Control Theory: Vol. II Analysis and Design of Space Vehicle Flight Control Systems*," Spartan Books, New York, 1970.

19 Hargraves, Charles, Forrester Johnson, Stephen Paris, and Ian Rettie, "Numerical Computation of Optimal Atmospheric Trajectories," AIAA Atmospheric Flight Mechanics Conference, Boulder, CO, Aug. 1979.

20 Hyde, R. A., "*H-infinity Aerospace Control Design-A VSTOL Flight Application*," Springer, Verlag, 1995.

21 Luenberger, David G., "*Optimization by Vector Space Methods*," John Wiley & Sons, Inc., New York, 1969.

22 "Matlab Simulink," The MathWorks, Inc., Natick, MA.

23 "mu-Analysis and Synthesis Toolbox for use with MATLAB," The MathWorks, Inc., Natick, MA 1998.

24 O'Donneli, C. F., "*Inertial Navigation Analysis and Design*," McGraw-Hill Book Company, New York, 1964.

25 Ogata Katsuhiko, "*Modern Control Engineering*," Prentice-Hall, Inc., Englewood Cliffs, NJ 1970.

26 Ogata Katsuhiko, "*State Space Analysis of Control Systems*," Prentice-Hall, Inc., Englewood Cliff, NJ 1967.

27 Perkins, William R., Jose B. Cruze, Jr., "*Engineering of Dynamic Systems*," John Wiley & Sons, Inc., New York, 1969.

28 Press, William H., Brian P. Flannery, Saul A. Teukolsky, and william T. Vetterling, "*Numerical Recipes*," Cambridge University Press, Cambridge, 1987.

29 Sage, Andrew P., "*Optimum Systems Control*," Prentice-Hall, Inc., Englewood Cliff, NJ 1968.

30 Smith, J. L., "*Analysis of Inertial Navigation Systems*," Autonetics, Rockwell International, Anaheim, CA 1979.

31 Sutton, George P., Oscar Biblarz, "*Rocket Propulsion Elements*," John Wiley & Sons, Inc., New York, 2001.

32 Thomson, William Tyrrell, "*Introduction to Space Dynamics*," Dover Publications, Inc., New York, 1986.

33 Tou, Julius T., "*Modern Control Theory*," McGraw-Hill Book Company, New York 1964.

34 Tse, Francis S., Ivan E. Morse, and Rolland T. Hinkle, "*Mechanical Vibration*," Allyn and Bacon, Inc., Boston, 1963.

35 Wertz, James R., and Wiley J. Larson, "*Space Mission Analysis and Design*," Kluwer Academic Publishers, Dordrecht, 1991.
36 Wertz, James R., "*Spacecraft Attitude Determination and Control*," Kluwer Academic Publishers, Dordrecht, 1978.
37 Williams, S. J., " H-infinity for the Layman," Measurement + Control, Vol. 24, Feb 1991.
38 Xing, Guang Q. and Peter M. Bainum, "mu Analysis of Expendable Launch Vehicle Robustness," 46th International Astronautical Congress, Oslo, Norway, Oct. 1995.
39 Xing, Guang Q., Pater M. Bainum, and Feiyue Li, "Design of a Reduced Order H-infinity Robust Controller for an Uncertainty," Acta Astronautica Vol. 41 No. 2, 1997.
40 Yang, S. M., and N. H. Huang, "Application of H-infinity Control to Pitch Autopilot of Missiles," IEEE Trans. on Aerospace and Electronic Systems, Vol. 32, No. 1 Jan. 1996.
41 Zhou, Kemin and Pramod P. Khargonekar, "An Algebraic Riccati Equation Approach to H-infinity Optimization," Systems & Control Letters 11, 1988.
42 Zhou, Kemin, John C. Doyle, and Keith Glover, "Robust and Optimal Control," Prentice-Hall, Upper Saddle River, NJ 1996.
43 Benoit C., Gilles D. , Mauffrey S., "Aerospace Launch Vehicle Control: a gain scheduling Approach," [Journal Paper] Control Engineering Practice, Vol. 13 no.3, March 2005, pp.333-347. Publisher: Elsevier, UK.
44 deVirgilio MA, Kamimoto DK., "Practical Applications of Modern Controls for Booster Autopilot Design," [Conference Paper] AIAA/IEEE Digital Avionics Systems Conference. 12th DASC (Cat. No. 93CH3327-4). IEEE. 1993, pp. 400-12. New York, NY, USA.
45 Dong Hyun Kim, Jae Weon Choi, "Attitude Controller Design for a Launch Vehicle with Fuel-Slosh," [Conference Paper] SICE 2000. Proceedings of the 39th SICE Annual Conference. International Session Papers (IEEE Cat. No. 00 TH 8545). Soc. Instrum. & Control Eng. 2000, pp 235-40, Tokyo, Japan.
46 Oberoi S., Bandyopadhyay B., "Robust Control of a Laboratory Scale Launch Vehicle Model Using Fast Output Sampling Technique," [Conference Paper] 2004 5th Asian Control Conference (IEEE Cat. No. 04EX904). IEEE. Part Vol. 1, 2004, pp. 383-91 Vol. 1. Piscataway, NJ, USA.

47 Oberoi S., Janardhanan S., Bandyopadhyay B., " Output Feedback Control of Practical Launch systems," [Conference Paper] Proceedings of the 2004 IEEE International Conference on Control Applications (IEEE Cat. No. 04CH37596). IEEE. Part Vol. 1, 2004 PP 13-17 Vol. 1. Piscataway, NJ, USA.

48 Rosanian J. Saleh AR, Jahexd-Motlagh MR., "On the Design of Adaptive Autopilots for a Launch Vehicle," [Journal Paper]. Proceedings of the Institution of Mechanical Engineers, Part I (Journal of Systems and Control Engineering), vol. 221, no. 11, Feb. 2007, pp. 27-38, Publisher: Mech. Eng. Publications for IMechE, UK.

49 Sifer JF, Prouty SJ, Mak PH. "Advanced Concepts for Launch Vehicle Control," [Conference Paper] Proceedings. IEEE/AIAA/NASA 9th Digital Avionics Systems Conference (Cat. No. 90CH2929-8). IEEE. 1990, pp. 111-16. New York, NY, USA.

50 Wang Zhao, Wang Qing, Zhang Xin-juan, " An Adaptive Launch Vehicle" Mechatronics and Automation, 2007, ICMA 2007, International Conference, Harbin China, Aug. 2007.

51 McFarlane, D.C. and K. Glover, "Robust Controller Design using Normalized Coprime Factor Plant Description," Springer Verlag, Lecture Notes in Control and Information Science, vol. 136, 1989.

52 McFarlane, D.C. and K. Glover, "A Loop Shaping Design Procedure using Synthesis," IEEE Transaction on Automatic Control, vol. 37, no. 6, pp. 759-769, June 1992.

53 Vinnicombe, G., "Measuring Robustness of Feedback Systems," PhD dissertation, Dept. of Engineering, Univ. of Cambridge, 1993.

Index

(Blank Page)

Index

Index

Index

(Blank Page)

About the Author:

He was born in Seoul, Korea in 1940. He is a graduate of Kyung Gi H. S. After completion of his sophomore at the Engineering College of Seoul National University, he joined the Republic of Korea Army. Upon being discharged from the Army in 1962, he came to the United States to study Mechanical Engineering at Iowa State University in Ames, IA. After he finished his BS program at ISU, he continued his study at Purdue University for MS degree. After the master's degree, he joined the Honeywell, Inc in For Washington, PA. Later he went to Canada to study theology for one year, and came back to the United States. He earned his Ph.D. from the University of Connecticut, and joined the team that designed the Space Shuttle Boost Phase control system. During his decades of professional experience, he was involved in a variety of aerospace engineering projects such as Trident Nuclear Submarine Ballistic Missile Trajectory Error Analysis, MX Missile GN&C IMU Calibration Algorithm Development, SDI Anti-Satellite Tracking Algorithm Development (Kalman Filtering), International Space Station Attitude Determination, Delta III and IV Launch Vehicle Stage I/II Separation Analysis, and Launch Vehicle Fuel Tank Sloshing Analysis. He is a member of Pi Tau Sigma, Tau Beta Pi, and Sigma Xi. He started writing this book after he retired from the Boeing Company.

(Blank Page)

www.ingramcontent.com/pod-product-compliance
Lightning Source LLC
Chambersburg PA
CBHW030241230326
41458CB00115B/6622/J
* 9 7 8 0 6 1 5 1 6 5 1 1 0 *